Building Performance Analysis

Building Performance Analysis

Pieter de Wilde

Registered Offices
John Wiley & Sons, Inc., 111 River Street, Hoboken, NJ 07030, USA
John Wiley & Sons Ltd, The Atrium, Southern Gate, Chichester, West Sussex, PO19 8SQ, UK

Editorial Office
9600 Garsington Road, Oxford, OX4 2DQ, UK

For details of our global editorial offices, customer services, and more information about Wiley products visit us at www.wiley.com.

Wiley also publishes its books in a variety of electronic formats and by print-on-demand. Some content that appears in standard print versions of this book may not be available in other formats.

Library of Congress Cataloging-in-Publication data applied for

Hardback ISBN: 9781119341925

Cover design by Wiley
Cover image: Supphachai Salaeman/Shutterstock

Set in 10/12pt Warnock by SPi Global, Pondicherry, India

Printed in the UK

Contents

Endorsement by IBPSA

The International Building Performance Simulation Association (IBPSA) was founded to advance and promote the science of building performance simulation in order to improve the design, construction, operation and maintenance of new and existing buildings worldwide. IBPSA recognises the complexity of building performance and the many factors that influence this. This book addresses these issues in detail, unpacking the meaning of building performance analysis by considering its history and current practices. In doing so, it leads the reader to an appreciation of the fundamental importance of building performance analysis and the role it plays at all stages of the life cycle of a building, leading to an emergent theory of building performance analysis in Chapter 11.

Along this journey, the book mobilises an extensive quantity of relevant literature on this broad subject, making it an invaluable resource for students at all levels. Each chapter concludes with a list of activities that not only serves as a summary of the material covered but also provides an excellent basis from which to develop student projects and assessments.

The book provides a broad range of insights, food for thought and suggestions for how to approach your own building performance analysis. It is hoped that the book will go some way to elucidating the topic, equipping graduates with the knowledge and awareness required to specify, design, procure and operate high performance buildings that deliver high quality indoor environments and low energy consumption.

IBPSA is grateful to Professor De Wilde for the many hours he has devoted to bringing this book to fruition and commends it to anyone pursing a detailed knowledge of building performance analysis and its allied disciplines.

Professor Malcolm Cook
Loughborough University, UK
Chair of the IBPSA Publication Committee

The International Building Performance Simulation Association (IBPSA) makes every effort to ensure the accuracy of all the information contained in publications that it endorses. However, IBPSA, our agents and our licensors make no representations or warranties whatsoever as to the accuracy, completeness or suitability for any purpose of the content. Any opinions or views expressed in this publication are the opinions and views of the authors and are not the views of IBPSA. The accuracy of the content should not be relied upon and should be independently verified with primary sources of information. IBPSA shall not be liable for any losses, actions, claims, proceedings, demands, costs, expenses, damages and other liabilities whatsoever or howsoever caused arising directly or indirectly in connection with, in relation to or arising out of the use of this content.

Foreword

Ever since I was a young researcher in building simulation at TU Delft, I have been intrigued by the prospect of being able to support rational dialogues in building design projects, in particular to express unambiguously how we want buildings to behave or what goals we want to achieve with them. This inevitably invites the hypothesis that design can be managed as a purely rational fulfilment process in which clients precisely define their expectations (as requirements) and designers verify their creatively generated proposals (as fulfilment) against these expectations. It doesn't take much to realise that this can only be realised by introducing a set of objectively quantifiable measures, agreed upon by both parties. When expectations are not met, design adaptations or relaxation of client requirements could be negotiated. For many years I have taught a graduate course on this subject that I loosely labelled as 'performance-based design'. It was meant to whet the appetite of PhD students that walked in with vague notions about the next generation of building design methods and frameworks to support them. The course examined the literature in an attempt to cement the foundation of the central concepts such as performance, measurement and quantification. Then I showed how their operationalisation requires the development of a plausible worldview of buildings in which their system specification is expressed at increasing levels of resolution and as steps in an evolving design process.

Pieter de Wilde was one of the PhD students who was brave enough to voluntarily enrol in the course. He was looking for answers to his fundamental thesis research, only to find out that the course stopped far short of offering a methodology that could be mapped onto real-world design projects without some vigorous arm waving. For one there were still many missing pieces that could only be 'covered' by fuzzy connections. But above all, a unifying theory that gives building performance analysis a precise meaning in every application setting was and still is missing. The lack of a rigorous definition of generic tasks in building projects is one of the prime reasons why this situation persists. In the course I repeatedly stressed that the lack of a textbook that offers all relevant concepts and underlying ideas in one place is felt as another obstacle to attract the recognition the domain deserves. Some 15 years later, during a long drive through the English countryside, Pieter offered the idea to do something about this, and 3 years later, this resulted in the monograph that is in front of you. The road travelled in these 3 years has been as curvy and challenging as the drive through rural Devon, trying to avoid the sharp edges of the stone hedgerows and slowing down enough at blind corners. Fortunately Pieter's skills at the steering wheel kept me safe, and his skills at the keyboard proved to be an equal match for all narrow theories and blinding misconceptions that lay ahead.

I am very happy that this book got written. For one, it brings together the extensive body of work that has gone before, thus providing the first coherent account of the state of our knowledge in building performance, from fundamental concepts to operational measures, followed by their quantification in real-life cases. In organising the book along these three parts, the author has succeeded in taking the reader from a generic basis to operationalisation that gets ever more specific towards the later chapters. This approach is the perfect reflection of the fact that although the basis of performance concepts is generic, their application demands creative thinking and will always be case specific. The link between the two is realised by a broadening palette of multi-aspect building simulation tools of which the book provides a good overview. The central theme of the book is in the experimental and simulation based analysis of building performance, elegantly wedged between the fundamental concepts of performance and their operationalisation in specific case settings.

Students, developers and scholars in the field of building performance simulation, design management, performance-based design and rationalisation of building design will find this book useful. And although the ultimate solution for the purely rational design dialogue that I have been chasing remains elusive, this book provides a new and essential stepping stone towards it.

Professor Godfried Augenbroe
High Performance Buildings Lab
Georgia Institute of Technology
Atlanta, GA, USA
July 2017

Preface

Building performance is a concept that is used throughout industry, government and academia. It plays an important role in the design of new buildings, the management and refurbishment of the existing stock and decisions about the built environment in general. Yet there is no clear definition of building performance or unifying theory on building performance analysis available in the literature.

This book is an attempt to fill this void and to answer the following key questions:

1) What is building performance?
2) How can building performance be measured and analysed?
3) How does the analysis of building performance guide the improvement of buildings?
4) What can the building domain learn from the way performance is handled in other disciplines?

In order to answer these questions, the book brings together the existent body of knowledge on the subject. It combines findings from a large number of publications on aspects of building performance that all contribute in different ways. The book tries to unify this previous work, establishing a range of observations that underpin an emergent theory of building performance and building performance analysis. At the same time, the material makes it clear that there still is significant work to do: the theory does not reach beyond a conceptual framework. Operational building performance analysis still requires deep expertise by those carrying out the analysis, and existing tools and instruments only support part of the work. A design methodology that truly ensures performance of a building according to predefined criteria still remains to be developed.

In providing a working definition and emergent theory of building performance analysis, the book caters primarily to the building science community, both from industry and academia. It aims to support the many efforts to build better buildings, run more efficient design processes and develop new tools and instruments. The book will benefit senior undergraduate and graduate students, scholars as well as professionals in industry, business and government. Students engaging with this material will typically be those that are taking a course at MSc level in one of the many directions in architecture and building engineering, such as building performance modelling, environmental building design and engineering, high performance buildings, intelligent/healthy/low-carbon/sustainable buildings, building science and technology or building services engineering. While the text is intended to be self-contained, it will be helpful if such readers have developed a solid appreciation of building technology and the construction process, as well as building science. It will also be beneficial if students have been introduced to building simulation and physical experimentation. Research students and academics will have their own specific research interests but will benefit from a unified theory upon which to base their efforts. Extensive references are provided so that these readers can connect to the underlying foundations. It is hoped that professionals can use this material to reflect on the current way of handling performance in the field and that they will help to implement some of the ideas of this book in practice.

The book is structured in three parts. Part I provides a theoretical foundation for building performance. Part II deals with operational performance analysis, providing a conceptual frame that shows what deliberations and decisions are required to carry out an analysis and what tools and methods are available to help. Part III discusses how this analysis can impact on building practice. The book closes with an epilogue that presents an emerging theory of building performance analysis. A study of the complete book allows the reader to follow the underlying thought process and how it connects the many contributions that already have been made to aspects of the field. However, readers who prefer to start with getting an understanding of the emergent theory, or want to test their own ideas against this, may start by reading the final chapter and then explore the underpinning material as required. Non-linear readers may start at any chapter of interest. The main chapters all include a case study that demonstrates the complexity of building performance analysis in real practice; these cases are intended as challenge for readers to reflect on applicability of the emergent theory. Each chapter also includes six activities that encourage engagement with the material; these have been designed to be 'real-world' problems without a right model answer but instead should provide a basis for deep discussion within groups or teams. Key references are included in the references at the end of each chapter; a complete list and secondary references are provided at the end of the text.

This book is written to encourage dialogue about an emergent theory of building performance and its analysis. A website is maintained at www.bldg-perf.org to support communication on the subject.

Acknowledgements

This book is the result of more than twenty years of research in and around the area of building performance. In these two decades, many people have influenced my thinking about the subject. By necessity, not all of them can be listed, so these acknowledgements only name those who had a pivotal role in the emergence of this work.

I was introduced to building performance simulation during my studies at the TU Delft, starting with my graduate work in 1994 and continuing on this subject during my PhD project. My supervisor at that time, Marinus van der Voorden, thus laid the foundations of this effort. During my time as postdoc on the Design Analysis Integration (DAI) Initiative at GeorgiaTech, Fried Augenbroe provided deeper insights and guidance. My involvement in DAI also established invaluable connections with Cheol-Soo Park, Ruchi Choudhary, Ardeshir Mahdavi and Ali Malkawi, who influenced my subsequent career. My years with Dick van Dijk and the other colleagues at TNO Building and Construction Research had a stronger emphasis on industrial application and physical experimentation, giving me a more balanced perspective on the interaction between academia and practice. At the University of Plymouth, Steve Goodhew and colleagues expanded my view in a yet another direction, emphasizing the actual construction process and importance of the existing building stock. Yaqub Rafiq introduced me to genetic algorithms. Derek Prickett became a trusted voice on the practical aspects of building services engineering. Wei Tian, postdoc on my EPSRC project on the management of the impact of climate change on building performance, introduced me to parallel computing and the handling of large search spaces and the application of sensitivity analysis to make sense of the results. Darren Pearson and his colleagues at C3Resources gave me an appreciation of the worlds of monitoring and targeting, automated meter reading and measurements and verification; Carlos Martinez-Ortiz, the KTP associate on our joint project, introduced me to machine learning approaches. Sabine Pahl and other colleagues in the EPSRC eViz project not only provided me with a deeper understanding of the role of occupant behaviour in building performance but also made me realise that building performance analysis is a separate discipline that needs its own voice. My Royal Academy of Engineering fellowship brought me back to GeorgiaTech in order to learn more about uncertainty analysis; the discussions with Yuming Sun on the energy performance gap also helped shape my thinking. My work at Plymouth with my postdocs and students, notably Rory Jones, Shen Wei, Jim Carfrae, Emma Heffernan, Matthew Fox, Helen Garmston, Alberto Beltrami, Tatiana Alves, João Ulrich de Alencastro and Omar Al-Hafith, helped me see some of the complexities of building performance and advance my thoughts on the subject. The colleagues within the International Building Performance Simulation Association (IBPSA) have provided an excellent frame of reference ever since my first IBPSA conference in 1997; over the years some of them like Chip Barnaby, Malcolm Cook, Dejan Mumovic and Neveen Hamza have become trusted friends and references for my efforts. The same goes for the colleagues such as Ian Smith, André Borrmann, Timo Hartmann and Georg Suter that are active within the European Group for Intelligent in Computing in Engineering (EG-ICE) and for those active within the Chartered Institution of Building Services Engineers (CIBSE).

The specific idea to write this book on the subject of building performance analysis crystallised in October 2014 during a visit of my long-term mentor Fried Augenbroe, on the basis of a casual remark as we were driving to Bristol airport. Further momentum was gained a month later from a discussion with Ruchi Choudhary about real contributions to the field of building simulation during a visit to Cambridge University, leading to the actual start on this manuscript. My special thanks to both of them for setting me off on this journey of discovery. Thanks are also due for many people who provided input on elements of the text and helped with images, such as Joe Clarke, Wim Gielingh, Nighat Johnson-Amin, Gayle Mault and Ioannis Rizos.

Achieving the current form of the book has been helped by efforts from a group of trusted friends who proofread the material; this included Fried Augenbroe, Cheol-Soo Park, Georg Suter and Wei Tian. Feedback on parts was also obtained by MSc students at both Georgia Tech and Plymouth, which helped me to develop the material. Any remaining misconceptions and errors are my own responsibility. Further thanks go to Paul Sayer and the team at Wiley who managed the production of the work.

Finally, I would like to thank Anke, Rick and Tom for tolerating the long hours that were invested to realise this book. Without your love, support and endurance, this work could not have been completed.

Pieter de Wilde, Tavistock, UK
pieter@bldg-perf.org
www.bldg-perf.org

Endorsements

Many disciplines are concerned with aspects of building performance and its analysis. Surprisingly, little work exists that presents a comprehensive and systematic overview of this diverse and growing field. This timely book by Pieter de Wilde, a leading researcher and practitioner of building performance analysis, thus fills a significant gap. The book guides readers through a wide range of topics from theoretical foundations to practical applications. Key concepts, such as performance attributes, performance targets or performance banding, are introduced, as are the methods to measure and evaluate building performance. Topics of both scientific and practical relevance, including decision making under uncertainty or data collection and analysis for improved building operation and control, are reviewed and discussed. Readers will appreciate the comprehensive coverage of relevant research and standards literature, which makes the book particularly valuable as a reference. In summary, this book is highly recommended reading for both novices and experts who are interested in or want to learn more about building performance analysis.

Georg Suter
Vienna University of Technology, Austria

It sometimes is a challenge to write a book to describe the things we always talk about. Dr. de Wilde deals with the important topic of 'building performance'. This sounds easy, but actually the subject is very complex. Yet we must define the meaning of building performance before designing and constructing green buildings, low-carbon buildings or high performance buildings. After a thorough review of state-of-art research on building performance, this book presents an 'emergent theory' of building performance analysis. This book will play an important role in a deeper exploration of this fundamental topic.

Wei Tian
Tianjin University of Science and Technology, China

Over the last two decades, I have been involved in simulation studies of more than 20 existing buildings in the United States and South Korea, analysing the performance of double skins, HVAC systems (such as the example briefly introduced in Chapter 6 of this book), occupant behaviour, machine learning models for building systems and many others. However, it has never been easy to unambiguously quantify building performance of these cases. For example, how can we 'objectively' quantify the energy/daylighting/lighting/thermal comfort performance of a double skin system under different orientations and changing indoor and outdoor conditions? The performance of this double skin is dependent on design variables (height, width, depth, glazing type, blind type), controls (angle of blind slats, opening ratio of ventilation dampers usually located at the top and bottom of the double skin), occupant behaviour (lights on/off, windows open/closed), HVAC mode (cooling/heating) and so on. As this example shows, objective performance quantification of a double skin is not an easy task. Moreover, so far there is no established theory or set of principles to help us direct the analysis of building performance at different building and system scales. The general way we presently describe building performance is at best a 'relative' comparison to a baseline case. This book by Professor de Wilde attempts to fill this void and presents an emergent theory of building performance analysis. I have observed for several years how Professor de Wilde has worked hard to complete this invaluable book. I firmly believe that it will contribute as a foundation stone to the area of building performance studies and will support efforts in this field for many years to come.

Cheol Soo Park
Seoul National University, South Korea

At last, a book that answers the question 'what is building performance?' not by theory alone, but through analytics and impacts on building practice. Pieter de Wilde has crafted a comprehensive compilation of what building performance truly means – from its place in the building life cycle and its relationship to stakeholders – through systems, technologies and the unpredictable occupants who often have the most influence on how buildings perform. The book goes beyond the merely theoretical by demonstrating the analytics, tools and instruments needed to evaluate building performance in practice. The case studies are relevant and specific to the system or technology but also to the appropriate part of the building life cycle. By the end, Pieter de Wilde ties it all together through life cycle phase specific theories for evaluating building performance – design, operation and research. Well written, insightful and a pleasure to read.

Dru Crawley
Bentley Systems, USA

This is a long awaited primer for those studying performance, simulation and analysis of buildings. As a subject, building performance analysis borrows from a wide variety of viewpoints and disciplines. This book takes on the difficult task of consolidating these together and goes a step further in articulating the particular nuances of building performance. It is the first book on building performance that goes beyond current trends in research and instead reflects on its foundations, remit and reach. The book is sure to become an essential read for graduate students wanting to grasp the breadth of the subject and its roots. The clearly identified reading list and scenario exercises (activities) at the end of each chapter are fantastic; they help the reader go beyond the text and are particularly valuable for generating discussion sessions for graduate courses.

Ruchi Choudhary
University of Cambridge, UK

1

Introduction

Modern society is strongly focussed on performance and efficiency. There is a constant drive to make production processes, machines and human activities better, and concepts like high performance computing, job performance and economic performance are of great interest to the relevant stakeholders. This also applies to the built environment, where building performance has grown to be a key topic across the sector. However, the concept of building performance is a complex one and subject to various interpretations. The dictionary provides two meanings for the word performance. In technical terms, it is 'the action or process of performing a task or function'. It may also mean the 'act of presenting a play, concert, or other form of entertainment' (Oxford Dictionary, 2010). Both interpretations are used in the building discipline; the technical one is prevalent in building engineering, while the other one frequently appears in relation to architecture and buildings as work of art (Kolarevic and Malkawi, 2005: 3). But the issue goes much deeper. As observed by Rahim (2005: 179), 'technical articles of research tend to use the term "performance" but rarely define its meaning'. In the humanities, performance is a concept that implies dynamic, complex processes with changing values, meanings and structures (Kolarevic, 2005b: 205).

Whether approaching building performance from a technological or aesthetic perspective, buildings are complex systems. Typically they consist of a structure, envelope, infill and building services. Many of these are systems in their own right, making a building a 'system of systems'. All of these work together to ensure that the building performs a whole range of functions, like withstanding structural loads caused by people and furniture, protecting the occupants from environmental conditions, allowing safe evacuation in case of emergency, delivering a return on investment or making an architectural statement. Building performance thus is a central concept in ensuring that buildings meet the requirements for which they are built and that they are fit for purpose. Building performance plays a role in all stages of the building life cycle, from developing the building brief[1] to design and engineering, construction, commissioning, operation, renovation and ultimately deconstruction and disposal.

Different disciplines contribute knowledge on specific performance aspects of buildings, such as architectural design, mechanical engineering, structural engineering

---

1 In the United Kingdom the term *briefing* is used, whereas in the United States this is named *architectural programming*.

Building Performance Analysis, First Edition. Pieter de Wilde.
© 2018 John Wiley & Sons Ltd. Published 2018 by John Wiley & Sons Ltd.

and building science.[2] Other disciplines focus on specific systems, such as building services engineering or facade engineering, or are grounded in a common method, such as building performance simulation or the digital arts; in many cases disciplines overlap. The knowledge of all these disciplines needs to be combined into a building design, a building as a product and ultimately an asset in operation, which adds further complexities of interdisciplinarity, information exchange, management and control.

Building performance is a dynamic concept. The architectural performance depends on the interplay between the observer, building and context. The technical performance relates to how a building responds to an external excitation such as structural loading, the local weather to which the building is exposed and how the building is used. This often introduces uncertainties when predicting performance. Furthermore building performance needs to materialize within the constraints of limited and often diminishing resources such as material, energy and money. Challenges such as the energy crisis of the 1970s, the concern about climate change and the 2008 global financial crisis all contribute to increasingly stringent targets and a drive towards more efficient buildings and a growing interest in building performance.

Within this context, a large body of literature exists on building performance. Underlying principles are provided by generic books like, amongst many others, Clifford *et al.* (2009) in their introduction to mechanical engineering, Incropera *et al.* (2007) on fundamentals of heat and mass transfer, Stroud and Booth (2007) on engineering mathematics, Zeigler *et al.* (2000) on theory of modelling and simulation or Basmadjian (2003) on the mathematical modelling of physical systems. The application of these principles to buildings and to the assessment of building performance can be found in more specialist works such as Clarke (2001) on energy modelling in building design, Underwood and Yik (2004) on energy modelling methods used in simulation, Hensen and Lamberts (2011) on building performance simulation in design and operation and Mumovic and Santamouris (2009) on their integrated approach to energy, health and operational performance. Architectural performance arguably is covered by Kolarevic and Malkawi (2005) in their work on performative architecture. This is complemented by countless articles in peer-reviewed archived journals such as *Building and Environment, Automation in Construction, Energy and Buildings, Advanced Engineering Informatics, Architectural Science Review*, the *Journal of Building Performance Simulation, Building Research and Information* and *Design Studies*. Building performance is also a day-to-day concern in the construction industry and is of central importance to building legislation.

With the complexity of buildings, the many functions they perform and the multitude of disciplines and sciences involved, there are many different viewpoints and interpretations of performance. The many stakeholders in building, such as architects, contractors, owners and tenants, all view it from a different position. Even in academia, different research interests lead to distinct schools of thought on performance. An example is the work by Preiser and Vischer (2005), who provide a worthwhile contribution on building performance assessment from the point of view of post-occupancy evaluation, yet do not really connect to the aforementioned building performance modelling and

---

2 This discipline is typically named building science in the Anglo-Saxon countries, but *building physics* in continental Europe.

simulation domain. This lack of common understanding is problematic as it hinders the integration that is needed across the disciplines involved. It impedes the use of modelling and simulation in the design process or the learning from measurement and user evaluation in practice, since it makes it hard to sell services in these fields to building clients and occupants. The absence of a common understanding also means that building science and scholarship do not have a strong foundation for further progress and that the design and engineering sectors of the building sector are seen to lack credibility.

The discussion about building performance is further complicated by some intrinsic properties of the building sector. Some may consider building to be a straightforward, simple process that makes use of well-tested products and methods like bricks, timber and concrete that have been around for a long time and where lay people can do work themselves after visiting the local builders market or DIY[3] centre; however this risks overlooking some serious complexity issues. Architectural diversity, responding to individualist culture, renders most buildings to be different from others and makes the number of prototypes or one-off products extremely large in comparison with other sectors such as the automotive, aerospace and ICT industries (Foliente, 2005a: 95). Typically, buildings are not produced in series; almost all buildings are individual, custom-built projects, and even series of homes built to the same specification at best reach a couple of hundred units. This in turn has implications for the design cost per unit, the production process that can only be optimized to a certain extent and, ultimately, building performance. With small series, the construction sector has only limited prospects for the use of prototypes or the use of the typical Plan-Do-Study-Act[4] improvement cycles that are used in other manufacturing industries. Quality control programmes, modularization with standard connectors, construction of components in automated factories and other approaches used in for instance the automotive or electronic system industries are thus not easily transferred to construction as suggested by some authors such as Capehart *et al.* (2004) or Tuohy and Murphy (2015). Buildings are also complex in that they do not have a single dominant technology. While for instance most automobiles employ a metal structure, building structures can be made from in situ cast concrete, prefabricated concrete, timber or steel or a combination of these; similar observations can be made for the building shell, infill and services. Furthermore the construction industry is typically made up of many small companies who collaborate on an ad hoc basis, with continuous changes in team composition and communication patterns, which are all challenges for the dialogue about building performance. Of all products, buildings also are amongst those that undergo the most profound changes throughout their life; while changing the engine of a car normally is not economically viable, it is common practice to replace the heating system in a building, to retrofit the façade or even to redesign the whole building layout, with profound consequences on the building performance (Eastman, 1999: 27–30). Once buildings exhibit performance faults, these are often hard to rectify; there is no option of a product recall on the full building scale. Moreover, buildings, because of their fixed position in space, are not comparable with other products in terms of procurement strategies; for

---

3 Do It Yourself.
4 Sometimes named Deming cycle or circle, or Shewhart cycle.

instance, the decision on the purchase of a building also relates to facilities in the vicinity, not just the building itself. The supply chain of buildings also is different, with the clients who start building processes often selling the product on to other end users (Foliente, 2005a: 95–96).

Yet another complication arises from shifting approaches to performance measurement, driven by the rapid developments in the ICT sector. In the past, measurement of the performance of buildings was an expensive issue, requiring the installation of expensive specialist equipment. Computational assessment of building performance typically took place in a different arena, detached from the world of direct observation. However, the digital age has meant huge reductions in the cost of sensors; wireless technology reduces the need to put intrusive cabling into buildings, and increases in memory size make it easy to harvest data at high frequencies. As more data on building performance is harvested, it becomes obvious that performance predictions and measurement do not always agree, leading to phenomena like the 'energy performance gap' (Carbon Trust, 2011; Menezes *et al.*, 2012; CIBSE, 2013; Wilson, 2013; de Wilde, 2014; Fedoruk *et al.*, 2015; van Dronkelaar *et al.*, 2016). Some believe that the main reason for this energy performance gap is a lack of accounting for all energy use in a building such as ICT systems, plug loads, special functions and others (CIBSE, 2013). Others see issues with software, software users, building, commissioning, maintenance and recording (Wilson, 2013). Yet others hold that a key to improvement is a better understanding and representation of the energy-related occupant behaviour in buildings (Duarte *et al.*, 2015; Ahn *et al.*, 2016; IEA, 2016b). To bridge this gap, it seems obvious that some of the prediction and analysis tools used in the sector need to be revisited in depth (Sun, 2014). However, the different views of building performance also compound the debate and need to be addressed if prediction and direct observation are to become aligned. A common understanding of building performance is also a prerequisite to make sense of the large amount of data collected from buildings and to drive new analysis and management processes.

In spite of the interest of many in building performance and its importance in what clearly is a complex context, building performance remains so far a rather evasive concept. While the term building performance is used regularly in literature, there is a paucity of text that actually defines what it is; in most cases the meaning is left implicit. The generic concept of performance is far from limited to the building domain. Yet literature on the subject of building performance seems mostly restricted to discussions within the discipline, with only few authors looking towards other sectors. With further integration through concepts like machine-to-machine communication and the 'Internet of Things', it is important to bring the concept of building performance in line with the approaches in the other fields.

From an architectural stance, building design can be considered as the combination of three types of integration: physical, visual and performance integration. Here physical integration relates to the need for building components to connect and share space. Visual integration is combining the components in a way that creates the buildings' shared image. Performance integration then deals with sharing functions (Bachman, 2003: 4). In this structure, building performance can also be seen as a guiding design principle in architecture, similar to form making. In this context building performance covers a wide domain – from spatial, social and cultural to structural, thermal and other technical aspects (Kolarevic and Malkawi, 2005: 3).

The International Council for Research and Innovation in Building and Construction (CIB),[5] taking a technical view, defined the 'Performance Approach' to building as 'working in terms of ends rather than means'. Here 'ends' relates to desired technical attributes of a building such as safety or structural stability of load-bearing capacity; 'means' are actual systems and solutions. The CIB definition was originally positioned in the context of building legislation and how to define performance in building regulations (Bakens *et al.*, 2005). However, with the passing of time, many regulations are now performance based, and this definition has thus lost in importance and urgency; moreover a lot of the earlier fundamental thinking by CIB in the 1980s seems to be lost to the performance discourse. In the domain of standards, ISO 6241 (1984: 2) on 'the principles of performance standards in building' simply equals performance to 'the behaviour (of a product) related to use'.

Even so, only very few authors actually define building performance:

- Williams (2006: 435) notes that building performance is a complex issue. Listing a range of items that buildings need to accommodate (people, equipment, processes, places, spaces, image, convenience, comfort, support systems, costs, income, profitability), he then defines building performance as 'the contribution made by a building to the functional and financial requirements of the occupiers and/or owners and the associated physical and financial characteristics of the fabric, services and finishes over time'. Williams identifies three key facets of building performance: physical performance, functional performance and financial performance.
- Almeida *et al.* (2010) define building performance as the behaviour of buildings as a product related to their use; they note that performance can also be applied to the construction process (for instance, interaction between parties) and services (such as the performance of an asset in support of business).
- Corry *et al.* (2014) define building performance as 'delivering functional intent of each zone in the building while accounting for the energy and cost of delivering this functional intent'.
- An interesting view of looking at building performance is provided by Foliente *et al.* (1998: 16), who draw the attention to the opposite of performance: non-performance, which they define as the failure of meeting a specified performance level.

Key figures in the domain mostly leave the concept undefined. Clarke (2001: ix–x) emphasizes the complexity of buildings and the large search spaces required for analysis, as well as the different interacting physical domains, and then focusses on the benefits of building simulation and how this can be integrated into the design process. Preiser and Vischer (2005: 6) do not directly define building performance but list the priorities of building performance as health, safety, security, function, efficiency, work flow, psychological, social and culture/aesthetic. They also note the interplay between performance and the scale of any performance evaluation and the relation to occupants (individuals, groups or organizations). Hensen and Lamberts (2011: 1–14) build up the need for models and tools from a discussion of sustainability challenges, user

---

5 CIB is an abbreviation of the French version of the name, 'Conseil International du Bâtiment'.

requirements and the need for robust solutions; they mention high performance and eco-buildings, but do not define building performance. In terms of building performance simulation tools, they emphasize that these are multidisciplinary, problem oriented and wide in scope. Augenbroe, arguably a leading thinker on the role of simulation in performance-based building, approaches performance as central to a stakeholder dialogue and dissects that discussion into an interplay between building functions, performance requirements, performance indicators, quantification methods and system attributes (Augenbroe, 2011).

It is also interesting to note the position of some international organizations on building performance:

- The International Building Performance Simulation Association (IBPSA, 2015) has as its mission 'to advance and promote the science of building performance simulation in order to improve the design, construction, operation and maintenance of new and existing buildings worldwide'. IBPSA's vision statement mentions the need to address performance-related concerns, to identify problems within the built environment and to identify the performance characteristics on which simulation should focus, yet it does not provide a definition of building performance.
- The American Society of Heating, Refrigerating and Air-Conditioning Engineers (ASHRAE, 2015) provides annual handbooks that are a key reference in this area. Yet their composite index across the handbook series, which does mention many topical areas such as building information modelling (BIM), performance contracting and performance monitoring, does not have an entry on building performance.
- The Chartered Institution of Building Services Engineers (CIBSE, 2015a) publishes the *CIBSE Guide A: Environmental Design* (CIBSE, 2015b). This opens with a section on quality in environmental design, which discusses key criteria such as thermal, visual and acoustic comfort, health, energy efficiency and greenhouse gas emissions. By focussing on quality assurance in buildings, this guide sidesteps the definition of building performance; however, the guide goes on to define legislation including the Energy Performance of Buildings regulations and discusses performance assessment methods (PAMs) as a key approach to select appropriate calculation methods to assess quality.

Standards typically address only aspects of the overall building performance, yet can provide interesting indirect insights. For instance, BS EN ISO 50001 (2011: 3) defines energy performance as 'measurable results related to energy efficiency, energy use and energy consumption'. It notes that these measurable results can be reviewed against policy, objectives, targets and other energy performance requirements.

Williams (2006: 435) and Cook (2007: 1–5) associate building performance with building quality. However, Almeida *et al.* (2010) note that 'quality' is a systems attribute that is hard to define; it is often taken to mean the absence of defects. It is related to a range of theories and approaches such as quality control, quality assurance, quality management, quality certification and others. Gann *et al.* (2003) agree, stating that 'design quality is hard to quantify as it consists of both objective and subjective components. Whilst some indicators of design can be measured objectively, others result in intangible assets'. Other authors, such as Loftness *et al.* (2005), use the term 'design excellence' rather than performance or quality.

Not having a proper definition of building performance also leads to misunderstanding, fuzzy constructs and overly complex software systems. This is especially the case where building performance is used in the context of a wide view of building sustainability, in the difficult context of building design or as part of larger ICT systems; see for instance Bluyssen (2010), Todorovic and Kim (2012), Becker (2008), Geyer (2012) or Dibley *et al.* (2011). Some authors such as Shen *et al.* (2010) promise systems such as 'fully integrated and automated technology' (FIATECH), which is based on a workflow that includes automated design in response to user requirements, followed by automated procurement, intelligent construction and ultimately delivering intelligent, self-maintaining and repairing facilities; clearly such systems are a good long-term goal to drive developments but require a deeper understanding of performance to become feasible. This has lead to a situation where the building industry is sceptical of the work in academia and prefers to move at its own pace and develop its own guidelines, standards and systems. This situation where building performance is, by and large, an undefined concept in both building practice and industry, and where the term is used without a clear frame of reference and common understanding, needs addressing. A clear definition and theoretical framework will strengthen the position of that part of the building sector that provides services, products and buildings in which performance is important; it will also provide a foundation to move scholarship in this area to a next level.

The purpose of this book is to explore and bring together the existent body of knowledge on building performance analysis. In doing so, it will develop a definition of building performance and an in-depth discussion of the role building performance plays throughout the building life cycle. It will explore the perspectives of various stakeholders, the functions of buildings, performance requirements, performance quantification (both predicted and measured), criteria for success and performance analysis. It will also look at the application of the concept of building performance in building design, building operation and management and high performance buildings. The following key questions drive the discussion:

1) What is building performance?
2) How can building performance be measured and analyzed?
3) How does the analysis of building performance guide the improvement of buildings?
4) What can the building domain learn from the way performance is handled in other disciplines?

In answering these questions, the book will develop a theoretical framework for building performance analysis.

1.1 Building Performance: Framing, Key Terms and Definition

Performance is of interest to many disciplines, such as engineering, computer science, sports and management. As noted by Neely (2005), some of the most cited authors in performance measurement come from rather different disciplines, such as accounting, information systems, operations research and operations management. Consequently there is a wide range of literature dealing with context-specific applications of the term such as structural performance, algorithm performance, athletic performance and financial performance. While a full coverage of the performance concept across all fields is impossible, the following gives an overview of some of the interests and approaches from outside the architecture, engineering and construction (AEC) sector, thus providing context and a wider frame of reference for the discussion of building performance:

- In *electronics*, performance typically relates to a system (for instance, a smartphone) or the components of a system (for instance, a transistor). In general the main performance targets are 'better' and 'cheaper'. Within devices, electronic engineers talk of analogue and digital performance of components (Guo and Silva, 2008).
- In *human resources management*, academic and job performance of individuals are key. This is typically measured across a range of factors such as verbal, numerical and spatial abilities, as well as knowledge, personality traits and interests (Kanfer *et al.*, 2010). However, team performance depends on the interaction between tasks, team composition and individual performance. Tasks typically have two key dimensions: speed and accuracy. Deep studies are undertaken to explore the role of incentives to make teams work faster and smarter, with tension between competitive and cooperative reward structures (Beersma *et al.*, 2003).
- In *organizations*, organizational performance is related to the workflow, structures and roles and skills and knowledge of the agents of the organization (Popova and Sharpanskykh, 2010).
- In *manufacturing*, the drive towards higher efficiency leads to more measurement, control and process improvement. Key aspects are the identification of key performance indicators and benchmarks (measurement) and monitoring, control and evaluation. An important enabler to achieve higher efficiency is ICT, which can lead to better process execution, resource planning, intelligent control and advanced scheduling. Standardization is another key enabler for better manufacturing performance (Bunse *et al.*, 2011).
- In the *medical sector*, performance of healthcare is typically measured by means of health and quality of life questionnaires, physical and psychological tests, costs and duration of treatment (van der Geer *et al.*, 2009). In healthcare it has also been noted that if performance is reviewed to steer the actions of employees, it is important that these employees have control of the performance variation and can manage the relation between actions and outcomes (*ibid.*).
- In the *performing arts*, the performance of for instance musicians is known to be related to various human tasks such as listening, reading and playing (Sergent *et al.*, 1992).
- In *social science*, measurements are undertaken to compare the economic, social and environmental performance of countries. Here the indicators used are for instance the Human Development Index (HDI), which takes into account the gross domestic product, life expectancy at birth and adult literacy rate. Other indicators have a more detailed view and might include such aspects as income inequality, carbon emissions or gender bias (Cracolici *et al.*, 2010).
- In *sports*, performance analysis is concerned with recording, processing and interpreting events that take place during training and competition. It covers

technical, tactical and behavioural aspects of both individuals and teams (Drust, 2010). Performance analysis in sport is considered to be a difficult undertaking, covering biomechanics, notational analysis (which covers movement patterns, strategy and tactics), motor control and human behaviour, so that one-dimensional analysis of raw data can easily lead to misunderstanding (Hughes and Bartlett, 2010).

- In the *tourism* sector, different offers are compared using Tourism Destination Competitiveness (TDC) studies. TDC looks at different aspects of competitiveness, but while it uses exhaustive lists of indicators, there is still some concern about completeness. One way to develop TDC is to review it by means of Importance–Performance Analysis (IPA), which basically positions efforts in four quadrants along an axis of importance and competitiveness, thus allowing to define where resources need to be sustained, increased, curtailed or remain unchanged (Azzopardi and Nash, 2013). Taplin (2012) gives a good example of application of IPA as applied to a wildlife park.
- In *transport and logistics*, management uses key performance indicators to measure and improve the overall process; the usual objectives are to decrease cost and to improve efficiency and effectiveness (Woxenius, 2012).

With all these different disciplines taking their own approach to performance, there clearly is a need to establish a clear definition of key terms. The following section reviews terminology that sets the scene for an initial definition of building performance at the end of this paragraph.

As mentioned in the introduction, the word performance has two meanings: in technical terms, it is 'the action or process of performing a task or function' and in aesthetic terms it is the 'act of presenting a play, concert, or other form of entertainment'. Within the technical interpretation, performance can be taken to relate to an object, such as a building, car or computer; alternatively it can relate to a process, such as manufacturing or data transmission. Within the literature, two generic disciplines cover these areas: systems engineering and process management. Systems Engineering is broadly defined as 'An interdisciplinary approach and means to enable the realization of successful systems. It focuses on defining customer needs and required functionality early in the development cycle, documenting requirements, then proceeding with design synthesis and system validation while considering the complete [design] problem' (INCOSE, 2016).

The area of (Business) Process Management is defined as 'A disciplined approach to identify, execute, measure, monitor, and control both automated and non-automated business process to achieve consistent, targeted results aligned with an organization's strategic goals' (ABPMP, 2015).

It must be noted that the relation is not one to one: systems engineering is concerned not only with systems but also with the process of creating and managing these systems, whereas process management also relates to the product/outcome of the process.

A system can be defined as a set of interacting elements that, together, accomplish a defined objective. The elements may include products, processes, people, information, facilities and others (INCOSE, 2015: 5). Systems exhibit behaviour, properties and functions, which are characterized by emergence and complexity. Most systems interact with other systems and their environment (SEBoK, 2014: 65). In a slightly different wording, systems consist of *components*, *attributes* and *relationships*. Here components are the operating parts of the system, attributes are properties of the components, and relationships are the links between components and attributes (Blanchard and Fabrycky, 2011: 17). Systems normally sit in a hierarchy; the components that make up a system can be named a subsystem. The designation of system, subsystem and component is relative; a reason for defining systems is to understand and handle complexity. Similarly, there are different classifications of systems, such as natural and human made, physical and conceptual, static and dynamic or closed and open (*ibid*.). Thinking in systems helps scientists, engineers and designers to think about the world by defining categories, guiding observation and measurement and supporting the development of models and generic laws (Weinberg, 1975: ix–xii).

There are many reasons for analyzing the performance of systems. On a high level, these include an interest in for instance (Montgomery, 2013: 14–15):

1) Factor screening or characterization – to find out which factors have most impact on the performance.
2) Optimization – to find the parameter values and system configurations that result in the sought performance.
3) Confirmation – to verify that a system performs as is expected.
4) Discovery – to establish the performance of new systems, combinations and so on.
5) Robustness – to study how system performance changes in adverse conditions.

In the context of systems engineering, performance is defined as a 'quantitative measure characterizing a physical or functional attribute relating to the execution of a process, function, activity or task. Performance attributes include quantity (how many or how much), quality (how well), timeliness (how responsive, how frequent), and readiness (when, under which circumstances)' (INCOSE, 2015: 264).

In different words, performance is an attribute of a system that describes 'how good' a system is at performing its functional requirements, in a way that can be measured (Gilb, 2005: 382). Gilb gives a slightly different classification of performance types, discerning quality (*how well* a system performs its functions), resource saving (*how much resource* is saved in relation to an alternative system) and workload capacity (*how much work* a system can do). Performance relates not only to the physical design of a system but also to the particular use of a system. As exemplified by Hazelrigg (2012: 301), 'the performance parameters such as acceleration and top speed of a car depend on its physical design. However, another performance parameter might be the lifetime of the engine. This will depend on the maintenance of the engine, such as the frequency of oil changes, the conditions under which the vehicle is driven, and manner in which it is driven. These items are a function of the use of the product, not of its physical design'.

As a consequence, performance requirements should include a description of the conditions under which a function or task is to be performed (SEBoK, 2014: 292).

A function of a system is a 'characteristic task, action or activity that must be performed to achieve a desired outcome' (INCOSE, 2015: 190). There are two kinds of functions: (i) functions that relate to the requirements the system has to meet and therefore relate to an 'outer environment' and (ii) functions that are intertwined with the actual design of the system; these relate to an 'inner environment' and are partly a consequence of design choices. As stated by Simon, 'The peculiar properties of the artifact lie on the thin interface between the natural laws within it and the natural laws without …. The artificial world is centered precisely on this interface between inner and outer environments; it is concerned with attaining goals by adapting the former to the latter' (Simon, 1996: 113).

In order to analyze performance, 'how well' a system meets the functional requirements, one needs to compare the measured performance with clear criteria. Different words are used in this context, such as goal, target and objective. The Systems Engineering Body of Knowledge defines a goal as 'a specific outcome which a system can achieve in a specified time' and an objective as 'a longer term outcome which can be achieved through a series of goals'; this can be extended with the concept of an ideal, which is 'an objective which cannot be achieved with any certainty, but for which progress towards the objective has value' (SEBoK, 2014: 115). A target can be defined as a performance requirement defined by the stakeholder, which is to be delivered under specified conditions (Gilb, 2005: 430). In most cases there are multiple criteria, and often these criteria conflict, resulting in a need for trade-off decisions (SEBoK, 2014: 414). Augenbroe (2011: 16) considers the notion of a criterion to be central to the whole process of performance analysis: a criterion is closely interrelated with the experiment that is required, the tool(s) that must be used and the way in which data is collected and aggregated into a performance statement while also defining what is required.

The concept of measurement is crucial to performance analysis of systems. Measurement is the process that collects, analyzes and reports data about products developed or processes implemented; this allows the demonstration of the quality of these products and the effective management of these processes (INCOSE, 2015: 130). Measurement is often governed by industry standards and policies and sometimes by laws and regulations. Data analysis and reporting typically includes verification, normalization and aggregation activities, as well as the comparison of actual data against targets (SEBoK, 2014: 406).

Analysis can be encountered at different stages of a project; different categories of analysis are estimation analysis, feasibility analysis and performance analysis. Estimation analysis is carried out during the initial planning stage and is based on projections to establish objectives and targets. Feasibility analysis aims to establish the likelihood of achieving objectives and targets; it provides confidence in assumptions and ensures that objectives are reasonable. It might also include a check with past performance of similar projects and technologies. Finally, performance analysis is carried out during development and operation in order to check whether objectives and targets are being met (INCOSE, 2005: 42–43).

On a fundamental level, the analysis of building performance can be approached through four routes:

1) Physical testing, either in laboratory conditions or under 'live' conditions.
2) Calculation, mostly in the form of computer simulation.
3) Expert judgment, depending on the insights of professionals.
4) Stakeholder assessment, capitalizing on the insights of occupants who know a specific building best.

It is interesting to note that ISO 7162 (1992), still actual on content and format of standards for performance evaluation of buildings, only mentions categories 1–3, but excludes category 4.

Quantification of performance is useful, but when doing so it is important to remember the context and not to get blinded by numbers. As phrased by Cameron (1963),[6] 'not everything that counts can be counted, and not everything that can be counted counts'. In some areas of management and policy, making quantifications sometimes becomes obsessive, leading some to comment that measurement and regulation are leading to an 'audit society' (Neely, 2005).

Traditionally, construction management has focussed on the key factors of cost, time and quality, sometimes named the 'iron triangle' where trade-off between these three factors is required (Atkinson, 1999) and where poor performance leads to time delays, cost overruns and quality defects (Meng, 2012). Recent work indicates that the emphasis in construction management is now shifting to a wider range of issues such as safety, efficient use of resources and stakeholder satisfaction (Toor and Ogunlana, 2010) and specific studies are taking these individual issues further – see for instance Cheng *et al.* (2012) on the interaction of project performance and safety management or Yuan (2012) on waste management in the social context of construction.

In the arts, the word performance mainly appears in the context of the performing arts such as dance, theatre and music. Here a key aspect is the involvement of artists who use their bodies and voices. It is less associated with other types of arts such as literature and visual arts. Performing art typically involves a creative process that develops an underlying source or text into a specific production. Here a director, playwright, scenographer and others use their own creativity and interpretation to define what will be presented to the audience (Féral, 2008). In the resulting production, there is a second creative process, where actors interpret their roles and interact with the audience, the stage and objects or props (Lin, 2006). In the communication with the audience, visual, auditory and verbal stimuli are of importance (Cerkez, 2014). In the arts, performance lives next to rhetoric. Both of these are concerned with communication, but performance sets itself apart by having some form of 'embodiment' and attempting to 'enchant' the participants and audience (Rose, 2014). In musical performance, overall quality, technical skills and individuality are all key aspects of a performer's expression (Wöllner, 2013). The notion defining performance in the arts is not uncontested, as exemplified by Bottoms (2008) who makes a case for staying with 'theatre' as visual and time-based art forms with specific social–cultural contexts. Counsell and Wolf (2001: i–x) present a number of ways to analyze artistic performance by looking at aspects such as decoding the sign, politics of performance, gender and sexual identity, performing ethnicity, the performing body, the space of performance, audience and spectatorship and the borders of performance.

The aesthetic notion of performance in the field of architecture is still under development. Some work showing progress in performative architecture or architecture performance can be found in Leatherbarrow (2005), who explores how buildings perform through their operations and how this concept of performance interrelates actions,

---

6 Sometimes related to Albert Einstein, either as quote or sign in his office, but not verified; see, for instance, Blyth and Worthington (2010: 84).

events and effects, or in a wider sense in Kolarevic and Malkawi (2005). Kolarevic (2005b: 205–208) himself writes that architecture typically takes place on a spectrum between 'blending in' and 'standing out'. Recent architecture sometimes takes the standing out position, with the building performing in its context, which acts as a stage. Sometimes there even are active interactions with occupants, dynamically changing light patterns and other movements and reaction to create movement and action. Hannah and Kahn (2008) discuss the tension and interplay between performance and architecture in a special issue of the *Journal of Architectural Education*. Schweder (2012) explores avenues such as 'architect performed buildings', 'buildings that perform themselves', 'bodily performance in architectural time', 'rescored spaces' and 'its form will follow your performance'. Hann (2012) discusses performative architecture as move from 'form follows function' towards a mixture of both 'form is a consequence of actions and events' and 'events and actions are shaped by form'. Hensel (2013) describes in his book on performance-oriented architecture how the concept of performance may even transform the complete notion of architecture and the built environment. Dwyre and Perry (2015) discuss architecture and performance in terms of a contrast between static and permanent qualities versus temporal and impermanent ones, with architecture and landscape design starting to take up more dynamics and movement since the start of the 21st century.

Based on these key terms, building performance can be defined as follows:

> Building performance relates to either a building as an object, or to building as construction process. There are three main views of the concept: an engineering, process and aesthetic perspective. The engineering view is concerned with how well a building performs its tasks and functions. The process view is concerned with how well the construction process delivers buildings. The aesthetic view is concerned with the success of buildings as a form for presentation or appreciation.

This position on building performance is summarized in Table 1.1. This initial take on building performance will be developed into a theoretical framework that defines in more detail what building performance is, and how it can be operationalized, in the remainder of this book.

While the definition of performance as being something that the building actively does is logical, it is important to keep in mind that most buildings are immovable artefacts. In most cases the concept of action involves interaction with occupants such as humans entering and experiencing the building (Leatherbarrow, 2005: 10). Taking this

Table 1.1 Building Performance Views.

	Building Performance		
View	Engineering	Process	Aesthetics
Definition:	Action or process of performing a task or function	Action or process of performing a task or function	Form for presentation or appreciation

further, two views of building actions are important. One concerns the active actions and operation of buildings, such as that of exterior surfaces, screens, doors, furnishing and building services; most of these actions concern the adjustment to foreseen and unforeseen conditions. A second view concerns the more passive action that the building needs to take to stay as it is, in terms of reacting to ambient conditions such as climate and gravity. While this second view of 'action' concerns something that is more resistance towards forces and events, buildings actually are subject to serious loads in terms of the weather, (mis)use by occupants and alterations (Leatherbarrow, 2005: 13).

1.2   Performance in the Building Domain

In spite of the lack of definition of building performance, the concept has implicitly been around for a long time. As long as humans are concerned with shelter, performance will have been of importance. Emerging humanity will have selected caves to dwell in based on performance criteria such as protection from the elements, access and stability. Similarly, primitive dwellings must have been constructed with a focus on keeping the inhabitants safe from the weather and wild animals. But after some development, early humans have also constructed some formidable buildings such as the Stonehenge monument depicted in Figure 1.1 (3000 BC–2000 BC) or the Great Pyramid of Giza (2580 BC–2560 BC). Neither of these has reached the modern age with historical records of their full purpose and leave archaeologists to discuss the construction process and meaning of details; however both have fascinating astronomical alignments that may point to these buildings performing roles as solar clock or stellar representation.

Figure 1.1 Stonehenge Monument, Wiltshire, UK.

Both are on the UNESCO World Heritage list, demonstrating sociocultural importance; both remain impressive in terms of the effort and organization that must have gone into their construction, especially with the means available at that time, and whatever detailed functions these buildings may have had, they have been made with a quality that has allowed them to endure more than four millennia and can thus be said to be early 'high performance buildings'.

A full history of architecture and construction is beyond the scope of this work; however a range of major cultures left the world a fascinating built legacy – see the buildings, cities and infrastructure created by the likes of for instance Mesopotamia, Egypt, India, China, Greece, the Roman Empire, the Christian middle ages in Europe and the Pre-Columbian societies of the Americas. In many cases the buildings followed typical architectural styles, such as Ancient Persian, Ancient Egyptian, Karnataka Architecture, Song Dynasty Architecture, Doric/Ionic/Corinthian Order, Romanesque/Gothic/Baroque or Mesoamerican and Maya. Many of these styles prescribe form, construction methods and materials. The construction of these buildings involved complex design, planning and coordination of large workforces. How building performance was incorporated in their design will often remain a question.

1.2.1 Development of the Notion of Building Performance

Amongst the oldest documents in archives are legal codes; these often relate to buildings. It should thus come as no surprise that the earliest and often quoted example of building performance in building regulations (Bakens *et al.*, 2005) stems from the oldest code of law in the world, dating back to about 1754 BC: the Hammurabi Code by the King of Babylon. This states in § 229: 'If a builder has built a house for a man and has not made strong his work, and the house he built has fallen, and he has caused the death of the owner of the house, that builder shall be put to death' (Johns, 1903).

Another crucial source from antiquity on design and construction of buildings, and the oldest work written by someone from the same period, are the ten books on architecture by Vitruvius,[7] named *De architectura*. This work includes the first deep discussions of how buildings should meet user requirements (Foliente, 2000; Becker, 2008). Vitruvius states that buildings must possess three key qualities: *firmitas, utilitas* and *venustas*. The standard English translation of these Latin terms gives them as *durability, convenience* and *beauty* and adds the explanation by Vitruvius that 'durability will be assured when foundations are carried down to the solid ground and materials wisely and liberally selected; convenience, when the arrangement of the apartments is faultless and presents no hindrance to use, and when each class of building is assigned to its suitable and appropriate exposure; and beauty, when the appearance of the work is pleasing and in good taste, and when its members are in due proportion according to correct principles of symmetry' (Morgan, 1960: 17).

Other translations name firmitas, utilitas and venustas as *strength, utility* and *beauty* or as *firmness, commodity* and *delight*; they can also be interpreted by a focus on Build

---

7 Marcus Vitruvius Pollio, a Roman architect, civil engineer and military engineer, and author.

Quality, Function and Impact (Gann *et al.*, 2003). While Vitruvius is not directly speaking of building performance, this is clearly implied in the three key qualities. It is interesting to note that the ten books of *De architectura* span a wide field, covering amongst others urban design (Book I), building materials (Book II), architectural design principles for temples (Books III and IV), civil buildings (Book V) and domestic buildings (Book VI) and decoration in terms of pavements and plasterwork (Book VII). Beyond this, Vitruvius also covers underlying building science and services in terms of water supply (Book VIII), astronomy and solar access (Book IX) and machines and building services (Book X). Included in the work are discourses about architectural education, structural engineering, physics and music, and acoustics (Morgan, 1960: vii–xii). Another interesting point about Vitruvius is that in many cases *De architectura* gives prescriptions on how to build in specific detail and solutions; for instance on the layout of a city in respect to winds (*ibid.*, 26), relative dimensions of a theatre (*op. cit.*, 148) or foundations of houses (*op. cit.*, 191). Such prescriptions are one way to ensure building performance, staying with solutions that are known to work.

When looking at the role of performance in building, it is important to note that for a long time and in many cultures, building design and construction was a craft, with the know-how of the trade being passed on from master builders to apprentices. Moving forward from Vitruvius and through roughly two millennia, this remained the generic case. But the industrial revolution, which started in the United Kingdom and roughly took place from the mid-18th to mid-19th century, meant a change in manufacturing and production processes. Key developments that impacted on construction and buildings were, amongst others, availability of new building materials such as iron and steel, the invention of Portland cement, gas lighting and new production processes for glass. But the industrial revolution also changed the construction sector from a craft-based undertaking into an industry with different production processes and approaches. An interesting account of many of these changes, and how they impacted on the built environment, is contained in *At Home – a short history of private life* (Bryson, 2011). The industrial revolution also was one of the drivers towards changes in architecture and the development of modernism, functionalism and determinism (Braham, 2005: 57). In terms of performance, it is interesting to note the famous statement by the American architect Louis Sullivan, made in 1896, that 'form follows function' and the impact this had on architectural design.

The industrial revolution gave rise to the emergence of host of new disciplines in the building domain, most notably structural engineering and a new field dealing with heating, cooling, ventilation and lighting.[8] Specialist in these areas quickly organized themselves, founding organizations that have been dealing with building performance for over a century. With the start of the industrial revolution in the United Kingdom, it is not surprising that many new associations were founded here, such as ICE, the Institution of Civil Engineers (1818), ImechE, the Institution of Mechanical Engineers (1847), and IstructE, the Institution of Structural Engineers (1908). CIBSE, the Chartered Institution of Building Services Engineers in the United Kingdom, has roots in the Institution of Heating and Ventilating Engineers (1897) and the Illuminating Engineering Society (1909). CIBSE still publishes guides, application manuals and technical memoranda.

---

8 In the United Kingdom, this field is typically named building services engineering.

In the United States, ASCE, the American Society of Civil Engineers, was founded in 1852; this covers many domains as evidenced by the 33 academic journals still published by ASCE to this day, which cover such diverse fields as architectural engineering, structural engineering and urban planning. The American Society of Heating, Refrigerating and Air-Conditioning Engineers (ASHRAE) traces its history back to 1894. ASHRAE still publishes, amongst others, influential handbooks and standards. In the southern hemisphere, the Institution of Engineers Australia (IEAust) dates back to 1919, with a similar coverage as ASCE. The German VDI (Verein Deutscher Ingenieure) was founded in 1856 and covers a range of disciplines, including construction and building technology. The profession of architecture continued and its professionals also organized themselves, leading to the foundation of the likes of the RIBA, the Royal Institute of British Architects (1834), AIA, the American Institute of Architects (1857), BDA, Bund Deutscher Architekten (1903) and others. In other countries, organizations were founded only much later; for example, the Union of Chambers of Turkish Engineers and Architects was only created in 1954.

In the early 20th century, after World War I, the concept of building performance became more prominent. In the United States, the Division of Building and Housing developed a publication titled *Recommended minimum requirements for small dwelling construction*, which was published in 1922 and is often considered to be the first modern model building code (Zingeser, 2001). Another key publication was the *Recommended practice for arrangement of building codes*, published by the National Institute of Standards and Technology (NIST) (Gross, 1996; Foliente, 2000). Obviously, World War II directed attention elsewhere, and afterwards the prime concern was rebuilding. In the 1960s the concept returned to the fore, not just within the engineering disciplines but also as a guiding principle in architectural design. Braham (2005: 57) mentions a special issue of the magazine *Progressive Architecture* that appeared in 1967 on the topic of 'performance design'. This positioned performance design as a scientific approach to analyze functional requirements, stemming from the developments of general systems theory, cybernetics and operations research and combining psychological needs, aesthetic needs and physical performance. This enabled radical new designs, such as the Centre Pompidou in Paris – built between 1971 and 1977 – which makes building services, and thereby to some extent building performance, into the key design feature.

As with many disciplines, the rise of the personal computer starting in the 1960s also had a profound impact on work in building performance. Right from the start in the 1960s and 1970s, researchers at the University of Strathclyde's ABACUS unit already promoted the use of digital performance assessment tools as a guiding principle for building design (Kolarevic, 2005a: 196). The first group completely dedicated to the study of building performance as a subject in itself was probably the Building Performance Research Unit (BPRU), again at the University of Strathclyde (United Kingdom). BPRU studied appraisal of building performance in the context of design and was an independent unit from 1967 to 1972. Findings were presented in a book that appeared in 1972, describing the interrelation between design decision making and performance, the interaction between various performance aspects (physical, psychological, economic), the use of computing tools and application in practice. The work has a strong basis in systems theory and covers a range of issues such as lighting, sound, thermal comfort, limited use of resources and costs. Interestingly there also is significant attention for spatial elements and organization, as well as bounding spatial elements, compactness and circulation patterns, topics that feature less in later work on the subject (Markus *et al.*, 1972).

In the 1960s, another new building discipline came to prominence: building science, typically named building physics in mainland Europe. This is an applied science that mainly studies thermal, lighting and acoustic performance of buildings. First handbooks that appeared at the end of the decade are *Handbuch der Bauphysik* (Bobran, 1967), *Thermal Performance of Buildings* (van Straaten, 1967) and *Architectural Acoustics* (Lawrence, 1970). Obviously, the importance of building science – and accordingly building performance – increased with the energy crisis of the 1970s, turning the field into a domain studied across the globe.

Yet another discipline concerned with building performance emerged in the 1960s: environmental psychology. Generally this field came into being as response to dissatisfaction with the built environment of the time (Gärling, 2014). Originally environmental psychology focussed on the impact of the human environment on people's well-being; more recently this is looking to change people and human behaviour in order to preserve the environment (*ibid.*). Environmental psychology views building performance from a range of viewpoints, such as individual choice, consumption, sacrifice, values and attitudes, education, motivation, incentives and lifestyle (Stern, 2000). These play a role in efforts to deal with global challenges such as climate change, human population growth and the use of finite resources (Sörqvist, 2016). Stern (2000) however warns against a tendency to put too much emphasis on psychological interventions and points out that it is important to position the role of human actions in a wider frame; in some cases there are more effective interventions in other domains.

Early computers led to further changes in the construction sector by the evolution of digital building models. From the development of the first computer-aided design (CAD) by Sutherland in 1963, there was progress in geometric modelling, discipline-specific analysis models and central models shared by a range of applications. Of course many of these specific analysis models dealt with various aspects of building performance. In the early days the shared cross-application models were named building product models; the seminal book on the subject is the work by Eastman (1999), which includes a full chapter on their history. Pioneering work on product modelling was carried out in the context of the European COMBINE Project (Computer Models for the Building Industry in Europe), with stage I (1990–1992) focussing on the development of an Integrated Data Model (IDM) and stage II (1992–1995) exploring the development of this IDM into an Intelligent Integrated Building Design System (IIBDS). Early on, COMBINE demonstrated the importance of the process dimension of information exchange between various stakeholders in the building performance dialogue (Augenbroe, 1994, 1995; Clarke, 2001: 311–316). COMBINE had a lesser-known follow-up in the United States through the Design Analysis Integration (DAI) Initiative, which ran from 2001 to 2002 and again had a strong emphasis on process (Augenbroe *et al.*, 2004). Another noteworthy effort combining computer models with building performance was the SEMPER project carried out at Carnegie Mellon University from 1996 to early 2000s (Mahdavi *et al.*, 1997a). Around 2002 the key term for work on digital building models became Building Information Modelling (BIM); with the increasing move towards digitalization, this became a key change for the building industry in the first decades of the 21st century. For a recent overview, see

Eastman *et al.* (2011). It is noted that BIM is becoming a regular data carrier for new buildings; however there are challenges in capturing the existing building stock and legacy buildings (Volk *et al.*, 2014).

Internationally, CIB (International Council for Research and Innovation in Building and Construction) started efforts on the subject of building performance by launching Working commission W60 on the *Performance Concept in Building* in 1970. Later a range of CIB Task Groups focussed on related aspects: Task Group TG36 dealt with *Quality Assurance*, TG11 with *Performance Based Building Codes* and TG37 with *Performance Based Regulatory Systems*. Many of these were active over a long period. Through W60, CIB supported a range of conferences together with RILEM (Reunion Internationale des Laboratoires et Experts des Materiaux) and ASTM (American Society for Testing and Materials) in Philadelphia in 1972, Otaniemi in 1977, Lisbon in 1982 and Tel Aviv in 1996 (Foliente, 2000).

In the early 1980s, the CIB Working Commission W60 published their seminal Report 64 on 'Working with the Performance Approach to Building' (CIB Report 64, 1982). It opens with the famous and often-quoted line: 'The performance approach is, first and foremost, the practice of thinking and working in terms of ends rather than means'. The report is a position statement that was developed by the Working Commission over a decade and builds on earlier CIB Reports and W60 Working papers. It describes the meaning of the performance approach, especially when contrasted with prescriptive requirements and specifications. It discusses who might benefit from the performance approach and what these benefits might be, the knowledge base required, how to establish performance requirements, how to predict and measure performance and how to evaluate the suitability for use, and it concludes with a discussion of application at various levels (whole building, component, design, manufacture, regulations and standards). Underlying the work is a drive to develop consistency in the building domain, combined with the promotion of innovation in the sector. As quoted from the report: 'In essence ... the performance approach is no more than the application of rigorous analysis and scientific method to the study of functioning of buildings and their parts.... However it does break new ground by attempting to define unified and consistent methods, terms and documentation, and by subjecting all parts of the building to systematic scrutiny' (CIB Report 64, 1982: 4).

It is interesting to note that within the W60 report, significant attention is paid towards applying the performance approach on component and product level (CIB Report 64, 1982), whereas most recent texts on building performance tend towards holistic performance assessment of complete buildings or even address the district and city levels. In terms of the development of the building performance field, CIB Report 64 contains a bibliography that gives a good historical perspective. It is noted in the report that 'this bibliography is limited to major national and international publications on the performance concept and its application. Individual articles and conference papers are not listed' (CIB Report 64, 1982: 26); however the bibliography already covers 92 publications from 17 different countries as well as international organizations like CIB itself. Many of these references are extensive standards that consist of several parts or volumes, showing the considerable interest and progress on the subject. Out of these 92 publications, only 11 date from before 1970, with the oldest one being conference proceedings from the National Research Council of the United States on

'Performance of Buildings' dating back to 1961 (CIB Report 64, 1982: 26–30). CIB Report 64 contains important thinking by the experts of the time. Unfortunately it appears to have had a limited circulation only and is seldom cited in recent work.

The CIB Proactive Program on Performance-Based Building Codes and Standards (PBBCS) ran from 1998 to 2001. This was a networking platform for furthering the earlier work done by CIB on the subject, establishing the state of the art and setting the agenda for new initiatives (Foliente *et al.*, 1998: 5–6). In 2001 this was followed up by a European Thematic Network named PeBBu (*Performance-Based Building*), which ran from 2001 to 2005 (Almeida *et al.*, 2010) and was coordinated by CIB. PeBBu was funded by the European Union (EU) Fifth Framework Programme. This network brought together over 70 organizations with an interest in the subject, facilitating information exchange and dissemination of knowledge. As a network, the main activities of PeBBu were to promote performance-based building; however the project also included activities that mapped research in the area, and it developed a compendium of knowledge on the subject (Jasuja, 2005: 19–20). In parallel to this EU PeBBu, there was also an Australian counterpart – AU-PeBBu, which started in 2003 (Jasuja, 2005: 28–29). Both PeBBu networks aimed at moving the performance approach as defined by CIB towards wider application through engagement with a variety of stakeholders such as policymakers, regulators, building officials, investors, developers, owners and owner–occupiers, architects and designers, engineering professionals, specialist consultants, product manufacturers, project managers, contractors and builders, facility managers, service providers, users and tenants, ITC professionals, researchers and educators (Bakens *et al.*, 2005; Augenbroe, 2011: 16). PeBBu developed scientific reports in nine domains: (i) life performance of construction materials and components, (ii) indoor environment, (iii) design of buildings, (iv) built environment, (v) organization and management, (vi) legal and procurement practice, (vii) building regulations, (viii) building innovation and (ix) information and documentation. Domain iv on the built environment positions building performance within the urban context (Jasuja, 2005: 10–12, 31). Some work resulting from PeBBu was published in the journal *Building Research and Information* (Jasuja, 2005: 104). It must be stressed that PeBBu was mainly a networking and dissemination project; most underlying thinking stems from CIB Report 64.

The proliferation of computers meant that the building science discipline was able to advance quickly and move from traditional calculations to computer simulation; the history of building performance simulation is outlined by Augenbroe (2003: 6–10) and Clarke (2001: 3–5). The year 1985 saw the emergence of an entity that was initially known as the Association for Building Energy Simulation Software (ABESS). This developed into IBPSA, the International Building Performance Simulation Association, which was formally founded in 1987. IBPSA organizes a biannual conference named 'Building Simulation'; it has regional affiliates of various levels of activity across the globe, such as IBPSA-USA, IBPSA-England, IBPSA-China, IBPSA-Netherlands + Flanders and many others. In the United Kingdom, there also was an entity named the Building Energy Performance Analysis Club (BEPAC), which acted as a predecessor to a regional affiliate; BEPAC existed from 1985 to the mid-1990s.

In the 1980s the building industry in many countries was faced with pressure from government, clients and increased international competition to improve building quality and construction speed and reduce costs. At the same time a range of deep studies into the performance of actual buildings (case studies) emerged. Often these found issues with energy efficiency and indoor air quality. The relation between buildings and ill health became a subject of study and gave rise to the use of the term of 'sick building syndrome' (Cohen *et al.*, 2001). These developments led to the emergence of the new discipline of Facilities Management[9] (Cohen *et al.*, 2001). Starting from a simple basis in building maintenance, service and cleaning, Facilities Management grew to the profession that 'ensures the functionality of the built environment by integrating people, place, process and technology' and is concerned with performance in each of these domains (Atkin and Brooks, 2009: 4). Beyond the performance of buildings and building systems, Facilities Management is also concerned with the performance of the processes that take place inside and around the building, such as change management, in-house provision and outsourcing and workplace productivity.

With Facility Management addressing the performance of buildings in use, there also re-emerged an interest in the handover of buildings at the end of the construction stage. This is typically named building commissioning, allegedly a term rooted in shipbuilding, where a ship undergoes a process of testing before it goes into operation as commissioned vessel. The idea of building commissioning has been around for a long time. Already in 1963, the Royal Institute of British Architects (RIBA) Plan of Work of included a stage for feedback, where the architect was to return to the building to assess the success of the design and construction (Bordass and Leaman, 2005). This was later dropped but brought back in a revision of the Plan of Work in 2013, which reintroduced a review of buildings in use, post-handover and closeout. At the end of the 1960s, CIBSE published the first edition of their Commissioning code A, which was regularly updated and still is available to the current day (CIBSE, 2006). A recent development is the application of commissioning throughout the building usage, which is named continuous commission and abbreviated as CC (Liu *et al.*, 2003). In the United Kingdom, the 'Soft Landings' process also includes the design and construction stages, thus also involving the actors that produce the building in the operational performance (Way and Bordass, 2005). Since 2016 the UK Government requires centrally funded construction projects to be delivered through a Government Soft Landings (GSL) process, which ties in with a requirement to use Building Information Modelling (BIM) of these projects. While Soft Landings is promoted as an open-source framework, unfortunately some aspects are commercialized by the Building Services Research and Information Association and the Usable Buildings Trust; for instance, the guide on how to produce soft landings and some checklists are only available via the BSRIA bookstore.

The 1980s also saw the concept of sustainability gain traction, with as notable moment the publication of the United Nations World Commission on Environment and Development (WCED) report *Our Common Future* (Brundtland *et al.*, 1987). Where the effects of the energy crisis of the 1970s had worn off, this renewed the interest in the environmental performance of buildings. It sparked interest in a range of concepts, such as sustainable buildings, eco-buildings, bioclimatic and autarkic buildings and, more generally, green buildings (Roaf *et al.*, 2003). Around the end of the millennium, the broader interpretation of sustainability, which augments environmental concerns with economic and social issues, led to an expansion of the aspects typically taken into account in assessing building performance beyond the traditional energy efficiency,

---

9 Equivalent to Facility Management.

health and environmental aspects (Lützkendorf and Lorenz, 2006). There is no final definition of sustainability, as the concept is still under development (Mann, 2011b; Smythe, 2014). As exemplified by Hrivnak (2007), there are many issues to consider when applying sustainability to buildings, with a conflict between 'pure' and 'relative' sustainability and issues of where to position the system boundaries. As in other industries, the lack of definition leads to 'greenwash' – the use of token systems and interventions to promote buildings as sustainable, without actual intent to make true on the image invoked. The use of solar panels on buildings that are otherwise of mediocre construction specification is a prime example. As such, most attempts to define and appraise sustainability in construction – such as the appraisal method for infrastructure projects by Ugwu *et al.* (2006) and Ugwu and Haupt (2007) or the planning model for sustainable urban planning by AlQahtany *et al.* (2013) – have a rather transient nature.

The building sector in the United States started using the term High Performance Buildings (NYC DDC, 1999); initially this mainly concerned non-domestic buildings, but ultimately the concept was also applied to homes (Trubiano, 2013). Especially in the United Kingdom, the concept of Zero Carbon Buildings rose to prominence, amongst other things leading to the establishment of the Zero Carbon Hub in 2008. For a time there were plans by the UK Government to require all new homes to be Zero Carbon by 2016; however this plan was abandoned in 2015. More recently the focus has returned to energy efficiency, and the system boundary and grid connection is being taken into account, leading to the use of the term Net-Zero Energy Building (nZEB); for a deeper discussion, see Pless and Torcellini (2010). An overview that lists some of the many definitions and exemplifies the confusion in this domain is provided by Erhorn and Erhorn-Kluttig (2011).

The United States introduced a Government Performance and Result Act in 1993, which also impacted commissioning and management of constructed assets. Hammond *et al.* (2005) describe how this was implemented by the US Coast Guard. They highlight the importance of measurement in relation to accountability of governmental organizations, especially those with a military role; here performance is related to organizational strategy, scope and mission assessment, operations and logistics, and tactics. Further work looking at the performance of the construction industry, beyond building performance, was sparked in the United Kingdom by a range of publications such as the 1994 Latham Report and the 1998 Egan Report *Rethinking Construction*. These reports are not without criticism; for instance, Fernie *et al.* (2006) discuss some of the issues with the underlying work, warning for the need to distinguish cause and effect, the need to ensure that measurement captures the wider context and issues with sample selection for representing 'best practice'. However, both reports led to a strong focus on time and cost of production, as well as some interest in waste and defects but possibly also to less interest in the design quality of the resulting buildings (Gann *et al.*, 2003).

Another approach towards building performance was developed under the title of Post Occupancy Evaluation (POE). In the present time, POE is often defined as a human-centred framework for building performance evaluation (BPE), with a strong emphasis on end-user requirements (Burman, 2016: 59). There may be some confusion about the name as some work in the POE area is completely technical and based on hard technical measurements; other efforts are actually user perception studies, whereas a third category combines both methods. There are different claims regarding the background of POE. Preiser and Vischer (2005: 4) suggest that POE is founded in cybernetics, whereas others emphasize a background in environmental psychology (Cooper, 2001). User feedback already played a role in the work by the Building Performance Research Unit at Strathclyde in the late 1960s but became much more prominent through the Probe (Post-occupancy Review of Buildings and their Engineering) project in the United Kingdom. Starting in 1995s, Probe studied a series of 20 buildings, combining walk-through surveys, energy surveys, discussions with occupants and management and pressure tests. Among other things, Probe found that buildings were overly complicated and often failed to address fundamentals first and reported 'poor airtightness, control problems, unintended consequences, a dearth of energy management, a tendency for systems to default to "on", and a pathological trend for information technology and its associated cooling demands' (Bordass *et al.*, 2001b). Probe typically studied buildings 2–3 years after completion (Cohen *et al.*, 2001).

In the humanities and social disciplines, the 1990s brought a development named the 'performative turn'. Grounded in intellectual theory from the 1940s to 1950s, the performative turn emphasizes the interaction between human behaviour, actions and practice and their context; this means that performance depends on both action and the context in which the action takes place. The performative turn started in areas such as literature and theatre but then expanded to the arts, including architecture (Hensel, 2013: 17–21), leading to the developing concept of performative architecture as described by Kolarevic and Malkawi (2005).

1.2.2   History of Building Codes, Regulations and Rating Schemes

The words *building regulations* and *building code* are used to indicate the requirements for a building imposed by government. Laws and regulations are put in place by the government to make sure buildings meet a range of basic building performance requirements. These laws and regulations aim to ensure the health and well-being of those that cannot influence the design and construction process themselves and benefit society at large. As such, performance as imposed by the government represents a minimal building performance; typically higher performance can be achieved by setting higher ambitions. Laws and regulations are closely related to building standards, which set rules and processes for the activities and processes that relate to construction. Rating schemes are similar but often voluntary. The history of building regulations, standards and rating schemes gives a unique view on how the concept of building performance developed, dating back all the way to the Hammurabi Code of 1754 BC. Therefore, they are discussed in this separate section.

In the early days, building performance regulation often developed in relation to fire incidents, such as the burning of Rome in 64AD, which led to rules that required the use of stone and masonry. There are also fragments of Greek and Roman laws that indicate requirements for buildings to be inspected during construction (Holt *et al.*, 2007). Historically codes, regulations and standards overlapped; in modern times building codes typically refer to standards where technical issues are concerned (Foliente, 2000). The Mayor of London in the United Kingdom put in place, as early as 1189, regulations known as the 'Assize of Buildings', which addressed issues around boundaries including passage of light, sanitation and rainwater discharge and encouraged the use of stone to reduce fire risk (British History Online, 2015). Spain produced the 'Laws of the Indies',[10] developed in the 16th century, which set out the rules for the development of towns and missions for the overseas territories of the Spanish Crown. The Great Fire of London in 1666 drove the development of the Rebuilding of London Act 1667, which regulated distances between houses and their heights, width of walls, and empowered surveyors to enforce the act. Original French building legislation developed from Roman laws into a range of feudal laws. There was a major step change with the introduction of the 'Code Civil' in 1804 with book III of the code dedicated to property; it must be noted that due the geopolitical status of the time, this code applied to many countries under French influence such as Belgium, the Netherlands, Poland, Italy, Spain and Portugal. In the early 1960s, CSTB launched the French *Agrément* system, which decided on approval of systems and techniques in France (Becker, 2008); this was later replaced by a system named *Avis Technique*. Building laws in Germany initially developed at the state level, such as the Prussian Code of 1794; during the Weimar Republic, there was a discussion about a National code, but this did not materialize; the same happened during the Third Reich. Only after formation of the Federal Republic of Germany in 1949 did work start on the 'Bundesbaugesetz' that was put in place in 1960. In the United States, the Building Officials and Code Administration (BOCA) introduced a National Building Code (NBC) in 1915, but in spite of its name, this mainly covered the East Coast and Midwest; other codes by the International Conference of Building Codes (ICBO) covered the West Coast, while the Southeast had building codes by the Southern Building Code Congress International. These were later replaced by the International Building Code (IBC) by the International Code Council, first published in 1997 but taking hold from 2000 onwards. Chinese law has a long tradition but was completely reworked following the revolution of 1911. Initially building matters were covered by the State Council, Ministry of Construction and both local and regional governments. The current national building regulations in the United Kingdom were introduced with the Building Act of 1984; the detailed requirements are covered by a range of 'Parts', such as Part B: Fire Safety, Part E: Resistance to the passage of sound, and Part L: Conservation of fuel and power. For an overview of coverage of the related 'Approved Documents', see Table 1.2. In 1997 China put in place a national Construction law that covers building quality and safety issues; with various revisions this is still in place. Modern building codes typically address issues like fire risk, building access and evacuation, structural stability, energy provision and sewerage and drainage.

For a long time the guarantees towards good building performance were based on experience and know-how, as can be handed down from master craftsmen to apprentices. Such know-how is best captured by prescriptive regulations, laws, codes and standards (Becker, 2008). Towards the end of the 20th century, the regulations and standards in many countries became performance based rather than prescriptive (Augenbroe and Park, 2005). The fundamental differences between prescriptive and performance-based buildings codes and standards are discussed in the seminal paper by Foliente (2000). Prescriptive building codes describe solutions that are acceptable. In other words, they define the parts that may be used in a building. These parts have

---

10 Leyes de Indias in Spanish.

Table 1.2 Overview of the Approved Documents in the UK Building Regulations 2010.

Part A	Structure	A1 Loading
		A2 Ground movement
		A3 Disproportional collapse
Part B	Fire safety	B1 Means of warning and escape
		B2 Internal fire spread (linings)
		B3 Internal fire spread (structure)
		B4 External fire spread
		B5 Access and facilities for the fire service
Part C	Site preparation and resistance to contaminants and moisture	C1 Site preparation and resistance to contaminants
		C2 Resistance to moisture
Part D	Toxic substances	D1 Cavity insulation
Part E	Resistance to the passage of sound	E1 Protection against sound from other parts of the buildings and adjoining buildings
		E2 Protection against sound from within a dwelling
		E3 Reverberation in the common internal parts of buildings containing flats or rooms for residential purposes
		E4 Acoustic conditions in schools
Part F	Ventilation	F1 Means of ventilation
Part G	Sanitation, hot water safety and water efficiency	G1 Cold water supply
		G2 Water efficiency
		G3 Hot water supply and systems
		G4 Sanitary conveniences and washing facilities
		G5 Bathrooms
		G6 Kitchens and food preparation areas
Part H	Drainage and waste disposal	H1 Foul water drainage
		H2 Wastewater treatment systems and cesspools
		H3 Rainwater drainage
		H4 Building over sewers
		H5 Separate systems of drainage
		H6 Solid waste storage
Part J	Combustion appliances and fuel storage systems	J1 Air supply
		J2 Discharge of products and combustion
		J3 Warning of release of carbon monoxide
		J4 Protection of building
		J5 Provision of information
		J6 Protection of liquid fuel storage systems
		J7 Protection against pollution

(Continued)

Table 1.2 (Continued)

Part K	Protection from falling, collision and impact	K1 Stairs, ladders and ramps
		K2 Protection from falling
		K3 Vehicle barriers and loading bays
		K4 Protection against impact with glazing
		K5 Additional provisions for glazing in buildings other than dwellings
		K6 Protection against impact from and trapping by doors
Part L	Conservation of fuel and power: new dwellings	L1A Conservation of fuel and power: new dwellings
		L1B Conservation of fuel and power: existing dwellings
		L2A Conservation of fuel and power: new buildings other than dwellings
		L2B Conservation of fuel and power: existing buildings other than dwellings
Part M	Access to and use of buildings	M1 Access and use
		M2 Access to extensions to buildings other than dwellings
		M3 Sanitary conveniences in extensions to buildings other than dwellings
		M4 Sanitary conveniences in dwellings
Part N	Glazing: safety in relation to impact, opening and cleaning	N1 Protection against impact
		N2 Manifestation of glazing
		N3 Safe opening and closing of windows, skylights and ventilators
		N4 Safe access for cleaning windows
Part P	Electrical safety: dwellings	P1 Design and installation of electrical installations

performance attributes that are known to satisfy the requirements of the legislator. In contrast, performance-based codes only prescribe the overall performance that is required of the building; it is left to the design team to specify the parts and to demonstrate that these parts provide the required performance. As pointed out by Gross (1996) and Foliente (2000), performance-based codes are actually not new. In fact, the Hammurabi Code itself is performance based (no body shall be killed by a building) and leaves it open how the building is to achieve that aim. Foliente (2000) lists three main problems with prescriptive building codes: they can act as a barrier to innovation, they might hinder cost optimization and they might hinder international trade. However, prescriptive codes might also have some advantages in terms of being easier to apply, check and enforce. One of the first countries to introduce performance-based building regulations was the Netherlands, which put these in place in 1991. The Dutch Government Buildings Agency subsequently introduced performance-based procurement and tendering (Ang *et al.*, 2005). Another early adapter of a performance-based building code was New Zealand, which introduced this in 1992. Here implementation issues led to a review of the code between 2005 and 2008 and more weight for the 'Acceptable Solutions' that supplement the code. The experience in New Zealand demonstrated that training of all stakeholders is crucial in successful introduction of performance-based regulations (Duncan, 2005). The leading disciplines in performance-based regulations were structural engineering and fire engineering, and the fields of project initiation and construction were leading in implementing performance-based codes and standards (Foliente, 2000).

Back to building performance legislation in general, in the European Union, the following Pan-European work is of importance: the Construction Products Directive (originally introduced in 1988), the work on EN Eurocodes (emerging since 1990) and the Energy Performance of Buildings Directive (originally introduced in 2003). The 1988 Construction Products Directive encouraged national legislation to formulate functional and performance requirements, while leaving the technical solutions to the market, with the aim of encouraging innovation (Ang *et al.*, 2005). The EN Eurocodes complement the Construction Products Directive, especially focussing on structural stability and fire safety (Gulvanessian, 2009). The Energy Performance of Buildings Directive, published in 2003 and implemented in 2006, requires that the member states of the EU have in place a system of building energy certificates, a system for boiler inspections and a system for inspection of air-conditioning systems (Olesen, 2005; Raslan and Davies, 2012). The EPDB was 'recast' in 2010 to set more strict targets and with the aim towards net-zero energy buildings (Janssen, 2010). However, most building regulation is still left to the national level with significant differences in organization and technical regulation remaining (Pedro *et al.*, 2010). It must be noted that the actual impact of legislation may be more moderate than hoped by politicians and those developing them; for instance, Oreszczyn and Lowe (2010) present a graph that shows a very slow reduction of household gas consumption between 1920 and the late 1990s, with an unlikely steep decent required to achieve zero consumption in the years beyond 2010 – which, in fact, has far from materialized by 2016.

Concerning standards, the International Organization for Standardization (ISO) was founded in 1947. ISO started work on building performance in 1980 when it published standard ISO 6240:1980 on contents and presentation of performance standards in building. This was followed 4 years later by ISO 6241:1984 on performance standards in buildings, which sets out how such standards are to be prepared and what factors are to be considered. From there ISO has developed a wide range of standards that pertain to building performance, such as thermal requirements (ISO 6242-1: 1992), air purity requirements (ISO 6242-2: 1992), acoustical requirements (ISO 6242-3: 1992), area and space (ISO 9836: 2011) and others. Note that many ISO standards find their way into national European regulations and are combined with Euronorms, for instance in Germany as DIN EN ISO, the Netherlands as NEN EN ISO and the United Kingdom as BS EN ISO. Generic quality systems developed by ISO such as standard ISO 9000 and ISO 9001 have been found to be applicable to the construction industry. However, research such as Landlin and Nilsson (2001) suggests that the building industry neglects the innovation and learning perspectives.

In the 1990s, the mandatory codes and regulations were supplemented with a range of voluntary rating schemes. Most of these have an environmental background. Examples are BREEAM (Building Research Establishment Assessment Methodology), introduced in the United Kingdom by the Building Research Establishment in 1990; Passivhaus, developed by Lund University and the Institut für Wohnen und Umwelt in 1990; LEED (Leadership in Energy and Environmental Design) by the US Green Building Council and introduced in 1994; MINERGIE, a Swiss rating system launched in 1994 and upheld since 1998 by the Minergie Association; NABERS (National Australian Built Environment Rating System) by the Australian Office of Environment and Heritage in 1998; CASBEE (Comprehensive Assessment System for Built Environment Efficiency) by the Japan Sustainable Building Consortium in 2004; and QSAS/GSAS (Qatar/Global Sustainability Assessment System) by the Gulf Organisation for Research and Development in 2009. A recent addition is the WELL standard, which specifically links building performance to human health and well-being. Many of these rating schemes are available worldwide and are in competition. An overview of selected rating systems is provided in Table 1.3. The way in which building performance is handled in rating systems is very diverse; sometimes credits may be awarded on the basis of extensive quantification of performance, as in the case of energy use; at the same time credits may be obtained by simply being located close to a railway station or for employing certified personnel during the design stage, which are not building performance aspects in a strict sense.

Some key points in this brief history of performance in the building domain are depicted in Figure 1.2.

1.2.3 Selected Recent Developments in Building Performance

Establishing the state of the art in building performance analysis is the subject of this book and the subsequent chapters. However, to set some context for the following discussion and to guide the reader, this paragraph mentions a (personal) selection of some of the current trends and developments. This is not intended as a full state of the art, but is a subjective selection of works and developments that are worth highlighting. As mentioned, building performance can be analyzed using physical testing, calculation/simulation, expert judgment and user assessment.

Recent advances in digital technology have resulted in an exponential growth of the amount of data that is measured in buildings. Automated Meter Reading (AMR) makes it possible to collect data on for instance indoor temperatures, electricity use and water consumption at high frequency; however due to the continuous use of buildings, such data tends to quickly turn into large data that needs proper analysis to be of use. Guidance on how to measure the performance of actual buildings in use is provided by the International Performance Measurement and Verification Protocol (IPMVP), a protocol for the monitoring of energy and water use by buildings. It defines standard terms and best practice and aims to support measurement and verification (M&V) while acknowledging that these M&V activities typically need to be tailored to each specific process (Efficiency Valuation Organization, 2014a: iv). The IPMVP makes generic recommendations in terms of making sure that accuracy of measurements should be balanced against costs. Furthermore, it encourages work to be as complete as possible (considering all effects of an intervention), the use of conservative values when making estimates, and efforts to ensure consistency across different projects, staff, measurement periods and both demand reduction and energy generation projects. Work done as per IPMVP should be relevant and transparent (Efficiency Valuation Organization, 2014a: 2).

Table 1.3 Overview of voluntary rating schemes.

Scheme	Developed/maintained by	Website
BEAM (Building Environmental Assessment Method)	BEAM Society	www.beamsociety.org.hk
BREEAM (Building Research Establishment Assessment Methodology)	BRE (Building Research Establishment)	www.breeam.org
CASBEE (Comprehensive Assessment System for Built Environment Efficiency)	JSBC (Japan Sustainable Building Consortium) + JaGBC (Japan Green Build Council)	www.ibec.or.jp/CASBEE/english/
DGNB System	DGNB (Deutsche Gesellschaft für Nachhaltiges Bauen)	www.dgnb.de/en/
HQE (Haute Qualité Environnementale)	ASSOHQE (Association pour la Haute Qualité Environnementale).	www.assohqe.org
LEED (Leadership in Energy and Environmental Design)	USGBC (US Green Building Council)	www.usgbc.org/leed
MINERGIE	Minergie Association	www.minergie.ch
NABERS (National Australian Built Environment Rating System)	Office of Environment and Heritage	www.nabers.gov.au
Passivhaus	Passive House Institute EU CEPHEUS project	http://passiv.de/en/ www.passivhaus.org.uk www.phius.org
QSAS/GSAS (Qatar/Global Sustainability Assessment System)	GORD (Gulf Organization for Research and Development)	www.gord.qa
WELL	IWBI (International WELL Building Institute)	www.wellcertified.com

On the level of building components, there is a large amount of performance test procedures and a corresponding body of knowledge on detailed aspects. For instance, façade sound isolation can be measured according to ISO 1628-3, with authors like Berardi (2013) describing issues around instrument positioning. Similarly, the fire hazard of materials used in building can be tested in fire test rooms according to ISO 9705, with Li *et al.* (2012a) discussing specific work on curtain materials. Wang *et al.* (2013) describe fire testing at a larger level, addressing work on continuous reinforced concrete slabs. With regard to thermal performance, a guarded hot box

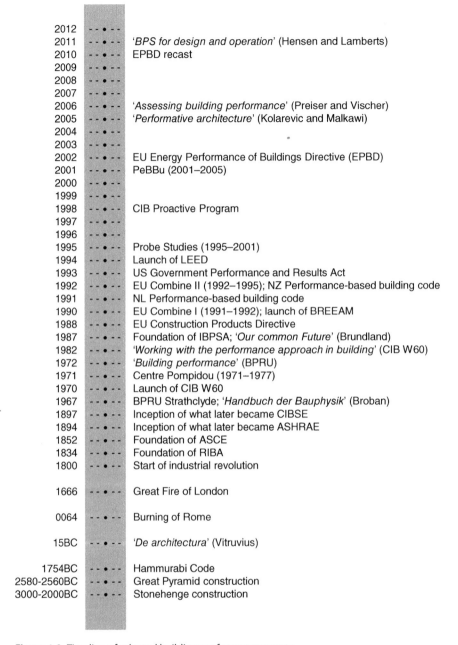

2012	
2011	'BPS for design and operation' (Hensen and Lamberts)
2010	EPBD recast
2009	
2008	
2007	
2006	'Assessing building performance' (Preiser and Vischer)
2005	'Performative architecture' (Kolarevic and Malkawi)
2004	
2003	
2002	EU Energy Performance of Buildings Directive (EPBD)
2001	PeBBu (2001–2005)
2000	
1999	
1998	CIB Proactive Program
1997	
1996	
1995	Probe Studies (1995–2001)
1994	Launch of LEED
1993	US Government Performance and Results Act
1992	EU Combine II (1992–1995); NZ Performance-based building code
1991	NL Performance-based building code
1990	EU Combine I (1991–1992); launch of BREEAM
1988	EU Construction Products Directive
1987	Foundation of IBPSA; 'Our common Future' (Brundland)
1982	'Working with the performance approach in building' (CIB W60)
1972	'Building performance' (BPRU)
1971	Centre Pompidou (1971–1977)
1970	Launch of CIB W60
1967	BPRU Strathclyde; 'Handbuch der Bauphysik' (Broban)
1897	Inception of what later became CIBSE
1894	Inception of what later became ASHRAE
1852	Foundation of ASCE
1834	Foundation of RIBA
1800	Start of industrial revolution
1666	Great Fire of London
0064	Burning of Rome
15BC	'De architectura' (Vitruvius)
1754BC	Hammurabi Code
2580-2560BC	Great Pyramid construction
3000-2000BC	Stonehenge construction

Figure 1.2 Timeline of selected building performance events.

experiment, standardized through ISO 12567-1, can be used; Appelfeld and Svendsen (2011) discuss the analysis of a ventilated window using this approach. Exploration of the impact of real outdoor conditions on façades in a semi-controlled experiment is studied using the EU PASSYS test cells (Wouters *et al.*, 1993), with Alcamo and De Lucia (2014) describing the modification of these cells to meet further requirements.

For theory on the use of calculation and simulation to assess building performance, the work by Augenbroe is seen by many to be leading the field. Building on the work of the EU COMBINE project, Augenbroe has provided a range of publications on building performance resulting from the work with his students at TU Delft and Georgia Tech. The insights into the use of building models, fundamental work in uncertainty and risk analysis (de Wit and Augenbroe, 2002; Heo *et al.*, 2012), a broader view on the use of knowledge in the construction industry (Kamara *et al.*, 2002) and the experience of teaching an MSc programme in High Performance Buildings have been integrated into a chapter in the book by Hensen and Lamberts (2011); see Augenbroe (2011: 15–36). This provides a deep review of the position to date on the role of building performance simulation in performance-based building and especially performance-based building design. A key in this work is the need to have a dialogue between stakeholders about the objective specification of performance measures. Augenbroe views building simulation as a 'virtual experiment' and approaches design as a choice from a range of alternatives, where systems theory helps to support multi-criteria decision making. Most of this theory applies not only to building simulation but also to the other assessment approaches.

Expert assessment has a long tradition in the construction industry. It is the key approach in assessing and handling risk in construction (Yildiz *et al.*, 2014). It also plays an important role in construction litigation, where courts pay special attention to the opinion of professionals in establishing why a building failed (Lindsey, 2005). Expert opinion is also used in advanced analysis efforts of rapidly developing and changing fields, such as the prediction of the home networking market (Lee *et al.*, 2008), as well as for complex areas such as the vulnerability assessment of buildings towards earthquakes (Dolce *et al.*, 2006). Professionals used for expert assessment are typically specialist working in academia and research institutes, ensuring that they are on the forefront of developments.

For building occupant or stakeholder assessment, the book by Preiser and Vischer (2005) presents a good overview of building user surveys; the appendices contain generic checklists as well as examples of detailed occupant questionnaires (Preiser and Vischer, 2005: 212–228 and 232–234). In terms of actual studies, the Probe project (Bordass *et al.*, 2001b) is the most prominent application. Probe studies consist of a number of stages. Stage 1 involves establishing an agreement for undertaking a Probe study. Stage 2 collects data in advance of a first visit by means of a pre-visit questionnaire (PVQ). Stage 3 is a first site visit, which includes an interview with the host, walk-around the building, informal discussions with stakeholders and staff, review of specifications, system control settings, initial spot measurements and readings. Stage 4 consists of initial analysis and the development of a draft report. Stage 5 is second site visit, with the aim to address some issues in more depth and to discuss preliminary findings with the stakeholders. Stage 6 comprises a Building Use Studies (BUS) occupant survey. Stage 7, which actually runs throughout the Probe study, is energy analysis based on meter readings as well as billing data; this is based on the Energy Assessment Reporting Method (EARM) and Office Assessment Method (OAM). Stage 8 is a pressure test to check air leakage of the building. Stage 9 results in the final Probe report. Some buildings went on to a Stage 10, where results were published in the CIBSE Building Services Journal (Cohen *et al.*, 2001). After working in this area for many years, the people behind Probe find that there still is surprisingly little detailed information about the measured performance of modern buildings available. Leaman *et al.* (2010) suggest that this may be due to poor results, which are not published for obvious reasons; unfortunately this leads to a lack of learning and improvement. The BUS methodology and survey is available through a network of partners named, unsurprisingly, BUS methodology. While this helps to maintain quality of the surveys, this also means that the process is not open source and open to general external scrutiny. Soft Landings, the methodology for ensuring a good handover from construction to use stage, supports current efforts in this area.

While the previous paragraphs outline key exponents of building performance assessment, the overall context of building performance is highly dynamic. There are various trends in building science and beyond that are impacting on the field. One recent development is a shift of interest beyond the system boundary of individual buildings towards studies at district and urban level. An obvious extension is from buildings to district heating systems; as an example of ongoing work in this area, Steer *et al.* (2011) report on the control settings for such networks. At district level, Tian *et al.* (2015) have studied correlations between building stock variables in the analysis of the thermal performance of university campus buildings. Orehounig *et al.* (2015) show how considerations at neighbourhood level lead to novel concepts such as an 'energy hub'. Stossel *et al.* (2015) present a study on the development of a composite environmental quality index for cities, while Yigitcanlar and Lönnqvist (2013) have explored the measurement of knowledge-based urban development performance. Kontokosta and Tull (2017) demonstrate the use of machine learning to model the thermal performance of the building stock of New York, which consists of 1.1 million buildings. Azadi *et al.* (2011) have applied the concept of performance to green urban spaces. Reinhart and Davila (2016) give an overview of the field of urban energy modelling, albeit leaving out some of the thorny issues of data input and the associated uncertainties as presented by Choudhary (2012). Obviously, the analysis of the built environment at district or urban level does not solve the problems that are still left in designing, modelling and understanding individual buildings, as there is no guarantee that errors at this single building scale will be cancelled out by the higher number of units at the urban scale.

Clarke (2015) presents a vision for the development of building performance simulation that includes a critique of the lack of a shared vision for a beneficial end goal for the discipline. He mentions the need to spend further efforts towards positioning performance analysis in the design process, abstracting building performance design problems, and to develop performance criteria, metrics and performance assessment procedures. Some of the issues are elaborated further in the paper by Clarke and Hensen (2015); this suggests that 'high integrity representation of physical processes', 'coupling of different domain models' and 'design process integration' should be the ultimate goals. Interestingly, the authors do not include a critique of the lack of definition of the concept of building performance itself. Also of interest are the 'ten questions' papers launched by the journal *Building and Environment* (Blocken, 2015), which discuss particular subjects in a way that 'provide younger researchers directions for future research'. These articles show some of the frontiers in building performance analysis that are currently being explored, such as work in pollen concentrations, allergy symptoms and their relation to indoor air quality (Bastl *et al.*, 2016), thermal environment and sleep (Lan and Lian, 2016) and hybrid computational–physical analysis of wind flow in the built environment (Meroney, 2016).

At various levels, ranging from systems to full cities, there is an interest in 'smart'. Kwon *et al.* (2014) present a system that fuses data from a range of sensors to detect elevator users even before they call the elevator, thus improving scheduling. Lin *et al.* (2014) provide an example at the building level, discussing how smart systems can improve the response to an earthquake. A general overview of the prospects of smart power grids, combining microgrids, a high-voltage national grid, storage, distributed generation and interaction with the energy users, can be found in Amin (2013). Lombardi *et al.* (2012) take a holistic look at smart city performance, incorporating governance, economy, human capital, living and environment. O'Grady and O'Hare (2012) present a discussion of how ambient intelligence may impact citizens of a smart city and some of the issues and technology involved. McLean *et al.* (2015) provide an example of a case study that explores the social and political implications of introduction of a smart energy grid.

The interaction between buildings and humans is another area where a lot of efforts is being invested. This fits in a much wider trend where there are high stakes in predicting human behaviour, such as sales, unemployment and healthcare. In these, the Internet is becoming an important tool, helping to establish what people are presently doing but also what they are likely to do in the future (Goel *et al.*, 2010; Chandon *et al.*, 2011). Humans play a role as clients, as explored by Hoyle *et al.* (2011) who present a study into the understanding of customer preferences in the automotive engineering domain. In other domains, humans are controlling systems. In automation, there is a long tradition of research into human computer interaction (HCI), which is built around the idea that humans control the computer; more recently this is developing into new models of collaboration between humans and machines (Hoc, 2000). When it comes to buildings, humans have a complex, tri-way relationship with performance. First of all, the simple presence of humans creates loads that buildings have to respond to, for instance in terms of structural loads resulting from body weight or excitation from walking, or in terms of heat and moisture emissions. Secondly, human beings actively operate buildings, changing control settings, opening and closing windows and blinds as well as a range of other systems. Thirdly, buildings are in place to meet human needs, so human perception is a key factor in judging the final performance of buildings. Seminal work in the area is the chapter by Mahdavi (2011) who reviews this interaction in the context of building simulation. The relation between lack of building performance and occupant complaints is discussed by Goins and Moezzi (2013). Webb *et al.* (2013) show how studies of human behavioural change can be linked to the energy us of households and thus to building performance. In general, for performance analysis to become an integrated part of the building cycle, it must be useful to the people that actually work with and inside these buildings rather than imposed from above (Bordass and Leaman, 2005). Further work on the interaction between occupants and the thermal performance of buildings is ongoing in the International Energy Agency Annex 66 on the definition and simulation of occupant behaviour in buildings (IEA, 2016b). At the same time, Stern (2000) warns against a bias that expects too much of psychological interventions in human–environment interactions. Kim (2016) provides evidence of some of the limitations of overly detailed occupant behaviour models in building performance simulation.

Advances in data analysis, as well as parallel and cloud computing, also make for a change in context. O'Neill *et al.* (2013) present advances in building energy management using advanced monitoring and data analytics for US Department of Defense naval station buildings; Hong *et al.* (2014) provide a similar analysis of data collection and analysis for the retrofit of the head office of a financial institution in California. Mathew *et al.* (2015) show how the gathering of energy use data from a group of over 750 000 buildings leads to 'big data' issues in terms of data storage, cleansing and analysis. Cloud computing and related developments such as Software as a Service (SaaS) are changing the concepts used in software and IT hardware design and purchasing (Armbrust *et al.*, 2010). Cloud computing is also starting to have an impact on academia; typical analysis tools like Matlab are already being tailored to work in a cloud computing environment (Fox, 2011). Ventura *et al.* (2015) describe how parallel and cloud computing support complex nonlinear dynamic analysis in structural building engineering, whereas Zuo *et al.* (2014) show how parallel computing can be used within building daylighting simulations. Barrios *et al.* (2014) present a tool for the evaluation of the thermal performance of building surfaces that is designed to run in the cloud. The PhD thesis of Obrecht (2012) focuses on the use of parallel computing to support airflow analysis of buildings. Beach *et al.* (2015) explore the relation between building information models, data management and cloud computing.

Developments on the Internet of Things (Kortuem *et al.*, 2010; Dijkman *et al.*, 2015) also relate to buildings and bring their own inherent challenges such as privacy risks (Weber, 2015). Qin *et al.* (2015) describe data flow and energy use pattern analysis for a smart building. Palme *et al.* (2014) present a practical application to classroom access control in schools, while Uribe *et al.* (2015) demonstrate an implementation that manages the acquisition, storage and energy transfer in an energy-efficient building. Caragliu and Del Bo (2012) conducted an econometric study into the relation of smart city attributes and economic growth.

1.3 Outline of the Book

This book brings together the current knowledge on building performance and its analysis. It discusses the concept of building performance in depth, explores how building performance can be measured and analyzed and how such an analysis of performance can be used to improve buildings, and explores how other disciplines can help to improve the field. The book aims to illustrate the rich and complex context in which building performance analysis takes place, which means that setting up a meaningful analysis effort typically requires a deep dialogue between the various stakeholders. It deliberately makes regular references to other areas in order to break the isolationist approach that sometimes dominates building science.[11] From this, an emergent theory of building performance analysis will be developed.

The remainder of the book emphasizes the engineering view of building performance being a concept that captures how well a building performs a task or function. It attempts to cover building as both a process and an object; it aims to integrate the aesthetic view

---

11 Testing whether external concepts are applicable in the context of building performance is a job that can only be done by experts in this area.

of presentation or appreciation in this approach by seeing this as one of the key functions of a building. Dedicated comments about the other interpretations of building performance are included where relevant.

The book is structured in three main parts. Part I provides a theoretical foundation on building performance, Part II explores assessment and Part III deals with impact. Part I starts from a wider view on building performance, zooming in to user needs and requirements. Part II builds up the fundamentals of an approach for building performance analysis and develops this approach into a conceptual framework for working with building performance. It explores what is needed to carry out analysis efforts, combining criteria for performance, performance measurement and quantification and operational building performance analysis. Part III discusses how building performance analysis impacts building design and construction, building operation and management and high performance buildings. The final chapter (epilogue) summarizes the emergent theory of building performance analysis. In other words, Part I introduces building performance, Part II explores building performance analysis and Part III deals with the application of building performance and its analysis. Figure 1.3 describes this structure in a graphical format.

While the book develops an emergent theory of building performance analysis, it emphasizes the significant work that is required to implement this theory in daily practice by discussing case studies in the main chapters of Parts I, II and III. The case studies highlight the complexity of real buildings and provide an indication of the future work that will be required to operationalize the theory.

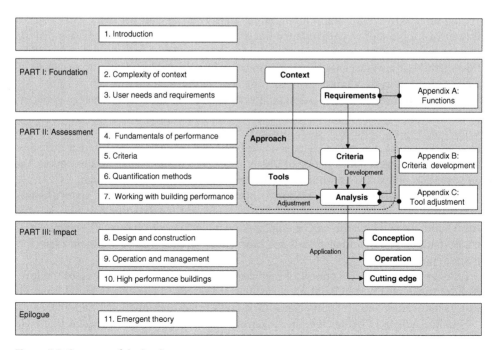

Figure 1.3 Structure of the book.

Within this structure, detailed contributions are as follows:

Chapter 2 starts by positioning building performance in its complex context, thus providing the basis for deeper discussion in the subsequent chapters. It discusses the building life cycle, the main stakeholders in building, building systems and the interaction between the fields of architecture and engineering. It then moves on to review some of the deeper challenges to the industry in terms of building performance, followed by general approaches to ensure building performance as well as some of the specific tools available.

Chapter 3 covers needs, functions and requirements. It explores the different world views of the various stakeholders, the corresponding building functions and their respective functional requirements. This is complemented with a discussion of how buildings typically meet these functional requirements through a range of systems and subsystems.

Chapter 4 deals in depth with the central concept of building performance and discusses the different attributes of performance and yardsticks to measure it. The chapter covers the experiments, observations and performance measures that are important for quantification.

Chapter 5 introduces performance targets and criteria, which are needed for the analysis of the performance that has been measured and which allow establishing 'how good' any score is. To do so the chapter reflects on goals, targets, ambitions, constraints, thresholds, limits, benchmarks and baselines. It also introduces performance banding.

Chapter 6 covers performance quantification. The chapter covers the four main approaches that can be used to quantify performance: calculation and simulation, monitoring and measurement, expert judgment and stakeholder evaluation.

Chapter 7 covers working with building performance and presents a conceptual framework for building performance analysis. It explores how performance criteria need to be developed for each specific case and matched to an appropriate quantification method. It also discusses the adjustments of methods to the specific situation, as well as iteration in the analysis process.

Chapter 8 returns to building design; it reviews how building performance concepts as established in Chapters 2–6 can be applied in a design context. This chapter covers some challenging issues like decision making under uncertainty and visualization; it focuses on virtual analysis, as there is no real building that can be analyzed at this stage apart from potential precedents or mock-ups. The chapter also briefly covers the construction phase and how buildings materialize from the original design.

Chapter 9 covers building operation, control and management. Here the focus shifts to performance analysis of real objects in use, looking at the data that can be harvested from buildings and how this can be employed to manage performance. The chapter also covers fault detection and diagnosis and how this feeds into the development of performance contracts.

Chapter 10 reviews the concept of high performance buildings. It discusses the forefront of the application of building performance analysis in construction and how this fosters innovation and emerging concepts such as smart and intelligent buildings.

Finally, Chapter 11 brings together all strands of the book in an emergent theory of building performance analysis.

1.4   Reflections on Building Performance Analysis

Building performance is an important concept, but so far the term has been left mostly undefined. This is an undesirable situation: it means the industry is working with a vague value proposition, while academics lack a strong foundation to move forward on the subject. It is important to fill this void and provide a working definition of building performance and establish an emergent theory of building performance analysis.

The current position on building performance is not surprising, given the complexity of the field. A deep understanding of building performance requires a broad knowledge base, which covers the domains of architecture, construction and science; typically it requires further insights in specific areas such as architectural design, engineering, building technology, construction, physics, material science, systems theory, computing and mathematics. To complicate matters, there also is a division between industrial practice and building science, with practice emphasizing the need to know how things work in the real world, while science often feels the industry is locked into proceeding with business as usual – a classical situation of conflicting views between 'boots in the mud' and 'ivory tower'. There are very few people who have the full overview of all aspects and can provide a unified view on building performance.

So far, development of theory on building performance has mainly taken place from within the discipline. However, there are other fields that all have their own interest in performance and that have made some progress in furthering the subject. It thus seems worthwhile to explore the adjacent fields, especially those of systems engineering and process management, to find out what concepts may fit within the building context. But the filtering of what external concepts can be integrated into a theory of building performance analysis should be left to experts in this specific discipline, who appreciate the uniqueness of building design and engineering, construction and building operation. In general, the building performance analysis field should be inquisitive and be neither xenophilic (Clarke and Hensen, 2015) nor xenophobic.

While there is no unified theory on building performance analysis, there have been a lot of contributions to aspects of the field. Journals like *Building and Environment, Energy and Buildings, Automation in Construction* and the *Journal of Building Performance Simulations* have published literally thousands of articles on related efforts. An attempt to filter this work and see what joined-up picture emerges is overdue. This book hopes to provide a solid starting point in this direction. A review of the history of the concept of building performance analysis as an area of scholarship demonstrates that this specific field can be tracked back for about 50 years, to the late 1960s, but the roots of building performance go much deeper and reach all the way to shelters built by early humans. It is important to build on this legacy, including the pioneering work by BPRU and CIB Report 64 and the contributions of many others, so as not having to reinvent the wheel again and again.

Ultimately, building performance analysis is an applied science. So, as pointed out by Hensen and Lamberts (2011: 3), one not only needs to ask how the field helps to support the production of desired results and products, but also has to strive for deep understanding and appreciate the inner workings. In this context it is worth remembering that while building performance analysis focuses on buildings, which are material objects, buildings in the end are created to serve humans and their activities.

1.5   Summary

This chapter introduces the concept of building performance analysis and its importance in the architecture, engineering and construction sector. It discusses the technical interpretation of performance, as in performing a function, and the aesthetic view of presenting and entertaining; it combines this with building as both an artefact and process. Buildings are shown to be complex systems, both in terms of the many systems involved and the long life cycle, which results in many disciplines having an interest in and interaction with the area of building performance. The architecture, engineering and construction industry itself adds further complications by its make-up and collaboration practices and the highly individual products it creates. Yet building performance is not immune into the rapid developments in ICT, and new developments in digital metering and measurement are presently driving a change in how building performance is analyzed.

While the term building performance is used frequently, there is no clear definition of the term. This is not uncommon in other fields. However, the absence of a common understanding impedes progress in both the building industry and the related academic fields. The lack of a unifying theory on building performance limits the credibility of the disciplines, a problem that is further exacerbated by casual use of concepts from other domains. The problem of the 'performance gap' also indicates a tendency to over-promise, with the ultimate product of the construction industry, buildings, regularly failing to meet the expectations of its stakeholders.

Building performance lives in a context of other disciplines, such as electronics, human resources, sports, manufacturing and others that all have their own interpretation of performance. Two fields that provide a deeper view are systems engineering and process management. Key concepts that help to better handle performance are systems, functions, criteria, goals, objectives and measurements. On a fundamental level, the main approaches for building performance analysis are physical testing, calculation/simulation, expert judgment and stakeholder assessment.

Based on the introductory discussion, this chapter then defines building performance as a tripartite concept that can relate to an engineering, process or aesthetic perspective. The engineering view is concerned with how well a building performs its tasks and functions. The process view is concerned with how well the construction process delivers buildings. The aesthetic view is concerned with the success of buildings as an object of presentation or entertainment.

This initial definition of building performance is followed by a brief history of the building performance, starting from shelter for emerging humanity, the impressive Neolithic monuments, and Vitruvius all the way to the present. Milestones in this history are the following:

- The industrial revolution and foundation of associations that represent specialisms.
- The post-war interest in performance as architectural driver as well as subject in itself.
- Development of building performance in relation to computers and ICT from 1960s onwards.
- The relation with the emergence of the field of building science/physics, again from 1960s onwards.

- The link with Computer Aided Design and Building Information Modelling, and notably the EU COMBINE project of the early 1990s.
- The work of CIB on building performance through Working Commission W60 on the Performance Concept in Building (1970 onwards) and the PeBBu project (2001–2005).
- The foundation of IBPSA, the International Building Performance Simulation Association in 1987.
- The emergence of Facility Management as a separate field.
- The interest in sustainability from the 1980s.
- The rethinking of construction industry processes from the late 1990s.
- The continuous work on user surveys and Post Occupancy Evaluation, and especially the Probe Project in the United Kingdom.
- The performative turn in the humanities since the 1990s.
- The emergence of diverse concepts such as High Performance Buildings, Zero Carbon Buildings and Net-Zero Energy Buildings that have appeared around the turn of the millennium.

Another area where developments are worth discussing is the field of building regulations, standards and rating schemes. From the Hammurabi Code of 1754 to modern performance-based regulations, this gives a good insight in some of the developments in performance thinking. European developments such as the Construction Products Directive, Eurocodes and the Energy Performance of Buildings Directive are presently shaping and driving developments. Also of note are the ISO standards, which include building performance since the 1980s. Voluntary rating schemes, such as BREEAM, LEED, NABERS and others, complete this overview.

Finally, a brief selection of more recent work in building performance is presented to help set the scene for the following chapters. Discussed are advances in monitoring and measurement, with a focus on AMR and the IPMVP, and test methods for building components as often defined in ISO standards. A theoretical framework on role of simulation in performance-based building is provided by Augenbroe (2011) and is mostly applicable to other analysis methods as well. In both the juridical context and prediction, expert assessment still plays an important role. For evaluation of building performance by stakeholders, POE is a key approach, with the Probe studies and the descendant BUS methodology representing important exponents of the field. Changes in building performance analysis are currently driven by expansion of the system boundaries towards inclusion of the district and urban level. Further development takes place in the field of smart systems, both at building component and whole building level, via smart grids, and smart cities. Another area driver of change is increased interest in the role of occupants and the introduction of elements of human computer/machine interaction into the building sector. And like in many fields, the progress on data analysis, cloud and parallel computing and the emergence of the Internet of Things are impacting building performance analysis.

The remainder of the book sets out to define building performance and establish how this can be measured and analyzed and how it can guide the improvement of buildings. The findings will be combined into an emergent theory of building performance analysis.

Recommended Further Reading:

- *Working with the Performance Approach in Building* (CIB Report 64, 1982) for generic introduction to performance-based building.
- *The role of simulation in performance based building* (Augenbroe, 2011) for a leading theory on the use of performance analysis in making decisions in building design and beyond.
- *Developments in performance-based building codes and standards* (Foliente, 2000) for a good overview of the emergence of performance-based codes at the end of the 20th century.
- *Assessing Building Performance* (Preiser and Vischer, 2005) for the wider view on post-occupancy evaluation and stakeholder surveys.

Activities:

1 Make a list of your top-three favourite buildings and your three most despised buildings. Then analyze what it is that makes you like or dislike these buildings, and express this in terms of building performance.

2 Find examples of specific buildings that perform well in terms of
   A Acoustical quality
   B Efficient use of water
   C High workload capacity in handling numbers of people going through the building
   D Responsiveness towards the difference between working week and weekend
   E Readiness to cope with an earthquake.

   Explore the design of these buildings and whether any special systems are in place to make the building perform well in this specific area. Review the design and engineering process of the building; where did the interest in this specific performance aspect originate from, and what was done by whom to ensure that the building would perform in this aspect?

3 Discuss the relations between a well-managed building construction process and the resulting building. Does proper quality assurance processes during construction guarantee that the final building performs well? Why or why not?

4 Identify a building that performs in the aesthetic interpretation of the word by making a creative statement and communicating with its context and the observer. Explore what mechanisms are used for communication and what attributes the building has to enhance this communication.

5 Review your national building regulations regarding the protection against noise hindrance of commercial aircraft for residential properties, and find out whether these are prescriptive or performance-based regulations.

6 Ask some colleagues or friends to give a definition of building performance. Contrast your findings with the discussion of the concept given in this chapter.

1.6   Key References

Augenbroe, G., 2011. The role of simulation in performance based building. In: Hensen, J. and R. Lamberts (eds.) *Building performance simulation for design and operation.* Abingdon: Spon Press.

Bachman, L.R., 2003. *Integrated buildings: the systems basis of architecture.* Hoboken, NJ: Wiley.

Bakens, W., G. Foliente and M. Jasuja, 2005. Engaging stakeholders in performance-based building: lessons from the Performance-Based Building (PeBBu) Network. *Building Research & Information,* 33 (2), 149–158.

Becker, R., 2008. Fundamentals of performance-based building design. *Building Simulation,* 1 (4), 356–371.

Blanchard, B.S. and W.J. Fabrycky, 2011. *Systems engineering and analysis.* Upper Saddle River, NJ: Prentice Hall, 5th edition.

Blocken, B., 2015. New initiative: 'ten questions' paper series in building & environment. *Building and Environment,* 94, 325–326.

CIB Working Commission W60, 1982. CIB Report 64: Working with the Performance Approach to Building. Rotterdam: CIB.

Clarke, J., 2001. *Energy simulation in building design.* Oxford: Butterworth-Heinemann, 2nd edition.

Clarke, J., 2015. A vision for building performance simulation: a position paper prepared on behalf of the IBPSA Board. *Journal of Building Performance Simulation,* 8 (2), 39–43.

Cohen, R., M. Standeven, B. Bordass and A. Leaman, 2001. Assessing building performance in use 1: the Probe process. *Building Research & Information,* 29 (2), 85–102.

Counsell, C. and L. Wolf, 2001. *Performance analysis: an introductory coursebook.* Abingdon: Routledge.

Efficiency Valuation Organization, 2014. *International performance measurement and verification protocol: core concepts.* Washington, DC: Efficiency Valuation Organization.

Foliente, C., 2000. Developments in performance-based building codes and standards. *Forest Products Journal,* 50 (7/8), 12–21.

Foliente, G., 2005a. Incentives, barriers and PBB implementation. In: Becker, R. (ed.) *Performance based Building international state of the art – final report.* Rotterdam: CIB.

Foliente, G., 2005b. PBB Research and development roadmap summary. In: Becker, R. (ed.) *Performance based building international state of the art – final report.* Rotterdam: CIB.

Gilb, T., 2005. *Competitive engineering: a handbook for systems engineering, requirements engineering, and software engineering using planguage.* Oxford: Butterworth-Heinemann.

Hazelrigg, G.A., 2012. Fundamentals of decision making for engineering design and systems engineering. http://www.engineeringdecisionmaking.com/ [Accessed 21 December 2017].

Hensel, M., 2013. *Performance-oriented architecture: rethinking architectural design and the built environment.* Chichester: Wiley.

Hensen, J. and R. Lamberts, eds. 2011. *Building performance simulation for design and operation.* Abingdon: Spon Press.

IBPSA, 2015. International building performance simulation association homepage [online]. Toronto, Canada. Available from www.ibpsa.org [Accessed 26 November 2015].

INCOSE, 2005. Technical Measurement. San Diego: Practical Software & Systems Measurement (PSM) and International Council on Systems Engineering, Report INCOSE-TP-2003-020-01.

INCOSE, 2015. *Systems engineering handbook: a guide for system life cycle processes and activities.* San Diego, CA: Wiley.

ISO 6241: 1984. Performance standards in building – principles for their preparation and factors to be considered. Geneva: International Organisation for Standardization.

ISO 7162: 1992. Performance standards in building – content and format of standards for evaluation performance. Geneva: International Organisation for Standardization.

Jasuja, M. (ed.), 2005. PeBBU Final Report – Performance Based Building thematic network 2001-2005. Rotterdam: CIB.

Kolarevic, B. and A. Malkawi (eds.), 2005. *Performative architecture: beyond instrumentality.* New York: Spon Press.

Leaman, A., F. Stevenson and B. Bordass, 2010. Building evaluation: practice and principles. *Building Research & Information,* 38 (5), 564–577.

Markus, T., P. Whyman, J. Morgan, D. Whitton, T. Maver, D. Canter and J. Fleming, 1972. *Building performance.* London: Applied Science Publishers.

Montgomery, D.C., 2013. *Design and analysis of experiments.* Hoboken, NJ: Wiley, 8th edition.

Morgan, M., 1960. *Vitruvius: the ten books on architecture.* English translation. New York: Dover Publications.

Neely, A., 2005. The evolution of performance measurement research. *International Journal of Operations & Production Management,* 25 (12), 1264–1277.

Preiser, W. and J. Vischer, eds., 2005. *Assessing building performance.* Oxford: Butterworth-Heinemann.

SEBoK, 2014. The Guide to the Systems Engineering Body of Knowledge, v.1.3. Hoboken, NJ: Body of Knowledge and Curriculum to Advance Systems Engineering/ Trustees of the Stevens Institute of Technology.

Part I

Foundation

2

Building Performance in Context

The previous chapter defined building performance as a concept that captures how well a building performs its tasks and functions, how well the construction process delivers buildings or the success of a building as presentation or entertainment. It also mapped some of the complexity around building performance. This complexity is often under-estimated, both from outside the building discipline and from within. Yet this is a key reason why concepts and approaches from other disciplines such as engineering and computing do not fit seamlessly into the domain of building performance analysis, leading to a waste of efforts as deplored by Clarke and Hensen (2015). This chapter reviews this context in more depth, thus providing a basis for the deeper exploration of building performance that will follow in the remainder of the book.

A first issue to consider is the complexity of the building life cycle. Buildings are known for their longevity; most buildings have a life expectancy of around 80–100 years. However, the service life of building systems and components of buildings has a wide range; roughly decoration lasts 5–7 years, building services 10–15 years, the building envelope around 25 years and the structure at least 50–75 years, whereas the building site is timeless (Blyth and Worthington, 2010: 49). Looking at specific systems there often is detailed research on service life; for instance Vieira *et al.* (2015) report around 20 years for rendered facades, Garrido *et al.* (2012) 20–25 years for paint coatings, while Silva *et al.* (2012) found values in the range of 60–90 years for stone cladding. This means that maintenance, renovation and replacement play an important role in managing building performance. The importance of the temporal context of building performance is exemplified by Preiser and Vischer (2005), who dedicate specific chapters to six key phases of the building delivery and life cycle, covering strategic planning, programming/briefing, design, construction, occupancy and reuse/recycling. Yet there are very few, if any, studies that track building performance over the full life cycle. Typical Life Cycle Assessment (LCA) still includes just one single deterministic analysis per life stage. Tracking over a period of time is usual in monitoring work, but mostly limited to a couple of years at best, with work like Thiel *et al.* (2014) being the exception rather than the rule. The inclusion of longitudinal changes is only done in very few simulation studies such as de Wilde *et al.* (2011). A complex and related issue is how society should deal with existing buildings and whether complete replacement by new buildings or renovation and refurbishment is the better option to improve performance on the key environmental, social and economic aspects (Power, 2008).

The complexity of the many stakeholders in building performance is already high-lighted by CIB Report 64 (1982), which lists the community, corporations, government,

Building Performance Analysis, First Edition. Pieter de Wilde.

building occupants, clients, designers, builders, product manufacturers and insurers as parties that all have a different interest. The actual stakeholders in most buildings change over time; there are different parties involved in the various life stages, but even within one stage actors often change – for instance when a building is being sold or a new management team brought in. The stakeholders mentioned before are, in most cases, just top-level names that point to more complex organizations with more or less formal internal structures and decision makers, communication channels and authority (Lizarralde *et al.*, 2011). Apart from the issue of keeping track of these many stakeholders, the individual actors involved and their roles, there sometimes are also issues with stakeholder involvement; for instance Kersten *et al.* (2015) discuss the issue of engaging residents in building renovation projects. Further complexity arises by the interactions between building ownership, relations between these stakeholders and responsibility for performance. Note that there also are some significant international differences here; for instance the discipline of building surveying is common in the United Kingdom, Australia and New Zealand, but does not exist in continental Europe or the United States.

Yet another issue is the complexity of buildings when viewed as a system. The Sweets building product catalogue (Sweets, 2016) provides an overview of well over 40 000 different products that can be included in buildings; many of these products are actually systems in themselves that consist of multiple components and parts. Often significant numbers of the same product can be used in a single building; for instance a simple house may contain over 5000 bricks. Moreover, many construction products – especially those made of wood, metal and concrete – are subject to a large degree of customization, yielding an infinite number of possibilities to combine these products into a building as a system. But systems are more than just a collection of components and parts; for instance Augenbroe (2011) identifies the need to understand the building as a system that provides specific functions. There are different ways of defining systems; as Weinberg (1975: 51) reminds us, 'a system is a way of looking at the world', a 'mental tool to aid in discovery' (*ibid.*, 54), so the systems view of building also needs further exploring. Again, this complexity of system build-up and ways of viewing it is often underestimated, for instance by some who aim to use optimization in the building design process.

The interest in building performance as a topic in itself is caused by a range of external factors, such as environmental concerns and client satisfaction. Environmental pressures and climate change are often quoted. However, too many authors seem to oversimplify the drivers for building performance to a statement of the percentage of final energy use attributed to the built environment (typically said to be in the order of 30%–40%), as well as the related greenhouse gas emissions. Yet much deeper arguments for looking at building performance can be provided; for instance Oreszczyn and Lowe (2010) give a detailed review of some of the challenges in energy and buildings research. Occupant health, comfort and productivity are further issues to consider. Regarding the building process, documents like the UK's Latham Report highlight issues in procurement and contracts in the construction industry, identifying some operational practices as adversarial, ineffective, fragmented and not serving clients (DoE, 1994); also in the United Kingdom the Egan Report is famous for highlighting the need to reduce construction time, construction cost and the number of defects in construction projects (DTI, 1998). Building performance challenges thus should not be oversimplified, either.

   A special context for the discussion of building performance is the relation between the fields of architecture and engineering. Probably because of historical reasons as well as some overlap between the fields, there is some tension between these disciplines, with regard to the leadership on design and responsibility for the performance of the building (Herbert, 1999) though some publications on building performance like Kolarevic and Malkawi (2005) are carefully aiming to provide a cross-party view. A related issue that warrants extra attention is the design process, as this is where buildings and their performance emerge. Again this is a complex area prone to oversimplification. As an indication, one may compare and contrast the waterfall, spiral and V-process models stemming from the systems engineering world (Blanchard and Fabrycky, 2011: 50–51), the classic work on 'how designers think' by Lawson (2005), 'design thinking' by Cross (2011) or the many issues of the journal *Design Studies*.

   Current approaches and tools to ensure building performance also provide a background to how the concept is perceived. They relate to what Weinberg (1975: 228) calls the 'Systems Triumvirate', which asks: (i) Why do I see what I see? (ii) Why do things stay the same? (iii) Why do things change? Frameworks and tools influence what we can observe and are instrumental to understanding of the concept. Before exploring such approaches in depth, it is important to study the perspective they provide on building performance from a more generic stance.

   By exploring the issues of life cycle, stakeholders, buildings as systems and challenges in building performance, this chapter provides a common understanding of the wider context that sometimes hinders the progress in building performance analysis. The chapter is mainly descriptive in nature as it provides a common basis for subsequent discussion. Critical reflection is needed whenever these concepts appear in the context of building performance, and care should be taken not to oversimplify these issues.

2.1   Building Life Cycle

In the context of systems, performance is an attribute that is of importance throughout the whole life cycle of that system. At the inception stage, various stakeholders define the requirements that the future system will have to meet; these requirements often take the form of a functional description that includes levels of performance that are to be met. The design process then focusses on developing a system description that will meet these performance requirements. Construction and manufacture actually realize the system; once the system is operational, it can be tested in order to see whether it indeed performs as intended. The system then goes into use; during this phase performance might degrade due to wear and tear, while maintenance and interventions can help restore performance to original levels. At the end of its life, systems are deconstructed or disposed of; this allows a review of overall performance throughout the life cycle but also comes with some particular end-of-life requirements such as disposability. This life cycle applies to all man-made systems, including buildings. INCOSE gives an overview of different views of the life cycle depending on the context, not only discerning the generic life cycle but also looking at more specialist systems such as the high-tech, military and aviation/space industry, which have more distinct phases for issues like user requirement definition, verification, internal and external testing, deployment, deactivation and others (INCOSE, 2015: 29–32).

However, buildings are capital assets with a long lifespan, which makes them different from many other products. A book like *A street through time: a 12 000-year walk through history* (Noon *et al.*, 2012), while aimed at children, gives a vivid account of the coming and going of buildings in the city fabric, the different functions these buildings may have over time and consequently different perceptions of building performance over time. As a consequence of the long life cycle, there also is a greater distance between those that design and construct the building and the occupants/users, and more complex relationships between the stakeholders. The long lifetime of buildings and the various stages of the building life cycle lead to the use of a range of disjoint performance assessment methods and the creation of a vast amount of performance data that is complex and context dependent. Moreover this data is hard to manage: there are issues with handover between the different phases of the life cycle, data overlaps and redundancies as well as missing data and generally a lack of integration (Gursel *et al.*, 2009).

Discussions focussing on the building life cycle tend to hone in on different areas and partly follow divisions of work in the industry. Architectural discourse focusses on the design phase, building and construction research looks mainly at the production process, and facility management publications logically cover the use phase. The full life cycle, including the demolition and deconstruction phase, is mostly covered in environmental work that deals with Life Cycle Analysis or in more economical studies that look at Life Cycle Costing. Lützkendorf and Lorenz (2006) discuss some of the complexities involved in combining the different building life cycles with various performance aspects and the related assessment tools.

Probably one of the most authoritative ways of viewing the building process is the RIBA Plan of Work. With the architectural emphasis on design, this clearly focusses on the front end of the building life cycle, but it also includes construction and use. First released in 1963, it has been revised many times. For a long time, the plan named work stages by letters; this naming still features in many older publications. The current version, released in 2013, discerns seven key stages and uses numbers rather than letters. Naming is as follows: 0: Strategic Definition, 1: Preparation and Brief, 2: Concept Design, 3: Developed Design, 4: Technical Design, 5: Construction, 6: Handover and Close Out and 7: In Use (RIBA, 2013). These stages do not explicitly name building performance, but it is clearly implied; for instance there is reference to project and quality objectives, as well as explicit 'sustainability checkpoints'. A similar naming has been issued by the American Institute of Architects.

The actual construction process includes a range of activities that range from site investigation, site preparation, setting out, excavation, construction of foundations, construction of the superstructure, fitting of windows and doors, installation of services and partitions, to painting and fit-out. A difference can be made between traditional methods of construction and more modern approaches such as off-site construction and modular construction. Work on performance and the construction process roughly falls in two categories. The first consists of efforts that focus on the performance of the construction process itself, with an emphasis on costs, time, waste reduction and health and safety of construction workers and the parties involved; see for instance Mawdesley and Al-Jibouri (2009) or Ogunlana *et al.* (2003). The second category consists of efforts that address the quality of the end product; this category can be subdivided in itself into efforts where quality is primarily 'freedom from deficiencies and defects' and efforts where quality relates to ensuring the user needs and requirements are satisfied

(Gann and Whyte, 2003). For a work that looks at building 'quality' as measured in terms of number of defects and rework needed, see for instance Mills *et al.* (2009).

A specific stage in the building life cycle sits in between the construction and occupancy phases and is concerned with the handover of the building from the construction team to the client. This stage is named commissioning; it is a key stage in terms of performance: this is where clients and their agents check whether the building performs as expected. The efforts invested in commissioning vary significantly; they range from a quick 'walk-through' with visual inspection to note any visual defects all the way to a structured long-term process to ensure the building is operating as designed; in some cases this may overlap with a significant part of the occupancy stage under the label of 'continuous commissioning'. Commissioning should also include the testing and checking of any performance monitoring systems that are present in the building (Gillespie *et al.*, 2007: 10). Commissioning can also be applied to existing buildings; sometimes this is named existing building commissioning (EBCx), retrocommissioning (RCx) or recommissioning (ReCx). In all cases it involves the investigation, analysis and optimization of building performance through better operation and maintenance (California Commissioning Collaborative, 2011: 7).

Facility management by nature looks towards the operation of buildings but has expanded in scope to cover a larger part of the building life cycle such as facility planning, outsourcing, human resources management, service-level agreements, health and safety, workplace productivity, building intelligence, facility management providers, supplier relationships, specialist support such as ICT, contract management and financial control and benchmarking (Atkin and Brooks, 2009). Building performance is core business for facility management (Douglas, 1996); for a recent discussion, see for instance Talib *et al.* (2013) on the application to a healthcare facility. In other disciplines beyond building, the in-use stage is subject to product life cycle management (PLM) that continuously monitors performance of products throughout their life cycle (Djurdjanovic *et al.*, 2003). In some cases, product life-cycle management may also apply to systems used in buildings.

While there is a tendency to focus on the design and construction of new buildings, it must be kept in mind that most buildings still have a limited lifespan. For some of the systems that are part of the building, this is even more important; while the structure may be in place for hundreds of years, systems like heating and ventilation systems these days typically have a lifetime of only something like 10–25 years. This makes it important to look at the performance of these systems, as well as at their expected lifetime. Furthermore building design needs to take into account the management resources that will be available later in the building life cycle (Bordass *et al.*, 2001a). There are two fundamental approaches for estimating the service life of buildings and their parts, components and subsystems. The first is the use of testing, exposure and experience to establish the service life based on measurements; the second is to estimate the service life by means of a reference value (Jasuja, 2005: 51). Service life can also be predicted by models (Trinius and Sjöström, 2005). At the end of their life cycle, both the building and its systems reach end-of life (EOL) and need to be demolished or deconstructed, and to be disposed of or may be recycled.

Efforts in Life Cycle Analysis in the building sector often are an extension from studies of energy consumption during building use towards embodied energy, but modern methods cover the full lifetime of the building from initial planning to deconstruction; for examples, see Bribián *et al.* (2010), Ramesh *et al.* (2010) or Malmqvist *et al.* (2011). Beyond the performance of the building, this work also needs to take into account the performance of processes such as the mining and quarrying of materials, production and manufacture, transport to the site, demolition and ultimately disposal or recycling. Interestingly, efforts seem to focus on exemplary buildings in an urban context, with less interest in traditional methods and rural areas (Cabeza *et al.*, 2014). For examples of work on Life Cycle Costs, see Han *et al.* (2014) or Kovacic and Zoller (2015).

While the view of the building life cycle as a number of stages or phases is useful for discussion, it is important to note that performance typically changes over time. At the beginning of the building life, building performance does not emerge in one go; typically there is a gradual process of components and systems being put in place, connected, and controls being implemented. Once a building is in use, performance normally degrades, due to material degradation, ageing, accidents and similar. At the same time maintenance and renovation may bring performance back to an initial level; in some cases this might even improve on the original performance. Predicting these effects is difficult. While in general older buildings will be harder to maintain, leading to an acceleration of performance degradation, at some point buildings and system get a historic value, and serious efforts will be made to maintain them at all costs (CIB Report 64, 1982: 20). Yet many views of building performance take this as a constant attribute, which can be measured at any point in the life cycle.

2.2 Stakeholders

A stakeholder can be defined as 'any entity (individual or organization) with a legitimate interest in a system' (INCOSE, 2015: 52). Identifying the stakeholders and their requirements is a key issue when exploring what functions a building needs to provide and hence what performance is needed. However, as worded perfectly in an earlier version of the *Systems Engineering Handbook*, this is not always straightforward: 'One of the biggest challenges … is the identification of the set of stakeholders from whom requirements should be elicited. Customers and eventual end-users are relatively easy to identify, but regulatory agencies and other interested parties that may reap the consequences of the system-of-interest should also be sought out and heard' (INCOSE, 2003: 61). This is especially true for buildings with their long lifespan, which may be subject to a significant change of use. The 'actor' perspective complements the 'hard system' view (de Bruijn and Herder, 2010).

CIB Report 64 (1982) already points out the range of stakeholders in building performance, identifying the community, corporations, government, building occupants, clients, designers, builders, product manufacturers and insurers. It notes that these parties might have different interests in performance, with some having a stake in many or all aspects, whereas others might only relate to specific ones. The report also makes the specific point that some stakeholders only get involved with building performance once, like a once-in-a-lifetime client, whereas others are engaged with various projects and can learn across these projects (CIB Report 64, 1982: 7). Later authors, such as Hitchcock (2003: 6), Becker (2005: 58) and Frankel *et al.* (2015: 10–12), give similar overviews, although the specific names and groupings may vary slightly. While all these stakeholders clearly have an interest in building performance, Sexton and Barrett (2005) note that not all of them have the required capacity, capability and motivation to work towards innovation and improved performance. Emphasizing the need for engagement, Bordass *et al.* (2001c) write that a successful building, in terms of overall performance, requires a good client, a good brief, good teamwork, specialist support, a good/robust design, enough time and budget, appropriate specifications, a contractor who builds with attention to detail, good control, post-handover support, independent monitoring and vigilant management.

Clients and developers are the stakeholders that initiate a building design and construction process. The interest of this stakeholder in building performance depends on a range of factors. Frankel *et al.* (2015: 10–12) note that the owner/developer category covers a spectrum that ranges with the duration of ownership. Regular developers are typically short-term owners and often make assumptions on the needs of future tenants. Owner/developers have a long-term interest and typically have a closer control over various performance aspects throughout the building life cycle. The California Commissioning Collaborative (2011: 14–15) supports this view, noting that owner/occupiers typically have full control over the facility and benefit directly from high performance and efficiency gains. Where facilities are let, who benefits from operational improvements and investments in the building depends on the arrangements about who pays the utility bills. Where the party paying for capital investments is not the party who pays utility bills, there may be a lack of motivation to invest, which sometimes is called the 'split incentive' problem. However, at the core of the client and developer interest in performance is a value proposition, where investment in building performance must be balanced by the requirements of the future building occupant; the value proposition may be a straightforward offer for a modest price but may also be a high quality offer at the top of the market. Clients and developers may invest their own money to fund the project or may rely on external investors such as banks to provide finances. Szigeti and Davis (2005: 30) point out that most clients need time and effort to develop a clear understanding of what they actually need in terms of building performance; typically this requires iteration and deep interaction with the design team.

Building occupants or users are the stakeholders that will actually inhabit the building. Because of this, they have a vested interest in the health and safety aspects of the building, as well as the comfort conditions provided. Foliente *et al.* (1998: 20) note that building stakeholders can be ordered at different levels: society, organization and individual. When using the term building occupants, this typically includes not only the permanent occupants but also any visitors, maintenance personnel, neighbours, owners and non-human occupants such as pets (CIB Report 64, 1982: 10). Regulatory provisions are typically in place to ensure adequate health and safety and comfort conditions are met. Becker (2008) points out that while building end users form the largest group of stakeholders, they typically are anonymous before the building is actually occupied and hence are not explicitly present or represented during all stages that precede occupancy. In other cases, users may be the client and developer and come with very specific performance requirements for the facility; for instance, consider the needs of a horticultural or zoological exhibition, a manufacturing plant or an intensive care unit in a hospital, and their direct impact in the design and development process.

The design team is the collective of construction professionals that defines and shapes the building, consisting of experts like architect, architectural technologist, structural engineering, building services engineer, building physics consultant and quantity surveyor. The way in which these experts interact varies significantly; sometimes the 'team' is comprised of one person who integrates all knowledge; sometimes there are many people from different companies involved who may even collaborate on a global scale. Sometimes there is a strict hierarchy in the organization; in other cases there is an intensive collaboration without clear boundaries. This is reflected by McCleary (2005: 217) who notes that where some see different tasks for different actors within the design process, others emphasize that 'the design and construction of buildings is the task of a multi-professional team'. By defining the building, the systems and materials used, the design team makes fundamental decisions regarding the performance of this building. However, the engagement of design team members with performance varies. Structural engineering is probably one of the building domains at the forefront of applying a performance-based approach to design (Becker, 2008); others like architecture do not yet have a unified approach and are still in the process of defining their way of moving forwards (Hensel, 2013: 16). In Europe, efforts are underway to teach integrated design – in terms of a cross-disciplinary approach that creates better integration of the work of architects, engineers and other consultants – at postgraduate level (Brunsgaard *et al.*, 2014). A related emerging concept is that of Integrated Project Delivery (IPD), which is defined as 'a project delivery approach that integrates people, systems, business structures, and practices into a process that collaboratively harnesses the talents and insights of all project participants to optimize project results, increase value to the owner, reduce waste, and maximize efficiency through all phases of design, fabrication and construction'. Key ingredients to IPD are multiparty agreement, shared risk and reward and early involvement of all parties (Kent and Becerik-Gerber, 2010; Uihlein, 2016).

Actual building construction takes place by range of companies. The size and legal structure of these companies again is very diverse, ranging from sole traders and small partnerships to large industrial conglomerates. These purchase materials from a wide range of traders and suppliers, all the way from aggregate for concrete to highly specialized façade systems. Being enterprises, all these companies provide products and services as a business; they balance building performance and customer satisfaction against creating a profit for their owners and shareholders. Lee and Rojas (2014) approach construction project performance from the managerial perspective, with a focus on staying with project plans. Love *et al.* (2015) discuss process performance while focussing on public–private partnerships (PPPs), with a focus on ensuring the delivery of value for money for the public sector. With the many actors in the construction value chain, performance plays a role in dealing with suppliers, adding value and delivering to clients. Key aspects that need to be managed are product quality, on-time delivery, customer satisfaction and prevention of waste (Bronzo *et al.*, 2012). Building performance gets an additional interest where buildings and building parts are innovative and fall outside the conventional everyday practice (Sexton and Barrett, 2005). In general, building firms are mainly interested in the process side of building performance, whereas suppliers will look towards the performance of individual products rather than the performance of the building as a whole. For a critical review of the collaboration and project performance in the sector, see for instance Meng (2012) or Fulford and Standing (2014).

Clearly facility managers have a direct responsibility for building performance. Note that in small buildings this role may actually be taken by the owner/occupier. In larger cases, facility management can actually be split over various domains; for instance there may be a specific subcontract to an energy management company that tracks the energy use of the building and intervenes where needed to make sure all energy systems work optimally.

A special stake in building performance is held by insurance companies. These have a financial interest in preventing claims due to problems with the construction process, as well as any faults and defects in the completed product. Where parties make use of performance contracts and these may be subject of litigation, it becomes extremely important to ensure that insurers have a good understanding of what is being agreed and how this can be verified.

The performance of buildings concerns the outside world at a range of different levels. This goes from the occasional visitor to the building, who has an interest in his or her own safety, health and well-being, all the way to the government, which needs to manage the performance of the built environment at a national level in the light of the interest of the general public. Some of these stakeholders may have very specific performance concerns; for instance a neighbour may be concerned with noise curtailing of a specific facility, or a municipality may have specific interest in the rainwater run-off behaviour in order to minimize flooding risk. Some stakeholders like the fire brigade will look at performance of the building under extreme conditions, such as accessibility for rescue teams and others in case of smoke and fire.

As will be obvious from this list, buildings thus have a range of different stakeholders and decision makers. There are many relations, interdependencies and contracts between these stakeholders. Just as an example, the building owner will have contracts with tenants, a facility management company and needs to meet the local town regulations. In construction, there are typically complex relations between a main contractor and subcontractors; there are special collaboration structures for construction projects, such as public-private partnerships (PPP). Sometimes interests may conflict, such as in the office retrofit situation discussed by Miller and Buys (2008). Typically these various stakeholders do not have the same needs in terms of performance information. Supplying all stakeholders with all information often creates a problem of managing all data (Marques *et al.*, 2010). Moreover, during the life cycle the stakeholders and experts involved with that building change, leading to loss of project information and the context under which data was generated or collected (Gursel *et al.*, 2009). Interaction between stakeholders often takes the form of a project, carried out by an organization that in most cases is temporary and transient. Most projects have inherent uncertainty and are strongly impacted by the project environment. Project managers have to deal with a number of people and companies and need to coordinate large amounts of data (Marques *et al.*, 2010). An important issue in communication and information exchange between various stakeholders is the use of a shared common knowledge base. Only with such a common base can a message sent by one stakeholder be understood by another. Development of a common knowledge base is called 'grounding' (Casera and Kropf, 2010); for building performance this grounding is still under development. Augenbroe and Park (2005) consider performance to be a key element in the dialogue between various stakeholders; they specifically stress the role of performance in the dialogue between the design team and client and in the dialogue between tenant and facility manager. In the context of such dialogues they observe 'asymmetric ignorance between communicating partners'. This concern about dialogue is supported by de

Bruijn and Herder (2010) who point out that decisions are often 'emergent' and the result of interactions and negotiations. Some R&D projects such as Shen *et al.* (2012a) specifically targets the communication between selected stakeholders.

The definition of categories of stakeholders, while useful for a general understanding of the various actors in the built environment, inherently carries the risk of stereotyping. However, there are always exceptions to every categorization. An example is Santiago Calatrava, who is both an architect and a civil engineer. Architects like Renzo Piano and Richard Rogers of the Centre Pompidou clearly manage to incorporate the design of building services systems in their design. On an organizational level, there is also crossover; for instance the case study of ECN Building 42 described by de Wilde (2004) involves a client organization with an in-house building physics department, thus blurring the lines between the traditional client and consultancy definition. Stakeholder roles may also shift over time; a typical example is a tenant in a rent-to-buy scheme. Training of various stakeholders, including specifying engineers and installation contractors, may have a role to play in achieving better performance evaluation of buildings (Gillespie *et al.*, 2007: iv). On a different level, the roles and interactions of stakeholders itself may be a target to achieve better building performance. Frankel *et al.* (2015: 18) believe that this requires a transition of the industry and marketplace to one that recognizes the value of building performance and where this issue is given importance within corporate culture and commitment. This might be brought about by public recognition and further comparison of buildings with their peers.

2.3 Building Systems

Performance relates to buildings and their constituent parts, as well as the occupants and processes that the building houses. Discerning between building parts, components and elements is in fact a way to distinguish building systems. As mentioned in the introduction, a system view is 'a way of looking at the world', a 'mental tool to aid in discovery' (Weinberg, 1975: 51–54). Gilb (2005: 426) defines a system as 'any useful subset of the universe that we choose to specify. It can be conceptual or real'. For Augenbroe (2011: 18) building design is about making functional requirements (needs) and aggregated technical systems (means) meet. This is a complex activity that requires deep knowledge of relationships of building systems and subsystems.

A seminal book on systems is *General System Theory* by von Bertalanffy (1968). Von Bertalanffy was a biologist, but his ideas on systems have been applied to education, psychology, sociology, philosophy, history and cybernetics. The original work of von Bertalanffy is often quoted in systems engineering to define systems; for instance the Systems Engineering Body of Knowledge or the INCOSE Systems Engineering Handbook emphasizes that 'a system is a set of elements in interaction' (SEBoK, 2014: 67; INCOSE, 2015: 5). In the original work, von Bertalanffy (1968: 19) contrasts systems theory with classical science, saying that a system is 'organized complexity with strong, non-trivial (nonlinear) interactions', whereas classical science often studies parts that can be isolated from their surroundings so that the behaviour of these parts can be worked out. Raphael and Smith (2003: 11) expand this view by defining three types of complexity: descriptive complexity, cognitive complexity and computational complexity.

Systems can be decomposed into subsystems, which again can be decomposed into a further level of subsystems or elements. This leads to system hierarchies. One key task for system engineering is to define the right hierarchy, without overdoing the decomposition; this would result in too great a span of control or excessive layers in the hierarchy. A 'rule of thumb' useful in evaluating this balance is that a system should have no more than 7 ± 2 elements reporting to it. Hierarchies with many more subordinate entities will suffer from too much complexity and become unmanageable (INCOSE, 2015: 7–8). Weinberg (1975: 62–67) notes that a systems view of the world creates order and often allows the mathematical handling of a set of systems or subsystems.

Gilb (2005: 47) links systems back to performance by stating that 'a system can be described by its set of function attributes, performance attributes, resource attributes and design attributes. All these attributes can be qualified by conditions, which describe the time, place and events under which these attributes exist.'

A system view is a way of looking at the world; however in reality most systems interact with other systems. Within most systems there is the hierarchy of subsystems and elements. Outside the system boundary, there may be interaction with an operational environment and enabling systems (INCOSE, 2015: 11). For a building, the operational environment may be the town or the city; enabling systems may be the power grid or the maintenance system provided by a facility management company. Inside the building, the operational environment of for instance the communication system may be defined with structural elements such as reinforced concrete walls that limit the range of transmission.

Systems have *attributes*, which are an 'observable characteristic' or 'property' of the system (INCOSE, 2015: 6). Note that performance is a system attribute, but not every attribute is related to performance. For instance a car may have top speed and colour as attributes, where only the top speed is a performance attribute. Typically, attributes are represented by symbols named variables.

When applying systems thinking to buildings, there are various ways to decompose a building; there is no right or wrong as it all depends on the objective of the decomposition. This is already demonstrated by CIB Report 64 (1982: 5–6), which mentions three different views: one view that distinguishes systems, subsystems, elements, components, general products and materials; another view that distinguishes structure, envelope, space dividers/infill and building services; and yet another classification that focusses on spaces and discerns between spaces between buildings, open areas inside buildings, rooms, ancillary spaces and service spaces. The classification into construction/structure, envelope/façade, services/HVAC and infill/content is a common one and can already be found in classic work such as Markus *et al.* (1972: 11), Rush (1986: 10) and Bachman (2003: 19). Douglas (1996) orders these systems as 'shearing layers' that have furniture and equipment on the inside while ultimately connecting to a building site and surroundings. He notes that when moving through this hierarchy, the typical lifespan of the system or layer increases from a few years all the way to permanent. The California Commissioning Collaborative (2011: 4), with a specific view on building performance management, suggests that building systems are typically complex and highly interactive. Their performance typically degrades over time, requiring recalibration of sensors, adjustments in control or further maintenance and refurbishment. Contrary to most other products, buildings are fixed in place, making them more sensitive to their context and giving them a strong cultural dependency (Gann and Whyte, 2003).

Bachman (2003: 18) notes that, in the domain of architecture,[1] systems thinking is an important factor in efforts to progress design as a structured methodology. He also points out that in architecture there is a tendency to combine a functional view of systems with the physical elements that provide those functions. He suggests that this leads to a 'corrupted' use of the term systems as a form of 'hardware classification'. Bleil de Souza (2012) notes that engineers and scientists typically operate from a systems point of view, with a concern about how parts are organized and a focus on how these parts interact. For these actors the use of models to study the behaviour of systems is a powerful approach. Augenbroe (2011: 18) makes an explicit distinction between functional (aspect) systems and technical systems. In this approach, aspect systems relate to a view of building (sub)systems that are based on specific functions and requirements, such as providing shelter or safety. Technical systems relate to the aggregation of building elements into more complex systems, which in turn can be subsystems in a further hierarchy. Augenbroe also points out that the relationship between the technical subsystems typically is much more complex than just that of constituent parts; there are many functional relationships, assembly rules and constraints amongst different subsystems. Textbooks for undergraduate students in building technology such as Bryan (2010) or Riley and Cotgrave (2013) try to cover both aspects by discussing the many systems that may be used in a building, as well as the many functions these systems may have.

Bringing systems together in a building requires physical integration, visual integration and performance integration (Bachman, 2003: 32–34). Rush (1986: 13–14) notes that integration can take place between systems that are remote, touching, connected, meshed or even unified. In order to achieve this integration the design team must have a good grasp of the different components that make up the system and how these may relate to one other, which is a non-trivial task and requires processing of a significant amount of information (Bar-On and Oxman, 2002). When combining products into a building, there is a need for dimensional coordination in order to manage the interfaces with and connections to other systems. Dimensional coordination often leads to the prescription of standard sizes, profiles and so on. To some extent this is contrary to a pure performance-based approach where all solutions that meet the user needs are allowed; however, dimensional coordination has been proven to be very effective in both the construction and other industries (Becker, 2005: 72). In some cases, such as industrialized housebuilding, the architectural design focusses on a limited number of house types that customers can choose from; in some instances a certain amount of customization is designed into the types so that customers can make these houses fit their specific requirements (Wikberg *et al.*, 2014). This approach lends itself to extra efforts to ensure the performance of the main types on offer, whereas the customization aspect fits well with modularization and a systems approach. Good examples in this area are the Scandinavian prefabricated timber houses or more expensive variants such as the German 'Huf Haus' system.

Mumovic and Santamouris (2009: 3–6) describe the complexity of the built environment system, noting the many interactions between architectural statement,

---

1 Bachman discerns two philosophical positions on building design: design as a structured methodology and design as free creative exploration.

energy, health, operational performance and ultimately the human building occupants. They describe buildings as often being 'complex, bespoke systems that are difficult to control, with little feedback available on their real operation and actual
performance'. Allen (2005: 31) points out that most components of a building have
several functions and that the functioning of these components mostly is interdependent. The complexity also becomes evident in work such as that by Geyer (2012)
who has used the SysML (Systems Modelling Language) to model the building systems that relate to building energy efficiency. De Bruijn and Herder (2010) observe
that buildings are in fact sociotechnical systems that include both complex technical
systems and a network of interdependent human actors. There are many factors that
cause complexity, such as the interdependence of the subsystems, feedback loops,
non-linear relationships and systems working in parallel or sequential as well as
synchronous or asynchronous.

Building subsystems and components may serve more specific functions and sometimes can be tested under laboratory conditions – something that is difficult and costly
at the whole building level. As a consequence, the use of performance requirements is
much more common at the component level, such as for windows (CIB Report 64, 1982:
22). Neumann and Jacob (2008: 92) mention the use of Functional Performance Tests
(FPT), which represent standardized tests at the system or system component level.
In contrast to the observational approach that is mostly used for whole buildings, FPT
typically study the system or component by exposing it to a specific operational regime,
thus forcing it into specific operation points. An example may be the testing of a gas-
fired boiler using a dedicated test rig facility that allows the measurement of boiler
performance with regard to both heating and domestic hot water supply under typical
loading conditions. Data obtained from component testing can be used to predict how
the component will perform within a complete building using modelling and simulation
approaches using a process of component model calibration, scaling and replication
(Strachan, 2008). However, it is important to keep in mind that the performance of
many systems also depends on the interaction with the rest of the building; in such cases
the performance of the system can only be established long after the decision to use the
system is actually made (Thompson and Bank, 2010).

A general observation on building systems that serve specific functions is that there is
a need to make building systems, components and their control more understandable to
the occupants. It is advisable to avoid unnecessary complication (Bordass *et al.*, 2001a).

For novel and innovative systems, the maturity of particular technologies is of interest. Technology Readiness Levels (TRLs) have been defined by NASA to identify
where problems in terms of budget, schedule and performance might be expected to
occur (INCOSE, 2005: 54–55). TRLs cover nine categories, ranging from basic
principles only (TRL 1) to flight proven technology that has been applied in actual
missions (TRL 9).

A specific point of interest in discussing building systems is the notion of system
architecture. While architecture has a specific connotation in the building domain, it
has a different one in systems engineering. INCOSE (2015: 64–65) distinguishes
between system architectures, which conceptually define a system (in terms of what
elements are part of the system), and the characteristics (attributes, design features and
parameters) of these elements themselves. For Blanchard and Fabrycky (2011: 100), the
system architecture is a solution that deals with both the requirements and the system

structure. The Systems Engineering Body of Knowledge defines system architecture as the organizational structure of a system in terms of both system components and interrelations (SEBoK, 2014: 202).

2.4   Building Performance Challenges

Before exploring the performance challenges that buildings face, it is worth to pause for a moment and to review a few peculiarities of the built environment: scaling from single building to the built environment, role in the economy and complexity and emergent behaviour.

An important point in considering building performance challenges is that buildings are omnipresent and thus play a role in many important aspects of human life. While individual buildings might play only a minute part in global issues, the large multipliers that apply often scale things up to a major contribution. A clear example is the impact of buildings on energy use. While the impact of single buildings is almost negligible, globally the building sector is responsible for roughly one third of the total final energy consumption and more than half of the electricity demand; three quarters of this energy use is consumed by households, turning residential buildings into a major area of concern (IEA, 2015: 400). A similar argument can be made for the material resources used in buildings; in general construction plays a role in major environmental issues such as global warming, ozone layer depletion, ecosystem destruction and resource depletion (Ding, 2005). The scaling effect is also clearly present in problems with the safety of construction materials that can have major and long-lasting impacts on whole populations, as exemplified by the health impacts of water from lead pipes that happened in ancient Rome and that still occur occasionally in modern times (Delile *et al.*, 2014; Edwards, 2014).

Another issue to keep in mind is that in many countries, the construction industry is a major part of the economy. Large sums of money are tied up in built assets; the building sector is labour intensive and employs a significant part of the workforce. In times of economic downturn, such as the financial crisis of 2008, construction is often hard hit, as investment in new buildings requires confidence by decisions makers in the financial outlook. Yet governments may want to invest in construction to give impulses to an economy that is struggling; for instance the construction of homes typically creates jobs in construction but also has significant spin-off as new residents will buy furnishings and appliances. Construction thus plays an important part in things like a country's Gross Domestic Product (GDP), employment rate and trade balance (Dasgupta, 2007). Maintaining a healthy interaction between construction and economy is a delicate two-way challenge, as exemplified by the role of the US subprime mortgage crisis in the lead-up to the financial crisis of 2008 (Demyanyk and van Hemert, 2011). The construction sector has peculiar problems in being functionally and organizationally fragmented, with many different companies, a gap between design and construction and often a separation of client and occupant (Ang *et al.*, 2005). At the same time, like other industrial sectors, construction is subject to globalization and a changing business environment, with increasing foreign ownership, foreign affiliates and changing forms of procurement (Jewell and Flanagan, 2012).

While one may take buildings for granted because of our familiarity with the built environment, buildings actually are not simple products and systems. As mentioned at the start of this chapter, buildings are typically one-off products and seldom made in series. Buildings are probably the product that is most modified after handover to the client, sometimes in very fundamental ways. In the systems engineering world, a distinction is made between complicated systems and complex systems. Here complicated systems consist of many parts that interact in a certain way. Complex systems have 'emergent behaviour', which is not just a summation of the behaviour of the parts, often have subsystems that are operationally independent and sometimes are subject of evolutionary development (INCOSE, 2015: 8–9). Buildings clearly are both complicated and complex – consider the many parts that make up the load-bearing structure of a typical office or the complex interactions that take place in the heating and ventilating system of such a building.

Buildings are primarily constructed to provide humans shelter and comfortable living conditions. Yet not all buildings meet this challenge. Some buildings are overheating, cold, or have problems with the indoor air quality and lead to health issues designated with the terms sick building syndrome (SBS) or building-related illness (BRI); see for instance Redlich *et al.* (1997) or Crook and Burton (2010). Sometimes the lack of performance is due to the buildings themselves, due to poor design and construction; in other cases the lack of performance is due to a range of factors. For instance, in some countries part of the population lives in fuel poverty, leading to cold homes, poor health and excess winter deaths; the cause here often is a combination of poor quality dwellings, maintenance issues, family income, work status and others (Boardman, 2010).

An important challenge when looking at building performance is the deep interrelation between buildings and environmental issues. Construction and operation of buildings requires the use of resources such as land, water, non-renewable materials and energy. Buildings contribute to greenhouse gas emissions, surface water run-off, eutrophication, and might damage local ecosystems; on a larger scale buildings thus play a role in problems such as climate change, smog formation and acidification. Efforts to make buildings and the building process more environmentally friendly have been underway for almost half a century, ever since the energy crisis of the 1970s. However, the complexities of building systems and the interaction between different aspects mean that serious challenges remain. For instance, as heating systems become more efficient and operational energy use decreases, more emphasis is placed on embodied energy in creating buildings; or in a different example, energy-efficient buildings might encounter problems with indoor air quality and/or overheating. Some of the environmental challenges link back to other domains; for instance the use of fossil fuels is related to economic aspects (peak oil) and politics (dependence on supply by foreign countries). Issues like fuel poverty have a deep impact on people health and well-being (Boardman, 2010). In general, end users, companies and governments worldwide have to pay more attention to energy efficiency due to climate change, unsecured energy supply and rising energy prices, which lead to new and stricter regulations and changing customer behaviour (Bunse *et al.*, 2011).

While there has been significant attention for making buildings more energy efficient and how this relates to greenhouse gas emissions and preventing climate change, there has been less attention for climate change adaptation: ensuring that buildings can cope with the impacts of climate change that is already underway. Some early work in this direction is reported by Hasegawa (2004). A generic overview of this area is given by de Wilde and Coley (2012), who discern graduate climate change, extreme weather events, sea level rise, effects on ecosystems and environmental degradation as related challenges that may impact on both buildings and the occupants and key processes inside those buildings.

The building design process in itself is another challenging area that sometimes causes building performance issues. Creating a built environment that is of good design, is enjoyable, has value and improves quality of life is a goal that is hard to achieve. One of the obstacles in realizing such a built environment is that there is no well-developed understanding of building quality and how to measure it (Gann and Whyte, 2003). Some design teams set unrealistic expectations and publish very low energy use projections that are hard to meet later on (Hinge and Winston, 2009). In general building performance targets are becoming more demanding and now typically require not only appropriate designs but also involvement of owners, operators and tenants (Frankel *et al.*, 2015: 1).

The construction industry has a tendency to focus on process performance rather than product performance and quality; see for instance Han *et al.* (2008) and its focus on productivity and inventory management. The industry often reduces upfront cost and allows for deliberate value engineering and cost-cutting that may lead to lower building performance (Loftness *et al.*, 2005). In general, the sector is not good at learning from its completed products. Performance feedback is often kept confidential; findings from real buildings are at best provided as anecdotes or as general comments without reference to specific projects (Cohen *et al.*, 2001). Typically learning only takes place if there is a serious issue or failure (Bordass and Leaman, 2005). After handover, many building owners and operators lack the expertise to move beyond simple temperature regulation and do not manage buildings and systems at their optimum settings (Neumann and Jacobs, 2008: 3). Building performance is sometimes put at risk because of lack of follow-up through later stages, for instance when modifications are made. There also may be oversights due to the complexity of many aspects that need to be tracked (Spekkink, 2005: 44). Frankel *et al.* (2015: 5–6) also note the issue of resistance towards guaranteeing performance and pushback of accountability to the operational stage, issues with awareness of the public and industry and disclosure of benchmarking data (*ibid.*, 5–6). An additional problem is that for the construction industry, building performance is often equal to the absence of defects. However, there is still some way to go in this respect. Georgiou (2010) reports an average of 5 defects per home built post-1996 in Australia, with an upward tendency; key defects in this work are cracking, damp, leaks, workmanship and not meeting the regulations rather than deeper performance issues. The idea that performance is established by compliance with rules and the absence of defects can also be found in construction management. For instance, Chow and Ng (2007) list compliance with the design brief, compliance with legislation and adequacy of a cost estimate as 'quantitative indicators' for Consultant Performance Evaluation.

Beyond handover to the client, facility management and maintenance are increasingly important. Management of building performance, in terms of aspects like occupant comfort and efficient operation, now often is an issue in attracting and retaining tenants (California Commissioning Collaborative, 2011: 4). At the same time, economic pressures in commercial buildings lead to intensified space use and higher occupant density. The now prevalent ICT systems sometimes lead to noise problems caused by printers and computers (Leaman and Bordass, 2001). Moreover, in modern technology, the degradation of system performance is often invisible, which may lead to unexpected failure of systems (Djurdjanovic *et al.*, 2003).

Increasing accountability in the juridical system is also providing challenges in the area of building performance. In terms of contracts and responsibilities, the building industry over many years has been confronted by a change towards increasing protection of consumers and a tendency to take legal action where things do not go as planned (CIB Report 64, 1982: 23). Problems with indoor environmental quality, acoustics and building integrity are leading to costly litigation and sometimes require significant efforts to put right (Loftness *et al.*, 2005). Moreover, in legislation there is a trend to strengthen the responsibility of the actors in the building design process, ensuring they meet the requirements of professional liability and duty of care as well as increased expectations in terms of warranties of end users (Almeida *et al.*, 2010). At the same time, as building codes are becoming more ambitious, they also tend to become harder to enforce (Frankel *et al.*, 2015: 4).

Performance requirements for buildings are not always static. The emergence of new threats like the terrorist attacks of 9 September 2001 in the United States have had permanent consequences for the operation of airports and other public buildings worldwide, whereas the more recent attacks in Paris, Brussels, London and Manchester are once more putting the safety measures in railway stations, stadia and theatres under review. Developments like the spread of contagious diseases such as severe acute respiratory syndrome (SARS), Ebola and the Zika virus have had serious impacts on the requirements placed on hospitals and the built environment in general. Some performance issues emerge only with the progression of science; for instance the problems related to asbestos or particle emissions of certain engine types were initially unknown. Even within one and the same building, and looking at the same requirements, building performance is typically dynamic and changes over time; moreover it is subject to interventions and the impact of management factors (Frankel *et al.*, 2015: 19).

Understanding and managing the role of building occupants also remains a challenge. In many areas, such as energy and water efficiency, building performance is strongly related to occupant behaviour. It has been suggested that behavioural change might allow further efficiency gains to be made; however this requires further understanding of the knowledge, attitudes and abilities of occupants (Gill *et al.*, 2010). Corry *et al.* (2014) point out that there also is a lack of integration between the building operation domain and soft data sources (e.g. information held by human resources, social media etc). This leads to the existence of 'untapped silos' of information, which limits the understanding of the data that actually is analysed.

The role of technology in current society also challenges building performance. The generic desire to upgrade to the latest model of mobile phone, tablet or automobile creates pressure on the technical disciplines, including construction (Rahim, 2005: 179). Yet due to the long lifetime of many building systems, these typically lag behind. A good example is Building Automation Systems (BAS), which originally are designed for operational control, not for data capture, storage and analysis. Most building automation systems require the installation of additional interfaces and systems if advanced data management is expected (Neumann and Jacobs, 2008: 1). In theory, the tracking of the performance of buildings in terms of metrics that capture occupant comport and use of resources may allow the operator to identify and remedy issues before complaints are raised (California Commissioning Collaborative, 2011: 53). However, this will not materialize without additional investments. Leaman *et al.* (2010) make the case that building evaluation is real-world research, requiring a multidisciplinary approach and fieldwork, with a focus on solving problems, actionable results and provision of value to the client.

A special kind of challenge is extreme events, such as natural disasters, industrial accidents and acts of terrorism. Sometimes these are seen as focussing events, which gain attention in the media and lead to efforts to prevent repeats from happening, or at least in mitigating the impact if a repeat occurs (Birkland, 2009). The evaluation of building performance under extreme events requires dedicated approaches that allow the study of specific 'what if' scenarios (Thompson and Bank, 2010). In some cases, extreme events are related, such as an earthquake that causes fires or a storm resulting in both high winds and flooding. Li and van de Lindt (2012) review the risk of multiple hazards in terms of high wind, earthquake, snow and flood in a study that combines probabilities, two levels in terms of construction standards and performance limit states. The recent events in the Northern England (2015), South India (2015) and Central Europe clearly show the challenge of dealing with flood risks and their impact on buildings (Kreibich and Thieken, 2009). Beyond models that predict simple water levels, work is now undertaken to predict the impact that floods have on buildings and building stock (Schröter *et al.*, 2014). The challenge of windborne debris that result from cyclones, hurricane, tornados and typhoons is driving research in the development materials and products that help buildings to perform well when subject to such extreme conditions (Chen and Hao, 2014; Chen *et al.*, 2014). Some relations that play a role in the wake of extreme events such as flooding and the subsequent link between moisture damage to buildings, exposure of occupants to these buildings and their health are complex and still subject of deep studies (Haverinen *et al.*, 2003). In some cases moisture may be the result of regular use conditions, such as surface and interstitial condensation; in other cases this can be related to extreme events, such as flooding. In this context He *et al.* (2014) studied the impact of flood and post-flood cleaning in Brisbane; Hasegawa *et al.* (2015) describe the long-term effects of the tsunami in Japan. While areas like building fire safety performance have a substantial body of knowledge on dealing with technological aspects such as building materials and fire protection measures, this is another area where the 'soft' aspects such as occupant characteristics still need further consideration; see for instance the work by Park *et al.* (2015). Building performance in terms of allowing the safe evacuation of occupants in case of emergency is an area where understanding of human behaviour and crowd movement is an active research field and where new design approaches are currently emerging (Pelechano and Malkawi, 2008). Specifically, crowd modelling is helping to evaluate exit strategies and the response to different emergency scenarios (Sagun *et al.*, 2011).

The theory of building performance itself also still requires further development. The dialogue about building performance between different stakeholders suffers from different approaches to express expectations (client demands) and actual provisions (supply by the design and construction sector). This leads towards many disconnects in the communication about performance throughout the industry. Performance-based building seeks to address this problem by providing frameworks for the systematic definition of stakeholder needs and actual fulfilment of these needs by buildings (Augenbroe, 2011: 16). At the same time, this concept of performance-based building design is now believed to be hampered by differences in responsibilities amongst various stakeholders; an example is the discussion that ensues if a building was designed for operating 50 hours per week, but tenants actually are operating the facility for 80 hours per week (Frankel *et al.*, 2015: 8). In general building performance is often hard to judge, even by experts, and typically can only be established after deeper investigation (Bordass *et al.*, 2001c). There might even be an element of 'cultural resistance' against the quantification of building performance by people who hold that performance is subjective and hence cannot be measured. Augenbroe and Park (2005) assert that this position of 'subjective quality' is now being replaced by 'objective utility', but this does yet seem to have pervaded the full building sector. And ultimately, actually carrying out performance measurement in building practice is still challenging. Fernie *et al.* (2006) point out that not all aspects can be measured and that the practice of measurement tends by its very nature to emphasize what can be measured. Performance measurement is seen as important by both academia and practice, but there appears to be a sparsity of empirical research and uptake in operational practice. Studies of the use of performance measurement in practice typically require an 'action research' methodology. Barriers to such work might be business organization, process maturity, employee engagement and organizational culture (Farris *et al.*, 2010).

Underlying science for performance assessment also is still under development. As an example, in the field of building materials, there is a lot of work being done on the assessment of external and internal impacts of these materials. The external side typically covers emissions, waste, natural resource depletion and toxicity during manufacturing, transport and construction, whereas the internal side looks at discomfort and well-being. A common approach is Life Cycle Assessment (LCA). An early overview of integration of these approaches into computation assessment of building performance assessment is presented by Mahdavi and Ries (1998), but gathering the data for this type of analysis is still challenging, and assessment methods are still under development (Assefa *et al.*, 2007). Different approaches may be needed here to differentiate between local and global depletion of resources (Habert *et al.*, 2010). Extra complexity is added to Life Cycle Assessment by the need to review land use as part of the impacts under consideration (Allacker *et al.*, 2014). Similarly, the quality of indoor air, in relation to occupant health, continues to be an area of work. An example is the detection, identification and relation to a source of airborne particles inside buildings, as discussed by for instance Gudmundsson *et al.* (2007). With the relation between indoor and outdoor air, many potential sources of particles and different ventilation systems, this is a complex field. As in other areas, the role of occupants is also gaining attention here, this time in terms of cleaning habits, ventilation behaviour, purchases of products and appliances and similar (Loftness *et al.*, 2007).

In spite of efforts to the contrary, some buildings do not fully perform as intended, leaving a significant discrepancy between predicted and measured performance. This mismatch is sometimes named a performance gap. An 'energy performance gap' has been noted since the mid-1990s, as reported by Norford *et al.* (1994), Bordass *et al.* (2001a, 2001b) and Scofield (2002), and is further discussed in more recent work such as Turner and Frankel (2008), Menezes *et al.* (2012), Stoppel and Leite (2013), de Wilde (2014) and CarbonBuzz (2016). In many cases, actual measured energy use is higher than predicted; however the deviation can go in all directions – in some areas performance might be better than expected, in others it might be worse and in yet others the building might just work differently from what was intended (Cohen *et al.*, 2001). Similar performance gaps can also be identified in other areas, such as daylighting, indoor air quality and acoustics.

Frankel *et al.* (2015: 2), reporting on a series of recent workshops with industry leaders in the United States, provide a good overview of some of the current challenges to achieve good building performance. These include the need to engage occupants in achieving building performance, a lack of recognition for building operation and facility management in some organizations and insufficient understanding of the complexities of building performance by policymakers. They suggest it would be good to establish a better link between real estate valuation and the performance of this real estate. A move towards 'integrated design', with better attention for building performance, also requires clear responsibilities for specific aspects of performance, which can be challenging in complex projects. They also point out that there is need to better understand how performance may change over the building lifespan, that there are issues with building data capture and storage, and that work is needed in scaling performance studies beyond the analysis of individual buildings.

2.5 Building Performance Context in Current Practice

A real building that demonstrates the challenges that the complex context poses towards building performance analysis is the Bank of America Plaza in Atlanta, United States. This building is a commercial skyscraper that offers a floor area of $122\,000\,m^2$, spread out over 55 storeys. It has a height of $312\,m$, and at the time of writing (2016), it ranked number 71 amongst the tallest buildings in the world and number 11 amongst tallest buildings in North America. It is the highest building on the Atlanta skyline, providing a landmark that is recognizable as far away as the Atlanta international airport, 12.5 miles away. The building is designed in modern Art Déco style; the top is formed by an open lattice steel pyramid that ends in an obelisk-like spire that is covered in gold leaf and lighted at night. See Figure 2.1. Beyond office space, the building offers a gallery café, conference and dining rooms, retail space, a health club, banking facilities, underground parking and a postal services distribution area.

The history of the building exemplifies the complexity of the many stakeholders involved with such a high-profile building. Originally, it was intended as headquarters for the Citizens & Southern National Bank, but already during construction ownership changed as Citizens & Southern National Bank merged with Sovran Bank; this in turn was acquired by North Carolina National Bank. The merged company took the new name NationsBank, and upon completion the building was named NationsBank Plaza. NationsBank in turn bought BankAmerica in 1997 and rebranded as Bank of America; this led to the building being renamed into the current Bank of America Plaza. In 2002 the building was sold to real estate firm Bentley Forbes from Los Angeles; however this company lost the Bank of America Plaza to foreclosure in 2012, when CW Capital Asset Management became the owner. In 2016 the building was purchased by Shorenstein Properties of San Francisco. Facility management was carried out by Parkway Realty Services before this was taken on by Shorenstein. Office floor space is rented out to a

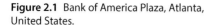

Figure 2.1 Bank of America Plaza, Atlanta, United States.

range of tenants, which now include Bank of America, law firms Troutman Sanders LLP and Hunton & Williams LLP and the CDC Foundation.

The building was developed by Cousins Properties together with the various banks that ultimately became Bank of America. The architectural design was provided by Kevin Roche John Dinkeloo & Associates (KRJDA), with CBM Engineers doing the structural plans and Environmental Systems Design Inc. carrying out the mechanical, electrical and plumber design. The main contractor was Beers Construction. In terms of the life cycle, construction started in 1991 and the project was completed in 1992. Construction of the building took place on a fast-track schedule; work was completed in 14 months, on time and on budget. Dating from 1991/2, the building is now 25 years old; a first retrofit cycle was completed in 2014. For this retrofit, advice on environmental issues was provided by the Sustainable Options LLC consultancy. Internal spaces were redesigned by interior architects Hirsch Bedner Associates.

Regarding systems complexity, the load-bearing structure of the skyscraper consists of a composite frame. It employs two super columns at each edge of the tower; this creates an interior that has no local columns and offers uninterrupted floor space. Above the lobby, the building has a curtain wall that mixes red granite with glass; given the height of the building, a good view from the offices is important. Sealants used on the façade are from Dow Corning. Vertical access to the building is provided by 24 Otis high-speed elevators; these have card access, while the overall status is controlled by a console in the lobby. The spire at the top of the building contains cooling towers and other mechanical equipment, as well as the elevator engine rooms. Overall access to the building is controlled by way of key card access and a turnstile; the building also has dedicated security personnel. Visitors are requested to use a web-based pre-authorization system and to bring photo ID. Fire and safety systems include emergency power supplies, which can use either the local Georgia Power service or backup emergency generators. Other systems include emergency lighting, fire alarms, sprinkler systems with fire jockey pumps, stairwell pressurization and smoke exhaust and safety controls for the elevators. Part of the retrofit included upgrades to the HVAC system, installation of more water-efficient systems and new electric vehicle charging points. This HVAC system now includes chillers, cooling towers, more than 400 VAV units and new Honeywell controls. The water systems upgrade, which includes low flush and flow fixtures, new water-saving toilets and urinals, repair of leaks and continuous monitoring, has reduced indoor water use by half; a cooling tower condensate recapture system has also been installed.

In terms of building performance challenges, the original layout from the early 1990s reflects the focus of the United States on economic growth in a time of relatively low fuel prices. The building was well received, winning the Building Owners and Managers Association (BOMA) International prize for The Outstanding Building of the Year, International Award for buildings over 1 000 000 square feet and also being named BOMA Southern Regional Building of the Year. However, raising costs of resources, increased concerns about environmental impact and difficult market conditions due to the economic downturn of 2007 caused a rethink during the 2014 retrofit cycle. After this work was completed, the Bank of America Plaza won a range of accolades for environmental performance. Amongst these was the recognition as 'Midtown EcoDistrict Luminary' at the Green level, awarded by the local Midtown Alliance of Atlanta, in recognition of high environmental performance in terms of energy, water, waste and transportation. The project was also given an 'Outstanding Water Project' award, as well as the 'E3' Liquid Asset status by the Metro Atlanta Chamber of Commerce. The current owner, Shorenstein, mentions that their management strategy includes the continuous evaluation of building performance, combined with continuous improvements to the facility. Efforts include energy efficiency, water efficiency, waste and e-waste, integrated pest management, green cleaning and tenant outreach and aim to meet 'green' construction standards. Key results include certification with the US Environmental Protection Agency (EPA) EnergyStar scheme, with a 2016 score of 76 – meaning that the building is more efficient than 76% of comparable buildings. Moreover, the building was certified as LEED (Leadership in Energy and Environmental Design) Silver in 2015, scoring 50 credits out of the 110 that are available under the Operations and Maintenance (O + M) for existing buildings scheme. Shorenstein markets the building as 'Atlanta's Tallest High Performance Green Building'. Ambitions include a further reduction of energy and water consumption by 20% in 2020.

Building management is provided through a local office in the building. Maintenance engineers are available round the clock to deal with emergencies. Building management includes a cleaning service, covering vacuum cleaning of carpets, dusting and trash collection. Tenants can require maintenance through an online system. Occupant safety is a key concern to the building management; specific processes are in place to deal with a range of issues such as elevator malfunction, suspicious packages, bomb threats, civil disturbance, fire, severe weather and toxic spill or exposure. Special rates are available for building tenants to undertake first aid training. Where tenants want HVAC operation outside regular opening hours, this can be requested from the management. Window washing and maintenance is provided by the specialist company HighRise Systems Inc.

2.6   Reflections on the Complexity of the Context

Building performance in itself is a difficult subject, but it also lives in a complex context: buildings have a long and often interesting life cycle, involve a multitude of stakeholders and consist of many systems, components and parts that all have their own properties and lifetime. It is very easy to underestimate this complex context, especially since buildings are so familiar. Yet at the same time many buildings are unique to some degree and have their own specific settings; this makes it hard to generalize findings and observations. But keeping in mind the contextual complexity is of crucial importance when working with building performance.

It is crucial to keep in mind that performance is a time-dependent building attribute; it varies as a building takes form during the design and subsequent construction process, and it changes under different loadings during the life cycle. Interventions, from a simple change of control settings to expensive extensions and retrofit programmes, are likely to cause step changes in performance. The fact that various building systems and component have a different lifetime makes this issue even more significant. As a consequence, assessments of building performance should always indicate the point in time for which they are valid. General statements of building performance should be tested to check whether indeed they are valid for the whole building life cycle. The temporal context cannot be ignored.

Recent work often states the need for buildings to be energy efficient and reduce greenhouse gas emissions, often in the context of climate change and quoting statistics that find that the built environment is responsible for something in the order of 30–40 percent of national energy consumption in most countries. This may be true and is indeed an area of concern, yet it is important to remember that there are other environmental problems such as depletion of resources like water, material and local ecosystems and other challenges such as population growth, maintaining economic prosperity or safeguarding health and safety. Highlighting the issue of thermal performance is fine, but at the same time the general coverage of building performance should be seen to be broader. Simplifying it to one single aspect only does not do the concept of building performance justice.

As building performance is recognized as an area of importance, different stakeholders claim ownership and attempt to lead on achieving it. This sometimes results in tensions, for instance between the different actors in a design team or between practitioners and academics. Once more, there is complexity, as many actors will face different tasks, have a slightly different understanding of performance and will use different tools that influence the way they think. Yet in the end, it should be realized that many parties have a role to play in creating and operating buildings that perform well, and that teamwork is the best way to move forwards. Where stakeholders are discussed, these are often reduced to stereotypes, with simple attributes, such as in the statement that 'architects think visually' or 'engineers employ calculations'. As always, stereotyping is dangerous; actual people and organizations cover a wide spectrum and attributes.

The practical example of the Bank of America Plaza shows the contextual complexity of the performance analysis of a real building. In thinking about building performance, one thus should take extreme care in considering this context. Referring back to Weinberg (1975: 228), it is wise to ask: 'Why do I see what I see? Why do things stay the same? Or why do things change?' The context has a deep impact on the observations made and how building performance is perceived.

2.7   Summary

This chapter positions building performance in a wider, complex context. Performance is a concept that lives on the interface between a building as a system and the functions of that building; however, a range of aspects frame our perception of performance. The chapter starts by looking at the temporal perspective, discussing the relation between building performance and the building life cycle. It then moves on to the organizational and human perspective, addressing the relation between building performance and stakeholders. The next perspective is that of systems, reviewing different ways to look at the different parts that make up the building and how they relate to one another and the performance of the whole. Finally, the chapter explores the different challenges that drive the interest in building performance and that provide urgency to the development of the concept and how this is handled.

Performance is something that changes over time and hence needs to take into account the various building life stages: inception, design, construction, commissioning, use, renovation and refurbishment and ultimately deconstruction and disposal. It is important to keep in mind that many buildings have a long lifespan and may be subject to significant change of function at some point in their life. An important view of the building life cycle is the RIBA Plan of Work. Most of the life stages can be divided into more detailed steps that all have their own dynamics and relation to building performance. Performance degradation with the passing of time and the impact of maintenance are aspects that deserve particular attention.

Building performance is of interest to a wide range of stakeholders: clients and developers, occupants and building users, the building design team, construction companies and their employees, facility managers, insurers, occasional visitors and the general public. Each of these stakeholders has different interests that need to be balanced in an attempt to make buildings meet all requirements and aspirations. At the same time, it is important to note that categorization should be handled with care as it may lead to stereotyping; in real life actors may move from one category to another and change their priorities over time.

A system view of buildings can help in understanding how parts, components and other entities work together to deliver building performance. It is important to keep in mind that there may be different ways of seeing a system. A classic view of buildings discerns building structure, services and contents. Often, systems can be decomposed into subsystems, leading to a hierarchy of systems. Building design typically works in the opposite direction, combining systems into a whole; this requires physical integration, visual integration and integration of performance. Building systems are mostly complex, bespoke systems with non-linear relationships and interdependence between subsystems and are related to human actions. Building components can sometimes be analysed in laboratory conditions, but care needs to be taken when interpreting how the findings from such tests scale and integrate in the performance of an actual building as a whole.

There is a wide range of drivers and challenges that feed the interest in building performance, such as human comfort, health and well-being, environmental issues, imperfections in the design and construction process, facility management and extreme events. Some of these issues are even more pertinent due to the large scaling factors from single buildings to the built environment in general and the role of the construction sector in national economies. A further factor that requires a good understanding of building performance is accountability. The role of occupants and technology in buildings is changing and puts further pressure on the concept. Yet the theory of building performance itself also needs further development to meet these challenges.

Recommended Further Reading:

- *An Introduction to General Systems Thinking* (Weinberg, 1975) for a seminal discussion of systems theory, as applicable in almost every conceivable domain ranging from computing and electrical engineering to archaeology and composition.
- *The Sciences of the Artificial* (Simon, 1996) for the classical introduction to complex systems, the process of design and analysis.
- *Integrated Buildings – The Systems Basis of Architecture* (Bachman, 2003) for a good overview of application of systems theory that does both the architectural and engineering view justice.
- *Fundamentals of performance-based building design* (Becker, 2008) that discusses the various stakeholders in building performance and provides a good insight into designing for functional requirements and *Performance-based design* (Kalay, 1999) to link this to architectural theory.
- *RIBA Plan of Work* (2013) as the industry standard reference for the various stages of the building process.

Activities:

1   Consider your current accommodation, whether this is a student room, apartment or family home, which we will call 'the place'. Where is the place in its life cycle, and are there any significant milestones on the horizon? Who has a stake in this place? What systems can you discern that make up the place? Are you aware of any performance challenges that the place faces?

Figure 2.2 Life stages of a human.

2   Identify all stakeholders in a modern penitentiary facility. What are the key interests of these stakeholders, and what conflicts between these interests can you identify?

3   Find an off-the-shelf system that is used in different buildings, and then study how this system performs in these different contexts. What can be learned from the interactions?

4   Make a schematic drawing of all the building services system in a building. Base the drawing on a section of the building, with the services overlaid. Use the schematic to show how key utilities like water, gas and electricity are routed through the building and how the relevant components are interconnected. Annotate with dimensions, materials and any standards and norms that apply.

5   Analyse whether you have ever been in a situation where external, non-building-related factors impacted your perception of a facility. What were the factors, and what was their impact on how you experienced the performance of that building?

6   Examine Figure 2.2, and think of 'person performance' for each of the seven life stages. Then explore how you would measure that performance in more detail. Is the same approach applicable for all life stages, or would you suggest using a different one for each stage?

2.8   Key References

Allen, E., 2005. *How buildings work: the natural order of architecture.* Oxford: Oxford University Press, 3rd edition.

Atkin, B. and A. Brooks, 2009. *Total facilities management.* Chichester: Wiley-Blackwell, 3rd edition.

Augenbroe, G., 2011. The role of simulation in performance based building. In: Hensen, J. and R. Lamberts, eds., *Building performance simulation for design and operation.* Abingdon: Spon Press.

Augenbroe, G. and C. Park, 2005. Quantification methods of technical building performance. *Building Research & Information,* 33 (2), 159–172.

Bachman, L.R., 2003. *Integrated buildings: the systems basis of architecture.* Hoboken, NJ: Wiley.

Bayazit, N., 2004. Investigating design: a review of forty years of design research. *Design Issues*, 20 (1), 16–29.

Becker, R., 2008. Fundamentals of performance-based building design. *Building Simulation*, 1 (4), 356–371.

von Bertalanffy, L., 1968. *General systems theory – foundations, development, applications.* New York: George Braziller Ltd.

Blanchard, B.S. and W.J. Fabrycky, 2011. *Systems engineering and analysis.* Upper Saddle River, NJ: Prentice Hall, 5th edition.

Bleil de Souza, C., 2012. Contrasting paradigms of design thinking: the building thermal simulation tool user vs the building designer. *Automation in Construction*, 22, 112–122.

Blyth, A. and J. Worthington, 2010. *Managing the brief for better design.* Abingdon: Routledge, 2nd edition.

Bolger, F. and G. Rowe, 2015. The aggregation of expert judgment: do good things come to those who weight? *Risk Analysis*, 35 (1), 5–11.

California Commissioning Collaborative, 2011. *The building performance tracking handbook.* California Energy Commission.

CIB Working Commission W60, 1982. CIB Report 64: Working with the Performance Approach to Building. Rotterdam: CIB.

Cross, N., 2011. *Design thinking: understanding how designers think and work.* London: Bloomsbury.

Douglas, J., 1996. Building performance and its relevance to facilities management. *Facilities*, 14 (3/4), 23–32.

Farrell, R. and C. Hooker, 2013. Design, science and wicked problems. *Design Studies*, 34 (6), 681–705.

Frankel, M., J. Edelson and R. Colker, 2015. Getting to outcome-based building performance: report from a Seattle Summit on performance outcomes. Vancouver/Washington, DC: New Buildings Institute/National Institute of Building Sciences.

Gilb, T., 2005. *Competitive engineering: a handbook for systems engineering, requirements engineering, and software engineering using planguage.* Oxford: Butterworth-Heinemann.

IEA, 2015. *World energy outlook 2015.* Paris: International Energy Agency.

INCOSE, 2003. *Systems engineering handbook: a guide for system life cycle processes and activities.* San Diego, CA: International Council on Systems Engineering.

INCOSE, 2015. *Systems engineering handbook: a guide for system life cycle processes and activities.* San Diego, CA: Wiley.

Kalay, Y., 1999. Performance-based design. *Automation in Construction*, 8 (4), 395–409.

Kent, D. and B. Becerik-Gerber, 2010. Understanding construction industry experience and attitudes towards integrated project delivery. *Journal of Construction Engineering and Management*, 136 (8), 815–825.

King, D., 2010. *Engineering a low carbon built environment: the discipline of building engineering physics.* London: The Royal Academy of Engineering.

Lawson, B., 2005. *How designers think: the design process demystified.* London: Architectural Press and Routledge, 4th edition.

Leaman, A., F. Stevenson and B. Bordass, 2010. Building evaluation: practice and principles. *Building Research & Information*, 38 (5), 564–577.

Macmillan, S., J. Steele, P. Kirby, R. Spence and S. Austin, 2002. Mapping the design process during the conceptual phase of building projects. *Engineering, Construction and Architectural Management*, 9 (3), 174–180.

Meng, X., 2012. The effect of relationship management on project performance in construction. *International Journal of Project Management*, 30 (2), 188–198.

Mumovic, D. and M. Santamouris, eds., 2009. *A handbook of sustainable building design & engineering*. London: Earthscan.

Preiser, W. and J. Vischer, eds., 2005. *Assessing building performance*. Oxford: Butterworth-Heinemann.

Redlich, C., J. Sparer and M. Cullen, 1997. Sick-building syndrome. *The Lancet*, 349 (9057), 1013-1-16.

RIBA, 2013. *RIBA plan of work 2013*. Bristol: Royal Institute of British Architects. Available from www.ribaplanofwork.com [accessed 31 January 2016].

Rittel, H. and M. Webber, 1973. Dilemmas in a general theory of planning. *Policy Science*, 4 (2), 155–169.

Rush, R., ed., 1986.*The building systems integration handbook*. Boston, MA: The American Institute of Architects.

SEBoK, 2014. *The Guide to the Systems Engineering Body of Knowledge*, v.1.3. Hoboken, NJ: Body of Knowledge and Curriculum to Advance Systems Engineering/Trustees of the Stevens Institute of Technology.

Sexton, M. and P. Barrett, 2005. Performance-based building and innovation: balancing client and industry needs. *Building Research & Information*, 33 (2), 142–148.

Simon, H.A., 1996. *The sciences of the artificial*. Cambridge: MIT Press, 3rd edition.

Sweets, 2016. Sweets™ Product Catalogs for Building Products [online]. New York: Dodge Data & Analytics, Inc. Available from https://sweets.construction.com [Accessed 26 June 2016].

Trinius, W. and C. Sjöström, 2005. Service life planning and performance requirements. *Building Research and Information*, 33 (2), 173–181.

Turner, C. and M. Frankel, 2008. *Energy performance of LEED for new construction buildings*. White Salmon, WA: New Buildings Institute.

Weinberg, G., 1975. *An introduction to general systems thinking*. New York: Dorset House Publishing, Silver Anniversary Edition.

Weinberg, G., 1988. *Rethinking systems analysis and design*. New York: Dorset House Publishing.

de Wilde, P., 2014. The gap between predicted and measured energy performance of buildings: a framework for investigation. *Automation in Construction*, 41, 40–49.

de Wilde, P. and D. Coley, 2012. The implications of a changing climate for buildings. *Building and Environment*, 55, 1–7.

3

Needs, Functions and Requirements

The previous chapter discussed the context of building performance in terms of life cycle, stakeholders, systems and the many challenges that drive the interest in performance. This chapter returns to the definition of building performance as a concept that expresses how well a building performs a task or function and explores these tasks and functions and their definition in further depth. This approach applies to the engineering, process and aesthetic views of performance.

Definition of stakeholder *needs* and *requirements* is an important activity at the very start of system design and development. It typically responds to a high-level problem or opportunity, which initiates the actual design and development process, such as a client wanting to invest in a new development or needing new facilities. The activity includes identification of the stakeholders, elicitation of stakeholder needs, identification of the operational scenarios that the system is likely to encounter, a translation of the stakeholder needs into requirements and formal description and management of the requirements (INCOSE, 2015: 52–57). Typically capturing stakeholder needs is not a trivial matter, as stakeholders may have only a generic idea of what they want. Needs of different stakeholders may conflict, and in many cases significant analysis is required to get the full list of needs. Requirements are a refinement of the needs that take into account constraints on potential solutions, which identify the critical qualities and levels of performance that a system must have to be acceptable to those stakeholders (*ibid.*, 54). Both needs and requirements are a thing that is needed or wanted; requirements may also be imposed by law and thus be compulsory. Other texts use slightly different wording for the same things; for instance Blanchard and Fabrycky (2011: 71–72) use problem definition and need identification for phrasing the real needs and then follow that up by a needs analysis that leads to the requirements of a system. The Systems Engineering Body of Knowledge sees a progressing of real needs, perceived needs and expressed needs to carefully selected and retained needs that become the requirements (SEBoK, 2014: 275). To some extent the definition of requirements is supported by ISO standards that cover the quality, evaluation, modelling, management and vocabulary of the field (Schneider and Berenbach, 2013).

Formally, a *requirement* can be defined as a condition or capability needed by a stakeholder to solve a problem or achieve an objective or as a condition or capability imposed by a contract, standard or legislation (Pohl and Rupp, 2015: 3). A requirement typically takes the form of a statement of this stakeholder need (Hull *et al.*, 2011: 6). A *function* relates to needs and requirements in that 'a function is a characteristic task, action, or activity that must be performed to achieve a desired outcome. A function may be

Building Performance Analysis, First Edition. Pieter de Wilde.
© 2018 John Wiley & Sons Ltd. Published 2018 by John Wiley & Sons Ltd.

accomplished by one or more system elements comprised of equipment (hardware), software, firmware, facilities, personnel, and procedural data' (INCOSE, 2015: 190). At the same time, functions are generally understood to be an activity that is natural to the purpose of a person or thing, such as bridges performing the function of providing access across water. In mathematics, a function is a relation or expression that involves one or more variables, such as the function $y = bx + c$. In computing, a function is a basic task of a computer, especially one that corresponds to a single instruction of the user. Some authors distinguish between functional requirements and non-functional requirements, where a *functional requirement* relates to action that the system must carry out, whereas a non-functional requirement relates to properties, qualities or attributes that the system must have (Robertson and Robertson, 2013: 10).

In the context of buildings, the identification of client needs, aspirations and desires is typically covered by the processes of strategic planning and briefing.[1] A seminal text on briefing is *Managing the brief for better design* (Blyth and Worthington, 2010); further insights can be gained from Schramm (2005: 29–38) and Marmot *et al.* (2005: 39–51). The RIBA Plan of Work (2013) starts with Stage 0 – *Strategic Definition*, which aims to identify the client's business case, strategic brief and other core requirements; this is followed by Stage 1 – *Preparation and Brief*, which develops project and quality objectives, aspirations, budget and constraints. Roughly this corresponds with first identifying needs and then specifying requirements. ISO 6241 (1984: 2) gives only a very brief definition of building stakeholder requirements as a 'statement of need to be fulfilled by a building'. As defined by CIB Report 64 in more detail, *user requirements* are conditions and facilities to be provided by the building for a specific purpose but independent of its context; *performance requirements* define conditions and facilities to be provided by the building for a specific purpose, taking into account the specific context. An example given is the requirement for thermal comfort in winter – from the users perspective, this can be stated as a minimum temperature, possibly with an allowance for some unmet hours; in terms of building performance, what is required depends on outdoor conditions, and under some conditions more is needed from the building than in others (CIB Report 64, 1982: 10). The identification of user needs may stem from a hierarchical, top-down process, based on various stakeholders, activities and schedules that are expected to be hosted by the building (Becker, 2005: 68). However, just asking stakeholders what they need is insufficient, nor is the compilation of lists of needs (Blyth and Worthington, 2010: 65). Moving forward, these needs – which are mostly formulated in user language – need to be translated into more formal performance requirements, which specify exactly what properties the building must have to meet these needs. This step is essential as only the technical requirements allow the comparison of 'needs' and 'solutions', but making the right translation from needs to requirements typically involves communication between the various stakeholders (Foliente, 2005b: 109) and requires the management of expectations (Blyth and Worthington, 2010: 64). However, as buildings are highly unique products, this work is typically complex. Augenbroe (2011: 16–17) points out the needs to engage various stakeholders early in the design process, and to ensure that there is a dialogue between these stakeholders and that conflicts are resolved while at the same time dealing with

---

1 Architectural programming in the United States.

knowledge asymmetry between parties and ensuring transparency in the demand–supply roles. Furthermore, defining needs and requirements is never fully complete. Becker (2005: 8) states that the generic target for performance-based design can be defined as 'satisfying most user needs most of the time in all building spaces'. She suggests that a distinction can be made between *essential needs* as defined in building regulations and *optional needs*, which vary between projects and stakeholders (*ibid.*). In later work, Becker (2008) suggests a terminology of user needs (UN) to represent the objectives or relevant performance attributes in non-technical terms; performance requirements (PR) as expressing the user needs as functional statements and performance criteria that express the needs as a quantifiable factor with a clearly defined target value. The formal description of user needs and requirements are typically described in documents, named Statement of Requirements (SoR). The content of these documents changes from project to project and may also change over the lifetime for one single project. They thus are dynamic documents that evolve with the building (Szigeti and Davis, 2005: 24).

Requirement specification is common to many disciplines. A common approach is found in requirement engineering (Pohl and Rupp, 2015), which provides insights and tools that are also applicable to buildings. Amongst others, requirement engineering provides support for the description of requirement in words as well as by means of dedicated models; it is seen as an integral part of systems engineering (INCOSE, 2015: 57–64).

Building functions and requirements can be handled at a general level, as in 'Total Building Performance', and in the context of specific buildings and project. The general level is of importance for building comparison and benchmarking and for appraising the built environment at higher levels of aggregation. The specific functions and requirements play a prominent role in individual building design and performance analysis projects, and are more sensitive to the views and perceptions of actual stakeholders. This yields an interesting tension between attempts at capturing performance through a generic performance framework, as evidenced in for instance the BREEAM and LEED rating systems, and highly individual and tailored approaches such as the dialogue between building functions, building systems and performance as described by Augenbroe (2011: 18–22).

Exploring the areas of requirement specification, building functions, stakeholder need and world views, and finally building performance requirements, this chapter provides a knowledge base for defining 'building functions', which is a prerequisite for working with performance as a concept that expresses how well the building performs these functions.

3.1   Requirement Specification

In various domains, such as software development, use is made of a dedicated Requirement Engineering (RE) process to elicit user requirements. In most cases requirement engineering involves collaboration between various partners to agree on a final set of requirements for the system. However, this is not an easy process due to the fact that stakeholders often are not fully sure about their needs, and because there often are differences in viewpoint, mental models and expectations between different parties; the identification of inconsistencies and ambiguities as well as conflict resolution is an important part of the process (Laporti *et al.*, 2009). As emphasized by Aurum and Wohlin (2003), requirement engineering is a process that is rich in decision making on the typical strategic, tactical and operational levels and that involves the gathering of intelligence, the design of solutions and the making of choices. This in turn leads to the stages of problem identification, development and selection (*ibid.*). According to Carrillo de Gea *et al.* (2012), requirement engineering in general involves the following:

- Elicitation – Identifying stakeholders, eliciting needs, capturing these needs, structuring the needs found, cross-checking with checklists and other sources, and documenting the outcomes.
- Analysis – Decomposing high-level requirements into more detail, feasibility studies, identification of conflicts, establishing whether any requirements are unclear, incomplete or ambiguous, and resolving any problems.
- Specification – Documenting the requirements and enabling storage and retrieval.
- Modelling – Representing data flow, relationships and others as required and where needed describing requirements using specific specification templates and languages.
- Verification and validation – Comparing requirements with plans, templates and other procedures.
- Management – Handling the maintenance and change of requirements and ensuring traceability of relationships and changes.

There are different types of requirements; good and generally corresponding classifications are provided by Pohl and Rupp (2015: 8), Gilb (2005: 37–38) and the SEBoK (2014: 292), who discern:

1) *Function/functional requirements*, which relate to 'what the system has to do, the essence and fundamental functionality' (Gilb, 2005: 37).
2) *Performance requirements*, which express 'how good' the stakeholders want the system to be; these can be subdivided into quality requirements (how well), resource saving requirements and workload capacity requirements (how much).
3) *Design constraints*, which impose certain details on the system and thus limit the solution space.
4) *Condition constraints*, which cover other constraints such as legal aspects and R&D resources available for a project.

Hull *et al.* (2011: 6–7) point out that requirements are typically presented in the form of a statement, which implies the use of language, although other options are available. Requirements may concern a physical system or a process and convey functional, operational and design characteristics, as well as constraints. Gilb (2005: 39) warns that requirements should be concerned with the ends, goals and needs but that sometimes solutions or means come in amongst the requirements. These should only be accepted if these are conscious constraints; in all other cases they are solutions and should be excluded from the requirements. A good set of requirements should be complete, consistent, feasible and clearly bounded (SEBoK, 2014: 296). Furthermore, the requirements should be unambiguous, measurable or testable and related to the acceptability of the system for the stakeholder (Hull *et al.*, 2011: 7). Typically requirements include identification of a business need, project objectives, functional needs, non-functional needs, quality requirements, acceptance criteria, assumptions and constraints; they may also include organizational statements such as required training and impact on the organization (PMBOK, 2008: 110).

Elicitation of requirements starts with identifying the sources for finding needs. Pohl and Rupp (2015: 19) list stakeholders, documents and systems that are in operation already. For the gathering of requirements from stakeholders, techniques like interviews and questionnaires may be used; other options are brainstorming, asking stakeholders to change perspective and using analogies. The *Project Management Body of Knowledge* adds focus groups, facilitated workshops, Joint Application Development or Design (JAD) sessions, further group creativity techniques such as Delphi studies, affinity diagrams and nominal group techniques and also emphasizes the different options to make group decisions (PMBOK, 2008: 107–108). Documents that serve the formulation of requirements typically relate to existing legacy systems or competition; using them may allow the reuse of existing requirements or improving over these. Systems in operation can be observed, not only for what goes well but also to find ineffective processes and issues for improvement. INCOSE (2015: 60) recommends relating future systems to the environment in which they will be used, as this will likely help with the identification of specific requirements. Interviews and workshops are generally considered to be the most effective need elicitation technique. Some specific approaches that may support requirement specification are Cooperative Requirements Capture, Joint Application Design, Viewpoint Oriented Requirements Definition, Quality Function Deployment (QFD),[2] Voice of the Customer (VOC), House of Quality (HOQ) and Collaboration Engineering. Pacheco and Garcia (2012) give an overview of specific stakeholder identification methods. Azadegan *et al.* (2013) describe a process that can be used to facilitate the elicitation process in workshops with stakeholders that helps to generate ideas about requirements and then clarifies, reduces, organizes and evaluates these towards consensus. Laporti *et al.* (2009) show how, for products with a strong process dimension such as software, stories can be transformed into scenarios and ultimately use cases. Carrillo de Gea *et al.* (2012) give an overview of over thirty dedicated software tools that aim to support the Requirement Engineering process. Meth et al. (2013) show how automation, and especially the analysis of natural language documents, can support the identification of requirements. Problems may appear if insufficient attention is paid to the analysis of stakeholder requirements, as this may lead to having to revisit these requirements later. Similarly, it is important to do a deep review of various system modes and operation scenarios. Incomplete sets of requirements are also a danger, as they will lead to a need to redesign the system to incorporate the missing ones. Lack of method in verification and insufficient traceability also cause issues (SEBoK, 2014: 297). In the context of one-of-a-kind complex systems, there may be a need to pay additional attention to negotiation between stakeholders and system supplier (Ratchev *et al.*, 2003). Méndez Fernández and Wagner (2015) note the following factors as needing attention in requirement engineering: communication, prevention of incomplete requirements, traceability, use of common terminology and ensurance of clear responsibilities. There also is a need to watch for moving targets and process stagnation. Some researchers, such as Tsai and Cheng (2012) or Aguwa *et al.* (2012), suggest the application of statistical analysis to the outcomes of interviews and/or panel discussions with experts in order to identify priorities and reach consensus.

---

2 Note the general issues about QFD and similar approaches put forward by Hazelrigg (2012: 495).

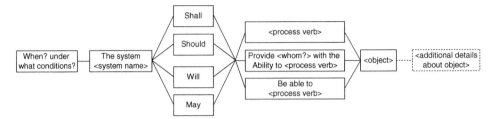

Figure 3.1 Requirement wording template. Image courtesy of Pohl and Rupp/Rocky Nook.

Requirements, once identified, can be captured in different ways. A common way is the use of natural language, which has the advantage of ease of communication but may introduce ambiguities. Robertson and Robertson (2013: 393–471) provide a complete template for a full requirement documentation, including project drivers such as stakeholders and purpose, function requirements, non-functional requirements and other project issues. One way to help the formulation of individual, specific requirements is the use of 'requirement templates' (Pohl and Rupp, 2015: 55–57) or 'requirement boilerplates' (Hull *et al.*, 2011: 84–88). An example is shown in Figure 3.1.

A second option to capture requirements is the use of models. One popular option is the use of the Unified Modeling Language (UML), developed by the Object Management Group (OMG, 2016). UML has four main types of models: a use case diagram, which describes what functions are provided to the stakeholder; a class diagram, which describes the relationships between system and context; a sequence diagram, which shows a process; and a state diagram, which captures events as well as the triggers and consequences of these events. Examples of the various UML diagrams are given in Figures 3.2–3.5 for different use cases, infrastructure classes, passenger handling activities and escalator states at an airport. Note that there are alternatives to UML, a classic option being the IDEF (Integration DEFinition) family of tools that also enables process modelling, information and data modelling, the capture of design intent and others. Another alternative is SysML, a tool based on UML but with less focus on software development. In many cases, natural language and diagrams can be combined to get the best of both worlds (Pohl and Rupp, 2015: 35–36). Requirement formulation in natural language may be supported by the use of a glossary (*ibid.*, 45–46) and the use of templates (*op. cit.*, 53–57). Gilb (2005: 39–43) adds that the requirements need to be developed in such a way as to make clear, in a quantified manner, what will be considered success and failure against these requirements and how these requirements relate to the overall project survival and failure. Gilb also suggests keeping a separate record of requirements that are desired but nice to have rather than must have. INCOSE (2015: 60) recommends to link requirements to standards wherever available. A deep study of requirement rationales may help to reduce the number of requirements, remove bad assumptions, identify design solutions that have slipped in amongst the requirements, and improve communication with the stakeholders (SEBoK, 2014: 294).

The UML Use Case diagram depicts the key interactions between a stakeholder and a system. In terms of identifying building requirements, it may help to identify what key activities will take place in the building; interactions may include not only direct interactions with the building but also activities that are to be hosted within the building.

Drawing UML Use Case diagrams can help in identifying space requirements and logical arrangements, without a preconceived requirement in terms of rooms, which would mix design into the requirement specification. Figure 3.2 shows some of the interactions between a passenger and an airport facility. Note that not all situations have such a strong logic. In many cases the user may have several options; there may also be exceptions and cases of misuse or negative scenarios (Robertson and Robertson, 2013: 139–143).

The UML Class diagram is helpful in identifying relations between entities. In terms of requirement engineering, it should mainly be used in terms of identifying relationships with, and dependencies on, the context of a building. However, Class diagrams can also be used to decompose a system, which may again lead to inadvertently mixing design solutions with the requirements. In the airport example of Figure 3.3, the focus should be in identifying requirements for a new terminal building; when designing a complete airport, care should be taken when using such a schedule, as there may be other solutions (such as integrating the control tower in the terminal) that would be precluded if the relationships become requirements.

The UML Activity diagram is particularly good in identifying parallel processes and synchronizing the completion of related sets of activities. This type of diagram is similar but more detailed than the Use Case diagram. Again scenarios may help identifying what processes belong together and thus support organizational logic of building spaces and design of building systems. The more technical the process, the more effective these diagrams; for instance in Figure 3.4 the physical separation of different groups of passengers (arrived from outside, security cleared and preboarded) relates directly to the corresponding activities. Similar logic will apply in, for example, a bank with vaults or a hospital with operations taking place in an operation theatre that needs to be clean.

Finally, the UML State diagram is most helpful for a system with clearly identifiable states. The example given in Figure 3.5 is that of an escalator and represents only one single building system. However, a similar diagram could be drawn to show the processes and system states that are involved in starting up or shutting down the full airport terminal or moving it to a special state – for instance, one could identify a 'fire alarm' status and how this impacts on the status of sirens, elevator operation and position, fire escape doors, exhaust ventilation and so on.

Note that all UML diagrams presented in this section are basic; the UML format has many more options that can be used to add specific details and further annotations in the various diagram types. Also, it must be stressed that in current AEC industry practice, this explicit level of modelling interactions and relations is very rarely done in the context of defining needs and requirements.

The definition of requirements needs to strike the right balance between being complete and manageable; each requirement will lead to system design activities as well as activities to check that the resulting solutions actually meet the requirements. This means that having many requirements may lead to unmanageable complexity, especially when it is noted that each requirement may involve a range of factors. INCOSE (2003: 91) suggests that, as a rule of thumb, something like 50–250 functional/performance requirements per specification would be appropriate. Note that these are functional requirements only; requirements in the physical or environmental areas would be in addition to this set.

Once requirements are identified, they need to be verified and validated. Hull *et al.* (2011: 13) point out the need to trace the interrelations between requirements, as this

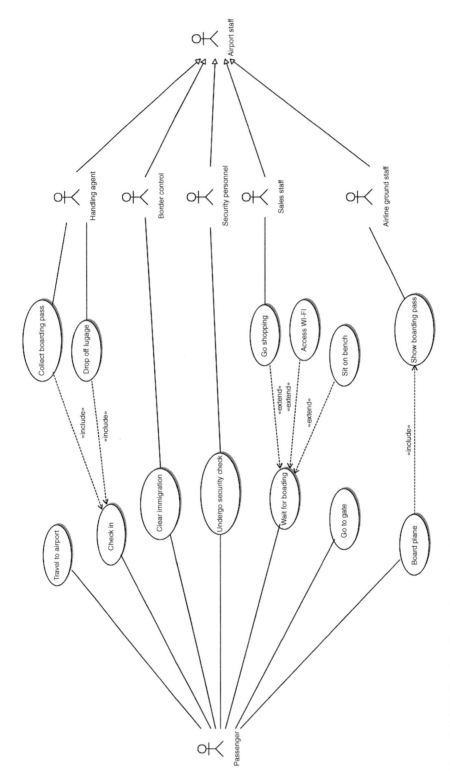

Figure 3.2 UML Use Case diagram for an airport.

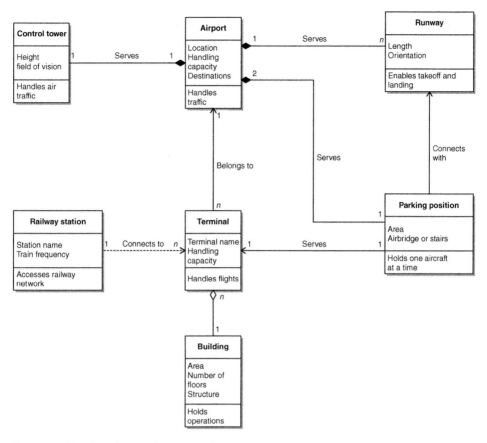

Figure 3.3 UML Class diagram for airport infrastructure.

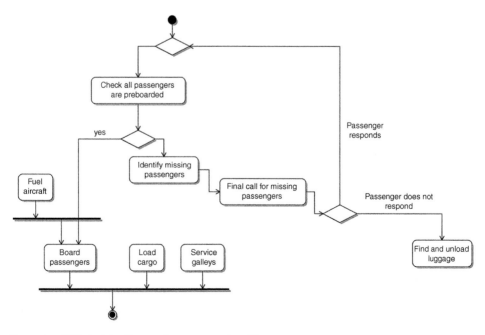

Figure 3.4 UML Activity diagram for passenger handling.

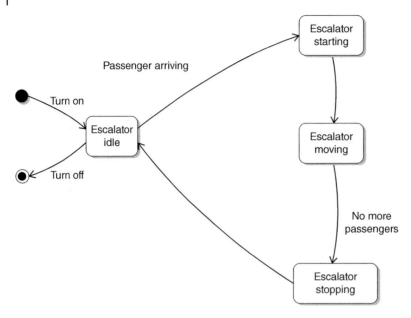

Figure 3.5 UML State diagram for escalator.

provides a range of benefits such as better confidence in meeting all objectives, ability to capture the impact of any changes that are made, greater accountability and organization and progress tracking. Pohl and Rupp (2015: 94–104) provide the following suggestions to handle requirements verification and validation: the involvement of the correct stakeholders, the separation of the identification and correction of errors and conflicts, the use of different viewpoints to analyze requirements, the change from one documentation type to the other, the use of artefacts such as prototypes and the repetition of the validation process. Even so, in requirement engineering it is well known that requirements may conflict. Shehata *et al.* (2007) have developed a taxonomy for identifying requirement interactions in software systems; this is based on a distinction between systems, behaviour and resources and both static and dynamic relationships. Some of the conflicts in the taxonomy include interaction between requirements, leading to ambiguous situations, or the overriding and blocking of requirements, where one requirement cancels out or delays another. Other issues may lead to conflicts because of the overuse of resources, degradation of behaviour and even damaging of systems (*ibid.*). Pohl and Rupp (2015: 105–106) discern data conflicts, with roots in information issues; conflict of interests, related to differences in goals; conflicts of values, such as ideas and cultural differences; relationship conflicts; and structural conflicts. Zambrano *et al.* (2012) state that interactions between requirements are a well-known problem in requirement engineering, especially in terms of traceability, dependencies and links to aspects and concerns. They list the key interaction problems between requirements as conflict, dependency, reinforcement and mutual exclusion (*ibid.*).

A final stage is the management of requirements. This is important, as requirements are used throughout the system life cycle and may well change along the way. Pohl and Rupp (2015: 111–122) suggest to define a range of attributes for each requirement that include a requirement identifier, name, brief description, version, author, source, stability of the requirement, potential risk and level of priority. Requirements are key to the development of a project work breakdown structure (WBS) and the management of project cost, schedule and quality (PMBOK, 2008: 105). User needs and requirements may feed into approaches such as the Balanced Scorecard, Performance Prism or Performance Measurement Matrix (Chytas *et al.*, 2011); however these are not completely without criticism due to responsiveness to dynamic change and evolution (*ibid.*), actual use in practice (Soderberg *et al.*, 2011) and underlying principles (Hazelrigg, 2012: 493–511; Watts and McNair-Connolly, 2012).

The architectural practices of briefing and programming are only partly aligned with the wider theory on requirement engineering and functional analysis. While there is a significant body of knowledge on briefing and architectural planning, this does not often quote the work in systems engineering. In the industry, there are two main approaches to briefing. One school of thought holds that the brief must be frozen after the process and then feeds into the design; another school of thought sees the brief as an evolving document that keeps changing during the design process (Yu *et al.*, 2007). There is a generic suggestion that the process should be made more efficient (Latham report, see DoE, 1994; Kamara *et al.*, 2001). Different approaches are applied to briefing to achieve this and to improve the requirements that are defined, such as Quality Function Deployment (QFD), agent and knowledge-based approaches, Total Quality Management (TQM) and Concurrent Engineering (Kamara *et al.*, 2001). Hansen and Vanegas (2003) have suggested the use of automation to develop design briefs that are better linked to the rest of the building life cycle. Yu *et al.* (2005) suggest the construction of a value tree and the use of weighting factors to reach consensus. There is a range of software applications that aim to support briefing; for an overview, see for instance Chung *et al.* (2009). However, most of these tools seem to stem from an academic research background. Blyth and Worthington (2010: 66–67) argue that a brief should 'set out performance requirements which are a statement about the measurable level of function that must be provided for an objective to be met'; to achieve this performance requirements should address the objective, be precise, unambiguous, measurable, operational and best be formulated in a positive manner.

3.2 Requirement Types

Only part of the stakeholder needs and requirements relate directly to building performance. As discussed, key categories are functional requirements, performance requirements, design constraints and condition constraints. Figure 3.6 represents these four main categories, together with further subcategories derived from Gilb (2005: 37–38) and the Systems Engineering Body of Knowledge (SEBoK, 2014: 292).

The first category of function(al) requirements specifies what it is that the system must do; however this does not say anything about how well that function is provided (Gilb, 2005: 54). Functional requirements appear to be prevalent in construction briefs, possibly due to the large diversity and uniqueness of buildings. As an example, a building requirement may be to provide a view to the outside world; however this requirement does not specify any details that allow to measure how well that function is provided, other than a binary test of whether the function is provided or not.

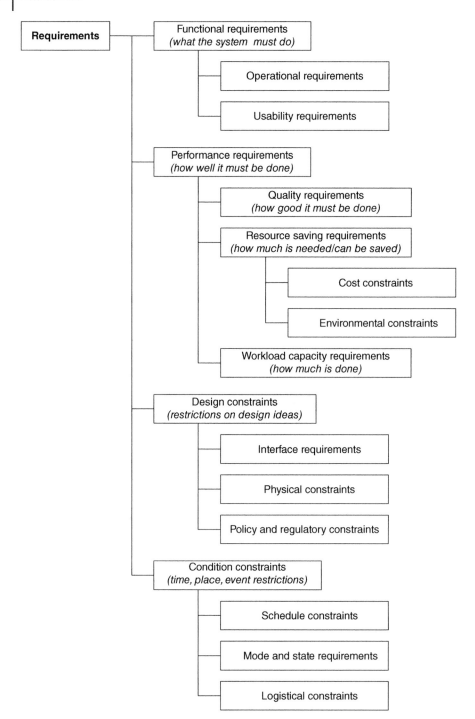

Figure 3.6 Overview of main requirement types.

Performance requirements are concerned with how well a system carries out a function. Gilb (2005: 54–57) discerns three subcategories: quality requirements that specify how good the system performs its functions, resource saving requirements that specify how much of some sort of means can be used or must be saved, and workload capacity requirements that specify how much the system delivers with regard to a function. An example of a quality requirement for a building may be that certain rooms should have a reverberation time that is optimal for clarity of speech. A typical resource saving requirement is the reduction of energy use; however also the volume of material used in construction and the funds required are resources. An example of a workload requirement is the handling capacity of passengers per annum for an airport. Note that there is a significant body of literature on the area of Quality Assurance in projects; see for instance the sections on project quality management, planning for quality, performing quality assurance and quality control in the *Project Management Body of Knowledge* (PMBOK, 2008: 189–214).

Design constraints are restrictions towards the design solutions, such as regulations that apply (Gilb, 2005: 57). There may also be a need to interface with other systems, which again reduces the option space. Furthermore, there may be limitations towards space available or materials that can be used, which yet again limit the range of design options. A building requirement may for instance relate to a specific plot of land that is available to the client, which comes with size and orientation restrictions.

Condition constraints are restrictions that address the factors of time, place and event and thus pertain to the system life cycle (Gilb, 2005: 57–58). For instance, they could include requirements that stipulate when a building is expected to be operational and when not; or they could require that a building is to be linked well to public transport facilities.

Some authors and research groups again have different classifications. For instance some systems engineers define a high-level requirement of effectiveness, which is decomposed into performance, availability and survivability. Here performance relates to how well the function is performed, availability to how often the function is there when called for, and survivability to the likelihood that the system will be able to use the function when needed (SEBoK, 2014: 117). Other terms used in the analysis of system effectiveness are usability, dependability, manufacturability and supportability; each of these comes with a specific way of looking at the system. Which factors are most relevant typically depends on the overall aim of the system and the priorities that are set by the stakeholders (*ibid.*, 320). Within systems engineering, specific attention is paid to ensure that systems are designed for operational feasibility. This leads to consideration of reliability, maintainability, usability (human factors), supportability (serviceability), producibility and disposability, and affordability (life cycle cost). These factors come on top of a design aim for competitiveness, which implies meeting customer expectations in a cost-effective way (Blanchard and Fabrycky, 2011: 375).

To make them operational, many requirements need an elaboration beyond pure identification. For instance, a building project typically has a cost constraint, but this only becomes meaningful once it is specified in terms of an available budget – for instance, 'the project shall cost in between £1.2m and £1.6m'. Hull *et al.* (2011: 112–114) observe that this elaboration can take two forms: stakeholders may define a situation or scenario where the system must demonstrate that it meets the requirement, or they may set a numerical value for some experiment or achievement that must be met.

Robertson and Robertson (2013: 279–301) present this as fit criteria; they recommend the use of numbers, decision tables, graphs and use cases. Pohl and Rupp (2015: 22–23) distinguish 'dissatisfiers', 'satisfiers' and 'delighters'. Dissatisfiers are requirements that must be fulfilled to prevent stakeholder disappointment; often these are subconscious requirements and expectations. Satisfiers are requirements that the stakeholders explicitly ask for; meeting these requirements is what they expect. Finally, delighters are requirements that the stakeholders did not directly ask for but give an added value beyond expectation. Gilb (2005: 128–135) discusses these criteria from a slightly different angle. At a basic level he discerns between fail and survival levels, which define whether a system will be acceptable to the stakeholder or not. Within this bandwidth there is a smaller target area defined by what the stakeholder would expect the system achieve and the possibility of doing better than this expectation. There also is the wish level, which would be the best one could hope for. All of these can be related to achievements of other or previous systems that can act as reference and that give information on the past, present, trends and records.

3.3   Functional Requirements

The introduction to this chapter already mentioned the definition of a function as *a task, action or activity that must be performed to achieve a desired outcome*. Identifying building functions therefore is crucial if performance is taken to be the expression of how well a building performs these functions. Further insights into system functions can be gained from the following sources:

- Ackoff (1971) in his classical text formulated that 'The function(s) of a system is the production of outcomes that define its goal(s) and objective(s). Put another way, suppose a system can display at least two structurally different types of behaviour in the same or different environments and that these types of behaviour produce the same kind of outcome. To function, therefore, is to be able to produce the same outcome in different ways'.
- Gilb (2005: 360) offers the following definition: 'A function is 'what' a system does. A function is a binary[3] concept, and is always expressed in action (to do) terms (for example, *to span a gap* and *to manage a process*)'.
- The Systems Engineering Body of Knowledge defines a function as 'an action that transforms inputs and generates outputs, involving data, materials, and/or energies. These inputs and outputs are the flow items exchanged between functions. The general mathematical notation of a function is $y = f(x, t)$, in which y and x are vectors that may be represented graphically and t = time' (SEBoK, 2014: 300).
- Simon (1996: 5) relates the function of a system to a purpose or goal but notes that whether the system fulfils that purpose or goal depends on interaction between purpose, character of the system, and the environment around the system.
- Van Ostaeyen *et al.* (2013) define function on a generic level as the intended purpose of a system. They note that function also is a link between user needs (subjective) and physical systems (objective).

---

3 Gilb does not explain in his book what is meant by a binary concept; however, INCOSE (2015: 6) points out that processes are binary in that they can be in one of two states: *idle* or *executing*.

Functions are intentional and related to stakeholder needs. These needs define what functions the system must perform (INCOSE, 2003[4]: 74). The function of a system must not be confused with the behaviour of that system; the function of a system relates to 'what the system can be used for or is asked to do in a larger system context' (SEBoK, 2014: 176), whereas the behaviour of a system is the range of actions through which a system acts on its environment (*ibid.*, 115). Behaviour is not necessarily related to a goal. The notion of behaviour is related to system events and state changes (*op. cit.*). The Systems Engineering Body of Knowledge expands on behaviour by stating that 'to have a function, a system must be able to provide the outcome in two or more different ways. This view of function and behaviour is common in systems science. In this paradigm, all system elements have behaviour of some kind; however, to be capable of functioning in certain ways requires a certain richness of behaviours' (SEBoK, 2014: 116). Functions also relate to the context of the system; they define what the system should be able to do when operating in its intended operating environment (INCOSE, 2003: 75). Van Eck (2011) takes a slightly wider view than these systems engineering definitions and discerns three archetypical functions: functions that define the desired behaviour of a system, functions that define the desired effect of a system and functions that define the purpose of a system. In defining functions it is important to establish the boundaries of the system under development, so that it is clear which functions the system has to carry out and what is left to other systems in the environment (*ibid.*, 2003: 74).

When defining functions, it is important to use the right semantics. As noted by Gilb (2005: 360), a function expresses action and what a system does. INCOSE (2003: 343) recommends using an active verb and measurable noun in the definition of any function. Further care must be taken in handling the system function as a concept of what the system can do and what the system is actually doing; this relates to the *state* of the system such as 'operational', 'turned off', 'undergoing maintenance' and others as appropriate (INCOSE, 2015: 67). At the same time, there are also things that should not be included in a 'pure' function definition: how well a function is done is performance and how the function is achieved is a design solution (Gilb, 2005: 107).

The process of *functional analysis* aims to identify all functions that a future system will need to accomplish. This requires the consideration of various use scenarios for the system and possible modes of operation (INCOSE, 2003: 157). Functional analysis breaks main functions down into subfunctions; it also studies the inputs and outputs of the main function and subfunctions and how these interact. This leads to a definition of internal and external interfaces; the outcomes are often described by means of functional flow block diagrams (FFBD). Clear system boundaries are needed to position these interfaces (*ibid.*, 166). Analysis then should proceed to find out which functions are critical, unwanted or may need reinforcement. Tools that help functional analysis are the use of a Work Breakdown Structure (WBS), the Function Analysis System Technique (FAST) diagram, or flow diagrams (INCOSE, 2003: 343). An example of the use of FAST in the built environment is provided by Du *et al.* (2012) who have used this methodology to analyze the functioning of fire equipment.

---

4 It is interesting to observe that the 2003 version of the Systems Engineering Handbook (INCOSE 2003) contains several references to functions that have been dropped from the more recent version (INCOSE 2015). Due to the interest in functions, the older reference is cited where appropriate.

Many functions are complex; in that case it is possible to identify a hierarchy of functions. The work involved in taking a function apart into subfunctions is named *functional decomposition*. Functional decomposition is a step-by-step approach towards identifying logical grouping of subfunctions that belong together, looking at various relationships in terms of temporal or data flow, control logic, typical state transitions and similar. This can be compared with a top-down approach to problem solving. When undertaking functional decomposition, the aim is to identify all sub-functions that are needed to carry out the overall function; this requires a deep analy-sis of all potential environments and scenarios that the system may encounter. Tools that may help in carrying out functional decomposition are, amongst others, behav-iour diagrams, context diagrams, functional flow block diagrams, functional flow diagrams, timelines, control/data flow diagrams, data dictionaries and entity relation-ship diagrams (INCOSE, 2003: 85 and 159–162); see for instance the UML diagrams discussed in the previous section and the options provided by the IDEF range of tools. When undertaking functional decomposition, the analyst should pay attention to logical sequences, so that subfunctions are recognized as part of higher-level func-tions. Similarly, it is good practice to define function inputs and outputs and to make sure that it is clear how functions interface with both higher-level and lower-level functions. It is important to ensure consistency, making sure that the full function can be traced from start to finish (*ibid.*, 157). The resulting set of functions and subfunc-tions is named a *functional architecture*; this defines what the system needs to do in order to complete its mission (SEBoK, 2014: 300). In this context Simon (1996: 128) emphasizes the interaction between systems and subsystems: 'The several compo-nents in any complex system will perform particular sub-functions to contribute to the overall function. Just as the "inner environment" of the whole system may be defined by describing its functions, without detailed specification of its mechanisms, so the "inner environment" of each of the subsystems may be defined by describing the functions of that subsystem, without detailed specification of its sub-mechanisms'. Even so, there may be various understandings of key functions and how they can be decomposed so discussion and dialogue remain important to create a common notion (Eckert *et al.*, 2012); there is no deep consensus on how to represent functions in the engineering field (Erden *et al.*, 2008; van Eck, 2011). In mathematics, functional decomposition is related to the exploration of parthood relations for functions and the theory of mereology (Vermaas, 2013). Erden *et al.* (2008) position recognition as a process that moves into the opposite direction of decomposition. Functional decom-position is part of the wider field of *function modelling* that relates human needs, system functions and the physical system in a context of roles, deployment and inter-actions with the environment. Key concepts that need to be taken into account are the system, the process and the function (Erden *et al.*, 2008). Functional models help designers to think about problems without being bound by physical solutions (Nagel *et al.*, 2011).

The functional analysis of a system, and the definition of functions that are to be performed by that system (functional allocation), describes what the system will do. However, in the context of need and requirement analysis, they should not define design solutions (INCOSE, 2003: 158); here system functions should be kept implementation independent so that they can clearly be distinguished from design solutions (*ibid.*, 75;

Gilb, 2005: 83). Interaction with the outside world can be negative (maintaining the system, trying to counteract impact on the system) or positive (responding to impact, reacting in a way that is desired). Furthermore, the interaction can be synchronous and regular or asynchronous and irregular (SEBoK, 2014: 116). In defining what is needed, it may be beneficial to combine a system decomposition with technical measures that can be used to quantify performance at each level; in systems engineering this would be realized by developing a 'parameter tree' that maps to a Work Breakdown Structure or WBS (INCOSE, 2005: 34–35).

It is important to remember that at some point of the design process, the requirements lead to design and that at that point the general, system-independent functional needs start to relate to actual systems and components. In other words, here 'what the system should do' becomes 'what the specific system actually does' (Gilb, 2005: 84–85). In the end, all functions are carried out by systems or subsystems, such as parts, equipment, software, facilities or human beings (INCOSE, 2003: 157).

This meeting of functional requirements and technical solutions is the core of the GARM (General AEC Reference Model) by Gielingh (1988, 2005), which describes the meeting of needs in terms of Functional Units (FU) and means in terms of Technical Solutions (TS); see Figure 3.7. Gielingh's GARM model is based on the idea of an inverted tree that depicts the functional needs of the client as well as the structure of a product, where each need can be decomposed into lower-level needs, to a point where technical solutions can be found or procured (Gielingh, 2005: 79).

In many cases, a given function can be provided by different systems. At the same time, many systems have more than one function (van Ostaeyen *et al.*, 2013). Once there is an actual system, undesired and incidental functions may arise that were not intended by the stakeholders and designers (Eckert *et al.*, 2012); there may also be instances where the system fails to provide the functionality and thus malfunctions. The actual system also provides for a different type of decomposition: it can be taken apart into a hierarchy of systems and subsystems, or, the other way round, a system is an

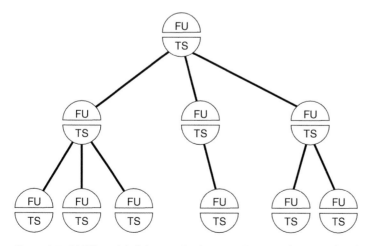

Figure 3.7 GARM model of the meeting between Functional Units and Technical Solutions. Image courtesy of Wim Gielingh.

assembly or aggregation of technical systems (Augenbroe, 2011: 18). A different way of expressing this is to talk about a part-whole hierarchy of technical systems and a means-end hierarchy of functions (van Ostaeyen *et al.*, 2013). Vermaas (2013) gives an interesting mathematical discussion that shows that functional decomposition does not necessarily map to subsystems in a one-to-one relationship. The initial split between what the system should do and later return to what an actual system does applies not only to functions but also to requirements in general; see the discussion in Hull *et al.* (2011: 93–136) on requirements in the problem domain and requirements in the solution domain. INCOSE (2005: 28) lists some of the issues that can only be assessed once a specific system has been designed, such as physical size and stability, functional correctness, supportability, maintainability, efficiency, portability, usability, dependability and reliability. Finding systems or technical solutions that meet the requirements is the domain of design; it can be supported by strategies like TRIZ, which explicitly links to user needs and problems (Mann, 2011a).

In order to turn functions into functional requirements, the function must be expanded with information about specific conditions such as time, place and event (Gilb, 2005: 85) and a definition of what is considered the difference between successfully providing the function or not. Functions are also playing a role in novel enterprise approaches such as the Product-Service Systems (PSS) concept. Here manufacturers move from offering systems to offering services; this leads to concepts such as performance contracts (van Ostaeyen *et al.*, 2013).

3.4 Building Functions

Building functions define what a building has to do. In general they reflect what occupants and other stakeholders expect of a building. Specific needs towards a certain function can be expressed in terms of levels of functionality; a range of functions and related levels can be captured by a functional profile (BS ISO 15686-10: 2010: 17). For some products such as batteries, the key functions are very clear: for instance Bae *et al.* (2014) discuss the role of key battery parameters on the maximum capacity and voltage. Compounding issues like battery temperature are well known and allow the authors to directly discuss these without going into much detail. This is not the case for buildings, which come in a great variety and with many functions. Consequently there are different approaches for thinking about building functions. One important group aims to identify a general structure of functions for most buildings; another limits itself to more narrowly defined categories of buildings.

The notion of building functions is firmly related to building design and architecture through the statement by Louis Sullivan that 'form follows function' from 1896. A more recent and elaborate theory is the Function–Behaviour–Structure model of John Gero and his colleagues at the University of Sydney (Gero, 1990; Rosenman and Gero, 1998; Gero and Kannengiesser, 2003, 2007). This FBS model positions function as variables that describe what an object is for, behaviour as variables that describe what the object does and structure as variables that describe what the object actually is.

Approaches aiming towards a generic view of building functions often build on high-level theories such as Vitruvius' *firmitas*, *utilitas* and *venustas* or Maslow's

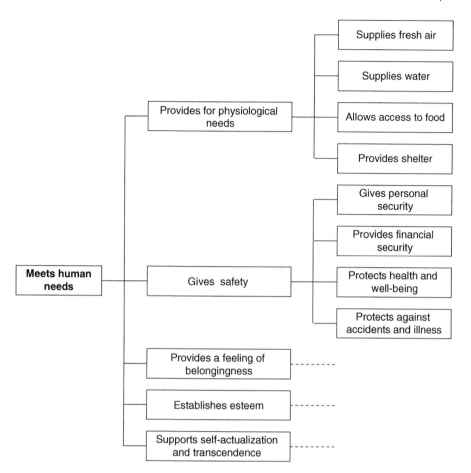

Figure 3.8 Maslovian hierarchy of building functions.

'hierarchy of human needs'.[5] A straight example of a building function hierarchy according to Maslow is presented in Figure 3.8. Typically such efforts are not presented as a dedicated functional decomposition study in themselves, but are embedded in theories such as 'total building performance' (Hartkopf *et al.*, 1986a, 1986b; Hartkopf and Loftness, 1999) or the performance priorities hierarchy of Preiser and Vischer (2005: 5). The total building performance approach uses six 'fundamental building performance mandates': thermal comfort, acoustic comfort, air quality, lighting comfort, spatial comfort and building integrity; these can be presented as building functions by adding a verb like maintain or provide. Each of these six areas can be viewed from different perspectives: physiological, psychological, sociological and economic. The combination

---

5 Maslow (1943) presents a classical theory of human need and motivation that is presented as a pyramid, where lower levels need to be met before higher levels can be addressed; it distinguishes physiological needs as the most basic level and then progresses via safety, social and esteem needs to the top level of self-actualisation.

yields 24 fields where detailed functional aspects can be defined, such as 'sense of warmth' on the intersection between thermal comfort and psychological aspects. Evaluation of these aspects allows the assessment of suitability, reliability and flexibility. While this theory is over 20 years old, it still finds application; for instance Low *et al.* (2008) relate total building performance to buildability; Mahbub *et al.* (2010) used it as a wider framework to position studies of building acoustic performance and hence sound-related functions; Oyedele *et al.* (2012) describe a recent application of this framework to a case study building in Singapore. Preiser and Vischer distinguish three levels of priority covering nine key aspects: (i) health, safety and security, (ii) functional, efficiency and work flow and (iii) psychological, social and cultural and aesthetic. Again, strictly speaking these are presented as performance aspects but can be turned into functions by adding a verb. The Preiser and Vischer view is influential due to the wide use of their *Assessing Building Performance* book. In project management, a common high-level view is the use of the 'iron triangle' that balances cost, time and quality (Marques *et al.*, 2010).

Allen (2005: 32) gives a long list of explicitly defined building functions, all positioned on the same level and without making use of a specific organizational structure. He gives the following building functions: controls thermal radiation, controls air temperature, controls thermal qualities of surfaces, controls humidity, controls air flow, provides optimum seeing and visual privacy, provides optimum hearing and acoustic privacy, controls entry of living creatures, provides concentrated energy, provides channels of communication, provides useful surfaces, provides structural support, keeps out water, adjusts to movement, controls fire, gets built, is maintained, provides clean air, provides clean water, removes and recycles waste. Allen warns that many of these functions are interconnected and that many building components relate to several of these functions (*ibid.*, 254). They are positioned on equal footing, without decomposition and over-arching high-level functions.

Others approach the decomposition of the functions of a generic building following a systems perspective. An clear example of this view is Douglas (1996) who builds the key functions of buildings around a view of buildings as a series of layers and uses this to discern three primary building functions: (i) to enclose space, (ii) to act as a barrier and modifier to climate and (iii) to provide protection and privacy. Douglas notes that the functions and performance of the building as a whole need to be separated from the functions of elements, components and materials. Many works on architectural design follow a similar route; for instance Bachman (2003) presents building design as an integration of systems. In his view performance results from 'meshing' or 'mapping' the functions of building components (*ibid.*, 4). Undergraduate textbooks on construction often also follow this approach, discussing the dominant technologies for parts of the building together with the main functions of these parts; see for instance Riley and Cotgrave (2013) or Bryan (2010).

There is a range of authors who implicitly define building functions, mixing functional decomposition and system-based definitions. For instance, CIBSE Guide A defines the building design process as an activity that transforms client needs and functional requirements into an environment that meets building users expectations; it describes initial cost, whole life cost, fire safety, noise and vibration, water and waste, regulations and climate as constraints that must be met. These constraints can be phrased as building functions by saying that the building 'must meet' the demands of each of these factors. At the same time, Guide A defines geometry, aesthetics, envelope/façade, building system and structure, building services, lighting and other systems as the mechanisms and systems that will allow the building to deliver these functions (CIBSE, 2015b: 0–2). Hartkopf *et al.* (1986a), while starting from a top-down functional approach to building functions, discuss the relations between these primary functions and building systems in significant depth. In the context of defining intelligent buildings, Wong *et al.* (2005) define 'Quality Environment Modules' in the areas of environmental friendliness, space utilization and flexibility, cost-effectiveness, human comfort, working efficiency, safety and security, culture, image, construction process and health and sanitation; each of these 'Modules' is then defined in terms of functional requirements, functional spaces and technologies.

Explicit exploration of the relationships between systems, system states and processes can be found in the work on domain ontologies. Ontologies can exist not only on a philosophical level (where they are concerned with what the categories of concepts are that exist in a scientific domain) but also on a more pragmatic level (where they are concerned with what the types, relationships and properties of entities are that can be used in something like a computer program). El-Diraby (2013), one of the authors often cited on the subject, has defined a domain ontology for construction knowledge (DOCK), which contains taxonomies of system entities with their attributes, as well as the interrelationships between these entities and the relation with their environment. Actions, processes, projects and events are represented in another dimension, discerning purposeful events, accidents, natural events, activities and tasks; together system and action give a structure that defines functions. Turk (2006) offers an ontology that is tailored to construction informatics – in other words, the role that ICT can play in the industry; while not directly concerned with the functions of building themselves, this provides an interesting insight into the range of approaches available to digitally manipulate and analyze them.

Most building codes and legislation are also based on an underlying classification of building functions. A good example are the UK's Approved Documents, which are named 'Part A' to 'Part P'; these deal with functional aspects such as structural stability, fire safety, contaminants and moisture, toxic substances, ventilation, sanitation, drainage and waste disposal, fuel storage and combustion, protection from falling/ collision/impact, conservation of fuel and power, access, glazing safety and cleaning and electrical safety. Similarly, environmental rating schemes like BREEAM and LEED are based on what the development team of these schemes believed to be key functions of buildings, albeit within the focus of environmental or sustainability performance. Environmental rating schemes are highly dependent on the concept of sustainability, which in itself is still under development with new views and aspects being added over the last decade; see Mann (2011b) for a fascinating visual overview.

Another view of building functions starts from the observation that the different categories of buildings, such as homes, offices, hospitals, factories, airports, prisons, schools or cinemas, all provide different top-level functions. Here the key function of a home is to provide a safe and comfortable living environment for some sort of household; the key function of an office is to provide a productive working environment for the employees of a company; a hospital is to provide suitable facilities for the diagnosis, treatment and recovery of patients; and so on. As the categories of these building types are close to the main function, these are often interchanged. In mixed-use buildings they take on an important role in designating parts of the building; see for instance the use of functions (societal functions, production, healthcare, education, financial, office, infrastructure, other) in Nilsson *et al.* (2013) or the similar functions (domestic, education, hospital, industrial, office and sport) in van Dronkelaar *et al.* (2014). This view of building functions is also expressed in Zhong *et al.* (2014), who explore building functions in an urban context and define it as a concept that reflects the daily activities that take place in a building.

Some norms, such as ISO 15686-10 (2010), take a similar view and define function as 'the purpose or activity of users and other stakeholders for which an asset or facility is designed, used or required to be used'. This is further developed into a comparison of functional performance requirements with serviceability, where the later is the 'capability of a facility, building or constructed asset, or of an assembly, component or product thereof, or of a movable asset, to support the functions for which it is designed, used, or required to be used' (*ibid.*).

Figure 3.9 shows functional decomposition for a very basic building, a bus shelter, in line of working with a main building function. This hierarchy clearly includes some of the functions that have been encountered before but is tailored to the priorities for the specific building type at hand.

Rather than to aim for a generic functional structure that applies to all buildings, such efforts look at a narrow building typology and the functionality that is to be provided in that specific case. As an example, Horonjeff and McKelvey (1994: 181–214) discuss some of the functions that need to be considered in airport design and planning, where some of the factors like giving access to ground transport, meeting an existing aeronautical demand and being free of surrounding obstructions are highly specific. Clements-Croome (2006) in his book *Creating the productive workspace* discusses many of the factors that play a role and that might be used to create a dedicated functional hierarchy for office buildings.

In organizational studies, an awareness of why organizations exist is a prerequisite to measuring their performance. For instance, a commercial organization may exist to realize a maximal profit; a not-for-profit one may exist to protect nature. Goal, function and performance measurement are thus interlinked (Popova and Sharpanskykh, 2010). When dealing with building performance, it is thus important to consider the goal of

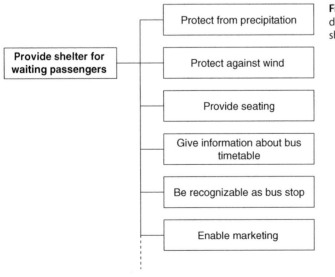

Figure 3.9 Functional decomposition of a bus shelter.

the client and future building occupants and what they see as the main function of the building, and how this contributes to their organizational objectives.

Most discussions of building functions seem to focus on primary aspects like provision of space, maintaining comfort (thermal, acoustical, visual, air quality) and structural integrity; environmental aspects are also prominent. Other disciplines put more emphasis on the operational feasibility of products. For instance Blanchard and Fabrycky (2011: 375–652) include separate chapters on each of the following subjects: reliability, maintainability, usability (human factors); logistics and supportability, producibility, disposability and sustainability; and affordability. INCOSE (2015: 226–241) features reliability, availability, maintainability, resilience, system safety, security and usability. INCOSE (2003: 13) also includes supportability and dependability as a high-level term to cover reliability, availability, maintainability and supportability. Apart from sustainability, in general, these factors are getting only limited attention in the building domain.

As already emphasized by Markus *et al.* (1972: 1–12), people play a key part in buildings – they are crucial to the financing of building initiatives, carry out the design and construction work and are the prime stakeholders. People interact with buildings by modifying them to fit their needs, by operating controls and by adapting to the building peculiarities. As a consequence, the discussion of building functions is likely to encounter many different requirements and external pressures, which may change over time. Leatherbarrow (2005: 8) warns that building functions, even at the whole building level, often change throughout their lifetime, such as in office to residential conversions. This change of function is hard to predict and a major critique of the 'functionalist view'. Becker (2005: 67) therefore recommends conducting an in-depth review of the known as well as the foreseeable needs, plus the expected and possibly disturbing events. This will give a more complete overview of required building functions throughout the building's lifespan. Overall, this gives extra importance to issues like flexibility, adaptability and robustness in a building context.

The development of a decomposition of building functions thus faces the following challenges:

1) Definition of the starting point for the decomposition is difficult, as this is very dependent on the context. For a specific building category or stakeholder, there may be a main function that is of prime interest, but in most cases there are various competing 'root' functions.
2) Branches of a functional decomposition may run into fields that are still under development and where competing views exist. A clear example is the branch that deals with sustainability and where different opinions are held on what needs to be included and in what relation these aspects stand: environment, society, economy and culture.
3) A proper decomposition separates functionality without duplication. However, some building decomposition branches may reconnect. For instance, a high-level view may separate the functions of 'maintain thermal comfort' and 'be energy efficient'. Further decomposition may then find that 'supply heating' is a subfunction of maintaining thermal comfort, whereas 'efficient supply of heating' may be a subfunction of being energy efficient – which leads to functions that are almost identical.
4) Hierarchies and order may not be as obvious as it seems – for instance should a function like 'maintain a good indoor air quality' be a subfunction of 'ensure health and safety', 'maintain good indoor environmental quality' or something else? There are different ways to view this and no good or wrong answer.

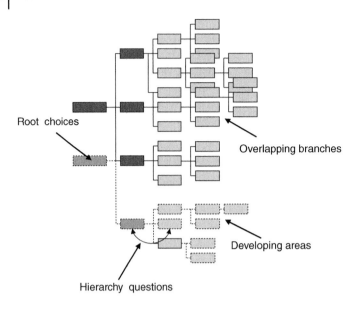

Root choices

Overlapping branches

Hierarchy questions

Developing areas

Figure 3.10 Functional decomposition challenges.

This is visually represented in Figure 3.10.

Given the wide span in building categories, stakeholders and occupant needs, the notion of a final authoritative overview of building functions is unachievable. However, Appendix A provides a list of some building functions assembled from the literature on the subject as guidance for anyone requiring an initial overview of many of the possible functions that may be identified.

3.5 Stakeholder World Views

The different stakeholders in a building project (architect, structural engineer, etc.) all have their own focus area and thus seem to experience different worlds. Some of these world views may be fundamentally different; for instance an architect and a contractor may have very different ideas about a new building under construction. Where it comes to building performance, Bleil de Souza (2012) suggests that there may be similar differences in the world view of architects and engineers or physicists. A key challenge in building design is to combine these different perceptions into a holistic view (McCleary, 2005: 219–220). Unsurprisingly, the previous discussion of building functions and stakeholder needs and requirements thus reflects a range of different viewpoints. This is typical for most systems; for instance Hull *et al.* (2011: 65–67) introduce the concept of viewpoint oriented requirements definition (VORD). This is yet another hierarchy; it can be used to create a logic structure to identify stakeholders and then list the requirements per stakeholder category. An example of viewpoint classes, mapping to the building performance stakeholders discussed in Chapter 2, is presented in Figure 3.11. In VORD, direct viewpoints interact with the system; indirect viewpoints do not interact directly but have an 'interest' of some sort. Viewing the needs and requirements from

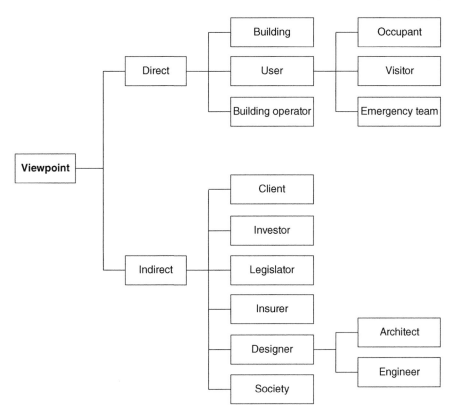

Figure 3.11 Hierarchy of building viewpoints.

different perspectives contributes to the completeness of any overview; however as pointed out by Pohl and Rupp (2015: 115–188), such perspectives can also be used to manage requirements and filter out those that are of importance in a specific context. Robertson and Robertson (2013: 473–477) suggest the development of a stakeholder map, followed by stakeholder analysis that looks at their role, goals, constraints and a range of other factors.

The review of different occupant and stakeholder world views and their needs and requirements helps to identify the totality of needs and requirements that a building must meet. Meeting all needs and requirements may be subject to trade-off and balancing conflicting ones. For instance, project development has two sides: it provides economic and social benefits, typically contributing to the standard of living and working and enhancing the quality of life, but at the same time it requires land that is then lost to nature or agricultural use; it causes pollution of land, water and air; it generates noise; it consumes significant resources; and it uses large amounts of energy (Ding, 2005). In order to handle such issues, the overall process often works with ranking and prioritizations of requirements. It may also be helpful to discern different stakeholder requirements as mandatory, optional and nice to have (Pohl and Rupp, 2015: 118–122).

The individual views of the stakeholders of one and the same building can be largely different. As an example, consider the key functions that will be important for two stakeholders in a new office project: the developer (see Figure 3.12) and a prospective tenant (see Figure 3.13). An initial thought on integrating different viewpoints of building stakeholders may be to use these views for a top-level hierarchy and then to continue by defining requirements and their decomposition for each view. However, this will lead to conflicts down the hierarchy, as various functions will have a range of stakeholders; for instance a requirement such as 'provide thermal comfort' is likely to be viewed differently by designers, clients, investors, legislators and building occupants.

Capturing and balancing the requirements of the building stakeholders, taking into account their different world views, is the essence of the briefing process. Collinge and Harty (2014) describe briefing as a social process that develops ideas as well as design solutions, where different stakeholders have their own interpretations and understandings. Kamara *et al.* (2001) stress the importance of clearly and unambiguously stating the requirements of the clients, which they detail as 'the persons or firms responsible for commissioning and paying for the design and construction of a facility, and all the interests they represent'. Yu *et al.* (2007) describe briefing as the stage in which the

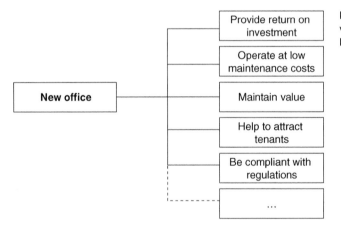

Figure 3.12 Stakeholder views on a new office: Developer perspective.

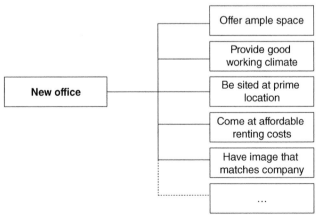

Figure 3.13 Stakeholder view of a new office: Tenant perspective.

values of client, future occupants, architect and society are collected and goals and needs are defined; interestingly they also add 'uncovering facts about the project'. In an earlier paper, they also note that briefing is sometimes made more complex by the involvement of 'multi-headed' clients, which they describe as an organization or group of stakeholders with different goals and needs. Further complexity is caused by a demand–supply gap between occupant clients and paying clients (Yu *et al.*, 2005). Chung *et al.* (2009) list some of the problems with briefing as a lack of common direction in the client organization, an overemphasis on financial consideration, unstructured approaches to collect client requirements and conflict between stakeholder needs; they also note a tendency to move towards solutions before requirements are properly defined. Many of these same issues are also raised in the earlier work of Barrett *et al.* (1999). Chandra and Loosemore (2010) have studied the interactions between various stakeholders in a hospital redevelopment process; they note significant cultural differences between different groups such as clinicians, planners, contractors and consultants and observe that iterations and development of mutual understanding are needed to converge towards shared goals. With the variety of stakeholder views, consideration of some factors may help the efficiency of the briefing process: a clear management structure, well-defined responsibilities, good facilitation, selection of the proper stakeholders, team commitment, honesty, openness and trust, and open and effective communication all play a role in defining client requirements (Tang and Shen, 2013). The quality of a brief may also depend on the expertise and background and, hence the viewpoint, of the brief writer (Kelly *et al.*, 2005).

3.6 Building Performance Requirements

Working with building performance requirements follows in a tradition of work that aligns with the history of the performance concept and the work of CIB W60, PeBBu and others as described in Chapter 1. Huovila (2005: 12–18) provides an overview of various historic sources that list stakeholder requirements, ranging from Vitruvius to CIB efforts. The seminal work in CIB Report 64 makes a distinction between building user requirements and building performance requirements, which they distinguish as follows:

- *User requirements* define conditions and facilities to be provided by the building for a specific purpose but independent of its context.
- *Performance requirements* define conditions and facilities to be provided by the building for a specific purpose, taking into account the specific context.

In general, occupant requirements relate to physiological, psychological, sociological and technical aspects. These are primarily defined in qualitative terms and are goals or objectives rather than performance targets. Performance requirements refine the need in quantitative terms (Gross, 1996). An example given is the requirement for thermal comfort in winter – from the occupant perspective, this can be formulated by stating a minimum temperature, possibly with an allowance for some unmet hours; in terms of building performance, what is required depends on outdoor conditions, and under some conditions more is needed from the building than in others (CIB Report 64, 1982: 10).

User requirements are derived from user needs (UN), which sometimes emerge in building briefs as goals and objectives. However, it is important to note that occupant needs may include needs that go beyond the building; for instance they might include a need for proximity to schools. Furthermore, there may be other drivers for requirements that come from something else than occupant needs, such as societal concerns about the environment. Typical building stakeholder needs have been defined in various contributions from the 1970s onwards and include spatial characteristics, indoor climate, acoustics, illumination, moisture safety, energy efficiency, serviceability, accessibility, health and hygiene, comfort, structural safety, fire safety, security, ease of operation and maintenance, durability and sustainability. Most of these needs and expectations are related to future building occupant activities and occupancy schedules. Needs of quality of air and water, protection from noise, traffic and access to sun and a view are needs of the general public. Costs are a main issue for owners and developers, while building regulations ensure that duty of care is not neglected (Becker, 2008). Occupant needs may relate to physiological, psychological, sociological and technical aspects. They are primarily defined in qualitative terms and are goals or objectives rather than performance targets (CIB Report 64, 1982: 10). Stakeholder needs should identify what stakeholders, spaces and timeframes they apply to; often this includes some uncertainty and is specified as 'most building occupants', 'most of the time' (Becker, 2005: 68).

User needs are typically formulated as qualitative, colloquial statements, which can be understood by all stakeholders. This leaves a need to elaborate the user needs in more technical and specific terms for the design team; this is what causes the difference between user needs and performance requirements. Building performance requirements and specifications should be devoid of value judgments; whether something is good, acceptable or similar is something that the decision maker should decide in the specific context in which a component or building is to function (CIB Report 64, 1982: 18). Performance requirements are not only dependent on stakeholder needs but also relate to the building spaces and components. As a consequence the translation of user needs to performance requirements starts at the level of the whole building and then follows decomposition into spaces, parts and systems and materials (Becker, 2008). An example is the realm of building acoustics, where one user need is to allow the occupant continuous and undisturbed sleep. Functional statements bring this generic objective to the realm of specific spaces within the building and looks at noise levels and their duration that are allowed; they will also cover the potential sources of noise. The next level down will address noise reduction levels of partitions (*ibid.*).

Deru and Torcellini (2005: 6) stress the need to start any effort towards building performance for a specific project with a vision statement, which then can be split into a number of underlying performance objectives or goals. In doing this, one must keep in mind that performance requirements may relate to the whole building or to parts thereof, such as specific systems or spaces (CIB Report 64, 1982: 5–6). One also needs to take into account that emphasis on what is important is likely to change as a project progresses. At the inception of building design, the prime concerns are issues such as space requirement and construction costs; other aspects like functionality, environmental impact, energy efficiency, adaptability, durability and occupant health, comfort and productivity are mostly considered more informally at this point but become key issues later in the life cycle (Hitchcock, 2003: 3).

To attain desired performance, it is useful to have a framework that makes sure all important elements of the key functions are addressed and subsequently measured (Muchiri *et al.*, 2011). Yet there is no single generally accepted formal framework or classification of building functions. A classic framework is provided in ISO 6241 (1984: 6), which lists user requirements pertaining to building stability; fire safety; safety in use; water/air/gas tightness; hygrothermal, visual, tactile, dynamic and hygiene aspects; the suitability of spaces for use; durability; and economy. Almeida *et al.* (2010) describe a range of performance-based frameworks that give such a structure, including the NKB Model (Nordic), IRCC Model (Inter-Jurisdictional Regulatory Collaboration Committee), PSM (Performance System Model), Systems Approach, ASTM Standard on Whole Building Functionality and Serviceability (WBFS) and ISO 15928 Series. All of these are frameworks that help the translation of a broad goal into operational requirements. Prior and Szigeti (2003) not only cover the same frameworks but also relate this to work by the US General Services Administration, the British Airport Authority, CIB, the European Commission and the PeBBu network. Hitchcock (2003: 14–24) reviews a range of performance frameworks that list different approaches to user needs and building attributes, related standard and models, including the CIB Compendium of Building Performance Models, the ASTM Standards on Whole Building Functionality and Serviceability, the ICC (International Code Council) Performance Code for Buildings and Facilities, the LEED Green Building Rating System and US DOE High-Performance Buildings Metrics Project.

Most voluntary rating schemes (see Table 1.3) have a background in environmental issues and will highlight the corresponding building requirements; however over the years, these schemes have grown to provide a good general coverage of requirements that relate to urban connectivity, site use and operation, provision of a good indoor environment, efficient use of resources like energy and water, minimizing of emissions and disposals as well as building construction and operation. The fact that these rating systems cover a range of aspects is also their weakness, with critics pointing out that this means that core issues do not get sufficient attention and become diluted (Campbell, 2009).

In the United Kingdom, the British Council for Offices gives the fundamental occupier needs for offices as delivering productivity, good value for money, and prospect of best place to work (BCO, 2009: 25). Within these overall needs, occupier priorities (ranked from most important to least important) are comfort, temperature, lighting, length of commute, noise, security, public transport access, car parking, kitchen facilities and access to shops and leisure (*ibid.*, 26). The US National Institute of Building Sciences provides an online Whole Building Design Guide (WBDG, 2016), which presents a range of key requirements in the form of the following design objectives: accessibility, aesthetics, cost-effectiveness, functional and operational needs, historical preservation, productivity, security and safety and sustainability. It also points out specific goals and objectives in terms of a range of factors such as adaptability, maintainability, scalability and others. Further information is provided for a range of building types: ammunition and explosive magazines, archives and record storage buildings, armouries, aviation, community services, educational facilities, federal courthouses, healthcare facilities, land port of entry, libraries, office buildings, parking facilities, research facilities, barracks and warehouses.

Further frameworks to help with the identification of building requirements are maintained in the industry. Obviously, such frameworks have a commercial value as they help to identify the unique services of the companies that develop them. A couple of them are discussed to give a flavour: Perkins + Will, a global architectural firm, has been active in charting key issues in building design (Cole *et al.*, 2012). In their *Wall Chart* of 2008, they distinguished 12 main fields: beauty/inspiration, material life cycle, water, ecosystem restoration, daylighting/circadian rhythm, resilience, material health, health and well-being, energy, community connectivity, nature/biophilia and environmental quality. Over the years this developed into a scheme that supports production functions, regulatory functions and habitat functions, and at a lower-level energy, water and material flows. The updated version of 2011 strives for regenerative capabilities. Arup, the global engineering firm, has a similar guide named *SPeAR2* (Sustainable Project Assessment Routine). This is used for specific projects and specific places, ensuring each project responds to its context. AECOM, another engineering firm, offers a six-step process that aims at reviewing building performance against business objectives and sustainability in the form of the AECOM *Climate Resilience Framework*. Atkins makes use of a scheme that emphasizes the future proofing of cities, with attention for buildings, water, transport, environment, economy, energy, waste, community, safety and security, health and social care, housing and education. BuroHappold make use of a *Living City* chart that, amongst others, mentions environmental quality, protection and stewardship, agriculture and food, waste and recycling, utilities and infrastructure, mobility and inclusion, health and well-being and governance. CH2M uses a *Materiality Chart* to guide their work in design, design-built, consulting and the area of project, construction and programme management; this chart reminds users of the broad aspects that go into good projects, from education and purchasing to environmental issues, innovation, economic performance and compliance.

Researchers also note problems with some of the requirements. Gann and Whyte (2003) point out the inherent difficulty of measuring social and political aspects, which often play a role in building. Hinge and Winston (2009) note that in the overall discussion of building performance and the corresponding requirements, energy efficiency and the related cost savings might be overemphasized. The function of 'being sustainable' is a complex one as the definition of sustainability is still under development. For instance Alwaer and Clements-Croome (2010) suggest that sustainable buildings should score well on environmental, social, economical and technological factors and make the case that there is a close relationship between sustainability and intelligence.

Some of the generic tools for requirement engineering, such as Quality Function Deployment, the Design Structure Matrix, and aspects of Multi Criteria Decision Making,[6] are also used for building requirements. However, there is also dedicated software such as the Finish tool EcoProp, which is developed purely to support the systematic management of building requirements (Huovila, 2005: 20–39).

Of this myriad of frameworks, the following are presented in more detail to convey a flavour of the contents. The ASTM standard on Whole Building Functionality and Serviceability is a widely cited document that lists about a hundred performance aspects that will appear in individual building requirement specifications; see Table 3.1 (Szigeti *et al.*, 2005). The Total Building Performance framework by Hartkopf *et al.* (1986a, 1986b) is another classic and comprehensive overview, summarized in Table 3.2.

---

6 The wording multi-criteria is in fact a tautology and should be multi-criterion; however multi-criteria has become the accepted term in much of the literature.

Table 3.1 Overview of building functions and related aspects covered by the ASTM WBFS Standard.

Function	Aspects
Support office work	Provide rooms, training facilities, generic space, storage space, interview rooms; space for printers and copiers; space for shipping and receiving goods and mail
Enable meetings and group effectiveness	Provide meeting and conference space, room for informal meetings and interaction and space for groups
Provide suitable sound and visual environment	Maintain privacy and speech intelligibility; minimize distraction and disturbance; control vibration; provide lighting while controlling glare; allow adjustment of lighting by occupants; provide distant and outside views
Provide suitable thermal comfort and indoor air	Control temperature, humidity, indoor air quality, fresh air supply; offer occupant control; provide operable windows
Enable typical office ICT equipment	Provide space for computers and related equipment as well as power, telecommunication network access and cooling
Allow changeability by occupants	Enable partition wall relocation; minimize disruption in case of change; provide illumination, HVAC and sprinklers in a way that enables change
Integrate building features in layout	Integrate HVAC, lighting and sound in space layout; minimize space loss
Protect occupants and assets	Control access, provide internal secure zones, provide vaults and secure rooms; ensure cleaning systems are secure in terms of cleaning and maintenance at all times; keep garbage secure; provide key or card access
Protect facility	Protect zone around the building, site and parking from unauthorized access; provide surveillance; protect perimeter and public zones; enable protection services
Enable work outside normal hours or conditions	Enable operation after hours or outside normal hours, during loss of external service, and enable continuity of work during breakdown of services
Provide image and identity	Provide good exterior appearance, lobby, internal spaces, image of spaciousness, good finishes and materials, identity, relation with neighbourhood, historic significance
Offer amenities for staff	Provide food, shopping, day care, exercise, bicycle racks and seating
Provide special facilities and technologies	Offer conference facilities, translation, satellite connection, computing, telecom centre
Enable access and wayfinding	Provide access to public transport, enable visitors, vehicular entry, wayfinding, internal movement, circulation
Provide key structures	Provide floors, walls, windows, doors, roof, basement, grounds
Be manageable	Be reliable, easy to operate, easy to maintain, easy to clean; have a low-energy use; provide controls; have a suitable service life
Enable operations and maintenance	Ensure occupant satisfaction; provide operations and maintenance strategy as well as info on resource consumption
Support cleanliness	Enable cleaning of exterior and public areas, cleaning of offices, cleaning of toilets and washrooms, special cleaning, waste disposal

Table 3.2 Overview of Total Building Performance Criteria.

Spatial performance: Individual space layout, aggregate space layout, convenience and services, amenities, occupant factors and controls

Thermal performance: Air temperature, radiant temperature, humidity, air speed, occupant factors and controls

Indoor air quality: Fresh air, fresh air movement and distribution, mass pollutants, energy pollutants, occupant factors and controls

Acoustical performance: Sound source, sound path, sound receiver

Visual performance: Ambient and task levels, contrast and brightness ratios, colour rendition, view/visual information, occupant factors and controls

Building integrity: Structural loads, moisture, temperature, air movement, radiation and light, chemicals, biological agents, fire, natural disasters, man-made disasters

---

It structures the view of building functions and requirements by linking design excellence to the human senses, differentiating between thermal, indoor air, acoustic, spatial and visual quality and building integrity. In achieving design excellence, mechanical, chemical and physical properties play a role, as well as physiological, psychological, socio-cultural and economic factors.

With these long lists of building functions and performance aspects, there is a question of how to handle this number of issues. In cases where a decision involves a range of performance aspects, excessive elaboration of criteria and aspects might be counterproductive as it makes it more difficult to compare and contrast the options (CIB Report 64, 1982: 18). Yet while a focus on a small subset of performance aspects helps to deal with conflict between aspects and will improve performance in the selected areas, it might also introduce a risk that overall performance will suffer (Jain *et al.*, 2011). In the generic literature of performance, Neely (2005) states that it has long been known that there needs to be a balance between various performance measures. Use of inappropriate indicators and poorly designed incentive schemes are partly blamed for a short-term culture in some businesses and leading to unintended consequences. Frameworks such as the 'performance pyramid', 'performance measurement matrix' and 'balanced scorecard' aim to strike the right balance (*ibid.*). It is likely that the same holds true for building performance, yet for the manageability of efforts, there still will be a desire to limit the number of performance aspects studied to those that are crucial to the intended building use. CIB Report 64 (1982: 22) suggests this may be as few as 3, but where time and resources permit this could be as many as 25.

As with other systems, user needs, performance requirements and performance criteria for buildings are all positioned within the demand side; they do not provide design solutions (Becker, 2005: 73). In practice, there often is a lack of connection between the practical needs and constraints of the building occupants and operators at the one hand and the intentions of the design team on the other (Hinge and Winston, 2009). It has been suggested that some poor definition of things like 'design quality' might be helpful to some degree in that this supports openness, innovation, invention, speculation and ingenuity (Gann and Whyte, 2003). However, ultimately it is important to establish the stakeholder requirements well and to refer back to them later in the design and building life cycle. It is also important to remember that buildings are complex, and not to miss this complexity in the formulation of requirements and subsequent building performance analysis. A building is more than just a set of parts, components and elements that have been defined in the design process in response to separate requirements and then combined in a construction process; such a view would overlook things like social and aesthetic aspects or the interactions between building subsystems. Similarly, a building is also not just a design that is realized and then experienced; such a view would overlook technical constraints and problems (Leatherbarrow, 2005: 9).

A final comment on building performance requirements is that these must be viewed in the context of specific interests. In some cases this context may be the development of a specific building. Here requirements may range from relatively generic ones that can be reused, such as those formulated by the British Council for Offices (2009), to highly individual requirements such as those for a sensitive medical laboratory facility that justifies the application of full requirement engineering approaches such as those described by Gilb (2005) or Robertson and Robertson (2013). In other cases the interest may be in a group of buildings, such as in portfolio management, studies of districts or sectors or rating systems; in such situations there is a need to find a common underlying framework of requirements so that various buildings can be compared for benchmarking and cross-case analysis.

3.7 Building Needs, Functions and Requirements in Current Practice

An actual project that exemplifies the challenges of handling stakeholder needs, functional requirements, performance requirements and constraints in real practice is Terminal 3 of the Beijing Capital Airport in China; see Figure 3.14. This building was developed in order to cope with rapid expansion of air traffic to Beijing in the first decennium of the 21st century, following China becoming a member of the World Trade Organization (WTO) in December 2001 and in preparation for the Olympic Games in Beijing in 2008. Terminal 3 was designed in 2003 and 2004 in response to a design competition and took 4 years to construct. The building was formally opened in February 2008, with operations starting in March. Upon completion, this was the largest airport terminal building in the world constructed in one single stage; when including the Ground Transportation Centre, this was the first building to go beyond a floor area of 1 million m^2. Stakeholders in the project included the Beijing Capital International Airport Company (client), Foster and Partners, Arup and NACO (design team) and the Beijing Urban Construction Group (contractor). Consultancy work in the domains of structural engineering, mechanical and electrical engineering and fire engineering was carried out by Arup; further consultants included Davis Langdon (quantity surveying), Michel Desvigne (landscaping), Speirs and Major (lighting), Design Solutions (retail), BNP Associates (baggage handling), Reef UK (façade maintenance) and Schumann Smith (architectural technical specifications).

The stakeholder needs for Terminal 3 were defined in the context of the design competition, which called for a modern gateway to China to be designed between the eastern runway of the existing facility and a new 3rd runway to be constructed further to the east. The new terminal was required to help the airport to raise the existing capacity

Figure 3.14 Beijing Capital Airport Terminal 3. Image © Nigel Young/Foster+Partners.

of 27 million passengers per annum to a projected 60 million in 2015.[7] Generous commercial space was requested in order to support airport operations. The brief asked for integration of the terminal into mass public transport systems and inclusion of a new rail link to the city centre of Beijing. The design was to respond to the local climate conditions, which comprise cold winters and hot, humid summers. The budget was set at US$ 650 million.

The winning design by the combination of Foster and Partners, Arup and NACO was selected in November 2003 and developed into a full preliminary design by March 2004; to enable this rapid development, Foster and Partners opened an office in Beijing, allowing intensive collaboration with the client and other local stakeholders. As part of the design process, the original brief was evolved into a full programme of requirements, which covers the wide range of workload capacity, quality, resource saving and timeliness and readiness issues that an airport needs to meet. It is worth pointing out that only a few key requirements, such as required passenger handling workload capacity, stem from the brief; many others are closely intertwined with specific design concepts and could only be established during the design process. For instance the requirement for capacity of a people mover only becomes an issue once it is decided that this system is to be used and when an approximate design is available that allows to identify crude passenger flows. By necessity, this paragraph can only provide a high-level overview; briefing documentation often consists of hundreds of pages, with parts being confidential. However, the overview should convey the complexity of a real brief

---

7 In reality, Beijing Capital Airport handled 53.6 million passengers per year in 2007, 55.9 million in 2008 and 65.4 million in 2009; since then numbers have continued to grow.

and the difficulties in tracking the drivers of building performance in a project of this order of magnitude.

Many requirements are functional requirements, which describe what the building needs to do, but which do not specify how well this needs to happen (although indicative words such as 'comfortable' may be included). The main functional needs for Terminal 3 are as follows:

- Provide convenient access for passengers, airport and airline staff, visitors and others. This needs to accommodate different modes of transport such as by private car, rental car, taxi, bus or rail and the 'modal split' between these systems. Each system has further requirements in terms of parking, loading/unloading, waiting and circulation.
- Offer facilities to process passengers, which include ticketing, check-in, baggage drop-off, baggage collection, immigration control and security checks.
- Accommodate existing airlines and aircraft that serve Beijing to use the building, taking into account the sizes and capacities of these aircraft while anticipating future industry developments (changes in airlines serving Beijing and the future use of other and novel aircraft types).
- Provide facilities that support passenger flows through the building (circulation space), including waiting areas, lounges, holding spaces for gathering the passengers of specific flights to ensure controlled boarding, routes for efficient deplaning and a layout that separates departing and arriving passengers.
- Include a baggage handling system that forwards checked-in baggage to the appropriate aircraft, moves arriving baggage to an appropriate collection point or carrousel and can also handle intraline and interline luggage that needs to go from one plane onto another.
- Display flight information to passengers and visitors, as well as wayfinding information throughout the facility.
- Support aircraft operations, maintenance and the various airline administrative functions that go with this. Amongst these is a functional requirement for Air Traffic Control (ATC) to have an unobstructed view over aprons, taxiways and runways.
- Enable aircraft handling in terms of aircraft parking spaces, taxiways, jetways for passengers to board, refuelling facilities and provision of ATC. Individual parking positions require electrical power supply, external air-conditioning systems that can be connected to planes and aircraft taxi guidance systems.
- Provide typical passenger amenities, such as bars and restaurants that offer food and beverages, restrooms, stores and retail outlets, offices for banks and insurance companies, internet access, public access, power outlets for passenger devices, vending machines, hairdressers, luggage wrapping and entertainment facilities.
- Support airport security by offering sterile areas beyond security screening checkpoints.
- Enable governmental control in terms of immigration, customs, agriculture and public health services of departing and incoming passengers.
- Provide facilities for disabled people or people with special needs, transfer passengers, VIPs and VVIPs.
- Accommodate the needs of the airlines operating out of the terminal, which includes provisions for airline ground staff, airline flight crew, catering and cabin service and maintenance departments.
- Offer facilities for air traffic control, police, fire brigade, medical and first aid personnel and the media/press.

Many of these functional requirements need to be detailed through performance requirements; this leads to a lengthy document that cannot be reproduced here. However, the most important performance requirement is about the workload capacity of the building and is specified as that it should be able to handle 7000 international passengers per hour and 7360 domestic passengers per hour for 2015, while allowing for 500 000 aircraft movements per year. This workload capacity in itself requires efficient operation of the building in terms of handling aircraft, passengers and crew, luggage and air freight. It also requires a good surrounding infrastructure to allow the efficient manoeuvring of aircraft around the facility on stands, apron and taxiways. For passengers, time spent at the airport is directly related to workload capacity; this time is carefully balanced not only to allow for efficient passenger processing but also to enable passengers to interact with retail facilities to generate airport revenue. For Terminal 3, the time from check-in to departure is 30–45 minutes for domestic flights and 60–75 minutes for international flights. Boarding times are given as 10–15 minutes for domestic as well as international departures. Minimum connection times for Terminal 3 are set as follows: domestic transfers, 50 minutes; international transfers, 60 minutes; domestic to international transfers or vice versa, 120 minutes.

The performance of an airport terminal in terms of quality needs to anticipate the various issues that feature in passenger satisfaction surveys, such as good orientation, comfort (thermal, visual, audio) and convenience. Key issues are providing a design with minimal walking distances for passengers, limitation of passenger processing time, overall occupancy levels and the risk of congestion. Obviously these aspects are closely interrelated with a specific design. An important issue for passengers is the reliability of the luggage delivery; this is ensured by setting a target for a low luggage mishandling rate; for Terminal 3 the requirement is less than 2.5‰. Safety concerns are a special quality requirement; an airport terminal needs to provide sterile areas that cannot be entered without extensive security check and for which the operator can guarantee that only people with the right paperwork and without dangerous items have access. Note that part of the security is independent on the building but predominantly rests on the operator, and that security risks may change over time; processes and systems such as Threat Image Projection (TIP) and the Screener Proficiency Evaluation and Reporting Systems (SPEARS) address operational issues. Other quality demands for airport terminals, for instance in terms of provision of thermal comfort, lighting, power supply, sanitation and drainage are not different from the wider building stock and follow typical national and international performance requirements.

Performance in terms of the saving of resources relates to different aspects. First and foremost, the airline industry is one that operates with small financial margins, so economic and financial viability are of utmost importance. Aircraft operation is a contributor to greenhouse gas emissions, typically in the order of 5%–10% of national carbon emissions; short taxi routes and efficient handling may help reduce this effect while also having a positive effect on local air quality and noise levels. For the building itself, there are the typical resources of time and money involved in design and construction, and subsequently operation. Another key performance requirement that drove the design and construction of Terminal 3 was the need to have the facility operational in time for the 2008 Olympic Games. Furthermore, the building must be able to adapt to future requirements beyond the initial planning horizon, for instance in terms of the accommodation of both existing and future airline fleets.

The final design and actual building of Terminal 3 in fact consists of three connected volumes: T3C, T3D and T3E; this allows for taxiways in between the volumes that create shorter routes for planes and service vehicles. Terminal 3 has 72 jetways and remote parking bays, which together serve 150 gates. The building includes an area of $162\,000\,\mathrm{m}^2$ for retail, a further $12\,600\,\mathrm{m}^2$ for duty-free shops and $7200\,\mathrm{m}^2$ for convenience service areas; furthermore it has car parking facilities for 7000 vehicles.

A key unifying aspect of the building is the single roof, which in form refers to a Chinese dragon; under the roof the building minimizes the level changes that passengers have to navigate while minimizing walking times and enabling quick transfers. The final design emphasizes operational efficiency, passenger comfort, a welcoming and uplifting atmosphere, sustainability, natural lighting and open views to the outside. Easy orientation is important in a large building like this and is supported by the form of the roof. This roof has an area of $360\,000\,\mathrm{m}^2$, with skylights oriented to the southeast that allow solar gain from the morning sun but block it out during warmer periods of the day and year. The external cladding covers an area of $275\,000\,\mathrm{m}^2$.

In terms of technical systems, the building includes an integrated environmental control system with low-energy consumption and carbon emissions; it has separate drinking water and non-potable water systems. It employs 243 elevators, escalators and moving walkways, as well as a high-speed automated people mover (APM). Furthermore it has an automated luggage transfer system that can handle 19 200 pieces of luggage per hour from the 292 counters, with luggage speed of 10 m/s and consisting of 15 km of transport belts, 66 carousels and 200 cameras. For passenger comfort, each row in the waiting area has electrical outlets to power and charge personal electronics. The design strongly features the traditional Chinese colours of red and yellow; a Feng Shui expert was employed to ensure the arrival area is designed in such a way as to be calming and inviting.

The design of Beijing Airport Terminal 3 has won a number of international prizes: the Condé Nast Traveller Award for Innovation and Design, category Favourite Airport (2007); The Emirates Glass LEAF Awards, Best Structural Design of the Year (2008); Travel + Leisure Design Awards: Best Transportation (2009); Architects Journal AJ100 Building of the Year Award (2009); and the Royal Institute of British Architects (RIBA) International Award (2009).

3.8 Reflections on Building Performance Requirements

The first step in analysing building performance is to identify the needs, functions and requirements that a building has to meet. Part of this is activity is covered by the process of briefing.[8] However, it is worthwhile to look across to the field of requirement engineering that is used in other disciplines. Briefing appears to be another theory that is mainly developed from inside the construction sector, yet requirement engineering gives a range of good pointers for requirement elicitation and for formally presenting requirements in either graphical format (such as UML diagrams) or words (boilerplates), whereas even the seminal work of Blyth and Worthington (2010) on briefing remains rather generic. Also, note that briefing mainly takes place in a design context; however

---

8 In this section, the term briefing is taken to also represent architectural programming.

an investigation of requirements may also take place when assessing the performance of an existing building; here the process may be similar but briefing may be felt to be the wrong terminology.

In the academic world, the process of briefing is sometimes underestimated. Some scholars expect clients to come up with a detailed brief for architects and engineers to work with. In reality, establishing the needs and requirements is hard work, which goes hand in hand with design development. Often the brief is developed by the architects and engineers themselves rather than supplied by the client. On a related note, it is surprisingly hard to obtain actual briefs; this is probably due in part to commercial and competitive sensitivities.

In the discussions of needs, functions and requirements, functional requirements are often confused with performance requirements. But functional requirements describe what a building needs to do, while performance requirements define how good the buildings needs to provide those functions. Note that key functions are often implied by building type names, such as office, hospital, train station or factory. However, definition of the functions that are needed does not say anything yet about actual performance levels that are required; this will be covered in the next two chapters.

Some scholars have tried to develop a definitive overview of all building performance aspects that may be studied or considered in designing projects. However, while these efforts may yield worthwhile checklists of aspects to review, it is unlikely that one definitive overview can ever be created: buildings are bespoke products, and the organisation and priorities in specific projects will thus show significant variations. Accordingly, overviews of what performance aspects need to be analyzed need to be tailored to the actual project at hand.

In thinking about performance requirements, it is useful to discern quality requirements (how good), workload requirements (how much) and resource saving requirements (how much is needed/can be saved), and possibly the additional requirements of timeliness and readiness. This high-level view guards against simplification of performance to just one single aspect and is a structure that can be learned and remembered, whereas long lists of all the various performance aspects that may or may not play a role are best used as checklists only.

Thinking about building functions and requirements helps to prepare the ground for analysis of performance, as well as for design. But in both analysis and design, it is important to keep in mind that the mapping between functions/requirements and systems/components is not necessarily one to one; a single building system such as a front door may provide many functions (provide access, prevent heat loss, allow daylight access, keep out noise), or a single function may require a series of systems (maintaining thermal comfort typically requires a building envelope as well as a heating/cooling system). Models such as GARM, which depict the matching of 'functional units' with 'technical solutions', may lead to a classification trap where each system is seen to provide one function, and functional decomposition just becomes a hardware classification. The case study of Beijing Airport Terminal 3 demonstrates the myriad of needs, functions and requirements that play a role in a real building design and construction project; obviously the management of all of these is a highly complex and difficult challenge.

At this stage the initial definition of building performance in Table 1.1 can now be expanded to include the most important attributes for each performance view. See Table 3.3.

Table 3.3 Building Performance View with performance attributes.

	Building performance		
View	**Engineering**	**Process**	**Aesthetic**
Definition:	Action or process of performing a task or function	Action or process of performing a task or function	Form for presentation or appreciation
Attributes:	1) Quality (how well) 2) Resource saving (how much) 3) Workload capacity (how many, how much) 4) Timeliness (how responsive, how frequent) 5) Readiness (when, which circumstances)	1) Cost 2) Time 3) Quality 4) Safety 5) Waste reduction 6) Customer satisfaction	1) Creativity 2) Interpretation 3) Communication 4) Embodiment 5) Enchantment 6) Movement

3.9 Summary

This chapter explores stakeholder needs and requirements as a key starting point for the exploration of how a building functions and, ultimately, performs. Stakeholder needs are typically formulated in user language. Requirements formulate these needs more precisely and in technical terms while also taking into account constraints, critical qualities and levels of performance that are needed. Functions are important in this, since functions define the characteristic tasks, actions or activities that a system must perform to achieve a desired outcome. General theory on stakeholder needs and requirements is found in systems engineering and more specifically in requirement engineering; application to the building domain is available in the principles of briefing and architectural programming.

Requirement engineering (RE) involves the elicitation of stakeholder needs, analysis of these needs, specification of requirements in words and/or models, verification and validation and ultimately management of requirements. There are different types of requirements: functional requirements, performance requirements, design constraints and condition constraints. Requirement elicitation needs to be based not only on working with all stakeholders through processes like interviews and workshops but also on the review of (legal) documents and existing systems. Formal requirements can be formulated in natural language using requirement templates or boilerplates. A popular technique for modelling is through Unified Modeling Language (UML) use case diagrams, class diagrams, sequence diagrams and state diagrams; alternatives are integration definition (IDEF) and SysML models. Requirement analysis identifies any conflicts, dependencies, reinforcements and mutual exclusions; it establishes the ultimate priorities. Requirements need proper identification and management to ensure traceability and updating as a development process progresses.

Performance requirements come in different categories: quality requirements, resource saving requirements and workload capacity requirements. Other stakeholder needs and requirements do not relate directly to building performance. Functional requirements specify what a building must do; however they do not state how well this must be done. Design constraints and condition constraints impose restrictions, for instance as required by legislation and regarding time, place and event. There also are other requirement classifications that emphasize system effectiveness, availability, survivability and operational feasibility. To become operational, most requirements need the definition of fail and success levels.

A function is something that the building does, in relation to a goal or objective. Semantically, functions should be defined by an active verb and a measurable noun. Functions can have various states, such as 'idle' and 'operational'. Functional analysis aims to identify all functions that a future system will need to provide. Functional decomposition identifies subfunctions that belong together in a top-down approach. Decomposition is often used to identify technological solutions that can be combined to develop a larger system that meets higher-level functional needs.

There is a range of approaches towards building functions. Some aim to develop a universal framework, using theories like Maslow's hierarchy of human needs or a generic building system perspective to come up with concepts such as 'total building performance'. Building code and legislation also are mostly rooted in an underlying classification of building functions. Work on ontologies for the building domain also relates to building systems and their functions and relationships. Other approaches take a more narrow view, exploring the functions that pertain to specific categories of buildings such as offices only, or even limit themselves to one single building project. Generic challenges in developing a functional decomposition of buildings are (i) the identification of the starting point/root, (ii) coverage of areas that are still under development, (iii) preventing duplication and overlap and (iv) hierarchical questions.

The different stakeholders have different world views, which lead to different requirements and priorities. These must be accounted for in the requirement analysis process through approaches such as viewpoint oriented requirement definition. The traditional building briefing or architectural programming covers part of these issues but is seen to leave room for improvement.

User needs and requirements are key concepts in the seminal work on the performance approach for buildings by CIB Report 64 (1982) and the later PeBBu project (Jasuja, 2005). Typical building user needs that are often cited include spatial characteristics, indoor climate, acoustics, illumination, moisture safety, energy efficiency, serviceability, accessibility, health and hygiene, comfort, structural safety, fire safety, security, ease of operation and maintenance, durability and sustainability. The General AEC Reference Model (GARM) (Gielingh, 1988, 2005), which relates Functional Units (FU) to Technical Solutions (TS), is often cited to explain how requirements lead to building (system) development. Frameworks that provide an overview of many of the possible building requirements and related performance aspects are the ASTM Standard on Whole Building Functionality and Serviceability, the Total Building Performance theory and the structures that underlie most of the voluntary rating schemes like BREEAM and LEED. Many actors in building design and engineering practice have their own proprietary overviews, schemes or wallcharts.

Dealing with stakeholder needs and requirements, and subsequently with building performance, should be based on a balanced approach; there is a danger that focus on a limited set of aspects may lead to a skewed view. In the end, finding the right set is context dependent and requires careful consideration by the analyst and project stakeholders. In some cases, such as individual projects, a specific and highly individual approach may be appropriate; other contexts such as benchmarking and portfolio management may require a more generic approach.

Recommended Further Reading:

- *Requirements Engineering Fundamentals* (Pohl and Rupp, 2015) for a practical self-study book on elicitation, documentation, validation and negotiation and management of requirements.
- *Mastering the Requirements Process* (Robertson and Robertson, 2013) for a deep discussion of requirement specification, including an extensive template for specifying requirements.
- *How Buildings Work – The Natural Order of Architecture* (Allen, 2005) for developing a good understanding of the interrelations between building functions and building systems.
- *Managing the Brief for Better Design* (Blyth and Worthington, 2010) for a good understanding of the briefing/architectural programming in the building sector.
- Overviews of the *ASTM standard on Whole Building Functionality and Serviceability* (Szigeti *et al.*, 2005) and the *Total Building Performance* framework (Hartkopf *et al.*, 1986a, 1986b) for a comprehensive overview of many of the functions that buildings perform and how these may be ordered.

Activities:

1 Obtain a real building design brief that is given by a client to an architect or design team. Review the requirements within this brief, and identify the following:
   A Functional requirements
   B Constraints
   C Design solutions that may have entered the requirement statement.

2 Install a UML editor on your computer and experiment with the different UML diagrams by:
   A Drawing an UML Use Case Diagram for the interactions between patient with a broken leg and a hospital/medical facility
   B Developing an UML Class diagram for the some of the entities that make up a elevator
   C Drawing a UML Activity Diagram for the people flow diagram for a large supermarket
   D Developing a UML State Chart for the heating system of your home.

3 Develop a functional decomposition for an indoor swimming pool.

4 Go online and search for the most detailed description of a modern, lightweight hiking tent you can find. Analyze what the key functions of this tent are and how the tent provides those functions. Then consider all the parts of the tent; what are the functions of these parts, and how do they help the tent as a whole to perform its function?

5 Explore the requirements that would apply to a penguin enclosure in a zoo and to a display of tropical orchids in a botanical garden. Pay special attention to the temperature and relativity humidity ranges that these animals and plants need; compare and contrast these with the typical comfort ranges for human visitors.

6 Investigate how a client may formulate performance requirements regarding aesthetics for a new building.

3.10 Key References

Allen, E., 2005. *How buildings work: the natural order of architecture.* Oxford: Oxford University Press, 3rd edition.

Blanchard, B.S. and W.J. Fabrycky, 2011. *Systems engineering and analysis.* Upper Saddle River, NJ: Prentice Hall, 5th edition.

Blyth, A. and J. Worthington, 2010. *Managing the brief for better design.* Abingdon: Routledge, 2nd edition.

CIBSE, 2015. *CIBSE guide A: environmental design.* London: The Chartered Institution of Building Services Engineers, 8th edition.

Cole, R., P. Busby, R. Guenter, L. Briney, A. Blaviesciunaite and T. Alencar, 2012. A regenerative design framework: setting new aspirations and initiating new discussions. *Building Research & Information*, 40 (1), 95–111.

Collinge, W. and C. Harty, 2014. Stakeholder interpretations of design: semiotic insights into the briefing process. *Construction Management and Economics*, 32 (7/8), 760–772.

Eckert, C., A. Ruckpaul, T. Alink and A. Albers, 2012. Variations in functional decomposition for an existing product: experimental results. *Artificial Intelligence for Engineering Design, Analysis and Manufacturing*, 26 (2), 107–128.

El-Diraby, 2013. Domain ontology for construction knowledge. *Journal of Construction Engineering and Management*, 139 (7), 768–784.

Erden, M., H. Komoto, T. van Beek, V. D'Amelio, E. Echavarria and T. Tomiyama, 2008. A review of function modeling: approaches and applications. *Artificial Intelligence for Engineering Design, Analysis and Manufacturing*, 22 (2), 147–169.

Gero, J., 1990. Design prototypes: a knowledge representation schema for design. *AI Magazine*, 11 (4), 26–36.

Gero, J. and U. Kannengiesser, 2007. A function-behaviour-structure ontology of processes. *Artificial Intelligence for Engineering Design, Analysis and Manufacturing*, 21 (4), 379–391.

Gielingh, W., 2005. Improving the Performance of Construction by the Acquisition, Organization and Use of Knowledge. Ph.D. thesis, Delft University of Technology.

Gilb, T., 2005. *Competitive engineering: a handbook for systems engineering, requirements engineering, and software engineering using planguage*. Oxford: Butterworth-Heinemann.

Hartkopf, V., V. Loftness and P. Mill, 1986a. Integration for performance. In: Rush, R., ed., *The building systems integration handbook*. Boston, MA: The American Institute of Architects.

Hartkopf, V., V. Loftness and P. Mill, 1986b. The concept of total building performance and building diagnostics. In: Davis, G., ed., *Building performance: function, preservation and rehabilitation*. Philadelphia, PA: American Society for Testing and Materials.

Hinge, A. and D. Winston, Winter 2009. Documenting performance: does it need to be so hard? *High Performance Buildings*, 18–22.

Hull, E., K. Jackson and J. Dick, 2011. *Requirements engineering*. London: Springer.

INCOSE, 2015. *Systems engineering handbook: a guide for system life cycle processes and activities*. San Diego, CA: Wiley.

Nagel, R., R. Hutcheson, D. McAdams and R. Stone, 2011. Process and event modelling for conceptual design. *Journal of Engineering Design*, 22 (3), 145–164.

Neely, A., 2005. The evolution of performance measurement research. *International Journal of Operations & Production Management*, 25 (12), 1264–1277.

van Ostaeyen, J., A. van Horenbeek, L. Pintelon and J. Duflou, 2013. A refined typology of product-service systems based on functional hierarchy modeling. *Journal of Cleaner Production*, 51, 261–276.

Pacheco, C. and I. Garcia, 2012. A systematic literature review of stakeholder identification methods in requirements elicitation. *The Journal of Systems and Software*, 85 (9), 2171–2181.

Pohl, K. and C. Rupp, 2015. *Requirement engineering fundamentals*. Santa Barbara, CA: Rocky Nook Inc.

Popova, V. and A. Sharpanskykh, 2010. Modeling organisational performance indicators. *Information Systems*, 35 (4), 505–257.

Robertson, S. and J. Robertson, 2013. *Mastering the requirement process: getting requirements right*. Upper Saddle River, NJ: Addison-Wesley (Pearson Education), 3rd edition.

Rosenman, M. and J. Gero, 1998. Purpose and function in design: from the socio-cultural to the techno-physical. *Design Studies*, 19 (2), 161–186.

SEBoK, 2014. The Guide to the Systems Engineering Body of Knowledge, v.1.3. Hoboken, NJ: Body of Knowledge and Curriculum to Advance Systems Engineering/Trustees of the Stevens Institute of Technology.

Szigeti, F., G. Davis and D. Hammond, 2005. Introducing the ASTM facilities evaluation method. In: Preiser, W. and J. Vischer, eds., *Assessing building performance*. Oxford: Butterworth-Heinemann.

Zambrano, A., J. Fabry and S. Gordillo, 2012. Expressing aspectual interactions in requirements engineering: experiences, problems and solutions. *Science of Computer Programming*, 78 (1), 65–92.

Part II

Assessment

4

Fundamentals of Building Performance

The first three chapters of the book have provided a foundation for working with the concept of building performance. Chapter 1 defined building performance as related to either a building as an object or to building as construction process. It points to three main views of the concept: an engineering, a process and an aesthetic perspective. Chapter 2 positioned building performance in wider context, reviewing the building life cycle, stakeholders, systems and challenges that drive the interest in performance. Chapter 3 explored the stakeholder needs and requirements that underlie the interest in building performance and the different functions that buildings perform. Building on these foundations, this chapter focusses on the key area of building performance itself and explores the quantification of building performance, the experiments and metrics used, and building performance measures as well as signs of performance problems.

Even with this extensive background, building performance remains a complex concept. Various authors have tried to dissect building performance into core elements and their interrelations, with key contributions being the work of Report 64 by CIB W60, Vanier *et al.* (1996), Foliente *et al.* (1998) and Augenbroe (2011). This has led to a general view with increasingly fine detail, although these sources often use slightly different terminology. An overview and mapping of terms is provided in Figure 4.1.

Report 64 by CIB W60 positioned building performance as dependent on three main aspects: *user needs*, the *context* in which the building has to provide for those needs and the *prediction method* (or possibly model) used to measure the performance (CIB Report 64, 1982: 9). Different authors have expanded this view. Vanier *et al.* (1996) identified five key aspects: (i) function requirement, (ii) performance requirement, (iii) functional element, (iv) agent or stress and (v) factor or use. In this view, a distinction is made between the functional needs of the users and the performance requirement, which is positioned as measurable concept that allows the determination of whether or not the user needs are met. The context is expanded to include systems and elements similar to the Technical Solutions introduced by Gielingh (1988, 2005), as well as agents (stress) and factor (use). Agents represent the actual loads, stresses and similar, which work on the elements or systems and play a role in how well these can meet the performance requirements. Factors represent the usage of the building, which is related to the level of performance that is required. In their view, Vanier *et al.* do not include the method or model used to measure performance. Foliente *et al.* (1998: 30) also distinguish five aspects, discerning objective statements, quantified performance criteria, agents and events, attributes of the building and performance parameters. Finally Augenbroe (2011: 19–20) integrated the previous frameworks

Building Performance Analysis, First Edition. Pieter de Wilde.
© 2018 John Wiley & Sons Ltd. Published 2018 by John Wiley & Sons Ltd.

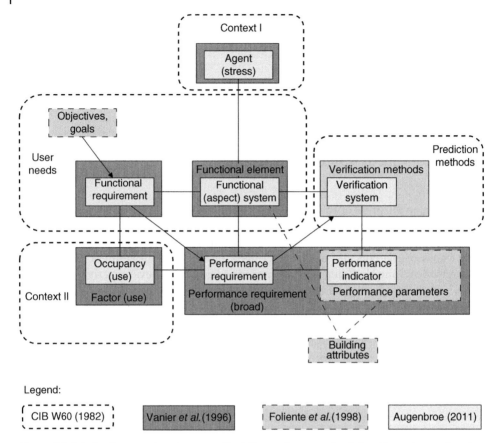

Figure 4.1 Relation between key elements of building performance (as defined by various authors).

using the following seven key aspects: (i) goal and stakeholder needs, (ii) functional requirements, (iii) performance requirements, (iv) performance indicators (PIs), (v) verification or testing methods, (vi) functional systems and (vii) agents that act on the systems.

Within these views of performance, the issue of quantification and measurement is crucial to the operationalization of the concept. INCOSE (2015: 264) defines performance of any system as a 'quantitative measure characterizing a physical or functional attribute relating to the execution of a process, function, activity or task. Performance attributes include quantity (how many or how much), quality (how well), timeliness (how responsive, how frequent), and readiness (when, under which circumstances)'. As expressed in the famous quote by Lord Kelvin (Thomson, 1889: 73), 'when you can measure what you are speaking about, and express it in numbers, you know something about it; but when you cannot measure it, when you cannot express it in numbers, your knowledge is of a meagre and unsatisfactory kind: it may be the beginning of knowledge, but you have scarcely, in your thoughts, advanced to the stage of science, whatever the matter may be'. Gilb (2005: 139) considers measurement as central to the use of performance and resource attributes of all systems. Note that measurement is involved in establishing quantity, quality,[1,2] timeliness and readiness. In the specific context of buildings, Augenbroe (2011: 15) points out that an agreement on how to quantify performance is crucial in the dialogue about stakeholder needs and how these needs are met. But this quantification is deeply interwoven with the user needs, building functions, performance requirements, indicators or measures that are used, quantification or observation methods, the system and subsystems, and the load and stress that act on the building and/or building system.

The purpose of measurement is 'to collect, analyze and report objective data and information to support effective management and to demonstrate the quality of products, services and processes' (INCOSE, 2015: 130). Measurement requires the definition of the base measures, derived measures, indicators and measurement frequency, followed by actual data collection and gathering, storage of this data in a repository, data analysis and verification. Documentation of findings, reporting to stakeholders and recommendation on possible actions is also included in most measurement processes (*ibid.*). Typically, the design of experiments to quantify building performance requires deep knowledge about the performance aspect at stake, the physical processes involved and the experimentation approach – whether that is by means of simulation, experimentation under laboratory conditions or monitoring in use (Augenbroe, 2011: 16). In measuring performance, it is also important to establish a level playing field. For instance, when measuring the performance of water production plants, one needs to take into account the quality of the water resource that these plants use; if this is not done, then plants using good water are advantaged over those using low-quality water. In the end, it is the water quality gain that is of interest in terms of assessing the plant performance, not just the end product (Igos *et al.*, 2013). Similarly, the measurement of thermal building performance should not penalize buildings that are subject to harsher climate conditions and that may have to work harder to achieve similar results as buildings in a milder climate. Assessment of building performance and contrasting of requirements, predictions and actual measurements is recommended at most stages throughout the building life cycle (BS ISO 15686-10: 2010). While measurement expresses performance in terms of numbers, a next step is to assign values to these numbers. Here measurement methods, which quantify performance, move into the domain of normative methods, which designate some outcomes as permitted/desired and other outcomes as not permitted/undesirable.

The expression of performance presently involves a wide range of terminology. Many authors use the term Key Performance Indicator (KPI). Originally, KPIs are used in business administration to define and measure organizational goals. Over time, the idea of KPIs has also been applied in the building sector, for both processes and buildings. In general, KPIs represent performance on some quantitative scale convenient for managers; this is not necessarily something that can be physically measured (Becker, 2005: 83). Some authors apply KPIs to building performance in the wider sense (Shohet *et al.*, 2003; Bertelsen *et al.*, 2010; Kesic, 2015). Mahdavi *et al.* (1997b) use a more specific term, 'aggregate space-time performance indicators', in the context of building performance modelling and analysis. Augenbroe (2011: 17) uses a more generic

---

1 For a deep philosophical discussion about quality, see Pirsig (1974).
2 Embedded in the measurement of quality is the comparison with other products or systems; a quality project is rated 'good', whereas others may be 'average' or even 'bad'.

performance indicator (PI) as concept to quantify building performance and for the representation of simulation and actual measurement outcomes. However, other authors use different terms. Hand (1998: 212) discusses different categories of 'performance assessment data', including principal performance data, derived performance data, and parameter distribution data. Hitchcock (2003: 3) and Deru and Torcellini (2005: 5) use 'performance metric', which is derived from data analysis. The systems engineering domain does not use one single term, but a range of 'technical measures' that differentiate between specific uses (SEBoK, 2014: 408). At the client side, Measures of Effectiveness (MOEs) are used to measure success in a way that is independent of a specific system. Measures of Performance (MOPs) are used to quantify the overall performance of a given system or product. Technical Performance Measures (TPMs) relate to elements or subsystems; typically the quantification of MOPs requires aggregation of TPMs. Finally, Key Performance Parameters (KPPs) capture some critical performance aspects of a system, which is likely to have an impact on success or failure of the overall system (INCOSE, 2005: 6–7). In their more generic *Theory of Modeling and Simulation*, Zeigler *et al.* (2000) differentiate between objectives, outcome measures and outcome variables. Here outcome measures allow the assessment of how well the objectives are being met while they are computed from output variables of the models.

While the emphasis in building performance rests on quantifying how well a building meets its objectives, it is also worthwhile to consider an important prerequisite: to perform, a building must be relatively free from defects. Unfortunately, this is not always the case (Georgiou, 2010; Park *et al.*, 2013). Some defects are detected and issues corrected, which in turn causes project schedule and cost overruns; others, especially those hidden from visual inspection, may remain and impact on the performance of the building throughout its life cycle. Hence it is important that building performance analysis also covers some of the indicators of hidden issues, so that these can be put right wherever possible.

Thus far, there is no unified theory on building performance analysis that is accepted across the building domain that can be used to check how novel ideas and suggestions fit. As a first step, Augenbroe (2017) suggests three axioms of building performance that can be used to critically assess contributions for basic validity:

Axiom 1: Building performance can only be observed through a real or virtual experiment that exposes a building to a load or excitation.

Axiom 2: Building performance does not aggregate. No component or part of a building has a performance, unless that component or part can be subjected to a test in which it is isolated from the building. The overall performance of a building cannot be derived from performance of its components and parts.

Axiom 3: Current building information model (BIM) systems contain insufficient information to support building performance analysis.

This chapter starts with a deeper exploration of building performance as a crucial concept at the interface between functions and systems. It then proceeds to discuss quantification of performance and the theory of experiments and metrics that are central to this quantification. Specific attention is paid to building performance measures and the way performance is expressed, as well as some of the indicators that allow the detection of defects and malfunctioning.

4.1   Performance: The Interface between Requirements and Systems

Previous sections explored building performance as a concept that expresses how well a building performs a task or function. This expression takes place on the interface between the actual building and its subsystems, components and elements on one side and the notion of tasks and functions on the other. Performance is thus something that exists on the interface between *building* and *function*.[3] This view is shared by both the building research and systems engineering worlds. On the building side, the pioneering work by BPRU describes this as a complex interaction between building systems at the one end and building objectives at the other. They position two layers in between: one for environmental systems and one for human activities (Markus *et al.*, 1972: 4). The link between objective and system was further developed in the Netherlands, where Gielingh produced the GARM (General AEC Reference Model), sometimes informally called 'hamburger model'. This uses two opposing half disks that represent Functional Units (FU) on top meeting Technical Solutions (TS)[4] at the bottom, with performance levels, validation methods and proof of performance at the interface between the two (Ang *et al.*, 2005). Also see Figure 3.7 and corresponding text. In a design context, the Functional Units represent the objectives and goals that stem from the user need and often can be decomposed over a number of levels. The Technical Solutions represent systems or subsystems that may meet the demands. On the interface between Functional Unit and Technical Solution a check takes place to evaluate the match (Szigeti and Davis, 2005: 11–12). Augenbroe (2011) has a similar view, positioning performance at the meeting place between building functions and building systems. Kalay (1999) approaches the same thing from the famous architectural proclamation by Louis Sullivan that 'form follows function', where form represents the building and its system. He goes on to define performance as the interpretation of behaviour, which results from the interaction between building form, building function and context, thus explicitly adding a third dimension. Kalay also gives context a wide coverage, including physical, cultural, social and other factors. Foliente *et al.* (1998: 15–16) also point out the need to integrate views along a number of axes in order to handle the performance concept in building. They differentiate between product hierarchy (from elements and systems to complete buildings), knowledge development (from research to application in industry), the construction process (from inception to occupancy) and over different performance aspects or attributes. On the Systems Engineering side, Simon (1996: 8–9) already notes that in many studies and observations, it is convenient to distinguish between an inner environment (system) and an outer environment (in which the system operates). The success of the system often depends on the goal and the interaction between the system and the outer environment. This leads Simon to say that 'description of an artifice in terms of its organization and functioning – its interface between inner and outer environments – is a major objective of invention and design activity' (*ibid.*, 9).

---

3 The aesthetic notion of performance can be incorporated in the view that performance is something that measures how well a building performs a task or function by defining a function for the building along the lines of 'to entertain', 'to be representative' or similar.
4 Confusingly, the paper by Ang *et al.* (2005) replaces Functional Units with Functional Concepts (FC) and Technical Solutions with Solution Concepts (SC); it is recommended to use the original terms by Gielingh.

INCOSE (2015: 68–69) notes that a key feature of systems is the interaction between system elements. To study the working of a system in depth, one has to look at the 'interface' between these elements. As pointed out by INCOSE, interface comes from the Latin words 'inter' and 'facere' and thus means 'to do between things'. Interfaces can act two ways, as input or output; the 'doing' requires an objective or function but also some form of physical elements to actually make something happen. The Systems Engineering Body of Knowledge (SEBoK) phrases the same issue as follows: 'As functions are performed by physical elements (system elements), inputs/outputs of functions are also carried by physical elements; these are called physical interfaces. Consequently, both functional and physical aspects are considered in the notion of interface' (SEBoK, 2014: 287).

This interface features in a range of seminal publications. In the GARM or 'hamburger' model, it is the connection between the Functional Units and Technical Solutions (Gielingh, 1988). In the Function-Behaviour-Structure model, one encounters function in a similar guise; the system side is represented here as structure and design description (Gero, 1990). The ASTM Standard for Whole Building Functionality and Serviceability presents a similar interface in the comparison of required functionality and rating of serviceability (Szigeti and Davis, 2002). Augenbroe (2011: 18–21) positions performance at the meeting point between top-down functional decomposition and bottom-up assembly of building systems. In systems engineering, the classical 'V' process model represents requirements and functional decomposition on the left side of the V, whereas system combination and integration are positioned on the right side (Blanchard and Fabrycky, 2011: 51). In the context of Product–Service Systems, van Ostaeyen *et al.* (2013) describe functional hierarchy modelling; this is based on a 'teleological[5] chain' where demands are decomposed into functions and an interface where this meets a whole-part decomposition. The original GARM model by Gielingh (1988, 2005) was further developed in the PeBBu project context, where the interface became a meeting place between performance requirements and performance specifications (Becker, 2005: 109); see Figure 4.2. Here the Functional Unit on top now contains a translation from the functional user need on the left to a more technical performance requirement on the right. The Technological Solution on the bottom sees a transition from the specification of a specific technical solution on the left to the performance of that solution on the right. The key to working with the performance concept now becomes a matter of comparing and matching the required performance with the specified performance of a candidate technical solution.

When considering the interface between actual building systems and the notion of function, one point to keep in mind is that the buildings and building systems are material entities, whereas stakeholder needs and requirements are abstractions that live in the subjective realm (Erden *et al.*, 2008). In other words, the building is an object that interacts with its environment and thus displays some sort of behaviour; however in essence this behaviour is independent of stakeholder objectives and intentions. The stakeholder needs and requirements are abstract, mental concepts. The concept of a system function connects the two; functions can be seen a mental concept that represents a desired system role or behaviour. However, the system may also have unintended

---

5 Teleology: the explanation of phenomena by the purpose they serve rather than by postulated causes.

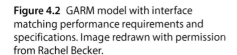

Figure 4.2 GARM model with interface matching performance requirements and specifications. Image redrawn with permission from Rachel Becker.

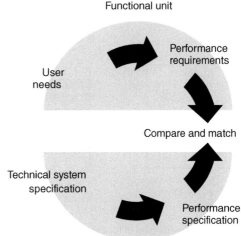

roles or behaviours (*ibid.*). Performance in terms of defining well the systems performs a function connects both sides: there is both the quantified performance of the system, which lives in the objective/material realm, and required performance, which resides in the subjective/mental realm.

In Gero's Function–Behaviour–Structure model, functions sit in the subjective domain (Gero, 1990; Gero and Kannengiesser, 2003) and are linked to the actual system via behaviour variables that describe what the system actually does. Gero and Kannengiesser make a further distinction between expected behaviour (what is needed) and behaviour derived from structure (what the system actually does); this maps to a performance requirement and a performance specification.

The following concepts are used in Gero's model: function (F), expected behaviour (B_e), structure (S), behaviour derived from structure (B_s) and design description (D). Moving between these concepts, eight design activities (seven transformations and one comparison) can be defined, see Table 4.1. While the model is developed to describe and guide design, it is obvious that the analysis of functions (goal, intentions) and the stepwise development of a corresponding system structure that provides these

Table 4.1 Fundamental processes in the Function–Behaviour–Structure model.

$F \rightarrow B_e$	Transformation of functions into expected behaviour
$B_e \rightarrow S$	Synthesis of a structure that will exhibit the intended behaviour
$S \rightarrow B_s$	Analysis of the actual behaviour of the structure
$B_s \leftrightarrow B_e$	Comparison of actual and expected behaviour
$S \rightarrow D$	Documentation of the structure
$S \rightarrow S'$	Development of new or modified structure
$S \rightarrow B_e'$	Modification of expected behaviours
$S \rightarrow F'$	Choice of new or modified functions

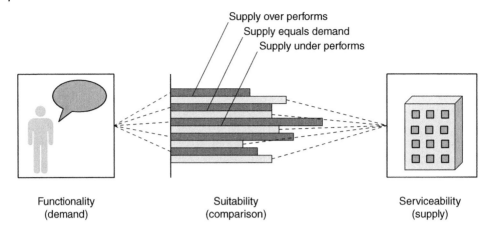

Figure 4.3 Comparison of functionality and serviceability. Image redrawn with permission from Szigeti and Davis.

functions is key. To check whether the structure provides the target functionality, the process requires the analysis and comparison of expected and actual behaviour or, in other words, of performance.

Note that the Gero model is still under development and discussion. Some of the issues at stake are complex and partly philosophical; they relate to design thinking and the reasoning of how needs and requirements influence the form of a future product. For a deeper discussion of the Function–Behaviour–Structure model, as well as some critiques, see Dorst and Vermaas (2005), Vermaas and Dorst (2007), Erden *et al.* (2008) and Galle (2009).

The ASTM Standard for Whole Building Functionality and Serviceability (Szigeti and Davis, 2002) uses yet another terminology for the demand and supply side, preferring 'required functionality' over performance requirement or expected behaviour and 'rating of serviceability' over performance specification or behaviour derived from structure. The comparison of both sides in this work is depicted in Figure 4.3. In addition to GARM and F-B-S, the ASTM standard provides an explicit depiction of a range of demands, pertaining to different functions or performance aspects, and a basic comparison, differentiating between supply that matches demand, underperforms or overperforms. Note that the way demands are depicted is deterministic, with one target value only – whereas in reality the target typically will be a range between an upper and lower threshold. Similarly, real performance may include some variation and uncertainty.

Gilb describes a more complex four-way interface. In his view, a system has four types of attributes: design, function, resource and performance. The design attributes describe the systems and processes that make up the main system such as a building. The function attributes describe what the building does. The resource attributes describe what is needed to keep the system (building) running, such as time, maintenance, money and fuel. Finally, performance attributes describe Gilb's three categories of performance: quality, resource saving and workload capacity. Each attribute type comes with a hierarchy of different subtypes; the interface between all of these then is a many-to-many relationship between these four categories of system attributes (Gilb, 2005: 105).

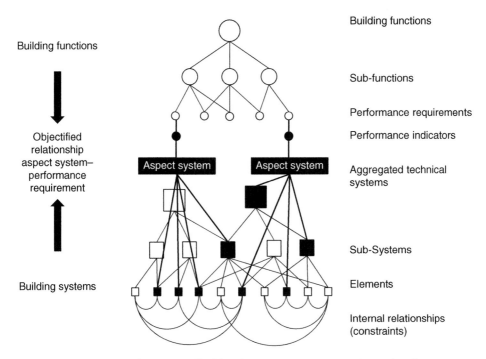

Figure 4.4 Interaction and mapping of building function, system aggregation and performance. Image courtesy of Augenbroe.

Augenbroe (2011: 21) has mapped three of Gilb's attributes in his depiction of the meeting between building function decomposition, technical system aggregation and performance as reproduced in Figure 4.4. Augenbroe (2011: 17) points out that the distinction between supply and demand is not always clear in the construction industry; for instance many clients need help in developing their brief and requirements. This help is often provided by the architects and consultants that develop the building design that will later meet this brief. Figure 4.4 shows some of the many-to-many relations mentioned by Gilb, with a range of systems and subsystems working together to meet a certain function and performance shown as an attribute of such a combination. A similar tripartite relation between function, building and performance is described in Pati and Augenbroe (2007), with specific examples for courtroom buildings.

Throughout his work, Augenbroe also mentions another issue: that performance is typically quantified on the basis of observation (in either reality or a model) that then yields observed states, which in turn are aggregated to obtain performance indicator values (Augenbroe, 2002; Augenbroe, 2003: 6; Pati *et al.*, 2009; Augenbroe, 2011: 17). This concept is also discussed by Zeigler *et al.* (2000: 3–4) who differentiate between the structure and behaviour of systems. For Zeigler *et al.*, state and state transition mechanisms are part of the system structure; behaviour is observed by a relation between system input and system output (*ibid.*). However, it must be noted that a significant function of buildings is to resist change and loads; therefore the observation of building states may be more central to building performance than that of more generic systems that may emphasise system input and output.

Linking several types of concepts with many-to-many relationships is a principle that is at the core of the Performance Framework Tool (PFT) developed at IRUSE in Ireland (Curry *et al.*, 2013; Corry *et al.*, 2014). This uses the idea of 'scenario modelling', where a scenario links a range of related building performance aspects, building (sub)systems, performance objectives and metrics that are of relevance in a certain context; these scenarios are called into action by a trigger event (O'Donnell, 2009: 51–54). This Performance Framework Tool includes a performance ontology that captures and interrelates various data streams about building performance, including measurement, simulation and others using semantic web[6] technology. A distinction is made between functional metrics and aggregation metrics, where functional metrics are for evaluation, whereas aggregation metrics capture and process a range of input values though processes such as calculation of counts, sums, products, deviation, average, minimum, maximum and variance (Corry *et al.*, 2015). To arrive at performance metrics, the framework identifies data sources and formulae, which are equal to Augenbroe's observed states and aggregation.

4.2 Quantifying Performance

The key to working with the concept of building performance is quantification and measurement. Quantification and measurement allow the stakeholder to express what is needed and then to check whether the building meets those needs. Quantifiable targets also help the building designer, as they give clear criteria to judge the failure or success of a building design – or actual building – in meeting the stakeholder needs. On the interface between needs and solutions, the numbers that result from measurement and quantification are the link between demand and supply and form the basis of comparison and matching.

Measurement is a process where one 'ascertains the size, amount of degree of something by comparison with a standard unit or with an object of known attributes'. Quantification, a closely related concept, is the process that 'expresses or measures the quantity of something' and that translates observations into numbers (Oxford Dictionary, 2010). As pointed out by Foliente (2000), quantification of building performance can be based on actual physical testing, calculation or a combination of the two. Further alternatives for performance analysis are expert judgment and stakeholder assessment, but these need to be handled with care as they have a subjective element by nature and do not naturally lead to numbers. Gilb (2005: 375) defines measurement as 'the determination of the numeric level of a scalar attribute under specific conditions, using a defined process, by means of examining a real system'. This requirement for real systems makes the term measurement less applicable to the expression of performance that is based on calculation or computation; however the latter is covered by quantification. So one may measure the performance of a real building but quantify the performance of a virtual building by way of simulation.

---

6 Semantic web: a standard for the Internet that aims to enable data sharing and reuse across application, enterprise and community boundaries.

While in daily life measurement and quantification can be relatively simple actions – such as measuring the length of something using a tape measure or measuring the weight of something using scales – the professional version of the process is mostly more elaborate. SEBoK (2014: 406) defines the activity of performing measurement as 'focussing on the collection and preparation of measurement data, measurement analysis, and the presentation of the results to inform the decision makers. The preparation of the measurement data includes verification, normalization, and aggregation of the data as applicable. Analysis includes estimation, feasibility analysis of plans, and performance analysis of actual data against plans'. According to INCOSE (2005: 6), 'Technical measurement is the set of activities used to provide the supplier and/or acquirer insight into progress in the definition and development of the technical solution, ongoing assessment of associated risks and issues, and the likelihood of meeting the critical objectives of the acquirer'. There is also variation across disciplines. For instance, ISO 15939:2007 defines key issues to take into account when measuring in a software or management context. Here measurement is concerned with capturing 'base measures' by way of some sort of process or method; these are aggregated using a function to arrive at 'derived measures'. Here the standard moves over to data analysis and uses further analysis to arrive at 'indicators', which in turn can be interpreted to yield 'information products' (Abran *et al.*, 2012). However, mathematical processing of measured values does not always yield meaningful numbers (*ibid.*).

The expression of performance in terms of numbers is typically captured by some sort of parameter or value. As indicated in the introduction to this chapter, current literature shows a whole range of terms: Key Performance Indicators (KPIs), Performance Indicators (PIs), Performance Assessment Data, Performance Metrics and others. In organizational research, a performance indicator reflects the state or progress of the organization (or a unit or individual within it) in a quantitative or qualitative manner (Neely *et al.*, 1997; Neely, 2005; Popova and Sharpanskykh, 2010; Pritchard *et al.*, 2012: 19). Systems engineering uses more detailed terms such as Measures of Effectiveness (MOEs) and Measures of Performance (MOPs) as defined by INCOSE (2005: 6–7). Beyond these high-level categories of indicators, measures and parameters, subcategories can be used to indicate a specific focus or use. For example, in healthcare a difference is made between process indicators and outcome indicators. Here process indicators relate to issues like the patient condition over time; outcome indicators may look at how the process leads to patients actually taking the drugs they have been prescribed and, ultimately, patient recovery (Reeves *et al.*, 2010). A slightly different way of looking at this is by discerning 'leading indicators' and 'lagging indicators'. Leading indicators are used to monitor the performance of tasks and functions that will lead to result. Lagging indicators are used retrospectively to check whether results have been achieved. Leading indicators typically relate to the process; lagging indicators relate to performance targets and benchmarks (Muchiri *et al.*, 2011). At the same time, one needs to be extremely careful in selecting indicators. In healthcare, it has been observed that teams developing performance indicators for tasks with a higher uncertainty developed more process indicators than outcome indicators. It was also found that these process indicators had more of a problem-solving focus. In contrast, teams developing performance indicators for tasks with low uncertainty focussed more on indicators for outcomes and procedures (van der Geer *et al.*, 2009). As an example of a specialist naming of indicators, the area of road transport uses Safety Performance Indicators (SPIs) that are defined as the measures for the road safety performance that reflect the operational conditions of the road traffic system. In this domain, SPIs are used in addition to accident and casualty statistics, especially to understand what factors such as alcohol lead to accidents and how they relate to the severity of these accidents (Assum and Sørensen, 2010). They also are used to understand the safety of a countries vehicle fleet or to measure the impact of road safety measures (Christoph *et al.*, 2013). Van Ostaeyen *et al.* (2013) introduce Functional Performance Indicators (FPI) in order to assess how well a system functions or, in other words, performs its intended purpose; these FPIs are used to discern variables that relate to performance from those that describe behaviour and structure. In the building domain, another typical distinction is between indicators that predominantly focus on a spatial distribution, those that predominantly study how some aspect changes over time and those that combine both spatial and temporal aspects.

In many cases, the quantification of performance requires expertise of the appropriate domain. Some of the underlying domains are generic, such as mechanical engineering, material science or thermodynamics; however these domains have conventions and well-tested approaches to study phenomena of interest. For instance, material science regularly considers mechanical strength, stiffness and toughness to compare various materials. These can be applied to relatively common situations, such as reinforced concrete beams and columns; yet the same approaches are highly useful when looking at a much more specialized area such as high performance fibres; here common knowledge helps to focus on typical stress–strain curves and strength–stiffness plots that could for instance be used to compare various samples of Kevlar and carbon fibres (Koziol *et al.*, 2007). In road safety studies, statistics help to identify key factors that often play a role in road accidents and the severity of their outcomes, such as vehicle age and weight (Christoph *et al.*, 2013); these are thus crucial for the appropriate measurement of the performance of some interventions such as barriers. In other cases, more specialized expertise is required. For instance studies on energy-storing devices often use 'Ragone plots', which map power density against energy density (Gogotsi and Simon, 2011). Obviously, for quantifying building performance, the expertise needed depends on the specific issue at hand; however in general care should be taken when encountering thermodynamics, lighting, acoustics and vibrations, moisture and airflow, and chemical and similar issues. Foliente (2005b: 111–112) notes that many performance criteria for buildings show some overlap between purely technical and sociological aspects. Where the technical aspects typically are well supported by the natural sciences, the sociological aspects require human response studies, an area where further work is still needed. Human response studies may take the form of laboratory or field studies and can take place in more or less controlled environments using careful research design and statistical approaches to explore expectation levels.

Measurement is a key tool for empirical science (Luce, 1996); the science of measurement is named metrology. For an introduction to metrology and measurement, see Bewoor and Kulkarni (2009). Inherent in the definition of measurement is the comparison with a standard unit or with an object of known attributes; this implies the use of a measurement scale. Fundamentally, there are four measurement scales: nominal, ordinal, interval and ratio. A nominal[7] scale only assigns names to elements of a set.

---

7 From nomen: Latin for name.

An ordinal[8] scale ranks elements of a set, but does not provide any information about the differences between different ranks – number one is not necessarily twice as good as number two. In an interval[9] scale, the difference between positions is equal; for instance the difference between a temperature of 5°C and 10°C is the same as between 15°C and 20°C. Finally, a ratio[10] scale has a true zero, such as in the Kelvin temperature scale. While metrology is an established field, there are still many issues in this area that always need to be considered in research and development; for instance there are complex issues to take into account when defining the number of categories on a nominal or ordinal scale and what these mean to the stakeholders (Viswanathan *et al.*, 2004; Lai *et al.*, 2010). Gilb (2005: 137–164) provides a deep discussion of scales and how to tailor them to interest in various performance issues.

Measurement as a process that ascertains the size, amount or degree of something must be very specific in defining what attributes of a building are being observed. The domain of metrology has agreed on a range of definitions of key terms that are listed in the *International Vocabulary of Metrology*[11] (JCGM 200:2012). This defines *measurand* as a quantity that is intended to be measured (*ibid.*, 17), while *quantity* itself is the property that can be expressed as a number and a reference. This reference then can be a measurement unit, a measurement procedure, a reference material or some combination of these (*op. cit.*, 2). These definitions underpin the International System of Units (SI) and many units that are derived from the SI System. However, authors that deal with (building) performance rarely, if ever, use the term measurand. Instead, common terms are metrics, measures and indicators.

Whether performance is measured by way of performance indicators, metrics, measures or parameters, the definition must be unambiguous and understandable to all stakeholders. In some cases this can be relatively straightforward, such as a room temperature or electricity consumption of a specific system; however due to the complexity of buildings, the task of identifying these attributes is often non-trivial; see for instance the performance index class diagram by Gursel Dino and Stouffs (2014) or the performance framework ontology by Corry *et al.* (2015). In other cases the attribute itself can be complex and require decomposition; see for instance the example provided by Gilb (2005: 154) in terms of the 'quality' of a system, which in the example is decomposed into availability, adaptability and usability.

Furthermore, measurement must capture the specific conditions under which the observation took place. In a controlled experiment these can be fully defined by the experimenter, as in a classical falling object experiment where the moment of release is captured and then the movement of the object over time is traced, or in a more complex car crash experiment where a vehicle is made to hit a barrier at a given speed. To be of use in measuring performance and stakeholder needs, these conditions need to map to user-validated mission scenarios. These scenarios may represent either typical or nominal conditions, or extreme 'rainy day' situations where loads and stress on the system are maximized. An example of performance evaluation where scenarios play a key role is that of the evaluation of fire risks; for proper evaluation of more complex

---

8 From ordinalis: Latin for sequence.
9 From intervallum: Latin for 'space between ramparts'.
10 From ratio: Latin for reckoning.
11 Often abbreviated as VIM, from French for Vocabulaire International de Métrologie.

(multifunctional) buildings, one needs to define for instance the initiating event, location of the fire, fire severity, exposure, impact of any fire protection systems and domino effects (Nilsson *et al.*, 2013).

Another requirement for measurement and quantification is the use of a properly defined process. Such processes are often laid down in technical standards, such as the British and European Standards like that for facility performance benchmarking (BS EN 15221-7:2012). Further standards are provided by International Organization for Standardization (ISO). Some of their older standards give guidelines on writing standards themselves; among these, ISO 6241 (1984: 2) prescribes that performance standards should state both the performance requirements for which the standard is applicable and the method(s) that can be used to assess the performance. The statement of the requirement should include the stakeholder need and the loads or agents acting on the building. The American Society for Testing and Materials International (ASTM) provides standards for a wide range of building and building system/component tests in the United States. Further standards are available from the Netherlands Standardization Institute (NEN), the German Institute for Standardization (DIN), Standards Australia and others. However, there are also cases where no standards are available; in such cases measurement and quantification need to follow the generic principles of good science and aim for a process that is repeatable, valid and accurate.

Again, different wordings are used for process definitions. The seminal Report 64 of CIB W60 (1982: 14) coins the term of Performance Test Methods (PTMs) for those tests that aim to represent behaviour in use, with complex conditions. It then distinguishes between PTMs for research purposes, PTMs for system design and development and PTMs for quality control. CIB Report 64 goes on to stress the importance of test conditions in PTMs, which need to represent the conditions under which a building or building system will be used, but which may vary for different regions and countries (*ibid.*).

It then goes on to show some real-life stresses on buildings and laboratory tests that replicate these stresses, such as the use of a water spray and ventilator to replicate driving rain, a swinging load to replicate a falling person or a tapping machine to replicate knocking or stepping on a floor (*op. cit.*, 15). However, the term seems to have been mainly limited to the CIB partners; it appears for instance in the PeBBu report by Becker (2005: 75) but does not seem to have gathered more general traction.

Another term that relates to the assessment process is that of Performance Assessment Method (PAM). This has its origin in the International Energy Agency Annex 21 on 'Calculation of energy and environmental performance'. It defines a Performance Assessment Method as 'a way of determining a desired set of data indicative of a particular aspect of building performance using a predictive computer program' but then expands this definition to measurement and quantification in general by observing that 'since we are concerned in many cases with using the results to inform design decisions the above simple definition has been extended to incorporate the interpretation of results in design or other terms' (IEA, 1994: iii). According to the original source, a PAM combines a purpose, program (software tool) and method. To ensure that assessments can be carried over between different programs and can be repeated, use is made of a performance assessment method document (PAMDOC) that, among others, describes the process for carrying out a performance analysis (*ibid.*). Performance Assessment Methods are still mentioned in recent guides such as CIBSE Guide A

(CIBSE, 2015b: 0-8, 0-10–0-11) and CIBSE AM11 (CIBSE, 2015c: 18); in the CIBSE context a PAM is taken to be a 'means of documenting a QA procedure in building energy and environmental modelling. PAMs will be produced by experienced modellers. PAMs must be developed for using a particular calculation method, normally enshrined in software', and lists the purpose, applicability, output, models and sub-models used, context, zoning, building and operation that is assumed in the assessment process (*ibid.*).

Finally, there is the concept of an Analysis Function (AF) as defined by Augenbroe and his colleagues. An Analysis Function is defined as a formal expression of a reusable analysis procedure, which can be used to identify the relevant information that is needed for doing an analysis, whether the analysis is based on a virtual experiment (simulation) or a physical experiment. In the context of simulation, an Analysis Function defines what information needs to be sought from a building information model. A key issue is that Analysis Functions should be defined in such a way as to be understandable to all stakeholders in the design analysis dialogue while being software neutral. An Analysis Function is defined for a specific performance aspect, a specific building (sub) system and a specific performance measure or indicator. It defines the process that is needed to map building variables, experimental variables and control variables to an unambiguous measure of performance (Augenbroe *et al.*, 2003: 5–6; Augenbroe *et al.*, 2004; Augenbroe and Park, 2005; Pati *et al.*, 2009).

Consistent quantification of performance is seldom straightforward. Even where the domain is technological and the underlying science well developed, there usually is a range of underlying assumptions to be made and decisions to be taken about performance expression. For instance, when accounting for energy generation, transmission and distribution, it is hard to identify the actual conversion factors (Hinge and Winston, 2009); assumptions for these factors need to be clearly expressed in order to allow comparison of any performance assessments. Non-technical performance aspects require even more diligence. For instance, Spekkink (2005: 32) expresses the view that architectural and cultural aspects cannot be captured by hard, measurable performance indicators; Gilb (2005: 162–163) counters that one 'can and should always define a scale of measure for any system critical variable' and that 'defining a scale of measure is a teachable practical process', giving examples for further difficult attributes such as environmental friendliness (*ibid.*, 141–144).

Quality control of experiments uses the terms verification and validation. In general, in science and technology, the process of verification is used to establish that a product or process meets some sort of specification and works according to that exact specification; for instance when verifying that some software code does precisely what one expects that software code to do, without errors. The process of validation looks at how well a model represents the real world, taking into account that a complete agreement is very hard to reach (Zeigler *et al.*, 2000: 367). In terms of computing, it can be said that verification is about 'solving the equations right', whereas validation is about 'solving the right equations' (Oberkampf and Roy, 2010: xi). Applied to experimentation in general, verification then is about making sure that the experiment and measurement are carried out in the correct way, whereas validation relates the work to the underlying objectives, such as whether an experiment indeed provides and answer to a stakeholder question or need. In the context of buildings and systems engineering and design, 'the purpose of the verification process is to provide objective evidence that a system or system element fulfils its specified requirements and characteristics' (INCOSE, 2015: 83), while 'the purpose of the validation process is to provide objective evidence that the system, when in use, fulfils its business or mission objectives and stakeholder requirements, achieving its intended use in its intended operational environment' (INCOSE, 2015: 89).

Verification compares the response of a system with a reference that gives an expected result. If the expected and observed results match, then the system is verified for that action (SEBoK, 2014: 347). Verification may involve inspection, testing, analysis, demonstration and simulation; sometimes verification can also involve processes of analogy/similarity and sampling. Inspection involves measurement and observation using the human senses to check whether a system has the defined properties, as defined in proper documentation, without subjecting the system to any excitation. Testing is where a system is subject to excitation, and the responses are observed and measured. This may involve a range of inputs or combination of inputs as in a scenario. Testing is used to establish more complex system attributes such as operability, supportability and performance. Testing often involves the use of dedicated test equipment and instruments to obtain accurate readings that can be analysed. In the context of verification, system analysis may be used as a separate process where actual testing is difficult; here there is no actual excitation, but the theoretical response is checked or simulated (INCOSE, 2015: 86–87). Verification may yield verification reports but also reports of issues and non-conformance that subsequently require change requests to the system (SEBoK, 2014: 350).

Validation involves a demonstration that a system meets a series of stakeholder requirements, can fulfil its functions in the intended environment and for a range of operational scenarios and, more in general, is ready for use. Validation of a system is sometimes required by a client before a change of ownership from supplier to customer. Validation may also lead to certification, where the product comes with written assurance that this product performs according to a certain standard (INCOSE, 2015: 93–94).

4.3   Experimentation and Measurement

In essence, the process for measuring or quantifying performance as defined by a Performance Test Method, Performance Assessment Method, Analysis Function and similar can be viewed as a measurement protocol. Augenbroe has, over many years, championed the notion of building performance simulation as a virtual experiment that maps a range of input variables, environmental and control variables, and system properties to a set of observable states. Aggregation of these observable states then leads to performance quantification in the form of a performance indicator. The structure underlying this notion of performance analysis is applicable to real, virtual and hybrid experimental set-ups. It also allows for different aggregation methods that can meet different needs (Malkawi and Augenbroe, 2003: 6; Augenbroe and Park, 2005; Augenbroe, 2011: 17).

Formally, an experiment is a series of tests, whereby purposeful changes are made to the inputs and variables of a process or system. Responses of the system and system output are observed in order to understand the relation between changes and effects. Careful planning of experiments helps with the subsequent analysis of results, which often involves statistical analysis (Wu and Hamada, 2002: 1; Montgomery, 2013: 1).

Experiments can lead to different levels of knowledge. At a first level, experimentation establishes to what inputs or stimuli a system of interest responds; this allows the observer to identify what variables must be measured over time. At a second level, experimentation yields a series of paired input and output variables, indexed over time, which can be studied to establish relationships between input and output. At the third level, experimentation allows reproduction of behaviour, provided this starts from the same initial state. At further levels, experimentation yields insights in how inputs affect the system states and state transitions and how these state transitions affect output (Zeigler *et al.*, 2000: 13).

There are different types of experiments: one can discern experiments that are conducted to help the understanding of some physical process, experiments that help to determine or improve the parameters used in existing models, experiments that support the design and analysis of systems, and experiments that are carried out to validate models and software (Oberkampf and Roy, 2010: 372–273). Montgomery (2013: 14–15) lists experiments to characterize systems and identifies factors of interest for optimization, confirmation and discovery, and to establish robustness of a system or process. In the specific area of building, CIB Report 64 (1982: 14) makes a difference between experiments that serve research purposes or quality control, as well as between tests at component or full building scale. Performance measurement and experimentation can be applied not only to products and production processes but also to design and engineering itself. In this case the objective is to manage the resources such as know-how and personnel in order to control costs, quality, flexibility, lead times, deadlines, etc. (Girard and Doumeingts, 2004).

INCOSE strongly focusses on experimentation as testing, in the context of system verification, thus making sure that a system has the required characteristics. This verification may also include other approaches, such as visual inspection, gauging and measurement, demonstration of how a system responds to stimulation or excitation or the use of models and mock-ups (INCOSE, 2015: 86–87). Older versions of the handbook define tests in more detail, distinguishing four basic test categories:

1) 'Development test: conducted on new items to demonstrate proof of concept or feasibility.
2) Qualification test: conducted to prove that the system design meets its requirements with a predetermined margin above expected operating conditions, for instance by using elevated environmental conditions for hardware.
3) Acceptance test: conducted prior to transition such that the customer can decide that the system is ready to change ownership from supplier to acquirer.
4) Operational test: conducted to verify that the item meets its specification requirement when subjected to the actual operational requirement' (INCOSE, 2003: 131).

The basic elements of an experiment are depicted in Figure 4.5. This picture expands the traditional ICOM (Input, Control, Output, Mechanism) model by expanding the controls to controllable and uncontrollable factors. Furthermore, it details the traditional representation of one central process by explicitly relating this to (a) system(s) of interest, which provides desired function to the process, while being subject to operation by that process; the system has (an) initial state (s) as well as changed states that may be observed over time and may have further abstracts mechanisms supporting it as well as the process. Classical texts like Zeigler *et al.* (2000: 4) or

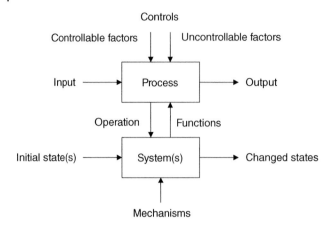

Figure 4.5 Basic elements of an experiment.

Malkawi and Augenbroe (2003: 6) often focus on the system only, with either a focus on input/output (Zeigler) or observable states (Augenbroe), which may relate to a more general system interest or the focus on buildings as more 'passive' systems that do not have production as main focus. Others like Montgomery (2013: 3) focus more on process and production. Figure 4.5 splits these to make it explicit that experiments need to observe both the process (input and output) and the states of the system. The systems and processes that are of interest in an experiment are named 'experimental unit' (Wu and Hamada, 2002: 8; Montgomery, 2013: 69), 'unit of analysis' (Fellows and Liu, 2008: 99) or sometimes 'unit of assignment' (Groat and Wang, 2013: 317).

The process of experimentation involves a series of key activities. The first of this is the analysis of the problem; this relates closely to the discussion in the previous chapters of the stakeholders in buildings and their performance requirements. The second is the selection of response variables, as well as the drivers of the experiment in terms of factors, levels and ranges. Typically these can be classified as design factors, factors that are held constant and factors that are allowed to vary. Some factors may not be of interest but still be intrinsic to the experiment; these are sometimes named nuisance or noise factors. The next activity is the choice of the experimental design in terms of sample size, the order of the experimental runs and the application of statistical principles to ensure that the results from the experiment can be analysed in an efficient manner. A key issue here is a move from one-at-a-time variation of factors towards the change of several factors at once in what is called factorial designs. This is followed by the actual doing of the experiment; it is good practice to include a few test runs, as well as to have rigorous processes in place to ensure that the measurement plan is followed. Observation during the experiment is an area that should be handled with due diligence (Weinberg, 1975: 228; Weinberg 1988: 43–60). After observation has been done and data has been collected, results from the experiments can be statistically analysed; often this analysis also employs graphical representation of the data, as well as the development of an empirical model that describes the relation between the experimental variables. Finally, analysis results should lead to conclusions, be documented and linked back to the original objectives of undertaking the experiment (Montgomery, 2013: 14–21). Good science requires experiments to be based on a correct sampling, to be repeatable (giving similar results if the same specimen is tested with the same test/apparatus) and to be reproducible (giving the same results if a specimen is tested with different tests/apparatus) in order to be accepted by the scientific community (CIB Report 64, 1982: 16).

One key aspect in experimentation is that typically, the experimenter is interested in establishing the impact of a specific factor or set of factors on the process or system under consideration. In order to get the best insight into the impact of that factor, it needs to be handled different from the other variables that impact the experiment; this is named 'treatment' and turns that key factor into an 'independent variable' (Groat and Wang, 2013: 316). It may also be useful to study the experimental unit in a situation where treatment is absent; this is often named a control or control group (*ibid.*). At the same time, other factors must be kept constant or eliminated; this is called experimental control (Fellows and Liu, 2008: 101). Sometimes an experiment is defined specifically to test a hypothesis or theory; the new hypothesis is named H_1 or sometimes H_A, whereas the opposite is named the null hypothesis or H_0 (Montgomery, 2013: 36; Streiner, 2013: 213–215).

Where the object of study in an experiment is a system, there often is the aim to get some sort of response from that system in order to establish differences. To ensure a response, a load or input is imposed on the system; this is termed 'system excitation'. Where there is a range if loads that often come in a specific order, these may be represented as a mission or scenario (Blanchard and Fabrycky, 2011: 80–82). Scenarios can be related to the use cases and activity diagrams used in requirements engineering. It can also be useful to group certain excitations in a 'swimming lane' diagram that captures which loads belong to a particular sequence (INCOSE, 2003: 172). It is important that testing conditions and excitations reflect the conditions that are expected once a building or component is in real use. Defining these conditions is a non-trivial effort, often requiring fundamental research. It is important to recognize that the conditions that are appropriate in one case might not be transferable to another (CIB Report 64, 1982: 14). Becker (2005: 8) observes that 'in many areas of building performance there is a recognized lack in dose-effect data' – which can be interpreted as that it is unknown what the proper system excitation would be to get a response, and that available excitation-response sets are still insufficient and incomplete. Where scenarios involve a long time span, such as the ageing of building products, time and cost constraints may prevent the use of a natural test, and experiments may be devised that speed up the process (CIB Report 64, 1982: 4).

Another important aspect of experimentation is observation. As emphasized in Figure 4.5, this should cover input/output pairs, system states and experimental factors. The interaction between input, output and state is exemplified by Zeigler *et al.* (2000: 16–18) via the case of forest fire. Here one may study the impact of lightning (input) on the development of smoke (output); however the impact of the lightning bolt may depend strongly on whether the vegetation is wet, dry or already burnt (state), while other factors such as wind may also play a role in the fire development. The relation between input and output can often be captured by means of mathematical equation, leading to an empirical model. In many cases the relation may be linear (Vesma, 2009: 8–14); however there can also be more complex relations, and thus it is important to establish what the correct form is, using knowledge of the underlying phenomena (Tingvall *et al.*, 2010). An important issue in observation is the handling of time. Zeigler *et al.* (2000: 14–17) emphasize that input, output and states are measured in relation to a time base. Depending on interest, one can then distinguish between longitudinal studies, where observation takes place over a longer interval, and cross-sectional studies, where a range of observations is related to one point in time (Fellows and Liu, 2008: 100). In the context of measuring building performance, Neumann and Jacob (2008) add an intermediate position and discern three types of observations in relation to time: spot measurement, which consists of only one single measurement at a particular point in time; short-term measurement, which is conducted over a limited period of time; and long-term measurement.

Observation often takes the form of measurement. What is being measured relates back to the original information need and decisions to be made. It is thus important that the observations can be communicated to all stakeholders, and that they help to identify any problems and anomalies and reflect the project objectives (INCOSE, 2015: 130). Just collecting observations for the sake of data collection does not make sense and is a waste of time and effort; it is thus important not to be tempted by the possibilities of instruments and data loggers. Instead, observation should aim to be useful and require minimal effort in collection, and most of all, it should be repeatable and captured in a logical, regular manner (INCOSE, 2015: 131). There is a danger in people trying to find a single measure or small set of measures that they expect to be applicable across all projects they conduct, as there normally is no one-size-fits-all measure or measurement set (SEBoK, 2014: 409).

Proper measurement does not rely on accidental observation; instead it takes place according to a well-defined measurement plan that allows systematic observation and is followed by evaluation of the findings. This measurement plan thus must refer to the stakeholders and their needs, then proceed to define what direct and indirect measures will be taken and what instrumentation is required, and specify the measurement frequency, how the measurements will be stored, what – if any – thresholds will be observed, and who will review the data. The actual measurement activity consists of the gathering, processing, storing, verification and analysis of observations. During this process the observations start as pure data but are turned into information through processing and analysis. Documentation and review of the measurements is an important stage; typically it includes data cleansing[12] and post-processing, for instance by presenting data as graphs and tables. Actors in the measurement process should be prepared to redo some measurements if data is found to be missing or not usable (INCOSE, 2015: 130–135); very few measurements are a single-pass activity (SEBoK, 2014: 409).

Measurement requires that an appropriate infrastructure is in place. Measurement System Analysis (MSA) provides an approach to review this infrastructure (SEBoK, 2014: 439). Typically, measurement requires properly calibrated instruments, trained personnel to conduct the measurement and analysis, and a suitable measurement environment; experimenters must understand why the measures are needed and how the raw data is turned into information, preventing inappropriate measurement and use of data (SEBoK, 2014: 409). Good measurement is mostly based on the use of well-defined testing and measurement procedures that are often laid down in standards, regulations and company procedures.

---

12 Data cleansing: the process of identifying incomplete, missing and incorrect records in a set.

It is important to note that the outcome of experiments and measurement is not always a single number; in many cases the outcome is actually a relation between factors of interest. A good example is the relation between static pressure difference and air-flow (CIB Report 64, 1982: 20). In such cases the result of the experiment is in fact a graph or sometimes a mathematical approximation of the measurement outcomes.

Most experiments involve some degree of uncertainty and experimental errors. This may be due to small variations in the system set-up or the process that the system is subjected to, limitations to the observation process and instruments and other factors that are beyond control of the experimenter. These are dealt with by applying statistics to the experiments. There is an area of statistics that specifically deals with 'design of experiments' (DoE), ensuring that these yield appropriate data that can be statistically analysed. This is achieved through three principles (Wu and Hamada, 2002: 8–11; Montgomery, 2013: 12–13):

- Randomization – Selecting any materials and other aspects in such a way that there is no systematic error in the observations.
- Replication – Repeating the experiment while taking care to make sure this is not just a repeated measurement of the same object/process.
- Blocking – Reducing or eliminating factors or parameters that are not of interest.

There are uncertainties that lead to risk in achieving building performance, where risk is a combination of probability and consequences (Ale, 2009: 4–5). Applied to the context of system design, risk becomes 'a measure of the potential inability to achieve overall program objectives within defined cost, schedule, and technical constraints'. The probability translates to the uncertainty in achieving the desired outcomes and the consequences in what not meeting those outcomes means for the system and the stakeholder (SEBoK, 2014: 395).

Zeigler *et al.* (2000: 27–29) use the concept of an experimental frame to guide observation or experimentation with a system. This experimental frame defines what variables are of interest, which is related to the objective of the observation. For most systems, a range of experimental frames can be defined, varying from simple to complex. An experimental frame can also be seen as an externality to the experiment, which generates input for the experiment, monitors that experiment is conducted properly, observes and analyses the outcomes.

In an ideal situation, experiments are carried out with as much control over the variables as possible. Such maximal control is provided in a laboratory. However, sometimes the use of a laboratory is not feasible due to for instance the size of the subject or the costs associated with lab testing. In other cases, there may be an interest in doing the experiment in a real-life context that includes social, industrial, economic and other factors; such experiments in a dynamic live context are sometimes named quasi-experiments (Fellows and Liu, 2008: 24; Groat and Wang, 2013: 322–327) or field studies. Note that in a laboratory context there still may be factors that are imposed on the experiment, such as gravity. Due to their size, there are only few experiments that involve tests on a complete building under laboratory conditions – the exceptions typically being tests like how buildings respond to crucial threats like fires or earthquakes. As a consequence, most whole building tests are carried out *in situ*, with only semi-controlled experiments. However, laboratory testing of building systems, like boilers or facades, is common. Physical experimentation typically is expensive, so where there is a choice the simpler option should be chosen. In some cases insights from more complex tests can be used to develop simpler approaches that are sufficient for follow-on tests, such as the development of full-scale fire test into simpler spread-of-fire tests on small material samples (CIB Report 64, 1982: 16).

The methodology for conducting experiments in live buildings needs careful consideration and must be developed based on the specific assessment goal. As per previous discussion, when one wants to establish the relationship between an independent and a dependent variable, in theory a randomized controlled experiment should be conducted. This random selection would split the sample in two groups. One of these will act as control group, whereas the other will be subject to manipulation of the independent variable. However, in real facilities, this is very hard to establish. For instance, in hospitals one might want to study the relation between lighting levels and errors in dispensing medication; however it is almost impossible to control and eliminate the other nuisance factors that might cause variation in the dependent variable. But in some cases, experiments may be helped by the use of naturally existing conditions, such as the presence of rooms with and without a view. This makes it imperative that researchers explore the specific situation and make maximal use of existing conditions in their semi-controlled experiment (Joseph and Hamilton, 2008).

4.4 Building Performance Metrics, Indicators and Measures

As discussed in the opening paragraph of this chapter, there is a range of terms that are used to capture the outcomes of measurement and quantification: performance metric, performance indicator and performance measure. As noted by Keeney and Gregory (2005), 'the terms performance measure, criterion and metric are often used as synonyms'. However, the terms can also be defined more in detail. For instance, Labate (2016) makes the following distinction:

- *Metrics* are used to represent how performance changes over time or in different dimensions. The word is also used as general term to capture a measurement method, the values obtained through measurement and calculated or combined values.
- *Measures* are the numbers and quantities that represent performance; these are directly observable and can be expressed as a number or value (which represents magnitude) and a unit of measure (which represents meaning).
- *Indicators* are combined values that are used to measure performance, achievement or the impact of a certain change.

Note that none of these terms has been defined in the vocabulary of metrology (JCGM, 2012).

Unfortunately, the field of building performance does not adhere to these definitions, instead using the terms as synonyms. An example in the field of building commissioning is the report by Neumann and Jacob (2008), which uses measured data, measured values, performance indicators, performance metrics, baselines, signatures and control parameters all in a similar vein and hence is hard to follow for any reader. Similarly, Bluyssen (2010) uses indicators to discuss building performance as well as occupant well-being and health; although the latter are undoubtedly important, they are a proxy for building performance. Yet whatever name is given to performance attributes, it must help to assess the value of various alternatives and be defined in a way that is unambiguous, comprehensive, direct, operational and understandable (SEBoK, 2014: 415). This paragraph reviews building performance metrics, indicators and measures as used in the building performance literature one by one, ordering concepts and providing an overview. It starts by discussing metrics, as a generic term and most closely related to experimentation, followed by indicators and measures. For each concept, particulars and the most common context in which they are used will be discussed.

4.4.1 Performance Metrics

A metric is a system or standard of measurement. The word metric relates to the metric system, which in turn refers to the well-known unit of length. In business, metrics are defined as a set of figures that measure results; in mathematics and physics, metrics are a topological function based on distances or on quantities analogous to distances. In the context of systems engineering, Gilb (2005: 376) points out that a metric is a system attribute, which is specified in relation to a specific scale of measure. Scales of measure are selected in such a way that measurements can be translated to numbers.

The term performance metrics is used in a range of domains in and beyond engineering and building. Lee (2015) discusses the use of performance metrics as drivers of quality in healthcare, where these help to measure reliability, safety, efficiency and patient experience; at the same time he warns that it is important to keep in mind what the overall objective is and to remember that all aspects are intertwined, even if they are measured separately. Melnyk *et al.* (2004) use metrics in the context of operations management, noting how these play a role in control, communication and improvement in business. In the world of ICT, Vigo and Brajnik (2011) discuss the use of automated computation of website/page accessibility metrics; aspects that play a role in these metrics are validity, reliability, sensitivity, adequacy and complexity. Gleckler *et al.* (2008) discuss the use of performance metrics to compare climate models, although their text also uses measures and variables. They point out the need to study the correlation between different metrics, as well as the sensitivity to observational uncertainty, spatial scale and the domain considered. Hölttä and Koivo (2011) discuss the use of metrics in the web-forming processes in the paper, plastic film and steel industries, where these are used in the context of control loops. This work mentions the use of filtering and comparison of metric values with an upper and lower threshold value. Metrics are also used in order to capture the productivity and impact of individual academic researchers, such as through the *h*-index[13] and i10-index,[14] although this is not without criticism (Pijpers, 2006; Macilwain, 2013).

In the building domain, Hitchcock (2003: 3) uses performance metrics for the expression of building performance objectives while using a dynamic, structured format to define quantitative criteria. According to Hitchcock, metrics must either be predicted or measured in relation to the passing of time and must allow the evaluation of the meeting of particular performance objectives. Often it is useful to study a hierarchy of

---

13 An *h*-index of *y* means that a scholar has published a number of *y* papers, each of which has been cited in other papers at least *y* times.

14 The i10-index represents the number of papers of a scholar with 10 citations or more.

metrics to capture a high-level objective. Further, Hitchcock distinguishes between performance metrics that are used for benchmarking and performance metrics that are used to express the outcome of a more specific assessment (*ibid.*, 10). Deru and Torcellini (2005: 5) define performance metric as a 'standard definition of a measurable quantity that indicates some aspect of performance'. They mention the following characteristics as essential: metrics need to be measurable or directly based on measurements; metrics require a clear definition of how they are measured; metrics must indicate progress towards achievement of a performance goal; and metrics must address specific questions about performance. The California Commissioning Collaborative (2011: 51) defines a metric as 'a key performance indicator which may be compared to historical or expected values'. They note that metrics can relate to the performance of subsystems or the whole building. In their view, an important characteristic of a metric is that it should capture more than simple status of data points; it should combine data to give a deeper meaning. Examples of metrics mentioned by the California Commissioning Collaborative are system efficiency, such as *kW per tonne of cooling delivered*, or thermal comfort when measured as *percentage of operational time that a building zone is within a predefined temperature range*. Use of the term metric seems to be rather inconsistent in the domain; for instance, Swarup *et al.* (2011) use metrics to describe project attributes that act as input to a study of success rather than to measure the outcomes of their evaluation.

Hitchcock (2003: 4) notes that performance metrics can have many forms and that there is little standardization of these metrics. Furthermore, the preferred metrics might even change over time within a single project. An example given is the performance of chiller efficiency using a single Coefficient of Performance value, multiple points on a load curve or load surface, and sometimes a mathematical function representing such a curve or surface. Furthermore, Hitchcock differentiates performance metrics on the basis of metric calculation variables and database filtering variables. Here metric calculation variables relate to concepts such as the building size, building location, fuel type and occupancy. Database filtering variables are concepts such as building use, HVAC system, special design features and ownership (*ibid.*, 27). This situation is not unique to the building sector. For instance, Joseph and Hamilton (2008) observe that even for an important field as healthcare, there still is a lack of standard performance metrics and tools. While most organizations in this discipline use metrics such as length of stay, number of falls and patient satisfaction, there is significant variation in how these are measured. A similar challenge exists for capturing the environmental variables in medical facilities, even though something like indoor air quality is crucial for operation theatres and wards. Hammond *et al.* (2005) point out that 'each metric by itself only provides a narrow perspective of a facility'. The California Commissioning Collaborative (2011: 54) takes the view that most metrics are building specific and that comparison of metrics for different buildings should be undertaken with caution. They suggest that building operators learn, over time, what is normal for a specific facility and what constitutes an anomaly. Deru and Torcellini (2005: 5) distinguish between 'tier 1' metrics, which are based on annual and monthly data, and 'tier 2' metrics, which are based on hourly or sub-hourly measurements. In their view performance indicators aggregate tier 1 metrics at a higher level and are mainly used by policymakers. Neumann and Jacob (2008: 16) use the term performance metrics to capture the outcomes of evaluations and calculations that combine general building data such as area, tariffs and zoning with measured values of use of utilities, such as electricity and water consumption; however they seem to consider performance metrics and performance indicators to be synonyms.

In an effort towards classification of building metrics, Then (2005: 85–86) discerns five types of building performance evaluation metrics: *Economic metrics*, which relate to the alignment of building assets with strategic business directions; *functional metrics*, which relate to the creation of assets that are 'fit for purpose'; *physical metrics*, which relate to the operational aspects and asset management; *service metrics*, which relate to the quality perception of end users; and *environmental metrics*, which relate to the impact on users, the community and the environment. Hitchcock (2003: 25) suggests the standardization of energy-related performance metrics on the basis of the US Department of Energy (DOE) High-Performance Buildings Metrics Project as reported by Deru and Torcellini (2005). Hitchcock also promotes the embedding of building performance metrics in the IFC. The more recent report by the California Commissioning Collaborative (2011: 53) gives an overview of five recommended specific HVAC systems metrics: occupant comfort index, cooling plant efficiency, heating plant efficiency, fan system efficiency and outside air ventilation. While these are well defined, they relate to specific targets such as the percentage of hours that a zone is within a target bandwidth, which may be different for specific buildings; the definition of zones; and the number of persons in a building.

Building performance metrics are mostly captured using engineering concepts and units, such as voltage (V), frequency (Hz) or temperature (K, °C or °F), and typically defined in the list of symbols and units in books and papers. One issue with units is the continued use of both the International System (SI[15]) and Imperial (IP) systems in parallel, which means that even on this basic level for something as fundamental as energy, there already is a choice between Joules (J) and British Thermal Units (BTU); however there are even further options to express this in the energy use in the built environment, such as kilowatt-hour [kWh] and tonnes of oil equivalents [toe]. These can be converted into each other but require careful specification to prevent errors and miscommunication.

Hitchcock (2003: 14–24) notes that most performance frameworks like the CIB Compendium, ASTM Standard on Whole Building Functionality and Serviceability and others do not include quantifiable metrics, the only exception being the US Department of Energy High-Performance Metrics Project, which is the subject of his report. This work is expanded in later work from Lawrence Berkeley National Laboratory (LBNL): in their technical guide for energy performance monitoring of buildings, Gillespie *et al.* (2007: 22–34) list a range of performance metrics that can be measured. At a basic level they list metrics such as outdoor air temperature, main electric power consumption, energy use intensity, gas heat rate, total water flow and zone temperatures. At an intermediate level, they recommend the use of metrics such as duct static pressure, HVAC electric power, the power of specific component such as chillers and air handling units, lighting electric power, electric plug loads, supply and return temperatures and boiler output. At an advanced level, they recommend to link building metrics such as plant operation hours, gas flow to specific systems and boiler efficiency to collected outdoor weather and building use parameters.

---

15 SI: from French: Système Internationale d'Unités.

Performance metrics also feature strongly in a recent report by the New Buildings Institute of the United States on 'getting to outcome-based building performance' (Frankel *et al.*, 2015), although the report also incidentally uses the word indicator. The report notes that energy performance metrics seem to dominate the discussion on building performance (*ibid.*, 23). A widely used metric for building energy performance is Energy Use Intensity (EUI). In the United States, this is typically measured in kBtu/sf/year, or sometimes kWh/sf/year, whereas elsewhere the use of kWh/m^2 per year is common (Neumann and Jacob, 2008: 16[16]). This metric is applicable to all fuel types and normalizes to building size. However, when using EUI one should still take into account differences that might occur due to variation in outdoor climate, building use, occupancy and special technology (Frankel *et al.*, 2015: 22). Bordass *et al.* (2001b) note that energy use per square metre of floor area across a sample of 20 buildings shows a large range and that this variation becomes even higher if studied as energy use per occupant. Other performance metrics that should be considered concern occupant comfort, health and satisfaction, economic efficiency, productivity and resource optimization and building functionality and resilience (*op. cit.*). Energy Use Intensity (EUI) requires only annual energy consumption and building floor area and thus is relatively easy to compute from limited data. According to the California Commissioning Collaborative (2011: 26), this makes EUI a good indicator for the comparison of buildings in a portfolio, with historical data, or a peer group. However, in practice even something as straightforward as building floor area may lead to dilemmas; for instance there remain decisions to be made about how to deal with voids (such as in staircases and atria) or inclusion of parking decks. O'Brien *et al.* (2017) stress the deep interaction between occupant behaviour and building performance and suggest that this ought to be reflected in a group of specific occupant-centric performance metrics. Wang *et al.* (2017a) present metrics to measure the performance of future buildings, trying to look a century ahead. Hitchcock (2003: 4) also points out that performance objectives and associated metrics of a building change and evolve over time and that it is important that one can track these changes in an archiving system.

4.4.2 Performance Indicators

An indicator is a thing that indicates a state or level; in many cases this means a gauge or meter of a specified kind. In British English the word indicator is also used for a flashing light on a vehicle that shows it is about to change lanes or turn; in chemistry an indicator is a compound that changes colour and indicates a pH value, the presence of particular substances or chemical change.

Performance indicators are used across a wide range of disciplines, but especially in business administration, where Key Performance Indicators (KPIs) are used to define and measure organizational goals. Performance Indicators are also pervasive in government in attempts to balance objectives and costs and improving efficiency and efficacy (Blyth and Worthington, 2010: 86). In the context of systems engineering, performance indicators are linked to the provision of services,[17] especially in terms of service

---

16 Neumann and Jacob (2008: 16) call EUI 'annual specific consumption'.
17 Services in general include systems that focus on human activities and development, flow of things and governance; examples are healthcare, education, retail and trade. See SEBoK (2014: 554–555).

availability, reliability and performance. The Systems Engineering Body of Knowledge makes a distinction between critical success factors (CSFs) and Key Performance Indicators, where CSFs are the elements of the service that are most important to the customer, whereas the KPIs are values or characteristics that can be measured in order to establish whether or not the service delivers the CSFs (SEBoK, 2014: 408). Performance Indicators are also used in standards and norms. For instance, ISO defines an indicator as a 'measurable representation of the condition or status of operations, management, or conditions' and a key performance indicator as an 'indicator of performance deemed by an organization to be significant and giving prominence and attention to certain aspects' (ISO 14031:2013). The concept is applied at various levels; for instance BS EN ISO 50001:2011: 3 defines an energy performance indicator (EnPI) as a 'quantitative value or measure of energy performance, as defined by the organisation', adding that EnPIs can be expressed in various ways, including simple metrics, ratios or more complex models.

In engineering, performance indicators are often used to indicate a parameter of interest. Tsitsifli and Kanakoudis (2010) develop a performance indicator to predict the risk of pipe failure in relation to pipe characteristics, product carried in the pipe, and operation process. Spyridis *et al.* (2016) use performance indicators in the context of dimensioning and appraisal of tunnel linings. Gernay and Franssen (2015) present a fire resistance indicator that includes not only the impact of peak temperatures but also the effects that result from the cooling phase. Messer *et al.* (2011) use a process performance indicator (PPI) to measure the level of model refinement needed in integrated product and material design, in their case applied to blast-resistant panels. Maghareh *et al.* (2016) define a predictive performance indicator to study the sensitivity of a real-time hybrid simulation towards phase discrepancies between an actual system and a computation.

One technical domain where performance indicators seem to be a prevalent term is that of road safety, where Safety Performance Indicators (SPIs) are widely used; this domain provides some insight into well-established use in a specific field. Road SPIs establish a link between interventions in the road system (actions) and final outcome (casualties in crashes); as such they are used in predictions as well as in the measurement of final outcomes. They typically are an intermediate measurement of a specific safety aspect, such as 'the proportion of car occupants using seat belts' or 'the proportion of the traffic volume travelling on divided roads'. In this domain these SPIs all have a well-established relation to the overall measure of casualties (Tingvall *et al.*, 2010). For road SPIs, European research recommends that these performance indicators should reflect operational conditions (the factors that actually cause crashes and injuries) but be more general than a direct measure of any specific safety interventions. In other words, SPIs need to reflect road safety, not just the implementation of an intervention. At the same time, they need to focus on specific risk factors, preventing overlap between different aspects. Finally, it is seen as important that SPIs enable their users to track the development of road safety over time, allowing assessment of the impact of interventions (Christoph *et al.*, 2013). For road SPIs, it is noted that it is difficult to collect data. In theory, one would want to collect data on the general road population, whereas data available in many countries is biased by measurements only taking place in case of an accident (Assum and Sørensen, 2010).

Performance indicators are used to study how well objectives are being met in a range of other fields, such as business administration, project management, commerce, manufacturing and healthcare. In all these domains, performance indicators can be used to identify a gap between desired and actual performance. By doing so they help the implementation of any strategies and initiatives aimed at improvement. Also, they help focus staff attention on particular areas of interest (Muchiri *et al.*, 2011).

Tsai and Cheng (2012) give an overview of KPIs for e-commerce and marketing of products targeting the elderly, listing financial, customer, internal process and learning and growth indicators; their list includes, among others, sales growth rate, market share, customer willingness to purchase, customer satisfaction, delivery speed, customer complaints and employee satisfaction. In healthcare, performance indicators are used to measure the quality of the care that is provided. However, they are also used as an instrument that links staff achievement with payment. Overall healthcare quality in the United Kingdom is monitored by the balanced Quality and Outcomes Framework that consists of over 130 indicators; for instance it measures blood pressure among patients as an indicator of coronary heart disease, haemoglobin and cholesterol levels as indicators of diabetes, and thyroid function as indicator for hypothyroidism (Reeves *et al.*, 2010). Anema *et al.* (2013) give an example of a process-related performance indicator from the medical domain and show how numerators and denominators feed into an indicator, given as mean score (and sometimes with SD) – for instance one can count the number of patients treated for a specific condition within a certain time (numerator), as well as all patients with that condition (denominator), and then use the quotient as performance indicator for how well patients with this condition are being helped within the set timeframe. They point out that the reliability of health performance indicators is dependent on the completeness, accuracy, consistency and reproducibility of the underlying data. Yet even in areas such as intensive care or breast cancer care, there appear to be ambiguities regarding the definition of this underlying data. This can lead to different interpretations that in turn lead to significantly altered indicator scores, which makes these indicators useless for transparency and pay-for-performance programmes (*ibid.*). Muchiri *et al.* (2011) give a summary of leading performance indicators for the maintenance process, listing category, measure/indicator, units, a description and recommended targets as aspects that need to be defined for each indicator. One example is the indicator of 'improvement work' that is needed. This has the unit of percentage; this is calculated by dividing the number of man-hours used for improvement and modification by the total number of man-hours available, while it is recommended to limit this to no more than 5% or 10%. In the context of organizations and management, Lohman *et al.* (2004) describe how a Performance Measurement System (PMS) uses performance indicators and standardized metrics. Popova and Sharpanskykh (2010) present a framework for modelling the performance of enterprises; as examples of performance indicators in this domain, they list the 'average correctness of produced plans' or the 'time to create a new short-term plan after all operational data is received'. Bunse *et al.* (2011) give an overview of various energy efficiency indicators used in the manufacturing sector; these are mostly amounts of energy used per timeframe, per amount of money spent or per unit produced.

There is an emergent field in the literature that discusses the definition and content of performance indicators at a more generic level. Popova and Sharpanskykh (2010) discuss the modelling of organizational performance indicators; they suggest that a proper definition of a performance indicator should take the form of a 'performance indicator expression' that includes the following attributes: *name, definition, type* (continuous or discrete), *timeframe* (evaluation interval), *scale* (measurement), *min/ max value, source, owner, threshold* and *hardness* (soft: not directly measurable; hard: quantifiable). A performance indicator expression may also include a goal definition, which can be defined by a mathematical statement containing the symbols $>$, $\geq$, $=$, $<$ or $\leq$, such as $PI28 \leq 24\,h$ or $PI08 = high$. The inclusion of goals leads to a status of the PI that can be that of achieved (ceased), maintained (avoided) or optimized (maximized, minimized, approximated). This PI status can be evaluated for different points in time (*ibid.*). Pritchard *et al.* (2012: 124) suggest that performance indicators to be used in their ProMES (Productivity Measurement and Enhancement System) should be captured using an Indicator Information Form that captures the *objective* to which the indicator relates, a short *indicator name*, the *rationale for selecting* that indicator, a description of *how the indicator is calculated*, *data sources* for the indicator and *responsibility* for data gathering, analysis and reporting.

Performance indicators come in many different forms. At a high level, key performance indicators are a subset of all the indicators that can be defined; they are selected in such a way as to represent the fundamental performance of an organization at a reasonable cost of measurement and monitoring (Popova and Sharpanskykh, 2010). Key performance indicators are regularly used in construction management; see for instance Skibniewski and Gosh (2009) who discuss how KPIs relate to enterprise resource planning (ERP), putting special emphasis on the relations between these indicators and the dimensions of time and knowledge. Another useful categorization is to distinguish between process indicators and outcome indicators, where the process indicators measure how well a process is executed, whereas outcome indicators measure the impact of that process on the quality of a product. Van der Geer *et al.* (2009) describe how both of these are used in a medical context and relate to task uncertainty; they stress that it is important that actors should be able to have an impact on the indicator (one may be able to follow the correct procedure, but cannot always guarantee treatment results). Performance indicators are often contained implicitly in existing documents such as company mission statements, job descriptions and others. Extracting these indicators from this documentation is another task for domain experts (Popova and Sharpanskykh, 2010).

Alwaer and Clements-Croome (2010) use KPIs in a building design context, in the context of priority setting for building assessment efforts. However, there is no consensus about this approach; for instance Bordass and Leaman (2005) state that there is an issue with the use of KPIs that they see as too generic and not close enough to what clients really want.

Some authors also list a range of criteria for performance indicators. A good example consists of the criteria given by Pritchard *et al.* (2012: 119–122), which requires that indicators must:

- *Cover all objectives, and cover each objective completely*; it is noted that missing an objective or performance aspect may lead to skewed efforts and that this can have significant negative consequences.
- *Be valid*; this means that the indicator needs to capture the actual objective and should not be a proxy that may lead to misinterpretation.
- *Be controllable by the unit/team*; measurement of something that is outside the control of the actors is unlikely to help improve the process or product.

- *Be cost effective to collect*; data collection and analysis requires time and money and hence should be carried out efficiently.
- *Be understandable and meaningful*; complex indicators that are not understood by actors in the process do not motivate or lead to change.

At the same time, it is noted that development of good indicators is a difficult, non-trivial task that requires deep understanding of the process, product and objectives (*ibid.*, 121). While the criteria that performance indicators should be measurable, controllable and valid sound like fair requirements, implementation in a real practice is often complex, with subjective interpretations, uncertainties and sensitivities; an example is the measurement of performance in healthcare (van der Geer *et al.*, 2009).

Anema *et al.* (2013), reporting on the comparison of hospitals' quality of care, stress that performance indicators need to be *reliable instruments*, especially if the results of measurements and comparisons are published in the public domain and interpreted by lay persons. This is a non-trivial issue, especially when the actual data gathering process is 'self-reported' by for instance the different hospitals. Even a formally defined process for calculating a performance indicator may still require some interpretation, such as where data needs to be cleaned to deal with exceptions; for instance an indicator for 'duration of medical care' may be influenced by how one deals with patients that decide they want to postpone an intervention or those who are unfit to undergo treatment, which are factors outside the influence of the hospital and thus not fair to include in measurement of their quality of care. Muchiri *et al.* (2011) present the development of performance indicators for the maintenance process in manufacturing. Their work reflects the fact that development of indicators requires a *deep understanding* of the maintenance function and how to measure the performance of maintenance work in relation to such issues as plant life, cost, health and safety, as well as overall manufacturing strategy. For this specific context they discern *leading indicators* as well as *lagging indicators*, where leading indicators relate to the actual maintenance process, whereas lagging indicators relate to the results of the maintenance process.

While there is huge variation in specific detail, there are a number of common types for performance indicators (Pritchard *et al.*, 2012: 121–123):

- *Output indicators* capture the amount of work done.
- *Efficiency indicators* relate the output to the resources required to generate that output (such as time, money and manpower).
- *Meeting objectives indicators* measure whether a process, unit or system is meeting the goals that have been set.
- *Readiness indicators* capture the demand that has been met within a given time.
- *Prevention indicators* measure the occurrence of some sort of unwanted incident or even the occurrence of issues that may lead to an incident.
- *Resource indicators* capture whether adequate resources are at hand to cover demands.
- *Customer satisfaction indicators* measure what the user/customer think of a service or product, either in positive way (satisfaction) or by measuring the opposite (complaints).

In healthcare it is recognized that performance in chosen areas will ultimately reach a maximum, beyond which no further improvement can be made. There thus comes a point where performance indicators can be retired. Criteria for such a decision ought to review the average rate of achievement, recent trends in achievement rate, extent and trend in achievement rate, average rate and trend in exception reporting and extent and trend in variation of the exception rate (Reeves *et al.*, 2010). This is reflected by the observation of Skibniewski and Gosh (2009) that key performance indicators in construction have their own life cycle.

Single performance indicators are not always independent of other indicators. There may be a positive or negative correlation between indicators, where a change in the one indicator corresponds with a change in another. In other cases one indicator may be an aggregation of others, where both indicators measure something similar but for instance on a different temporal or spatial resolution. Where possible any causal relationships between indicators must be identified (Popova and Sharpanskykh, 2010).

The interaction between individual performance indicators leads to the need to look at performance evaluation systems. Tsai and Cheng (2012), quoting Brown (1996), suggest that, at evaluation system level, the *fewer indicators, the better*; measurements should *link with success factors*; indicators should cover *past, present and future*; design of indicators should *reflect the interest of customers, shareholders and other stakeholders*; evaluation should be *diffused from top to bottom* to ensure consistency; indicators can be combined to contribute to better and more *comprehensive evaluation*; indicators should be *adjusted to changes* in the environment or organization; and indicators *should reflect research and goals of a specific organization*.

Bunse *et al.* (2011) have studied the use of energy performance management in the manufacturing sector, where one would expect to see a decent degree of process control. In spite of the general guidance mentioned earlier, even in this field there are various issues with measuring performance; they conclude that only few key performance indicators are suitable for use at the plant and process level and that there is a lack of standardized indicators. Moreover, they conclude that there is no widely applicable set of benchmarks for plants, processes and equipment. They suggest that machines should communicate their energy use profile, so that better comparisons will be possible, and that there still is a need for better benchmarks. Figure 4 of their paper summarizes this in a strong table and contrasts this with literature. Edwards (2008) compares injury prevention indicators in the health and building sectors and notes that indicators in the building domain typically are oriented towards business, building life cycle and asset management, whereas indicators in the health domain are oriented towards injury prevention and human behaviour. It is noted that while there is literature that spans both the technical and sociological domain, there are not many indicators that combine these fields. An example of such an indicator would be the number of injuries sustained by occupants of buildings with a certain design feature such as stairs without handrails.

However, what information is relevant depends from project to project. As a consequence, it is difficult – if not impossible – to develop a list of performance indicators that covers projects in general. Factors that cause this need for a tailored approach include project size, uniqueness, complexity and stakeholder viewpoints. The use of a set of key performance indicators aims to focus discussion on the core objectives. However, establishing how performance data is aggregated into such key performance indicators, in such a way that one can understand their meaning without knowing all details, remains a challenge (Marques *et al.*, 2010). Muchiri *et al.* (2011) observe that much of the literature proposes lists of key performance indicators but does not provide a methodological approach for selecting or deriving the indicators that can be used in specific contexts. Popova and Sharpanskykh (2010) stress the need to formally capture the relationships between various performance indicators and the need to use a formal modelling method. They also note that the process of selecting or defining performance indicators is not trivial, as it is case specific and requires an understanding of various information sources and deep expertise. For instance a performance indicator for 'customer satisfaction' typically involves subjective choices. One way to deal with these is to define a 'soft' indicator and allow for subjective evaluation; another approach is to try to cover this by other indicators – in the case of customer satisfaction, one could for instance combine the percentage of returning customers, on-time deliveries and number of complaints. In the medical sector, it has been suggested that the aggregation of data into performance indicators is better handled by a central coordinator. This would ensure that the underlying data is really comparable and that all calculations are done the same way. Similarly, it has been suggested that performance indicators should be defined in a way that ensures maximal use of standard information and terminology (Anema *et al.*, 2013).

In ProMES, Pritchard *et al.* (2012: 118–119) emphasize that the selection of performance indicators should not take place directly, but needs to be related to an accurate and complete list of project objectives. Performance indicators then are put in place to measure how well each of these objectives is achieved. They recommend developing performance indicators by way of holding a series of workshops or brainstorm sessions, where iteration takes place not only to prevent the process from getting to a premature stop but also to get consistency among the indicator set. The outcome should be a set of indicators that is complete and accurate and that maps to the objectives. A number of 5–15 indicators is recommended. Popova and Sharpanskykh (2010) hold that the relationships between various performance indicators should be made explicit by modelling, so that a formal framework is available for analysis and verification. The relationships to be captured include causal relations, positive and negative correlation and aggregation.

The term performance indicator (PI) is widely used in the building sector. The PeBBu project state-of-the-art summary (Becker, 2005: 6) mentions that this is a term that has been coined in this field in the late 1990s and then goes on to suggest that performance indicators 'must be quantifiable, well understood, and preferably amenable to computational analysis in order to enable performance prediction during the generation of design solutions. Statistical data is needed on relations between effect of the physical factors and health, comfort, human response, perception of building performance and satisfaction' (*ibid.*, 70). Indeed, performance indicators are promoted in the UK Egan Report, which states that the construction industry should adopt 'quantified targets, milestones and performance indicators' (DTI, 1998: 14) and that the Task Force wants to see the industry start using targets as it has no accepted performance indicators (*ibid.*, 15). Yet not everyone in the industry subscribes to this approach; there are some who suggest that predicting, planning and measuring for the future may be misplaced and prefer to put their trust in the professional judgment of experts. Also, it is noted that some areas of construction have more advanced indicators than others and that in some cases the focus on measurement and indicators carries the risk that teams focus on those areas that have easy-to-quantify indictors while neglecting more difficult issues (Gann and Whyte, 2003). This is also emphasized by Bordass *et al.* (2001c) who warn that there is a risk that performance indicators become an end in themselves, distracting from the needs and requirements that stakeholders have in buildings under various contexts, and that this may lead to things such as high occupant density in spite of dissatisfaction of occupants.

Performance indicators feature clearly in the EU Combine Project, where it was realized that a conceptual integrated data model (IDM) of a building requires the inclusion of a schema to 'provide a way to communicate the expert evaluations among actors in a specific design context, i.e. linking a performance result to a specific request thus giving the essential meaning to the performance indicator' (Augenbroe, 1995: 30). In the later Design Analysis Interface project, the concept is defined in more detail: 'a performance indicator is defined as an objective measure of the performance of a building (sub) system. Its quantification is based on an unambiguously defined experiment from which the output state can be observed by using a real or virtual experiment (usually a virtual experiment conducted through simulation). Each performance indicator comes with "metrics" that exactly define the aggregation steps of the observed states that lead to a objective evaluation of system performance, i.e. a quantified performance indicator' (Augenbroe *et al.*, 2003: 5). Work in the context of facility management by the same team adds the constraint that performance indicators should be based on first-order physical principles in order to prevent an interpretation bias and misinterpretation by various stakeholders. This idea has been used to develop a building performance assessment toolkit that bases itself on normative calculations for energy, lighting, thermal comfort and maintenance, all taking into account the current state of the building, the environmental conditions and the actual usage. Calculations are based on the appropriate norms and standards that are available (Augenbroe and Park, 2005). In further work, a distinction is made between 'hard' and 'soft' performance indicators. The hard indicators come with a well-defined calculation process that is applicable to different types of buildings; the soft indicators are based on user satisfaction and environment-behaviour studies such as Post-Occupancy Evaluation and are comparable for buildings and spaces with similar critical functions (Pati *et al.*, 2009). Ultimately, Augenbroe (2011: 19) positions performance indicators as 'quantifiable indicators that adequately represent a particular performance requirement. A performance indicator, by definition, is an agreed-upon indicator that can be quantified specifying an agreed-upon (and sometimes standardized) measurement method'.

As a full description of all uses of performance indicators in buildings goes beyond the scope of this book, some selected examples are given in three categories: building design, project management and urban studies.

1) For building design, performance indicators are used by many authors to capture the progress of design towards performance goals and targets. Chesné *et al.* (2012) distinguish between performance indicators that describe what already has been achieved and potential indicators to assess remaining prospects. In the United Kingdom, a Design Quality Indicator 'toolkit' has been developed, which consists of a conceptual framework, a data gathering tool and weighting mechanisms, with the intent to support building design. The DQI is supported by the Construction Industry Council. The DQI includes building performance through the concept of build quality but also looks at other aspects such as building function and impact. It is described as a tool for thinking, bringing together various aspects into an aggregate indicator that can be used as a tool for informing design decisions, by a

range of stakeholders (Gann *et al.*, 2003). Chew and De Silva (2004) report that in later work, the DQI sometimes is used in conjunction with two other indicators: a Workmanship Quality Indicator and a Maintenance Indicator. An attempt at a wide framework towards the use of (key) performance indicators in the building domain is the Sustainable Built Environment or SuBETool by Alwaer and Clements-Croome (2010). Their work aims to support the assessment of buildings and building designs by giving stakeholders the option to select performance indicators from pre-existing sets. They make a distinction between mandatory indicators, desired indicators, inspired indicators and inactive indicators and also differentiate between different levels of building resolution by means of a micro, meso and macro scale. Consensus about the importance and ranking of a set of environmental, social-cultural, economic and technological indicators has been based on interviews and the use of the analytical hierarchical process (AHP).[18] In presenting their tool, the authors list a set of criteria from literature that their performance indicators should meet. First of all, 'an indicator system should provide a measure of current performance, a clear statement of what might be achieved in terms of future performance targets and yardstick for measurement along the way'. Furthermore, indicators should be representative and support the making of design decisions, usable by the stakeholders without being overly complex, flexible and applicable to a range of buildings and situations and usable in different phases of the building life cycle, reflecting critical issues, and should not constrain the design process. An indicator system should be easy to use and scientifically valid (*ibid.*).

2) In the context of construction project management, an example of the use of performance indicators is provided by Navon (2007) who uses project performance indicators (PPI) to monitor labour, equipment, materials management and safety. Lee and Rojas (2014) list the following indices as being crucial to project management: Schedule Performance Index (SPI), Cost Performance Index (CPI), Number of Reworks (NOR), Number of Unchecked Items in Checklist (NUI), Number of Incidents or Accidents (NOIA) and Number of Safety Violations (NSV). They present a visual tool, Activity Gazer, which helps to track performance development over time for each of these indicators. Love *et al.* (2015) look at performance from a process management point of view; they use both the term indicator and measure. They provide a long list of key performance indicators for measuring the performance of a project; however their list mixes various concepts such as 'client satisfaction' (performance aspect), 'feasibility study' (project step or activity) and 'legal, commercial and technical and engineering structures' (project views and system levels). Tam *et al.* (2006) develop performance measurement indicators (PMIs) for environmental management in construction organizations using three main categories: regulatory compliance, auditing activities and resource consumption. Of these three, the compliance and auditing link back to the view that lack of issues/defects equals performance. Yeung *et al.* (2009) develop a Partnering Performance Index (PPI) based on seven weighted key performance indicators to measure the performance of partnering projects.

---

18 Note the comments of Hazelrigg (2012: 499–503) on the analytical hierarchical process.

3) At the urban level, ISO/IEC JTC 1 (2014: 60–70) reviews the concept of Smart Cities and in order to operationalize the concept provides an overview of existing indicators for smart and sustainable cities. This overview covers a wide range of aspects, such as education, fire and emergency response, health, recreation, safety, waste, transportation, energy, finance, governance, planning, social equity, civic engagement, biodiversity, ICT access and innovation. The long list of indicators contains for instance student/teacher ratio, percentage of students completing their studies, response time for fire departments, average life expectancy, number of homicides in relation to population size, percentage of population with access to certain services, debt service ratio and voter participation in elections. On the urban level, ISO 37120:2014 is concerned with performance indicators on a city scale. Ugwu and Haupt (2007) use key performance indicators in the context of infrastructure delivery. In their work in the context of urban development of disused areas, Laprise *et al.* (2015) discern between three types of performance indicators: those linked to normative values, those linked to measured values and those linked to reference situations; in other words, they point out that the indicator may be there to measure progress towards a goal, the status of something without a direct value judgment, or to measure the performance in relation to some sort of benchmark.

Attempts are made to integrate various building performance indicators in one overarching framework. An example is the CREDIT Performance Indicator Framework (Bertelsen *et al.*, 2010), which covers different building system levels, the design, operation and use phases and different performance aspects including environmental, social and economic performance; in total this system comprises 187 indicators.

CREDIT Indicators are intended to be used for the specification of requirements in the briefing phase, for guidance and comparison in the design phase and for assessment and measurement after completion. CREDIT Indicators are designed for criteria that include relevance, objectivity, accessibility, readability, measurability and sensibility and have been gathered from 28 case studies in the Nordic and Baltic countries. Many of the indicators listed measure a building in price per some sort of unit/system or classify performance in a Likert scale classification. However, even though the system is over 5 years old, it does not seem to enjoy a wide uptake.

It is interesting to note that the term performance indicator is sometimes wrongly used in construction papers to describe system attributes that, strictly speaking, are properties but not related to performance. Along similar lines, Wong *et al.* (2008) use 'system intelligent indicators' to capture whether building systems have certain features, such as connectivity with a fire detection system, or not.

Some literature also uses the concept of a performance index rather than a performance indictor. For instance, Gursel *et al.* (2009) define such an index in the context of an information exchange framework as an information object type that is typically related to building components. In their view, a Performance Index can be one of five main categories: requirement, specification, measurement, inspection or correction. Of these, they consider requirement and specification to be design indices; inspection, measurement and correction are assessment indices. Chua and Chou (2011) use the term index to represent the heat gains and cooling energy of commercial buildings in the tropics, specifically via the Overall Thermal Transfer Value (OTTV) and Envelope Thermal Transfer Value (ETTV). Ye *et al.* (2014) introduce an energy saving index (ESI) for windows, which is based on calculating an energy saving equivalent and comparing that to the working of an 'ideal' material. Another index is the Zero Energy Performance Index (zEPI). This index equals zero net annual energy use as a value of 0, while a value of 100 is equal to the measured baseline from 2001 in the United States (Frankel *et al.*, 2015: 23). Rodrigues *et al.* (2015) discuss the use of geometry-based building indices, although the examples they list (shape coefficient, south exposure coefficient, relative compactness, window–wall ratio, window–floor ratio and window–surface ratio) are normally seen as design characterizations rather than performance measurements.

4.4.3 Performance Measures

A measure is a standard unit used to express size, amount, degree, extent or quality. The word is also used to indicate a means of achieving a purpose or an instrument marked with standard units and used for measuring such as a container, rod or tape.

As with metrics and indicators, performance measures are used in a range of fields. Stryer and Clancy (2003) discuss the role of performance measures in the quality management and accountability of the health sector. Evans (2004) and Moers (2006) provide two examples of the use of performance measures in the domain of organizational management. Ornelas *et al.* (2012) review the use of performance measures for the ranking and selection of mutual funds in the world of finance. Chen *et al.* (2015) describe the use of performance measures in the comparison of corporate performance in strategic management research. Jang *et al.* (2012) describe the use of performance measures to quantify human system interaction in the critical tasks of nuclear power plant control. On the engineering end of the spectrum, Kang (2011) describes the definition of a performance measure to capture the robustness of residual vibration control in undamped or underdamped systems.

Hansen and Vanegas (2003) use Design Performance Measures (DPM) in the context of construction, especially briefing. They note that cost-based measures are most prevalent in the literature but suggest the use of other measures that cover flexibility, aesthetics, engineering performance, environmental friendliness, accessibility, constructability and maintainability. Dainty *et al.* (2003) have investigated the use of performance measures for the activity of project management itself, listing factors that relate to team building, leadership, mutuality and approachability, honesty and integrity, communication, learning, understanding and application, self-efficacy and external relations; interestingly they name these measures 'indicators' in the main body of their text. Love *et al.* (2015) use 'key measures' in their paper on public–private partnerships in construction, listing cost savings, cost overruns, percentage on time, technical efficiency and on budget as important ones used by industry, the national audit office and researchers. Ozorhon *et al.* (2011) use a multidimensional performance measure to assess international joint ventures in construction, which covers project performance, performance of the management and partner performance. In a study on the impact of earthquakes, Zareian *et al.* (2010) use a ground motion intensity measure (IM) for excitation and an engineering demand parameter (EDP) for an estimation of probability of building collapse.

Neely *et al.* (1997) provide findings of a deep literature on the development of performance measures, with a focus on operations and production management.

They note that performance measures are traditionally seen as 'a means of quantifying the efficiency and effectiveness of action', 'a means of tracking' used in planning and control and part of a 'feedback control loop'. Tsang *et al.* (1999) point out that performance measures can be considered at three levels: as individual indicators, as a set of indicators that forms a performance measurement system (PMS), and at the level where the performance measurement system interacts with its environment. They note a difference between diagnostic measures, which are used to guide operations, and strategic measures, which are dealing with higher-level goals and objectives. Furthermore, they point out that 'what is measured gets done', so that there is an interaction between the process observed and the measurement. One can also distinguish between technical measures, which relate to a product or process, and service measures. The latter typically need to balance effectiveness with efficiency, which often are opposing objectives. Service measures focus on what is important to the customer; this means they often relate to outcomes, are forward-looking and need to be timed well (SEBoK, 2014: 408).

Neely *et al.* (1997) also provide a long list of criteria for the design of performance measures and suggest that performance measures should be derived from strategy, be simple to understand and provide timely and accurate feedback. They should be based on quantities that can be influenced, or controlled, by the user alone or in cooperation with others. Performance measures must reflect the 'business process', that is, both the supplier and customer should be involved in the definition of the measure. They should relate to specific goals or targets, be relevant, be part of a closed management loop, be clearly defined, have visual impact, focus on improvement, be consistent and maintain their significance as time goes by; they should provide fast feedback, have an explicit purpose and be based on an explicitly defined formula and source of data. They must employ ratios rather than absolute numbers and use data that are automatically collected as part of a process whenever possible. Measures should be reported in a simple consistent format, be based on trends rather than snapshots, provide information, be precise and exact about what is being measured and be objective and not based on opinion (*ibid.*).

INCOSE (2015: 130–131) suggest that the best measures are designed to only require a minimal effort to collect and that where possible one should aim for measures that are repeatable, that are straightforward to understand by all involved and that typically can be presented based on an underlying regular pattern such as a daily, weekly or monthly basis. They should support management and decisions, project needs, timeliness, performance requirements, product quality, effective use of resources, cost management and the meeting of relevant standards. The Systems Engineering Body of Knowledge defines performance measures indirectly, by stating that 'technical measurement is the set of measurement activities used to provide information about progress in the definition and development of the technical solution, ongoing assessment of the associated risks and issues, and the likelihood of meeting the critical objectives of the acquirer' (SEBoK, 2014: 407). Where measurement data is aggregated into performance measures, the following approaches can be used (which all should be defined clearly): algebraic relationships; techniques such as trend analysis, pairwise comparisons, objective matrix analysis and operations research analysis; and expert judgment, where needed combined with statistical approaches (INCOSE, 2005: 39).

To define performance measures, Neely *et al.* (1997) recommend the use of a record sheet that requires the following inputs:

- Title, which is self-explanatory and makes clear what the measure does, and its importance.
- Purpose, describing the rationale that underlies the measure.
- Relates to, which links the measure to business objectives.
- Targets, giving an explicit level of performance that is to be achieved.
- Formula, specifying how the performance will be measured.
- Frequency, specifying when the measure should be recorded and reported.
- Responsibility, identifying who will collect and report the measure.
- Source of data, ensuring a consistent data gathering process.
- Follow-up, defining who is to take action on the measure and in what way.

Some domains within the systems engineering realm, such as software and aeronautic engineering, use specific terms to differentiate between different aspects of performance. Specifically, they discern (INCOSE, 2005: 6; SEBoK, 2014: 408):

- Measures of Effectiveness (MOEs) – These capture the performance from the point of view of the client/customer, independent of a particular system or solution. They are closely aligned with the overall objective and mission.
- Key Performance Parameters (KPPs) – These capture the performance of the most critical and essential aspects.
- Measures of Performance (MOPs) – These capture the particular performance of a specific system or solution.
- Technical Performance Measures (TPMs) – These capture the performance of a subsystem or element within the overall system or solution.

These are not independent; typically MOEs evolve into KPPs and MOPs and then into TPMs (SEBoK, 2014: 408). Interestingly, these terms have thus far not been taken up in the building domain.

The use of these different categories of measures helps the dialogue between different parties in the system development and acquisition process. Measures of Effectiveness allow clients to compare alternatives, test whether solutions are sensitive towards assumptions, define and evaluate key performance and can be used to define which solutions are acceptable. MOEs are solution independent but do represent the intended operational environment, so that the client/customer can evaluate how well the solution will work under these conditions. MOEs help to operationalize the success or achievement of a mission or objective and capture how well any potential solution achieves these. Key Performance Parameters capture and quantify those performance issues that are critical to the client. KPPs are a subset of the MOEs that are critical to success, and where failure to satisfy the requirements may lead to rejection of a solution, or to re-evaluation or termination of a project. Together, MOEs and KPPs help to detail the demand side of system performance (INCOSE, 2005: 10–12; INCOSE, 2015: 133–135).

Measures of Performance are attributes of specific systems; they can be compared with the Measures of Effectiveness to establish whether a particular solution meets the client's requirements. MOPs capture the physical and functional behaviour of a specific system. MOPs, like MOEs, factor in the intended operational environment, so they pertain to specified testing and/or operational conditions. Technical Performance Measures specify further detail of the performance of a given system element or subsystem. TPMs are used to assess design development, monitor progress towards objectives and assess compliance to performance requirements or technical risks. Moreover, they provide visibility into the status of important project technical parameters, which in turn enables effective management, thus enhancing the likelihood of achieving the technical objectives of the project. TPMs are used during system development with an aim to capture design progress, status in terms of meeting requirements and risk. Together, MOPs and TPMs detail the supply side of system performance (INCOSE, 2005: 10–12; INCOSE, 2015: 133–135).

Working with MOEs, KPPs, MOPs and TPMs requires some care to keep things manageable. As MOEs are the top-level measures that drive the other categories, it is recommended that they are kept independent; a typical number used in industry seems to be about 2–12 MOEs. Each MOE might use anything from 1 to 10 MOPs to provide information on the meeting of core requirements by a given system, detailing things like system affordability, technical performance, suitability and risk. Typically there are again more TPMs, often 2–400 per project (INCOSE, 2005: 30–31). Some concepts that are captured by TPMs may relate to product stability, product quality (be measured in terms of functional correctness, supportability and maintainability, efficiency, portability and usability and dependability and reliability) and technology effectiveness. Typical candidate TPMs that would be applicable in these areas would be useful life, structural load bearing capacity, coverage area, accuracy, power performance, emissions, noise transmission and others; some of these have a broad applicability across different domains, and others are specific to certain disciplines (INCOSE, 2005: 50–53). To support the working with these concepts, INCOSE (2005: 48–49) provides a checklist that is intended to help a project team to keep track of all relevant considerations such as stakeholder involvement, traceability, risk management and scheduling.

4.5 Handling and Combining Building Performance

The previous paragraph covered performance metrics, indicators and measures as terms that capture and express building performance. However, they pertain to individual performance aspects and leave open how these aspects are handled as a set. There often is a desire to combine the measurement of performance aspects to a higher level through a process of aggregation; this is similar to moving towards the original 'stem' in a functional decomposition tree. However, at the same time, it is recognized that it is important to maintain the relevance and meaning that is available at the aspect level. It is important to remember that measurement and quantification are a way to get objective statements about what a building does in a specific situation. However, the judgment of exactly how a stakeholder perceives those numbers remains by definition subjective. As stated by Gilb (2005: 382) for systems engineering in general, 'System performance is an attribute set that describes measurably "how good" the system is at delivering effectiveness to its stakeholders. Each individual performance attribute level has a different "effectiveness" for different stakeholders and, consequently, different value or utility'. An alternative is to stay with different performance measures, but to bundle these in a performance measurement system (PMS). Lohman *et al.* (2004) describe such a system in an actual case in the process industry; they suggest that the development of such systems often is more of a matter of coordinating existing efforts than designing a new system from the start. Lam *et al.* (2010) describe a similar project in the context of building maintenance and facility management. Wong *et al.* (2008) aggregate a range of intelligence indicators into a system intelligence score (SIS) for buildings. Ren *et al.* (2013) follow a similar approach in their development of collaborative design performance measurement (CDPM).

On the subject of aggregation, CIB Report 64 (1982: 20) says that 'whenever several attributes are considered, the comparison of products or solutions against a performance specification requires a measure of judgment. It can, however, be useful to use numerical methods to combine the separate performances into a single index of overall worth or quality'. The typical process of combining performances is to first of all normalize the measurements, so that they are all expressed on a similar scale; subsequently they can be combined into one overall measure, where needed using weighting factors that express the relative importance of each aspect. Aggregation needs to be handled with some care, especially where the (subjective) weighting factors are designed to represent group preferences.[19] In management science, the balanced scorecard method is one approach that aims to ensure a full coverage of all performance aspects from a top-down perspective (Kaplan and Norton, 1992; Neely *et al.*, 1997; Pritchard *et al.*, 2012: 208–209). Various processes for setting weights, or 'ranking', are available for use in a building decision context; see for instance Hopfe (2009: 67–71) or Iwaro *et al.* (2014). Weighting is also key to attain an overall index in for instance a rating scheme like BREEAM, which includes an explicit weighting where 'credits' obtained are weighted as follows: management 12%, health and well-being 15%, energy 19%, transport 8%, water 6%, materials 12.5%, waste 7.5%, land use and ecology 10% and pollution 10% (BRE, 2011: 23). These weightings have been selected consciously and may include a deliberate imbalance in order to provide an incentive to pay more attentions to specific factors (Lee *et al.*, 2002b). One question that must be asked of performance indicators in frameworks and aggregation is whether or not they are independent variables, which can be modelled as a single concept. In road safety statistical analysis is used to establish such independence (Tingvall *et al.*, 2010). It is noted that in various disciplines there are efforts attempting to develop a measurement set that covers different levels of detail, catering for different actors and their needs; however it is becoming more generally accepted that it is not possible to have one universal list of criteria to ensure the performance across projects, as most projects are unique to some extent (Marques *et al.*, 2010). In Practical Software and Systems Measurement (PSM), a distinction is made between the planning, performing, and evaluation of measurements. This approach links information categories, measurable concepts and a set of candidate measures to support project teams in identifying their information needs and priorities. Also in the software domain, the Goal, Question, Metric (GQM) approach encourages measurement at the conceptual level (Goal), operational level (Question) and qualitative level (Metric). In this approach, each goal is decomposed into several questions that are then linked to metrics that address the question at stake and provide insight into the goal achievement (SEBoK, 2014: 404–405).

Many authors suggest depicting the performance for different indicators or measures in one graphical format such as a 'performance profile' (CIB Report 64, 1982: 20) or 'radar chart' (Augenbroe, 2011: 27–31). In ICT systems, it is possible to keep track of a

---

19 See Hazelrigg (2012: 227–243), and his discussion of Arrow's Impossibility Theorem.

large set of numbers; this may circumvent the need to use weightings such as in the Performance Framework Tool (PFT) for the continuous measurement and monitoring of building (Corry *et al.*, 2014).

A prime example of the fundamental questions related to performance aggregation is the use of costs as an indicator. In one approach, all preferences represent some sort of value judgment, and hence any ranking can ultimately be monetized. But an alternative approach is to present the decision maker with technical performance data and the cost and allow that decision maker to make their own trade-off between cost and performance.

In themselves, quantification and measurement are technical processes. However, they are embedded in a much more complex sociotechnical context that imposes external constraints. As listed by INCOSE (2005: 22),

1) Measurement and quantification must reflect the information needs of the project and stakeholders; these are seldom static but rather tend to change over the duration of a project.
2) Measurement and quantification methods must be understandable to the decision makers, so that they appreciate what the measurements mean and can relate this to the options they have.
3) Measurement and quantification must be used in an effective manner, playing an actual role in decisions and choices.

Yet handling performance is not always straightforward. For instance, there is a risk that project managers start to focus on project performance only and thus prioritize project cost and scheduling status over product performance (INCOSE, 2015: 135). It is thus important that the right balance is achieved. As recommended by SEBoK (2014: 392), 'do not neglect harder things to measure like quality of the system. Avoid a "something in time" culture where meeting the schedule takes priority over everything else, but what is delivered is not fit for purpose, which results in the need to rework the project'.

Given the importance of the performance of buildings and building products, many clients in the industry require that products and buildings are certified or tested by an independent third party or at the very least that a third party guarantees that testing and verification methods are valid (Foliente, 2000). Examples of test providers are independent laboratories such as the Building Research Establishment (United Kingdom), Fraunhofer Institute (Germany), TNO Building Research (Netherlands), the Lawrence Berkeley National Laboratory, National Renewable Energy Laboratory or Pacific Northwest National Laboratory (United States), CSRIO (Australia) and BRANZ (New Zealand). In cases where product testing in laboratory facilities is not feasible, the use of international codes and standards can be mandated, such as ISO standards.

4.6   Signs of Performance Issues

The main discussion in this chapter is about performance and how to capture and express this. However, it is also worthwhile to look at the opposites of performance: failure and non-performance. Defects can sometimes be observed directly, as in a crack in a wall or a gap under a window frame; in other cases there may be signs of non-performance such as mould growth, which is an indication of a ventilation and moisture excess problem. However, absence of defects is only a precondition for performance. Absence of defect does not guarantee anything beyond that the product is 'as specified'. If this specification is not right, there may still be insufficient performance, for instance because of the use in environmental conditions that are not anticipated or operation of a 'perfect' product with inappropriate control settings.

The detection of defects is mostly a visual task. However, there are some techniques that can be used to help the process. For instance, there are various structural health assessment techniques such as the use of acoustic emissions (Behnia *et al.*, 2014), the impact-echo method (Lee *et al.*, 2014), ultrasonics and electric resistivity (Shah and Ribakov, 2008). One of the ways to identify thermal defects is the use of infrared cameras. These help the user by providing an image that represents the infrared radiation emitted by buildings, which, when interpreted correctly, allows to spot areas with excessive heat loss, moisture content or ventilation losses (Taylor *et al.*, 2014; Fox, 2015; Fox *et al.*, 2015).

Georgiou (2010) presents a defect classification system for houses in Australia, where a building defect is defined as where 'a component has a shortcoming and no longer fulfils its intended function'; the main categories proposed are cracking, damp, drainage, external leaks, incompleteness, internal leaks, regulation conflict, structural adequacy, window sill gaps and overall workmanship. Ahzahar *et al.* (2011) use a wider definition, where a defect is linked to 'non-conformity with a standard of specification', listing examples such as structural defects, faulty wiring, inadequate ventilation, issues with soundproofing or inadequate fire protection systems; as causes for such defects, they suggest design errors, manufacturing flaws, defective or improperly used materials and lack of adherence to the design by the contractor, as well as external factors such as ageing, land movement, vermin and even corruption in the industry. Forcada *et al.* (2014) note that in construction terms like error, fault, failure, quality deviation, snag and non-conformance are used as synonyms to defect; they go on to define a defect as a term that covers any shortcoming in function, performance, statutory or user requirement in any part of the building. They then order defects according to the following categories: impact on functionality, detachment, flatness and levelness, incorrect installation, misalignment, missing part(s), soiled, stability/movement, surface appearance, tolerance error, water problems and a mixed bag of 'others'. Based on analysis of over 3000 defects, they conclude that poor workmanship is the main cause of defects (*ibid.*). Josephson and Hammarlund (1999) take a wider view and identify causes for defects as stemming from the client, design, production, material and maintenance; they also note that most defects involve a chain of events that link defect cause, defect occurrence and defect manifestation but also may involve corrective action. Park *et al.* (2013) define a defect as some part of the construction process that needs redoing due to being incorrect; in their view the main impact relates to time and cost of rework rather than loss of performance. Other work focusses on the defects in one specific type of building system; for instance Chew and De Silva (2004) explore the typical defects that occur in traditional facade systems, such as cracks, corrosion, dampness, staining, biological growth, sealant and joint failures, efflorescence, blistering, peeling and flaking, discoloration and delamination. Georgiou (2010) points out that previous studies indicate an average of over 2 defects per house in the 198–1996 timeframe, which has increased to over 5 defects in a sample collected for 1006–2010. Josephson and

Hammarlund (1999) report high numbers varying between 280 and 480 defects for large projects such as a museum, school, fire station and shopping centre.

Preventing defects is a hard task; most efforts in industry so far are remedial rather than preventive. Research in the field looks at identifying the number and types of defects, the reasons for occurrence of these defects, management of defect information and knowledge, and the development of systems that prevent defects during construction (Park *et al.*, 2013). Clearly, maintenance and inspection play a role in prevention; for instance Wang and Su (2014) describe the inspection of HVAC ducts in order to identify gaps and cracks, settled dust and contamination, rust and other debris. Here rust and cracks are typically defects that may develop over time, whereas dust and debris relate to cleanliness of the system.

Health issues of occupants, such as asthma, Legionnaires' disease and the more generic sick-building syndrome (Redlich *et al.*, 1997; Burge, 2004), are consequences of building malfunctioning rather than causes. Hartkopf *et al.* (1986a: 274–307) note the following further signs of non-performance or 'stress' on offices: weak definition of workspaces, poor accessibility, lack of privacy and/or identity, missing services, missing amenities, mould, dust build-up, cracking, lack of acoustical ceilings, brought-in lamps/heaters/coolers, covered windows, fading, yellowing, sloping floors, worn carpets and dented columns.

4.7 Building Performance in Current Practice

The Akershus University Hospital in Nordbyhagen, Norway, is a state-of-the-art facility that shows the current practice of handling performance in the development of a new project. This hospital is located in the municipality of Lørenskog near Oslo. The building discussed here was opened in 2008 and replaced an earlier facility. The building is designed by C.F. Møller Architects, in collaboration with Multiconsult, SWECO, Hjellnes COWI, Interconsult and Ingemansson Technology, and was constructed by NCC Norge. The hospital itself aims to be the most patient focussed and friendly hospital in Norway; the building is seen as one of the most modern hospitals in Europe.

Being a university hospital, Akershus offers medical care to patients while also training medical staff; it is affiliated with the University of Oslo and falls under the Southern and Eastern Norway Regional Health Authority. A Research Centre was established at the hospital in 2000 and publishes around 250 scientific papers per year. The hospital provides about 50 000 day treatments, as well as 60 000 overnight stays per year; it serves around 200 000 outpatients with 6000 members of staff, 650 of which are medial doctors. The hospital has 22 operation theatres that allow it to carry out 21 000 operations per annum. Medical specialties include, among others, neurology, paediatrics, children habilitation, physiotherapy, surgery, orthopaedics, anaesthesia, obstetrics, gynaecology, clinical chemistry, immunology and radiology.

The core function of a hospital is to provide medical and surgical treatment as well as nursing for sick or injured people. To support this, hospital buildings provide facilities to medical personnel, patients, visitors and a range of other stakeholders. Well-designed and constructed hospitals contribute to patient well-being and may help to shorten patient recovery time in comparison with less welcoming hospitals, many of which are legacy buildings that have emerged from organic growth. This means that there is an interaction between building performance and organizational performance; there is an interesting crossover where an attempt is made to apply the health concept of evidence-based medicine to hospital buildings under the banner of evidence-based design.

To live up to its ambition of being a modern and patient-friendly facility, Akershus combines modern architecture and ICT with state-of-the-art medical approaches. On the architectural side, it fits in with the Scandinavian approach (Burpee and McDade, 2013) of a horizontal and perforated building with good access to the outdoor environment. On the technological side, it employs Smart Cards that control parking, work clothing, access to buildings and rooms, priority in elevators, network access and printing, and home office solutions. Akershus Hospital has IP phones, over 400 laptops and 3800 stationary terminals, an automated drug management system with robot 'pill pickers' and delivery of drugs by a combination of automated guided vehicles and tube transport. Heating and cooling is partly provided by a ground-heat exchange system with 350 energy wells of 200 m depth each. Waste disposal is supported by a stationary vacuum system. Medical state of the art is supported by the embedded research centre and university links.

Obviously a building like Akershus Hospital fulfils many functions, so a full functional decomposition is beyond the scope of this example. At the top level, the building supports medical care and academic research; this can be decomposed into functions for medical staff, patients, visitors, researchers and other stakeholders. Support for each of these groups can be decomposed in further detail; for instance researchers will need to be provided with library support; doctors will need access to patient records, specialist medical equipment such as MRI scanners and provisions for surgery instruments; patients require a place to wait for care and overnight facilities if they are inpatients. The building combines a wide range of systems that provide these functions, for instance by offering rooms, ICT systems and specialized facilities such as HVAC systems that ensure a sterile environment in operating theatres. As noted, there is no one-to-one match between the functional decomposition and the technical systems. A good example is a key feature of the hospital: the glazed indoor street that connects the various departments of Akershus. To begin with, this glazed indoor street is used by all occupants of the hospital and thus meets transport and orientation functions across all different user groups. In terms of patients and visitors, this is a central axis that facilitates navigation. The use of glass allows excellent daylighting and outside views, which is known to support patient well-being and recovery; at the same time it allows for minimal use of artificial lighting that reduces electricity use as well as the cooling load. Solar control is inherent in such a system to prevent overheating and is provided by interstitial blinds in the triple glazing units. The glazed street provides further functions by hosting receptions, polyclinic wards and information boards; it also includes a pharmacy, florist, café and church, catering for further needs of the hospital users. See Figure 4.6.

The performance of a hospital building is closely related to that of the performance of the organization that uses the building to provide medical care. This is related to a process and is commonly measured in terms of medical KPIs such as admission rate, patient waiting time, patient readmission rate, referral rate, inpatient mortality rate, harm events occurrence, percentage of medication errors, time between symptom onset and hospitalization, number of preterm births, bed and room turnover rate, average number of patient rooms in use at one time, patient satisfaction scores, patient–staff ratio, media presence (split in positive and negative mention), average insurance claim processing time and cost, claim denial rate and average treatment charge. University research is also increasingly measured in terms of KPIs; here key indicators are for instance postgraduate student completion rates, number of scientific publications coupled with journal impact factor, volume of research funding obtained, impact in terms of commercial spin-off companies and number of patents. Some of the medical KPIs, such as patient mortality rate, are highly sensitive, whereas there is significant discussion about the appropriateness of trying to measure scholarly activity.

Akershus Hospital can also be measured in terms of conventional building performance measures. For instance, the Energy Use Intensity of the buildings on the site on average is 117 kBtu/f^2 or 369 kWh/m^2 per year; taking into account the efficiency of generation and transport as well as the effect of the ground-heat exchange, the source Energy Use Intensity can be calculated at 273 kBtu/f^2 or 861 kWh/m^2 per year. Overall geothermal energy covers 85% of the heating demand, which equals 40% of overall energy consumption (Burpee and McDade, 2013).

The design of Akershus Hospital was won several prizes: Best International Design in the Building Better Healthcare Awards (2009), Design & Health International Academy Award in the category of Best Large International Health Project (2015) and Highly Recommended in the European Healthcare Awards (2016).

Figure 4.6 Akershus Hospital main thoroughfare.
Image © Jorgen True, provided by courtesy of C.F. Møller.

4.8 Reflections on Working with Building Performance

The essence of building performance lies in the comparison between demand (what users need and want) and supply (what a building provides). However, beyond this basic notion, things quickly become more complicated. As established in the previous chapter, performance can take a number of forms and relate to quality (how good), workload (how much), use of resources (how much is needed/can be saved), timeliness and readiness. The quantification of performance is not necessarily absolute; good performance can also be established in comparison with other options that are deemed to have average or bad performance. The assessment of how good the building meets these demands also depends to loading or excitation of the building, behaviour of building and building systems and the way observation takes place. Models like those by Vanier *et al.* (1996), Foliente *et al.* (1998: 30), Gero (1990) and Augenbroe (2011: 21) represent this same message in different ways while also using a slightly different terminology.

Building performance is not a static attribute. As discussed before it changes over time; it is highly dependent on the context, loads that work on the building, control settings, occupant behaviour, system ageing and degradation, maintenance and refurbishment. Because of this, performance assessment efforts need to carefully define the experimental conditions under which performance is being observed. Unfortunately, it seems the building sector is not doing a very good job at capturing these conditions; this may be due in part to the complexity that is involved. But as a consequence, only few performance assessments as reported in the literature can be properly replicated, whether that is in the domain of monitoring, simulation, expert assessment or stakeholder evaluation.[20] This lack of replicability is a serious impediment for progress in building performance analysis as a domain of science, as well as for industrial development. Moreover, building performance analysis often seems to focus rather quickly on the quantification of some aspect of performance, without proper definition of a hypothesis (H_1/H_A) and null hypothesis (H_0). The uptake of formal design of experiments (DoE) in line with statistical studies, randomization of input factors, efforts at replication and efforts at blocking factors is also rather limited.

Beyond the proper definition of analysis conditions, the drivers of the analysis may also have impact on how performance is viewed. Yet why a specific analysis is done is often left implicit. Key reasons for undertaking building performance analysis are to provide a proof of concept and to demonstrate that a building meets requirements and goals or targets, as part of an acceptance test or for operational verification. These drivers also need to be captured properly; again this is often overlooked and taken for granted.

Current practice in specifying building performance is to use rather shorthand statements, for instance by reference to a quantity like Energy Use Intensity (EUI), without specifying how this is measured or calculated, or under which conditions; yet it is known that an EUI for heating differs with climate conditions and occupancy length. The state of the art for the capture and specification of experimental details is

---

20 Challenge to readers who disagree: select a journal article on building performance of your choice, and try to replicate the work. It will soon become obvious that there are many modelling assumptions and measurement details that are not mentioned.

represented by the work on PTMs, PAMs and AFs. However, these are to be considered starting points and need further work if the field is to become more mature. Work in this area also needs to include more fundamental studies of how one may excite buildings and observe changes in state and input/output behaviour and how one deals with the reality that some factors cannot be controlled – leaving only semi-controlled experiments as an option to move forwards.

The case study of Akershus Hospital demonstrates the intricate interaction between stakeholder needs, building functions and performance requirements on the one side and design solutions and actual performance on the other. Full traceability of how these interact is far beyond current capabilities of the building design and engineering industry to capture design decisions and information upon which these decisions are based.

As should be clear from the discussion, performance is a building attribute that depends on a range of factors that relate to building behaviour. Yet it is easy to confuse other building attributes, such as dimensions or material properties, with performance attributes. While some building properties can be highly complex, such as the average U-value of a composite wall, a performance attribute typically requires some form of excitation that entices behaviour, observation and ultimately quantification of this behaviour.

Finally, the terminology that is in use to talk about building performance is relatively immature and can be a source of confusion: metrics, indicators and measures are all mixed. Systems engineering has more detailed terms that may be helpful but also may be overly complex. Some researchers in the building domain use systems engineering performance terms without fully appreciating their meaning. For instance, Geyer and Buchholz (2012) use measures of effectiveness (MOEs) where they talk about the performance of an actual design solution, so they should be using measures of performance (MOPs) if they wanted to stay with systems engineering terminology. It is suggested that the way forward is to reserve Key Performance Indicators for business and process management. Building engineering could then focus on Performance Measures, and develop Performance Measure Expressions (PMEs) to specify exactly what is meant by each individual PM. In time, this would allow an easy transition to the wider view of measures of effectiveness, key performance parameters, measures of performance and technical performance measures as used in systems engineering.

4.9   Summary

This chapter addresses the core concept of building performance, how performance is quantified through experiments, and the terminology used to capture the concept; it also looks at signs of performance issues. Building performance links the goals and needs of stakeholders, functional requirements and performance requirements with actual systems and sub-systems (technical solutions) that meet these requirements and needs. How well these systems do this is captured in by a metric, measure or indicator. This is quantified through a testing method for a given load/stress and use. Measurement is crucial for the operationalization of building performance, providing objective data about the quality of products, services and processes.

Building performance expresses how well a building performs a task or function. This concept thus lives on the interface of performance requirements that define the tasks and functions that are needed and technical solutions that meet these requirements. Different models have been made of this interface, including the GARM or 'hamburger' model (Gielingh, 1988), the Function–Behaviour–Structure model (Gero, 1990) and the ASTM-WBFS mapping of functionality and demand. More complex multidimensional mappings that link design, function, resource and performance are described by Gilb (2005: 105) and Augenbroe (2011: 21).

Quantification of building performance requires measurement, which typically needs domain expertise. This expertise drives what behavioural or state attributes are being observed and compared with a measurement scale. Factors to take into account are the conditions under which the quantification takes place and the process that is used – where possible based on a norm or standard. In the building science literature, terms like performance test method (PTM), performance assessment method (PAM) and Analysis Function (AF) are used to describe measurement processes in various contexts. For any quantification effort, verification and validation are also of importance; verification ensures that the experiment and measurements are carried out correctly, whereas validation is concerned with whether or not the experiment and measurements fully relate to the issue (stakeholder requirement) at stake.

Experiments are tests where changes are made to inputs and variables of a process or system, followed by the observation of the responses to these changes. Proper observation of input, output, initial states, changed changes and controllable as well as uncontrollable factors is the key to good experiments. This leads to knowledge about what stimuli make a system respond, pairs of input and output variables that can be studied, and ultimately how inputs affect state transitions and output of a system. Experiments can take place in the real world, in a virtual environment and in hybrid set-ups. Experiments help to demonstrate the feasibility of a concept and to meet requirements and can establish that a system is fit for use in an intended operational environment. In some cases customers require an acceptance test before a system or product changes ownership.

Measurement should be based on systematic observation, according to a well-defined plan that defines what instrument are to be used, as well as what parameters will be observed, what the measurement frequency will be and how data will be stored, analysed and reported. Good measurement takes into account uncertainties and quantifies the potential experimental errors.

There is a range of terms to capture the outcomes of the measurement and quantification of performance: performance metrics, performance indicators and performance measures. At a more detailed level, one finds key performance indicators (KPIs), measures of effectiveness (MOEs), key performance parameters (KPPs), measures of performance (MOPs) and technical performance measures (TPMs). In many of these there is a relation to engineering concepts and units. For performance indicators, it has been suggested to formally define these using a 'performance indicator expression' (Popova and Sharpanskykh, 2010); similar work has been suggested for measures (Neely *et al.*, 1997).

The quantification of building performance for individual aspects can sometimes be aggregated into higher-level concepts. However, when undertaking such aggregation, one introduces subjective weighting factors that should be handled with care. The use of performance profiles or radar charts circumvents this issue by visualizing all information in a single normalized figure. Even so, in the end it is important to ensure that all measurements relate to the actual needs of the project, are understandable by the stakeholders and/or decision makers and relate to actual decisions and choices.

While the focus of this work is on building performance, it is worth to realize that there also is a realm of failure and non-performance. Defects, where a building, system or component does not meet its specifications, may seriously compromise performance. However, the absence of defects is only one precondition for performance; other factors like improper use or control can also result in unsatisfactory results. At the other end of the spectrum, the impact of lack of building performance can sometimes be observed through occupant complaints, health and visual clues in the building.

Recommended Further Reading:

- *Competitive Engineering* (Gilb, 2005) for a deep discussion of performance, scales of measure and performance quantification.
- Publications on the Function–Behaviour–Structure model by Gero and his colleagues (Gero, 1990; Gero and Kannengiesser, 2003).
- *Evidence-based Productivity Improvement* (Pritchard *et al.*, 2012) and *Modeling organisational performance indicators* (Popova and Sharpanskykh, 2010) for a good overview of the state of the art in performance measurement in organizations and manufacturing.
- *Technical Measurement* (INCOSE, 2005) for the leading publication on Measures of Effectiveness, Key Performance Parameters, Measures of Performance and Technical Performance Measures.
- *Design and Analysis of Experiments* (Montgomery, 2013) and *Experiments* (Wu and Hamada, 2002) for theory on the set-up of experiments while taking into account subsequent statistical analysis.

Activities:

1 Select any paper from a building related academic journal that carries the word 'performance' in the title. Critically appraise this paper using the material discussed in this paper. Does the paper define building performance? Does it use metrics, indicators or measures? Does the paper describe what, if any, test was used to quantify performance?

2 Study the functions of an 'outhouse', making a functional decomposition of the requirements these buildings have to meet. Then define a range of performance measures that allow quantification of how well a specific outhouse meets its functions. When in need of inspiration, refer to *Toilets: a Spotter's Guide* (Pickard, 2016).

3 Set yourself a specific situation (dwelling, occupant) where a stakeholder needs a doorbell (chime). Then carefully define an experiment, short of actual installation, that allows this stakeholder to test some available systems and select the one best suiting his or her requirements.

4   Select an advanced and complex building system such as a lift/elevator, revolving door or air source heat pump. Collect product information from competing manufacturers of this system. Then analyse that information – what subset pertains to plain system properties, and what subset is concerned with system performance? Do the manufacturers provide any information on the experimental conditions that were applied during the testing process?

5   Define a set of performance metrics to measure how well a 'Measure of Performance' (MOP) captures the performance of a specific system.

6   Search for a journalistic article that discusses complaints about an office or school building. Then analyse the complaints listed, and identify which ones relate to building performance and which ones are actually not about the building. Critically appraise how clearly one may distinguish these root causes.

4.10   Key References

Alwaer, H. and D. Clements-Croome, 2010. Key performance indicators (KPIs) and priority setting in using the multi-attribute approach for assessing sustainable intelligent buildings. *Building and Environment*, 45 (4), 799–807.

Augenbroe, G., 2011. The role of simulation in performance based building. In: Hensen, J. and R. Lamberts, eds., *Building performance simulation for design and operation*. Abingdon: Spon Press.

Augenbroe, G., P. de Wilde, H.J. Moon, A. Malkawi, R. Choudhary, A. Mahdavi and R. Brame, 2003. *Design Analysis Interface (DAI) – Final Report*. Atlanta, GA: Georgia Institute of Technology.

Augenbroe, G., P. de Wilde, H.J. Moon, A. Malkawi, 2004. An interoperability workbench for design analysis integration. *Energy and Buildings*, 36 (8), 737–748.

Blanchard, B.S. and W.J. Fabrycky, 2011. *Systems engineering and analysis*. Upper Saddle River, NJ: Prentice Hall, 5th edition.

California Commissioning Collaborative, 2011. *The Building Performance Tracking Handbook*. California Energy Commission

CIB Working Commission W60, 1982. *Working with the Performance Approach to Building – CIB report 64*. Rotterdam: CIB.

Corry, E., P. Pauwels, S. Hu, M. Keane and J. O'Donnell, 2015. A performance assessment ontology for the environmental and energy management of buildings. *Automation in Construction*, 57, 249–259.

Deru, M. and P. Torcellini, 2005. *Performance metrics research project – final report*. Golden, CO: National Renewable Energy Laboratory.

Erden, M., H. Komoto, T. van Beek, V. D'Amelio, E. Echavarria and T. Tomiyama, 2008. A review of function modeling: approaches and applications. *Artificial Intelligence for Engineering Design, Analysis and Manufacturing*, 22 (2), 147–169.

Foliente, G., R. Leicester and L. Pham, 1998. *Development of the CIB proactive program on Performance Based Building codes and standards*. Highett: CSIRO.

Gero, J., 1990. Design prototypes: a knowledge representation schema for design. *AI Magazine*, 11 (4), 26–36.

Gero, J. and U. Kannengiesser, 2003. The situated function-behaviour-structure framework. *Design Studies*, 25 (4), 373–391.

Gielingh, W., 2005. Improving the Performance of Construction by the Acquisition, Organization and Use of Knowledge. Ph.D. Thesis, Delft University of Technology.

Gilb, T., 2005. *Competitive engineering: a handbook for systems engineering, requirements engineering, and software engineering using planguage*. Oxford: Butterworth-Heinemann.

Gursel Dino, I. and R. Stouffs, 2014. Evaluation of reference modeling for building performance assessment. *Automation in Construction*, 40, 44–59.

Hitchcock, R., 2003. *Standardized Buildings Performance Metrics – Final Report*. Berkeley, CA: Lawrence Berkeley National Laboratory.

IEA, 1994. *Annex 21: Calculation of energy and environmental performance of buildings. Subtask B: appropriate use of programs. Volume 1: executive summary*. Watford: Building Research Establishment.

INCOSE, 2003. *Systems Engineering Handbook: a guide for system life cycle processes and activities*. San Diego, CA: International Council on Systems Engineering.

INCOSE, 2005. *Technical Measurement*. San Diego, CA: Practical Software & Systems Measurement (PSM) and International Council on Systems Engineering, Report INCOSE-TP-2003-020-01.

INCOSE, 2015. *Systems Engineering Handbook: a guide for system life cycle processes and activities*. Hoboken, NJ: Wiley.

Kalay, Y., 1999. Performance-based design. *Automation in Construction*, 8 (4), 395–409.

Markus, T., P. Whyman, J. Morgan, D. Whitton, T. Maver, D. Canter and J. Fleming, 1972. *Building Performance*. London: Applied Science Publishers.

Marques, G., D. Gourc and M. Lauras, 2010. Multi-criteria performance analysis for decision making in project management. *International Journal of Project Management*, 29 (8), 1057–1069.

Montgomery, D.C., 2013. *Design and Analysis of Experiments*. Hoboken, NJ: Wiley, 8th edition.

Muchiri, P., L. Pintelon, L. Gelders and H. Martin, 2011. Development of maintenance function performance framework and indicators. *International Journal of Production Economics*, 131 (1), 295–302.

Neely, A., 2005. The evolution of performance measurement research. *International Journal of Operations & Production Management*, 25 (12), 1264–1277.

Neely, A., H. Richards, J. Mills, K. Platts and M. Bourne, 1997. Designing performance measures: a structured approach. *International Journal of Operations & Production Management*, 17 (11), 1131–1152.

Oberkampf, W. and C. Roy, 2010. *Verification and Validation in Scientific Computing*. Cambridge: Cambridge University Press.

van Ostaeyen, J., A. van Horenbeek, L. Pintelon and J. Duflou, 2013. A refined typology of product-service systems based on functional hierarchy modeling. *Journal of Cleaner Production*, 51, 261–276.

Pati, D. and G. Augenbroe, 2007. Integrating formalized user experience within building design models. *Computer-Aided Civil and Infrastructure Engineering*, 22 (2), 117–132.

Popova, V. and A. Sharpanskykh, 2010. Modeling organisational performance indicators. *Information Systems*, 35 (4), 505–257.

Pritchard, R., S. Weaver and E. Ashwood, 2012. *Evidence-based productivity improvement: a practical guide to the productivity measurement and enhancement system (ProMES).* New York: Routledge.

SEBoK, 2014. *The Guide to the Systems Engineering Body of Knowledge, v.1.3.* Hoboken, NJ: Body of Knowledge and Curriculum to Advance Systems Engineering/Trustees of the Stevens Institute of Technology.

Simon, H.A., 1996. *The sciences of the artificial.* Cambridge: MIT Press, 3rd edition.

Wu, C. and M. Hamada, 2002. *Experiments – Planning, Analysis, and Parameter Design Optimization.* Chichester: Wiley, 2nd edition.

Zeigler, B., H. Praehofer and T. Kim, 2000. *Theory of modeling and simulation – integrating discrete event and continuous complex dynamic systems.* London/San Diego, CA: Academic Press, 2nd edition.

5

Performance Criteria

The previous chapter developed building performance in terms of a measurable concept on the interface between performance requirements and solutions that meet these requirements. Theory of experimentation was discussed in some detail, as were different terminology and indicators of non-performance. However, one key element for defining building performance is still missing: that of criteria to judge the assessment, which is the subject of this chapter. As phrased by Blyth and Worthington (2010: 86), 'whether or not a building is successful will depend on the criteria used to judge success. The ability to measure the success of a building project relates to the yardsticks introduced in the brief, against which it can be measured'. Criteria are important and might have been discussed as a first element of building performance; however they are covered here as this allows building on the earlier chapters on context, functions and requirements and the deeper exploration of building performance.

A criterion is a 'principle or standard by which something may be judged or decided', which typically involves the comparison of observed behaviour with a goal ('an aim or desired result') or target ('an objective or result to which efforts are directed'). Criteria allow one to establish how well a system such as a building performs.

Gilb (2005: 324) uses the word ambition as a term that specifies a target performance level. Such an ambition must state to what requirement it is linked, as well as 'a notion of the level being sought'. This notion of the level being sought requires the use of a scale of measure; it typically relates to a past track record, trends in the area of this performance aspect and possibly state-of-the-art achievements or so far unsurpassed records (*ibid.*, 117–118). Blanchard and Fabrycky (2011: 206) name this ambition an 'aspiration level'; they observe that the aspiration level can not only be a desired level of achievement but also some undesirable level that is to be avoided. Embedded in this positioning of ambition is a review of appropriate references. These may be limited to one system and historical performance data, where one aims to improve on past performance; alternatively one may compare with a wider group of peers and look at the performance of others to set the level that is desired. In the systems engineering domain, INCOSE (2015: 54) uses 'validation criteria' for measuring how well a solution matches the measures of effectiveness (MOPs) and other requirements that have been defined; in other words, validation criteria can be used to show how a solution scores in terms of customer and user satisfaction. They thus quantify the match between an actual value and a goal or target; this quantification can be used to evaluate effectiveness, efficiency and compliance of a solution. This in turn helps to guide decisions in a design/engineering

Building Performance Analysis, First Edition. Pieter de Wilde.
© 2018 John Wiley & Sons Ltd. Published 2018 by John Wiley & Sons Ltd.

process or in system operation. While goals in terms of championships are only on the fringe of building performance interests, there are some areas where such a status acts as a big motivator for technological development. An example is the efforts to develop algorithms for playing chess or go, which have been tested against reigning world champions (Kroeker, 2011).

The Systems Engineering Body of Knowledge highlights the importance of criteria in analysing alternative solutions and, ultimately, decision making. The assessment criteria are defined as 'required properties and behaviour of an identified system situation,' typically based on a description of a best or ideal system (SEBoK, 2014: 180). It is noted that criteria can be expressed as either absolute or relative, typically on a scale; boundaries may be put in place to limit the options taken into account. In decision-making theory, criteria are often captured by means of an objective function (Hazelrigg, 2012: 45–53). Objective functions capture preferences; to do so they must understand and maximize the objective of the decision maker. Preferences can be written as $a > b$, which means that 'the decision maker strictly prefers object a to object b' and then handled according to rules of mathematical set theory (French, 1988: 62; Jordaan, 2005: 10–12). In the context of search and optimization, the objective function (which represents the performance of one specific solution towards the ideal) is augmented with a fitness function, which captures how well a population of solutions does on average (Raphael and Smith, 2003: 176–177). Goals and targets also play an important role in motivation. Locke and Latham (2002) describe this in the context of organizational psychology, discussing 35 years of research on the relation between goal setting and task motivation. They describe how goal difficulty and goal specificity drive task performance, impacted by factors such as choice, effort, persistence, strategy and moderated by commitment, goal importance, self-efficacy, feedback and complexity. Jung *et al.* (2010) show how individual performance feedback and explicit goals strengthen the performance of employees working collaborative via an ICT environment, whereas Kleingeld *et al.* (2011) report that group performance may require group-centric goals rather than individual objectives. Klotz *et al.* (2010) demonstrate how goal setting may create unintended anchoring bias in the design of energy-efficient buildings.

An initial knowledge base on criteria for building performance was already described by CIB Report 64 (1982). Amongst others, this identifies the need to review a range of performance aspects when making a choice between two alternatives. It notes that setting single performance values is seldom satisfactory, since this may lead to anchoring bias; it also is at odds with the fact that performance might not necessarily increase linearly, exhibiting diminishing returns at some point of the curve. The same may be true of the value of increasing performance beyond a certain level (*ibid.*, 18). CIB Report 64 suggests the use of banded levels of performance while differentiating between performance measures that represent better performance through decreasing values, increasing values, logarithmic values, authoritative limit values or peak level values (*op. cit.*). The use of performance bands prevents too much focus on precision of a performance prediction method, which might be misleading. Instead, decision makers can just focus on comparing the bands (CIB Report 64, 1982: 18). This approach has found its way into many frameworks; an example is EN15251 for indoor environmental criteria, which groups measurements into four categories: I - high expectation levels for sensitive and fragile people, II - normal expectation levels for new buildings and renovations, III - acceptable levels that have an extra allowance for existing buildings, and IV - values outside these categories that should only be accepted for a limited duration (Olesen, 2007). Some of the criteria in EN15251 assign these categories on the basis of a percentage of time that design values are being met; for instance in the area of indoor temperatures, category I requires that the indoor temperature is within a specified range for 90% of the time, where the relation between outdoor temperature and acceptable indoor temperatures is specified in a graph. Foliente (2000) defines building performance criteria as 'a statement of the operative or performance requirement', giving examples of a floor having to have sufficient strength to support the anticipated load and not vibrating or sagging in such a way that this results in discomfort for the occupants. In both cases, this implies a relation between some predefined load and system behaviour. Foliente then proceeds to discern two types of performance criteria: *technology-based criteria*, which deal with the performance of buildings and products under controlled testing conditions, and *risk-based criteria*, which deal with the performance of buildings and products in actual use.

Other goals and targets in the building sector are defined more loosely, such as the goals of Zero Energy Buildings (ZEB) and/or Zero Carbon Building (ZCB). Marszal *et al.* (2011), Kibert and Fard (2012) and Heffernan *et al.* (2015) give overviews of the various interpretations and details of these concepts and problems with proper definition, with McLeod *et al.* (2012) and Heffernan (2015: 72–73) expanding on the specific definition and understanding of the term zero carbon homes in the UK housing market. There is ongoing discussion on the aspects that should be included in these concepts such as on-site versus off-site energy production (Berggren *et al.*, 2013; Mohamed *et al.*, 2014), the link with embodied energy and material use (Hernandez and Kenny, 2010), the relation between building and neighbourhood (Koch *et al.*, 2012; Sartori *et al.*, 2012), and the sociotechnical embedding (Pan, 2014). A risk of some confusion is caused by the use of the abbreviation nZEB for both 'net zero energy buildings' and 'nearly zero energy buildings' (Visa *et al.*, 2014). This multitude of papers demonstrates the need for clear definition of criteria in the building industry.

Building performance criteria are needed for each of the performance requirements as discussed in Chapter 3: for quality requirements they provide a goal for how good a function must be done, for resource saving requirements a goal for how much of a resource may be used or must be saved, for workload capacity requirements a goal of how much capacity is to be provided, for timeliness requirements how responsive the building must be and for readiness requirements when and under what circumstances certain building functions must be available. Examples would be criteria for how much vibration is allowed under a certain type of wind load; criteria for the amount of energy, water or materials used in operating and constructing a building; or criteria for the handling capacity of a facility. The criteria define what level of performance is needed and the frame of reference for setting those levels. Criteria may include thresholds, where some state change takes place, and constraints that exclude options with certain unwanted characteristics.

Criteria are sometimes also used to describe a wider context, where they are equivalent to requirements and needs. For instance, Blanchard and Fabrycky (2011: 126–127) talk about design criteria relating to functional capability, interoperability, sustainability, reliability, maintainability, usability and safety, security, supportability and serviceability, producibility and disposability and affordability in general, without specifying an ambition level for each of these.

This chapter provides a framework for handling performance criteria by discussing goals, targets and ambitions as an expression of aspirations, benchmarks for comparison with peers, bands of performance and finally constraints, thresholds and limits that help manage the evaluation space.

5.1 Goals, Targets and Ambitions

A goal can be defined as an aim or desired result; the word is also used in the context of sport, where it indicates a line or surface that a ball needs to cross to score a point. A target is something similar and used to identify an objective or result towards efforts are directed; however a target also has a military connotation in that it can be used to designate a person, object or place selected as the aim of an attack. However there is a fine difference that these other uses help to explain: a goal typically is something that is either achieved or not, like a goal in a soccer game; see Figure 5.1. A target is something that allows for measurement in how close one gets, as in getting close to the centre in shooting or archery; see Figure 5.2. Other mental images for a goal are the destination at the end of a journey of some kind, whereas the notion of a target evokes the images of radar screens and crosshairs.

According to the dictionary the words aim and objective are synonyms for goal; however, in the context of management, aim and objective are often taken to mean strategic, longer-term ambitions, whereas goals and targets identify steps to be taken to get towards the objective. But as use of these words varies, it is important to try to find out their specific use in particular contexts or manuscripts. There are many texts that give suggestions on how to properly define a goal or objective; a well-known suggestion is to set SMART objectives (Specific, Measurable, Attainable, Realistic and Timed); lesser known are PURE (Positively stated, Understood, Relevant and Ethical) and

Figure 5.1 Goal, with binary scoring option.

Figure 5.2 Target, with gradual scoring scale.

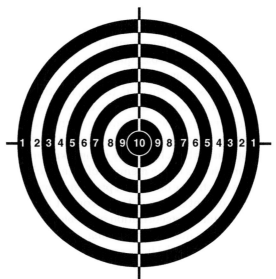

CLEAR (Challenging, Legal, Environmentally sound, Agreed and Recorded) objectives (Krogerus and Tschäppeler, 2011: 21). To make the situation more complex, there often exists a hierarchy of main goals and sub-goals. These can be mapped by a metamodel that uses AND and OR relations to express some of the underlying relations (Popova and Sharpanskykh, 2010). Whether one uses goals or targets, it is important to ensure that criteria are kept independent to prevent double counting, in line with the earlier discussion of building functions and functional decomposition. Where this is not possible, it is better to work with one single criterion. Moreover, care must be taken to ensure that goals and targets reflect the actual needs of the stakeholder and system under development (INCOSE, 2003: 210).

Goals and targets are sometimes represented by a mathematical statement using the symbols $>$, $\geq$, $=$, $<$ or $\leq$. Such a definition is appropriate for a well-defined performance quantification in terms of performance measures or indicators as discussed in the previous chapter; Popova and Sharpanskykh (2010) use the term 'performance indicator expression' for such a mathematical statement. Examples may be PI28 ≤ 24 hours or PI08 = high (*ibid.*). However, goals and targets are not always static; for instance the performance in terms of ventilation provided is a function of the static pressure difference. CIB Report 64 (1982: 20) points out that such behaviour cannot be represented by single numbers; instead they suggest the use of representative curves to compare 'compound data'. Augenbroe (2011: 15) warns that raw performance data typically is not suitable for the comparison of competing design alternatives; in most cases there is a need for a more advanced aggregation of this data to make an objective comparison, and specifying this aggregation should be part of setting up the criteria. In manufacture, there often is conflict between objectives, such as in the trade-off between due date, inventory level, product quality and machine utilization (Jain *et al.*, 2011).

Goals and targets are often used in computing, for instance in optimization efforts and agent-based simulation. The level of detail of these goals and what can be done with them vary. Some models allow users to change and remove goals over time, whereas in other models they are fixed and completely embedded in computer code (Choi, 2011). Moreover, objectives sometimes conflict; there is no reason to assume that general tenets like the iron triangle of cost, quality and time do not hold in the field of building performance. Some researchers have tried to quantify the degree of conflict between objectives, see for instance Shi *et al.* (2016), but this is unlikely to solve the underlying issue. In mathematical terms, the trade-off between conflicting goals is embedded in the idea of Pareto fronts, where improvement towards one goal requires that the performance against the other goal must be allowed to decrease (Lozano and Villa, 2009). A goal can sometimes be achieved by using a known skill or process. Finding the right solution requires access to the memory of actual solutions to achieve the goal in question, and this is exactly the domain of problem solving and skill learning. In other cases, there may be a concept from a different context that may be applied; finding solutions along that line requires categorization of problems and inference (Choi, 2011).

In the context of organization modelling, a goal is 'an objective to be satisfied describing a state or development'. The description of a goal requires a goal name, definition, priority, evaluation type, horizon, ownership, perspective, hardness and negotiability (Popova and Sharpanskykh, 2010). Goals and performance measurement are naturally linked; where possible this relation should be defined explicitly (*ibid.*). Goals can either be self-set, assigned externally or negotiated (Miles and Clenney, 2012).

Goals can be hard (measurable) or soft (more difficult to establish if met). Hard goals can be labelled as satisfied, undetermined or failed; soft goals can be labelled as satisfied, weakly satisfied, undetermined, weakly denied or denied (Popova and Sharpanskykh, 2010). The status of a goal may also change over time: a goal can be achieved (ceased), maintained (avoided) or optimized (maximized, minimized, approximated) at different points in a process (*ibid.*). On a fundamental level, the hard, measurable goals can be stated as 'the more, the better', 'the less, the better' or 'the more centre, the better'. Such goals can be described by preference functions, with the most desired situation shown in a target interval, possibly flanked by allowable intervals (Inoue *et al.*, 2012).

In complex situations there may be a whole number of goals, with some goals being assigned priority and given the label of top-level goals (Choi, 2011). Popova and Sharpanskykh (2010) take this concept of goals and sub-goals further by defining a 'goal pattern'. A goal pattern relates to a set of performance measures, with formal goals being expressed for each of these measures. The goal pattern defines which sets are favourable and which sets are unwanted; they also may add information about the priority of one performance aspect over others. Goal patterns should also add a temporal dimension and define when these values are desirable. In this context, goal patterns may be (i) achieved/ceased, (ii) maintained/avoided and (iii) optimized/approximated (*ibid.*). The factor time is an important one, as performance expectations might change over the buildings' life cycle. In some cases there is a stable level that remains similar; in other cases expectations go up (such as the increased requirements in terms of energy efficiency), while in other cases it is accepted that older buildings have a lower performance (CIB Report 64, 1982: 21).

The concept of goals is widespread in organizational and management science. Here goals are often classified using the categories as low, high, increasing and do-your-best structures. In production and operations management, one extremely important goal is on-time operation. This goal can be captured by different performance metrics such as 'average flow time' or 'percentage of deadlines achieved'. Care must be taken to identify the metric that really reflects the need of the operation (Doerr and Gue, 2013). In terms of task performance, setting goals enhances the work done by 'directing attention, mobilizing effort, increasing persistence, and motivating strategy development'. Similar effects have been found at team level (Hoegl and Parboteeah, 2003).

There is a body of knowledge on goal-setting theory; however this mainly pertains to the domains of psychology, human behaviour and management. In general the underlying thought is that, in order to achieve an optimal level of performance, a goal should be set that is challenging yet attainable. The consensus seems to be that high goals, set at the 90th percentile, work best (Welsh and Ordóñex, 2014). An example of such goals is those that are set for a negotiation process (Miles and Clenney, 2012). While goals are an important driver for performance of individuals, the link between the two is often moderated by other factors. For instance, research on a sample of 145 software development teams shows a direct relation between goals and both effectiveness and efficiency, but this link is moderated by the quality of the teamwork (Hoegl and Parboteeah, 2003). The use of performance goals may also have a negative side; where they are used consecutively, they may lead to resource depletion and contribute to unethical behaviour (Welsh and Ordóñex, 2014). Goals can lead some individuals to become hyper-motivated. However those not reaching a goal may be demotivated, stressed and may get a lower self-esteem; for this group not having a goal at all may be better (*ibid.*). Hyper-motivation may also have unintended side effects, such as a competitive and individualistic mindset (*op. cit.*).

Gilb (2005: 118–124) points out that a target typically consists of a range of numbers on a scale that are considered desirable, rather than a single value. Within this target range, the levels that form the target represent what is needed to achieve success. However, one may also go beyond the target and enter a part of the scale that Gilb names 'stretch'; this stretch level challenges to exceed the goal and perform even better than expected. Even further out is a part of the scale that is labelled 'dream'; this is to identify some optimal theoretical areas that may be achievable in the future. Next to these desirable areas, Gilb also recognizes 'fail' areas, which mean that success has not been achieved; beyond these performance can get even worse and one risks going beyond 'survival' levels, where the total system will fail (*ibid.*). Another view of looking at performance targets is to aim for 'the smaller the better' or 'the larger the better'. There are also performance targets where the main concern is to achieve a specific target and where deviation from target is to be prevented; in the context of a business process, examples of the latter are for instance providing the right ingredients of food, consistency and acceptance rate (Lim and Zhu, 2013). Examples of details that may need definition when setting a target include the percentage of change that is the ambition, the baseline period to which change is compared, use of absolute or normalized data, any interim targets, responsibility for the target and monitoring procedures (Robinson *et al.*, 2015). When expressing a target in the form of numbers, it is also imperative to consider the appropriate measurement scale and whether the target is to be expressed using a nominal scale, ordinal scale, interval scale or ratio scale.

The general process for setting targets involves getting support by the decision makers for working with targets, selecting the actual target, deciding on system boundaries, defining a reference baseline, setting the actual target and deciding on the length of the commitment (Rietbergen *et al.*, 2015). Targets should not only be technically achievable but also desirable. To assess what is a good target therefore often requires subject knowledge and expert assessment (Ruiz *et al.*, 2015). Rietbergen *et al.* (2015) list the main target types as absolute or volume targets, physical efficiency targets, economic targets and emission targets. Targets do not necessarily have to stay fixed over time; for instance in computing there are specific approaches that ask decision makers to provide feedback along the way about the priorities, allowing them to update preferences and set new targets along the road (Lozano and Villa, 2009). It must be kept in mind that decision makers often have to deal with a range of performance aspects, which all have their own metrics and targets; some of these may conflict (Jain *et al.*, 2011).

The definition of targets is a non-trivial task. To be of use, targets must be significant to stakeholders, comparable amongst peers, and must have some level of ambition (Rietbergen *et al.*, 2015). Targets that are based on external benchmarks need to be adapted to the unique circumstances of any specific situation (Jain *et al.*, 2011). Even in advanced production processes in industry, there often is a lack of guidance on how to set performance improvement targets (*ibid.*). When defining targets, it is worth noting that the variation of many variables follows the bell curve of a normal distribution. It is common to describe the form of a distribution by reporting the mean and the 5% and 95% percentiles,[1] or the mean and standard deviation (SD) where the standard deviation is defined so that 68% of the values sit within one SD of the mean, 95.5% within two SDs of the mean and 2.3% in each tail. About 99.7% sit within three SDs of the mean (Norman and Streiner, 2003). Allowing only for very small tails is not uncommon in manufacturing processes, where the target may very well be a 100% pass rate and where a drop to 'only' 99% would mean a serious problem (Pritchard *et al.*, 2012: 20–21). In many disciplines, people are interested in how the main population behaves and less so in the tails of a distribution. Yet this is not the case for engineering design; here there is more concern about the behaviour of extremes such as high temperatures, weak components and the like. As a consequence engineers are more interested in the tails (O'Connor and Kleyner, 2012: 47).

Percentiles are regularly used to define targets for product design. For instance one could define a target that those potential drivers who fall within the 5th–95th percentile in body length must be able to reach the pedals of a car, as well as being able to see out of the window. Where health and safety issues are involved, typically much more stringent targets are used such as 1th–99th percentile (Nah and Kreifeldt, 1996). Where the prediction of outcomes is at stake, another way of handling variation and probabilities is the use of confidence intervals (Norman and Streiner, 2003: 27–28). Typical values used are 60%, 90% and 95% confidence intervals (Smith, 2011: 52–53). The confidence intervals calculated for performance predictions in thermal building performance during early design stages are not as high, with Rezaee *et al.* (2015) reporting values of below 50%.

Where there is range of targets, the actual performance can be plotted on a target plot. This is similar to a radar plot, but here the centre of the plot represents the optimal performance, while the outer circle equals zero performance. In a target plot, achieved values are represented as impacts on the target circle. The technique can be used in any

---

1 A percentile is a statistical measure that indicates what percentage of observations fall under this value; the 50th percentile is the median in a normal distribution.

domain. An example in the medical sector is given by Stafoggia *et al.* (2011); however application in the building domain seems to be extremely limited. In such cases it may also be worthwhile to use Data Envelopment Analysis (DEA) and Statistical Process Control (SPC) charts for performance measurement and target setting and for identifying best practice (Jain *et al.*, 2011).

The way that data will be analysed often has an impact on how this data is collected. This also is the case for setting targets. Some techniques such as data envelopment analysis (DEA), which is used in Operations Research (OR), have implicit assumptions about targets. DEA assumes that lower input and higher output levels indicate a better performance. However, there are situations where this is not the case and where the actual objective is to stay close to a certain target and minimize deviation (Lim and Zhu, 2013). In other cases, such as greenhouse gas emissions, less output is better. When using DEA, this means that the target must be redefined to meet the technique (Sueyoshi *et al.*, 2010).

Targets are used in many disciplines. A good example is mineral exploration, where targeting is related to the direction of efforts to identify ore deposits; related activities are conceptual targeting, target modelling, target ranking and ultimately target identification. These efforts typically start at a global scale and then hone in on the provincial, district and ultimately project level. Prediction and uncertainty are intrinsic to all steps. Results that help direct searches are often presented in hierarchical and Venn diagrams. In hierarchical diagrams, information is gathered on a larger area; smaller areas nested within are then explored in more detail, until a target area is found. In the Venn diagram approach, information on a range of critical parameters is used to find the areas where properties of interest overlap (Hronsky and Groves, 2008). In the military sector, target identification is an important aspect of the use of radar and sonar. One technique used is to compare signals received to a reference library; typically the process also involves the use of filters to pick up relevant signatures (Carroll and Rachford, 2011). Targets are also used in the medical literature in the context of identifying illnesses and agents. Traditionally the approach was to find a single drug for each target. Recognizing the complexity of some situations, medicine is now moving to a 'multi-target, multi-drug' situation, which is named polypharmacology (Vitali *et al.*, 2013).

In a corporate context, targets are also increasingly linked to financial incentives for personnel and business units. One term used in this context is pay-for-performance (P4P); obviously the size of the incentive in relation to the unit of assessment is of key importance in such programs. Most P4P systems score performance in bands and tie quality to a financial bonus (Kirschner *et al.*, 2012). In this way, performance targets are used as an incentive. The linking of a monetary reward to achieving the target capitalizes on the concepts of extrinsic and intrinsic motivation, where the latter represents the fact that doing good work and meeting targets provide their own inherent rewards (Rablen, 2010). The target level that is set might have an impact on the effort put into achieving this target and getting the reward set. However, this is not endless; at some point it will become attractive to bear extra risk for non-achievement rather than to invest effort in reaching a target that is seen to be unattainable (*ibid.*). The acceptance of uncertainty and tails in target setting may lead to some hard ethical questions. In much of society, the willingness to accept risk as a part of daily life is disappearing. However, creating systems that do not pose any risk is extremely expensive, if possible at all.

A further complication is the division of ethics into two main streams: *deontological ethics*, which is about 'doing the right thing' regardless of the consequences, and *teleological ethics*, which takes a utilitarian view and judges actions in reference to results. A deep discussion of how this relates to environmental regulation for toxic substances is provided by Simon (2010). Issues to be considered in this context are the dose–response data, precautionary principle, conservatism in target setting and individual susceptibility, which all need to be balanced in the setting of a reasonable maximum exposure (RME).

Within the ProMES system for productivity measurement and enhancement, the concepts of simple goals and targets have been replaced with so-called contingencies that describe the amount of some indicator that is desired in relation to other process-dependent parameters. A key idea that underlies these contingencies is the fact that typically there is not always a linear relationship where more of an indicator is better; there are often general expectancies that push indicator values to some skewed distribution (Pritchard *et al.*, 2012: 20–24). In mainstream economic decision-making literature, the same idea is represented by the concept of utility, which measures preferences of decision makers.

The underlying drivers for setting goals and targets, as mentioned by Gilb (2005: 324) and Blanchard and Fabrycky (2011: 206), are ambitions and aspirations. These drivers have been explored in some depth in the management field. Shinkle (2012) describes three frameworks that are concerned with organizational aspirations: strategic management, behavioural theory and strategic reference point theory. Shinkle points out that most work in this area has an explanatory nature but that there is a paucity of measurement and documentation of aspirations and comparative studies. In social science research, ambition has been described as 'a commonly mentioned but poorly understood concept' (Judge and Kammeyer-Mueller, 2012).

The design of green and sustainable buildings is generally accepted to depend on the ambitions of the stakeholders and whether or not they embrace these issues (Hojem *et al.*, 2014). This ambition can be impacted by learning about new technology or knowledge becoming available to the actors; ambition also is helped by enthusiasm, social learning and interpersonal trust (*ibid.*). Ambition and motivation may stem from a range of factors. Often environmental concerns or monetary arguments are cited as key issues that drive building performance (Russell-Smith *et al.*, 2015b), but others like corporate reputation or employee values also play a role (Pellegrini-Masini and Leishman, 2011).

High-level drivers and motivation are often based on a range of considerations. By way of example, Joas *et al.* (2016) describe over 16 factors that contribute to the German policy to increase the share of renewable energy in the power mix, which includes the typical concern about climate change or conservation of exhaustible resources but also goals like job creation, cost reduction, global leadership and decentralization. Policy ambitions may not always be realistic; for instance Özcan and Arentsen (2014) describe how policies to increase the use of energy from biomass may be out of step with the actual potential of a region.

The relation between targets or goal setting and ambitions is complex; for instance institutional environmental targets may be set very high due to political pressure and then fail to result in actual performance improvements (Robinson *et al.*, 2015). Some of the targets in building performance are driven by political ambitions, such as the reduction of greenhouse gas emissions. Strachan (2011) discusses the interaction between the 'twin challenge' in energy policy, decarbonization and security of supply, and the development of models and tools to handle decisions in UK policymaking, quoting rigorous debate on policies and what measures best meet the challenge. Yet there is evidence that the impact of political targets on something like the reduction of gas consumptions by dwellings is relatively limited (Oreszczyn and Lowe, 2010).

The process for setting targets may be driven by certification requirements, such as those that are tied to environmental rating schemes. Certification may require the reporting of actual performance, disclosure of findings and formal auditing by external experts. Often certification requires that quantifiable targets are put in place and that improvement takes place over time (Rietbergen *et al.*, 2015). Target levels in certification may be voluntary, fixed by the certification agency (imposed) or open to negotiation (*ibid.*).

Just as in any other human endeavour, goals, targets and ambitions play an important role in the building sector. However, the study of these concepts in this field is difficult due to the generic use of the terms, with many authors using them as synonyms. At the same time, it is interesting that a discussion of goal and targets seem to be missing where one expects them. In their book on *Assessing Building Performance*, Preiser and Vischer (2005: 5–7) discuss performance levels, user's needs and priorities; the book has a chapter on strategic planning (*ibid.*, 29–38) with an example on option rating and another on briefing (*op. cit.*, 39–51), but the words 'goal' and 'target' do not feature in the index. Blyth and Worthington (2010) in *Managing the Brief for Better Design* discuss managing expectations and levels of performance (*ibid.*, 59–75), but again 'goal' and 'target' do not feature in the index. Atkin and Brooks (2009) on *Facilities Management* discuss best value (*ibid.*, 7), service-level agreements (*op. cit.*, 101–117) and various other relevant fields, including human resource management, yet again their index does not mention goals or targets.

It thus appears that actual building performance goals and targets are mostly hidden within subject-specific contributions. For instance, a deep discussion of thermal comfort criteria for buildings can be found in Nicol and Humpreys (2007, 2010) or de Dear *et al.* (2013). A discussion about (day)light in rooms is provided by Nabil and Mardaljevic (2005), Kittler (2007) and Mardaljevic *et al.* (2009), whereas examples from the field of room acoustics are provided by Cerdá *et al.* (2009), Bradley (2011) and Augusztinovicz (2012). Exploration of these articles makes it clear that an understanding of the goals and targets in each of these fields requires specific domain knowledge: in the area of thermal comfort about comfort temperatures, operative temperatures, free-running temperatures, adaptation, temperature drift, the Fanger model (Predicted Mean Vote and Predicted Percent Dissatisfied) and heating seasons; in the domain of daylighting about illuminance, irradiance, sky conditions, daylight factors, daylight coefficients, daylight autonomy and field of view; and in the domain of room acoustics about reverberation times, sound decay, sound clarity, sound intelligibility or vibration propagation.

It should therefore not be a surprise that some papers that explicitly discuss building performance targets use specific terms. In calculating their energy saving index (ESI) for windows, Ye *et al.* (2014) relate an actual energy saving value to the 100% saving target that would be achieved by a hypothetical ideal material. In construction project management, a technique known as target value design (TVD) is used to manage costs; it can be combined with life cycle analysis in order to include some environmental impacts (Russell-Smith *et al.*, 2015a). Lee *et al.* (2015a) give an example of how energy efficiency targets and allowable costs for a building retrofit can be defined on the basis of an analysis of project cost risks and operational practice risk. Even at an institutional level, targets can be contentious. As an example, Robinson *et al.* (2015) have studied self-set carbon emission targets against actual achieved emission reductions for a group of universities in the United Kingdom. They note discussion on the key performance indicators and how these are calculated in order to get fair figures across different institutions, with key issues being operating hours, endowment and the efficiency of floor space, making targets hard to compare.

To support the working with goals and targets, INCOSE (2005: 13–14) employs the following concepts:

- *Achieved to date/current actual*: performance quantification that can be plotted over time to see how a project moves in terms of meeting objectives and milestones, thus contrasting the achieved/actual performance with the target or goal. The quantification can be based on either a measurement or modelling and simulation.
- *Current estimate*: the expected performance at the end of the development process, which typically is an extrapolation of achieved-to-date values.
- *Planned value or target*: the performance that is the objective at a given point in time; note that the final target might be different from that at an intermediate point in time of the development process.
- *Planned performance profile*: the development of performance towards the final objective.
- *Tolerance band*: an area of deviation from the performance profile that is deemed acceptable. Typically moving outside the tolerance band requires intervention to limit risks to the overall performance.
- *Milestone*: a point in time where a performance is quantified; typically this is linked to review and test moments.
- *Threshold*: the minimal required value of a performance indicator.

5.2   Benchmarks and Baselines

Where goals and targets are defined in reference to some sort of existing performance, use is made of benchmarks and baselines. A benchmark is a 'standard or point of reference against which things may be compared or assessed'. The origin of the word comes from the discipline of surveying (geological analysis), where surveyors used a mark that was cut in a building or wall, allowing them to reposition levelling rods in exactly the same place; see Figure 5.3. Since the 1970s, the term became widespread in industry in the context of comparing the performance of one process, machine or unit with the performance of others within a peer group in order to check or improve production (Pérez-Lombard *et al.*, 2009). It appears that benchmarking was first introduced to buildings in the context of energy benchmarking. Closely related is the term baseline, which is a 'minimum or starting point used for comparisons'. This word has a strong link with construction: in drawings it is the line of reference from which an object is defined.

Figure 5.3 Traditional surveying benchmark, cut in stone (Devon, UK).

For buildings typically the baseline is the ground level, indicated as base level of 0.00 m, with positives values going up and inverted level (IL) going down. In more general systems engineering, a baseline is a set of expectations against which a design configuration or alternative may be assessed (Blanchard and Fabrycky, 2011: 53–54). The related verb is baselining; in many cases this relates to the comparison of the observed performance with historical track records (Nikolaou *et al.*, 2011).

Building performance benchmarks can take different forms. The California Commissioning Collaborative (2011: 25) discerns two main approaches towards benchmarking: the comparison of performance of a building against the historical performance or track record of that same building, or the comparison of performance against the performance of a group of similar buildings or 'peers'. Deru and Torcellini (2005: 8) list a number of reasons why the comparison with a benchmark may be useful: in order to compare actual performance with the design intent, to compare the performance of a specific building with that of others, to support building rating, in the context of an economic analysis, or to establish records that can be used to find trends. Where performance is compared to other buildings, there is a choice of different levels of benchmarks: those that represent average (typical), above average (good) and excellent (best practice) found amongst a group of products or processes (Pérez-Lombard *et al.*, 2009). Frankel *et al.* (2015: 26) suggest that in the ideal case, data from existing buildings should be used to define targets and objectives for design of new projects. Predictive models can help assess assumptions and progress throughout these projects. After completion, at least 2 years of post-occupancy evaluation should be undertaken to monitor actual achievements and make any necessary improvements. Benchmarking can also be applied to construction processes and projects, rather than to buildings themselves; for instance Yeung *et al.* (2009) use it to compare the partnering performance in construction projects in Hong Kong.

Like building performance analysis in general, the definition of benchmarks has to consider the range of performance criteria involved in a specific situation: benchmarks can be created for single performance aspects, such as energy performance, or across a range of aspect, such as energy, comfort, acoustics, indoor air quality and others. One approach towards benchmarking with multiple aspects is the use of multi-criteria decision making, such as the Technique for Order of Preference by Similarity to Ideal Solution (TOPSIS) as presented by Wang (2015).

When setting building performance benchmarks, one has to consider variation of the building stock. Some issues to take into account are location, year of construction, building size/floor area and main utilization of the building, which are all recognized as important discriminants for many performance aspects (Neumann and Jacob, 2008: 15). In most cases it is appropriate to define a set of buildings that are similar to the building being studied. Within this set, the aforementioned issue of grouping according to performance remains. When looking at such a group or portfolio, the California Commissioning Collaborative (2011: 27) recommends looking at the best and worst performing buildings. Often the best buildings may have practices that can benefit other buildings in the set; the lowest performing buildings are prime candidates for interventions. However, Hinge and Winston (2009) suggest that the use of actual median performance values might be better as benchmark than the use of some theoretical targets set by legislators, or only the front runners, as this prevents a focus on 'playing the system'.

In the context of building project management, Ahuja *et al.* (2010) distinguish between four types of benchmarking: *internal*, which compares business units in the same company; *competitive*, which compares two businesses in the same sector; *industry*, which looks at several companies in a sector in general; and *strategic*, which aims at long-term developments for a company in context. In line with business process performance studies, benchmarks in this field are typically captured by a set of performance indicators that cover a range of aspects. Performance for each of these indicators is measured, after which learning, process intervention and monitoring can take place (*ibid.*). A similar approach is described by Lam *et al.* (2010), who list time, cost, quality, functionality, safety and environmental friendliness as the KPIs that need to be benchmarked.

Within the literature, some authors distinguish other categories of benchmarks. Chung (2011) discerns between public benchmarking systems and internal benchmarking systems, depending on who has access to the data. He notes that public benchmarking can be used as an incentive, creating pressure on some stakeholders to improve the performance of their buildings. Nikolaou *et al.* (2011) differentiates between benchmarks based on the underlying data, which they split into large statistical data sets that represent the overall stock or segments thereof, and data for specific building types that represent either good or typical practice. Pérez-Lombard *et al.* (2009) discuss benchmarking in the specific context of building energy performance certification. Gao and Malkawi (2014) mention hierarchical benchmarking as a type of benchmarking that starts at the whole building level and then moves down to the level of subsystems.

In physics, sometimes one can provide analytical solutions to specific problems, which then can be used as benchmark for studying the accuracy of either measurements or numerical approximations; for an advanced application see Skylaris (2016). In computing, numerical benchmark solutions may be used, provided that their numerical accuracy can be demonstrated (Oberkampf and Roy, 2010: 244). In the building industry, benchmarks are mostly established on the basis of data collection from existing buildings; however one may also use simulated performance data to create a benchmark set for situations where measurements are not available (Lee and Lee, 2009; Chung, 2011; Gao and Malkawi, 2014). The creation of benchmarks from simulation can be based on data from fully transient methods, simplified methods and hybrid approaches such as Bayesian networks (Li *et al.*, 2014b). Federspiel *et al.* (2002) point out that existing data is always limited by the performance of the buildings included, which may be inefficient to some point, especially for some specialized type of buildings like laboratories; model-based data is not restricted in this sense. Nikolaou *et al.* (2009) add that there are some countries such as Greece where measured data is not available and where the use of a Virtual Building Dataset (VBD) based on simulation is the only viable alternative. Lee (2010) provides an example of benchmarking in conditions where the cooling of buildings is dominant. Shabunko *et al.* (2014) report on benchmarks for an energy-rich economy, Brunei Darussalam; obviously such values are only applicable to similar situations. Pérez-Lombard *et al.* (2009) note four key steps in the building benchmarking process: the gathering of data for a significant set of buildings that will act as frame of reference, the gathering of data for a specific building being studied, comparison of the data from the single building to that of the larger set and finally recommendations for actions to improve the single building.

Data for benchmarking may be obtained from a range of sources, such as automated measurement devices and professional surveyors, or can be self-reported by building owners and occupants (Hsu, 2014). Chung (2011) points out that the definition of set of 'similar' buildings in building energy benchmarking is not a straightforward task, as it needs to account not only for differences in building attributes such as age, number of floors, material and so on but also for differences in climate conditions, energy efficiency incentives and occupant behaviour. Dealing with these differences requires normalization of results to create comparable data. Hinge and Winston (2009) make the case for the further collection of actual, measured performance data, with random samples, to create better benchmarks.

In the field of energy performance building, arguably the oldest building benchmarking domain, different countries have their own data sets. In the United States, data on non-domestic building energy performance is available from CBECS, the commercial buildings energy consumption survey, and for domestic buildings from RECS, the residential energy consumption survey (EIA, 2016). In the United Kingdom, the ECON19 guide on energy benchmarking for commercial buildings (Action Energy, 2000) has long been the main benchmark for offices. The guide gives typical energy profiles for four main types of offices: naturally ventilated/cellular, naturally ventilated/open-plan, air-conditioned/standard and air-conditioned/prestige. A minor revision of the guide appeared in 2003. ECON19 defines energy benchmarks as 'representative values for common building types, against which one can compare a building's actual performance'. Values for the ECON19 benchmarks were 'derived from surveys of a large number of occupied buildings and include all energy uses, not just building services' (*ibid.*). The Probe studies benchmarked buildings, mainly in the educational domain, against a broader data set, which allows a direct comparison with its peers (Leaman and Bordass, 2001).

However, building benchmarks do not always need to be on the whole building level. An example is the use of benchmarks for cleanliness, which can be applied to individual surfaces in rooms. In hospitals the measurement of cleanliness is important, especially with the increasing risks posed by superbugs like MRSA. One approach that is being used is visual inspection, but obviously that does not relate directly to the presence of microbes. An alternative is the detection and measurement of adenosine triphosphate (ATP), a chemical that is produced in every living cell and hence acts as an indicator that there was something alive on a surface. ATP levels can be identified with bioluminescence; ATP benchmark levels that identify unacceptable levels of soil have been defined from multiple experiments and statistics (Mulvey *et al.*, 2011).

Once the data set on which benchmarks will be based has been selected, this data needs processing. Tools that model the relation between input and output parameters from observed data pairs are sometimes named data-driven or inverse models (Zhang *et al.*, 2015b; ASHRAE, 2017: 19.2).

Mathematical techniques used to derive benchmarks include simple normalization, regression analysis, data envelopment analysis and stochastic frontier analysis (Chung, 2011); another slightly wider classification is to discern bin methods, multiple linear regression, support vector regression, Gaussian process regression, artificial neural networks and decision trees (Li *et al.*, 2014b). Principal component analysis and principal component regression may also be used (Wang *et al.*, 2014), as may be support vector approaches (Zhang *et al.*, 2015b). Gao and Malkawi (2014) suggest the use of clustering techniques (specifically K-means clustering) in order to allow benchmarking against a group of buildings that has the same characteristics and use that can be selected from a large dataset. Yalcintas (2006) provides an example of the use of artificial neural networks (ANN) in the context of benchmarking, where the ANN is used to predict the energy performance of a building; results are compared with linear regression to determinate whether that predicted performance requires intervention.

One specific analysis technique frequently used in benchmarking is the aforementioned data envelopment analysis (DEA), from the fields of economics and operations research. Data envelopment analysis works by identifying a 'production frontier' or 'best practice frontier' for a data set, so that one can establish how the performance of elements of that set relates to the best peers in that set. In order to compare the performance of a specific case to the benchmark as described by the data envelopment analysis, the distance to the efficient frontier is computed. There is a range of techniques for doing this such as the directional distance function, Farrell efficiency measure or Russell efficiency measure (Lozano and Gutiérrez, 2011). Data envelopment analysis allows the consideration of large set of parameters and aspects, which may help a decision maker in a complex situation (Jain *et al.*, 2011). Furthermore, data envelopment analysis allows the inclusion of a range of driving factors in the analysis. Application to building energy benchmarking is presented by Lee and Lee (2009) and Kavousian and Rajagopal (2014), with the latter explaining the principles of frontier models in some detail.

The quality of benchmarks based on statistical analysis of measured data depends on the quality of that underlying data, which involves data cleansing – assessment of that data to remove outliers, correcting or removing corrupt records and dealing with missing data points – as well as the handling of uncertainties (Hsu, 2014). An important issue to bear in mind is that the view of what is quality data depends on the use; what may be excellent for one objective may not be appropriate for another (*ibid.*). For benchmarking to be of use, the products or processes that are compared must be similar.

For buildings this typically requires that they are of the same type; when comparing energy use or thermal comfort, they must also be subjected to a similar climate. Where possible, an option to filter for peers with similar attributes and use should be provided (Pérez-Lombard *et al.*, 2009). In any case, most external benchmarks need some form of adjustment to a specific situation before they can be used to define targets (Jain *et al.*, 2011). There are different approaches to look at data quality. One approach is the internal comparison between different categories within the same data set. Another is to split the data in a set into different subgroups in order to identify systematic differences between these subgroups. Control groups can also be used that bring in exogenous data; this is taken to a higher level by using external validation (Hsu, 2014).

As in many areas, there are dedicated programs for benchmarking. Examples of early software tools that were designed specifically to support building energy performance benchmarking are Energy Star and Cal-Arch (Matson and Piette, 2005). More recently a wide range of programs has become available to benchmark utility use, such as EnergyIQ, FM Benchmarking, MSC, WegoWise and others. Many of these combine functionalities for benchmarking with more advanced analysis options, moving into the domains of performance tracking, system automation, and fault detection and diagnosis (California Commissioning Collaborative, 2011: 71).

A baseline is a 'minimum or starting point used for comparisons'. In the context of building analysis and benchmarking, the concept is usually taken to identify more than one single point; the construct of a line is extremely useful to represent the relation between some performance aspects and driving factors. Examples would be the energy use of a facility in relation to a usage pattern that changes for different days of the week or that is strongly dependent on outdoor air temperature. Rajasekar *et al.* (2015) reinforce this notion of a benchmark being a simple point of reference; in the context of thermal performance in relation to climate conditions, they use baselines to create 'dynamic benchmarks'. Building energy performance baseline models are typically fitted to data from a 'training period'; afterwards the model can then be used for a 'prediction period' (Granderson and Price, 2014).

For analysis of the impact of energy-saving measures, a baseline must be established that captures the typical energy use before the introduction of these measures. A baseline is also the starting point for advanced control and fault detection and diagnostics (FDD) approaches. Baselines are a key tool to quantify the improvement after intervening in the systems or operation of a building. As one cannot measure both the pre-intervention and post-intervention state at the same time, baselines are needed that ignore or standardize the impact of any noise factors and single out the impact of the intervention (Zhang *et al.*, 2015b).

There is a range of models that are often used to develop baseline models from existing data: mean-week models, which just take into account the time of each day of the week; change-point models that account for relation between outdoor temperatures and HVAC setpoints; and day-time-temperature regression models that cover these key factors. Some of these models are commercial and proprietary models, whereas others are academic and more widely available (Granderson and Price, 2014). Zhang *et al.* (2015b) explore the baselines developed via change-point regression models, Gaussian process regression models, Gaussian mixture regression models and artificial neural networks. Srivastav *et al.* (2013) list the following approaches for establishing an energy use baseline: variable-base degree day models, regression models, artificial neural networks and generalized Fourier series models. An advanced way of capturing a baseline while accounting for uncertainties is the use of Gaussian Mixture Regression (*ibid.*).

Baselines are prominent in the International Performance Measurement and Verification Protocol (IPMVP), which, interestingly, does not refer to benchmarks in any way. In the IPMVP, baselines are related to the installation of an energy conservation measure (ECM). Before the installation of the ECM, actual energy use is measured during a 'baseline period'; measurements done after installation take place during a 'reporting period'. In terms of the IPMVP, the extrapolation of use patterns observed during the baseline period to the reporting period yields an 'adjusted baseline'; savings can then be calculated by subtracting the consumption measured during the reporting period from the adjusted baseline (Efficiency Valuation Organization, 2012: 7). The IPMVP highlights the importance of selecting the baseline period in such as way that it represents the full spectrum of operation modes of the building, covering the whole operational cycle, with minimum and maximum energy consumption, under known conditions, directly before intervention (Efficiency Valuation Organization, 2014a: 3); see Figure 5.4.

With the increased temporal resolution of building energy metering, more robust baseline models can be created than the traditional ones that were based on monthly data (Granderson and Price, 2014). At the same time, there are still questions about the accuracy of baseline models, the best interval to use for developing the model, ownership of data and baseline models, and others (*ibid.*).

In terms of application of baselines in building performance studies, Zhang *et al.* (2015b) use the concept in their inverse modelling studies on the quantification of the savings obtained by installing energy conservation measures. In their papers on calibration of building energy simulation programs, Reddy *et al.* (2007a, 2007b) use baseline information from a wide range of other buildings as benchmark for 'preliminary data proofing' and screening of the quality of simulation outcomes before engaging in deeper analysis. At the same time, they also discuss parameter baseline values that they define

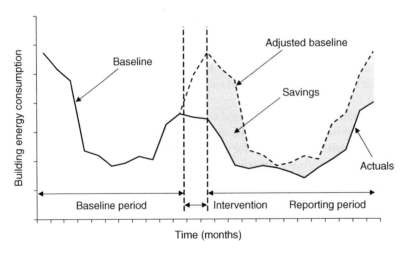

Figure 5.4 Baseline principles.

as parameter values before calibration. In industry, the use of baselines is a common feature of advanced energy information systems; in these systems baselines represent historical data to display what the energy use would have been without a certain intervention (California Commissioning Collaborative, 2011: 46).

5.3 Constraints, Thresholds and Limits

Constraints, thresholds and limits are important in the definition of criteria because they represent special points of interest on the scale(s) used to measure building performance. The words are related and partly overlap. Formally, a constraint is a limitation or restriction; a threshold is 'a level or point at which something starts or ceases to happen or come into effect'. Interestingly, in terms of buildings, a threshold also is the term used for the component that forms the bottom of a doorway and that is crossed when entering into a room or house. A limit is 'the point beyond which something does not or may not pass or extend, a terminal point or boundary, a restriction on the size or amount of something'. Limits are commonly used to indicate things like the speed limit on roads, drink limits when driving or weight limits for bridges and vehicles. These concepts also play important roles in the natural and formal sciences. For instance, in physics the laws of thermodynamics apply to all man-made systems and will put constraints on what systems are feasible, preventing the development of a perpetual motion machine, whatever the stakeholder goal may be[2] (Bryson, 2003: 107). Phase transitions such as the freezing and boiling point are natural thresholds for materials used in construction and encountered in buildings. Limits are also an important concept in mathematics where series converge to a definite value (Stroud and Booth, 2007: 780–789).

Many constraints, thresholds and limits in the area of building performance are rooted in human perception thresholds; for instance Perez *et al.* (2010) discuss the experiments to study vibrotactile thresholds, Loffler (2008) explores the perception of contours and shapes through the human visual system and Zhang *et al.* (2011) present work on air temperature thresholds for occupants in air-conditioned buildings. Understanding these thresholds is part of the domains of human physiology and psychophysics. Note that there can be both upper and lower limits and thresholds; for example, the human eye cannot see below a certain light level, but gets blinded beyond another. Similarly in for instance a restaurant setting, too much noise from talking and music can be annoying, but similarly a silence where one hears all others eat is not comfortable. Inside the borders that define an area of human comfort and well-being, there may also be an area that acts as a stimulus, such as a well-lit environment (Bluyssen, 2010).

Planning, building and safety regulations provide minimum requirements, which are defined in terms of thresholds and limits. However, requirements can be set beyond those required by codes and regulations (Bakens *et al.*, 2005). Becker (2005: 79) lists a generic 10-step process for establishing minimal performance levels from basics, which includes the definition of potential occupant and activity combinations for the building,

---

2 First law: one cannot create energy. Second law: some energy will always be wasted. Third law: one cannot reduce temperatures to the absolute zero. Bryson quotes Dennis Overbye (1991) expressing this as 'you can't win, you can't break even, and you can't get out of the game'.

identification of generalized loads that may adversely affect building performance and ultimately definition of when performance is deemed to be unsatisfactory or failing. This process demonstrates that constraints and limits not only apply to the performance of the building but also to the load that the building is expected to take. Another point made by Becker is that one may identify whether there are any allowed accepted percentages of stakeholders who remain dissatisfied; a target of 100% satisfaction is not always achievable or needed.

Thresholds and limits play an important role in structural engineering, where it is important to establish critical limit states for processes such as buckling of beams and columns (Honfi *et al.*, 2012; Chuang *et al.*, 2015). They are also of crucial importance in fire safety engineering, where building evacuation time is a prime concern and limits are imposed by regulations (Gupta and Yadav, 2004; Pursals and Garzón, 2009). Chow and Chow (2010) discuss the issue of limits to heat release rates of fires in residential high-rise buildings in the Far East; they relate this to the amount of combustibles that are stored in a typical residential flat. They also note that the new supertall buildings have problems meeting the mandatory evacuation time of 5 minutes. Laprise *et al.* (2015) discern between a limit value (V_L) as a minimum required score, an average value (V_A) for representing usual practice, a target value (V_T) to express ambition, and best practice value (V_B) to represent high performance.

Thresholds and limits need to be compared with care; for instance comparison between different countries may run into complications by slight differences in underlying standards. In other cases, it is the limits themselves that cause issues; for instance, studies of the impact of alcohol on road safety need to deal with different definitions of when a driver is considered to be impaired by alcohol, with different allowable thresholds of 0.0–0.9 g/l in place across Europe (Assum and Sørensen, 2010).

The use of thresholds and limits also may introduce a risk that people who are close, but not fully, at the goal start behaving dishonestly. This is especially the case where the goal is stated (and possibly rewarded) as an all-or-nothing goal (Welsh and Ordóñex, 2014); in such cases a gliding scale may be more productive.

For Gilb (2005: 341) a constraint is something that restricts a system or process. To qualify as a constraint, it must be explicit and intentional. Not meeting the constraint leads to some form of loss or penalty. In his work on decision making, Hazelrigg (2012: 79) defines a constraint as 'something that the decision maker does not get to decide about', such as the laws of physics. He believes that a constraint that is imposed to make sure a certain result is achieved is not a true constraint; this view may conflict with the notion of design constraints by Gilb. Hazelrigg (2012: 42) also points out that a constraint is a restriction; the introduction of constraints never improves the potentially best performance. Within that situation, there are two types of constraints. *Inactive* constraints take away some of the options, but not the optimum. *Active* constraints make it impossible to attain the optimal solution. Constraints may have their origin in a range of factors, such as limited resources (finance), regulations or competition. Not meeting the constraints leads to partial or complete project failure. At the same time, each project has constraints, so the best thing is to see constraints as design and engineering challenges (Gilb, 2005: 69). Constraints may relate to various parts of the life cycle; there may be specific design or engineering constraints or operational constraints; other constraints are important throughout the whole system life (*ibid.*, 70).

Some constraints are binary and can be defined in terms of 'must' or 'must not'. Other constraints can be scalar and defined in terms of lower and upper levels (Gilb, 2005: 341–342). Raphael and Smith (2003: 135), basing themselves on the number of variables that are included, discern between unary constraints, binary constraints, ternary constraints and N-ary constraints for functions with one, two, three or n arguments.

Gilb (2005: 118–133) identifies two ways of violating constraints. The first way is where the system misses some required level or condition, while it can still operate. In this case the system fails to meet the constraint but may still be feasible. The second way is where a system misses a required level that leads to total system demise and thus affects the survival of the system. Both failure and survival can depend on either lower or upper levels. The use of constraints to repair an objective function that is not valid may lead to dangerous results. A classic example is an objective function that states 'more is better' for something edible like chocolate. However, this is only true to certain point; yet setting a limit such as 'more is better, but only to the point of one pound a day' means that the constraint will impose the solution rather than the preference and optimization will always return the limiting value (Hazelrigg, 2012: 46–47).

Constraint-based reasoning is a branch of computer-aided engineering that exploits constraints to find solutions; often these solutions can be found on a frontier that describes where the constraint is met exactly, such as on a Pareto front (Raphael and Smith, 2003: 133). Constraint-based reasoning may aim to find a solution that meets all constraints but may also aim to identify all options in an option space that are ruled out by the constraints – or the other way round, all options that satisfy the constraints given (*ibid.*, 136) – thus reducing the search space.

A threshold is a point at which something starts to happen. The word is used to indicate the beginning of a runway and the bottom component of a door. In psychology and medicine, it is the point where some stimulus starts triggering a response. In construction, it is often used to define the border between what is desirable or not; for instance BS ISO 15686-10:2010: 5 defines a threshold level as a 'number indicating the level of functionality which, if not provided, would significantly or completely impair the ability of users to carry out their intended activities or operations'. In electrical systems, protective devices are used to prevent circuits from overload and short-circuiting. The common options are fuses and circuit breakers that open the circuit if the current increases beyond a given threshold value (Stein *et al.*, 2006: 1185–1187). Thresholds also play a role when deciding whether two data instances are similar. In information science, this is established by defining a similarity function and then setting a threshold value that establishes whether the two instances match or not (dos Santos *et al.*, 2011). Algorithms in the related domains of pattern recognition and information retrieval are typically measured in terms of precision and recall.

Thresholds that identify the difference between dissatisfying and satisfactory performance are often based on statistical data on human health and comfort. Other thresholds are based on data on human responses and perceptions (Becker, 2005: 70). Care must be taken when setting thresholds to account for the context in which these thresholds will be used; what is perfectly acceptable in one situation may be not in a different one (CIB Report 64, 1982: 19).

In the field of human perception of light, it is well known that the human eye contains two types of photosensitive cells: cones and rods. The cones allow humans to perceive colour but require higher lighting levels. Rods only pick up light intensity; they are 10–100 times as sensitive than cones and produce the image humans experience at night when they still can see but do not perceive colour such as at night (Geebelen, 2003: 83–94). In terms of hearing, humans can perceive pitch (frequency), loudness (sound pressure level) and timbre (combination of frequencies). Sound pressure level perception ranges from 0.20 mPa (0 dB) at the threshold of hearing to 20 Pa (120 dB) at the threshold of pain (Mommertz, 2008: 9). The impact of sound also relates to the duration of exposure, which is expressed in the long-term average sound level L_{eq}. (*ibid.*, 41). In the United Kingdom, action is required where personal noise exposure exceeds 80 dB (McMullan, 2012: 201). Where humans are exposed to too much noise, a distinction is made between effects in terms of temporary threshold shift (TTS) and permanent threshold shift (PTS); the effects of TTS normally recover, whereas the effects of PTS do not (McMullan, 2012: 187).

The perception of thermal comfort has been studied for a long time. A seminal publication in this area was *Thermal Comfort* by Fanger (1970), which establishes a steady-state heat balance to assess thermal equilibrium and expresses thermal comfort as a Predicted Mean Vote (PMV). The PMV is a function of air temperature, mean radiant temperature, relative humidity, air speed, metabolic rate and clothing. Thirty years later this was challenged by 'adaptive' thermal comfort models that also consider physiological acclimatization as well as behavioural and psychological adjustment (de Dear and Brager, 1998). Most adaptive models are based on empirical studies and express how acceptable indoor comfort temperatures relate to outdoor temperatures. Beyond comfort, the thermal environment may also venture into areas that start posing a risk to life. When the body core temperature of a human being becomes lower than 35°C, that person experiences hypothermia, which is to be treated as a medical emergency. Often this is caused by being in cold outdoor conditions for a long time or falling into cold water; however it may also be experienced in a poorly heated building. At the other end of the spectrum, body temperatures of over 38°C lead to hyperthermia; again this can become a medical emergency in the form of a heat stroke.

For indoor contaminants, thresholds are specified by means of the concentrations of these contaminants in the indoor air. What values are appropriate requires an understanding of how these concentrations relate to human health. This is a challenging area; not only are there many contaminants (over 80 000 chemical compounds have been listed), but moreover the direct experimentation on human health is mostly unethical. Additionally, the impact of these contaminants depends on the duration of the exposure. Upper limits or thresholds are often given in ppm, ppb, mg/m^3 or μg/m^3 (Awbi, 2003: 49–50). For some contaminants such as carbon monoxide, there is clear consensus on threshold levels; for others such as ozone, there still is a debate about appropriate limits (Morrison *et al.*, 2011).

Similar to thresholds, limits can be used to indicate the borders at the edge of a numeric scale. In this way, limits indicate when a performance enters the range of failure, and beyond that, where survival is at stake (Gilb, 2005: 374). Foliente (2005b: 111–112) notes the relation between parameters and acceptable limits in performance criteria. The performance parameter is objective and can either be measured or calculated. The acceptable limits are related to user expectations and may be subjective in nature. Limits can have multiple levels based on the user preferences in terms of quality and cost; this is captured by 'performance bands' that define classifications. Limiting factors are often stated in such a way that most, but not all, occupants of a building will remain satisfied; typical boundaries are 70% and 90% (Becker, 2005: 75).

The chance that a building fails, or that it does not meet the specified performance levels, is typically a matter of risk. This leads to the need to establish acceptable risk levels. Note that even when not defined specifically, such risk levels can also be implied and for instance be hidden in safety factors (Foliente *et al.*, 1998: 16). To cope with risk, requirements are sometimes adjusted to the severity and probability of events impacting on the building. An example is the requirements for earthquake resistance in structural engineering: here events with lower impact but higher probability should not impair serviceability, but for events with higher impact and lower probability the only requirement is the prevention of total collapse (Becker, 2005: 70). Given that providing full protection against earthquakes is never possible, the discussion here focuses on establishing risk and identifying 'how safe is safe enough'. This in turn requires an investigation of earthquake probability and consequences (Lin, 2012). Limits are used to discern between classes of consequences; typically the impact of an earthquake on building structures is classified as slight, moderate or extensive damage or complete collapse. The earthquake engineering community uses probabilistic calculations to establish the exceedance of limit states of the building in order to discern what the impact of a certain quake will be (*ibid.*).

An example of limits and constraints in the field of thermal environment is shown in Figure 5.5. Note that the thermal comfort band is relatively small; thresholds are drawn as straight lines, representing that control setpoints for building heating and cooling systems are often single values. Human survival limits are far away but not intangible. The figure also demonstrates the relation between the exposure duration and exposure level.

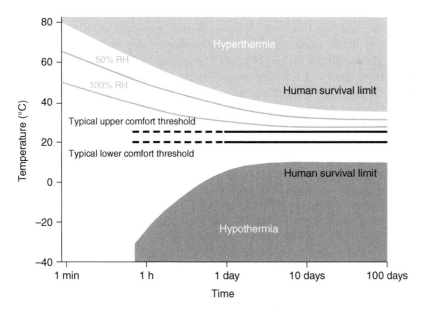

Figure 5.5 Thermal comfort and survival thresholds and limits. Redrawn from Krieger (1957) – with kind assistance of Karl Tate.

5.4   Performance Categories and Bands

The performance of products is often grouped in a class, label or bands. Such bands may be based on the use of numerical indicators such as strength levels, leading to a description such as 'flammable', 'fire retardant' or 'fire resistant'; however one can also group by simple properties such as weight, size and colours (Becker, 2005: 72). In other words, bands may be based on a nominal scale, ordinal scale or even an interval scale. A typical example is the energy labels that are applied to electronic consumer goods. A key issue in the use of bandings is the determination of the boundaries of each band (Nikolaou *et al.*, 2011). Beyond the use of such bands to inform consumers, bands may also be used to get an overview of the status of a design or product within a professional context. Here, if a range of performance measures is used, one can define a 'performance measure status' in terms of being within expected bandwidth or being in need of attention. The need for taking action on some performance aspects can be reflected by traffic light indicators – green: ok; amber: requiring monitoring; and red: requiring intervention (INCOSE, 2005: 45).

The use of performance bands has advantages and disadvantages. A positive aspect of bands is that they present a simple and relatively coarse approach, which is easily understood. CIB Report 64 (1982: 18) points out that many performance test methods have some degree of inaccuracy and that the use of banding prevents attempting evaluations that are 'accurate to the fourth decimal place'. For products, banded performance sometimes can be part of a certification process that guarantees a certain level (*ibid.*, 23). At the same time the use of bands lead to the introduction of borderlines between quality and performance categories, which may lead to discussion. It is also important to note that 'criteria to set the scale are subjective and, perhaps, closer to policy decisions than to technical analysis. Thus, there is a great disparity between different scales' (Pérez-Lombard *et al.*, 2009).

Bandings have a natural link to Likert scales as used in many perception questionnaires and surveys, typically using a 5- or 7-point scale (Fellows and Liu, 2008:171). A clear example is the Predicted Mean Vote (PMV) by Fanger (1970), which has seven categories: −3 (cold), −2 (cool), −1 (slightly cold), 0 (neutral), 1 (slightly warm), 2 (warm) and 3 (hot). Note that the PMV banding has a neutral position in the middle; other bandings such as product energy efficiency have a clear direction towards one side of the spectrum.

CIB Report 64 (1982: 19) recommends that banded levels should typically use three main bands, with an extra band at each side for extreme cases – giving five performance bands in total. They suggest naming these bands at the middle of the alphabet using (K), L, M, N, (O). This prevents too much of a bias at one end of the spectrum and allows for addition of other bands if needed later. Nordin *et al.* (2009) describe the development of what they name a 'category ratio scale' for environmental annoyance such as noise and air pollution. They stress the need to have reliable and valid comparisons when using subjective categories and point out earlier research that found that the use of simple numbers to represent categories may be skewed. As an alternative one can use a standardized scale with descriptors, with a seven-point scale being suggested as optimal. Wang *et al.* (2014) suggest the use of six performance bands: (i) *very bad*, for an object that is dysfunctional and need replacement; (ii) *bad*, if the object needs rehabilitation to get back to a satisfactory condition; (iii) *poor*, if the object needs major repairs; (iv) *fair*,

Figure 5.6 Typical EPBD performance bands.

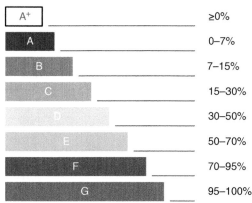

A⁺	≥0%
A	0–7%
B	7–15%
C	15–30%
D	30–50%
E	50–70%
F	70–95%
G	95–100%

if the object needs medium repairs; (v) *good*, where degradation has started and maintenance is needed; and (vi) *excellent*, if the object is like new and only needs preventive maintenance.

In the area of whole building energy use, the European Energy Performance of Buildings Directive (EPBD) sets limits to energy use by defining maximum allowed values for an Energy Performance Index (EPI); these limits are then used to grade buildings and to award bands from A to G (Pérez-Lombard *et al.*, 2009). The bands underlying the EPBD ratings are based on a cumulative distribution function graph that maps energy efficiency to a reference case. Buildings that use between 0% and 7% of the reference case are labelled as class A; 7%–15% are class B; 15%–30% class C; 30%–50% class D; 50%–70% class E; 70%–95% class F; and 95%–100% class G (*ibid.*); see Figure 5.6. Interestingly, research indicates that in Germany, the impact of Energy Performance Certificates on the purchase of properties is limited; while there is a generic awareness of the certificate, it ranks behind factors such as location, price, amenities such as garden, terrace or parking space, the size and condition of the building and the construction method used (Amecke, 2012). Similar results are found for the Netherlands (Murphy, 2014). Two recent developments in the EPBD banding are of special interest. In England, the original energy bands are now complemented by an environmental or CO_2 rating. This adds extra information but makes the labels less transparent to consumers. Moreover, it represents a case of dual representation of operational energy use. The use of classes A–G, contrary to recommendations by CIB Report 64, has caused a need for an additional class of buildings that are actually net energy producers; thus a special class of A+ has had to be incorporated. It is interesting to note that in most representations the energy producing category A+ is still projected along the same positive X-axis as the other categories, rather than pointing in the negative direction. Moreover, the spacing between the various categories typically is kept the same, ignoring the underlying limits.

In the BREEAM and LEED rating system, certification levels are also banded and related to the percentage of points scored across a range of aspects; see Tables 5.1 and 5.2. The position of the borders between the levels is relatively arbitrary, and no deep underpinning can be found in the literature. Work like that of Mansour and Radford (2016) may help to establish opinions of groups of building stakeholders, but this needs to be used with care when setting up weighting systems: one needs to keep in

Table 5.1 Banding for BREEAM certification.

BREEAM	
Score (% of available credits)	**Certification level**
<30	Unclassified
≥30	Pass
≥45	Good
≥55	Very good
≥70	Excellent
≥85	Outstanding

Table 5.2 Banding for LEED certification.

LEED	
Score (number of points obtained)	**Certification level**
<40	Uncertified
≥40	Certified
≥50	Silver
≥60	Gold
≥80	Platinum

mind that ultimately decisions are made by individuals and that there is a body of knowledge that questions group decision making (Hazelrigg, 2012: 227–243).

Chow and Ng (2007) show the use of banding in construction management using performance levels of poor, average, good, very good and excellent in their evaluation of the performance of engineering consultants. In this case, rather than predefined thresholds, they use the intersections of best fit lines in distributions gathered from a survey to set class limits.

5.5 Criteria in Current Practice

A building that exemplifies the challenges of juggling performance criteria in practice is the Davies Alpine Greenhouse at Kew Royal Botanic Gardens in the United Kingdom. This greenhouse was constructed as part of redevelopment of Kew Gardens that followed the registration as UNESCO World Heritage Site in 2003. It replaced an older structure that needed to make way for new laboratory space. The new Alpine Greenhouse was the first new greenhouse at Kew in over 20 years, needing to be attractive to visitors and meeting 21st-century expectations. The building was opened in 2006. It was designed by architectural practice WilkinsonEyre in collaboration with Dewhurst

Macfarlane & Partners (structural engineering), Atelier Ten (building services engineering) and Green Mark International (glasshouse engineering). Quantity survey-ing was carried out by Fanshawe; groundwork contractors were Kilby and Gayford.

The general brief for the Alpine Greenhouse requires a building for the display of alpine plants to the public, which fits in with the other innovative, high-quality glasshouses at Kew, such as the famous Palm House and Orangery. It asks for good accessibility, including wheelchair access, and low energy use. Architecturally, the building is to compete with contemporary projects such as the biomes at the Eden Project in Cornwall and the glasshouse of the National Botanic Gardens of Wales. The new Alpine house replaces an older greenhouse, which was kept cool by constant use of fans while shading was provided by whitewashing the glass; alpine plants have been kept at Kew Gardens since 1887.

Beyond these high-level needs, the key requirement for the new greenhouse is to provide suitable conditions for growing alpine plants in the UK climate. In general terms, alpine plants are dormant during a long and cold winter; they come to live during a rapid spring with high moisture levels and high light intensity and have a short grow-ing season of around 180 days. Growing on mountains, the plants typically are used to a thin atmosphere, meaning large amounts of sunlight. Experience from botanic gar-dens in the United Kingdom is that these plants can be grown under local conditions, provided they have extra ventilation over their leaves.

A first point of reference for the desired conditions is the definition of alpine climate; this represents the weather in mountainous areas that stretch between the tree line and the snowline. A well-used system for climate classification is the Köppen climate sys-tem; this considers alpine climates to be part of the Polar climate (Group E), with the mildest category defined as one where all months of the year have average climates between 0°C and 10°C. This exemplifies the design challenge, as in Surrey, England (which in itself has a temperate oceanic climate), only half of the year (November–April) fits in that range. Another way of defining the alpine climate is by biotempera-ture, which is an average of all temperatures with values below freezing adjusted to zero, as plants will be dormant below the freezing point. In the alpine zone, the biotempera-ture will be between 1.5°C and 3.0°C. Again, unsurprisingly the Surrey conditions are much milder, with the typical biotemperature calculated at 10.3°C. Unless active cooling (refrigeration) is used, which is at odds with the brief for an energy-efficient buildings, the local climate conditions thus become a constraint for the design and ultimate building performance.

Given that typical greenhouses have a higher indoor temperature in comparison with the outdoor situation, the decision was taken to set a design limit at 20.0°C. This is well above the average temperature of the warmest month in Surrey, which is 17.3°C in July; however it becomes a serious challenge for peak temperatures, with Surrey having 544 hours of temperatures over 20°C and a highest value that reaches 31.3°C.

The Davies Alpine Greenhouse meets this challenge by means of passive cooling. It consists of a glass building that creates a stack effect that moves air, which is pre-cooled while moving through a concrete underground labyrinth. This underground labyrinth combines heavy thermal mass with rippled surfaces to enable maximal heat transfer; it is pre-cooled with summer night ventilation and manages to produce air that is 7°C colder than typical ambient air on a warm summer day. The system has a small Air Handling Unit to support airflow. The stack effect is created by high-level vents in the

Figure 5.7 Davies Alpine Greenhouse, Kew Gardens, UK.

greenhouse, which are sized at about 20% of the floor area. Twin arches form the back-bone of the structure, which consists of an optimal glass area constructed from low-iron glass supported by stainless steel rods and clamps; silicon joints allow the structure to do without mullions, which would have caused significant shading. For solar control, each side of the glasshouse has retractable white polyester blinds that provide shading and prevent radiation losses at night; see Figure 5.7.

The Alpine house has won the following awards: RIBA Award (2006), Commendation for the Design Week Award 2006 and Civic Trust Award (2008); the building was shortlisted for the IstructE Award (2006) and the Mies van der Rohe Prize (2007).

5.6 Reflections on Performance Criteria

To work with building performance and to analyse whether demand and supply meet, one needs criteria that define what it is the building stakeholders need and expect, and what the building provides. These criteria are expressed in terms of goals, targets, aims, ambitions, thresholds, benchmarks and baselines – another area where there are many terms that look pretty similar but actually have different meanings. It thus is useful to be as precise as possible and where feasible to express demand in mathematical terms. This allows for experiments and measurement to unambiguously check whether or not the demand and supply match.

Most human needs are subjective to some degree. However, if these needs are trans-lated into exact computable criteria, then at least the check of whether the criteria are met is objective; any discussion is then about whether or not the criteria were set correctly and not about whether the building did meet the criteria. Still, criteria remain a complex area. Goals are binary and can either be achieved or not. Targets allow for a degree of success; this is especially suitable for needs where there is a diminishing return, as can be expressed in utility to the stakeholder.

Only few aspects of building performance can be observed directly; mostly the issues at the core of human needs require the collection of significant amounts of data that need aggregation to find out how the building is doing. For instance one can measure room temperatures, but a deeper evaluation of whether or not a building provides thermal comfort requires measurement over a range of conditions and integration over the different rooms in that building.

At present, many building performance criteria are formulated in a deterministic way – a building may achieve a certain energy efficiency, lighting level or sound isolation goal, or not. But increasingly criteria are now being formulated as targets that allow for some deviation and uncertainty, such as the criteria for adaptive thermal comfort. The prominence of binary goals may have its background in the fact that many performance criteria have their roots in building legislation, which enforces a lowest acceptable level and where the ultimate decision for the legislator is between accept and reject. Targets are more appropriate for handling aspirations to create buildings that perform well, giving direction as well as allowing for trade-off between various aspects.

To compare the performance of a building with its own track record (performance history) or that of other buildings (peers) requires baselines and benchmarks. Unfortunately data about wider sets of buildings is still surprisingly hard to obtain. There are some publicly available data sets, such as the ECON19 data in the United Kingdom or the CBECS data in the United States; however a lot of data is seen to be commercially sensitive and is not shared. Moreover sharing raw data has only limited use, as one needs once more to know exactly under what circumstances this data has been obtained to be able to interpret it. Comparison to other buildings is only useful if one can identify the true peers: buildings that have the same use and are subject to similar conditions.

The case study of the Davies Alpine Greenhouse at Kew Gardens shows the difficulties that are encountered when working with specific performance criteria. Even though the case study is a specific building with a clear focus on thermal aspects, the actual criteria and threshold values remain, to some extent, moving targets; they are dependent on what can be achieved with the design in the specific context.

5.7 Summary

This chapter explores the criteria that are necessary for the operationalization of the concept of building performance. Criteria take the form of goals and targets that allow a given performance to be judged against the aims and ambitions/aspirations of the stakeholders. Criteria allow the ordering of different alternatives in terms of preferences. Goals and targets are related to benchmarks as standards or points of reference. Typically they refer to the past performance of the building itself or to the performance of a group of similar buildings. Within such a peer group benchmarks may look at average/typical performance, above average/good performance or excellent/best practice performance. Another reference used in defining goals and targets is a baseline, which captures the standard case or situation as a starting point for making comparisons. Further points of interest in the definition of performance are constraints, thresholds and limits. Constraints are limitations or restrictions. Thresholds represent a level or point at which something starts to happen, while limits are terminal points and boundaries. Sometimes it is useful to group performance of objects in a limited number of bands; typical examples are the thermal comfort levels of the Predicted Mean Vote (PMV) as defined by Fanger or the different performance classes that are used in building energy performance rating and legislation such as the EPBD.

Goals and targets are used to define the aim or desired result. A difference is that a goal is either met or not; a target allows some sort of measurement of how close the result is to the ultimate aim. There is a large number of sources that discuss aim, objectives, goals and targets from a business point of view. In technical terms, goals and targets are often defined by means of a mathematical statement or 'performance indicator expression' that uses the symbols $>$, $\geq$, $=$, $<$ or $\leq$. It is important to note that performance objectives may change over time and that objectives within one and the same project may conflict. Goals may be hard (measurable) or soft (more difficult to observe). Where there are several goals, these may be combined into a 'goal pattern'. Goal setting is important in motivating stakeholders by directing attention, mobilizing effort, increasing persistence and motivating strategic development. Targets are mostly defined by way of numbers on a scale of measure. Within the scale one can be on target but also surpass demand by stretching the target towards the ultimate 'dream' values. Not achieving a target leads to failure and typically some kind of loss; some ranges of non-achievement threaten the survival of the system. Due to the probabilistic nature of production processes and uncertainty in use, statistical terms as percentile, standard deviation and confidence interval are often used in definition of targets. The drivers behind goals and targets are ambitions and aspirations of the stakeholders, which is a difficult area of research. In the building sector, goals and targets are used on a generic level; deeper understanding typically takes place at more subject-specific levels such as daylighting, room acoustics, thermal comfort and structural engineering.

Benchmarks provide a basis for evaluation or comparison; they relate to either past performance (history), a group of comparable buildings (peers) and represent the average, good or best possible performance for some sort of aspect. Benchmarking can be internal (one system/group), competitive (peer group) or industry-wide (whole sector). Benchmarking is also used for certification purposes. In physics, sometimes analytical or numerical solutions to some specific problems may act as theoretical benchmark. In the building sector, simulation models may be used to establish virtual data for comparison where there is insufficient measured data available. Typical data analysis techniques used to establish benchmarks are normalization, regression analysis, data envelopment analysis, bin methods, support vector regression, Gaussian process regression, artificial neural network modelling and the use of decision trees. Baselines capture the performance of reference behaviour, again for the purpose of comparison. In developing a baseline one aims to filter out the noise in order to obtain the typical performance. Baselines are often used to describe what the behaviour would have been without some sort of intervention; here the observed baseline is extrapolated to an adjusted baseline.

Constraints represent restrictions and limitations outside the control of the decision maker. Active constraints make it impossible to achieve some optimal performance. Thresholds are the points on a measurement scale where something happens and where a state changes; one of these is where a target moves from being met to being missed. Limits indicate edges of the scale; however they may be related to some percentage of occupants, or different levels of system excitation, accounting for the fact that perfect solutions are seldom viable.

In order to work with goals and targets, performance may be banded; such bands may be based on nominal scale, ordinal scale or interval scale. Banding is used in for instance thermal comfort assessment or in energy efficiency and environmental rating of buildings.

Recommended Further Reading:

- *Competitive Engineering* (Gilb, 2005) for an in-depth discussion of goals, targets, constraints and limits.
- *The International Performance Measurement and Verification Protocol* (IPMVP) for the leading industry work on baselines for building performance (Efficiency Valuation Organization, 2014a).
- The publications by Pérez-Lombard *et al.* (2009) and Zhang *et al.* (2015b) on benchmarking in the building performance context.

Activities:

1 Identify a building that has set a distinctive performance record, as well as one or two previous holders of the record in that same category.

2 Explore the contest criteria of the Solar Decathlon competition. Identify the underlying ambitions and aspirations of the competition organizers and analyse how these are translated into the main contests that make up the competition. Review how goals and targets are defined and what thresholds, limits and constraints are in place.

3 Develop a 'performance indicator expression' to set criteria for lighting levels in an office. Consider both visual performance and visual comfort. Underpin your work with references to relevant papers in journals such as *Lighting Research and Technology, Building Services Engineering Research and Technology, Energy and Buildings* or *Buildings and Environment*.

4 Install a simple domestic electricity monitoring system, with a clamp around the electricity supply and a display inside the house. Observe and register the electricity use in your home for the period of a day, week or month. What trends can be discerned? Use the data to establish your own electricity use baseline.

5 Explore the different approaches used in a fire alarm system. What properties of the building are being monitored in order to detect a fire? Find out what threshold values apply to each of these building properties. Review the risk for false alarms.

6 Set up a rating scheme, with goals, targets and thresholds, for students undertaking a course in building performance.

5.8   Key References

Ahuja, V., J. Yang and R. Shankar, 2014. Benchmarking framework to measure extent of ICT adoption for building project management. *Journal of Construction Engineering and Management*, 136 (5), 538–545.

Blyth, A. and J. Worthington, 2010. *Managing the brief for better design*. Abingdon: Routledge, 2nd edition.

California Commissioning Collaborative, 2011. *The Building Performance Tracking Handbook*. California Energy Committee.

Choi, D., 2011. Reactive goal management in a cognitive architecture. *Cognitive Systems Research*, 12 (3/4), 293–308.

Chung, W., 2011. Review of building energy-use performance benchmarking methodologies. *Applied Energy*, 88 (5), 1470–1479.

Deru, M. and P. Torcellini, 2005. *Performance metrics research project – final report*. Golden, CO: National Renewable Energy Laboratory.

Doerr, K. and K. Gue, 2013. A performance metric and goal-setting procedure for deadline-oriented processes. *Production and Operations Management*, 22 (3), 726–738.

Efficiency Valuation Organization, 2012. *International Performance Measurement and Verification Protocol: Concepts and options for determining energy and water savings, Volume 1*. Washington, DC: Efficiency Valuation Organization.

Efficiency Valuation Organization, 2014. *International Performance Measurement and Verification Protocol: Core concepts*. Washington, DC: Efficiency Valuation Organization.

Foliente, C., 2000. Developments in performance-based building codes and standards. *Forest Products Journal*, 50 (7/8), 12–21.

Foliente, G., 2005. PBB Research and development roadmap summary. In: Becker, R., ed., *Performance Based Building international state of the art – final report*. Rotterdam: CIB.

French, S., 1988. *Decision theory: an introduction to the mathematics of rationality*. Chichester: Ellis Horwood/Halsted Press.

Gao, X. and A. Malkawi, 2014. A new methodology for building energy performance benchmarking: an approach based on intelligent clustering algorithm. *Energy and Buildings*, 84, 607–616.

Gilb, T., 2005. *Competitive engineering: a handbook for systems engineering, requirements engineering, and software engineering using planguage*. Oxford: Butterworth-Heinemann.

Granderson, J. and P. Price, 2014. Development and application of a statistical methodology to evaluate the predictive accuracy of building energy baseline models. *Energy*, 66, 981–990.

Hazelrigg, G.A., 2012. Fundamentals of decision making for engineering design and systems engineering. http://www.engineeringdecisionmaking.com/ [Accessed 21 December 2017].

Hsu, D., 2014. Improving energy benchmarking with self-reported data. *Building Research & Information*, 42 (5), 641–656.

INCOSE, 2005. *Technical Measurement*. San Diego, CA: Practical Software & Systems Measurement (PSM) and International Council on Systems Engineering, Report INCOSE-TP-2003-020-01.

Jain, S., K. Triantis, and S. Liu, 2011. Manufacturing performance measurement and target setting: a data envelopment analysis approach. *European Journal of Operational Research*, 214 (3), 616–626.

Jordaan, I., 2005. *Decisions under uncertainty: probabilistic analysis for engineering decisions*. Cambridge: Cambridge University Press.

Krogerus, M. and R. Tschäppeler, 2011. *The decision book: fifty models for strategic thinking*. London: Profile Books.

Locke, E. and G. Latham, 2002. Building a practically useful theory of goal setting and task motivation. *American Psychologist*, 57 (9), 705–717.

Pérez-Lombard, L., J. Ortiz, R. González and I. Maestre, 2009. A review of benchmarking, rating and labelling concepts with the framework of building energy certification schemes. *Energy and Buildings*, 41 (3), 272–278.

Popova, V. and A. Sharpanskykh, 2010. Modeling organisational performance indicators. *Information Systems*, 35 (4), 505–257.

Pritchard, R., S. Weaver and E. Ashwood, 2012. *Evidence-based productivity improvement: a practical guide to the productivity measurement and enhancement system (ProMES)*. New York: Routledge.

Raphael, B. and I. Smith, 2003. *Fundamentals of Computer-Aided Engineering*. Chichester: Wiley.

Robinson, O., S. Kemp and I. Williams, 2015. Carbon management at universities: a reality check. *Journal of Cleaner Production*, 106, 109–118.

Shinkle, G., 2012. Organizational aspirations, reference points, and goals: building on the past and aiming for the future. *Journal of Management*, 38 (1), 415–455.

Smith, D., 2011. *Reliability, maintainability and risk*. Oxford: Butterworth-Heinemann, 8th edition.

Srivastav, A., A. Tewari and B. Dong, 2013. Baseline building energy modeling and localized uncertainty quantification using Gaussian mixture models. *Energy and Buildings*, 65, 438–447.

Welsh, D. and L. Ordóñex, 2014. The dark side of consecutive high performance goals: linking goal setting, depletion, and unethical behavior. *Organizational Behavior and Human Decision Processes*, 123 (2), 79–89.

Zhang, Y., Z. O'Neill, B. Dong and G. Augenbroe, 2015. Comparisons of inverse modeling approaches from predicting building energy performance. *Building and Environment*, 86, 177–190.

6

Performance Quantification

The previous chapters have expanded the notion of building performance as a measure that establishes how well a building meets user needs into a more extensive concept of what it is we want buildings to do and how that can be formulated as requirements. This has been related to quantification by means of experimentations and measurement and how performance can be captured through metrics, indicators and measures. Furthermore, a deeper background has been provided on goals and targets, benchmarks and baselines, as well as constraints, thresholds and limits. This provides the foundations to take the next step to analysis of building performance or turning back to the original starting point: towards the quantification of how well buildings fulfil requirements.

Quantification is the 'expression or measurement of the quantity of something', where quantity is an amount, number or other property of something that is measurable. To measure means 'to ascertain the size, amount, or degree of (something) by comparison with a standard unit or with an object of known size'. More in generic, it can also mean 'to assess the extent, quality, value or effect of something'. For a specific object or entity, it can indicate that something is of specified size or degree or reaches the required or expected standard. As a noun, a measure is 'a means of achieving a purpose, such as a cost-cutting measure; a standard used to express size, amount or degree; or an instrument such as a container, rod or tape marked with standard units and used for measuring' (Oxford Dictionary, 2010). In some cases, such as when establishing the length of an object, measurement can be a relatively straightforward process. However, measurement of performance typically involves a more complex context, where there is a need to manage a carefully designed experiment and where measurement involves the study of a range of observable states that may have to be aggregated into some overall metric. This requires an understanding of the scientific principles involved. However, measurement may also be bound by an industry and societal context and require knowledge of applicable laws, regulations and industry standards; at a lower level it may require the adherence to company procedures, standards, policies and directives and may have to respond to available infrastructure and project needs (INCOSE, 2003: 243).

In the context of systems engineering, Gilb (2005: 396–397) takes a more nuanced view of quantification. He limits quantification to the generic principle of 'articulating a variable attribute using a defined scale of measure and specifying one or more numeric levels on that scale'. This generic view sets quantification apart from measurement, since, as pointed out by Gilb, quantification can also be achieved via estimation,

Building Performance Analysis, First Edition. Pieter de Wilde.

feedback control and others. Gilb defines measurement as 'to determine the numeric level of a scalar attribute under specified conditions, using a defined process, by means of examining a real system' (*ibid.*, 375). This view of quantification fits well with the fact that building performance analysis can be approached through four main approaches: physical testing and measurement, calculation and simulation, expert judgment and stakeholder assessment. Apart from the technical quantification, these approaches also lend themselves to quantify process performance in the process management dimension, noting that in project management the focus is mostly directed towards effectiveness, efficiency and relevance (Marques *et al.*, 2010). Expert judgment and stakeholder assessment may – with care – also be applied to the aesthetic interpretation of performance.

Performance quantification through *physical testing and measurement* is often seen as the classic approach, with direct observation of the actual building under 'live' conditions. In some cases only a part of a building (a system or subsystem) may be subject to testing and measurements, which then may be done under the controlled conditions of a laboratory. Due the size, high cost and unique character of buildings, full laboratory tests on complete buildings are sparse; exceptions are mostly related to highly critical issues such as fire safety and resistance to earthquakes (Beji *et al.*, 2015; Li *et al.*, 2015; Wang *et al.*, 2015a; Ikeda, 2016). A term often encountered in the context of testing and measurement is *monitoring*; this relates to observations and measurements that take place over a period of time.

Performance quantification through *calculation* and *simulation* is another key approach for buildings. This has the advantage of enabling analysis without the need of physical access to a building; assessment can be done remotely and may also apply to virtual buildings that do not yet exist in reality, as often encountered in design projects. Formally, calculation is the 'mathematical determination of quantity or extent'. Simulation is similar but on a larger scale; it takes the mathematical equations into the computer (Shiflet and Shiflet, 2014: 376–377). Both calculation and simulation employ a mathematical model that consists of equations that describe the physical behaviour of a system (Basmadjian, 2003: 1). The main difference between calculation and simulation is an issue of what is still feasible through manual calculation; typically equations with up to 10 variables can still be solved by hand, whereas everything beyond that is best done through simulation. Note that building simulation can easily require the handling of 10 000 variables in a single model.

Performance quantification through *expert judgment* builds on the knowledge and experience of specific persons in an area of interest; often these experts are senior academics or consultants. De Wit (2001: 68–69) lists some of the principles that should be observed when making use of expert judgment, such as transparency (names of experts and tools used are made available for peer review), independence (experts have no vested interest in the assessment at stake), neutrality (methods that are used do not lead to bias) and empirical quality control (robustness study).

Performance quantification through *stakeholder assessment* has the advantage that it puts the building user in a central place; some argue that this is the ultimate way of assessment as buildings are created for their occupants (Preiser and Vischer, 2005: 10–11). However, it must be remembered that stakeholder assessment involves a subjective element; it involves the value that an individual attributes to something (Hazelrigg, 2012: 195). In that sense, stakeholder assessment presents a proxy of the actual building performance. The situation is further complicated by a generic practice where stakeholder assessment studies are often named post-occupancy evaluation (POE) studies.

In some cases, it makes sense to combine experimentation with computation. For instance, O'Neill *et al.* (2014) combine simulation and measurement for building monitoring and diagnosis of discrepancies. The Building Controls Virtual Test Bed (BCVTB) allows, amongst others, the testing of control hardware in virtual buildings (Wetter, 2011). Many papers such as Raftery *et al.* (2011) or Royapoor and Roskilly (2015) discuss the calibration of models with measured data. In the context of earthquake engineering, Maghareh *et al.* (2016) make the case for real-time hybrid applications.

Whatever the approach, CIB Report 64 (1982: 9) points out the need 'to always bear in mind the degree of simplification or uncertainty involved in practical data and methods, so that the apparent precision of a systematic, analytical approach does not mask the underlying approximations'. This again holds true across the different approaches to quantify building performance – for instance, in monitoring there may be a significant impact of the number, position and accuracy of the sensors used, whereas in building performance simulation a whole range of models with different accuracy and resolution can be used for a given analysis task. Assumptions and uncertainties also play a role in expert judgment and stakeholder evaluations.

In the world of systems engineering, further pointers are given for the measurement of system performance; however these recommendations apply to all forms of quantification, whether by means of measurement and monitoring, calculation and simulation, expert judgment and stakeholder assessment. The Systems Engineering Body of Knowledge recommends using consistent but flexible processes that are adapted to meet the specific information needs and characteristics of any particular process. Especially, it recommends that 'decision makers must be able to connect what is being measured to what they need to know'. Furthermore, it notes that 'measurement must be used to be effective' (SEBoK, 2014: 404). It is also good practice to base quantification on a formal strategy. Such a strategy should include (i) the definition of a repository that captures all quantifications that are produced; (ii) a formal quantification report that describes how the quantification was done, in terms of data collection/generation, analysis and evaluation, as well as the actual data; and (iii) an evaluation report that describes any findings about the quantification process and potential issues and improvements (INCOSE, 2003: 245).

Another category of building performance not covered by the above is that of the construction process performance. Here the typical issues of working to schedule, budget, preventing waste and maintaining health and safety are the prime concerns. This group of aspects can be measured as well. Data collection in this context is often empirical, by observing the state of the project against planning, cash flows, number of skips used or reported incidents amongst the workforce. However, calculation, expert judgment and surveys of project participants may also be used.

This chapter explores the four main approaches available for assessing the performance of a building as an object: physical testing and measurement, calculation and simulation, expert judgment and stakeholder assessment. As these are all extensive subjects, the emphasis is on positioning these approaches as tools within the emergent theory of building performance analysis. The chapter also briefly discusses the measurement of the construction process.

6.1 Physical Measurement

A direct measurement of building behaviour is the traditional approach towards assessment of building performance. A broad range of approaches and instruments has been developed to serve a multitude of interests by various stakeholders. For instance, the German Fraunhofer Institute for Building Physics showcases over 100 different measuring and test facilities on their website (Fraunhofer, 2016); in the United Kingdom, BSRIA[1] lists over 50 different product tests for various building services systems (BSRIA, 2016). In the Unites States, the National Renewable Energy Laboratory (NREL), National Institute of Standards and Technology (NIST), Lawrence Berkeley National Laboratory (LBNL), Oak Ridge National Laboratory (ORNL) and Pacific Northwest National Laboratory (PNNL) all offer extensive measurement facilities. As there is such a large number of existing techniques that are continuously being improved, and as new approaches keep being added, this paragraph cannot claim to give a definitive overview. Furthermore, many measurements have various levels of detail. The content of this section is thus to be seen as primer and a starting point for further investigation.

Physical measurement is based on the use of a range of instruments, such as anemometers, illuminance meters, motion sensors, flow meters, thermocouples and pyranometers. A brief overview of some widely used instruments is provided in Appendix D; for more in-depth coverage the ASHRAE *Handbook of Fundamentals* gives a detailed discussion of some of this equipment (ASHRAE, 2017: 37.1–37.40). It is important to keep in mind that instruments come in a wide range of qualities and prices. Sometimes measurement takes place in scale models; for an overview of applications, see Lirola *et al.* (2017).

A key issue in measurement is the accuracy of the instrument; calibration and adjustment help to reduce errors. Gillespie *et al.* (2007: 13), in the context of building monitoring, define accuracy as 'an indication of how close some value is to the true value of the quantity in question' and go on to say that the term can be used to refer to 'a set of measured data or to describe a measuring instrument's capability'. They define an error as 'deviation of measurements from the true value' (*ibid.*) Oberkampf and Roy (2010: 25) point out that accuracy can only be measured if a reference can be determined. The quality of measurement depends on the control of random errors and bias. As implied by the name, random errors simply mean there are deviations between measured values and true values but without a pattern; bias is a systematic error. In measurement of building performance, a typical source of random error is the quality of the measurement equipment; typical sources of bias are assumptions and analysis techniques (Efficiency Valuation Organization, 2014b: 1). Errors in measurement can be caused by the accuracy of the sensors, issues with data tracking, sensor drift and others. One way to limit these errors is regular – preferably planned – calibration of the sensors (*ibid.*).

The typical temperature loggers often used in building performance studies will have an accuracy of around ±0.2°C; however some may allow for variations as high ±0.6°C; these numbers can be significant when one is for instance trying to establish whether a room or building is overheating or not. For domestic electricity monitors, the accuracy

---

1 Building Services Research and Information Association.

of measured values is in the order of ±3 W; while this figure sounds high, in fact the difference is marginal when energy use is in the order of 2000 W to 3000 W. Gillespie *et al.* (2007: 14) note that not only sensors have accuracy but also any data acquisition and logging systems; they recommend a conversion accuracy of 0.1%. When selecting measurement equipment, other important factors to consider are the options for communications between devices from different vendors, data storage location and overall data storage capability; adding external storage and backup systems may increase costs (California Commissioning Collaborative, 2011: 43). Good measurement practice will repeat the test multiple times in order to get a measure for natural variation in observed values. Such repeated testing then establishes characteristic performance values, which can be expressed with some determined level of significance such as 0.05. This can be modified by adding a safety factor to set a performance requirement that must be met (CIB Report 64, 1982: 16). The quality of the measurement depends not only on the accuracy of the sensors but also on the exact installations of these sensors in the building, the data acquisition (logging) system, any data transfer (hardwired or wireless), calibration procedures used and analogue-to-digital conversions (Gillespie *et al.*, 2007: 1).

Measurement that takes place over a period of time is named monitoring; more detail on this activity, especially in the context of operating buildings, is provided in Chapter 9. However, for many measurement activities, the number and frequency of measurements is of importance. For the monitoring of energy use, Vesma (2009: 1) suggests weekly measurements as starting point; these give a higher temporal measurement than monthly data as typically used for billing, but less issues of data management as the typical high-resolution hourly or half-hourly data. Neumann and Jacobs (2008: 10) give an overview of data that is needed in the context of continuous commissioning of building services; they suggest capturing data on at least an hourly resolution. Their parameters cover energy use at different system levels, outdoor conditions (weather), indoor conditions and parameters that represent the system state. Such an approach is generally used in building monitoring projects. Bolchini *et al.* (2017) give an overview of some of the considerations that need to be taken into account when setting up a (thermal) monitoring campaign.

Many measurements have been standardized, with appropriate equipment and procedures described in dedicated norms, standards and protocols; others can be explorative and conducted in the context of research. For a good generic primer on developing novel experimental facilities, see Moore *et al.* (2009).

It is worth noting that not all quantities of interest can be measured directly. In the context of building performance and especially the saving of resources, the Efficiency Valuation Organization (2012: 7) points out that 'energy, water or demand savings cannot be directly measured, since savings represent the absence of energy/water use or demand. Instead, savings are determined by comparing measured use or demand before and after implementation of a program, making suitable adjustments for changes in conditions'.

6.1.1   Selected Physical Measurements and Tests

There are different ways to classify measurement approaches. One first distinction is between approaches that are typically used in the laboratory and approaches that are typically used in real or 'live' buildings; the latter is sometimes named 'on-site' or, using Latin, 'in situ'. Where measurements are done on actual buildings, some measurements may actually require the removal of samples from the building or require gaining access to spaces such as cavity walls that normally are hidden; other measurements do not require such activities and are classed as non-destructive techniques (NDT). Another distinction is between measurements that are part of the regular operation and control of the building, such as captured by the sensors of a building (energy) management system, and directed measurements that capture some aspect of building performance for a specific reason such as the investigation of faults or causes for complaints by occupants. Measurements may be categorized as whole building approaches or those that only deal with components and systems. Another categorization groups measurements per performance aspect, such as thermal, lighting, acoustical or indoor air quality. As all of these classifications are useful in their own way, this section presents a selection of some common test procedures in alphabetical order.

- **Air pollutants measurement**
  There are many pollutants that may impact the comfort and health of building occupants; identification typically takes place via physical, chemical and biological analysis. Measurement equipment requires an instrument to sample the air for pollutants; typically this is done with a membrane filter or precipitator. After a sample is taken, analysis may follow different routes, depending on the objective of the measurement. Particle numbers can be counted using microscopy, as well as specialist equipment such as an aerosol particle-size spectrometer (APS), condensation particle counter (CPC) or laser light scattering. Identification of gases or groups of gases requires specialist techniques such as chromatography. Odours may also be identified by human experts using an olfmeter apparatus. For some common pollutants such as CO_2, dedicated sensors have been developed, such as non-dispersive infrared (NDIR) and photoacoustic (PA) sensors. Typical measurement results include the number of particles per unit of air volume or mass of these particles per the same unit of air volume. For specific pollutants, their concentration is often expressed in parts per million (ppm); air quality is sometimes expressed in decipol (Awbi, 2003: 494–498). Examples of work using this approach are Stephens and Siegel (2012), who describe the use of air pollutant measurements to study the penetration of ambient particles into residential buildings in the United States, or Mentese *et al.* (2015), who explore the relation between indoor air pollution and respiratory health in Turkey. An example of work advancing the detection of hazardous volatile organic compounds (VOCs) is Leidinger *et al.* (2014). Marć *et al.* (2017) discuss the use of miniaturized test chambers to measure the emissions of floor coverings.

- **Building systems and component testing**
  Building components and systems, such as windows, facades and HVAC systems, can be subject to measurement before they are integrated into a building. Most manufacturers will have in-house facilities to test their own products. Measurement is also part of most product certification cycles, with independent third parties offering laboratory facilities and test processes that represent typical loads; often products are subjected to long stationary loads to identify standard behaviour, as well as repeated cycling to test durability. It is important to note that the impact of construction, weatherproofing and ageing might cause a product to perform different in real life

than under laboratory conditions (CIB Report 64, 1982: 16). Systems and equipment such as lifts and elevators, electrical circuits and systems, backup generators, emergency lighting, fire and smoke detectors, lightning protection, closed-circuit TV (CCTV) systems or public announcement installations all undergo serious testing by the manufacturer, as well as regular checks once installed in a building. There is a wide range of industry standards for such tests; for instance in the United Kingdom, Loss Prevention Standard (LPS) 1280 (BRE, 2014) provides testing procedures for smoke detectors. In many cases testing is mandated by law, and test certificates are to be held on file; this for instance applies to sprinkler, pressure and drinking water systems. Some examples of testing that are reported in the scientific literature are Schleibinger and Rüden (1999), who report on the testing of air filters in the laboratory. They compare both new samples and samples retrieved from operational HVAC systems using a specific chemical identification process (DNPH method). Raynor and Chae (2003) present similar work. Angrisani *et al.* (2010) report on experimental analysis of a small-scale polygeneration system. Carlon *et al.* (2015) report on the testing of a residential biomass boiler, albeit with data being used to improve models that represent this system. Demonstrating that even relatively simple systems may still need further analysis, Nilpueng and Wongwises (2010) report on an experimental investigation of a single plate heat exchanger. Van Den Bossche and Janssens (2016) discuss the results of airtightness and watertightness tests of windows and window frames.

- **Climate chamber measurement**
  Climate chambers, also known as environmental chambers, are used to create controlled indoor environmental conditions. This may be useful for a range of experiments and tests, varying from the reaction of construction materials and systems to human subjects on these conditions. The key set-up is of course the chamber itself; this can vary in size from a small compartment that holds sample materials (climate cabinet), actual room size, to larger volumes that can take larger building systems and assemblies. In many cases the chamber is constructed from stainless steel, ensuring high airtightness, is thermally insulated and has high quality seals on any openings. To maintain controlled indoor conditions, climate chambers have dedicated HVAC systems and controls. There is a high variation in specifications; typical systems will have a temperature range of 10°C–30°C and relative humidity of 10–90%; however there are also chambers that have a wider temperature range (−30°C to 60°C or more), whereas some chambers also reach higher relative humidity (up to 98%). Chambers will have air supply ducts, plenums, perforated floors, recirculation ducts and similar as required for specific applications. In larger rooms, extra fans may be needed to ensure good mixing of air. Furthermore, climate chambers need to be equipped with environmental data loggers; typically these measure temperature (T), relative humidity (RH) and carbon dioxide concentration (c). In some cases, use may be made of manikins to represent humans; advanced manikins may include systems to simulate breathing (by removing air from the room and injecting CO_2) and heat emission as provided by human bodies. For some situations, it is important to capture stratification in the room, leading to sensors being positioned at a number of heights, from floor level to ceiling. In other cases human subjects may be used to observe and assess conditions in the climate chamber. Test procedures and results obtained cover wide range; however it is always important to keep in mind that most climate

chambers need some time to reach the set conditions and for the indoor environment to stabilize. Some examples of the use of climate chambers are Wei *et al.* (2014b), who describe the use of an environmental chamber to study volatile organic component (VOC) and formaldehyde emissions from building materials. Tham and Pantelic (2010) use a climate chamber to study the performance of desk mounted fans and a desktop personalized ventilation air terminal device. Similarly, Dalewski *et al.* (2014) use a climate chamber to study ductless personalized ventilation (DPV) systems, a type of displacement ventilation (DV), while Hesaraki *et al.* (2015) use a climate chamber to study the thermal performance of low-temperature hydronic heating systems. Lan *et al.* (2011) studied thermal discomfort and its relation to perceived air quality, sick building syndrome symptoms, physiological responses and human performance using a climate chamber, with Leyten and Kurvers (2013) commenting on some aspects of this study, especially on the translation of experiments in a climate chamber to a free-running, naturally ventilated environment. Cui *et al.* (2013) use a climate chamber to study the impact of temperatures on the thermal comfort sensation, motivation and task performance of human subjects. Zhang *et al.* (2016) study the impact of exposure of human test persons to CO_2 at 5000 ppm, a typical exposure limit, on sensory discomfort, task performance and physiological responses. Liang *et al.* (2014) describe the use of large climate chambers that can be used provide a controlled outdoor environment for the testing of HVAC systems for road vehicles, which include simulation of high-speed wind and solar simulation. Latif *et al.* (2016) have developed a dual climate chamber that allows controlling hygrothermal conditions at two sides of a material being tested, similar to a hot box set-up but now adding control of the humidity. On the specific subject of manikins, van Treeck *et al.* (2009) have developed a multi-segment model and use this to validate advanced simulation for thermal comfort evaluation.

- **Co-heating test**
This measurement method establishes the heat loss of an unoccupied building under quasi stationary conditions. The instrumentation consists of electric resistance heaters to heat the building and ventilators to ensure an even temperature distribution throughout. Additional equipment includes temperature and humidity sensors and thermostatic control for the heaters. The test procedure is to use the heaters to achieve an elevated internal temperature that is roughly 10K higher than the outdoor temperature and maintain this for an extended period. In Western Europe, co-heating tests are typically undertaken during fall, winter and spring, with an internal temperature of around 25°C, and have a length of 1–3 weeks. During the test, heating energy Q_h and indoor and outdoor temperatures are captured, typically using a 5 or 10 minutes interval. Historical versions of the test included the use of the building services, which explains the name 'co-heating'. The data obtained from the test are used to compute an overall heat loss coefficient (*HLC*), based on the equation $Q_h = \text{HLC} \cdot \Delta T$. As ΔT will have varied over the test period, this typically is done via linear regression. Further detail on the co-heating process can be found in Bauwens and Roels (2014) or Farmer *et al.* (2016).

- **Durability tests**
Durability tests typically focus on the performance of materials and components and are mostly carried out in the laboratory. In general, they play a role in quality control and service life prediction of materials and components. Issues of interest are the risk

of corrosion, resistance to freeze–thaw cycles, response to solar exposure or durability in extreme conditions such as a marine environment. Equipment used varies highly with the type of testing. The generic process consists of the preparation of specimens that will be tested, conditioning of these specimens to reach the typical environmental conditions to which they will be subjected, repeated durability cycling that puts the specimen through the process it needs to endure, and ultimately inspection for durability. Durability testing may involve subjecting a sample to carbonation (exposure to water that contains dissolved carbon dioxide, as in 'acid rain'), exposing it to salt or salt water, freezing and thawing cycles, or possibly actual or simulated solar radiation. Typical approaches for the latter are the use of xenon light, fluorescent UV light or a carbon arc open flame. In some cases material ageing may be accelerated by artificial intervention, such as subjecting a specimen to a hydric environment, salt crystallization or another process. Steel, while a relatively well-known material, may still be studied under impact from aggressive environments such as those containing chloride. Wood, which is subject to decay, can be tested through exposure to moisture and a pure culture of selected fungi that may spread from feeder blocks that are in contact with the sample. A key aspect of durability testing is inspection after the cycling; the prime outcome of the tests is the count of the number of cycles that are conducted before a sample fails. Detailed inspection of specimens may involve visual inspection, sometimes using microscopy and regular structural tests but also specific approaches such as tests for the delamination of materials that are sensitive to that process. Inspection can also involve absorption/desorption tests, weighting using scales, or porosimetry to establish density, pore volume and size. Specialist work can employ ultrasound techniques to detect fractures inside a material and electromagnetic wave propagation measurements. Examples of durability testing are the work by Baroghel-Bouny *et al.* (2011), who review how a range of material properties, such as porosity, permeability and diffusion coefficient impact the durability of concrete with different supplements. Barbera *et al.* (2012) as well as Ludovico-Marques and Chastre (2012) study the durability of limestone under accelerated ageing; Duprat *et al.* (2014) study the durability of concrete structures under accelerated carbonation tests. Liu and Hansen (2016) have investigated the freeze–thaw durability of concrete under exposure to de-icer salt as common near roads. Youm *et al.* (2016) use rapid freeze–thaw cycles to study the durability of lightweight aggregate concrete. Sánchez de Rojas *et al.* (2011) describe the use of freezing and thawing cycles for assessment of the durability of clay roofing tiles. Shi *et al.* (2012) provide an overview of the durability of steel reinforced concrete in chloride environments. Villain *et al.* (2012) present the durability analysis of a concrete structure in a maritime environment (tidal zone). Singh *et al.* (2014) are developing accelerated tests to study the durability of framing timber when subject to decay. Ohnesorge *et al.* (2010) explore the bond durability of glue-laminated ('glulam') beams made of beech. Wolf (2008) reviews the test methods for the durability of wet-applied curtain-wall sealants. Johansson (2014: 17–20) discusses work to assess the durability of a specialist building component, a vacuum insulation panel, which comes with very specific properties and requirements and hence needs similarly specific test methods.

- **Evacuation experiments**
A complex area of analysis is the evacuation of buildings, as this involves the interaction between people, the building structure and environmental factors. As a

consequence, evacuation is often studied through experiments with human subjects in so-called 'fire drills'. It is well known that evacuation times depend on the physical characteristics of the building, physical and psychological characteristics of the evacuees and the characteristics of the fire or other emergency; for instance there are large differences between offices and hospitals and between people of different age, sex and health. Thus the selection of subjects for an evacuation experiment needs to be conducted with care. Evacuation has a clear behavioural and psychological dimension, as evacuation depends on social influences of neighbours, signal and information processing, the comprehension of evacuation plans and diagrams by the building occupants, and other factors. It may be delayed by actions such as having to wake up, collecting belongings, calling the fire brigade or even refusing to respond to an order to evacuate. Evacuation also involves human wayfinding from a known starting point, via intermediate points, to a final destination. This wayfinding requires the choice of an exit and is impacted by emergency signage and instructions. The equipment used in evacuation experiments typically consists of a system to give the evacuation cue, recording and observation systems such as video cameras, and stopwatches for timing. The cue can be provided by an existing building system, such as a fire alarm (sound or light), but this can also be a simple start sign by a human, or a more subtle cue like insertion of smoke into rooms. If the subjects are aware that they participate in an experiment, they may be asked to wear signs that help identification of individuals on any footage taken. In some cases they are also asked to wear masks or blindfolds to simulate low visibility. During the evacuation, conditions may be controlled, for instance by reducing lighting levels, introducing smoke, or obstacles in the evacuation path. Measurement typically establishes the premovement time (consisting of a recognition and response phase) and time needed to evacuate rooms, floors and the whole building. Other parameters captured or calculated may be the movement range of participants in a given time, average velocity, the velocity that is achieved in particular areas such as stairwells, density in key areas, flow rate at key points such as exits and stairs, and a generic space–time distribution. These can be used to establish the overall time needed for evacuation, or the required safe escape time (RSET), as well as the identification of evacuation bottlenecks. Some work in this area is provided by Shi *et al.* (2009b), who present a good overview of the many factors that play a role in emergency evacuation of buildings. Shields and Boyce (2000) report on a study of unannounced evacuation of four large retail stores in the United Kingdom. Fang *et al.* (2012b) studied the evacuation process in the stairwell of an eight-storey building using video recordings. Jeon *et al.* (2011) conducted evacuation experiments from an underground transport facility under different visibilities. Peacock *et al.* (2012) present evacuation data of eight office buildings in the United States. Ma *et al.* (2012) conducted evacuation experiments in the Shanghai World Financial Center, including one where participants had to transfer from stairs to elevators at a refuge floor. Huo *et al.* (2016) have observed the evacuation of a high-rise building via the stairs, studying both a phased and total evacuation of the building. Tang *et al.* (2008) present exploratory research into how humans read escape diagrams and plan their route, while Xie *et al.* (2012) study how people use the information provided by emergency signage during an evacuation. Nilsson and Johansson (2009) explore the role of social influence on evacuation behaviour in evacuation experiments in a cinema theatre. Andrée *et al.* (2016) blur the lines between physical and virtual experiments by studying the evacuation behaviour of people from an existing high-rise building in a virtual reality lab. Galea (2012) lists some of the challenges in the study of evacuation and pedestrian dynamics: the details of stair usage, impact of issues like age and gender on mobility on stairs, use of lifts in evacuations and group responses.

- **Experimental test cells or buildings**
The development of experimental buildings is not a standard measurement approach; instead this can be considered as prototyping. Yet many experimental buildings are closely monitored. Experimental buildings range from relatively simple temporary constructions to full-scale residential buildings. A classic example of experimental test cells is the PASSYS project (1986–1992), which developed high quality test facilities for solar components and facades across 12 countries in Europe. Each facility contains two to four identical test cells and a service room to the same specifications. PASSYS was later continued as the PASLINK network, and most test cells remain in place. For details, see Wouters *et al.* (1993), van Dijk and van der Linden (1993), Strachan (1993), Baker (2008) or Baker and van Dijk (2008). An example of a purpose-built temporary building is the HemPod, constructed by Shea *et al.* (2012) to test the hygrothermal properties of hemp line as a construction material, using the building for a co-heating test and comparing the findings with simulation results. Tong *et al.* (2015) describe the study of identical test buildings adjacent to a busy road that have been built for acoustical testing. Karlsen *et al.* (2015) use a test cell to study discomfort glare from windows. Brinks *et al.* (2015) use specially constructed test stands to measure the airtightness of light steel structures, whereas Ascione *et al.* (2016) have developed a test room to determine the performance of innovative building and HVAC components in a Mediterranean climate. At the other end of the spectrum, the buildings at the BRE Innovation Park in Watford, United Kingdom, are both a showcase for novel products and a test bed for the performance of novel technologies. A large series of experimental buildings has been created in response to the Solar Decathlon competition, organized by the US Department of Energy. After first events in 2002, 2005 and 2007, this is now a biennial event in the United States, with spin-off competitions in Europe and China. This is a special case of experimental buildings that require detailed measurements as part of the competition. Results are making their way into the scientific literature; for instance Osborne *et al.* (2010) discuss an evaluation of the Team Missouri Solar Decathlon 2009 entry, Hu and Augenbroe (2012) use the Georgia Teach entry to the Solar Decathlon 2007 competition as case study, Navarro *et al.* (2014) give an overview of the Solar Decathlon Europe (SDE) competition in 2012, Cronemberger *et al.* (2014) compare trends between SDE2010 and SDE2012 and Peng *et al.* (2015) discuss the performance of the Solark I house at the Solar Decathlon China 2013.

- **Façade test rig**
Building façades are a prominent and important part of buildings, and hence segments are often tested in special set-ups or test rigs. Here the façade segments may be subjected to outdoor conditions in a semi-controlled experiment where weather conditions are closely monitored; alternatively the test rigs may be positioned in a laboratory where conditions can be fully controlled. Experimental facilities almost always have a climatized box representing the indoor environment; if positioned outdoors the box may be able to rotate and tilt in order to control the angle of

incidence of sunlight. In laboratory conditions, the sun may be represented by a series of high intensity lamps such as xenon lamps. Wind may be simulated by a series of fans, or even a propeller driven by an aircraft engine. Pumps, tubes and nozzles may be used to spray water to represent rain. Measurement equipment typically includes thermocouples, heat flux sensors, humidity sensors and similar on the inside; often a weather station and pyranometer are used to capture climate conditions. Test procedures depend on the specific set-up and may range from a brief 5 minute test in the lab to an annual observation under outdoor conditions. The test may be stationary, requiring a stabilization period, under artificial cycling of varying conditions, or follow the natural outdoor conditions. Typically it captures the pressure difference ΔP between outdoor and indoor, rain/water spray intensity, solar radiation received and the regular parameters such as temperatures, relative humidity and heat flux. These can be used to classify outdoor conditions using indices such as the annual driving rain index (aDRI), directional driving rain index (dDRI) and others. Results can be used to determine the energetic performance of facades, as well as watertightness. A schematic drawing of a façade test rig to test for wind-driven rain is already found in CIB Report 64 (1982: 15). Recent examples of work in this area include Sahal and Lacasse (2008), who discuss water penetration tests for wall assemblies for different countries. Abuku *et al.* (2009) describe field measurements of wind-driven rain (WDR) obtained from a façade test facility and compare this with numerical simulations. Serra *et al.* (2010) describe the use of what they call an outdoor test cell to measure the thermal performance of a climate façade. Liu *et al.* (2014) study the impact of different control strategies on the performance of a glazed façade. Langmans and Roels (2015) discuss the study of full-scale test walls in an experimental facility in Belgium, which were subjected to tracer gas measurements, pressure measurements, wind speed measurements, and temperature and moisture monitoring. Grynning *et al.* (2013) study a glazing unit with four windows that include Phase Change Material (PCM) in a climate simulator, with xenon lamps supplying radiation to simulate the sun. Cuce *et al.* (2015) position façade test rigs inside an environmental chamber, studying thermal properties only. Meng *et al.* (2016) use a pneumatic cannon to test the impact of windborne debris on structural insulated panels. Blocken *et al.* (2013) provide a recent overview on research into wind-driven rain (WDR) and rainwater run-off from building facades, noting the complexity of predicting issues such moisture accumulation, frost damage, salt migration, discoloration and cracking. They suggest a strong need for further observational and experimental research in this area. Earlier work on the subject by the same authors can be found in Blocken and Carmeliet (2004). Pérez-Bella *et al.* (2013) discuss the role of exposure time in watertightness tests for facades, whereas Pérez-Bella *et al.* (2014) relate test methods to climatic records.

- **Fan pressurization test**
  This test method aims to establish the airtightness of the building envelope. The typical set-up consists of a fan (or set of fans) that is able to pressurize a building or room at a specified level and a manometer that measures air pressure. For practical reasons the fan is often positioned in a door that can be easily positioned in an existing door opening, leading to the colloquial name of 'blower door test' (BDT); see Figure 6.1. The test procedure employs the fan to create a pressure difference (ΔP) between the inside and outside; often a range of values is measured including 5, 10, 20, 30, 40 and 50 Pa.

Figure 6.1 Blower door test.

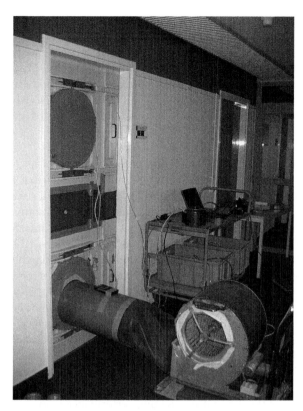

For each of these values, the airflow Q needed to maintain the pressure difference is measured. Data obtained from the test can be used to calculate an air leakage coefficient C. Metrics often used for comparison are the ACH_{50} value, which captures the required air change rate at 50 Pa, and the Q_{50} value, which captures the corresponding airflow in m^3/s. d'Ambrosio Alfano *et al.* (2012) provide an example of application to residential buildings in the Mediterranean region, Sinnot and Dyer (2012) in Ireland, while Chan *et al.* (2013) look at the US domestic stock.

- **Fire test (burn hall)**
  Fire tests are conducted to study issues such as initial flame spread in an enclosed space, the risk of occurrence of secondary ignition in interconnected spaces and general development of fires and smoke. For safety reasons these tests are typically conducted in a specialized laboratory. Equipment typically consists of a test room, sometimes with furniture and real content or more generic fuel packages. Fire may be ignited by means of a fuel pool fire or more specific causes (for instance when studying a fire resulting from an electric fault, this fault may be replicated in some detail). Dedicated equipment is used to measure temperatures, smoke development and air movement over time. Most of this equipment is specialized for the high temperatures in fires; for instance radiation is captured by Gardon gauges. CIB Report 64 (1982: 15) contains a figure that shows how the progress of a domestic fire through a door may be measured with different sensors on the door surface. Another common approach in fire test is Video Fire Analysis (VFA). Results from fire tests often include

histograms showing the spatio-temporal development of the fire and smoke and related parameters such as flame height, smoke layer height, temperature and fire velocity. Another crucial metric is the heat release rate (HRR). See for instance Beji *et al.* (2015). Note that due to the complexity of fire spread, fire experiments are also regularly used to support the development of fire models; see for instance Gutiérrez-Montes *et al.* (2008) or Xiao and Ma (2012). It is also worth pointing out that there may be many scenarios that can lead to fire; some of these can even be intentional (Nilsson *et al.*, 2013).

- **Full-scale lighting measurements**
  An area where full-scale experiments are common is that of lighting systems; one reason for this is the fact that criteria for glare and visual comfort are still under development, so the use of human subjects for assessment is common. Moreover, experiments may be better for the representation of a view to the outside, rapidly changing sky conditions and the impact of obstructions and reflections. The set-up typically consists of a test cell, the size of a regular office. Advanced experiments use two identical test cells, located side by side, where one cell is used to measure the impact of the sky conditions without interference, whereas the other then allows to do an identical measurement of the impact of some glazing or daylighting system of interest under exactly the same conditions. Equipment used includes illuminance meters, pyranometers and others as appropriate. Results typically include luminance levels (L), illuminance levels (E) and visual perception judgments of human occupants of the rooms. Some further work in this direction is provided by Li *et al.* (2010), who present measurements of a light-pipe system for a test facility in Hong Kong. Carletti *et al.* (2016) report on a combined analysis of the thermal and lighting performance of external venetian blinds. Lee *et al.* (2013a) use full-scale field measurements for a study of an advanced thermochromic window system, where behaviour of the system still needs to be identified and hence cannot be modelled. Kesten *et al.* (2010) use measurements in a full-scale mock-up to assess the performance of an artificial sky, providing an interesting contrast between the two approaches; Xiong and Tzempelikos (2016) present similar work.

- **Hot box measurements**
  The hot box method is a laboratory approach for the measurement of the thermal transmittance of building components such as windows, doors and façade panels. The set-up consists of two rooms or experimental chambers (boxes), with one kept at a high temperature, whereas the other is kept cold; the component being measured (specimen) is installed in between these two chambers. Thermocouples and heat flow meters are attached at specific positions of the specimen, as well as positioned strategically to measure the temperature of the adjacent air. See Figure 6.2. The key objective of the experiment is to measure the heat flow from the hot chamber, through the specimen, to the cold chamber. There are different designs to minimize the impact of heat loss in other directions. One is a 'guarded hot box', where the hot chamber is positioned in a larger chamber that is kept at the same temperature. Another method, named 'calibrated hot box', is to measure the heat flow through the other walls of the hot chamber, thus allowing to account for the heat flow that does not go through the specimen. The test procedure first allows for a stabilization period to ensure steady-state heat flow conditions are reached. Typically a time series is still measured,

Figure 6.2 Hot box measurement, showing sensors on specimen and in air void on cold side.

allowing the averaging of values and removal of outliers from the measurements. The thermal transmittance of the specimen, U_s, can then be found from the equation $Q_s = A_s \cdot U_s \cdot \Delta T_s$, where Q_s is the heat flow through the specimen, A_s the area and ΔT_s the temperature difference between hot and cold chamber, taking into account heat transfer coefficients at both surfaces (Asdrubali and Baldinelli, 2011). Applications of hot box measurements are described by for instance Martin *et al.* (2012) and Kus *et al.* (2013).

- **Indoor climate analysis**
  Occupant well-being is related to indoor air quality (IAQ) and, taking a wider scope, indoor environmental quality (IEQ), which includes other aspects such as lighting and sound levels. Typically a range of parameters is of interest; often these are measured in a concerted effort. Instruments often used to capture the thermal environment are a globe thermometer, dry-bulb thermometer, relative humidity sensor, anemometer and net radiometer. Sometimes thermocouples are added to directly record surface temperatures. Other aspects may be captured by adding a sound meter with analyser, an illuminance meter, and sensors to register CO and CO_2 concentration and the presence of particulate matter. Note that some manufacturers offer multifunctional instruments that can capture a range of these parameters. In most cases this equipment is connected to data loggers. Measurement may also be combined with collection of stakeholder evaluation by means of surveys. Results are the typical parameters such as temperature (T), relative humidity (RH), carbon dioxide concentration (CO_2), presence of volatile organic compounds ($VOCs$), and a count of

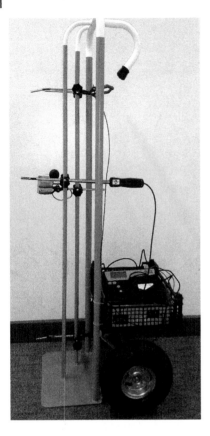

Figure 6.3 Thermal comfort measurement trolley. Image courtesy of Shen Wei.

particulate matter. For work in this direction, see for instance the noteworthy report by Newsham *et al.* (2012: 14–19) who have developed two dedicated sensor platforms that can be positioned at occupant work stations to collect data on a wide range of parameters. One platform takes the form of a trolley that contains 17 instruments; some of them capture the same parameter but at different positions. The second platform is less complex and contains 7 instruments to capture 7 distinct parameters. Application in 19 different office buildings is reported. A simple indoor climate analysis trolley is depicted in Figure 6.3. Wang (2006) reports on the use of an indoor climate analyser to study the thermal comfort in residential buildings in China. Della Crociata *et al.* (2012) report on the study of the IEQ of a hypermarket in Italy. Lee *et al.* (2012) study the IEQ in air-conditioned university rooms and correlate this to learning performance of students in Hong Kong; Yusoff and Sulaiman (2014) report similar work for Malaysia. Saha *et al.* (2012) study the indoor air quality in kitchens in India. Wang *et al.* (2015b) measured IEQ in airport terminal buildings in China. Maciejewska and Szczurek (2015) contribute a study on the required duration needed to capture indoor air quality.

- **Moisture measurement**
  Moisture can have different negative impacts on buildings – it may reduce thermal insulation, cause decay, lead to visible stains or contribute to mould growth.

Hygrometers allow monitoring of relative humidity inside rooms; however the measurement of moisture inside construction elements is a more complicated matter. In many cases such measurement requires some damaging of the construction, either to insert equipment or to take samples for analysis in the laboratory. Moisture content of construction materials can be obtained via gravimetric measurement, electrical resistance measurements or the use of hygrothermal sensors to establish the relative humidity. Most moisture meters make use of the fact that moisture content changes the electrical properties. A key aspect is the positioning of electrodes of such meters. The electrodes themselves may take the form of insulated or non-insulated pins that are inserted directly into the fabric being measured or attached to the surface of a sample of ceramic material or a wooden disc (which is in equilibrium with the surrounding building fabric). Some electrodes are simple metal pins or screws that can be left inside the material and that have a permanent position in the building structure. The water content in a concrete mix may be established using rapid heating, based on thermal response, or advanced technology such as ground-penetrating radar. A range of standard test methods exist for measuring the water ingress into materials like concrete: the Karsten-Tube test, contact angle measurement, infrared spectrometers, water-repellent meters and a sorption test method. Thermography may be used to detect temperature changes due to differences in moisture content. In a laboratory setting, moisture content can also be established using a precision balance to register the change of weight of a system. For materials like straw, moisture measurement may take place in the lab, using a sample removed from the construction, or in-situ: indirectly via the relative humidity (RH) in the wall, indirectly by measuring the moisture content of timber embedded in the straw or directly by measuring the electric resistance of the straw itself. Results of moisture measurement yield moisture content (w) and moisture diffusion coefficient (D), as well as sometimes a hysteresis effect[2] of construction materials. For further information, Saïd (2007) provides a literature review of different methods for the measurement of moisture in building envelopes. Brischke *et al.* (2008) present an in-situ study of moisture in the wood of a building and a pedestrian bridge. Kääriäinen *et al.* (2001) use microwave transmission to study the moisture content of concrete, wood and sand samples. Vrána and Björk (2008) present an experimental set-up that uses precision weighting to establish the moisture content in insulation materials when subject to certain temperature conditions. Carfrae *et al.* (2010) present an overview of approaches to measure moisture content of straw bale walls and go on to describe development of a moisture probe. Carfrae (2011) presents a full PhD thesis on the subject, with one section providing more detail on moisture measurement in general (*ibid.*, 82–85). Chen *et al.* (2012) describe the use of ground-penetrating radar to establish water content in fresh concrete mixes. Weisheit *et al.* (2016) assess techniques to measure water ingress into concrete.

- **Monitoring of buildings in regular use**
  Obviously, measurement in full-scale buildings that are in regular use is a key source for data on building performance; an in-depth discussion of this area will follow in Chapter 9. Here it is just mentioned that there is a whole range of measurement

---

2 A situation where the history has an impact on the state of a system; with moisture content of materials this may depend on whether the material is drying out or not.

processes available to capture things like the use of electricity, gas, water and heat, to establish temperatures, vibration levels, daylighting access, artificial lighting levels, sound levels and moisture transport in buildings, building structures and systems. If readings are taken repeatedly over time, measurement becomes monitoring. Buildings themselves may also be subject to evacuation tests, evaluation of occupant satisfaction and others as appropriate. A specific type of testing of buildings is the verification of whether all systems and components function as designed, which leads into the field of commissioning. Another thing that requires measurement in real buildings is the interaction between occupants and the building and its systems; this may involve the monitoring of occupant presence, interaction with control systems and things like the status of windows and doors (open or closed, angle, what percentage of opening). The equipment required for monitoring obviously relates to the aspects of interest; often specific meters are installed for the purpose. The level of metering equipment in buildings varies hugely, with some buildings having none, while others have detailed metering at subsystem level in place (Neumann and Jacob, 2008: 14). Often there already exist systems that may be accessed for monitoring purposes, such as utility meters and the sensors of a building (energy) management system (BMS or BEMS). An important issue in monitoring is to capture the loading of the building, for instance through the use of a weather station, observation of building occupancy and use and similar. According to Cohen *et al.* (2001), the integrity of the building fabric of a building is one of the few performance aspects of a building that can be measured objectively, when combining a range of techniques such as heat flux sensors and pressurization tests. To provide a flavour of the breadth of monitoring work, Corgnati *et al.* (2007) report on the monitoring of a building with an active transparent façade in Italy. Marsh (2008) reflects on the use of demonstration projects in the development of renewable energy systems. Nastar *et al.* (2010) report on measurements on a building in the United States that has been exposed to an actual earthquake. Zisis *et al.* (2011) study wind-induced pressures on a test house in Canada. Blocken *et al.* (2013) note the need for extensive observation to gain a better understanding of the complexities of rainwater run-off from buildings. Çelebi *et al.* (2013) provide an example of monitoring work that analyses the impact of ground and wind motion on a 64-storey high-rise building in California.

- **Scale model experiments for lighting**
  Architectural scale models, in combination with an artificial sky or sun, can be used to study solar access, daylight penetration into buildings and the impact of apertures, shading devices and advanced daylighting systems. The equipment required is, first and foremost, an artificial sky or sun simulator, although sometimes the real sky and sun can also be used. Illuminance sensors are used to capture lighting levels,[3] whereas cameras can be used to capture photographic images. Furthermore, this method requires a scale model that represents the building. Scales used are typically 1:10, 1:20 and 1:50 for devices, components and rooms; 1:50, 1:100 and 1:200 for daylight penetration into a building; and 1:200 or 1:500 for massing and shading studies. In making models, camera and sensor positions need to be considered, and access to these must be enabled. Models may also be equipped with observation openings.

---

3 The science of measuring light as visible to the human eye is also known as photometry.

Materials must be selected that appropriately represent the final finish, and care must be taken to prevent gaps and unwanted material transparency in the model. Measurements obtained from scale models are typically luminance levels (L) and illuminance levels (E). A quotient of indoor illuminance and outdoor illuminance is often used to calculate a daylight factor (DF), where $DF_j = Ei_j \ / \ Ee_j$ for each zone j. Bodart *et al.* (2007) discuss some of the issues that need to be taken into account when building models for use in scale studies of lighting, whereas Bodart *et al.* (2008) expand on photometric and colorimetric material properties. Kesten *et al.* (2010) as well as Thanachareonkit and Scartezzini (2010) report on work that compares scale models, a full-scale mock-up, as well as simulation studies. Saraiji *et al.* (2015) present work that uses scale models under a real sky. Dubois *et al.* (2007) describe the use of scale models to study the impact that coatings applied to glass have on human perception.

- **Sound measurements**
Building science has a range of specific tests that study the acoustical properties of materials, components and rooms. The principal instrument for sound measurements is obviously a sound meter that can be used to establish the environmental sound level in a room (N or NC value) and to measure airborne, impact or flanking sound (dB) as well as reverberation time (RT). Often the sound meter is connected to an external analyser, and specific cards or modes are used to arrive at the desired values. The sound that is measured may be naturally occurring, such as the noise from HVAC equipment, traffic or people walking on floors or stairs; however sound may also be artificially generated and for instance can be created by a system consisting of a noise generator, amplifier and loudspeaker. Depending on the needs, the loudspeaker may be a specialist omnidirectional version. Further sound sources can be a tapping machine (light impacts) or a bang machine (high impacts); this is already depicted in CIB Report 64 (1982: 15). In specialist laboratories, special rooms are used to establish material properties. A reverberation room, equipped with a loudspeaker and a microphone, is used to identify the sound absorption coefficient (a) of a material sample. A pair of reverberation rooms (sometimes know as transmission suite), with one room containing a loudspeaker and a microphone and the other room containing a microphone only, can be used to establish the sound insulation in terms of the sound reduction index (SRI) of a sample positioned in an opening between these two rooms. Anechoic chambers are used to study the relation between a sound source and receptor without interference of reflections or echoes; this is for instance of relevance when measuring loudspeaker properties. Typically such rooms or chambers are boxes that are isolated from the rest of the building structure to prevent structure-borne sound from impacting the measurement. Often several microphones are positioned in both the outgoing and ingoing chamber in order to obtain measurements of the sound field in the room. Structure-borne sound may be measured using accelerometers. The measurement process of sound is relatively fast, apart from measurements that need to average sound or noise over a certain period of time; in-situ there may be a need to schedule the measurement with the occurrence of the sound (such as traffic). Results include sound pressure levels (SPL), the relative loudness of a space (the noise criterion (NC) is used in the United States, whereas the noise rating curve (NR) is used in Europe) and reverberation times. More specific detail may be provided by calculating signal-to-noise ratio (SNR), equivalent sound levels (L_{Aeq}), or by creating a graph of the SRI over frequencies, measured by octave

band (typically 100–3150 Hz). For more details, see for instance Ismael and Oldham (2005), who use a scale model to study the impact of façade properties on the scattering of sound; Tang (2010) also uses scale models to study the screening effect of balconies on a façade. Januševičius *et al.* (2016) describe measurements to establish the sound reduction index of environmental friendly materials such as clay, straw and reeds. Scheck and Gibbs (2015) present a detailed study into structure-borne sounds resulting from lightweight stairs bolted to dwelling walls. Larsson and Simmons (2011) present the findings from a round-robin test[4] where structure-borne sound from a washing machine as well as tapping machine was measured in six different laboratories. Roozen *et al.* (2015) use laser Doppler vibrometry as an advanced method to obtain low frequency measurement of sound transmission and vibration. Schiavi *et al.* (2015) discuss the accuracies of impact sound reduction measurement of floor coverings in the laboratory. Bibby and Hodgson (2013) present a study of the acoustical performance of internal natural ventilation openings and silencers employing the two-room transmission set-up. Velis *et al.* (1995) and Sun *et al.* (2009) describe the development of anechoic chambers; Famighetti *et al.* (2006) study the acoustics of a reverberation chamber with fibreglass diffusers. Mahbub *et al.* (2010) describe a detailed acoustic evaluation of an office in Singapore, combining physical measurements with plan analysis, expert walkthrough and a stakeholder evaluation. Pinho *et al.* (2016) provide an example of acoustic field measurements, studying the reverberation time, sound isolation between rooms and sound isolation of the façade of schools in Portugal.

- **Structural testing in the laboratory**
  Structural testing is a common type of measurement, which is not surprising as this is closely intertwined with the safety and stability of buildings. Components and complete structures are subjected to load testing, impact testing, tensile testing, fatigue testing and others. Some of the aspects tested for include the analysis of the impact by a small hard body, impact by a soft large body, door slamming, crowd pressure, material fatigue and fracturing. Depending on what system or component is being tested, as well as the performance aspect of interest, a whole array of equipment is available, such as universal testing machines (UTM) for tensile and compressive strength, loading systems (actuators/rams), structural test rigs, testing tables (sometimes named 'shaking table'), loading jigs for bending and servo-hydraulic machines for static and dynamic loading. Structural testing typically takes place in a laboratory that has specific 'reaction floors' (and sometimes walls), which are designed to resist the high forces that may result from experiments. Sometimes special internal hoist facilities are in place to move heavy structural components around. Dedicated measurement equipment may include extensometers and strain gauges. Measurements may also be conducted by advanced techniques such as tomography, radiography, ultrasonics and thermography. Test procedures vary, but in general loads are applied at some point in time, and then either kept in place, slowly increased or cycled according to a predefined testing regime. Results include maximum loading capacity, such as the ultimate tensile stress (UTS), stress (σ) and strain (ε) relationships, and failure patterns/mechanisms. A large body of knowledge is available in the domains of solid

---

4 A round-robin test involves a range of laboratories doing exactly the same test independently in order to verify a test method, or to assess reproducibility.

and structural mechanics and material science that pertains to such tests; often these are specialist works that study a specific construction component or material. CIB Report 64 (1982: 15) gives general examples of dynamic and static loads (soft blow, knock, scrape) that may occur in actual buildings and experiments that may represent these conditions in controlled experiments. By way of example of the many specialist papers in this domain, Delhomme *et al.* (2010) have studied the effects of fatigue stress on anchoring bolts that attach metal components to a concrete structure. Narmashiri *et al.* (2012) conducted a failure analysis of carbon fibre reinforced polymer (CFRP) strengthened steel I-beams. Illampas *et al.* (2014b) present research into the response of adobe bricks to compressive loading. Kozłowski *et al.* (2015) present testing of the load-bearing capacity of timber–glass composite shear walls and beams. Xu *et al.* (2016) provide an example of the development of a fatigue test machine, which also shows the movement towards hybrid systems that combine testing of critical components with simulation of the remainder of the structure.

- **Thermography**
  Thermography, also known as infrared or thermal imaging, employs a special kind of camera that is sensitive to thermal radiation. Originally this required an expensive specialist camera; fortunately the price of these cameras is going down, with more affordable instruments becoming available in the market. Some manufacturers now also offer extensions to smartphones that allow these to capture infrared images at relatively low cost. However, the resolution of most products available on the market remains intentionally low, with high-resolution instruments limited to military use. Thermography can be used in different ways. A major difference is between active thermography, where a heat pulse is used to excite the building fabric, and passive thermography, which captures only 'natural' variations. When applied to buildings, further differentiation depends on the access to the building and the speed of the assessment – it can be limited to only external data capture but may also include observations from inside the building; furthermore it may range from a single image captured during a drive-by or a series of images captured during a longer session. When using thermography, outdoor conditions (which are outside of control of the thermographer) need to be observed, as wind, rain and previous exposure to direct sunlight may strongly impact results; similarly the distance to the building and angle of observation play important roles. Note that the instrument captures thermal radiation; using assumptions about surface emissivity, this can be converted to apparent temperatures. Infrared cameras do not allow direct observation of heat flow through the building fabric. Thermal images are used to identify areas of excessive heat loss but can also used for other purposes such as the identification of pipe and duct locations, delamination of materials or identification of faults in electrical systems. Balaras and Argiriou (2002) and Lo and Choi (2004) provide a broad introduction to the use of thermography for building defect detection. For more recent and in-depth reviews, see Kylili *et al.* (2014) or Fox *et al.* (2014). Ohlsson and Olofsson (2014) present work that explains some of the issues in establishing the heat flow rate through the building envelope from thermographic images. Barreira *et al.* (2016) discuss the use of thermography to study moisture content of materials. Spodek and Rosina (2009) discuss the application of thermography in historic buildings and conservation. Ham and Golparvar-Fard (2013, 2014) have combined thermography

with three-dimensional (3D) visualization to study thermal insulation and condensation problems. Fox *et al.* (2015) have explored the use of time-lapse thermography using a series of thermal images to study transient effects with the aim of detecting thermal defects such as moisture patches and missing insulation. Taylor *et al.* (2014) suggest the combination of thermography and computer simulation for a better identification of cold bridges; Bauer *et al.* (2016) present work to improve the understanding of thermal images by way of controlled laboratory studies using tiles and mortar. An example of a thermographic image is provided in Figure 6.4, showing a UK residence in winter, with heat loss around the lintel and through a vent over the window.

- **Tracer gas tests**
  The tracer gas technique (TGT) is used to establish air change rates and leakage from buildings. The equipment required consists of a gas bottle, mass flow controller and nozzle to supply the tracer gas, and some sort of detector that can establish the presence of the tracer gas. Sometimes tubes are used to insert gas or take samples at various locations at the same time. There is a wide range of tracer gases, such as CO_2, N_2O, helium and ethane. The process of the tracer gas test consists of injecting gas into the space and measuring the concentration over time. This allows for three measurement principles. The first injects a certain amount of gas, then stops the injection and measures how the concentration drops over time; this is referred to as the decay method. The second method establishes some constant rate of injection and then monitors how the concentration develops over time. The third method establishes what injection rate is needed to keep the concentration constant. In all cases, the experiment requires good mixing of all gases in the space. The key resultant of the measurement is a gas concentration (c) or a gas concentration change rate (dc).

Figure 6.4 Thermographic image of a domestic property in the United Kingdom. Image courtesy of Matthew Fox.

Results can be used to calculate the air change rate (AER or ACH), and, since it is known where measurements have been taken, iso-level contours in plans or sections. Tracer gas tests are also used to capture decay time, presence of leaks and the local mean age (LMA) of air or gas in the space; they can also be used to study the dispersion of pollutants. For more detail see Awbi (2003: 128–133). Some examples of tracer gas tests are Han *et al.* (2011), who use carbon dioxide, introduced in the supply air inlet, to study the local mean age of air in livestock buildings. Samer *et al.* (2011, 2012) present similar work. Lo and Novoselac (2012) have included tracer gas measurements in a study of cross-ventilation in an actual building, combining this with analysis of wind properties, façade pressures and airflow rates through window openings. Nikolopoulos *et al.* (2012) study the impact of the incidence angle of wind on natural ventilation of a building, using both tracer gas tests and numerical simulation. Afonso (2015) applies tracer gas studies to both domestic and non-domestic buildings and explores how this can be used within building zones, appliances such as refrigerators and in HVAC ducts, thus giving a good overview of application at different scales. Batterman *et al.* (2006) combine the tracer gas technique with the measurement of volatile organic compounds in homes, garages and vehicles. Shen *et al.* (2012c) present work that aims at finding optimal sampling positions, thus helping to enhance the tracer gas methodology.

- **Wind tunnel measurement**
A wind tunnel is an experimental device that creates an airflow moving past models of objects such as buildings; typically the airflow is recirculated. While automotive and aerospace application may require high-speed versions, wind tunnel use for the built environment is typically low-speed wind engineering. As unwanted turbulence would impact measurements, fans that propel the air are mounted downstream of the observation space, thus drawing in air over the model. Ducts are smooth, and vanes and dampers may be used to smooth out unwanted turbulence in the intake or recirculated air. The working area may be large, with sizes as large as $22 \times 4.5 \times 2.5\,\mathrm{m}$ reported in the literature; this working area needs access and a viewing port. Various equipment may be installed in the working area to measure air velocities and patterns, such as hot-wire or hot-film anemometers, pulsed wire anemometers and Irwin probes; alternatively more expensive systems may be used such as laser Doppler anemometry, particle image velocimetry (PIV) and flame ionization detectors (FID). Airflows may also be visualized using smoke, oil or fog/vapour in the airflow, or tufts on the model surface; another option is the use of scour techniques such as sand erosion. For a detailed description, see Blocken *et al.* (2016). Experiments consist of inserting a scale model in the wind tunnel and then creating a representative wind flow. Normally this means relating the wind speed at a meteorological station (U_{pot}) versus wind speed at the actual building site (U_0). For many conditions, it also requires to create an atmospheric boundary layer using vorticity generators and roughened surface, representing wind flowing over the Earth's surface. Realism may be increased by including buildings and other objects upwind of the test building in the model. Depending on the equipment used, results vary from visualized airflow patterns, streakline plots and identification of turbulence zones to measured wind speeds (v) at specific locations and heights, wind pressure on the building (P) and time series showing how these develop over time. As examples of specific work in this area, Law *et al.* (2004) use wind tunnel experiments to study how pollutants may move around low-rise industrial buildings. Mfula *et al.* (2005) study the impact of air pollution

sources on buildings in terms of concentration patterns at the surface, combining wind tunnel testing with tracer gas analysis. Bu *et al.* (2010) have used wind tunnel tests to study the ventilation of sunken courtyards in an urban setting, while Ji *et al.* (2011) use the approach to explore the impact of fluctuating wind direction on natural cross-ventilation. Another wind tunnel study on cross-ventilation, with a focus on the impact of opening positioning, is provided by Tominaga and Blocken (2016). Zhang *et al.* (2015a) use a climatic wind tunnel to study building evaporative cooling; the climatic wind tunnel is a wind tunnel that also allows the control of air temperature, humidity and solar irradiation. Blocken *et al.* (2016) evaluate the use of wind tunnel tests, as well as some computational fluid dynamic techniques, for the analysis of pedestrian-level wind around buildings.

6.1.2 Standards for Physical Measurement

A wide range of standards is available that specifies details on physical measurements of buildings and building components. A prominent source of such standards is the International Organization for Standardization (ISO); a selection of some of these standards is provided in Table 6.1. Further standards are provided by the American Society for Testing and Materials (ASTM), which now operates internationally; a selection of their standards is shown in Table 6.2. Some of the test methods that are provided by the European Committee for Standardization (CEN) are listed in Table 6.3. A small selection of test methods defined by the US National Fire Protection Association (NFPA) is provided in Table 6.4; a selection of standards from other organizations is listed in Table 6.5.

Further standards are provided by a wide range of bodies and institutions, such as the Air Tightness Testing and Measurement Association (ATTMA), the Chartered Institution of Building Services Engineers (CIBSE), the International Union of Laboratories and Experts in Construction Materials, Systems and Structures (RILEM), the International Commission on Illumination (CIE) and others. Many international standards by ISO and CEN are adapted by national standardization bodies, such as the British Standards Institution (BSI), German Institute for Standardization (DIN), Dutch Standardization Institute (NEN) and others; such standards may carry a double or triple coding such as BS EN ISO.

There are many papers that report on studies that follow specific measurement standards. For instance, Kalz *et al.* (2009) demonstrate the assessment of thermal comfort in twelve buildings in Germany on the basis of comprehensive monitoring and calculation of thermal comfort classes in according to the applicable EN 15251:2007 and EN ISO 7730:2005 standards. Berardi *et al.* (2014) use NFPA 286 for fire testing of novel fibre reinforced polymer wall panels as an alternative to regular ASTM E84 testing. Limbachiya *et al.* (2012) use standard BS EN 12390-3:2002 and standard ASTM C1260 in their studies of granulated foam glass concrete. Li *et al.* (2014a) discuss the classification of skies according to the standard sky distributions as defined through the CIE Standard Skies. Turner and Awbi (2015) include air leakage tests according to ATTMA Technical Standard 1 in their investigation of a hybrid ventilation system. Liu and Novoselac (2015) use the ANSI/ASHRAE Standard 55 to capture thermal comfort in their work on diffusers. Sameni *et al.* (2015) refer to CIBSE TM52 on overheating, as well as BS EN 15251, in their work on temperatures in flats built to Passivhaus standard in the United Kingdom.

Table 6.1 Overview of some ISO standards on building (component) measurement (selection).

Subject	Applicable ISO norm/standard
Fire resistance of glazed elements	ISO 3009:2003
Laboratory test of air permeability of joints	ISO 6589:1983
Thermographic (infrared) detection of irregularities	ISO 6781:1983
Mechanical tests for windows and door height windows	ISO 8248:1985
Calibrated and guarded hot box measurements	ISO 8990:1994
Adhesion/cohesion properties of sealants	ISO 9046:2002
Performance test for concrete floors: non-concentrated load	ISO 9882:1993
Performance test for concrete floors: concentrated load	ISO 9883:1993
Fan pressurization method	ISO 9972:2015
Measurement of room acoustic parameters – perf. spaces	ISO 3382-2:2009
Measurement of reverberation time – ordinary rooms	ISO 3382-2:2008
Measurement of room acoustic parameters – open plan	ISO 3382-3:2012
Determination of sound power and energy levels (noise)	ISO 3746:2010
Measurement of the vibration of structures	ISO 4866:2010
Measurement of fluid flow	ISO 5167-1/2/3/4:2003
In situ measurement of thermal resistance/transmittance	ISO 9869-1:2014
Measurement of impact sound insulation	ISO 10140-2:2010
Thermal transmittance of glass – guarded hot plate method	ISO 10291:1994
Laboratory measurement of flanking sound transmission	ISO 10848-3:2006
Measurement of radioactivity in the environment: radon	ISO 11665:2012
Ignitability of building products	ISO 11925-3:1997
Tracer gas dilution method for airflow rate determination	ISO 12569:2012
Determination of moisture content of building materials	ISO 12570:2000
Destructive testing of joints	ISO 12996:2013
Resistance of thermoplastics to elevated temperatures	ISO 13257:2010
Reaction-to-fire tests for facades (large scale)	ISO 13785-2:2002
Determination of declared thermal conductivity	ISO 13787:2003
Determination of water absorption by partial immersion	ISO 15148:2002
Measurement of sound insulation (field measurements)	ISO 15186-1:2003
Watertightness under dynamic pressure	ISO 158321:2007
Sampling strategy for determination of airborne asbestos	ISO 16000-7:2007
Determination of local mean age of air in buildings	ISO 16000-8:2007
Emission of VOCs from building products	ISO 16000-9:2006
Sampling strategy for moulds in indoor air	ISO 16000-19:2012
Field procedures for testing geodetic/surveying instruments	ISO 17123:2005

(Continued)

Table 6.1 (Continued)

Subject	Applicable ISO norm/standard
Airborne sound insulation	ISO 16283-1:2014
Impact sound insulation	ISO 16383-2:2015
Façade sound insulation	ISO 16283-3:2016
Resistance of glass to shock-tube loading	ISO 16934:2007
Determination of curing behaviour of sealants	ISO 19861:2015
Determination of water–vapour transmission properties	ISO 21129:2007
Determination of adsorption/desorption properties	ISO 24353:2008
Reaction to fire test for sandwich panel systems	ISO 13784:2014
Determination of bending behaviour of insulation products	ISO 13788:2012
Quasi-static reversed-cyclic test for timber structures	ISO 16670:2003
Tests for glass subjected to destructive windstorm	ISO 16932:2016
Measurement of conductivity of wet porous materials	ISO 16957:2016
Measurement of antibacterial activity on plastics	ISO 22196:2011
Determination of resistance to axial withdrawal of screws	ISO 27528:2009
Simplified seismic assessment of concrete buildings	ISO 28841:2013
Test methods for vibration emissions of power tools	ISO 28927:2009

As per July 2016, the ISO website lists 431 standards that pertain to buildings. http://www.iso.org/iso/search (building).

Table 6.2 Overview of some ASTM standards on building (component) measurement (selection).

Subject	Applicable ANSI norm/standard
Measurement of sound impedance by tube method	ASTM C384-04 (2016)
Measurement of absorption by reverberation room method	ASTM C423-09a
Test for volatility of channel glazing compounds	ASTM C681-14
Test method for weathering of elastomeric sealants	ASTM C793-05
Measurement of flame propagation for gaskets	ASTM C1166-06 (2011)
Strength test of panels for building construction	ASTM E72-15
Static load testing of truss assemblies	ASTM E73-13
Visual assessment for lead hazards in buildings	ASTM E225/E225M-13
Test method for wear testing of rotary window operators	ASTM E405-04 (2012)
Flexural tests on beams and girders	ASTM E520-04 (2011)
Test for frost/dew point of sealed insulation glass units	ASTM E546-14
Test for shear resistance of framed walls	ASTM E564-06 (2012)
Determining air change rate by means of tracer gas dilution	ASTM E741-11
Measurement of vibration-damping properties of materials	ASTM E756-05 (2010)

Table 6.2 (Continued)

Subject	Applicable ANSI norm/standard
Determining air leakage by fan pressurization	ASTM E779-10
Tests for metal railing systems and rails for buildings	ASTM E935-13e
Test method of deglazing force on fenestration products	ASTM E987-88 (2009)
Test for glass breaking probability under static loads	ASTM E997-15
Measurement of outdoor A-weighted sound levels	ASTM E1014-12
Test for measurement of speech privacy in open plan space	ASTM E1130-08
Practices for air leakage site detection	ASTM E1258-88 (2012)
Determination of load resistance of glass in buildings	ASTM E1300-12ae1
Determination of steady-state transmittance of windows	ASTM E1423-14
Outdoor sound measurement using digital analysis	ASTM E1503-14
Testing methods for bond performance of anchors	ASTM E1512-01 (2015)
Test for structural performance of sheet metal roof systems	ASTM E1592-05 (2012)
Serviceability of office for sound and visual environment	ASTM E1662-95a (2012)
Serviceability of office for location, access and wayfinding	ASTM E1669-95a (2012)
On-site electrochemical/spectrophotometric lead test	ASTM E1775-16
Measurement of the strengths of structural insulated panels	ASTM E1803-14
Determination of airtightness using an orifice blower door	ASTM E1827-11
Test for electromagnetic shielding effectiveness	ASTM E1851-15
Test method for impact by missiles and cyclic pressure	ASTM E1886-13a
Test for tensile breaking strength of exterior insulation	ASTM E2098/E2098M-13
Evaluation of water leakage of building walls	ASTM E2128-12
Water penetration by static water pressure head	ASTM E2140-01 (2009)
Test for accelerated ageing of electrochromic devices	ASTM E2141-14
Measurement of equipment-generated continuous noise	ASTM E2202-02 (2009)
Test for air permeance of building materials	ASTM E2178-13
Test method for determination of sound decay rates	ASTM E2235-04 (2012)
Measurement of argon concentration in sealed glass units	ASTM E2269-14
Test for transverse and concentrated load on floors/roofs	ASTM E2322-03 (2015)
Durability testing of duct sealants	ASTM E2342/E2342M-10 (2015)
Method for field pull test of in-place wall assembly	ASTM E2359/E2359M-13
Determination of the resistance of glass to thermal loading	ASTM E2431-12
Test for freeze–thaw resistance of exterior insulation	ASTM E2485/E2485M-13
Measurement of impact resistance of finishing systems	ASTM E2486/E2486M-13
Test for particulate matter emissions from wood burning	ASTM E2558-13
Determination of air leakage through windows, doors and walls	ASTM E2830-04 (2012)
Determination of forced entry resistance of windows	ASTM F588-14
Fire test response of mattresses and furniture	ASTM F1550-16

As per July 2016, the ASTM website lists 426 standards that pertain to buildings. https://www.astm.org/Standards/building-standards.html.

Table 6.3 Overview of some CEN standards on building (component) measurement (selection).

Subject	Applicable EN norm/standard
Testing of grout for pre-stressing tendons	EN 445:2007
Test methods for building lime	EN 459:2010
Testing of gypsum plasterboards	EN 520:2004
Test methods for water pipelines	EN 545:2010
Determination of U-value of glass by guarded hot plate	EN 674:2011
Determination of U-value of glass by heat flow meter	EN 675: 2011
Method to test the active soluble salt content of masonry	EN 772:2016
Test for flexural and shear resistance of lintels	EN 849-9:2000
Test methods for air permeability of windows and doors	EN 1026:2016
Test methods for water tightness of windows and doors	EN 1027:2016
Test methods for glass in buildings	EN 1036-2007
Method to test air tightness of joints in plastic piping	EN 1054:1995
Test for elevated temperature cycling of plastic pipes	EN 1055:1996
Test of resistance of glass against bullet attack	EN 1063:1999
Test methods for joints in timber structures	EN 1075:2014
Measurement of self-cleaning performance of glass	EN 1096: 2016
Test methods for panic exit devices in escape routes	EN 1125:2008
Test methods for temporary work equipment	EN 1263:2014
Test methods for cylinders for locks	EN 1303:2015
Fire resistance tests for non-load-bearing walls	EN 1364:2015
Fire resistance tests for load-bearing elements	EN 1365:2014
Tests for fire dampers in building services installations	EN 1366:2015
Tests for valves/hydraulic safety groups	EN 1487:2014
Test methods for hardware for sliding and folding doors	EN 1527:2013
Aerodynamic testing of dampers and valves	EN 1751:2014
Test methods for suspended access equipment	EN 1808:2015
Testing of air handling units	EN 1886: 2007
Test methods for level handles and knob furniture	EN 1906:2012
Testing of wind resistance of external blinds and shutters	EN 1932:2013
Testing of single-axis hinges	EN 1935:2002
Thermal measurements by guarded hot plate method	EN 1946-2:1999
Thermal measurements by heat flow meter	EN 1946-3:1999
Thermal measurements by hot box method	EN 1946-4:2000
Thermal measurements by pipe test method	EN 1946-5:2000
Assessment of resistance against earthquakes	EN 1998:2005
Determination of bending behaviour of insulation products	EN 12089:2013

Table 6.3 (Continued)

Subject	Applicable EN norm/standard
Test for maximum service temperature of insulation	EN 12097:2013
Requirements and tests for mechanically operated locks	EN 12209:2016
Test methods for pigments for colouring of cements	EN 12878:2014
Performance testing of residential ventilation systems	EN 13141:2014
Tests for electrically controlled exit systems on escapes	EN 13637:2015
Testing of apertures of letter boxes and letter plates	EN 13724:2013
Test methods for suspended ceilings	EN 13964:2014
Laboratory measurement of walking noise on floors	EN 16205:2013
Measurement of ventilation flow on-site	EN 16211:2015
Tests for protection against strangulation by internal blinds	EN 16433:2014
Laboratory tests for water-repellent products	EN 16581:2014
Acoustic test for drywall systems	EN 16703:2015
Determination of propensity to undergo smouldering	EN16733:2016
Test methods for strength of sheared connections	EN 16758:2016
Method to measure tobacco exposure	EN 16789:2016

As per July 2016, the CEN website lists 1006 approved and published standards for buildings. https://standards.cen.eu.

Table 6.4 Selected NFPA standards for fire protection testing of buildings and building products.

Subject	Applicable NFPA norm/standard
Integrated fire protection and life safety system testing	NFPA 4-2015
Testing of water-based fire protection systems	NFPA 25-2014
Tests of fire resistance of constructions and materials	NFPA 251-2006
Test for resistance against cigarette ignition	NFPA 260-2013
Test of flame travel and smoke of wires and cables	NFPA 262-2007
Measurement of ignitability of exterior wall assemblies	NFPA 268-2012
Establishment of toxic potency data in fire hazards	NFPA 269-2012
Test of thermal barriers	NFPA 275-2013
Fire test of individual fuel packages	NFPA 289-2013
Fire flow testing and marking of hydrants	NFPA 291-2016

Table 6.5 Selection of standards from other organizations.

Subject	Applicable norm/standard
Field performance measurement of fan systems	AMCA 203-90 (R2011)
Protocol for radon decay measurement	ANSI/AARST MAMF-2012
Measurement of ambient noise level in a room	ANSI/ASA S12.72-2015
Measurement, testing and adjusting of HVAC systems	ANSI/ASHRAE 111-1988
Method of test for residential thermal distribution systems	ANSI/ASHRAE 152-2014
Safety glazing materials – methods of test	ANSI Z97.1-2015
Laboratory measurement of sound insulation of elements	AS 1191-2002 (R2016)
Measurement of airborne sound transmission	AS 2253-1979
Laboratory measurement of room-to-room sound insulation	AS/NZS 2499:2000 (R2016)
Wind tunnel test procedures for buildings and structures	ASCE MOP 67-1999
Sampling and testing of mastic asphalt used in buildings	BS 5284:1993
Guidelines for measurement of areas in healthcare	CAN/CSA Z317.11-2002 (R2013)
Measurement of daylight	CIE 108-1994
Measurement of reflectance and transmittance	CIE 130-1998
Test of fire behaviour of building materials and elements	DIN 4102-1:1998
Measurement of strength of non-load-bearing clay blocks	DIN 4160:2000
Measurement of vibration emission	DIN 45669-1:2010
Measurement of the whiteness of building materials	GB/T 5950-1996
Test method for building sealants	GB/T 13477.16-2002
Test for smoke generation from burning materials	GB/T 16173-1996
Fire resistance test for structural elements	JIS A 1304:1994
Lab measurement of sound insulation of building elements	JIS A 1428:2006
Measurement of sound insulation in buildings	JIS A 1441-1:2007
Performance test of sorptive building materials	JIS A 1905-2:2015
Measurement of sound and vibration of HVAC systems	SMACNA SSVPG-2013

Where AMCA is the Air Movement and Control Association, ANSI the American National Standards Institute, AARST the American Association of Radon Scientists, ASA the Acoustical Society of America, AS an Australian Standard, NZS a New Zealand Standard, ASCE the American Society of Civil Engineers, BS a British Standard, CAN/CSA the Canadian Standards Association, CIE the International Commission on Illumination, DIN a German Standard, GT/T a Chinese Standard, JIS a Japanese Standard and SMACNA SSVPG stands for the Sheet Metal and Air Conditioning Contractors National Association Systems Sound Vibration Procedural Guide.

6.2 Building Performance Simulation

Computer simulation offers a second method to quantify building performance. This approach uses a computer program to imitate reality (Shiflet and Shiflet, 2014: 376). Simulation offers some unique opportunities for experimentation and measurement: it is a key approach that can be applied if the object of study does not yet exist, for instance during design; it allows testing of variants under completely identical conditions, which may be hard in real experiments; it often allows testing at a lower cost in terms of time, money and potential danger; and it may be the only feasible option at some scales of study, such as urban, regional and global levels (*ibid.*, 277). Augenbroe (2011: 15) positions building simulation as a 'virtual experiment', which is especially suitable to analyse and quantify the performance of two or more competing design alternatives.

In a general sense, simulation is reproduction or imitation of something. However, to differentiate from general animations, visualizations and representations, building performance needs to be based on the use of a computer-based mathematical model that accounts for fundamental physical principles and that follows good engineering practice. In building simulation this model represents some part of building reality. The model can then be exposed to a number of scenarios or experiments, where each experiment involves the monitoring of observable states, which in turn help to understand the performance of the model when subject to the scenario at hand (Augenbroe, 2011: 16). As a virtual experiment, the essence of setting up a valid simulation is similar to that of doing physical tests; it requires a carefully planned process where the experimenter manages the experimental set-up (building, systems, measurement points), the experimental conditions or system excitation, and adheres to a measurement protocol that establishes what data is captured (temporal and spatial resolution: where and when). The difference between engineering calculations and computer simulation is one of complexity. Calculations can be done by hand and may have in the order of 10 variables; computer simulation easily deals with 10 000 variables.

In order to imitate reality, computers execute a computational model, which is built on an abstraction of physical phenomena and entities (Zeigler *et al.*, 2000: 26; Malkawi and Augenbroe, 2003: 5; Hensen and Lamberts, 2011: 12–13). Simulation thus goes hand in hand with modelling: the specification of the system that is to be simulated in terms of 'a set of instructions, rules, equations or constraints' (Zeigler *et al.*, 2000: 29). Modelling and simulation belong to the domain of scientific computing; for a general overview and history, see Oberkampf and Roy (2010: 1–15) or Shiflet and Shiflet (2014: 3–6). The words 'modelling' and 'simulation' are sometimes interchanged in discussions; generally a model is an abstraction or description of reality, whereas a simulation is the execution of that model in the computer (INCOSE, 2015: 181).

Building simulation models come in many different forms, covering a wide spectrum of capabilities. They can run on the basis of a short list of inputs or require extensive modelling efforts; some stem from a research background and are mainly used in academia, while others are provided as commercial tools with full back-office support and dedicated training provisions. The underlying mathematical models may have different properties and be linear or nonlinear, static or dynamic, discrete or continuous and deterministic or stochastic. Models may cover different physical processes, such as heat and mass transfer, acoustics, light or structural behaviour. They may represent a small part of the building, such as a one-dimensional section through a wall, or a complex combination of geometry, materials and systems representing a full building. One common way of classifying building simulation models is by their temporal dimension and distinguishing between stationary, semi-dynamic[5] and transient models. A

---

5 Semi-dynamic models sometimes are also named quasi-stationary models.

look at thermal models shows how the study of dynamic effects leads to increased computational expenses. A stationary model of heat transport through a wall requires the mathematical solution of one set of equations. In a semi-dynamic model one may for instance use 12 sets of conditions to represent the 12 months of the year, which means the same equations need to be solved twelve times. By moving to transient conditions, often studied by looking at hourly values, this increases to 8760 steps per year.

Another way of categorizing simulation models distinguishes between black box, grey box and white box models. This world view stems from the world of system identification; in all three categories, the model links a series of inputs with outputs. In the black box model, the relation between input and output is based on machine learning that captures the correlation by learning from correlated data pairs; there is no knowledge of the internal workings of the system that causes this correlation. Typical techniques used are regression analysis and neural networks. In a grey box model, there is a certain insight in the internal workings of the system that allows to do predictions; however these models have unknown properties and relations that need to be estimated. In a white box model all governing principles are known, allowing for explicit modelling of the relation between input and output. Sometimes white box models are also known as glass models; these models typically represent the state of the art in a certain domain. In order to support the development of white box models, use can be made of predefined component models. Black box models may also be named implicit, with white box models taking the opposite term and being named explicit. See Figure 6.5; further

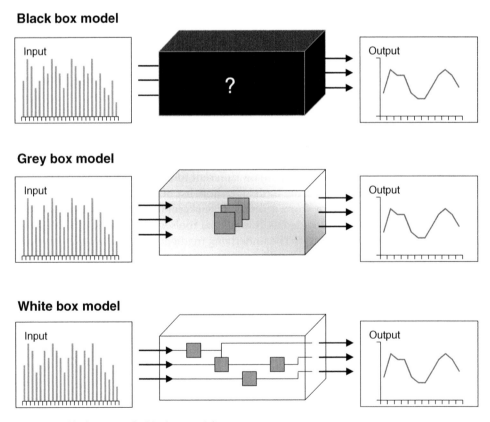

Figure 6.5 Black, grey and white box models.

discussion can be found in Zhao and Magoulès (2012), Fumo (2014) and de Wilde and Augenbroe (2018).

The process of modelling helps in managing complex situations as it simplifies the real world to the variables of interest. In doing so, modelling allows to analyse a situation that otherwise could not be handled and the making of decisions that could not be made when dealing with the full complexity (Marques *et al.*, 2010). This aspect of modelling applies not only to the building and building systems but also to the excitation of the building system that is studied: in real life, a building is subject to a complex pattern of external and internal excitation. In modelling these patterns are typically simplified in order to allow a practical evaluation of the building performance (CIB Report 64, 1982: 14).

Apart from appreciating the advantages of using models, it is also important to consider that there are times when using models and simulation is not appropriate. Basmadjian list a number of cases where it might be better not to go this route:

1) If an answer is needed within hours only, leaving no time for a serious modelling effort.
2) Where a simple and inexpensive experiment might provide the answer.
3) If the client is suspicious of theory and can be better convinced by physical evidence.
4) In cases where the system is too complex for meaningful modelling.
5) Whenever the answer is actually self-evident.
6) Where modelling will fail to provide any new or meaningful insights (Basmadjian, 2003: 3).

Oberkampf and Roy (2010: 51) point out that in terms of scientific computing, the concept of simulation involves a combination of certainty and uncertainty. This may be related to the complexity of the system that is simulated; while the principles of how components respond may be well known and captured, this may not hold for the combined overall system. However, simulation is also a process that can handle the fact that some issues have inherent randomness. Augenbroe (2011: 24) stresses that a model is always a design idealization. In real life, very few actual buildings behave like the corresponding building model. Differences are caused by model uncertainties, operation that is different from assumptions, variation in physical parameters, performance degradation of systems over time, maintenance interventions and others. As a consequence, the performance of the actual building is one element taken from a larger statistical distribution of predicted performances. It is also worth to bear in mind that using a detailed model is not necessarily a solution to this problem; the more complex models require more parameter estimations and thus increase the risk of bias in the underlying estimations (Augenbroe and Park, 2005).

While building simulation has seen significant development and is considered to be one of the main tools available to the building industry (Hensen and Lamberts, 2011: 3–9), several challenges remain. One issue is the 'performance gap' between predicted and measured performance (Carbon Trust, 2011; Menezes *et al.*, 2012; CIBSE, 2013; Wilson, 2013; de Wilde, 2014; Fedoruk *et al.*, 2015; van Dronkelaar *et al.*, 2016), which erodes the confidence of the industry and general public in the domain. Unfortunately this is not an easy issue to address, as there is a range of underlying factors, including fundamental model uncertainties, tool deficiencies and training issues, which are exacerbated by the complexity of the building design, construction and operation processes. In general, it is recommended to have simulation tasks done by experts, trained in both the software that is used and general calibration techniques.[6] Where possible models should be aligned with actual data on the loading of the facility and then be calibrated using metering data and utility bills. A good audit trail of the modelling and calibration effort is important, as well as keeping track of the software being used (Efficiency Valuation Organization, 2014a: 14). A related challenge is the modelling of occupant behaviour. This is a difficult field as it requires a prediction of occupant presence and activities, which is a complex matter. It is often stated that a mismatch between modelled and actual occupant behaviour leads to divergences that may be as large as 40% (Donn *et al.*, 2012; Duarte *et al.*, 2015); recent work in the context of the International Energy Agency (IEA) Annex 66 tries to explore some of these issues (IEA, 2016b). There also remains uncertainty about the thermophysical properties of materials and how this impacts simulation results (Prada *et al.*, 2014). Yet another challenge is the sharing of data between different digital environments, which links the simulation domain within the wider Building Information Modelling (BIM) world. A lot of efforts have been invested in creating interoperable models that exchange data; unfortunately many of these efforts suffer from the fact that this approach to interoperability does not contain the specific information that is needed to fully define appropriate digital experiments (Augenbroe, 2011: 25). Further challenges are inherent to the evolution of buildings and systems themselves; simulation of buildings that include large atria, unusual shapes and complex shading systems or that are partly underground remains difficult; radiant barriers and complex HVAC systems or innovative technology may push the envelope of simulation and may even require the development of new models (Efficiency Valuation Organization, 2014a: 15). Clarke and Hensen (2015), in a more general critique of the field of building performance simulation, list 'seven deadly sins' of simulation: macro-myopia, déjà vu, xenophilia, non-sustainability, failure to validate, failure to evaluate and failure to criticize. These seven points are excellent points to discuss progress in the discipline. However, some of them of them may be seen as controversial; for instance xenophilia in terms of too much attention for other disciplines and missing the core of building simulation may be a danger, but so is xenophobia and isolating the discipline from developments elsewhere.

General information on building performance simulation can be found in a series of books on the subject. *Design energy simulation for architects* (*Anderson*, 2014) provides a good overview of how simulation can support design and actually also covers other domains like daylighting and ventilation. *Building performance simulation for design and operation* (Hensen and Lamberts, 2011) is a seminal volume that provides guidance on many of the key topics relevant to actual simulation efforts. Further worthwhile books are *Energy simulation in building design* (Clarke, 2001) and *Advanced Building Simulation* (Malkawi and Augenbroe, 2003). Modelling of buildings and building systems is also discussed in significant depth in a dedicated section of the *ASHRAE Handbook of Fundamentals* (ASHRAE, 2017: 19.1–19.53) and in the CIBSE Application Manual 11 (CIBSE, 2015c). Well-known papers from the start of the millennium that capture the status of the field and ongoing developments, many of which are still continuing, are 'Trends in building simulation' (Augenbroe, 2002) and 'Building

---

6 In a rather controversial paper, Imam *et al.* (2017) question the modelling literacy of building simulation experts.

performance simulation: the now and not yet' (Spitler, 2006); accounts of the wider history of building performance simulation are provided by Oh and Haberl (2016) or Wang and Zhai (2016).

A range of professional organizations aims to advance the field of building performance simulation. First and foremost amongst these is the International Building Performance Simulation Association (IBPSA); other notable ones are Simulation for Architecture and Urban Design (SimAUD, part of the Society for Modeling and Simulation International), the Technical Council on Computing and Information Technology (TCCIT) of the American Society of Civil Engineers (ASCE) and the CIBSE Building Simulation Group.

6.2.1 Selected Building Simulation Tool Categories

There are many ways to group building simulation approaches and tools, which all have their own merits. This section follows a traditional approach by ordering simulation techniques according to the physical aspects that are simulated. The main domains are ordered alphabetically, from acoustics to whole building energy simulation. Beyond this, two extra categories that often miss from discussion of building performance simulation are included: fire simulation, which is sometimes seen as a specialism, and structural simulation, which seems to live rather independently in the domain of civil and structural engineering. Furthermore, there are a number of categories that do not fit the ordering by physical aspect; this concerns simulation with general engineering software, the emergent area of simulation using the Modelica modelling language, coupled simulation that combines domains and finally the area of tool interfaces, shells and environments. Information on a wide range of building performance simulation tools can be found through the online building energy software tools directory (IBPSA-USA, 2016).

- **Acoustical simulation**
  This field of building performance analysis deals with the modelling, simulation and assessment of sound inside and around buildings. It also covers unwanted sound, which is named noise. In most acoustical simulations that focus on indoor sound, the building is represented as a collection of planes that have specific absorption properties. Models that consider other properties such as mass and stiffness are used to study sound transmission through structures. The main modelling approaches underlying acoustical simulation models are geometrical models, wave-based models and diffusion equation models. The geometric models are based on the concept of sound travelling along lines or rays and are therefore also sometimes known as ray acoustics. Specific approaches in this category include image source models, particle tracing, ray tracing, pyramid tracing, cone tracing and radiosity models. Wave-based models use partial differential equations to describe how acoustic waves travel through any medium. Diffusion equation models describe how the sound energy spreads. The main simulation results typically take the form of the description of a sound field that is produced from one or more sound sources placed in the model. Recent work sometimes uses the term 'soundscaping' where the simulations are used to develop a positive sound field for listeners, as in deflecting noise. Another new

term is 'auralization', where simulation is not only used to predict sound but also to actually reproduce this sound so that listeners can judge how a room or facility may sound in the future.

Some of the acoustical software packages used in building simulation are CATT-Acoustic, Enhanced Acoustic Simulator for Engineers (EASE), Odeon, LMS VirtualLab Ray Acoustics, OTL-Suite/Room and LMS Raynoise. Acoustic simulation also is supported by the COMSOL Multiphysics software, which contains a dedicated acoustics module.

Good introductory overviews of modelling methods for building acoustics and their areas of application are provided by Hornikx (2015) and Mak and Wang (2015). Some further examples of scientific papers in this domain are El Dien and Woloszyn (2004), who use a pyramid tracing model to study the impact of road noise on the facades of high-rise buildings with balconies. Chatillon (2007) presents work that first empirically identifies the directivity of noisy machines using an anechoic chamber and then simulates the sound of these machines using commercial software to study how it propagates in industrial halls. Mahdavi *et al.* (2008) simulate the reverberation times and sound distributions in atria, which they contrast with actual measurements. Remillieux *et al.* (2012) present work on the modelling of impulsive sound, such as sonic booms from aircraft or explosions, around rather than inside buildings. Arjunan *et al.* (2014) develop a vibro-acoustical finite element model to predict the sound reduction index of a stud-based double-leaf wall. Navarro and Escolano (2015) review the use of the diffusion equation to deal with rooms, coupled rooms and rooms with objects. Hornikx *et al.* (2015) present a study into the acoustic modelling of sports halls, comparing and contrasting the diffuse field, geometric and wave-based methods and relating these to actual measurements. Alonso and Martellotta (2015) investigate the specific problem of modelling textile materials hung freely in space, such as tapestries and banners in halls and churches. Sequeira and Cortínez (2016) present the use of a simplified acoustic diffusion model to support the design of industrial buildings. An area that attracts special attention is the study of acoustics in churches and cathedrals. Examples of work in this field are Queiroz de Sant'Ana and Trobetta Zannin (2011) who simulate the acoustics of a modern church in Curitiba, Alonso *et al.* (2014) who simulate the cathedral of Seville or Álvarez-Morales *et al.* (2014) who present simulations based on geometric algorithms for the cathedral of Malaga.

- **Computational Fluid Dynamics (CFD)**
  While the name of the technique says fluids, in the context of building simulation, CFD is mainly concerned with the study of airflow. Where some aspects of ventilation and infiltration may be captured by simple nodal or zonal models (often named multizone models), CFD techniques are needed to study issues like airflow patterns (particle speed and direction), temperature distributions and dispersion of contaminants. This is especially so where there may be effects like buoyancy, stratification and turbulence. CFD simulation typically rests on the Navier–Stokes equations for fluid flows. A range of models is used, depending on the particulars of a specific situation: Reynolds-Averaged Navier–Stokes (RANS) equations, with variants Steady Reynolds-Averaged Navier–Stokes (SRANS) and Unsteady Reynolds-Averaged Navier–Stokes (URANS); Large Eddy Simulations (LES), which are fundamentally 3D and unstable; Detached Eddy Simulations (DES), which is a hybrid URANS/LES

model; and Direct numerical simulation (DNS). The study of fluid dynamics uses a number of parameters to characterize the status of the fluid and its behaviour. Amongst these are the Reynolds number (Re), which captures the ratio of inertial versus viscous forces inside a fluid and in which can be used to establish whether or not turbulence occurs; the Grashof number (Gr), which captures the ratio of buoyancy forces to viscous forces as an indicator of occurrence of natural convection; the Archimedes number (Ar), which indicates fluid motion due to differences in density; the Froude number (Fr), which represents the ratio between flow inertia and the field and which is used to define a speed/length relation; and the Schmidt number (Sc), which captures the ratio of viscosity and mass diffusivity in flows that have both momentum and mass diffusion processes. Knowledge of the related processes and their characterization is essential for selecting the appropriate modelling approach. The modelling requires the definition of the geometry of the object of interest, which may be a room or series of rooms indoors, but may also be a group of buildings or even a full urban environment subject to wind, HVAC systems and other systems/components. Extra complexity may arise from the inclusion of openings and stacks. The volume occupied by air is then divided into discrete cells; this process is called meshing. Mesh definition and resolution are important, as CFD is computationally intensive; often mesh definition requires a trade-off between desirable spatial resolution and temporal scales. The cells of the mesh may be tetrahedral, hexahedral or polyhedral. Mesh definition is sometimes supported by grid sensitivity analysis and can be based on the use of specialized mesh generation packages. CFD modelling also requires the definition of boundary conditions; this includes setting inlet properties, temperature effects (buoyancy) and definition of the aerodynamic roughness of the surfaces of the model. A critical issue in any CFD simulation is the choice of a turbulence model. A common turbulence is the $k-\varepsilon$ model; however this comes in a range of variations, such as the standard $k-\varepsilon$, RNG $k-\varepsilon$, (where RNG stands for Re-Normalisation Group), realizable $k-\varepsilon$ or low Re $k-\varepsilon$. Further turbulence models are the $k-l$ model, v^2f, LVEL, RSM and kω-SST (shear stress transport) models. Beyond this, the modeller must also decide on control and operation of openings, stacks and factors that impact airflow such as the regime of heat sources inside the building.

Some of the CFD software packages used in building simulation are ADREA-HF, FloVENT, PHOENICS, STAR-CCM+ and OpenFoam. The generic engineering analysis software ANSYS includes dedicated CFD tools, of which CFX and Fluent are often used for building simulation.

A starting point in the scientific literature is Srebric (2010), who gives a concise introduction to CFD simulation in an editorial of a special issue of the journal *HVAC&R Research* on the subject. Another good introduction is the paper by Durrani *et al.* (2015). Liu *et al.* (2004) use CFD to predict condensation on the surface of walls. Ji *et al.* (2007) and Ji and Cook (2007) describe the use of CFD simulations to study the natural ventilation, driven by buoyancy, of a space connected to an atrium. Wang and Chen (2007a, 2007b) present the coupling of a multizone network model with CFD in order to capture the effect of incomplete mixing in individual rooms. Huang *et al.* (2007) demonstrate the use of CFD to correlate wind loads and airflow on the structural behaviour of a tall steel building. Hanby *et al.* (2008) use detailed CFD simulation in order to obtain a benchmark for a simplified lumped parameter model of a double-skin facade with blinds. Cropper *et al.* (2010) combine a CFD model with the multi-segment human thermal comfort IESD-Fiala model for a detailed analysis of the interaction between temperatures, airflow velocities and the human response to the thermal environment. Blocken *et al.* (2011) give an overview of using CFD to study wind conditions around buildings and its impact on pedestrian comfort, wind-driven rain, convective heat transfer and pollutant dispersal. Defraeye *et al.* (2011) use CFD simulations for a deep exploration of the correlation between the wind speed and the convective heat transfer coefficient that is often needed in thermal studies. Gousseau *et al.* (2011) compare RANS and LES for the prediction of the dispersion of pollutants in an urban environment. Walker *et al.* (2011) report on the study of an open floor office building with atrium, which they studied using both CFD and a scale model, stating differences between 2.5% and 15% between both approaches. Mavroidis *et al.* (2012) use CFD to study the impact of atmospheric stability on the residence of pollutants in the wake of a cubical building. Ramponi and Blocken (2012) presents steady-state RANS simulation of a generic naturally ventilated building, validated by wind tunnel experiments, in order to give guidance on the setting of key CFD computational parameters such as the computational domain, grid resolution, atmospheric boundary layer definition, the turbulence model, discretisation scheme and convergence criteria. Balocco *et al.* (2012) use CFD to study the effect of sliding doors on airflow in an operating theatre, which is important for patient health and infection risk. Ai and Mak (2013) study the airflow and dispersion around buildings, focussing their research on the role of the inhomogeneous atmospheric boundary layer. Younes and Abi Shdid (2013) explore the complexity of modelling air leakage and infiltration through the building fabric using CFD, discussing the complexity of capturing the cracks and airflow paths. Durrani *et al.* (2015) compare LES and URANS models to represent observations conducted on a naturally ventilated (buoyancy-driven) enclosure such as an open floor office. Gloriant *et al.* (2015) use CFD simulation to study the thermal behaviour of a triple-glazed supply air window. Liu and Niu (2016) compare SRANS, LES and DES models with experimental measurements on the wind climate around an isolated high-rise building. Peng *et al.* (2016) report on an experiment where 19 teams used CFD to model a low turbulent flow and found significant variation amongst these teams. Perén *et al.* (2016) use CFD to study the ventilation of low-rise buildings with sawtooth roofs.

- **Heat and Moisture (HAM) models**
  While the thermal and hygric behaviour of buildings sometimes may be studied separately, there are many situations where the interaction between these two aspects needs to be considered. Typically this is the case where constructions are at risk of surface or interstitial condensation or where moisture may lead to a significant change in thermal properties of the fabric. Consequences of certain combinations may include mould growth, rot, mildew, cracks, swellings and increased energy loss. Models that consider both thermal and moisture are sometimes named hygrothermal or HAM models; interestingly, some authors also include airflow modelling to the work, noting that air is an important carrier of moisture into constructions and therefore considering HAM an abbreviation for Heat, Air and Moisture. Hygrothermal simulation is based on solving the fundamental heat and mass equations. In general terms these describe, using partial differential equations, that heat or mass entering a

system either leaves the system or is accumulated within the system while also taking into account generation and consumption, thus obeying the classic laws of conservation of energy and mass. On the hygric side, things are slightly more difficult due to a range of complex processes that must be accounted for: capillary motion (the ability of a liquid to flow in narrow spaces, sometimes even in opposition to gravity), diffusion (the tendency of molecules to spread out and intermingle), phase change (transition from water to vapour or ice and vice versa), adsorption (accumulation of a gas or liquid on a solid), absorption (molecules of a gas or liquid entering the volume of another material) and desorption (the opposite of adsorption and absorption). Computationally the simulation of heat, moisture and air transport may be difficult because the transport of each of these takes place at rather different time scales. Moreover, there are complex interrelations; for instance moisture diffusivity is temperature dependent, and one may have to distinguish between sorption and desorption curves. Before the spread of simulation, assessment of condensation risk was typically based on the use of the Glaser diagram or dew point method, where water vapour pressure and saturation pressure are drawn across a construction to identify areas at risk. Simulation tools also take into account capillary flow, sorption, vapour diffusion and changes in material properties. However, in hygrothermal simulations, the modeller needs to take great care in ensuring the relevant processes are included. Actual conditions that lead to moisture problems are typically complex and not captured well by standard values and approaches; most building assemblies deviate substantially from the design.

Some of the dedicated HAM simulation tools are Delphin, HAM-Tools, HAM-VIE, Match and WUFI. Due to the fact that HAM models combine various domains, this is also an area where there is a substantial application of generic engineering tools such as Matlab/Simulink and COMSOL Multiphysics.

Hens (2015) provides a good overview of the field of combined heat, air and moisture modelling, including a review of the history of the domain, underlying principles and physical laws, as well as a critical review of remaining challenges. A full PhD thesis on the topic is provided by Sasic Kalagasidis (2004). For further academic work, Künzel *et al.* (2005) explore how the hygrothermal behaviour of the building envelope interacts with the indoor environment inside a room, taking into account capillary action, diffusion and vapour absorption and desorption. Dos Santos and Mendes (2006) employ simultaneous simulation of heat and moisture transfer in order to predict ground heat transfer, air temperature and humidity for buildings in Brazil. Lengsfeld and Holm (2007) describe the development and validation of the WUFI software, with an emphasis on the impact of construction materials on humidity peaks. Janssen *et al.* (2007) explore the impact of atmospheric boundary conditions, in terms of wind-driven rain and vapour exchange, on hygrothermal simulations. Barbosa and Mendes (2008) combine a whole building hygrothermal model with HVAC system models in order to explore combined heat, vapour and liquid transfer; they find that leaving out the moisture aspect may lead to underprediction of the cooling load by 13%. Qin *et al.* (2009b) present an integrated model for predicting heat and moisture transfer at the building level. Tariku *et al.* (2010) have developed a holistic model that captures the heat, air and moisture exchange between the building indoor environment, fabric and services. Abahri *et al.* (2011) present a one-dimensional coupled heat and moisture transfer model for porous materials. Kong and Wang (2011) model heat and mass transfer in the context of the dry-out period of newly built houses, where the excess moisture may lead to increased energy use and damage by freezing in the winter. Antretter *et al.* (2013) have expanded the WUFI hygrothermal simulation tool with a module that allows including the impact of thermal bridges in the analysis. Portal *et al.* (2013) have developed a heat and moisture model into a combined simulation model for heat and moisture-induced stress and strain (HMSS) in order to manage the risk posed to historic sites and artefacts, linking HAM models to structural analysis. Künzel and Zirkelbach (2013) contribute work on the modelling of rainwater leakage in one-dimensional models, noting that the current hygrothermal models mostly assume a perfect assembly without gaps or cracks. Vasilyev *et al.* (2015) present a coupled numerical model for a multilayered wall that considers temperature, vapour pressure, water concentration in pores and vapour condensation and evaporation. Berger *et al.* (2015) present a reduced-order model for heat and moisture multizone modelling in order to overcome some of the computational load in typical hygrothermal models. Langmans *et al.* (2012) have expanded an existing heat and moisture model with a quasi-stationary airflow model that allows accounting for buoyant and externally driven air movement.

- **Lighting simulation**
  The area of lighting is well recognized as an important part of the simulation domain. This is especially true for daylighting, which can be challenging due to fast and large changes to outdoor conditions caused by the weather (clouds) and which also is a target for energy efficiency since the use of daylight does not require electricity. The field also has inherent complexity due to the interaction between the outdoor environment, indoor environment, occupant and system behaviour and a range of performance aspects that cover visual performance, visual comfort, psychology and perception. Daylight has a relation with the circadian rhythm and human well-being. Lighting also plays an important role in the human perception of objects and space and hence in assessment of buildings. As a consequence, there is a wide range of criteria for (day)lighting, which includes, amongst others, illuminance level of task areas and surroundings, uniformity of the light, colour rendition, contrasts, absence of glare and absence of flickering. For the more advanced aspects, these need to reflect individual needs, thus requiring thresholds that capture human diversity and variability. Classic lighting metrics include the daylight factor (DF), the horizontal illuminance level and the horizontal to vertical illuminance ratio. Lighting simulation may also have the objective of creating photorealistic renderings that take into account the laws of physics. The main underlying methods for lighting simulation are ray tracing and radiosity. Ray tracing, as the name suggests, follows the path of the light. Subcategories are forward ray tracing, which starts at the light source; backward ray tracing, which starts from the point of measurement or observation (such as the eye, a point of interest or the pixel on a screen) and then follows the path towards the light sources; and bidirectional ray tracing, which combines the other two. Radiosity methods work similar to thermal models of radiation exchange; they split the surface of entities of interest into 'patches' and then account for light exchange between these patches. In both ray tracing and radiosity, it is important to manage specular and diffuse reflections; surfaces that can be considered completely diffuse are also known as Lambertian surfaces. Ray tracing is generally considered to be a good approach for cases with direct illumination, shadows, specular reflection and refraction through

transparent materials; however this method becomes complicated and computationally expensive when dealing with diffuse interreflections. Lighting simulation first of all requires the modelling of the geometry; often this is created in a CAD or BIM environment. Surfaces in the model then need to be assigned their relevant properties, such as reflectance and transmittance. Where relevant, artificial light sources need to be positioned in the model and properties of the light source and luminaire defined, as well as operational parameters such as time of day and lighting levels where the light is on. In relation to daylight, blinds and shutters and their operation may also have to be included. For studies of daylight, another important modelling step is the selection (or definition) of a sky model; often reference is made to the CIE standard sky models (CIE standard overcast sky, CIE intermediate sky and CIE clear sky), the Perez all-weather sky and the Matsuura intermediate sky; in other cases measured lighting data may also be used. Simulation results can be used to obtain a range of performance metrics, such as the daylight autonomy (DA), which captures the percentage of occupied hours when a certain illuminance threshold is met by daylight only; the spatial daylight autonomy (sDA), which represents the percentage of an area that meets a minimum daylight illumination criterion for a specified fraction of operating hours; or the useful daylight illuminance (UDI), which uses thresholds of 100 and 2000 lux to identify when there is too little, enough or too much daylight. Simulation results can also be used to predict glare or glare avoidance; metrics used include the unified glare rating (UGR), daylight glare index (DGI) and daylight glare probability (DGP). Simple simulation results such as illuminance levels can also be used to develop contour maps; where appropriate these can be extended towards temporal maps, which indicate areas that receive certain amounts of light for a given period of time.

Lighting simulation tools include AGi32, Adeline, DElight, DIALux, DIVA, Inspirer, Lightscape, Lightsolve, Lightswitch, Mental Ray, Radiance, Relux, Zemax OpticStudio, 3ds Max and Velux Daylight Visualizer; there also are specialist tools to cover optical systems with lenses, light pipes and similar. Amongst these tools, Radiance is extremely versatile and well used; unsurprisingly there thus is a range of interfaces or preprocessors to Radiance, including Daysim, Desktop Radiance, LightSketch and Fener. Ecotect is also sometimes named in this context, but in fact is intended as an early design tool with access to a range of analysis engines.

For a further introduction to this area of simulation, Reinhart and Selkowitz (2006) give a brief overview of the field of daylighting in their editorial for a special issue of the journal *Energy and Buildings* on the subject and how this relates light, form and people. A deeper discussion, based on literature review, is provided by Ochoa *et al.* (2012). Mardaljevic *et al.* (2009) give a good overview of some daylighting metrics, while Andersen (2015) provides a deep review of the interaction between (day)lighting and human needs, covering the impact of daylight on health, visual comfort and perception of spaces. Other academic literature on the subject includes Ng *et al.* (2001), who report on the use of Radiance simulations to support the redesign of a museum in Singapore. Reinhart and Walkenhorst (2001) present a validation of the Daysim method on the basis of measurements in a full-scale test office. Glaser *et al.* (2004) present their visual interface, LightSketch, for the Radiance simulation engine. Reinhart (2004) describes the background of the Lightswitch model, which accounts for manual and automated control of lighting and blinds in offices. Mantzouratos *et al.* (2004) describe a workflow for illumination studies of buildings with historical and architectural importance like the monuments of Greece. Maamari *et al.* (2006) have conducted an experimental validation of a range of lighting simulation methods for two types of complex fenestration systems: a Serraglaze^{TM} element and a laser cut panel. Reinhart and Fitz (2006) present survey results on the use of daylighting in design, capturing the status in the first decade of the new millennium. Du and Sharples (2010) use Radiance to study the impact of different geometries and reflectance distributions on the vertical daylight levels in atria. Mahdavi and Dervishi (2010) explore the five options to generate sky models from a limited set of measurements and study how these options impact computed irradiance values for vertical surfaces. De Keyser and Ionescu (2010) present the modelling and simulation of a lighting control system that aims to keep illumination levels in a room constant. Reinhart and Wienold (2011), building on a review of dynamic daylight simulation capabilities, suggest the use of a daylighting dashboard to support the design of daylit spaces. Wong *et al.* (2012) simulate the use of remote source lighting (RSL) technology in lift lobbies in high-rise blocks in Hong Kong, adding experiments to validate the simulations. Khosrowshahi and Alani (2011) simulate the impact of ageing of a light bulb and collection of dirt and dust on a reflector and shade, and visualize the effects of these processes on the view inside a room using virtual reality modelling language (VRML). Kota *et al.* (2014) discuss the integration between BIM and daylighting simulation tools, focussing on the link between Revit and both Radiance and Daysim. Yu *et al.* (2014) assess the energy saving potential offered by the use of daylighting in an educational building in Nottingham based on simulations with Relux. Bueno *et al.* (2015) have developed a Radiance-based approach to assess complex fenestration systems. Saraiji *et al.* (2015) compare the results from an experimental set-up with simulations in DIALux and AGi32. Amundadottir *et al.* (2017) use Radiance to predict ocular light exposure in an educational building, exploring advanced performance aspects such as visual interest and gaze behaviour. Konis (2017) uses simulation to investigate the relation between building design, daylighting and the circadian rhythm.

- **Pedestrian movement (evacuation) models**
  The movement of humans inside buildings (occupants) and around them (pedestrians) is another distinct area within building simulation. Within this generic area of pedestrian movement, one can discern between two modes: normal movement and movement in an emergency situation; the latter is typically studied in terms of building evacuation. There are similarities and differences between both modes; however it must be noted that the evacuation mode always follows up on a normal situation due to some sort of incident.

  Normal pedestrian movement may be studied for a number of reasons; these include comfort, efficiency and the risk assessment of pedestrian safety such as the chance of trampling in crowds. How people move depends partly on the space layout but also on the activities; for instance there is a clear difference between individuals who wander freely or a group of people marching together. Other factors to be included are the impact of autonomy, individuality, time, function, appearance, age, sense of orientation and familiarity with the location. Interest may range from movement of individuals to that of large groups; crowds can be defined as a large number of people (typically more than 100), which is in the same place at a given time. Often crowd density and speed are classed in a concept termed 'level of service' (LOS), with higher classes having more risk. Pedestrian movement analysis may also help to find out which parts of a building are less used, allowing designers to reroute people or change the use of such areas. Pedestrian movement simulation needs to combine a range of issues, such as routings, speed and crowd density as well as physiologic and psychological issues. How humans move depends not only on their local knowledge and familiarity with the building or district, interaction with other people in terms of leaders, followers and supporters and relationships with the place, but also on personal characteristics such as health and mobility, well-being and incentive. Occupant density has a clear relation with occupant flow; typically walking speed goes down with higher numbers of people and less space, such as narrow corridors. Crowd movement also exhibits some emergent behaviour, such as tendencies towards lane formation, stop-and-go waves, pressure build-up or herding. Some characteristics may be regional, such as a tendency to walk on either the right or left side. Crowd motion may be unidirectional or multidirectional; further subcategories include patterns such as straight flow, flow around a corner, entering or exiting of spaces, parallel flow and crossing flow. There is a range of underlying models that is used in pedestrian movement simulation, such as cellular automata (CA), social force models, agent-based (activity-choice) models,[7] network flow models, fluid dynamics models and gas kinetic models. Some of these models are also used in other domains, such as traffic simulation. The cellular automata, social force models and agent-based models are sometimes named microscopic models, as they handle pedestrians as individuals; the other models are macroscopic in that they describe the movement of humans as a particle system that is governed by general rules and relations. Another way of looking at these models is to distinguish discrete event models (where one occupant moves from one location to another in a time step, with a spatial and temporal resolution) or continuous models where a flow is described by some sort of equation. Many models use a range of separate layers, such as a locomotion, steering and decision layer, in order to handle different aspects. There are also separate models to describe specific aspects of movement, such as queuing models; again these may be shared with other domains such as retail or computer networks. Others may be navigation graphs, routing and steering algorithms or Belief-Desire-Intentions (BDI) models for definition of psychological issues. Some models are 2 dimensional, whereas others are 3D. Underlying algorithms may be for instance Shortest Path Algorithms, A* Algorithms, or Ant Colonization Algorithms. The modelling effort that precedes pedestrian simulation typically requires a definition of the topography in order to set the navigable pedestrian space. In non-continuous models there will also be a need to define space discretisation and a mesh. Topology may be imported from CAD or BIM programs or drawn directly in the simulation tool. Specific attention must be paid to ensure a good representation of targets and obstacles. Furthermore, the modeller will have to define where pedestrians emerge in the model (source and position). Assumptions may have to be made regarding deceleration as a function of high density, lane formation in bidirectional flows and dynamics near bottlenecks or regarding dynamics in dense crowds. Sometimes modellers have to define navigation graphs

---

7 Agent-based models may be abbreviated as ABM, multi-agent systems as MAS.

with navigation points. In discrete models, there also has to be a decision on how many individuals can be contained in one grid cell, noting that such a cell can be as large as a room or corridor.

Some of the simulation tools for normal pedestrian movement are Aseri, Crowd-Z, Legion, Nomad, PedGo, SimWalk, VisSim (including VisWalk) and Witness. As in other domains, general engineering tools may also be used; an example is the application of the Stroboscope discrete event simulation software.

A well-known review of pedestrian behaviour models and good introduction to the field is the paper by Papadimitriou *et al.* (2009). Duives *et al.* (2013) give a broad overview of the state of the art in crowd simulation modelling. Nassar (2010) has developed an occupant flow model for use during the early design stages, where building circulation and public spaces are still taking form. Sagun *et al.* (2011) review the role of crowd simulation in designing buildings that are better at coping with emergencies. Kneidl *et al.* (2012) present work on the modelling of the movement of individuals that combines a microscopic model with a sparse navigation graph; further work in this direction is presented in Kneidl *et al.* (2013). Ma *et al.* (2013) combine an agent-based pedestrian model with a Geographic Information System (GIS), in order to deal with complex buildings. Song *et al.* (2013) discuss the development of grid-based spatial models for crowd simulation in micro-spatial environments such as indoor rooms. Kountouriotis *et al.* (2014) have developed an agent-based crowd model that couples individual parameters with a high number (thousands) of agents. Zawidzki *et al.* (2014) introduce the agent-based CZ model for crowd simulations on floor plans. Kim *et al.* (2013) have developed a simulation algorithm that focusses on pedestrian queuing for a cinema ticketing booth, comparing simulation results with experimental data. Yanagisawa *et al.* (2013) also look at queuing but address more complex situations, where one single queue leads to several service windows. Duives *et al.* (2013) write that existing models 'can roughly be divided into slow but highly precise microscopic modeling attempts and very fast but behaviorally questionable macroscopic modeling attempts'. Davidich and Köster (2013) warn that there is a disconnect between real-world observations and model results and present a calibrated model.

Evacuation behaviour is studied with the prime ambition of reducing risk to human life and health. It is a key aspect in Fire Safety Engineering (FSE); however there are also other threats such as terrorist attack or bombs, crowd panic or natural disasters such as earthquakes. All of these can be considered to be extreme events. A key metric that often is the objective of evacuation simulation is the estimated evacuation time of a building. Sometimes further detail is provided by distinguishing between the Required Safe Egress Time (RSET) and the Available Safe Egress Time (ASET), where RSET must be smaller than ASET. In comparison with normal pedestrian movements, evacuation comes with a distinct chain of events: after the incident, this must be recognized. Beyond that, the alarm will be raised. Occupants have to recognize the need to evacuate and then actually start to respond. This is followed by the actual evacuation. Beyond the evacuation, the building may become uninhabitable. Challenges to a safe evacuation may be things like panic and competitive egress. In case of fire, fumes are a major risk to human life due to toxicity, suffocation and reduced visibility; fumes are difficult to model as they occur in layers and build up over time. Stress may reduce the occupant awareness and capability to navigate.

Where buildings are large and complex, occupants may face additional challenges by not knowing the detailed escape routes and connectivity of spaces. It must be noted that in most cases, occupants only experience local environmental conditions; typically they do not know the exact status of the building and details of the incident that raised the alarm and led to order to evacuate. Contrary to regular pedestrian movement, evacuation has an increased risk of occupant health and mobility changing over time, for instance by getting hurt or having to deal with respiratory or visual problems. People in an evacuation may make hasty decisions based on incomplete information, and they may engage in interaction with the building and incident, such as when trying to put out a fire. Evacuation also requires occupants to select a route and speed as well as to decide how to deal with other people and objects. However, there is an increased risk of congestion, jamming of doors and opening and obstacles along the egress route. Exit knowledge may play a large role as well, with people tending to leave the building along the route they took on the way in. There are similarities between the evacuation of buildings and that of transport systems such as aircraft, trains and ships; some of the modelling approaches may be applicable across domains. Most simulations of building evacuation employ macroscopic models, focussing on the overall situation and mean egress speed values. However, coupled models may have to be employed where there is an interest in how individuals respond to low visibility conditions, obstacles and the like. This is also needed where behaviour such as pushing, crushing and trampling may be of interest. Another type of coupling is needed where there is an interest in how something like a fire evolves; this requires further specialist tools that are able to model flame spread, heat, soot density, smoke movement, presence of carbon monoxide and other aspects that may be of importance in a particular situation. Modelling for the simulation of building evacuations first of all poses the question of what incidents and scenarios need to be analysed. This is a non-trivial question, with many alternatives being possible. Once these scenarios have been selected, they need to be specified, including their aftermath in terms of things like fire spread and blocking of critical paths. The modeller will have to decide on whether to model occupants as individuals or as a group, and whether to employ a coupled model. Characteristics that define how occupants behave during the evacuation have to be set, as well as a relation of some of these characteristics to environmental conditions such as lower visibility. In some cases the modeller also has to define how panic propagates amongst the population of building occupants.

Simulation tools that focus on building evacuation are AIEva, BGRAF, Defacto, DBES, Escape, Egress, EgressPro, ESM, Evacnet, Evacsim, Exit, Exodus, F.A.S.T, Simsite, Simulex, Steps, Timtex and Wayout.

A seminal report on building evacuation models is the work by Kuligowski and Peacock (2005). Pelechano and Malkawi (2008) review the simulation of the evacuation of high-rise buildings, with a focus on cellular automata. A good overview of models for fire and smoke simulation, which includes a section on egress models, is provided by Olenick and Carpenter (2003). Further literature on the subject is for instance Shen (2005), who presents ESM, a network flow model that simulates evacuation under fire conditions. Shendarkar *et al*. (2008) include a belief-desire-intention model in an agent-based simulation of evacuations of urban areas in response to bomb attacks in public areas. Shi *et al*. (2009a) present a model that combines agent-based evacuation simulation with heat and mass simulation of fire and smoke. Tavares and Galea (2009) discuss work that addresses the question of what fire scenarios need to be simulated in the context of fire safety engineering, with special emphasis on providing safe exits. Yuan *et al.* (2009) present an approach that combines fine and coarse networks to simulate the evacuation of large and complex buildings. Zheng *et al.* (2009) review the use of cellular automata, lattice gas models, social force models, CFD models, agent-based models, game theory models and experiments with animals to study crowd evacuation. Oğuz *et al.* (2010) focus on the simulation of crowd escape in the urban outdoor environment in a context of computer graphics. Rüppel and Schatz (2011) have developed a serious game to study the impact of the building conditions on human behaviour in case of fire evacuation; here actual human decision making can be studied in a virtual environment. Borrmann *et al.* (2012) present a model that couples microscopic and macroscopic pedestrian dynamics in order to better predict evacuation times. Gelenbe and Wu (2012) review how sensor and communication networks may impact intelligent evacuation and emergency responses. Joo *et al.* (2013) include affordance theory in their modelling of human behaviour during evacuations, noting that predicting human behaviour in complex and uncertain situations is considered almost impossible.[8] Manley and Kim (2012) explore the inclusion of individuals with disabilities in evacuation simulations for rare but catastrophic events; Koo *et al.* (2012) study specifically how occupants with disabilities impact the evacuation of others. Wagner and Agrawal (2014) present an agent-based simulation approach for the evacuation of crowds from a concert venue in the case of a fire disaster. Li and Han (2015) have extended cellular automata with representations of familiarity and aggressiveness and studied their impact on evacuation efficiency. Thompson *et al.* (2015) warn that much data used in evacuation modelling is becoming dated and is not representative of current population characteristics. Tan *et al.* (2015) explore the inclusion of changes in spatial accessibility in an agent-based simulation.

- **Whole building thermal simulation**
Analysis of the thermal performance of whole buildings is a prominent area of building performance simulation. The domain emerged in the late 1960s and early 1970s together with the new discipline of building science/physics, spurred on by the energy crisis and the emergence of personal computing. Continuing concerns about energy supply, costs and greenhouse gas emissions have kept the momentum. Furthermore, whole building thermal simulation also is a key tool for the analysis of thermal comfort conditions and hence for the study of human health and well-being. In the literature, whole building thermal simulation is sometimes simply named building energy simulation and abbreviated as BES. Due to the prominence of the field, thermal models are relatively well developed; at the same time, the complex interplay between building fabric, occupant behaviour, HVAC systems and control systems remains challenging. Thermal building models use heat and mass balances that describe the three main heat transfer principles of conduction, convection and radiation. These three mechanisms occur in different forms and combinations in the five main energy flow paths that take place in buildings: transmission (heat flow through walls, floors,

---

8 In computational terms, this prediction problem is sometimes classed as non-deterministic polynomial-time or 'NP hard'.

roofs, windows and doors), ventilation (heat flow related to movement of air), solar radiation (heat entering the building through windows and other glazed surfaces), energy storage (where heat is added to, or removed from, building parts such as concrete walls and floors) and internal gains (heat produced by occupants, lights, appliances and the building services; negative in case of cooling). A full description of these energy flows is usually long and complex as it needs to account for the fact that most buildings have a multitude of rooms, each enclosed by several surfaces. Moreover most constructions are multilayered; some may be opaque, while others are transparent and allow solar gains. Air paths for ventilation and infiltration also need to be described. Building services may be simplified or included in significant detail, depending on the objective of the simulation. Internal gains and system controls typically require assumptions about occupant presence and preferences. The heat and mass balances describing this multitude of flows typically take the form of partial differential equations, which are solved using numerical approaches; the two most common methods are the finite difference method and the finite element method. Given the resulting overall complexity, most thermal models assume significant simplifications, such as full air mixing, one single temperature per zone and one-dimensional heat flow through surfaces. Some heat flow processes may be represented by simplified models such as the response factor method, transfer surface method or lumped capacitance method. Modelling of buildings for thermal analysis mostly starts with the definition of the building geometry, taking into account any zoning decisions (where individual rooms are modelled as one larger space). Walls, facades, roofs and floors will need to be modelled in terms of the order of layers, layer thickness and material properties of each layer such as conductivity, density and thermal capacity. For glazed elements, the modeller will have to set reflectance, absorption and transmission values. As mentioned before, building services may be modelled at various levels of detail; the simplest form is as instantaneous heat or cooling loads, but it is also possible to include detailed models of boilers, furnaces, distribution systems and others, including detailed operational regimes. Modelling of occupants also comes in various levels of detail; in any case it needs to represent the heat emitted by human bodies present in a room and comfort conditions that these humans require. The modeller will also have to assign weather conditions; typically this is done by making use of a measured, statistically representative or artificial file of weather data including outdoor temperature, relative humidity, wind direction and speed and solar irradiance; most weather files have an hourly format and cover a duration of 1 year. Weather ties in with the building location and the local climate; solar access modelling also requires correct orientation. Modelling has to include the definition of heat transfer coefficients, boundary conditions[9] and airflow within a building. Sometimes the modeller also needs to select spatial discretisation (for instance the number of sub-layers used to represent a layer of the building fabric) and temporal discretisation (time step). Given the full complexity of the model, most models require a selection of output parameters, for instance specific points in the model where temperatures are to be reported.

---

9 An adiabatic boundary condition is often applied to the symmetry plane of walls and floors that have similar temperatures at both sides, such as internal walls.

The most prominent whole building thermal simulation tools today are EnergyPlus, ESP-r, IDA-ICE and TRNSYS. Other tools, some of them legacy, are BLAST, BSim, DOE-2, Capsol, COMFIE, DeST, Ener-Win, HAP, PowerDomus, Sunrel and TAS.

Thermal building simulation is arguably the most mature of the subfields of building performance simulation. As a consequence, key information on this area is described in a number of books. *Energy simulation in building design* (Clarke, 2001) and *Modelling methods for energy in buildings* (Underwood and Yik, 2004) are fundamental works. *Building performance simulation for design and operation* (Hensen and Lamberts, 2011) contains excellent chapters on thermal load and energy performance prediction, as well as relevant sections on climate data, occupant behaviour, HVAC systems and building automation that all relate to thermal analysis. Another generic overview of the field is the chapter by de Wilde and Augenbroe (2009). Seminal papers in the area are Crawley *et al.* (2008), who compare the features and capabilities of twenty building energy simulation tools. Oh and Haberl (2016) describe the history of building energy simulation tools, with a focus on the United States; their work includes a genealogy chart that allows to trace some of the developments. A small selection only of the large body of literature on the subject is mentioned here. Crawley *et al.* (2001, 2004) give an interesting insight in the development of the EnergyPlus program. Sahlin *et al.* (2004) discuss the differential algebraic equations and general-purpose solvers used in the IDA-ICE simulation tool. Strachan *et al.* (2008) discuss the long history of validation of the ESP-r simulation program. Dols *et al.* (2015) describe ongoing work on the TRNSYS simulation tool, focussing on improving the handling of coupled modelling of heat transfer, airflow and contaminant simulation. Daum and Morel (2009) simulate how different blind control and lighting schedules may impact the energy use (both thermal and electric) of an office room. Dorer and Weber (2009) show how component models, in this case for micro combined heat and power (CHP) systems, can be integrated into whole building simulation tools such as TRNSYS. Zhou *et al.* (2014) give a more general comparison of the modelling of HVAC systems in simulation tools, describing some of the fundamental differences. Pang *et al.* (2012) present a framework to compare predicted and measured building performance, mainly in terms of power consumption, in real time, by linking EnergyPlus simulation, the Building Controls Virtual Test Bed, and actual energy management and control systems. Rysanek and Choudhary (2012) present a Matlab environment that supports an automated search for energy-efficient building refurbishment options within both commercial and custom simulation models. Zhao and Magoulès (2012) review a wider range of building energy consumption prediction tools, including engineering methods (which cover the traditional building simulation tools), statistical methods, neural networks, support vector machines and grey methods. Foucquier *et al.* (2013) present a review of models for building energy performance prediction, discerning physical models, statistical and hybrid methods; however this work clearly blends the borders between thermal and CFD models. Fumo (2014) builds on the two preceding papers but also adds a discussion of calibration and verification.

- **Fire simulation**
Fire simulation is a specialist area. Physical principles are partly overlapping with other building simulation fields, notably those of thermal analysis and airflow modelling. However, the overall process is different; over time the model needs to capture

ignition, fire development and flashover, the full fire and finally the collapse of the fire. Temperatures are mostly in a much higher range than used for heating and cooling analysis, with temperature gradients playing a role in the fire development. Construction materials and building content provide fuel for the fire, which gets depleted over time. Fire simulation also may cover the spread of smoke through a building. Some fire models actually cover the district or urban level.

Some of the dedicated fire simulation tools are ASET, Consolidated Model of Fire Growth and Smoke Transport (CFAST), Fire Dynamics Simulator (FDS), FireFOAM and SCHEMA-SI. There also are dedicated fire risk assessment tool, such as CRISP2 and FiRECOM.

Many authors present work that compares and contrasts numerical simulations with experimental results from fire tests. For instance, Chow and Zou (2009) look at the case of closed chambers; Chen *et al.* (2010) discuss fires inside a furnished office room; Yuan and Zhang (2009) study a compartment with solid fuel; Yang *et al.* (2011) review a shelf fire in a storehouse; Zhang *et al.* (2010) study fires of single burning items; Byström *et al.* (2012) model fire in a building compartment; Xiao and Ma (2012) analyse the fire behaviour of a laminated bamboo building; and Horová *et al.* (2013) look in detail at the travelling of fire along a surface. Björkman and Mikkola (2001) present a fire risk assessment of a timber frame building with a dedicated risk analysis tool that models fire spread, tenability of rooms and occupant response. Similar work, looking at the risk in multipurpose buildings, is provided by Chi *et al.* (2011). Shen *et al.* (2008) discuss how fire simulation can help the reconstruction of an arson fire. Muller *et al.* (2013) present a Petri net-based simulation model for fire growth and smoke spread that is used in forensics. Hua *et al.* (2005) model smoke propagation in a multistorey building. Qin *et al.* (2009a) focus their attention on the specifics of the spread of smoke in an atrium in the case of a fire, whereas Xia (2014) simulates the spread of fire smoke in the elevator shaft of high-rise buildings. Wang *et al.* (2008) employed a range of simulation models, including CFD and FDS, to study a backdraught fire in a townhouse in Taiwan. Lee and Davidson (2010) have developed a model to simulate fire spread after an earthquake, which not only captures the evolution of a fire in a room but also the spread from room to room and from building to building. Zhao (2011) uses cellular automata to model fire spread in urban areas. Hantouche *et al.* (2016) use simulation to explore the impact of fire on steel structures.

- **Structural simulation**
The impact of loads on building structures is another major field for the application of simulation; this is important for the analysis of the strength, stability and deformation of buildings. Interestingly, this domain seems to be mostly separate from traditional reports on building performance simulation, instead being left to civil and structural engineers. Overall this is another mature area, where a range of books covers the basic models and equations; see for instance the *Structural Engineering Handbook* by Gaylord *et al.* (1997), *Mechanics of Materials* by Gere and Goodno (2013) or *Structural Analysis* by Kassimali (2015). The main underlying physical concepts are compression, tension, shear, bending and torsion and how these lead to reactive forces, deflection and, ultimately, failure. Modelling approaches in this area often are based on finite element analysis (FEA), where a complex problem is subdivided in smaller parts that are then analysed using partial differential equations.

Modelling requires the definition of the building geometry and material properties; many modern tools allow importing of relevant data from BIM tools. Furthermore, the modeller needs to define the loads that act on the building. These are generally considered to come in the following categories: dead loads, which relate to permanent loads such as the weight of the structure itself; live loads, which relate to variable and dynamic loads such as those caused by people or furniture; and environmental loads, which are caused by snow, rainwater, wind, frost, thermal expansion and seismic activity. Other loads may include things like fire or the impact of explosions. More advanced structural analysis may include the impact of corrosion and fatigue on the performance of structures.

Structural simulation tools include Abaqus, Adaptic, ADINA, ANSYS Mechanical, Autodyn, Calrel, Comet, FEAP, LS-DYNA, midas FEA, MSC Marc, Nastram, OpenSees, OpenSHA, Perform-3D, SeismoStruct, SimScale, SolidWorks and VisualFEA; general engineering software is also widely used in the structural analysis domain.

In terms of academic literature, there is a significant number of papers that studies the impact of earthquakes on building structures. By way of example, Hayashi *et al.* (1999) have simulated the impact of an earthquake on real buildings and compared the results with actual observed damage. Polycarpou and Komodromos (2010) use simulation to study the effect of seismic pounding on the response of a building with adjacent structures. Liu *et al.* (2012) model the interaction between soil, wave propagation and clusters of buildings. Yang *et al.* (2012) evaluate the behaviour of high-rise buildings with concrete core walls; Lu *et al.* (2013) simulate the impact of extreme earthquakes on a high-rise building and analyse the potential collapse processes. Fallah and Taghikhany (2015) study the effectiveness of smart seismic dampers. Murray and Sasani (2016) simulate the impact of severe pulse-type ground motion on column shear-axial failure of pre-1970s reinforced concrete structures. Looking at other loads, Luccioni *et al.* (2004) have simulated the impact of a blast load on a building, including subsequent collapse. Kwasniewski (2010) simulates the progressive collapse of a multistorey building. Tsai (2012) presents the use of simulation to study the response of building frames to sudden loss of columns, which may result from an accident or malevolent human action and which may lead to progressive collapse; another contribution on simulation of the impact of blast load is the work by Kelliher and Sutton-Swaby (2012). Song *et al.* (2012) describe a BIM-based approach for the simulation and optimisation of the construction of structural frames. Deiterding and Wood (2013) discuss the simulation of the impact of blasts and shocks on structures and present a case study of a multistorey building. McAllister *et al.* (2013) report on the structural analysis of the collapse of the World Trade Center buildings in New York after the 2001 terrorist attack, which included simulation with ANSYS and LS-DYNA. Roca *et al.* (2013) use finite element analysis to study the construction process and subsequent deformation of masonry arches in Mallorca Cathedral. Bojórquez and Ruiz (2014) discuss the load factors for structural design that align with local building codes. Illampas *et al.* (2014a) combine finite element simulation and laboratory testing of an adobe masonry building. Siefert and Henkel (2014) present simulations to study the potential consequences of the impact of a commercial aircraft on a nuclear reactor building. Barazzetti *et al.* (2015) present an approach to capture the geometry and other relevant data from historic buildings, translate this to a BIM environment and then turns this into input for a finite element structural analysis. Brunesi *et al.* (2015) present a computational approach to study the fragility of reinforced concrete frames against progressive collapse when subjected to an extreme load. This work includes probabilistics in terms of geometry, materials and loads; it looks at the overall structure, connections as well as elements. Cui and Caracoglia (2015) simulate the impact of wind on tall buildings, combining the evaluation of forces, peak displacement, structural damage and intervention costs. Llau *et al.* (2015) present a method to simulate cracking of concrete. Petrone *et al.* (2016) provide another example of simulation-based studies of progressive collapse of reinforced concrete framed buildings.

- **General engineering software**
Since building simulation is in essence the use of computers to solve mathematical models of building behaviour, it is obvious that general engineering tools supporting the manipulation and solving of explicit equations can also be used for the specific discipline. There is a range of powerful tools and environments that offer a range of facilities, such as matrix computation, finite element analysis, data visualization and analysis. In general these tools allow full access to the equations that underlie (building) simulation, making them extremely flexible and versatile; however their use typically requires high levels of expertise and, in some cases, programming skills. As these tools are used across several disciplines, they tend to have a large professional user base and advanced capabilities. A typical approach is to develop models of building components and then to combine these into larger systems.

Some of the well-known tools in this category are Matlab, ANSYS, Strand7, COMSOL Multiphysics and Abaqus. Many of these have dedicated 'toolboxes' that help to apply them to specific areas such as mechanical engineering, fluid dynamics, electronics or heat and mass transfer analysis. For instance, Matlab comes with the Simulink environment, which offers access to a library of customizable models; this is popular in the design and analysis of control algorithms. ANSYS offers a range of tools for special domains, as well as suites or workbenches that give access to the full functionality. COMSOL Multiphysics has add-on modules for electrical, mechanical, fluid flow and chemical analysis, as well as multipurpose modules that support things like optimisation and interfacing with other tools. For some tools, further modules and components have been developed that specifically support building simulation; for instance Matlab/Simulink can make use of the SIMBAD (SIMulator of Building And Devices) component library and the International Building Physics Toolbox (IBPT) libraries and models; Matlab/Simulink and COMSOL Multiphysics both have access to the HAMlab model and tool collection.

Van Schijndel (2014) reviews the use of Matlab/Simulink to handle differential algebraic equations and ordinary differential equations within the building simulation context. Most other papers in the literature are about application. For instance, Ucar and Inalli (2008) simulate the thermal behaviour of seasonal storage tanks for solar heating systems using ANSYS. Zhang and Wachenfeldt (2009) use COMSOL for a detailed numerical study of the heat transfer in concrete walls that contain air cavities. Zanghirella *et al.* (2011) present a model for the simulation of active transparent facades that has been developed in Matlab/Simulink. Kumar *et al.* (2012) study the impact of tornado wind loads on wooden gable-roof buildings in the United States using ANSYS. Gerlich *et al.* (2013) present work that validates heat transfer calculations with COMSOL Multiphysics and also compare the results with simulation outputs of a range of other building simulation tools. Royon *et al.* (2013) use COMSOL Multiphysics to simulate the thermal behaviour of a new floor component with phase-change materials. Tomažič *et al.* (2013) use Matlab/Simulink to simulate a controller of an HVAC system in an office, whereas they use the Dymola/Modelica environment to simulate the thermal behaviour of the office and Matlab for the lighting. Barbason and Reiter (2014) use ANSYS Fluent for airflow simulation in a study of overheating of a two-storey house. Chen and Treado (2014) have developed a simulation environment that is based on HVAC modules within Simulink that aims to enable the study of different control strategies. Gustafsson *et al.* (2014) compare the simulation of advanced building systems in both Matlab/Simulink and TRNSYS, obtaining good agreement. Michalak (2014) has used Matlab/Simulink to automate the monthly quasi-steady calculations according to the EN ISO 13790 standard for the thermal performance of buildings. Sasic Kalagasidis (2014) reports on the modelling of phase-change materials (PCMs) using the International Building Physics Toolbox (IBPT) in Simulink. El Mankibi *et al.* (2015) use Matlab/Simulink to identify efficient design alternatives for an active multilayer living wall. O'Kelly *et al.* (2014) report on hygrothermal building simulations conducted with HAMlab. Dimitri and Tornabene (2015) use various approaches, including Strand7 modelling, to study the seismic capacity of masonry arches and portals. Lim *et al.* (2015) have used the Simulink toolkit within the general Matlab environment to model a district heating system with five buildings. Arrunda *et al.* (2016) study the bond between concrete and carbon fibre reinforced polymer laminates through finite element analysis with Abaqus. Lu *et al.* (2016) employ ANSYS to model the seismic performance of a reinforced concrete frame before and after the total collapse due to an earthquake. Rahman *et al.* (2016) use COMSOL Multiphysics to model a pressurized stratified thermal storage tank.

- **Modelica**
A specific general engineering modelling language that is getting significant traction in the building simulation domain is Modelica. This is discussed separately as it is a language rather than a tool; models built in Modelica still need to be simulated in an environment, which is provided by tools such as Dymola, MapleSim, OpenModelica, SimulationX or Vertex. Modelica is an object-oriented equation-based language. Moreover, Modelica uses acausal models; this means that the direction of information flow between sub-models is not predefined, as in traditional imperative software where there are strict rules about inputs and outputs of modules, and how these may be connected. This structure enables the development of larger systems from predefined component models; this is further helped by the use of graphical user interfaces (GUIs) and icons for components. There are different open-source libraries with building models for Modelica available; further work on development of models is conducted in the context of the International Energy Agency Annex 60 project (IEA, 2016a) and, since 2016, IBPSA Project 1.

A good background to the use of Modelica for building simulation is provided by Wetter (2009); this paper also introduces the 'Buildings' component library by Lawrence Berkeley National Laboratory. More detail on this 'Buildings' library can be found in Wetter *et al.* (2014). Another good overview of Modelica as applied to buildings is provided in Wetter *et al.* (2016); this paper compares and contrasts traditional imperative programming languages with the equation-based languages. Further academic papers on the topic are Felgner *et al.* (2006), who report on an early Modelica library of building systems developed at Kaiserslautern in Germany. Sodja and Zupančič (2009) discuss the transfer of a passive solar building model from Matlab/Simulink to the Modelica environment. Bonvini *et al.* (2012) describe how airflow models suitable for building rooms can be implemented in Modelica. Ali *et al.* (2013) optimize chilled water systems using Dymola/Modelica. Nytsch-Geusen *et al.* (2013) present the German 'BuildingSystems' library of complex energy buildings systems for Modelica. Jeong *et al.* (2014) have developed a process to transfer building information model (BIM) data into building energy models for Modelica. Kim *et al.* (2014) use Modelica to develop a simplified and reduced building envelope model for use in urban simulations. Li *et al.* (2014c) review the modelling of HVAC systems in Modelica. Fuchs *et al.* (2016) present an approach for modelling both buildings and districts within Modelica, thus enabling integrated analysis at the larger scale level that represent buildings, energy substations, pipes and energy plants. Sangi *et al.* (2016b) use Modelica for the thermal simulation of a micro CHP unit in the wider context of a thermo-economic analysis; authors of the same group also use this approach to model a high-temperature water circuit, a low-temperature water circuit and the related controls (Sangi *et al.*, 2016a). Perera *et al.* (2016) present building models in Modelica for inclusion in actual building energy management systems and compare the response speed as well as robustness to that of general Matlab models; they conclude that Matlab is faster but Modelica more robust. Zuo *et al.* (2016) describe the coupling of heat transfer, airflow and control when modelling HVAC systems and the building envelope in Modelica.

- **Coupled simulation**
Another special area within building simulation is coupled or co-simulation. There may be different reasons for coupled simulation; for instance one may want to study the interaction between different physical domains such as daylighting and thermal, or one may want to simulate a building component with one tool while using another tool to simulate the full building. In principle, there are four situations with regard to coupling. In the first, tools are fully stand-alone, and only results are combined. In the second, tools share input data but do not interact during the simulation; this is named interoperability. Tools may just share some data and use their own model or may run from a common shared model. The third situation with regard to coupling is where tools exchange information during the simulation; for instance a CFD code may be used to compute air temperatures that can be taken forwards by a thermal simulation code to compute heat transfer by transmission and radiation. The fourth situation with regard to coupling is complete integration; in such a case the different domains are fully integrated within one code, for instance when CFD capabilities are embedded in a thermal simulation program. Situations three and four are commonly considered to be proper co-simulation; it requires attention to convergence, stability and speed of the co-simulation. There may be one master simulation code, with the other acting as a slave; simulations can be carried out in parallel or in sequence. Interactions between the coupled tools may be handled and harmonized by middleware. Underlying modelling may have to ensure consistency; for instance, if in a thermal network a room is studied using CFD, then there is a need to connect the nodes of the

mesh with the overall thermal network. Importing one simulation into another may also be done using a functional mock-up unit (FMU), which packs models and equations for use in an appropriately designed master program; formats for this are defined in the Functional Mock-up Interface (FMI) standard.

Trčka *et al.* (2009, 2010) provide a good overview of co-simulation, focussing on the study of integrated and innovative HVAC systems. Wetter (2011) introduces the Buildings Control Virtual Test Bed (BCVTB), a middleware that enables the co-simulation between a range of different simulation tools such as EnergyPlus, Radiance, Matlab/Simulink and Modelica/Dymola. Arendt and Krzaczek (2014) discuss the coupling of transient CFD models and heat transfer in the building envelope at the level of equations and cover the handling of heat transfer coefficients as well as mesh mapping. Zhai and Chen (2006) provide guidelines for the coupling of thermal simulation with CFD. Further academic papers include the work by Citherlet *et al.* (2001) who provide a general discussion about the integration of various domains such as heating, lighting and ventilation in building performance simulation. Bartak *et al.* (2002) discuss integration of CFD and thermal simulation, specifically focussing of the extension of ESP-r. Zhai and Chen (2005) present coupling of EnergyPlus with CFD. Steeman *et al.* (2009) describe a coupled simulation model for heat and moisture transfer in air and porous materials. Fan and Ito (2012) use co-simulation with TRNSYS and ANSYS/Fluent to explore the energy saving potential of an energy recovery ventilator; Gowreesunker *et al.* (2013) use a similar set-up to evaluate the performance of phase change material heat exchangers. Goldsworthy (2012) has coupled thermal building simulation in TRNSYS with fire simulation in FDS. Feng *et al.* (2012) present the development of a fully integrated model for the simulation of heat, air, moisture and pollutant transfer. Beausoleil-Morrison *et al.* (2014) discuss the coupling of ESP-r and TRNSYS in order to capitalize on the strengths of each tool. Hafner *et al.* (2014) explore the communication between various discrete and continuous simulations models in the context of applying the BCVTB for the modelling of a complex factory. Nouidui *et al.* (2014) present an functional mock-up unit that enables co-simulation within EnergyPlus as master program. For instance, this allows the modelling of a novel HVAC system and its controls in Modelica and then to package this for simulation in EnergyPlus. Yao (2014) presents a co-simulation effort to explore the energy performance of window shades that employs EnergyPlus and the BCVTB in order to integrate a Markov chain method that represents manual control. Du *et al.* (2015) use co-simulation of TRNSYS and CFD to identify the optimal placement of a temperature sensor for an HVAC control system. Hong *et al.* (2016) use co-simulation in order to integrate an occupant behaviour model in thermal building simulation. Rouault *et al.* (2016) use co-simulation to model a novel latent heat thermal energy storage in Matlab, while a full house is simulated in EnergyPlus. Sangi *et al.* (2016b) use co-simulation with Modelica and Matlab/Simulink in order to model a building and a system controller.

- **Tool interfaces, shells and environments**
A final category of tools to discuss is a class of software that helps the use of other software, for instance by providing a graphical user interface (GUI), default values and preconfigured models. In many cases, tools in this category enable coupled

simulation, where the shell program acts as third-party interface. This type of tool is highly popular in academic education and some areas of building services consultancy. There also is significant academic work, as this is seen as one way to support the uptake of building simulation tools in actual building practice.

Examples in this category include Ecotect, IES-VE, Energy-10 and eQuest. The whole building simulation tool EnergyPlus was explicitly designed as a 'simulation engine' and from the start the creators of EnergyPlus anticipated the development of user interfaces; these are now provided by for instance DesignBuilder, OpenStudio and Sefaira.

Examples of work in the literature discussing the use of interfaces or shells are Reinhart *et al.* (2012) who have used DesignBuilder in an educational context and Sesana *et al.* (2016) who present work with students using Sefaira. Gupta and Gregg (2012) use IES-VE as simulation tool in academic research on the impact of projected climate change on homes in England. Ke *et al.* (2013) employ eQuest in a study on the prospects of energy performance contracting for office buildings. Peng (2016) employs Ecotect in a study of life cycle carbon emissions. Roth *et al.* (2016) present the measure feature of the OpenStudio environment, a way to simplify the modelling of energy-saving measures.

6.2.2   Validation, Verification and Calibration

An introduction to building performance simulation also should include a discussion of the principles of validation, verification and calibration, which are used to assure the quality of simulation efforts. These are common approaches in the field of scientific computing.

Verification is defined as 'the process of determining that a model implementation accurately represents the developer's conceptual description of the model and the solution to the model' (Oberkampf and Roy, 2010: 26). In more general terms, it can be explained as making sure that the computer code actually does what the software developer wants the code to do. In software engineering, verification deals with whether the product is built correctly (Pressman, 2005: 388). Oberkampf and Roy (2010) distinguish two subcategories: code verification, which is 'the process of determining that the numerical algorithms are correctly implemented in the computer code and of identifying errors in the software' (*ibid.*, 32), and solution verification, which is 'the process of determining the correctness of the input data, the numerical accuracy of the solution obtained, and the correctness of the output data for a particular simulation' (*op. cit.*, 34). Software engineering in general requires formal technical reviews, which should start at the software component level and then expand to cover the complete system and which should employ different testing mechanisms. The process should accommodate but not be limited to debugging. In some cases verification may benefit from the feedback from an independent test group (Pressman, 2005: 387).

Validation is 'the process of determining the degree to which a model is an accurate representation of the real world from the perspective of the intended uses of the model' (Oberkampf and Roy, 2010: 25). In terms of software engineering, validation deals with whether the right product is being built (Pressman, 2005: 388); in the context of building simulation, it is taken to ensure a good mapping between the physical processes and the rendering of these processes in the model. In general, there are three established approaches for validation; for best results these ought to be followed in parallel:

- Empirical validation – based on a comparison of simulation outcomes to measurements taken from a from real building or a dedicated test facility.
- Analytical verification – based on a comparison of simulation outcomes to a known analytical solution, typically for an isolated case.
- Comparative analysis – based on the cross-comparison of simulation outcomes of a range of simulation tools, preferably including tools that are in wide use and are generally accepted to represent the state of the art.

Validation leads to acceptance of a software program or recommendations for improvement. Data for empirical validation may stem from dedicated test facilities, such as the PASSYS test cells (Jensen, 1995), or can be collected from the monitoring of real buildings in operation. In some cases it may also be possible to augment measured data with synthetic data, for instance by adding direct normal solar radiation that correlates with measured global solar radiation (Song and Haberl, 2013). A general problem for the validation of building simulation programs is the complexity of actual operational conditions, which include the local weather, control settings and occupant behaviour (Ryan and Sanquist, 2012). Another difficult area to model is occupant behaviour. As an example, Tahmasebi and Mahdavi (2016) critically discuss differences between stochastic and non-stochastic window operation models that they tested in a calibrated EnergyPlus model of an office. The Building Energy Simulation TEST (BESTEST) procedure provides a range of case studies and corresponding benchmarks. These cases include diagnostic cases and mandatory qualification cases (Judkoff and Neymark, 1995; Neymark *et al.*, 2002). A classic contribution on validation of building energy simulation tools is the work by Lomas *et al.* (1997), who studied the interprogram variations between 25 tool/user combinations; findings of this work became part of the BESTEST methodology.

Calibration is 'the process of adjusting physical modeling parameters in the computational model to improve agreement with experimental data' (Oberkampf and Roy, 2010: 44). Calibration leads to the definition of parameter values in order to obtain an optimal match between model outcomes and empirical observations. Heo *et al.* (2012) distinguish between the calibration of observable model parameters (such as dimensions, control settings and system configurations), which they name *operational adjustment*, and the calibration of non-observable parameters (such as heat transfer coefficients), which they name *parameter estimation*. There are three main approaches to achieve calibration. The first is manual iterative calibration, which relies on user experience and a trial-and-error process to adjust the model. This is sometimes named 'heuristic calibration'. A more structured approach is achieved by using graphical and statistical methods. Finally, one may also automate calibration and use advanced machine learning techniques such as genetic algorithms to obtain the best possible fit between prediction and measured data (Reddy *et al.*, 2007a). While calibration only applies to individual models, expertise gained from the calibration process can be of use for later development of other models and, in a wider sense, for tool development in general. Samuelson *et al.* (2016) report on the calibration of design stage building energy models with actual weather data and revised internal loads and plug loads; they found this reduced an average underprediction of energy use from 36% to 7% only.

Where actual data is available, models can be calibrated. In general the process of calibration of thermal building models consists of the following steps:

1) Collect data about input parameters for the modelling effort, and ensure proper documentation.
2) Gather weather data from the calibration period, which can replace the standard weather used in most simulations.
3) Simulate the model and study the prediction of temperature and humidity.
4) Compare the simulation results with metered data, on either a monthly or hourly resolution.
5) Evaluate differences in simulated versus metered data using bar charts, time series graphs and scatter plots.
6) Revise the input data of step 1 where needed and repeat steps 3 and 4 to achieve a predefined calibration accuracy.
7) If needed, step down from high-accuracy hourly data to lower-accuracy monthly data if budget and time for calibration work is limited (Efficiency Valuation Organization, 2014a: 15–16).

Reddy *et al.* (2007a) recommend an iterative process, where a coarse grid search and ranking are used to identify promising parameter vectors, followed by a guided search to further refine these. Raftery *et al.* (2011) suggest a formal calibration process that includes the use of version control software, a range of iterative model revisions that brings modelling results close to measurement data, the capturing of how well revisions meet acceptance criteria and extensive process documentation. While the modelling of building geometry and the related parameters feature prominently in building simulation, other input data such as material properties and emission rates receive significant less attention. These are mostly reported in very simplified form only in papers on modelling, and the provenance of the numbers often is unclear (Hand, 2008).

A difficult issue is how close one requires simulation predictions and measurement data to correspond. Often this is expressed in statistical terms, with as criteria a Mean Bias Error (MBE) of 5% and a Cumulative Variation of Root Mean Squared Error (CVRMSE) of 20%. However, these values are mostly used in the context of energy use prediction; different values may be set for prediction of overheating or predictions in totally different domains such as acoustics, lighting or building evacuation. Another problem in calibration is the size of the parameter space, which is typically very large; often there also is a lack of empirical data, as there can be a multitude of values that become of interest in a calibration effort.

An influential document that gives guidelines for calibrated energy simulation of whole buildings is ASHRAE Guideline 14 (ASHRAE, 2002). This suggests a calibration approach that includes the following steps: planning of the calibrated simulation, data collection, model development and simulation, comparison of simulation results with measured data, model refinement, development of a baseline model that can be used for analysis of energy-saving measures, and reporting (*ibid.*, 2002: 32–44). Guidelines on when and how to use calibrated simulation in order to establish the performance of energy conservation measures are included in the International Performance Measurement and Verification Protocol (Efficiency Valuation Organization, 2014a). Seminal papers on calibration of building energy simulation papers are the work of Reddy *et al.* (2007a, 2007b). Further worthwhile contributions are the papers by Raftery *et al.* (2011), Coakley *et al.* (2014) and Sun (2014: 10–30). Strachan *et al.* (2016) includes an overview of the many International Energy Agency tasks and annexes that included substantial efforts on building energy tool validation.

Some examples of work that discuss the validation of simulation tools are as follows: Neymark *et al.* (2002) describe the application of the BESTEST diagnostics to the simulation of space conditioning equipment. Reddy *et al.* (2007b) demonstrate the process of calibration of building energy models in DOE-2 for two artificial and one actual office building. Judkoff (2008) reports on validation activities for the simulation of ground-coupled floor slabs, airflow, the interaction between shading, daylighting and internal loads, mechanical equipment and controls and buildings with double-skin façades. Beausoleil-Morrison *et al.* (2009) describe the use of program-to-program comparison in order to validate models of fuel cell micro-cogeneration systems. Kim and Park (2011) attempt empirical validation of the EnergyPlus simulation tool when used for the simulation of a double-skin façade by calibrating the model of this double-skin façade using extensive empirical data from a dedicated experimental set-up. Heo *et al.* (2012) use Bayesian calibration to deal with uncertainties in semi-dynamic models used for retrofit analysis. Tabares-Velasco and Griffith (2012) present diagnostic tests for the simulation of surface heat transfer processes. McNeil and Lee (2013) discuss the validation of Radiance modelling of optically complex fenestration systems, demonstrating the application of validation principles in the context of lighting rather than the prevalent thermal domain. Buonomano (2016) reports on the validation of a research tool named DETECt using BESTEST. Strachan *et al.* (2016) present validation studies that include one round of blind validation, where the modelling teams have to predict building performance without having access to calibration data, as well as a second round or remodelling, where modellers get access to empirical data that allows them to improve their models.

There is a large number of papers in the scientific literature that discusses model calibration for specific buildings. Pan *et al.* (2007) report on the calibrated simulation of a high-rise commercial building in Shanghai using DOE-2. Song and Haberl (2013) study the calibrated simulations of state office buildings in the United States, looking at the use of a combination of measured and synthetic data. Mustafaraj *et al.* (2014) present the calibration of a DesignBuilder/EnergyPlus model of a university building that is used as living laboratory for intelligent buildings and hence needs to be understood in significant detail. Kandil and Love (2014) present the calibration of an EnergyPlus model of a school with displacement ventilation and radiant thermal slab. Sun (2014: 101–136) offers advanced case studies of six buildings at a university campus in the United States, where measured data are compared with probabilistic simulation results. Ji and Xu (2015) discuss the calibration of a mixed-use building in Shanghai, noting the problem of disaggregating the electricity use of HVAC and plug load systems. Kim and Haberl (2015) present the calibration of a simple change point regression model for single-family residential buildings in the United States. Royapoor and Roskilly (2015) report on the calibration of an EnergyPlus model of a modern 5-storey office in the United Kingdom.

6.3 Expert Judgment

An expert is 'a person who is very knowledgeable about or skilful in a particular area'. Expert assessments are often used in complex situations where modelling is difficult or in cases where there is no or insufficient data available. Gordon and Gallo (2011) provide an example from the field of ecology, where experts are important in defining watershed conditions. Another area where expert judgment is regularly used is the evaluation of future products where, by nature of the novelty of the product, there is no hard data to underpin decisions (Ozer, 2008); experts thus often play a key role in managerial decision making in industry (Coussement *et al.*, 2015). Expert judgment may also be used to improve the predictions by models, in which case the expert judgment is combined with other data (Wilson, 2016); in modelling the use of expert judgment is a common approach to reduce uncertainty in complex predictions (Scholz and Hansmann, 2007). Expert judgment may also be hardwired in calculation and assessment methods like simulation; it is recommended that where possible any underlying assessments are clearly labelled as such (Kinney *et al.*, 2010). There are advanced mathematical models that deal with uncertainty and that benefit from expert elicitation, such as Bayesian model averaging (BMA) or credal model averaging (CMA); for a discussion and contrast of the two, see Corani and Mignatti (2015). It must be borne in mind that expert judgment studies typically are expensive, as they require significant time to run properly (Cooke and Goossens, 2008) and by their nature involve the participation of specialists.

Expert judgment also plays an important role in law; in such case the expert is typically referred to as expert witness. The expert may be called by various parties in a lawsuit: prosecution, defence, or judge and jury. The criteria for being an expert witness are the same as in other contexts and include having the right education, knowledge and experience. Keller (2015) discusses this position for the field of geomorphology, but the principles apply across most other disciplines. A key issue for expert witnesses is to focus on scientific research and use this come to their own conclusions; the line of reasoning then needs to be conveyed to the parties involved in the case. In forensic science, which deals with the analysis of incidents and similar, sometimes past events can be clarified by interviewing experts. A good process involves definition of the event of interest, asking the expert for a description of the event, constructing a timeline of the related decisions and events and where relevant identifying decision points, and finally probing these decisions. While information may be missing from the recollection, data obtained from this process is still seen to be valid (Alha and Pohjola, 1995). Expert judgment also plays an important role in the evaluation of student performance; in this context the term professional judgment may also be used. There is some discussion on how this compares with the use of checklists, faculty calibration and similar efforts that aim at making grading more objective; however there is evidence in the medical sector (dentistry) that expert judgment gives a better insight in the capacity of students to perform, and this capacity is something that must be inferred from observations (Chambers and LaBarre, 2014). An underlying cause may be that the evaluation of the performance of 'deconstructed competencies' misses some of the challenges of expert intervention in a real-life complex context (Mylopoulos and Regehr, 2011).

In principle, expert judgment is just one source of scientific data (Goossens *et al.*, 2008). An important aspect of it is that it does not require formal experiments or calculation procedures; as stated by ISO 6241 (1984: 3), 'a judgment or appraisal can permit the extent of satisfaction of performance requirements to be assessed on the basis of experience of similar cases and conditions or compliance with well-established solutions'. The development of 'expert systems' or 'artificial intelligence' in computing aims to replicate part of the human reasoning and judgment processes, combining a knowledge base of facts and rules with an inference engine that deduces new rules and facts.

According to the Project Management Body of Knowledge, expert judgment can be provided by groups or individuals with specialized training or knowledge (PMBOK, 2008: 77). However, human experts may not always be free from bias (Coussement *et al.*, 2015; Zhang *et al.*, 2017). But even if expert opinions may not be fully objective, they remain be valuable; this value is further increased if the experts can indicate which part of an assessment is based on hard facts and which part is subjective (Dror, 2013). Expert judgment relies on opinion and thus is subjective by nature; however that fact is compensated by the authority of the experts in their field. The key issue in ensuring that expert judgment is seen to be valid is the use of a transparent and well-documented methodology, which identifies the right experts and elicits the right information. It is important to prevent the expert assessment to be seen as having changing opinions, biased viewpoints or a focus on only one perspective, or to include overly conflicting judgments (INCOSE, 2015: 120). A place where uncertainties and bias become clearly visible is when experts are asked to estimate a parameter value or probability distribution; experts will always use intervals and point out that few things in nature have deterministic, clear-cut parameters (Utkin, 2006). In such cases, experts may tend to give conservative judgments; and it may become obvious that some experts might be more reliable than others. De Wit (2001) suggests that the credibility of expert judgments is strengthened by opening up all data, including the expert names and their contributions, to peer review. Furthermore, he states that experts should not have a stake in the outcome of the study, that methods should not risk a bias in the results and that where feasible any quantitative outcomes should be subjected to empirical quality checks. In general, predictions by experts may be improved by structuring the tasks given to experts, and by sharing the expertise and making sure experts interact (Ozer, 2008).

The process of asking experts for their opinion or assessment of qualitative values is sometimes named expert elicitation (EE). The general approach for expert elicitation is relatively straightforward: one needs to identify the experts, obtain their assessments, and where needed post-process these assessments and combine them if more than one expert was used. However, practical and methodological issues need to be considered in the process; for instance the questions that will be asked need careful preparation as they should be posed in a neutral way, without introducing a bias. One needs to decide whether it is practical to bring experts together in a panel session, which allows discussion and debate (Gordon and Gallo, 2011) but which may also lead to group think, or whether telephone or online methods should be used; these have the advantage of being anonymous but typically have a lower engagement (Khodyakov *et al.* (2011). Expert elicitation may have different goals: to represent the overall view of the scientific community, to represent a political consensus amongst different stakeholders or stakeholder groups, or to gain a rational consensus about what a group believes to be a representation of uncertainties in a certain domain (Cooke and Goossens, 2008).

Selection of experts is a key step in any form of expert assessment or judgment (Gordon and Gallo, 2011). In some cases, identification of experts may be based on peer nomination (Kinney *et al.*, 2010). Sometimes statistical procedures are used to support the selection of experts; this can be done by using expert calibration techniques if known variables are available for a test (Cooke and Goossens, 2008) or by a process where extremely high or low judgments are excluded (Scholz and Hansmann, 2007). Advanced mathematical techniques such as Bayesian aggregation of probability estimates are used in human reliability analysis and the prediction of human error probabilities and may be used to support the selection of reliable experts; see for instance the work by Podofillini and Dang (2013). In general however, the status of expert is typically obtained based on the track record and professional characteristics of an individual (Coussement *et al.*, 2015), such as membership of and status within professional organizations or universities (Carpio *et al.*, 2015). Expertise may also stem from working with complex systems, without formal training, leading to local workers having relational and causal insights that are not formally codified (Abernethy *et al.*, 2005). Some research in the social side of science and technology explores how professional expertise is achieved. For instance, Gendron *et al.* (2007) have studied how auditors become seen to have the expertise needed for their role in measuring government performance. In general, an expert needs to be confident about his or her work and will benefit from good communication skills (Keller, 2015).

The actual expert elicitation is of course highly dependent on the issue at hand. However, it is imperative that this is planned to the highest possible standard, as one is dealing with experts who may pick up methodological shortcomings. As expert judgment is known to lead to a subjective outcome, it is good practice to explicitly aim to capture and quantify any uncertainties and risks in these assessments (INCOSE, 2015: 119). It is also recommended to probe underlying beliefs and sources of evidence (Kinney *et al.*, 2010). Structured expert judgment (SEJ) aims to obtain expert assessment in a transparent, methodological manner. This process can be broken down in three main stages: preparation, expert elicitation, and post-elicitation. It focusses on very clear target variables, for which uncertainty is to be established. The rationale for the expert judgments is captured together with the assessment. Finally, outcomes are handled statistically (Cooke and Goossens, 2000; Goossens *et al.*, 2008).

Where a group of experts gives an opinion, there is a need to aggregate these opinions into one overall opinion.[10] The use of several experts helps to reduce the impact of errors in individual assessments, but there is debate in the scientific community on whether the opinion of these experts should be weighted and, if so, what basis there is to decide upon a weighting (Bolger and Rowe, 2015). One way of coming to an overall opinion is voting or ranking. This removes the need for performance measurement and data collection; however it runs the risk of strategic voting by some of the experts (Hurley and Lior, 2002). Mathematical models may be also be used to aggregate the individual judgments; however most methods assume that the individual judgments will be independent, which is not always the case in reality as experts may share the same education or have the same level of expertise (Wilson, 2016). Zhang *et al.* (2017) capture the judgment of a number of experts in a decision-making trial and evaluation laboratory (DEMATEL) and use optimisation and ranking to arrive at preferred design solutions. A difficult issue in the context of aggregation is how to deal with self-confidence ratings of experts (Scholz and Hansmann, 2007). Another way to combine opinions is to bring experts together in an 'expert panel'; this is often done where there is a broad group of stakeholders. However, there is a risk that such sessions lead to bias resulting from the personal interactions. In some cases such an expert discussion may be carried out online using structured process such as ExpertLens (see for instance the work by Khodyakov *et al.* (2011) in the medical quality improvement field) or

---

10 Note the comments of Hazelrigg (2012: 467) with regards to group decision making.

Delphi methods (for an application in the area of sustainable building, see Heffernan, 2015: 163–232).

Expert judgment is relatively common in risk and reliability engineering and in the medical domain. One example is the work by Mallor *et al.* (2008), who use expert judgment to assess the risk at a public sporting event where insufficient historical data is available to run statistical models and show how this can be integrated into a risk assessment tool for participants. Kinney *et al.* (2010) discuss the use of expert judgment in the context of the assessment of the benefits of regulatory controls of particulate matter to human health, which requires the definition of a concentration–response function as well as an estimate of uncertainties in the relationship. Berendonk *et al.* (2013) discuss expert assessment in the context of medical education, where expert assessment of trainee performance by senior staff is common. It has been found that the following aspects play a role in such expert assessments: the characteristics of the expert, the perception of the assessment task by the expert and the context of the assessment. Proper expert assessment requires that the assessor has domain specific knowledge, practical experience and credibility. Cooke (2013) discusses an example of using structured expert judgment (SEJ) in the context of climate change: the assessment of contribution of the melting of ice sheets to sea level rise. This text draws a distinction between expert informed modelling and the use of experts to gain deeper insight into the sources of uncertainty. Carpio *et al.* (2015) have used an expert panel assessment to review the strength and weaknesses of documents to certify energy efficiency in Spain. Expert judgment also plays an important role in the scenario planning process in industry; systematic scanning of expert opinions is used here to detect emerging threats and opportunities (Meissner *et al.*, 2017). Strachan and Banfill (2017) use expert judgment to rank retrofit measures for non-domestic buildings.

Expert judgment may play an important role in the building sector, but unfortunately there is a paucity of scientific literature that explores what defines an expert in this domain and how to elicit the best assessments from these experts. Yet Gann and Whyte (2003) point out the adaptivity of expert judgment as a key advantage in dealing with uncertainty in a design context. Blyth and Worthington (2010: 86) emphasize that many qualitative performance aspects of buildings, such as their image, character or success, are subjective and that especially such aspects require judgment. It is interesting to note that the architectural community indeed puts a strong emphasis on awards like the Royal Institute of British Architects (RIBA) Stirling Prize, American Institute of Architects (AIA) Gold Medal and Pritzker Architecture Prize; similarly the building services engineering community in the United Kingdom values the CIBSE Building Performance Awards.

Further evidence of the use of expert judgment in building performance is mostly implicit and contained under discussions of different subjects. Galiana *et al.* (2012), in the context of music hall acoustics, mention the risk of having experts use concepts and parameters that may not correspond to the perception of actual buildings; they conduct experiments with both groups in order to ensure that differences in perceptions are noted and taken into account. De Wit (2001: 68–93; 110–120) and de Wit and Augenbroe (2002) use expert judgment to assess uncertainties in building performance simulation, notably the ventilation rate of building spaces and room air temperature distributions. In this context, the expert judgment pertains to the uncertainty in a variable; the selection of experts thus needs to take place from the field to which the variable belongs, giving a clear definition of the domain. In itself, the uncertainty as given by the experts is an observable quantity that is clearly defined. The procedure used in this work is the one described in Cooke and Goossens (2000). Blyth and Worthington (2010: 89) point out that the Post Occupancy Review of Building Engineering (PROBE) studies strongly rely on experts gathering and analysing data from the buildings.

6.4 Stakeholder Evaluation

Yet another approach to evaluate the performance of buildings is through stakeholder evaluation. Here the occupants of the building are asked for their opinion on how well the building performs. Stakeholder evaluation is not unique to the construction domain; for instance it is used extensively in product and software design; it plays a key role in the study of human–computer interaction (HCI) where the key concern is to see how computer users assess input devices or on-screen representations. In the context of building evaluation, the terms stakeholder, occupant and user are often used as synonyms. In general the term stakeholder covers all parties who have an interest in a building, occupants are the people who live and work in a building, whereas user is more generic term that stems from product design and usage but can also be applied to buildings. A closely related concept to occupant evaluation is occupant satisfaction; however it is important to bear in mind that occupant satisfaction also has a relation to the expectations of that occupant. In other industries consumer satisfaction (CS) surveys are not just about the performance of the object under study, but focus especially on the attitudes and perceptions of human users. User satisfaction is of great interest in the areas of product design and ergonomics; see for instance the work by Han and Hong (2003) on CD players or Yun et al. (2003) on mobile phones. Another term that sometimes surfaces in user evaluation is affordance. Affordance expresses what an artefact offers to the user; in the context of product design, it is also taken to represent how well a user knows what to do with, and how to operate, the product (Hsiao et al., 2012) and thus is a measure of how easy it is to use that product. In some cases user feedback may be obtained from virtual prototypes; see for instance the work by Kim et al. (2011) on automobile interior design.

Stakeholder evaluation is seen by some as a key approach to evaluate building performance. Preiser (2001) states that 'the customer is king' and that this also applies to buildings. Leaman et al. (2010) take the position that ultimately, people are the best measuring instrument for observing building performance. CIB Report 64 (1982: 13) lists stakeholder evaluation as one option in a wider range of feedback processes that allow the assessment of building performance. This list includes panels with skilled or informed observers, the systematic study of complaints and issues about which occupants seek information, and direct surveying of what the building occupants do or perceive; further items on the list include reviews of records pertaining to the building, measurement of performance under actual use conditions, indirect assessment on the basic of statistics or indirect analysis on the basis of laboratory testing. Report 64 points out that these approaches are not to be seen as mutually exclusive alternatives, but that instead two or more methods may be used in combination (ibid.). Depending on the objectives of a study, occupant evaluation may take place in either a controlled chamber (laboratory) or an actual building (field test). Studies with occupant evaluation are common in for instance the area of indoor environmental quality. Gossauer and Wagner (2007) compare the two approaches with a focus on the study of thermal comfort. Blyth and Worthington (2010: 87–88) list the following techniques for gathering user feedback: interviews and focus groups, questionnaires, comparison against leading examples and observation by way of a building walk-through. Some areas of occupant evaluation, such as the perception of the view out of windows, are rather complex, and assessment processes for these areas are still very much under development (Matusiak and Klöckner, 2016). Another area where stakeholder studies are under development is statistical correlation; for instance Frontczak *et al.* (2012) and Candido *et al.* (2016) study the relation between occupant satisfaction and indoor environmental parameters using statistics; such work may help to rank performance aspects in an order of importance but needs to take extreme care to correctly capture contextual and occupant characteristics.

A key term in the context of stakeholder evaluation of building performance is post-occupancy evaluation (POE). There is a wide range of interpretations of post-occupancy evaluation; a selection of definitions is as follows:

- 'Examination of the effectiveness for human users of occupied, designed environments' (Zimring and Reizenstein, 1980).
- 'The act of evaluating buildings in a systematic and rigorous manner after they have been built and occupied for some time' (Preiser and Vischer, 2005: 8).
- 'An evaluation by people in buildings on the basis of questionnaires of various extent and generally in connection with physical measurements of different levels of detail' (Gossauer and Wagner, 2007).
- 'A process that involves a rigorous approach to the assessment of both the technological and anthropological elements of a building in use' (Hadjri and Crozier, 2009).
- 'A platform for the systematic study of buildings once occupied so that lessons may be learnt that will improve their current conditions and guide the design of future buildings' (Meir *et al.*, 2009).

In general, the aims of post-occupancy evaluation are to identify the occupant's perception of a building and to quantify their level of satisfaction; both of these involve a subjective element. Post-occupancy evaluation is also used to establish what issues may need improvement and to detect those end stakeholders that are not satisfied; in a wider sense it may be used to guide facility management and design processes (Jaunzens *et al.*, 2003). Post-occupancy evaluation studies often include explorations of productivity, occupant comfort and satisfaction, and sometimes aim to establish fitness for use of a facility.

The process of post-occupancy evaluation starts with a preparation phase, where a target building is identified, researchers/analysts ensure they have access to key stakeholders such as building owners, facility managers and actual occupants, and data collection is prepared. This is followed by actual data gathering. Most studies employ questionnaires amongst the stakeholders or use a series of interviews. Often use is made of Likert scale ratings to capture the occupant satisfaction with aspects such as thermal comfort. This may be augmented with data acquisition from the building systems and dedicated monitoring efforts, which can take significant time to capture the relevant operational conditions. Other data collection techniques may also be included, such as audits, expert walk-through or focus group discussions. Large amounts of data may result; often post-occupancy evaluation captures responses for many rooms and spaces in a building, and this may be further increased by the addition of time series on ventilation, natural lighting, artificial lighting, noise, temperature, relative humidity and similar. It may also include information on the building layout and condition, operational settings and use patterns, as well as opinions on the design quality, building management, ICT provisions, waste collection and similar. Where relevant, occupant characteristics such as age, gender, education, occupation, income, family size and ownership may also be added. Unsurprisingly then, the final phase covers in-depth analysis of all this data, where relevant through statistical analysis. Conclusions must be drawn on overall occupant perceptions, and feedback and suggestions for improvement formulated. The Usable Building Trust, an organization that promotes the use of post-occupancy evaluation, suggests to use a mix of approaches for a review of building performance, which includes audits, such as via the CIBSE TM22 Energy Assessment and Reporting Methodology; the use of discussion and review techniques such Learning from Experience Workshops; deployment of questionnaires, such as the Building Use Survey (BUS) Occupant Survey and CIC[11] Design Quality Indicators and Overall Liking Score; the inclusion of process support methods such as Soft Landings; and the development of a comprehensive strategy that combines all of the previous into a single approach (Bordass and Leaman, 2005). Given this extensive reach, post-occupancy evaluation studies are sometimes scaled according to the level of effort invested and rated indicative, investigative or diagnostic. Efforts that limit themselves to stakeholder perception and satisfaction are sometimes seen as mainly belonging to the social sciences (Mlecnik *et al.*, 2012).

Brief histories of post-occupancy evaluation studies are provided by several authors, such as Gossauer and Wagner (2007) and Hadjri and Crozier (2009). Several authors have described the development of post-occupancy research in the United Kingdom, the related PROBE (Post Occupancy Review of Buildings) studies and Building Use Survey (BUS) methodology, and the Soft Landings approach; see for instance Bordass *et al.* (2001a, 2001b, 2001c), Cohen *et al.* (2001), Cooper (2001), Bordass and Leaman (2005), Way and Bordass (2005) or Leaman *et al.* (2010). Developments in the field of POE in the United States at the start of the new millennium are discussed by Preiser (2001). An early version of what was later to become post-occupancy evaluation is the approach to the collection of building performance data described by Markus *et al.* (1972: 230–245). This includes amongst others the sourcing of data from local authorities, architects and building occupants, data on building use and crowding, acoustic measurements, details on the construction process and details on the organization using the building. Later projects that used post-occupancy evaluation were, amongst others, the Smart Controls and Thermal Comfort (SCAT) Project, the ProKlima study into sick building syndrome in offices in Germany, the EU HOPE project on Health Optimisation Protocol for Energy-Efficient Buildings or the Vital Signs Project on building performance at the University of Berkeley.

In spite of the popularity of post-occupancy evaluation, there are also some critical remarks to be made. First of all, the name may be confusing. Strictly speaking, post-occupancy evaluation does not imply stakeholder evaluation; it just signposts that

---

11 Construction Industry Council.

analysis takes place after a building has been occupied. For instance Mumovic *et al.* (2009) have studied the ventilation rates in recently built schools, but their work does not report on stakeholder evaluation of these buildings, yet they report their work under the name of post-occupancy evaluation. Moreover, the name post-occupant evaluation implies a focus on building occupants, which risks missing other stakeholders and users. Göçer *et al.* (2015) discuss the closing of the loop between post-occupancy evaluation and building design; their work also shows how the term post-occupancy evaluation overlaps stakeholder evaluation, measurement and monitoring and simulation and sometimes causes confusion. Furthermore, the version where post-occupancy evaluation does include occupant evaluation does not only relate to the performance of the building; such evaluations also depend on organizational aspects, weather conditions, the availability of control actions, time of day, expectations of the occupants and adjustments that occupants can make to their clothing or position in the building (Gossauer and Wagner, 2007). Lappegard Hauge *et al.* (2011) discuss the occupant experiences with different types of energy-efficient buildings based on literature review; they conclude that in general, post-occupancy evaluation would benefit from better capturing the social context of building occupants, system operation and occupant training. They also recommend that such studies should be longitudinal, broadened to include an assessment of aesthetics and should capture the reasons why occupants use the building in question. This limited representation of sociotechnical aspects is sometimes seen as restricting the usability of the results; see for instance Chiu *et al.* (2014) for a deeper discussion in the context of domestic retrofit. Yet another problem with conventional post-occupancy evaluations studies is that they are specific to a single case or at best a small group of buildings; findings are context specific and hard to generalize (Mansour and Radford, 2016). A different type of research, such as preference ranking, is needed to overcome this problem. An example of work in that direction is provided by Rebaño-Edwards (2007), who explore the relation between occupant lifestyle and perceptions of building quality, and even attempts to model these perceptions using a neural network approach. Nicol and Roaf (2005) compare and contrast post-occupancy evaluation studies with field studies in thermal comfort, highlighting that thermal comfort is dependent on time and context, something that may be overlooked in traditional post-occupancy surveys. Candido *et al.* (2016) list the following as 'methodological shortcomings' of current post-occupancy tools: a lack of contextualizing of the results, missing the combination of metered data and survey results, and not actually producing meaningful feedback to the key stakeholders. Their Building Occupants Survey System Australia (BOSSA) tool is an attempt to address some of these issues. All the same, post-occupancy evaluation at present is the main overarching approach that combines various building performance assessment methods.

There is a large number of stakeholder surveys in circulation, with many researchers developing their own questionnaire that fits their specific needs. For those looking for a starting point, a good resource is provided in the appendix of *Assessing Building Performance* by Preiser and Vischer (2005: 209–234); readers are encouraged to modify these for their own research purposes. The Building Use Studies (BUS) questionnaire is available against a licence fee, but with a special academic licence for those who wish to use it in a scholarly projects.

Some researchers in the domain of POE, such as Preiser and Vischer (2005), Meir *et al.* (2009) and Leaman *et al.* (2010), show a tendency to ignore the potential role of building performance simulation and expert judgment. This may be due to a view that measured data is the ultimate truth and that POE is about 'real world research' (Leaman *et al.*, 2010), but this exclusion is not helpful in developing an integrated approach towards building performance that makes use of all methods available.

There is a wide range of literature that covers building stakeholder evaluation, especially post-occupancy evaluation. Turpin-Brooks and Viccars (2006) report on a comprehensive post-occupancy evaluation of a sustainable office in the United Kingdom. Wagner *et al.* (2007) report on a summer field study on thermal comfort in office buildings in Germany. McCuskey Shepley *et al.* (2009) used post-occupancy evaluation to explore how staff members of an architectural firm in the United States experienced a new building. Lai and Yik (2009) study the perception of indoor environmental quality of high-rise residential buildings in Hong Kong. Zhang and Barrett (2010) evaluate five primary schools in the United Kingdom. McGrath and Horton (2011) use post-occupancy evaluation to evaluate the satisfaction of students with accommodation in a modular building built using volumetric construction. Choi *et al.* (2012) report on the post-occupancy evaluation of 20 Federal office buildings in the United States. Kansara and Ridley (2012) report on evaluation of a zero carbon city in Abu Dhabi. Mlecnik *et al.* (2012) present a wide review of post-occupancy evaluation studies into the occupant satisfaction of nearly zero energy houses in Germany, Austria, Switzerland and the Netherlands. Lenoir *et al.* (2012) evaluate a net zero energy building on La Reunion in the tropics. Newton *et al.* (2012) evaluate three schools built in Australia according to the same design template. Lai (2013) evaluates the perception of a board house at a university in Hong Kong. Collinge *et al.* (2014) report on the post-occupancy evaluation of a university building in the United States in order to augment a life cycle analysis framework. Leung *et al.* (2014) study the stakeholder satisfaction with care homes for the elderly in Hong Kong. Mundo-Hernández *et al.* (2015) evaluate the perceptions of the occupants of a factory building that has been converted into an art and design gallery in Mexico. Hassanain *et al.* (2016) describe the post-occupancy evaluation of a student cafeteria in Saudi Arabia. Pretlove and Kade (2016) present a post-occupancy evaluation study of seven new social housing dwellings in the United Kingdom, and Wongbumru and Dewancker (2016) have used post-occupancy evaluation to study occupant satisfaction with old and new public housing schemes in Thailand. A nice perspective is provided by Shen *et al.* (2012a, 2012b) who introduce the notion of a pre-occupancy evaluation, capitalizing on the possibilities of building information modelling and occupant activity simulation. They define pre-occupancy evaluation as 'the application of POE in the pre-construction, pre-occupancy or pre-project stage', in order to find out how well a design will meet the user requirements. Kuliga *et al.* (2015) explore how virtual reality may allow occupants to experience a building; they compare experiences in a real building, the virtual equivalent and a virtual equivalent with modifications.

6.5   Measurement of Construction Process Performance

There is a large body of knowledge on the measurement of process performance in the management literature; measurement is fundamental to management approaches such as the Balanced Scorecard (Kaplan and Norton, 1992), ProMES (Pritchard *et al.*, 2012) and Lean Systems Engineering (INCOSE, 2015: 203–207). Some management methods, such as the Six Sigma process, have been successfully applied in construction; for example, see Tchidi *et al.* (2012).

Key aspects of the building construction process are cost, time, product quality, safety, waste reduction and customer satisfaction. Traditional process performance measurement uses techniques such as the manual registration of work quality, time studies and recording of activity rating, crew balance, utilization rates and work sequence. Modern digital techniques, such as digital cameras and video, building information modelling, geographic information systems (GIS), barcodes and radio-frequency identification (RFID), 3D laser scanning and photogrammetry are augmenting the traditional approaches. One area that remains problematic is the difficulty of capturing what goes on inside a building, as this is often obstructed from the outside view and may be hidden inside layered construction.

In the field of construction management, accounting techniques such as cash flow analysis, the review of leading parameters and activity-based ratios, statistical process control (SPC) or earned value (EV) are used to measure and capture progress in terms of cost, schedule and scope (Al-Jibouri, 2003; Maravas and Pantouvakis, 2012; Aliverdi *et al.*, 2013). Measured project progress is often contrasted with the original planning and optimized with techniques such as the critical path method (CPM) or the program evaluation and review technique (PERT). Vanhoucke (2012) provides an example of how both simulated and empirical project measurement data can be used in process management. Product quality is predominantly measured through defect studies (Josephson and Hammarlund, 1999; Dong *et al.*, 2009; Park *et al.*, 2013); interestingly this captures the exceptions where the construction process did not produce what was intended. Construction process safety is also measured in terms of exceptions; here the common approach is to register near-misses and incidents by means of recorded injury rates and similar. Recent work is expanding this with upstream measurement of factors that influence safety, such as the percentage of personnel with safety certification or the results of drug tests amongst workers (Hinze *et al.*, 2013). Further attention is also directed towards the underlying drivers of unsafe situations, such as production pressure, risk perception and peer pressure (Guo *et al.*, 2015). Measurement of construction waste can have both a narrow and wide scope. In a narrow sense, waste coming from construction may be measured empirically, by counting or weighting skips; modelling the process that leads to waste is complex and typically takes the form of a mass balance that accounts for aspects such as the work breakdown structure, conversion ratios and wastage levels (Li *et al.*, 2016). In the wider sense, waste may be seen to include all emissions from the building site, such as sound, dust and water, and may even cover the efficient timing of employing resources such as personnel, equipment and materials. Customer satisfaction is typically measured in terms of user surveys using post-occupancy evaluation. A special category of process performance relates to the design process that precedes the actual construction work; there is some work that aims to measure design productivity and design quality (Dent and Alwani-Starr, 2001).

A selection of some of the work pertaining to process performance measurement in the literature includes the following: Navon (2007) presents work that aims at automated measurement of project performance rather than the typical measurement of the performance of a single building as a system. Techniques used in this context are barcodes and radio frequency identification (RFID), global positioning systems (GPS), video and audio technology and laser detection and ranging (LADAR). Cha *et al.*

(2009) review the different types of waste that may occur on a construction site and suggest a tool to measure the level of waste for these types. Cheng *et al.* (2009) use process operation time and customer satisfaction as the main drivers for construction process performance measurement in the context of process re-engineering and optimisation. Fernández *et al.* (2009) study the noise exposure of workers in the construction sector, which may have implications on their hearing. Ballesteros *et al.* (2010) study noise emission from a construction site in general, with a focus on the neighbourhood. De Marco *et al.* (2009) present a case study that demonstrates how project monitoring feeds into earned value analysis. El-Omari and Moselhi (2011) review the different approaches that are available to collect data on construction progress for site management and reporting. Gong and Caldas (2011) use automated analysis of video data to measure the productivity of construction projects. Roh *et al.* (2011) discuss the challenges of monitoring the progress of construction indoors and suggest the use of an object-based approach in combination with as-planned building information models in digital walk-through models. Pradhananga and Teizer (2013) describe the use of a global positioning system (GPS) to identify and track the activity of construction equipment such as loaders and excavators. Ochoa (2014) discusses the efforts to reduce variations between planned and actual construction times. Siebert and Teizer (2014) use an unmanned aerial vehicle to measure the process of excavation and earth movement on construction sites. Turkan *et al.* (2012) focus on the use of three dimensional laser scanning in order to track construction progress. Ajayi *et al.* (2015) review the different factors that lead to the large amount of construction waste that ends up as landfill and suggest management strategies that may be used to reduce this waste flow. Teizer (2015) explorers the prospects and challenges of using vision-based technology as applied to semi-automated or fully automated tracking of resources on (infrastructure) construction sites.

6.6 Building Performance Quantification in Current Practice

One building that has been the subject of an extensive performance quantification effort is the office that provides the headquarters of SK Telecom in Seoul, South Korea. This paragraph gives a short summary of these analysis efforts; the case study represents a state-of-the-art simulation study that identifies some of the practical constraints that researchers in the field have to overcome.

SK Telecom is a multinational and the largest wireless network carrier of the country. The building was designed by OMA Asia, a subsidiary of the Office for Metropolitan Architecture in collaboration with Junglim Architecture; further partners were Chang Minwoo Structural Consultants and Arup Engineering. Construction of the building took place between 2000 and 2003. The building consists of 33 storeys above and 6 storeys below ground; it has a height of 148 m and a floor space of 91 898 m^2. The building mainly provides office space for SK Telecom but also includes car parking, employee welfare facilities, an auditorium, dining rooms and multipurpose halls. Architecturally, the building is defined by a cranked shape and a blue curtain wall with individual sloping glass panels that create a unique texture; the window–wall ratio is 70%. In terms of building services systems, the building employs 3 boilers, 5 chillers, 1 ice storage system and 5 cooling towers, 47 air handling units, 122 fan coil units and 4 packaged air conditioners.

The SK Telecom Office was subject of advanced simulation in early 2012 in order to assess the prospects of a range of energy conservation measures. A thermal model of the building was developed by a team of researchers from Sungkyunkwan University in Seoul using the EnergyPlus simulation program and calibrated with measured data. While the geometry of the building was simplified significantly, the final model still employed 785 zones. See Figure 6.6. Monitored data of the building was available from the building energy management system, which employs 1692 measurement points throughout the facility. Close review of the measured data at component level has identified small discrepancies in this data, such as a mismatch between recorded fan airflow rate and fan electricity use; however, overall interest is at facility level.

Results from the initial model were compared with measured energy metering data. Electricity consumption prediction was deemed to be in good agreement; however the gas prediction was seen to require further calibration. After inspection of various data, changes were made to the indoor setpoint temperature and boiler efficiency curves, leading to substantial improvements. Results are presented in Table 6.6; differences between simulated and measured energy performance were expressed in terms of Mean Bias Error (MBE) and Coefficient of Variation of the Root Mean Squared Error (CVRMSE) in accordance with ASHRAE (2002) guidelines.

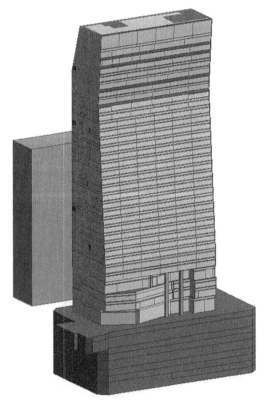

Figure 6.6 Geometry of the thermal model of the SKT Headquarters, Seoul, South Korea. Image courtesy of Sungkyunkwan University.

Table 6.6 Measured and simulated energy use of the SK Telecom building (2012 data).

	Measured	Model	
Electricity use	10 257 547 kWh	10 831 690 kWh	
MBE	—	4.0%	
CVRMSE	—	7.6%	
		Initial model	**Calibrated model**
Gas use	426 132 Nm^3	262 383 Nm^3	432 663 Nm^3
MBE		−38.4%	1.5%
CVRMSE		46.5%	14.0%

Data courtesy of Ahn *et al.* (2013, 2016), Sungkyunkwan University.

The research team has provided detailed comments on difficulties in gaining access to full details of the facility, partly because of security and privacy protocols but also because modifications to the original design may not have been recorded. They note that the modelling of zones is mainly left to subjective judgment, while properly representing the HVAC systems and their control is complex given the many systems in the real building and the many options within EnergyPlus. They report that internal loads and detailed control settings are extremely hard to capture even if a real building is available as reference.

Further details of the simulation efforts described in this section can be found in Ahn *et al.* (2013, 2016).

6.7 Reflections on Quantification Methods

The number and variety of building performance quantification approaches is huge. One may wonder whether these are all needed, but most of these approaches have been developed with a particular interest and scope in mind. The key challenge for doing a particular analysis is to find the right approach for the job. But practice often works the other way round: availability of specific tools may have significant impact on what analysis is actually conducted. It is important not to underestimate how tools and instruments may influence observations and courses of action taken.

Quantification approaches live in the four main silos of physical measurement, building simulation, expert judgment and stakeholder evaluation; within each silo are containers of expertise on specific methods, such as thermography or thermal simulation. There are some relations between silos and containers, but these are not one to one. For instance, experts in ventilation may combine wind tunnel experiments, CFD simulation and expert judgment. But typically experts in co-heating test have limited expertise in working with wind tunnels, although both are experimental techniques. Nor are there standard models available to do a virtual co-heating test, climate chamber experiment and durability test, or to use a virtual hot box. This lack of mapping makes it difficult to create further unity amongst the quantification methods.

It is interesting to note that there is an abundance of standards and norms for physical measurements, but only a few general guidelines such as CIBSE AM11 for building simulation; there is even less on expert judgment and stakeholder evaluation studies. This imbalance contributes to a lack of comparability between the different assessment pathways. The development of common assessment definitions that could be handed to experts in each of the four approaches would help to ensure that the same experiment is conducted, whether physically, virtually or mentally. This in turn would go some way to bridge the 'performance gap' that is mentioned in so many recent studies.

It must be noted that some evaluations are still very challenging. One example is the coupled simulation of a range of interacting physical phenomena such as thermal, moisture, ventilation and lighting. While existing tools allow some runtime interaction, real integrated simulation requires deep integration of simulation codes; simply providing different stand-alone tools with the same input data is not sufficient and only leads to limited insights. As a consequence, the use of BIM offers limited prospects towards achieving fully integrated analysis.

The area of expert judgment seems to have attracted relatively little attention in building performance assessment; further studies on the role of expert witnesses and their approach to performance quantification may yield interesting insights.

The term POE seems rather dangerous; it emphasizes an evaluation that takes place after a building has been occupied, and a limited sample of building occupants only, but is often used as synonym for stakeholder satisfaction or perception surveys. It would be better to talk about stakeholder perception studies and use different wording (such as in-use) to identify the time when a performance evaluation takes place. Note that the lifetime of buildings is lengthy, so post-occupancy and in-use may refer to any point in time in a period that may span several decades.

Given the complexity of buildings and their context, the description of real performance analysis efforts such as the study of the thermal performance of the SK Telecom Headquarters is a non-trivial task. It will be clear that the case study description in this chapter does not contain sufficient information to consistently replicate the simulations studies, nor do the full journal papers that contain further information.

6.8 Summary

This chapter gives an overview of the main approaches for the quantification of building performance: physical measurement, computer modelling and simulation, expert judgment and stakeholder evaluation; it also discusses some of the work on measuring the construction process. This quantification establishes how well a building fulfils its function(s) on the basis of established scientific methods. Some approaches such as expert judgment and stakeholder evaluation include a subjective component by nature but are the best approach available for dealing with missing information or complex situations where the performance depends on the interaction between the building and its occupants.

Physical measurement typically makes use of existing instruments such as thermocouples, flow meters, illuminance sensors and sound meters. These are applied in a range of specific measurement and testing approaches: air pollutants measurements, building systems and component testing, climate chamber measurements, co-heating testing, durability testing, evacuation experiments, use of experimental test cells or buildings, façade test rigs, fan pressurization measurements, fire tests in burn halls or chambers, full-scale lighting measurements and lighting measurements in scale models, hot box measurements, indoor climate analysis, moisture measurements, monitoring of buildings in regular use, sound measurements, structural testing in the laboratory, thermography, tracer gas testing and wind tunnel measurements. Many of these are described in dedicated standards from the International Organization for Standardization (ISO), the European Committee for Standardization (CEN), the American Society for Testing and Materials (ASTM) and others.

Simulation is a virtual experiment, where a computer is used to imitate reality; this approach is part of the wider field of scientific computing. Building performance simulation employs a computational model, based on abstraction of physical realities, which is executed in a computer. There are different ways of classifying simulation models; one prominent approach is to distinguish between stationary, semi-dynamic and transient models, and another is to differentiate between black box, grey box and white (glass) box models. Another view follows the different physical domains and thus differs between acoustical simulation, computational fluid dynamics, heat and moisture models, lighting simulation, pedestrian movement and evacuation models, whole building thermal simulation, fire simulation and structural simulation. General engineering software may also be used for building performance simulation. Special cases are simulation with Modelica, a general engineering modelling language, coupled simulation where different physical domains interact, and the use of tool interfaces, shells and environment that support the access to underlying simulation engines. As in all scientific computing, the concepts of validation, verification and calibration are important for quality control of building performance simulation.

Expert judgment is a way to assess building performance in cases where modelling is difficult or where these is no or insufficient data. It may be used in the context of computing, for instance in order to establish uncertainties in input parameters. Experts can also play a role in law, where they can be called to give their opinion as expert witness. Expert judgment is also common in the evaluation of student performance in education. A fundamental issue for expert judgment is the authority of the expert that makes the assessment, as well as transparency of the whole expert elicitation process; a recommended approach is structured expert judgment (SEJ). Where several judgments are used, care must be taken in bringing assessments together; this may be based on discussions between the experts in an expert panel or can be done using mathematical techniques. Expert judgments may not be very visible in the building performance literature, but are embedded in many other performance assessment approaches such as building simulation and post-occupancy evaluation.

Stakeholder evaluation makes use of the building occupants to assess building performance. As these occupant assessments are subjective by nature, this is a good approach to capture occupant perception and satisfaction. CIB Report 64 (1982) already discusses the use of stakeholder surveys; it notes that different quantification approaches are not to be seen as mutually exclusive and that good results may be obtained from combining methods. A common term in building performance evaluation is post-occupancy evaluation (POE). There are various definitions of this term, as well as measurement approaches that range from only a stakeholder survey to a comprehensive approach that combines interviews, questionnaires, measurement and monitoring and statistical data analysis. Depending on the level of effort, post-occupancy evaluation may be grouped into indicative, investigative or diagnostic studies. Post-occupancy evaluation is not without criticism. There is a tension between the name, which relates to a stage in the building life cycle, and the predominant use of stakeholder surveys; some researchers feel that post-occupancy evaluation belongs to the social sciences, whereas others promote the correlation between subjective opinions and hard measurement data.

Building process performance measurement is a different category of performance quantification. Key areas of interest are cost, time, product quality, safety, waste reduction and customer satisfaction. Some of these may be measured using traditional accounting approaches, augmented with new digital technologies. Others have their own approaches and peculiarities. For instance, the product quality resulting from the process is typically measured in terms of number of defects; process safety is captured in terms of near-misses and incidents. Some process aspects may be measured using general building performance quantification approaches, such as post-occupancy evaluation.

Recommended Further Reading:

- *Building Performance Simulation for Design and Operation* (Hensen and Lamberts, 2011) for a solid primer in building simulation, endorsed by IBPSA.
- *CIBSE AM11* on building performance modelling (CIBSE, 2015c) for further guidance on the use of simulation in the building services engineering context.
- *Assessing Building Performance* (Preiser and Vischer, 2005) for further backgrounds on post-occupancy evaluation and stakeholder surveys; note the appendix with model questions that may be reused in the development of new questionnaires.
- *ASHRAE Handbook of Fundamentals*, on measurement and instruments (ASHRAE, 2017: 37.1–37.40) for a good overview of the tools available for physical measurements in both experimental set-ups and actual buildings and systems.
- The paper on procedures for structured expert judgment by Cooke and Goossens (2000) and the paper by Keller (2015) on expert witnesses in a juridical context.
- The review on sensing and tracking on construction sites by Teizer (2015) for some insights into recent developments and open challenges in construction process performance measurement.

Activities:

1 Identify all the standards, norms and requirements for product certification that apply in your country to a complete front door unit, including door, door frame, side panels, letter box, security lock and others. Then distinguish between standards that give design specifications and those that define tests and measurements.

2 Explore the different empirical approaches that are available to measure the spread of smoke in a building with multiple rooms and floors. Compare and contrast this with computational approaches to simulate the same physical phenomenon.

3 EnergyPlus is primarily a whole building thermal simulation tool; however it has other capabilities that allow the coupled simulation between thermal and other domains. Explore what these other domains are.

4 Identify national experts who may be expert witness in litigation about the following situations (which are based on actual cases that attracted international press interest):

    A The collapse of a listed building after removal of a load-bearing wall from the basement.

    B Damage to a car parked in the street, resulting from solar radiation, which was concentrated by the façade of a building, with the glass acting as mirrors.

    C Massive costs overrun and opening delay of an international airport development.

5 Select an office building to which you have good access, such as a university campus building or one of your company offices. Then:

    A Collect as much information as you can about the building and its use, such as floor plans, materials and systems used, typical week and weekend use schedules.

    B Model the building in a simulation tool of your choice – this may be commercial software – or you can use EnergyPlus/OpenStudio, which is freeware. Use the simulation to predict energy use of the facility.

    C Compare the simulation results with metering or billing information as available.

    D Carry out a calibration of your model, changing parameter values to get a better alignment between simulation results and measurement data.

    E Explore the reasons why there remains a difference between prediction and measurement, even after calibration.

6 Conduct a stakeholder satisfaction survey in a building of your choice. You may use an existing questionnaire, such as the BUS Survey, or develop your own questions.

6.9 Key References

Andersen, M., 2015. Unweaving the human response in daylighting design. *Building and Environment*, 91, 101–117.

Anderson, K., 2014. *Design energy simulation for architects*. New York: Routledge.

Asdrubali, F. and G. Baldinelli, 2011. Thermal transmittance measurements with the hot box method: calibration, experimental procedures, and uncertainty analyses of three different approaches. *Energy and Buildings*, 43 (7), 1618–1626.

ASHRAE, 2002. *Measurement of energy demand and savings – ASHRAE Guideline 14-2002*. Atlanta, GA: American Society of Heating, Refrigerating and Air-Conditioning Engineers.

ASHRAE, 2017. *Handbook of fundamentals*. Atlanta, GA: American Society of Heating, Refrigerating and Air-Conditioning Engineers.

Awbi, H., 2003. *Ventilation of buildings*. London: Spon Press, 2nd edition.

Baker, P. and H. van Dijk, 2008. PASLINK and dynamic outdoor testing of building components. *Building and Environment*, 43 (2), 143–151.

Basmadjian, D., 2003. *Mathematical modeling of physical systems: an introduction*. Oxford: Oxford University Press.

Bauwens, G. and S. Roels, 2014. Co-heating test: a state-of-the-art. *Energy and Buildings*, 82, 163–172.

Blocken, B., D. Derome and J. Carmeliet, 2013. Rainwater runoff from building facades: a review. *Building and Environment*, 60, 339–361.

Blocken, B., T. Stathopoulos and J. van Beeck, 2016. Pedestrian-level wind conditions around buildings: review of wind-tunnel and CFD techniques and their accuracy for wind comfort assessment. *Building and Environment*, 100, 50–81.

Bodart, M., A. Deneyer, A. De Herde and P. Wouters, 2007. A guide for building daylight scale models. *Architectural Science Review*, 50 (1), 31–36.

Candido, C., J. Kim, R. de Dear and L. Thomas, 2016. BOSSA: a multidimensional post-occupancy evaluation tool. *Building Research & Information*, 44 (2), 214–228.

CIBSE, 2015. CIBSE AM11: Building performance modelling. London: The Chartered Institution of Building Services Engineers, 8th edition.

Clarke, J., 2001. *Energy simulation in building design*. Oxford: Butterworth-Heinemann, 2nd edition.

Cooke, R. and L. Goossens, 2000. Procedures guide for structured expert judgment in accident consequence modelling. *Radiation Protection Dosimetry*, 90 (3), 303–309.

Cooke, R. and L. Goossens, 2008. TU Delft expert judgment data base. *Reliability Engineering and System Safety*, 93 (5), 657–674.

Duives, D., W. Daamen and S. Hoogendoorn, 2013. State-of-the-art crowd motion simulation models. *Transport Research Part C*, 37, 193–209.

Durrani, F., M. Cook and J. McGuirk, 2015. Evaluation of LES and RANS CFD modelling of multiple states in natural ventilation. *Building and Environment*, 92, 167–181.

Fox, M., D. Coley, S. Goodhew and P. de Wilde, 2014. Thermography methodologies for detecting energy related building defects. *Renewable and Sustainable Energy Reviews*, 40, 296–310.

Goossens, L., R. Cooke, A. Hale and L. Rodić-Wiersma, 2008. Fifteen years of expert judgment at TU Delft. *Safety Science*, 46 (2), 234–244.

Hens, H., 2015. Combined heat, air, moisture modelling: a look back, how, of help? *Building and Environment*, 91, 138–151.

Hensen, J. and R. Lamberts, eds., 2011. *Building performance simulation for design and operation*. Abingdon: Spon Press.

Hornikx, M., 2015. Acoustic modelling for indoor and outdoor spaces. *Journal of Building Performance Simulation*, 8 (1), 1–2.

IBPSA-USA, 2016. Building Energy Software Tool Directory [online]. Available from http://www.buildingenergysoftwaretools.com [Accessed 15 February 2016].

Keller, E., 2015. Being an expert witness in geomorphology. *Geomorphology*, 231, 383–389.

Kesten, D., S. Fiedler, F. Thumm, A. Löffler and U. Eicker, 2010. Evaluation of daylight performance in scale models and a full-scale mock-up office. *International Journal of Low-Carbon Technologies*, 5 (3), 158–165.

Kuligowski, E. and R. Peacock, 2005. A review of building evacuation models. Washington, DC: National Institute of Standards and Technology, Technical Note 1471.

Langmans, J. and S. Roels, 2015. Experimental analysis of cavity ventilation behind rainscreen cladding systems: a comparison of four measuring techniques. *Building and Environment*, 87, 177–192.

Leaman, A., F. Stevenson and B. Bordass, 2010. Building evaluation: practice and principles. *Building Research & Information*, 38 (5), 564–577.

Mak, C. and Z. Wang, 2015. Recent advances in building acoustics: an overview of prediction methods and their applications. *Building and Environment*, 91, 118–126.

Malkawi, A.M. and G. Augenbroe, eds., 2003. *Advanced building simulation*. New York: Spon Press.

Mardaljevic, J., L. Heschong and E. Lee, 2009. Daylight metrics and energy savings. *Lighting Research and Technology*, 41 (3), 261–283.

Newsham, G., B. Birt, C. Arsenault, L. Thompson, J. Veitch, S. Mancini, A. Galasiu, B. Gover, I. Macdonald and G. Burns, 2012. *Do green buildings outperform conventional buildings? Indoor environment and energy performance in North American offices*. Ottawa: National Research Council Canada.

Nilsson, M., H. Frantzich and P. van Hees, 2013. Selection and evaluation of fire related scenarios in multifunctional buildings considering antagonistic attacks. *Fire Science Reviews*, 2 (3), 1–20.

Oberkampf, W. and C. Roy, 2010. *Verification and validation in scientific computing*. Cambridge: Cambridge University Press.

Ochoa, C., M. Aries and J. Hensen, 2011. State of the art in lighting simulation for building science: a literature review. *Journal of Building Performance Simulation*, 5 (4), 209–233.

Olenick, S. and D. Carpenter, 2003. An updated international survey of computer models for fire and smoke. *Journal of Fire Protection Engineering*, 13 (2), 87–110.

Papadimitriou, E., G. Yannis and J. Golias, 2009. A critical assessment of pedestrian behaviour models. *Transportation Research Part F*, 12 (3), 242–255.

Preiser, W. and J. Vischer, eds., 2005. *Assessing building performance*. Oxford: Butterworth-Heinemann.

Raftery, P., M. Keane and J. O'Donnell, 2011. Calibrating whole building energy models: an evidence-based methodology. *Energy and Buildings*, 43 (9), 2356–2364.

Reddy, T., I. Maor and C. Panjapornpon, 2007. Calibrating detailed building energy simulation programs with measured data – Part I: general methodology. *HVAC&R Research*, 13 (2), 221–241.

Reinhart, C. and S. Selkowitz, 2006. Daylighting – light, form, and people. *Energy and Buildings*, 38 (7), 715–717.

Saïd, M., 2007. Measurement methods of moisture in building envelopes – a literature review. *International Journal of Architectural Heritage*, 1 (3), 293–310.

Sasic Kalagasidis, A., 2004. HAM-Tools: An integrated simulation tool for heat, air and moisture transfer analyses in building physics. Ph.D. thesis. Gothenburg: Chalmers University of Technology.

van Schijndel, A., 2014. A review of the application of Simulink S-functions to multi-domain modelling and building simulation. *Journal of Building Performance Simulation*, 7 (3), 165–178.

Shen, W., Q. Shen and Z. Xialong, 2012. A user pre-occupancy evaluation method for facilitating the designer-client communication. *Facilities*, 30 (7/8), 302–323.

Shi, L., Q. Xie, X. Cheng, L. Chen, Y. Zhou and R. Zhang, 2009. Developing a database for emergency evacuation model. *Building and Environment*, 44 (8), 1724–1729.

Shiflet, A. and G. Shiflet, 2014. *Introduction to computational science: modeling and simulation for the sciences*. Princeton, NJ/Oxford: Princeton University Press, 2nd edition.

Srebric, J., 2010. Computational Fluid Dynamics (CFD) challenges in simulation building airflows. *HVAC&R Research*, 16 (6), 729–730.

Sun, Y., 2014. Closing the building energy performance gap by improving our predictions. Ph.D. thesis. Atlanta: Georgia Institute of Technology.

Teizer, J., 2015. Status quo and open challenges in vision-based sensing and tracking of temporary resources on infrastructure construction sites. *Advanced Engineering Informatics*, 29 (2), 225–238.

Trčka, M., J. Hensen and M. Wetter, 2009. Co-simulation of innovative integrated HVAC systems in buildings. *Journal of Journal Building Performance Simulation*, 2 (3), 209–230.

Underwood, C.P. and F.W.H. Yik, 2004. *Modelling methods for energy in buildings*. Oxford: Blackwell.

Wetter, M., 2011. Co-simulation of building energy and control systems with the Building Control Virtual Test Bed. *Journal of Building Performance Simulation*, 4 (3), 185–203.

Wetter, M., M. Bonvini and T. Nouidui, 2016. Equation-based languages – a new paradigm for building energy modeling, simulation and optimization. *Energy and Buildings*, 117, 290–300.

de Wilde, P. and G. Augenbroe, 2018. Energy modelling. In: D. Mumovic and M. Santamouris, eds., *A handbook of sustainable building design and engineering*. Abingdon: Taylor & Francis, 2nd edition.

Wilson, K., 2016. An investigation of dependence in expert judgment studies with multiple experts. *International Journal of Forecasting*, 33 (1), 325–336.

Zeigler, B., H. Praehofer and T. Kim, 2000. *Theory of modeling and simulation – integrating discrete event and continuous complex dynamic systems*. London/San Diego, CA: Academic Press, 2nd edition.

7

Working with Building Performance

Previous chapters have developed a foundation for working with building performance and have introduced the main quantification methods that may be used in building performance analysis. However, it should be clear from the previous discussion that actual performance analysis always takes place in a specific context and thus requires customization.

There are many issues that interact when conducting building performance analysis in practice. All of these revolve around the central concept of a (set of) appropriate performance measure(s). However, this set needs to be developed through hard work. Each analysis effort relates to a specific case, which has its own dynamics and mostly is unique. The case needs to be related to appropriate performance criteria, as well as to a suitable quantification method. Within each of these factors, there is further complexity. Specific cases all come with their own special interests and drivers, which define what performance aspects are to be evaluated. The context varies strongly in terms of stakeholders, point in the building life cycle and surroundings of the building. The building design itself is another factor, with most buildings being bespoke and unique. There may be an interest in specific systems that drives the performance analysis; similarly the analysis effort itself may highlight the importance of some systems. Loads and system excitation typically vary highly and are yet another factor that needs to be considered when setting up analysis work. On the other side of the equation, criteria interact deeply with the quantification method that is selected for the analysis work. Setting of criteria needs to consider what is desired (goal and target) as well as their operationalization in terms of specific limits and thresholds. The choice of quantification methods not only relates to selecting an underlying evaluation principle but also needs to consider the specific tools and instruments that can be used to work with these principles. See Figure 7.1. Obviously all factors interact; typically a number of iterations will be required to co-develop criteria and quantification methods that fit the situation.

The operationalization of building performance analysis is a core effort that requires expertise from those involved, applying knowledge about thermal behaviour of buildings (CIBSE, 2015a; ASHRAE, 2017), ventilation (Awbi, 2003), lighting (Boyce and Raynham, 2009) and related domains such as reliability engineering (Smith, 2011; O'Connor and Kleyner, 2012). Accounts about operationalization are mostly limited to two categories. Firstly there are some reports about the development of tools and support environments, such as Augenbroe and Park (2005) and Pati *et al.* (2009), or the EnergyPlus Engineering Reference (US Department of Energy, 2016). Secondly there

Building Performance Analysis, First Edition. Pieter de Wilde.
© 2018 John Wiley & Sons Ltd. Published 2018 by John Wiley & Sons Ltd.

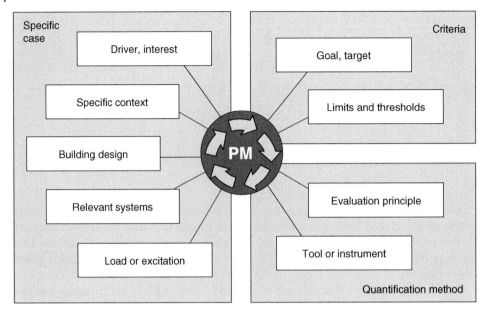

Figure 7.1 Interacting factors in Building Performance Analysis in practice.

are deep accounts of performance analysis of specific buildings, such as Lehmann *et al.* (2010) or Konis (2013). Support for operationalization of performance analysis is also provided by norms and standards, but the analyst will always have to adjust these to the specific case at hand.

Operationalization of building performance analysis will remain hard work that requires effort and expertise. This chapter explores working with building performance, employing a conceptual framework that shows what steps are required. It starts by presenting a number of relatively straightforward examples, which demonstrate the deep interaction between building, systems, stakeholders and life cycle for each analysis effort and point out which criteria and quantification methods may be used. The examples are intended as a sample from the infinite number of actual analysis contexts that may be encountered in practice, which show how performance criteria emerge from a specific situation but then need further development to allow actual performance quantification and analysis. They thus emphasize the need for stakeholder dialogue in order to tailor building performance analysis to the specific situation and context.

Following on the examples the chapter proceeds to develop the criteria for a selected number of performance aspects in further depth. The starting point here is given by single-performance aspects rather than the full complexity of a specific case, which mostly comes with multiple performance issues. Each performance aspect is briefly explored in term of the typical situations in which this is likely to be encountered and the typical scope in terms of system boundaries and time horizons that are used for analysis. For each aspect, some alternative tool-independent evaluation methods that may be used to conduct the analysis are discussed. Natural tendencies for setting targets and limits are presented. The discussion of each criterion is completed with reference to closely related performance aspects as well as some of the standards and norms that apply.

While development of criteria goes hand in hand with the choice of an evaluation method, actual analysis also requires the selection of specific tools and instruments – whether these are hardware for physical experiments, software for performance simulation, expertise for an expert judgment or methods of enquiry to obtain stakeholder feedback. These tools and instruments need to be adjusted to the case at hand but may also impact the criteria and evaluation process. As the number of actual cases and criteria for analysis is endless, and combined with a large toolset, this issue is discussed using one example for each of the key approaches towards performance quantification.

Like many issues in design and engineering, operationalization of building performance analysis may be helped by iteration, following the Deming cycle of Plan, Do, Study/Check and Act. This is especially advantageous where such iteration helps the stakeholder dialogue, ensuring that all parties understand the analysis process and allowing them to ensure that the results are fit for purpose.

This chapter develops a conceptual framework for operational building performance analysis. It uses a selection of building performance analysis cases, the development of analysis criteria and the further efforts required to adjust performance quantification methods to situation at hand. It also presents the benefits of iterative analysis. In doing so it prepares the ground for practical handling of the building performance concept.

7.1 Examples: Selected Building Performance Analysis Cases

With the high variation in building design, the many system alternatives, the different stakeholders and complex life cycle, most building performance analysis efforts relate to a unique situation. The following examples represent some typical cases that discuss the required effort to customize the analysis effort. This list of examples is not intended to be complete, nor are the approaches intended as a blueprint for specific analysis work. For a wide overview of the many building performance aspects and functions that may foster the development of analysis criteria, see Appendix A.

Design Choice: Gas-Fired Heating or Heat Pump with Photovoltaic Array

A common case that requires building performance analysis is the choice of a heating system, such as between a conventional gas-fired central heating and the combination of a heat pump with photovoltaic array. This specific choice is a frequent one in many European countries, where there exists an extensive gas supply network and gas heating is a common system but where the combination of a heat pump and PV is seen as a competitive 'green' alternative that reduces dependency from the gas supply. The systems involved in performance analysis of both options involve all heated rooms in a building, the two technical solutions under consideration and a heat supply system such as radiators and possibly underfloor heating. Stakeholders during the design process will be the architect, consulting engineers (HVAC, building physics and possibly electrical) and the project developer. As indicated in the title of the example, the case is presented as a design choice; however a similar choice may arise in a retrofit context. If the latter is the case, the situation involves the reuse of an existing supply system, which may or may not be retrofitted at the same time.

Typical performance aspects (requirements) to be considered in this case, and to be agreed through stakeholder dialogue, would be thermal comfort (quality), peak heat supply (workload capacity), energy efficiency (resource saving) and costing (resource saving). A further consideration may be heating availability, especially in terms of the PV-driven system, which depends on solar irradiation (timeliness).

Building performance analysis could support this choice in the following way:

1) Thermal comfort:
   One evaluation principle suitable for the analysis of thermal comfort is the use of the average comfort temperature during operational hours. A typical criterion would be to require $19°C \leq T \leq 23°C$, in line with typical comfort ranges. This way of quantifying thermal comfort requires an analysis method that provides temperatures for all rooms of the design for typical climate conditions, as provided by a whole building thermal simulation tool. A tool is needed that is able to model both systems under consideration; this would suggest the use of simulation tool that has a strong focus on building services, such as EnergyPlus or TRNSYS.

2) Peak heat supply:
   A good evaluation principle for the analysis of peak heat supply is the study of the heat load on a winter design day – which is highly design and climate dependent. The common criterion is to require that $Q_{system} \geq Q_{design\ day\ load}$; sometimes one may even decide to include a safety factor and set $Q_{system} \geq f \cdot Q_{design\ day\ load}$. Peak heat supply quantification requires a heat balance and could be established by a stationary calculation method; however since thermal comfort is already quantified using a whole building thermal simulation tool, it makes sense to use the same tool to establish peak heat load.

3) Energy efficiency:
   Energy efficiency may be evaluated through the total amount of energy to be obtained from the grid per year, expressed in primary energy. A typical criterion for this aspect is to aim for a primary energy consumption per year that is as close to zero as possible. Quantification of the primary energy consumption at design stage requires a method that accounts for conversion, transport losses and end use. This may be established from a semi-dynamic calculation, but since thermal comfort and peak heat load are already quantified using a whole building thermal simulation tool, it again makes sense to use the same tool.

4) Costing:
   Analysis of costs might include a review of procurement cost as well as payback period without consideration of feed-in tariff. An example of a criterion is to require that procurement cost should be $\leq 5\%$ of the total construction cost and that payback period should be ≤ 7 years. Quantifying these costs could be done by combining engineering energy consumption estimates with cost estimates using quantity surveying and accountancy software.

5) Heating availability:

An evaluation principle for analysis of heating availability is to analyze the period of occupied hours that heating system is not available. A heating system availability requirement could be a demand for heating during at least 99.5% of occupied hours. This availability could again be established using whole building thermal simulation.

System Sizing: Percentage of Glazing of an Office Building

For some systems, even once they have been selected, there are important options in terms of sizing. One example is the percentage of glazing that is used in a façade; in theory this may range from 0% (no window) to 100% (fully glazed) although in reality there mostly is some sort of window in most external walls, while even a fully glazed façade needs to make some allowances for structural systems and sealants. Highly glazed façades are popular in some architectural approaches. Performance analysis of variation of the percentage of glazing is mostly carried out at the level of a single room, allowing one to study the performance across a range of building physics aspects. Often these analyses are repeated for representative rooms at key orientations and positions in the building, allowing inference for the other rooms in the building. Where glazing levels are high, the analysis will include lighting systems, shading systems integrated in the façade or positioned adjacent to the glass, and HVAC systems; control settings will play an important role in the analysis. Typical stakeholders in a design decision about glazing percentage, and the related performance analysis, are the architect, building physics consultant and client. The decision about glazing percentage is often made early in the design process, as this may have consequences for the load-bearing structure of the façade and indeed the building itself.

Typical performance aspects (requirements) to be considered in deciding on the glazing percentage are view to the outside (quality), daylighting (resource saving), energy efficiency (resource saving) and architectural performance (aesthetics).

Building performance analysis could support the definition of a glazing percentage as follows:

1) View outside:

One possible evaluation principle to analyze the view to the outside is to analyze the Field of View (FOV) of the outside world from a well-defined sitting position in relation to the façade, measured horizontally. A criterion for the FOV may for instance be to aim for a minimum of 15°. Assessment of outside view is an area of building performance analysis that is still under development; there are no known tools that support FOV evaluation. Quantification will thus have to be done by the design team on the basis of detailed design drawings that include assumptions of the layout of the office space.

2) Daylighting:

A good evaluation principle to assess the daylighting situation in a room is the use of the spatial Daylight Autonomy (sDA). Assuming there will be some degree of artificial task lighting anyway, the criterion may be to require that 50% of floor space will be receiving at least 300 Lux for 50% of occupied hours. To quantify lighting levels, one needs a lighting simulation tool, with Radiance being a likely candidate.

3) Energy efficiency:

The evaluation of energy efficiency in this case needs to be defined closely to the glazing percentage that is of interest. A good evaluation principle would be to use a customized measure, such as the ratio of total energy use of the office (including heating, cooling and lighting loads) for each customized glazing percentage and the total energy use of the office assuming a typical glazing percentage of 16.7%. The criterion would be to find an alternative that has at least the same efficiency (ratio = 1) or a better efficiency (minimal ratio). Energy efficiency quantification of these ratios requires a building performance simulation tool that can capture the impact of glazing percentages on heating, cooling and lighting load, such as EnergyPlus.

Performance Analysis for Certification: Rating

A specific type of building performance analysis is the work that takes place in the context of certification for voluntary rating systems such as LEED or BREEAM. Rating applies to complete buildings, including all systems and components; some schemes may also include some of the landscaping and amenities around the building. Rating requires a client who sees the value of certification, a design team that is willing to put in the extra efforts required to achieve a good score and certified assessors on behalf of the rating scheme. Rating mostly takes place at the end of the design stage; however there are also options to rate existing facilities.

By way of example, BREEAM certification covers the following aspects, with aspect weighting in brackets: energy (0.19), health and well-being (0.15), innovation (0.10), land use (0.10), materials (0.125), management (0.12), pollution (0.10), transport (0.08), waste (0.075) and water (0.06).

In terms of analysis, the BREEAM system enables projects to achieve a number of credits in each of the categories as indicated. Some of these require a detailed performance analysis, such as for predicted energy use; in other categories points are awarded for issues that strictly speaking are not building performance, such as the employment of certain management principles or the distance to local public transport. However, the framework is rigid and based on assessment by certified assessors. Some examples, from a much longer list, are as follows:

1) Energy efficiency:

Based on calculation of the Energy Performance Ratio for New Constructions (EPR_{NC}) as compared with a baseline value; credits are awarded for better achievement.

2) Water use efficiency:

Based on assessment of water use using a dedicated BREEAM calculator; again credits are awarded for achieving improvement over a baseline value.

3) Waste reduction:

BREEAM awards credits for preventing demolition of existing construction on a building site and for preventing waste material going to landfill. Further credits can be gained by designing buildings in such a way as to limit the amount of waste generated during construction, which is measured in m^3 of waste generated for the construction of $100\,m^2$ of gross internal floor area.

4) Management:

Credits are awarded for sharing roles and responsibilities, handover and aftercare, early involvement with BREEAM professionals and for conducting a thermographic survey and airtightness test. Note that the credits are not awarded for building or construction process performance, just for undertaking some activities without quantifying how well these are done.

Overall credits are added up and then used to gain certification based on the following criteria: outstanding requires a score of ≥85%, excellent ≥70%, very good ≥55%, good ≥45% and pass ≥30%.

Portfolio Management: Prioritization of Interventions

Many building owners (governmental organizations, universities, corporations) have only limited resources and therefore need to decide carefully in what order they make interventions in their building portfolio. The prioritization may be driven by a range of factors; typically ensuring the stakeholders' process is fully supported is the main concern, followed by providing building quality and resource efficiency. The scale of this analysis is at building level. Typical stakeholder for this type of analysis is the local estate department or facility manager, although decisions may have to be approved by general management. This type of analysis is typical for a portfolio of existing buildings that are well into their operational stage, with some systems needing maintenance or even replacement.

Typical performance aspects to be considered in portfolio management are work productivity (workload capacity), occupant satisfaction (quality), energy use (resource saving) and water use (resource saving).

The following building performance analysis efforts can support prioritization of interventions:

1) Work productivity:

This assessment mainly looks at the process that takes place inside buildings and is highly dependent on the nature of the organization; one should bear in mind that the built environment is only one part of the infrastructure that supports the activities of the organization. In cases where the main process is production, one can measure the process output to capture productivity. Where the main process is administrative, the measurement becomes more complex; often some composite performance index is used that aggregates a number of KPIs. An example KPI may be the call abandonment rate for incoming phone calls. The criterion would be to keep abandonment rate as low as possible but setting a maximum allowance rate of 0.5%.

2) Occupant satisfaction:

Occupancy satisfaction typically will be measured using a survey. The criterion for occupant satisfaction is highly dependent on the actual stakeholder survey that is being employed. Where a 5 point Likert scale is used, allowing occupants to rate aspects as excellent, good, fair, neutral or poor, one may require that no aspect scores poor. The method to gather occupant opinions could be a bespoke POE questionnaire.

3) Energy use:

For existing stock, the obvious evaluation method for energy efficiency is by way of metered energy (electricity, gas) consumption. In order to be able to compare buildings of different size, this metered data can then be used to calculate the energy use intensity (EUI) by normalizing for floor area, giving a value in kBTU/ft^2 year or kWh/m^2 year. For prioritization of interventions, the buildings in the portfolio would be ranked, and those with highest consumption be targeted first.

4) Water use:

For the analysis of water use of existing buildings, the obvious evaluation method is again metered consumption. As with energy, this consumption can be normalized for floor area, yielding gal/ft^2 year or l/m^2 year. Prioritization of interventions would again be based on ranking and addressing buildings with high consumption first.

Fault Detection: Overheating

Some building performance analysis efforts may be following up from complaints; for instance a survey may have flagged up that offices in a certain building are generally perceived as being uncomfortably hot. In this case, the scale of the analysis will be at room level, covering the rooms in question. Stakeholders would be the occupants of the rooms as well as the facility management team. The analysis takes place in an existing building.

The obvious performance aspect to be considered in an investigation into overheating is overheating itself (quality). However, further aspects may be analyzed to find the cause and potential remedies, such as HVAC control and operation (responsiveness) and solar gains (responsiveness).

The following building performance analysis approaches might be used to investigate overheating:

1) Overheating:

The first step in responding to overheating complaints is to ascertain the temperatures in the rooms and analyze these to see whether there indeed is overheating. A common approach would be to deploy temperature loggers and capture indoor air temperatures over a period of time. If the problem is urgent, the measurement period may be short, such as 10 days of capturing temperatures at a 15 minute interval. If the issue is less urgent, one may take a longer view and capture two key seasons (the summer and an intermediate season), which requires 6 months of data; given the longer duration, one may opt for hourly readings. As a criterion, a limit of exceeding 28°C for no more than 1% of occupied hours may be used to establish whether a room is indeed overheating.

2) HVAC control and operation:

In case there indeed is overheating, a first check would be to see whether the HVAC system is providing chilled air as required. This may be achieved by monitoring supply temperatures and contrasting readings with control settings. This can be done through a portable temperature logger or possibly through an existing BAS sensor. Typically this would be done for a couple of days, with readings taken at a shorter time interval such as 5 minutes. As a criterion, one may require that supply air temperature is within a range of plus or minus 1°C of the setpoint value. Further fault analysis efforts will be required if supply air is not in the required range; one may also want to check the amount of air supplied.

3) Solar gains:

Assuming that the room is overheating and that the HVAC system functions as required, a prime issue to investigate is solar gains and how these are controlled. Analysis might monitor solar irradiation received inside the room, for instance at desk level, and compare that with the operation of any solar shading devices. Criteria are highly context specific; readings will depend on glazing percentage, room orientation, urban context and specifics of the shading device. However, with a system like controlled external blinds, one would expect less than 15% mismatch between control settings and actual readings.

Court Case: Partial Building Collapse

Building performance analysis can be serious where buildings are not deemed safe or, even worse, actually fail under some sort of load. For instance, a roof may collapse due to the load induced by severe rainfall. In such cases there may be injuries and even fatalities amongst occupants and the general public, the business process of the building operator may be halted, and there can be significant insurance claims. Stakeholders in this case will be victims, building owners and managers, as well as a range of legal actors. Typically court cases involve existing buildings, although some claims may address faulty design.

Typical performance aspects to be considered in such a context would be structural load-bearing capacity (workload), resistance to extreme events (readiness) and safety risk (quality).

The following building performance analysis is likely to be required:

1) Load-bearing capacity:

A first 'post-mortem' analysis of a building that has had a partial collapse will be a review of the original load-bearing capacity before the event. Typically this will be based on structural calculations using finite element analysis software. The pre-existing structure will be assessed to find out a safe design load, which is calculated from permissible stress, strain and safety margins. This safe load will be highly dependent on design detail, component sizing and materials. While there are general criteria such as maximum displacement of elements, the key criterion of safe load actually stems from the analysis rather than being defined upfront. A complicating issue may be the ageing and degradation of load-bearing elements over time.

2) Resistance to extreme events:

Related analysis efforts will explore what maximal loading had been anticipated for the building. In the case of collapse due to severe rainfall, this needs to include a review of predicted maximum rainfall, a review of capacity of the drainage systems and resulting structural loads under various scenarios. Findings will be compared with the established load-bearing capacity.

3) Safety risk:

In a court case, the ultimate questions relating to building performance are whether the building was prepared to cope with the extreme rainfall load and, if not, who is responsible for that mismatch. While the structural collapse mechanism may be obvious after the event, actual causes such as maintenance may be much more difficult to attribute. The main method to deeply evaluate the safety risks and whether the collapse could have been prevented is by employing expert witnesses, who comment on the handling of risks. The main criterion here becomes whether any party was negligent in the face of anticipated probable loads.

Research: Impact of Occupant Behaviour

Building performance analysis may also be explorative and driven by discovery. An example is the interest in the impact of occupant behaviour on the thermal performance of buildings, such as studied in IEA Annex 66. Research is driven by a desire to establish unknown relationships, often in the context of an ambition to develop better systems and approaches. In this specific case, there is an underlying expectation that better understanding may help to reduce building energy use. Parties in research are scientists, funding bodies and industrial stakeholders. In most cases research employs a range of techniques to study a range of situations, so there is not just one single building at stake but a set, although each element of the set may form a specific case study and analysis effort in itself.

Typical performance aspects that need to be analyzed in this context are energy use (resource saving) and thermal comfort (quality). Beyond that, there is special interest in the role of occupant behaviour as driving factor; this means that both performance aspects need to be correlated to observations of behaviour.

The following performance analysis efforts will support a study of the role of occupant behaviour:

1) Energy efficiency:
   The impact of occupant behaviour on energy use will be highly dependent on the specifics of any given building, so any variation is to be quantified in relation to a baseline for the same building. In the research context, both simulation and measurement in real buildings may be used; however both have drawbacks. To ensure maximal comparability, it is suggested that the impact of changes in occupant behaviour is related to percentual changes in heating, cooling and ventilation energy use per annum.
2) Thermal comfort:
   In the context of studying the impact of occupant behaviour on energy use, thermal comfort is a control variable that needs to be monitored. A realistic route would be to require the indoor air temperature to stay within a reasonable bandwidth, such as $19°C \leq T \leq 23°C$. This condition can be observed both experimentally and in building simulation software.

Added to these is the analysis of occupant behaviour itself. This can be split in occupant presence and occupant action, such as changing thermostat settings, lowering blinds or opening windows. In simulation care must be taken to properly document assumptions and how these are translated into simulation input. In measurement, these may be measured through the use of a range of sensors. Analysis then needs to establish the correlation between changes in these occupancy-related signals and the energy efficiency and thermal comfort inside the building.

7.2   Criterion Development

As demonstrated in the previous paragraph, the setting of criteria for building performance analysis requires more than just the selection of which performance aspects are of interest to the stakeholders. While performance aspects are a good point of entry, there typically is a choice to be made for a specific performance evaluation method, which has deep consequences for setting goals, targets and limits and thus for developing the criteria. Performance aspects themselves tend to have a typical occurrence that relates them to certain types of buildings and conditions where they are of increased relevance; they may also have a system and temporal scope that points the evaluation in a certain direction. The evaluation methods that are available to assess each building performance aspect tend to be a set of principles that undergoes natural selection. New methods are developed in response to evaluation needs, technical developments and in attempt to overcome shortcomings of existing methods; over the years some methods become well established, while others fade away. Evaluation methods are a first step towards performance quantification; however they need to be made operational by selecting appropriate tools (such as measurement instruments, software packages, experts or occupant surveys), which is discussed in the next section. However, evaluation methods already provide a good grip on performance measurement and allow the expression of goals and targets in terms of what values on a numerical scale are desirable and what limits and thresholds are to be used. The following examples discuss a range of typical performance aspects and how criteria may be developed for each of these aspects. Again, the list is not intended to be complete, nor are the criteria intended as blueprints for straight application. A general template that helps to consider the key issues that need attention in developing specific performance criteria is contained in Appendix B.

Acoustical (Auditory) Comfort   **Quality Requirement**

Acoustical comfort is the performance aspect that captures the satisfaction of occupants with the different manifestations of sound in buildings. It has a bearing on the ability to hear and speak, enjoying music, as well as on the control of noise (unwanted sound). Acoustic comfort relates to human hearing, which recognizes sound frequency (pitch), sound intensity (loudness), duration, timbre, sonic texture and spatial location. High sound pressure levels (SPL) cause pain, with 140 dB recognized as the threshold of pain. Extreme sound levels, or lower levels that are maintained for a significant duration, may cause reversible and ultimately permanent damage to the hearing system. Good acoustics enable speech intelligibility and music rendition.

What is appropriate in terms of acoustics depends highly on the function of a building and a room; there are different requirements for an office space, meeting room, church or auditorium. Interestingly the presence of human bodies and clothing changes the acoustics of a space. Some rooms have features that allow changing these properties.

The amount of flexibility allowed for acoustical comfort depends strongly on the activities of the occupants of a room; while strict margins may be required for a studio space or office for work that requires concentration, significant more allowances can be given for informal spaces.

Occurrence: Acoustical comfort is of importance in most buildings; obviously it is a key concern in rooms such as recording and broadcast studios, auditoria and theatres. It is also of special interest in educational spaces, where communication is important. Acoustical comfort may require special attention in rooms with hard surfaces such as concrete floors, walls and ceilings or a high percentage of glass.

Scope: The prime factor in assessing acoustic comfort is the performance of the room in which hearing is considered; however, in most cases, the adjacent spaces also play a role and may be a source of noise. Acoustical analysis can be relatively fast, given the high speed of sound, although an analysis across a range of frequencies obviously requires repeated measurements.

Methods: A first method for the analysis of acoustical comfort is the analysis of *reverberation time*, with indications of good values calculated by the *Stephens and Bate formula*. More advanced analysis typically covers a range of octave bands and relates the sound produced at a source location in the building to the sound received at other positions, including the effect of sound attenuation and absorption; this type of analysis employs a wider range of parameters such as early decay time (EDT) and sound pressure levels (SPL) and results in a sound field representation that shows significant variation in relation to the position in a room/building. Where duration is considered, a typical parameter is L_{Aeq}, which is an equivalent sound level using the A-weighted spectrum.

Criteria: Acoustical comfort criteria are highly dependent on room function. By way of example, guidelines for reverberation times may be as follows: classroom, 0.6–0.8 s; open-plan office, 0.8–1.2 s; restaurant, 0.8–1.2 s; and atrium space, 1.5–2.0 s. Absence of noise in bedrooms may be defined by requiring that $L_{Aeq} < 35$ dB between 23:00 and 7:00. See Figure 7.2.

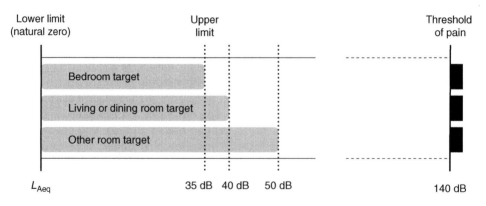

Figure 7.2 Sample upper limits for equivalent sound levels in a dwelling.

Related performance aspects: Noise, speech intelligibility, privacy
Relevant standards and norms: ISO 3382-1:2009; ANSI/ASA S12.60-2010

Indoor Air Quality **Quality Requirement**

Indoor air quality (IAQ) is a performance aspect that is concerned with minimizing the contamination of air with pollutants and hence is of high importance for occupant health and well-being.

IAQ is important for the occupants of the building, as it relates to their physical condition; however it is also important to those that run business processes in buildings as IAQ may seriously impact staff productivity. At a more generic level, the relation between IAQ and health is of interest to society, in the context of ensuring safety and managing healthcare budgets.

IAQ is a generic term; how strict it can be enforced depends highly on the type of pollutant. Some agents are well known for detrimental health effects, and in that case strict norms are in place. For other agents the relation to health is still object of investigation, and thresholds and limits remain to be established.

Occurrence: IAQ is a generic performance aspect that is of interest to all building occupants. However, it takes on extra weight in situations where the outdoor air is known to be contaminated, such as for buildings that are positioned along busy roads or industrial estates. IAQ also becomes extra important in medical facilities and in some manufacturing facilities where pollutants may negatively impact product quality.

Scope: The performance in terms of IAQ is typically analyzed on a whole building level. However, in detail there is an obvious correlation with the ventilation systems that are in place. Where air handling plant is used, there may be different zones or sectors that share the same systems; natural ventilation will relate to openings in the facade and air paths through the building.

Methods: A prerequisite for good IAQ is to ensure that there is sufficient ventilation, which can be established by measuring the air exchange rate of rooms and buildings. More in depth, the key to assessment of IAQ is the quantification of the concentration of any pollutants in the indoor air. The most appropriate way to express this concentration depends on the pollutant in case, as well as the danger it poses to health. Amongst the common pollutants in buildings are allergens, carbon dioxide, carbon monoxide, ozone, particulates, radon gas, tobacco smoke and volatile organic compounds (VOCs).

Criteria: Criteria are provided for limits of pollutants, such as a typical limit $150\,\mu g/m^3$ of suspended particulate matter. For some pollutants with known critical impact on human health, such as benzene, carbon monoxide, formaldehyde, naphthalene or radon, guidelines with threshold for maximum exposure to specified concentrations have been published by the World Health Organization (WHO). See Figure 7.3.

Pollutant	Averaging time	Lower limit (natural zero)	Upper allowable limit
SO_2	24 h	0 $\mu g/m^3$	365 $\mu g/m^3$
CO	1 h	0 ppm	33 ppm
O_3	1 h	0 pphm	12 pphm

Figure 7.3 Some maximum concentrations for selected gaseous pollutants.

Related performance aspects: Air temperature, relative humidity, air speed
Relevant standards and norms: ASHRAE Standard 62.1, CIBSE AM10

Thermal Comfort   **Quality Requirement**

Thermal comfort is the performance aspect that deals with the satisfaction of an occupant with the thermal conditions in a building. Human beings experience thermal comfort when they are in thermal equilibrium with their surroundings, thus feeling neither too warm nor too cold. Outside the thermal comfort zone, the human body will react: it will start shivering when cold and start perspiring when warm. Initially effects are mainly psychological, leading to dissatisfaction and loss of productivity. More extreme situations may lead to significant stress. Where thermal conditions change far beyond the comfort zone, they may become life threatening, leading to hypothermia when too cold and hyperthermia when too warm.

   Thermal comfort is of importance to all building occupants. However, there is significant variation in how thermal comfort is rated by individuals; some factors of influence are personal activity level, clothing, age, gender, health and regional expectations. Thermal comfort may also be an aspect to consider in buildings occupied by animals or plants, which may require very different thermal conditions from humans.

   Thermal comfort may allow for a small margin of negotiation. Occupants may be able to adapt to some discomfort by changing their clothing level or consuming hot or cold drinks. On balance, acceptance of discomfort is slightly higher in buildings where occupants have some control over the thermal conditions, for instance through openable windows or adjustable fan speed.

Occurrence: Thermal comfort is a performance aspect that needs to be considered in almost all buildings. However, it is particularly important for the weak, the very young and the elderly. Patients undergoing an operation may have reduced thermal regulation and require a very stable environment; the young and the elderly may not be able to respond adequately (changing clothing level, moving away from direct sunshine), and thus nurseries and care homes also require deep consideration of this aspect. Thermal comfort often is an issue in highly glazed spaces.

Scope: Thermal comfort is typically assessed per room or zone, since this is the thermal environment of a human body. The time horizon for assessment may vary from a single moment in time, which allows a deeper understanding of a specific situation and what causes discomfort, to assessment over a long period of up to a year, which allows a better understanding of the duration of periods of discomfort. Often analysis considers extremes, such as a heat wave or cold spell. Obviously, thermal comfort is directly related to the building fabric, internal gains, heating and cooling, as well as local climate conditions.

Methods: The most basic method for quantifying thermal performance is the use of a *comfort temperature*. In more detail, thermal comfort is dependent on the heat exchange between a human body and the environment; this is influenced by the metabolic rate, clothing, air temperature, mean radiant temperature, air speed and relative humidity. A well-known method is the *Fanger method*, which uses heat balance equations to calculate a Predicted Mean Vote (PMV) as well as a corresponding Predicted Percentage of Dissatisfied (PPD). Further effects may be considered in *adaptive models*, which relate an allowable operative temperature to the prevailing outdoor temperature. There are also further methods that consider the impact of radiation asymmetry and drafts. Duration of discomfort is often captured by means of degree hours, which multiply duration with severity.

Criteria: Thermal comfort relates to equilibrium, so the criteria often aim for a comfort band in the middle between ranges that capture conditions that are too cold and too hot. A target range for the comfort temperature is often given as 19°C–23°C for temperate climate regions, whereas warmer climates may allow 19°C–29°C. PMV is expressed on a typical Likert scale that rates –3 as cold, –2 as cool, –1 as slightly cool, 0 as neutral, 1 as slightly warm, 2 as warm and 3 as hot. Within this range comfort is often taken to be any value that falls within the range of $-0.5 < PMV < 0.5$. See Figure 7.4. For the duration of being outside the comfort range, there are often different accepted times for exceeding specific severity limits; for instance one could allow 5% of occupied hours to exceed 25°C but only 1% of occupied hours to exceed 28°C.

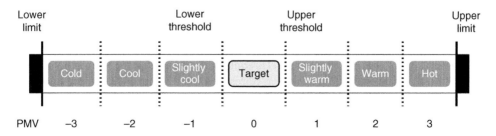

Figure 7.4 Typical thermal comfort target and criteria expressed using PMV.

Related performance aspects: Overheating, undercooling
Relevant standards and norms: ISO 7730; EN 15251; ANSI/ASHRAE Standard 55; CIBSE TM52

Visual Comfort   **Quality requirement**

Visual comfort is a performance aspect that covers the satisfaction of building occupants with the lighting conditions in a room or building. Traditionally, the focus is on ensuring sufficient light levels, management of contrast and prevention of glare. In more detail visual comfort requires a balanced brightness distribution, varying luminance levels, avoidance of discomfort glare, a uniform illuminance of and around any task areas, and flicker-free lighting. In most situations the best results are achieved by combining daylight and natural lighting.

Visual comfort is relevant to all building occupants. Lighting levels are known to affect the body clock or circadian rhythm and thus impact sleep patterns, coordination and productivity. Visual discomfort such as glare or flicker can have significant impact on well-being.

Visual comfort is still a subject of research; achieving visual comfort is mainly achieved by making good use of any daylight and preventing visual discomfort conditions.

Occurrence: Visual comfort conditions are again a factor of relevance to most buildings. Special attention is required in rooms that host activities that require high visual precision, such as libraries, manufacturing or repair of objects with small parts, or computer work. Extra attention may have to be paid to situations where there is direct sunlight or high contrast.

Scope: Visual comfort is mostly related to the field of vision of building occupants; where possible it thus makes sense to assess the aspect for specific locations such as working positions at desks in relation to windows, lights, computer screens and so on. Given the variation in natural light (fast changes due to climate conditions and continuous change in solar position throughout the day and year), both instantaneous and long-term analyses are required.

Methods: Initial assessment of lighting levels may be based on capturing *illuminance levels*, at work plane level, in a room. Uniformity is often captured by assessing *luminance ratios*. Further effects are captured through the use of a *Glare Index, Daylight Glare Index, Colour Rendering Index, Unified Glare Rating* and similar others. Flickering is typically captured by the amplitude, average and frequency.

Criteria: Typical illuminance levels required for visual comfort would be 100 lux for corridors, 150 lux for stores and plant rooms, 200 lux for entrance halls, 300 lux for teaching spaces, 500 lux for general offices and 1000 lux or more for electronic assembly workshops and similar. A typical luminance ratio between task and adjacent surroundings may be 1:3, while the ratio between task and remote surroundings may be 1:10, and between task and window 1:40. Some typical limits and thresholds for a Daylight Glare Index (DGI) are depicted in Figure 7.5.

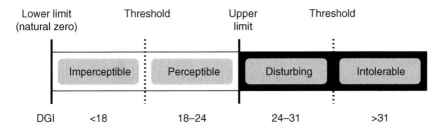

Figure 7.5 Typical limits and thresholds for DGI values.

Related performance aspects: Visual performance, glare prevention, daylighting
Relevant standards and norms: EN 12464-1, CIE 17, CIE 146:2002, CIE 147:2002

Productivity   **Workload Capacity**

Many buildings exist to facilitate some sort of business process. This can be the manufacture of items but may also be the provision of a service. To the client, productivity of this process is a core concern. Of course productivity is not just dependent on the building, but also relates to organizational and technical factors such as staff morale, reliability of machines and availability of parts and components. Yet assuming all other factors are under control, buildings and what they offer may make a difference towards productivity. Still, performance measurement in buildings needs to be closely correlated with the assessment of business and organizational performance. Key stakeholders for productivity are the building operator and staff working in the facility. Given the interaction with organizational and technical factors, there typically is significant room for negotiation about criteria.

Occurrence: Productivity may be considered in all buildings that host organizations, especially where these are commercial, governmental, educational and medical. Thus factories, offices, schools and universities, and hospitals are important categories of buildings where productivity may require assessment.

Scope: The appropriate scope for analyzing productivity depends on the focus of the assessment but typically relates to the facility as a whole or to business units. In some cases it may be reviewed at zonal level, for instance if one wants to establish a correlation between something like indoor temperature or CO_2 level and productivity.

Methods: Methods for the analysis of productivity are strongly related with the prime process, as well as the specific interests of the stakeholders. Examples are the measurement of *production volume, inspection pass rate, number of calls or cases handled, exam pass rate* or *patient readmission rate.* In many cases a single analysis view only gives a partial view of the overall productivity, so use is made of aggregation methods that cover a combination of aspects.

Criteria: Criteria depend highly on the purpose of the organization and the related measurement method. In general one wants productivity to be high; however quality also needs to be maintained. Figure 7.6 shows a range of productivity criteria that may be used to analyze the operation of a call centre.

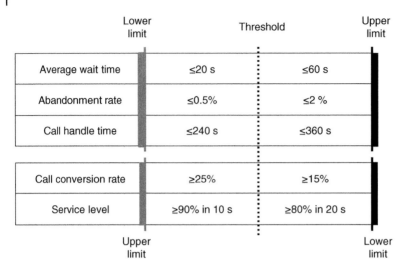

	Lower limit	Threshold	Upper limit
Average wait time	≤20 s		≤60 s
Abandonment rate	≤0.5%		≤2 %
Call handle time	≤240 s		≤360 s
Call conversion rate	≥25%		≥15%
Service level	≥90% in 10 s		≥80% in 20 s
	Upper limit		Lower limit

Figure 7.6 Call centre productivity criteria.

Processing Capacity   Workload Capacity

An important performance aspect for many buildings is their capacity to hold or process people. Physical limitations are seldom the key constraint; more often processing capacity is limited by health and safety risks or quality considerations. Capacity can be directly related to the potential to generate revenue, for instance where seats are sold. Stakeholders are facility operators and visitors, guests, passengers and similar. Processing capacity may be open to significant negotiation where comfort is at stake; room for negotiation disappears where capacity becomes constrained by safety risks.

Occurrence: Processing capacity is important in transport facilities such as airports, railway and subway stations and ferry ports. Capacity also plays a big role in sport and entertainment venues, restaurants and hotels. It is crucial for medical facilities, prisons and educational buildings. In some cases it can also be of high relevance to retail facilities.

Scope: Processing capacity is typically analyzed at the facility level. However, one may also study the processing capacity of rooms and even specific systems as part of the effort to assess how the overall facility performs; for instance one can study the capacity of stairs and escalators within a transport facility.

Methods: For some buildings, processing is directly related to *seat capacity*. For a sport venue, the processing capacity per year is simply the number of events per annum multiplied by the number of available seats at each event. The same calculations can be done for hotels, schools and similar. For technical systems such as elevators, processing capacity is measured in the form of the *handling capacity* (HC) of an elevator group, which depends on the number of cars, capacity and loading of each car and round trip time; typically HC relates to the percentage of the number of occupants of a building that can be served in 5 minutes.

Criteria: For basic capacity calculations based on number of seats, criteria are simply based on availability – although there often are underlying safety and quality assumptions, such as a decision not to allow standing guests

Table 7.1 Level of service definition for airports according to IATA.

Level of service (LOS)	Passenger flow	Delay	Comfort
A – Excellent	Free	None	Excellent
B – High	Stable	Small	High
C – Good	Stable	Significant but acceptable	Good
D – Adequate	Unstable	Just passable	Adequate
E – Inadequate	Unstable	Unpassable	Inadequate
F – Unacceptable	Halted	Blocked	Unacceptable

in order to prevent crowding and evacuation risks. Things become more complex if the number of people to be handled varies over time and certain levels of comfort and speed need to be maintained. Table 7.1 shows criteria for the Level of Service (LOS) that is defined for airports. A specific airport may require the LOS to be excellent during 95% of operating hours, with an allowance for LOS to go down to the level of good for only 5% of those operating hours.

Related performance aspects: Revenue generation, availability

Revenue   **Workload Capacity**

Many buildings are created by project developers, who invest in construction in order to make a profit. Revenue is the total income generated by any business activities; profit is calculated by subtracting expenses from the revenue. Revenue is another performance aspect that results from serious negotiation, which reflects the interaction between supply and demand; however it typically is also impacted by regulations, which cap prices, housing benefits and similar.

Occurrence: Revenue is an important aspect for all commercial building developments and operations. In the domestic sector, revenue is of importance for rented accommodation such as council houses, student accommodation and the private rental market.

Scope: The building scale that is most appropriate for assessing revenue depends on the business activity. Where the activity is construction and sale of buildings, it obviously makes sense to analyze at a whole building level. However, where business units, apartments or rooms are rented out, a higher spatial resolution is appropriate. Typical temporal scales are revenue after construction or rental income per year, month or week.

Methods: Revenue can be either assessed by way of *financial forecasting* or established from bank and accountancy statements that show *cash flows*. Revenue is directly measured. A more advanced economic measure is *Return on Investment (ROI)*, which gives a ratio between gains and cost of an investment.

Criteria: The criteria for what is acceptable revenue depends highly on the investment made; see Figure 7.7. It also often reflects general financial conditions, such as the stock market. A target may be to aim for revenue that is 10% larger than investments.

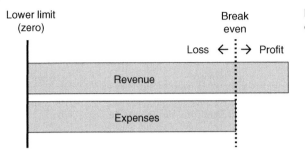

Figure 7.7 Balance between expenses and revenue.

Energy Efficiency   Resource Saving

Energy efficiency is an important societal concern, which needs to be considered in all buildings; it is one of the most deeply studied building performance aspects. The interest in energy use is driven by a range of environmental, economic and political concerns. At the level of single buildings, clients and tenants may strive for high-energy efficiency on the basis of individual motivators, while a degree of efficiency is mandated by governments. For some buildings, such as PassivHaus and similar, energy efficiency is a defining feature. Governmental mandates have to be met; going beyond the imposed criteria is an issue of negotiation.

Occurrence: Most buildings consume energy in order to maintain a comfortable indoor climate and lighting levels and to operate other building services. How much work the building needs to do to maintain these is of course highly dependent on the local climate conditions. The issue is of concern for virtually all buildings, apart from a few exceptions such as bus shelters or storage sheds. There also are some buildings that host energy-intensive processes; here the energy efficiency of the building itself is of lesser interest to the owner/operator. Energy efficiency is more important where buildings are remote and do not have access to regular supply via the electricity or gas grid.

Scope: Energy efficiency needs to be assessed at the whole building level while bearing in mind that the outcome of the analysis depends on the interaction between many elements. It is worth keeping in mind that within a building, some energy is used by the building itself (for heating, cooling, ventilation and lighting purposes); however, a certain amount of energy is used for non-building-related purposes (such as electricity for computers). This latter amount of energy, while delivered via the building and passing through building systems, is not really related to the performance of the building itself. In assessing energy efficiency, it is important to discern which uses are included in the analysis: heating, cooling, hot water supply, lighting and plug loads are the key loads, and all have their own particulars.

Method: Energy efficiency can be assessed by both *simulation/calculation* (especially during design) and *experimentally measured* (once the building is operational). Simulation options are discussed in Chapter 6 and include stationary, semi-dynamic and transient methods; individual methods can have more or less explicit representations of the underlying thermal processes along the black/grey/white box distinction. A special category is that of *normative methods*; normative methods are used in codes and standards to review whether or not buildings meet a certain energy efficiency criterion under assumed conditions. Normative methods can be contrasted to descriptive methods; confusion sometimes is created by the fact that the underlying calculation process can be identical. Measurement may capture the key energy supply (electricity, gas, oil, district heating) at the whole building level; however understanding of the specific use is helped significantly by sub-metering.

Criteria: Criteria for energy efficiency vary with the analysis objective. For comparison with peers, a common measure is Energy Use Intensity (EUI). For the management of a specific building that is in operation, it may be more interesting to set criteria in relation to the historical track record (baseline) and for instance aim for a 10% reduction in energy use. A specific balance between locally generated energy from renewable sources and energy use is required to establish whether a building is a net-Zero Energy Building (nZEB); see Figure 7.8.

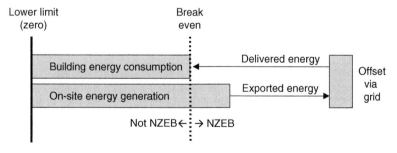

Figure 7.8 Criterion for establishing whether a building is a NZEB.

Related performance aspects: Energy generation; energy loss; thermal comfort
Relevant standards and norms: ISO 13790; ASHRAE 90.1; CIBSE AM11

Water Efficiency   **Resource Saving**

Water is increasingly recognized as another resource that needs to be used wisely. This is especially true in countries with limited rainfall. There is also concern about the use of drinking water for uses that not require that quality of supply, such as flushing the toilet.

Occurrence: Most buildings use water, again leaving out some very basic facilities like bus shelters and storage sheds.

Scope: The analysis of water efficiency often considers the capacity to collect rainwater, which involves analysis of roof size and drainage systems. Beyond this, the analysis considers the different uses of water: as drinking water, for toilet flushing, showering, bathing, laundry, dishwashing, outdoor use and others.

Methods: In principle the physics of fluid flow apply to water and drainage. However, in practice the assessment of water efficiency during building design is done using simple rules, such as relations between numbers of building occupants and standard numbers of water-using appliances required, which are then combined with typical daily water use figures. Calculations of water discharge are often based on the use of discharge units (DUs), which provide a simple way to account for amount and simultaneity. Sankey diagrams may be used to review flows of drinking water, rainwater, grey water (reusable for some purposes) and black water (sewerage) within the building. Metering of water use is mostly done on the supply side (water meter) only.

Criteria: Since water use is closely related to occupancy, a typical criterion is to require limitation of use based on standardized use, for instance less than 80 l/person/day. Individual design targets may try to better that value; see Figure 7.9.

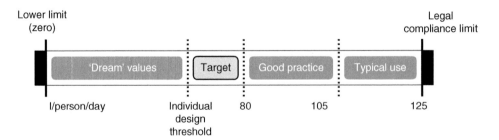

Figure 7.9 Water use (efficiency) target for a typical UK building.

Related performance aspects: Storm-water drainage capacity
Relevant standards and norms: EN 12056

Material Efficiency   Resource Saving

As natural resources are limited, there is an increasing interest in ensuring that buildings make efficient use of the materials employed. The use of materials also has direct implications for the costs of realizing construction projects. Furthermore, the production of parts, components and systems requires energy, as does the transport of these products to the construction site; in this context material efficiency is often quantified in terms of embodied energy, which is contrasted to operational energy, which is the conventional use of energy to operate a building. There are different approaches of handling the life cycle in the calculation of embodied energy. Stakeholders in material efficiency are the project developer, design team, contractor and the material and systems supply chain. Material efficiency is very much an issue of negotiation between the various stakeholders.

Occurrence: All building construction, without exception, requires materials, so this performance aspect is truly omnipresent. However, the importance given to the aspect varies; it may be given prominence in projects that value environmental issues, whereas it may be marginal in projects where other aspects such as safety, prestige and income generation are prime considerations.

Scope: Material efficiency needs to consider all the materials that are employed in creating a building. Often, whole life cycle considerations expand the scope to include sourcing of raw materials, fabrication of components and transport to the construction site. Sometimes it is expanded to include the product of excavation or demolition of previous buildings on a site.

Methods: The assessment of material efficiency requires an analysis of the amount of materials used in a specific building; in some cases this can be expressed in volume (such as concrete), while in others it is best handled in number of parts (such as bricks). For each volume or part, one then needs to consider the process of fabrication, transport and installation on the site and the detailed issues of interest to the analysis. In terms of material costs, a key underlying method is *quantity surveying*. Material efficiency in terms of dealing with loads requires methods from *structural engineering*. Embodied energy is typically assessed using *Life Cycle Analysis* (LCA).

Criteria: As buildings are highly unique, there is no generally applicable criterion; instead one typically develops an initial design and measures material efficiency as improvements gained over that reference. See Figure 7.10. This means that implicitly all quantification is normalized for the reference situation.

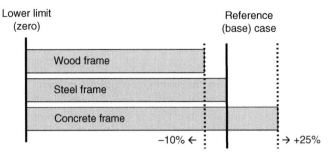

Lower limit
(zero)

Reference
(base) case

Figure 7.10 Material efficiency compared with a selected base case.

−10% ←   → +25%

Related performance aspects: Construction costs; operational energy efficiency
Relevant standards and norms: ISO 14014

Greenhouse Gas Emissions   Resource Saving

The interest in Green House Gas (GHG) emissions is driven by the concern about anthropogenic climate change, which leads to increasingly strict targets for various sectors such as transport, manufacturing and the built environment. This issue thus affects all buildings as well as construction processes. As performance aspect, the reduction of GHG emissions is classified as resource saving with a view on preserving the global climate and mitigating climate change effects.

Stakeholders who have to deliver reductions in GHG emissions are the building design and supply chain, owners and operators; beneficiaries will be society and humanity at large and ecosystems worldwide.

While policies for GHG reduction are increasingly stringent, the complexity of the issue sometimes creates room for negotiation. There also may be options to offset local emissions by buying 'carbon credits' in emission trade schemes.

Occurrence: GHG reduction is an issue for all buildings that operate some sort of system that uses energy. Note that energy may be supplied through a grid, with generators and thus actual emissions positioned away from the building.

Scope: GHG emissions are typically analyzed per building. However, further scrutiny of system boundaries may be required to review the impact of remote generation, conversion and transport of energy, especially for electricity as the underlying fuel mix is constantly changing. Temporal resolution is mostly per annum. A debate may be had about the need to include embodied energy in the analysis.

Methods: Measurement of GHG emissions is mostly done using computational tools and expressed in terms of carbon (CO_2) emissions. These typically start from heating and cooling needs inside the building; a range of factors is used to find the required primary energy to cover the demand, as well as generation, transmission and conversion losses. Where fossil fuel is used, a chemical formula is then needed to find out the amount of carbon dioxide produced at the source. Further conversion factors may be used to translate the relative impact of other GHG emissions into CO_2 equivalents. Beyond calculation, GHG emissions may also be measured empirically; however this typically requires sophisticated equipment such as gas chromatography.

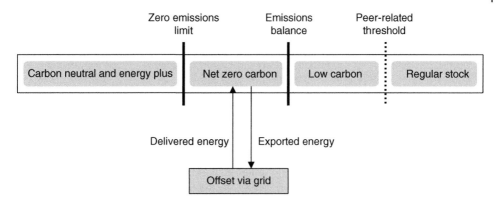

Figure 7.11 Different types of buildings that aim to manage carbon dioxide emissions.

Criteria: An often-used criterion for GHG emissions is Net Zero Carbon or NZC. In this case the building is assumed to generate energy on site, for instance through the use of renewable energy systems such as PV; it is allowed to export energy to the grid if production exceeds local demand and it can reimport this energy at times where local demand is larger than the production. See Figure 7.11.

Related performance aspects: Energy efficiency
Relevant standards and norms: ISO 14064

Burglary Resistance   **Readiness**

Security is an important issue for building occupants; this requires that the building is ready at appropriate times to resist forced entry while at the same time maintaining easy access for authorized entrants. Stakeholders are the building owners, occupants, law enforcement department, insurance companies and on the other side of the equation criminals. Resistance against burglary is a matter of making it difficult to gain unauthorized access; full protection against all eventualities is typically not possible within reasonable costs. The trade-off between protection and cost is a clear area for stakeholder discussion.

Occurrence: The performance aspect is applicable to almost all buildings, apart from a limited number that are left unsecured and open to everyone on purpose, such as bus shelters.

Scope: Overall burglary resistance concerns the whole building and all points of entry; typically resistance is deemed only as good as that offered by the weakest point. In general the analysis of the building performance thus focuses on doors, windows, skylights and similar and is reviewed at component level. However, this usually includes the door and window frames. Time is an important factor in assessing burglary resistance; typical timeframes range from 3–5 minutes (resistance to burglary by an opportunistic criminal) to 20–50 minutes (resistance to an experienced perpetrator who is determined to gain access).

Methods: Resistance against forced entry is mostly tested under laboratory conditions; it is extremely difficult to assess computationally. The process

Table 7.2 Burglary resistance class criteria according to the former EN 1630.

Resistance class	Method	Resistance time (min)
RC1	Kicking, pressing (vandalism)	—
RC2	Simple tools such as screwdriver, pliers (opportunist)	3–15
RC3	More tools: extra screwdriver, crowbar (opportunist)	5–20
RC4	Saw, hammer, electric drill (experienced perpetrator)	10–30
RC5	Extra power tools (experienced perpetrator)	15–40

is that lab personnel tries to gain access using increasingly destructive methods, starting from simple pressing and kicking against the element, using simple tools like a screwdriver, scaling up to a crowbar, then hammer and electric drill, and ultimately more powerful electric tools.

Criteria: A resistance class can be awarded for resisting the mentioned methods for a given time span. See Table 7.2.

Related performance aspects: Structural integrity; privacy
Relevant standards and norms: EN 1626

Access Control   Responsiveness

Ensuring that the right people have access to a building, while ensuring others are kept out, is important for many private, public and commercial organizations. Buildings need access control in order to maintain the safety of the occupants and contents. At the same time, access systems can also be a significant nuisance, creating delays when entering the building. A responsive system will recognize who is authorized to enter and who is not while causing minimal disruption to the movement of authorized occupants.

Stakeholders to access control are all occupants of a building, whether these are permanent or visitors; one may also identify a group of 'stakeholders' that the access control tries to prevent from entering the building, such as burglars.

Performance criteria for access control are typically seen as critical and non-negotiable. However there may be room for some flexibility during design of buildings that have zones with various sensitivities. For instance there may be differences between a ground floor with access for the general public, general circulation zones for staff circulation, and office space with more strict access control.

Occurrence: Access control is relevant for most buildings, except those that cater for the general public or that have functions that are deemed not to be at risk, such as storage of materials that are difficult to transport.

Scope: Access control focuses on the formal points of entry into buildings, zones and rooms; however the effectiveness also depends on maintaining the integrity of the perimeter of the building/zone/room.

Methods: Performance of access control will be primarily measured in terms of effectiveness. For instance this may be measured in terms of people gaining unauthorized access to a certain space, which can be expressed in a percentage where occupant numbers are large, or a count of exceptions in a highly controlled area. The performance of the access control system may also be quantified in terms of the rate of incorrect denials to enter the space or waiting time for people wanting to gain access.

Criteria: One way of quantifying correct working of access control is by measuring the False Alarm Rate (FAR) on systems that have that functionality and requiring a maximum FAR of 0.1% for high security areas, 0.3% for medium security areas and 1.0% for low security areas. Responsiveness to occupants can be established by processing speed, with systems that handle 30 or less persons per minute rated slow and systems that handle 31 or more persons per minute rated fast. See Figure 7.12.

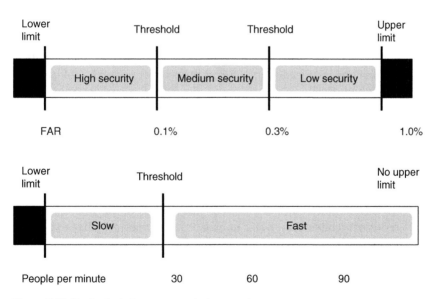

Figure 7.12 Dual criteria for access control responsiveness.

Related performance aspects: Burglary; fire escape
Relevant standards and norms: EN 6039-11-1: 2013, EN 60839-11-2:2015

Aesthetics   **Aesthetics**

Aesthetics is an important performance aspect in architecture and the built environment yet probably also one of the most challenging to include in any framework that attempts to set criteria and quantify performance due to the fact that the assessment of 'beauty' always involves a subjective judgment. In order to deal with aesthetics in a performance context, it is helpful to limit measurement and quantification to an objective assessment of relevant building attributes while refraining from judging 'beauty', since the latter also relates to characteristics of the observer (age, taste, training, etc.).

Key stakeholders in aesthetics are the client or project developer, architect and planning authorities, while the general public and architectural discipline at large have a significant interest.

Aesthetics often is subject of long and deep negotiations; it is key to how people perceive buildings and closely related to stakeholder satisfaction.

Occurrence:	Aesthetics is a concern in most building design projects, apart from some basic and industrial buildings where function may be completely dominant.
Scope:	Aesthetics pertain to building details (door handle, window), to the whole building, and even to urban level.
Methods:	Quantification of aesthetics may measure formal attributes, such as shape, order, proportion, scale, symmetry, complexity, colour or hierarchy. Examples of assessments that help to support the analysis of the aesthetics of a building may include the following:

  • Establishing whether a building is symmetric or asymmetric.
  • Checking whether rectangular forms and surfaces such as door openings and façade panels follow the proportions of the golden ratio.
  • Mathematically measuring the degree of visual entropy along orthogonal axes.

  Another approach would be to establish whether a building design represents a recognizable architectural style, such as postmodern, brutalist or Art Deco.

Criteria:	Criteria for building aesthetics are notoriously difficult and typically require negotiation between stakeholders. However, there are some universally accepted ones, such as the proportions of the golden ratio. In terms of architectural style, a human observer may recognize a style and assign value to it, without necessarily liking it.

7.3 Tool and Instrument Configuration

Active building performance analysis also requires interaction with the tools and instruments that are used. This is very much a two-way process: the development of criteria and choice of an evaluation method impose constraints on the tools and methods, but

Figure 7.13 Tool configuration.

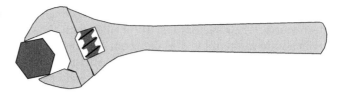

similarly those tools and instruments have their own inherent issues that will impact on the criteria and expected outcome. A simple visualization is the image of the wrench and the nut in Figure 7.13; although the wrench may be the right tool for fastening the nut, the quality of how well that is done depends on the adjustment of the tool to the specific job. And a deformed or damaged nut may make the use of the wrench more difficult. Similarly, some evaluation method may be the right approach to quantify performance, but ultimately how well that performance is quantified also depends on the tool that is used and how that tool is used. Two different aspects play a role: managing the intrinsic properties of the tool and ensuring that a good experiment is carried out in line with the discussion in Chapter 4.

In terms of tool configuration, it is worth remembering that experiments may (i) establish to what stimuli a system responds, (ii) capture a series of input–output pairs or (iii) reproduce system behaviour. To do this properly – which means that the analysis is reliable, valid and reproducible – requires that the experiment captures the system(s) of interest, the controllable as well as the uncontrollable factors that impact on the experiments, the excitation (input) and the system response (output). In other words, a tool or set of instruments must allow:

A) To capture the building and building systems of interest in terms of 'hardware'.
B) To apply an 'experiment' to that hardware in terms of subjecting it to a certain load or excitation or at least to register these if the experiment is only semi-controlled.
C) To measure how the building and building systems respond to the experiment by changes to states and output.

Given that there is an endless number of criteria, and a large number of tools and instruments, the following paragraphs discuss three selected examples to explore this interaction. These cover physical measurement, simulation and stakeholder evaluation. A general template that helps to consider the key issues in configuring tools and methods is contained in Appendix C.

Analysis with Hardware: Temperature Measurement

The use of small temperature loggers is common in various contexts, for instance to record overheating in different rooms of a building. There is a range of commercial products that can be used for the purpose. Yet real-life loggers impose practical constraints that impact the analysis. A first issue is logger accuracy, which varies across the different options. Once deploying loggers, an important practical issue is measurement location. From an evaluation point of view, ideally loggers must be positioned at a central location in the room/zone that is being monitored; however occupancy constraints may dictate a position on the periphery. The device should not be put in a position that may receive direct sunlight or be influenced through proximity to a radiator or similar heat source. These factors may be at odds with the desire to capture an average air temperature in the middle of the room. Another issue is logger set-up. Temperature loggers mostly come with dedicated software that allows the programming of the logger in terms of setting date and time, starting time and frequency of logging. This frequency may require a trade-off between temporal resolution and measurement duration due to limited storage capacity of the device. Analysts may also have to consider whether they will have easy access to the logger, allowing simple download, or whether the logger will be in a remote building or another situation that makes it worth to invest in some sort of wired or wireless data transfer. Furthermore, if devices are powered by batteries, measurement frequency is also related to lifetime of the battery. These considerations may impact on what criteria can be used; for instance, if temperature is only monitored on an hourly basis, then any limits that require a higher temporal resolution are not feasible. This constraint is not trivial; for instance a maximum or minimum temperature in a room may differ significantly from the on-the-hour measured value.

Data obtained from temperature logging is a straightforward time series. However, analysis may require some special operations, such as selection of a subset that corresponds to occupied hours, calculation of moving averages or calculation of weighted overheating hours. This in turn may require a correlation to other measurements; for instance one may want to know when the building was actually occupied or how indoor temperatures relate to outdoor temperatures. Where the interest focuses on a whole building, it is important that measurements are taken in different and representative spaces, where spaces are large or high multiple loggers may have to be used within a single space. In order to allow that kind of analysis, it is important to ensure that various data sets can be synchronized.

Analysis with Software: Whole Building Thermal Simulation

Thermal simulation is regularly used to study energy use and thermal comfort in buildings. There is a large number of software programs that may be used for the purpose. Built into these programs is a wide variety of physical models and assumptions and different mathematical approximations and solvers. While many programs offer some degree of flexibility, simulation tools still require specific input to be able to run and can only produce outputs that have been designed into them. In general terms, whole building thermal simulation tools require the definition of the building and systems, the climate conditions to which the building is exposed, the control regime, the occupancy schedule and the definition of the specific output they are to generate. Significant emphasis is often placed on geometry; however it may be useful to carefully consider zoning and simplification. Beyond the geometry, the materials that make up the building need to be defined; this is done through the specification of layers that make up wall, floor, roof, window and door segments of the model. However, models available in simulation tools may not cover all systems encountered in real buildings; for instance many tools only have a limited number of HVAC components that can be selected. It thus is important to ensure that a tool is found that maps the analysis interest. In many models, there are built-in assumptions; for instance there may be coefficients of performance that can be either fixed or customizable. Some common aspects remain surprisingly difficult; one such issue is the modelling of stairs and the associated airflows between the different floors of a building. The same goes for sunspaces, atria and courtyards. There also may be issues with geometry; for instance some tools expect every space to be bounded by six panes, which leads to constraints where the assessment pertains to a triangular section such as in a typical domestic attic space. Another area that may cause issues when setting up an analysis is the modelling of occupant behaviour and control. By way of example, in many cases one may assume that occupants open windows when a building gets too warm, but in terms of simulation, this requires a range of assumptions about when this human intervention takes place and how it impacts natural ventilation. Yet simulations that do not represent such effects may be seen as unrealistic and risk being discarded by some stakeholders. Weather data needs to be provided according to a specific format; if measured data on a specific site is to be used in simulation, it needs to be converted to the exact format that is understood by the tool. In some cases, one can only use climate data provided with the tool and will have to settle for the climate that is most similar to that of the actual situation. It is important to ensure that a simulation tool can export the data that is needed for the evaluation that is planned, such as air temperatures in spaces of interest or surface temperatures that are needed for assessment of thermal comfort (and/or radiative temperature), with the appropriate frequency. Usefulness may depend on mathematical modelling; for instance if a model is running on the basis of an hourly time step, capturing temperatures at a 10 minute interval is at best an interpolation, if possible at all. Some tools have advanced post-processing capabilities; for instance EnergyPlus can output PMV and PPD values or equivalent carbon emissions; other tools may require manual post-processing.

Analysis with Occupants: Stakeholder Satisfaction Survey

The use of surveys amongst occupants is a common method for capturing the satisfaction of occupants of a facility. As this approach is based on the opinion of human beings, it is important to tailor the survey to specific needs and to prevent bias. While there are good standard surveys, such as the BUS methodology, it is mostly beneficial to tailor the survey to the specific situation. A general survey will cover the common performance aspects and check for satisfaction across main performance areas; however more specific feedback can be obtained by asking about specific features or issues of the actual building under investigation. As in all surveys, care must be taken when phrasing questions and providing options. For instance, a Likert scale with an even number of points may force respondents to answer either positively or negatively, whereas providing a neutral position may lead to superficial responses. Two other issues are of key importance when administering a survey: the sampling of respondents from the total building population and the timing of the survey. In both cases one should aim to minimize the impact of organizational noise on findings; for instance conducting a stakeholder survey during a period of peak workload may say more about the organizational pressures than about the building.

7.4 Iterative Analysis

Like in most design and engineering processes, building performance analysis also benefits from iteration. Iteration allows the development of criteria and application of a performance quantification method, testing the results and then refining earlier decisions if needed. This is similar to the use of the Deming cycle of plan, do, check and adjust. Feedback iteration is acknowledged as important in systems engineering (Blanchard and Fabrycky, 2011: 57; INCOSE, 2015: 32–33). The deep interaction for each specific case between stakeholder interests, the unique context, the relevance of selected systems and components, design particulars, loads and excitation, the evaluation principles and corresponding tool/instruments and the developed goals and targets is complex, so it is always wise to evaluate the selected approach and allow for readjustments.

The following situations briefly discuss some selected cases where iteration has been found to be essential to the success of the analysis:

- Temperature logging in a building, whether through dedicated measurement kit or via the building energy management system, typically takes place over a longer period. In many cases the analyst sets up the monitoring and then waits for a significant time to capture the typical range of values over a season, peak values during a heat wave or minima during a cold spell. Faulty sensor readings or incorrect measurement set-up may mean the need to wait a full year to redo the measurement, so ensuring that the process works as desired is an important part of carrying out such an analysis efficiently.
- Initial measurement of the temperatures in a building may indicate that controls have been wired in the wrong way, with the thermostat in one room leading to heating or cooling in a different room and vice versa. In this case it is important that the wiring is corrected; further measurement will not provide any meaningful results beyond confirming what can be established through the first observations. Once control functions as intended, further studies may be carried out to study how well the heating and cooling systems work and whether the control settings may require fine-tuning.
- In the use of building performance simulation tools, many assumptions need to be made to create a model. Yet a simple oversight can render the simulation outcomes completely useless. By way of example, a lot of effort may be invested in modelling an office with an atrium space. Initial analysis of indoor air temperatures may point towards significant overheating of the atrium. At this point, the design team may mention that they anticipate the use of openings in the atrium roof and stack ventilation to ensure natural cooling. Without modifying the model to include this design feature, the analysis will not be accepted as valid by the design team.
- A building design project may start off by aiming for a net-Zero Energy Building. Initial analysis, using simulation, may show that this is not fully feasible unless significant investments are made in battery storage. Discussion with the client may then find out that the aim for Net-Zero Energy is to be relaxed to Near-Zero Energy, rather than investing resources in the procurement of sufficient battery capacity. This case shows change of criteria in response to initial analysis results.

7.5 Building Performance Analysis in Current Practice

One building that has undergone various rounds of performance analysis is the Roland Levinsky Building at the University of Plymouth in the United Kingdom. See Figure 7.14. The building is a good case study to demonstrate the need to fit performance analysis

Figure 7.14 Roland Levinsky Building, University of Plymouth, UK.

efforts to a particular context and need. The Roland Levinsky Building is a multifunctional building that houses the Faculty of Arts and Humanities, featuring lecture theatres, studio space, laboratories, staff offices, art galleries, a café and storage space. The building has 10 levels, offering a gross floor area of $12.660\,m^2$. It was constructed between 2003 and 2007. The architectural concept employs an atrium named 'crosspoints' that connects university and public space, with student and staff facilities provided in a low west and high east tower. The building has large glazed facades towards the South, providing a view over the city, as well as towards the North in order to create good lighting conditions for artists. This scheme is wrapped in a copper façade that morphs into the roof. In terms of building services, the building employs mechanical ventilation. The air handling system consists of variable speed fans and multiple air handling units, some of which are positioned in the basement (close to the lecture theatres), while others are positioned in a plant room on top of the east tower (close to the student and staff spaces). Cooling is provided by air-cooled chillers, whereas heating and hot water supply is through gas-fired condensing boilers.

The different rounds of building performance analysis had various drivers, taking place at different stages of the building life cycle:

- During design, extensive analysis was conducted by two building engineering companies involved in the project: Hoare Lea and Scott Wilson. Amongst others, these carried out structural engineering analysis, wind tunnel testing, concrete frame analysis, fire smoke analysis and thermal modelling. The emphasis in these efforts is on confirming system design choices and optimizing parameters under assumed future conditions.

- After 5 years of operation, the estate department of the university embarked on an ambitious carbon reduction programme. In collaboration with the provider of the building automation system, operation and control of the building was reviewed. Overall this programme has been delivering a reduction of 32% of carbon emissions in 2015, with further reductions planned by 2020 and 2030. These savings are verified by an external 3rd party M&V (measurement and verification) contractor using C3ntinel software for analysis.
- Between 2009 and 2011, the building was a case study for academic research into the impact of climate change on thermal building performance. A range of simulation studies was carried out to analyse the risks of increased overheating and changing energy consumption for heating and cooling. In these efforts, the focus was on predicting thermal comfort, overheating risk and impact on occupant productivity in relation to predicted future climate conditions while taking into account potential changes to the energy use of the building.
- Between 2012 and 2013, the building also was used as a case study for an academic investigation of the 'energy performance gap' between predicted and measured energy use. Here the focus of interest shifted to the impact of uncertainties on simulation and measurement results. This was later followed by studies that looked into detail in occupant presence in the building.

Obviously these analysis efforts are placed at various points in the building life cycle (during design, initial and current operation, as well as looking at the future) and involve a wide range of stakeholders (client, design team, facility management, building automation experts, M&V specialists and academic researchers). This brief overview demonstrates that a typical facility is likely to undergo various analysis efforts by different parties. While all four analysis efforts have a common thread around the energy use and thermal comfort of the building, the detailed interest and drivers behind the analysis vary: during design the scope is system design and dimensioning; in the operation review, the scope is optimal control; in the academic climate change study, the scope is resilience against change; and in the performance gap research, the scope shifts to the interaction between occupant presence and building performance.

In terms of development of analysis criteria, choice of evaluation method and configuration of tools and instruments to the task at hand, the following insights can be gained from the academic studies:

- The performance aspects of energy efficiency and thermal comfort are closely interrelated in these studies, as a warming climate may push cooling demand beyond existing system capacity.
- Energy efficiency has been closely linked to greenhouse gas emissions, as these are generally accepted to be a causal factor in anthropogenic climate change.
- The interest in exploration of performance under future conditions, combined with a need to quantify the impact of a range of factors, mandates the use of building performance simulation as the main quantification method.
- Selection of the building simulation tool EnergyPlus precedes the development of analysis criteria. EnergyPlus was selected for a number of reasons, including good reputation of the software in research contexts, access to input and output files enabling the automation of repeated simulation runs in distributed simulation, and expertise of the research team.

- For the analysis of thermal comfort conditions in the building, an adaptive model was selected that reflects different indoor setpoint values as well as comfort bands in relation to a warming outdoor temperature, as it is unrealistic to expect that the occupants of the building will judge the future conditions using exactly the same criteria as today. In detail, building performance was quantified using 'likelihood of thermal discomfort [%]' as defined by Nicol and Humpreys (2007).
- The EnergyPlus simulation engine calculates primary heat demand to condition spaces inside the building. This heating demand can be used to calculate greenhouse gas emissions by taking into account system efficiencies to arrive at natural gas and electricity consumptions. These in turn are converted into carbon dioxide equivalents on the basis of assumptions the mix of energy sources that supply energy to the electricity grid, calorific values and combustion products of fuel fuels and equivalent impacts of these combustion products on greenhouse gas effects.
- Actual metered data of the Roland Levinsky Building is reported in the common format of kWh; the estate department logs this data on an hourly basis. Metering includes gas and electricity use. Assumptions are required to disaggregate the overall use into energy that is used for space heating, hot water supply and plug loads, as there initially was no sub-metering (this has later been added to the facility).
- In both the study of adaptation to climate change and impact of occupant behaviour, the analysis of building performance relates to studying the impact of an independent variable (climate, occupancy) on dependent variables (energy use and thermal comfort). Of these two dependent variables, thermal comfort needs to be maintained; energy use is a function of this set-up and is studied without a specific criterion, although the general assumption is that 'less is better'.
- These analysis efforts highlight significant uncertainties in the performance quantification. In terms of climate change, use has been made of the UKCP 2009 probabilistic climate change projections, which by nature already reflect uncertainty. However, on the time horizons required for climate change prediction, there are further uncertainties that may have significant impacts, such as change of use of the facilities and retrofit of facades and building services. The impact of these uncertainties has been assessed by exploring a range of intervention scenarios. In terms of occupancy, significant discrepancies between assumed use (as reflected in the building automation setpoints) and actual observed building occupancy have been found.

The research efforts have included a large number of simulation results. Some representative simulation outcomes pertaining to the energy use of the building are presented in Table 7.3. Note that this is just a basic selection; hundreds of simulation runs have been carried out to study the impact of various climate variations for the UKCP 2009 predictions, as well as different use and retrofit scenarios. Overall findings indicate that in buildings in a temperate oceanic climate, which already employ an active cooling system, the onset of climate change can most probably be mitigated by retrofitting appropriate HVAC units, noting that the pace of climate change is relatively slow, whereas the building services have a lifetime of approximately 15 years only.

Further information about the analysis of the Roland Levinsky Building can be found in Tian and de Wilde (2011), de Wilde and Tian (2012), de Wilde (2014) and Amore *et al.* (2016).

Table 7.3 Energy consumption quantification of the Roland Levinsky Building.

Metered data

	Gas use (kWh/year)	Electricity use for HVAC (kWh/year)	Total energy use for HVAC (kWh/year)
Measured, 2011	8.81×10^5	1.20×10^5	10.01×10^5
Measured, 2012	6.68×10^5	1.29×10^5	7.97×10^5

Simulated data

	Heating (kWh/year)	Cooling (kWh/year)	Total energy use for HVAC (kWh/year)
E+ baseline	5.84×10^5	1.89×10^5	7.73×10^5
E+ low warming	3.87×10^5	3.86×10^5	7.73×10^5
E+ medium warming	3.54×10^5	4.22×10^5	7.76×10^5
E+ high warming	3.39×10^5	4.51×10^5	7.90×10^5

7.6 Reflections on Working with Building Performance

Actual building performance analysis is a complex and knowledge-intensive process. A conceptual framework depicting the main ingredients is presented in Figure 7.15. At the centre of this graph is the selection and development of performance measures. The starting point is a specific case, which is a unique building project that comes with an individual design and dedicated building systems and responds to the specific needs of the stakeholders. The requirements allow the identification of relevant performance aspects, which will typically relate to the main building performance aspects and functions. Criteria development then requires that an appropriate evaluation method is found and possibly modified to fit the context; it also requires that appropriate limits and thresholds are set. Beyond that, one also needs to complete the puzzle by identifying suitable tools or instruments to carry out the actual analysis. The case study of the Roland Levinsky Building gives a brief overview of how various interests drive different performance analysis efforts and may require a different analysis set-up for one and the same building.

The development of analysis criteria may be supported by providing access to available evaluation methods, typical limits and thresholds based in human physiology and generally accepted benchmark figures. Ultimately however the criteria needed for the assessment of the performance of a specific building will be unique. Consequently, analysis methods need to ensure they have the capability to adjust to the specific needs of a situation. A general template that helps to consider the main aspects that need attention during the development of criteria is provided in Appendix B. A checklist that reviews crucial issues of tool configuration is provided in Appendix C.

While the conceptual framework as developed in this chapter helps to think about building performance analysis, actual operationalization of this framework and any emergent theory still remains a serious challenge. As described in the previous

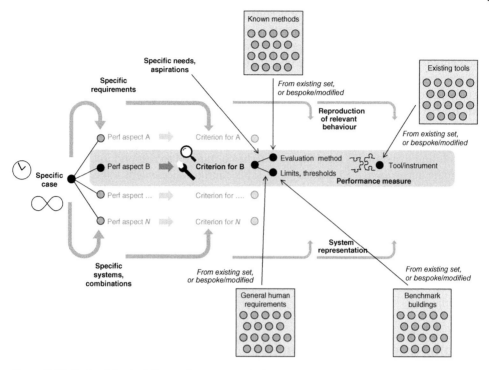

Figure 7.15 Customizing building performance analysis.

chapters, the Bank of America Plaza case shows the complexity of the changing context in terms of life cycle, stakeholders and systems. Beijing Airport Terminal 3 exemplifies the large number of needs, requirements and functions that need to be considered. Akershus Hospital presents the intricate interactions between requirements, design and actual performance. The Davies Alpine House shows how difficult it is to establish criteria and how targets are prone to change. The SK Telecom Headquarters provide a feel for the complexity of quantifying performance through simulation and measurement. Finally, the Roland Levinsky Building shows how different analysis needs all lead to different criteria and analysis procedures.

7.7 Summary

This chapter explores the deep customization that is needed to work with building performance. It presents a range of typical situations that show how most needs for analysis require the study of performance across a range of aspects while stressing that specific cases are mostly unique and come with their own building design, systems, drivers/interest and loads/excitations. The chapter then proceeds to explore in more depth how a number of performance aspects can be measured or quantified, discussing acoustical comfort, indoor air quality, thermal comfort, visual comfort, productivity, processing capacity, revenue generation, energy efficiency, water efficiency, material efficiency, greenhouse gas emission control, burglary resistance, access control and aesthetics. Note that this list is a small subset of a wider range of performance aspects as listed in Appendix A. Each aspect is discussed in terms of general importance, key stakeholders and likely room for negotiation on the subject, followed by a review of likely occurrence of the performance aspect, typical system and temporal scope of the analysis, some of the tool-independent evaluation methods that may be used to evaluate the performance for this aspect, and an example of some of the criteria that may be used. A template to support the development of performance criteria is provided in Appendix B.

Tools and instruments typically need to be configured to respond to the dynamics of a specific performance analysis at hand. This configuration needs to be done in such a way as to make the performance analysis a scientifically sound experiment, which is reliable, valid and reproducible. It requires a review of the building and systems under investigation (hardware), experimental conditions as applied (load/excitation) and measurement and data aggregation (observation). A checklist to support the configuration of tools and instruments is provided in Appendix C.

Development of a performance analysis approach and evaluation criteria, like most design processes, will benefit from iteration and the Deming cycle of plan, do, check and adjust.

Recommended Further Reading:

- The accounts by Lehmann *et al.* (2010), Konis (2013) and the case studies contained in CIBSE AM11 for examples of operationalization of building performance analysis in specific cases.
- The papers by Augenbroe and Park (2005) and Pati *et al.* (2009) for an example of the operationalization of building performance within a tool that targets a specific category of buildings.
- Technical manuals for measurement instruments or simulation tools, such as the *EnergyPlus Engineering Manual* (US Department of Energy, 2016), for an insight into prospects as well as constraints that stem from specific analysis instruments/tools.

Activities:

1   Develop a set of criteria for the performance analysis of a building-integrated photovoltaic (BIPV) system.

2   Develop a set of criteria for the performance analysis of a simple door handle. Explore how this performance could be assessed (i) in a laboratory, (ii) through computational analysis and (iii) in actual use.

3   Explore the evaluation principles that are available to analyze the performance of the fire resistance of a multilayered non-load-bearing partition wall.

4   Find an existing survey to capture the stakeholder evaluation of the lighting (both through daylighting and artificial lighting) for a room of your choice. Review the survey and adjust it to make it as specific as possible to the study of the lighting performance of that one specific room.

5   Explore what the range of building performance aspects is that one may study with a simple CO_2 data logger.

6   Study the history of ISO standards in general. What is the process that leads to standards, and how are they maintained and improved?

7.8   Key References

ASHRAE, 2017. *Handbook of fundamentals*. Atlanta, GA: American Society of Heating, Refrigerating and Air-Conditioning Engineers.

Augenbroe, G. and C. Park, 2005. Quantification methods of technical building performance. *Building Research & Information*, 33 (2), 159–172.

Awbi, H., 2003. *Ventilation of buildings*. London: Spon Press, 2nd edition.

Boyce, P. and P. Raynham, 2009. *The SLL lighting handbook*. London: CIBSE Society of Light and Lighting.

CIBSE, 2015. *CIBSE AM11: Building performance modelling*. London: The Chartered Institution of Building Services Engineers, 8th edition.

O'Connor, P. and A. Kleyner, 2012. *Practical reliability engineering*. Chichester: Wiley.

Konis, K., 2013. Evaluating daylighting effectiveness and occupant visual comfort in a side-lit open-plan office building in San Francisco, California. *Building and Environment*, 59, 662–677.

Lehmann, B., H. Güttinger, V. Dorer, S. van Vlesen, A. Thieman and T. Frank, 2010. Eawag Forum Chriesbach – simulation and measurement of energy performance and comfort in a sustainable office building. *Energy and Buildings*, 42, 1958–1967.

Pati, D., C. Park and G. Augenbroe, 2009. Roles of quantified expressions of building performance assessment in facility procurement and management. *Building and Environment*, 44 (4), 773–784.

Smith, D., 2011. *Reliability, maintainability and risk*. Oxford: Butterworth-Heinemann, 8th edition.

US Department of Energy, 2016. *EnergyPlus*[TM] *Version 8.7 Documentation: Engineering Reference*. Washington, DC: US Department of Energy.

Part III

Impact

8

Design and Construction for Performance

The preceding chapters have dealt with the foundations and assessment of building performance, exploring how this performance is related to user needs, discussing what approaches are available to quantify performance and exploring working with building performance. However, to make a difference in actual practice and for real buildings, this theory about building performance then needs to be implemented. Two broad areas are crucial to ensure that thinking about building performance actually has impact: building design and construction, and building operation and management. Design and construction is the subject of this chapter.

Design is a key activity in bringing systems into being, by an activity that decides on the look and functioning of something, especially by making a detailed drawing of it. Design is a key part of both architecture and engineering. In his seminal *Sciences of the Artificial*, Herbert Simon (1996: 111) states that 'Everyone designs who devices courses of action aimed at changing existing situations into preferred ones. The intellectual activity that produces material artifacts is no different fundamentally from the one that prescribes remedies for a sick patient or the one that devises a new sales plan for a company or a social welfare policy for a state'. Construction is the process of actually making the building, according to the plans that result from the design process. Both design and construction are extensive disciplines in themselves, to which complete textbooks and study programmes are dedicated, such as *How designers think* (Lawson, 2005), *Design Thinking* (Cross, 2011), *Architectural Design Procedures* (Thompson, 2002), *House Construction* (Riley and Cotgrave, 2013) and *Industrial and Commercial Building* (Riley and Cotgrave, 2009). In this chapter the focus is on how building performance can be given a place in the process that leads to buildings taking shape.

There is a myriad of approaches that have the ambition of embedding the consideration of performance in the building design process. This is explicit in performance-based building design (Becker, 2008) and the more widely cast performance-based building (Bakens *et al.*, 2005; Jasuja, 2005), or in high performance building design as per the professional certification offered by ASHRAE in the United States. However building performance also is a key ingredient of the many variants of sustainable building design, such as eco-house design, bioclimatic building design, green building design, Passivhaus design, whole building design and others; see for instance Roaf *et al.* (2003), Ward (2004), Butler (2008), Krainer (2008), Day (2015), Hopfe and McLeod (2015), Keeler and Vaidya (2016) or Kibert (2016). Hansen (2007: 35–64) gives an overview of a range of building design strategies such as self-sufficient, ecological, green, bioclimatic and low-energy design and maps how these relate to building performance aspects such

Building Performance Analysis, First Edition. Pieter de Wilde.

as the use of daylighting, ventilation and others; this work shows some of the subtle differences between these generic design approaches. Most books and papers on such design strategies provide a good overview of the need for energy efficiency, conservation of resources and other drivers for performance, and introduce a range of performance aspects – some of which may conflict, such as energy efficiency, occupant comfort and cost. Many also introduce building science in some detail, cover the generic stages of the design process, and many contain a significant discussion of case studies. The main suggestions provided for reaching good performance are a generic instruction to pay attention to building science and to make use of computational analysis, as expressed by for instance Becker (2008), Zuo *et al.* (2010), Sabol and Nishi (2011) and Zalok and Hadjisophocleous (2011). However, the fine mechanics of how to actually design a building that delivers good performance remain evasive. The best guidance is typically given in the use of simple rules, such as the Trias Energetica or Ladder of Lansink. The Trias Energetica suggests to (i) limit the need to use energy by prevention, (ii) use renewable energy sources where a demand is inevitable and (iii) where there is no other option, use fossil fuels but only as efficiently as possible. The Ladder of Lansink is a similar approach to deal with the use of materials (Entrop and Brouwers, 2010). Another similar design guideline is the 'fabric first' approach to energy-efficient buildings, which suggests to first address the building envelope before dealing with systems. Again, opinions on details may differ; for instance Chesné *et al.* (2012) are of the opinion that too much attention is paid to insulating buildings, which makes it impossible to fully exploit the potential of natural resources such as solar energy and outside air. The complexity persists when design is limited to subsystem design rather than whole building design, such as in the work on facades by Ünver *et al.* (2004), Aksamija (2013), Singhaputtangkul *et al.* (2014) or Huang and Niu (2016), the selection of energy conservation measures (Khazaii, 2016) or building retrofit (Juan *et al.*, 2010; Appleby, 2013; Ferreira *et al.*, 2013; Langston, 2013; Rysanek and Choudhary, 2013 or Shao *et al.*, 2014). Similarly, approaches for designing with a single performance aspect in mind, such as the work on daylighting by Kleindienst and Andersen (2012), do suggest how to analyse performance but suffer from similar issues where it comes to guide the actual design activities.

Where the generic design approaches seem to remain on a high level and do not provide a detailed roadmap on how to actually proceed with performance-based design, another school of thought focusses on small steps (by comparison to the whole design process) and tries to support performance-based design decisions. This approach underlies the work of Gero (1990) on the Function–Behaviour–Structure model, the Generic AEC Reference Model of Gielingh (1988), the ASTM Standard for Whole Building Functionality and Serviceability (Szigeti and Davis, 2002) and the work by Augenbroe (2011: 21) on connecting building functions with building systems via Performance Indicators. It connects building performance and building design with theory on decision making, which in itself is a substantial area of research with an extensive literature; see for instance French (1988), Keeney and Raiffa (1993), Klein *et al.* (1993), Simon (1997), Jordaan (2005) or Hazelrigg (2012). Decisions have three ingredients: a set of alternatives to choose from, expectations about the future and preferences for outcomes of the available choices (Hazelrigg, 2012: 9). At a high level, decision theory consists of two main categories: normative decision theory, which suggests how a decision is to be made to realize the best possible outcome, and descriptive decision theory, which studies how decisions are made in practice and accepting that not all decisions are rational or lead to the best possible outcome. Normative decision theory bases itself on mathematics and mathematical economics (Hazelrigg, 2012: 3–4) and assumes that decisions are made by an 'economic actor' who can deal with the world in all its complexity and who can select the best alternative from all options available (Simon, 1997: 118–119). The identification of the best alternative is often supported by mathematical optimization. However, in real life there mostly are limits to the information about the situation, time for making the decision and other constraints; to describe this situation, Simon (1996: 38) introduced the concept of 'bounded rationality' and the verb 'satisficing' in order to describe decision making that leads to choices that are 'good enough' rather than optimal (*ibid.*, 27). Another view of decisions is the theory of naturalistic decision making, which does not focus on decisions as events that can be isolated but instead approaches decision making as a continuous flow of actions that work towards a set of goals. Naturalistic decision making has been found to be evident in complicated scenarios, where the decision maker faces ill-structured problems, an uncertain and dynamic environment, shifting, ill-defined or competing goals, action/feedback loops, time pressure, high stakes, multiple actors and complex organizational goals and norms (Klein *et al.*, 1993: 6–7).

The process of designing for building performance is supported by a wide range of design tools that may range from simple objects, such as pencil and paper, to advanced computer programs. The performance analysis tools discussed in Chapter 6 are thus part of a wider spectrum. Zapata-Lancaster and Tweed (2016) report that both architects and engineers make extensive use of experience and rules of thumb. In engineering, computing supports a range of tasks, such as object representation and reasoning, database management, constraint-based reasoning, optimization and search, knowledge management, and geometric and graphic design (Raphael and Smith, 2003). In the wider building industry, modern digital tools support and enable the sketching of design concepts, three-dimensional (3D) rendering of various design stages, validation of the programme of requirements, the engineering of building services (HVAC) systems, development of material schedules, digital prototyping, documentation of proposals and contracts, archiving of changes to the design and many others, often sharing data via a common building information (BIM) standard (Eastman *et al.*, 2011: 7). Hitchcock (2004) points out the need to repeatedly share building performance information gained from different perspectives during the building design process and discusses the role that interoperability, performance tracking and simulation can play. Integration of computers in the design process may follow different paradigms. At one extreme is automated design, where the computer drives the full process; however the more common approach is computer-assisted design where human designers make the high-level design decisions. However, tools are definitively becoming more able at some design tasks, as is evident from the emergence of generative tools as described by for instance Chase (2005), Shea *et al.* (2005), Krish (2011), Granadeiro *et al.* (2013) or Kitchley and Srivathsan (2014). It appears that the literature on building performance tools is biased towards analytical tools such as those discussed in Chapter 6; however the other approaches and concepts also may have important roles to play.

Ensuring building performance during the design process cannot be just an internal process of tools and designers. Augenbroe (2011) has long championed the notion that performance should be the subject of a dialogue between various stakeholders in a building design project. Communication and negotiation thus become important issues in designing for building performance, taking into account the peculiarities of the construction industry (Dainty *et al.*, 2006). Zapata-Poveda and Tweed (2014) remind us that design for performance involves a dialogue that takes place in a social context that requires common frames of reference and interpretation. This leads to a need to explore the subjects of visualization of data and information, another field in itself as described by for instance the seminal work of Tufte (2001).

The final step towards creating buildings that perform is actual construction. This is the domain of construction management. There is a significant body of literature on work in Quality Assurance (QA) and the implementation of the related ISO 9000 series of standards in construction, as well as on the uptake of the wider Total Quality Management (TQM) that also includes the management and practices of the companies that carry out the construction process; see for instance Xiao and Proverbs (2002), Lau and Tang (2009), Harrington *et al.* (2012) or Kraft and Molenaar (2014). Unfortunately, in spite of these efforts, actual as-built quality of buildings is not always in line with expectations, with research often reporting quality levels of poor and average (Kazaz and Birgonul, 2005; Sommerville and McCosh, 2006). The problem of the 'performance gap' between predicted and measured building performance is further evidence of issues in translating design into buildings that perform as expected (Carbon Trust, 2011; Menezes *et al.*, 2012; CIBSE, 2013; Wilson, 2013; de Wilde, 2014; Fedoruk *et al.*, 2015; van Dronkelaar *et al.*, 2016). To ensure building performance, construction not only needs to link back to the fundamental issues of building physics but also needs to cover the procurement process and stakeholder management (Goodhew, 2016). Another related issue may be the difficulty of enforcing evermore complex building regulations (van der Heijden *et al.*, 2007).

This chapter explores performance-based design of buildings, paying attention to the design process, design decision making, tools that support design, visualization and communication, and the construction process, all of which need to interact in a proper manner to ensure that buildings are created that meet the expectations of all stakeholders.

8.1 Performance-Based Design

As already noted in CIB Report 64 (1982: 22), building performance can be considered at all stages of the building design project, from briefing to post-occupancy in order to drive the design as well as to define responsibilities for delegated tasks. However, it plays a special role in design, as this is where the building is conceived and many of the properties that will later impact on performance are set. Design is a complex, creative activity; to date there is no 'how to' book on the subject (Cross, 2011). Instead design remains a challenging research subject (Bayazit, 2004). Design plays an important role in both architecture and engineering. In general, architecture is the art and science that is concerned with the design and construction of buildings. Originally, it incorporated many other disciplines; the famous Vitruvius was an architect, civil engineer and military engineer. However, over time, architecture has evolved to become very focussed on visual appearance, leading the design process; in other disciplines this often is the other way round, with engineers and inventors driving the development process, assisted by industrial designers where required (Foliente, 2005a: 96). In contrast, engineering is the branch of science and technology concerned with the design, construction and operation of engines, machines and structures. The discipline is very broad and not just applicable to buildings; branches include civil engineering, chemical engineering, electrical engineering, mechanical engineering, naval engineering and many others. The use of mathematics and physical principles is central to engineering. Engineering disciplines typically focus on systems and components, applying laws and equations such as Ohm's law, Hooke's law, Newton's laws, Maxwell's equations, the Navier–Stokes equations, Knuth's compendia of sorting and searching algorithms, and Fitts' law of human movement (SEBoK, 2014: 25).

The key challenge in design is to figure out how new systems, such as buildings, ought to be in order to meet the goals of the stakeholders and to function properly (Simon, 1996: 4). Design is complex; it has been described as 'problem solving' or 'puzzle making'. In problem solving the designer starts with a requirement and tries to find a design that meets this requirement; in puzzle making the designer begins with a set of forms and shapes and modifies these to meet the requirements (Kalay, 1999). A different view has been defined in the seminal paper by Rittel and Webber (1973), who, in the context of planning, introduced the notion of 'wicked problems' that are hard to define and where it also is difficult to define when to stop searching for solutions. More recent work such as Farrell and Hooker (2013) suggests that there may be different nuances in between 'tame' and 'wicked' problems. Harfield (2007) argues that requirements are not equivalent to design problems and that a significant interpretation takes place by the designer; he also holds that solutions and problems coexist.

With the high stake in building performance, many authors use the term performance-based design for design processes that emphasize this aspect of buildings. Yet within this common frame, there are various interpretations. A straightforward definition is provided by Oxman (2006), who equals performance-based design to 'a process of formation that is driven by a desired performance'. Kalay (1999) provides a more complex view, which defines performance-based design as a process that aims to find a qualitative solution for a combination of building form, function and context and where the quality of the solution requires a multi-criteria and multidisciplinary evaluation of building performance. Others such as Nembrini *et al.* (2014) simplify this to the view that performance-based design is a design process that employs computer-based building performance analysis tools. In structural engineering, there is a generally accepted design process named performance-based seismic design (PBSD), which explores the response of buildings to seismic excitation to establish overall performance levels and risks in a number of cycles that include preliminary design and detailed design (Wei and Qing-Ning, 2012). Some of the differences between these viewpoints go back to the notion of performance itself. As already noted by Gross (1996), 'The performance concept itself means different things to different people. To some, it is a concept of qualitative aspirations for buildings without a systematic methodology for analysis and verification. For others, it is a concept which requires quantitative analysis and rigorous evaluation that at times discourages those who wish to use the concept when these tools are not available.' Yet even for those who see it as a concept that needs analysis and evaluation, an exact roadmap on how to identify and carry out performance-based design remains evasive. The European PeBBu project, which explored and promoted performance based building (PBB) and in doing so also considered design, concluded in 2005 that performance-based building design was still relatively unknown to European design professionals (Jasuja, 2005: 54). In the decade that has passed since, regulations such as the European Directive on the Energy Performance of Buildings and rating schemes such as BREEAM and LEED have had a significant impact on building design; however the emphasis is mostly on compliance and certification, while the process of how to achieve those still remains debated and, in some aspects such as tool selection, contended (Kokogiannakis *et al.*, 2008; Raslan and Davies, 2010). Performance-based design can be imposed on the design team by clients and building regulations; however, the design team itself can also be the driving force to opt for this approach (Spekkink, 2005: 37).

In essence, performance-based design focusses on ensuring that building design solutions meet the performance specifications. This requires work on ensuring that client and user needs are defined as performance requirements and keeping track of requirements and specifications required by the government. To do so, the design team should manage the involvement of client and other stakeholders during the design process and ensure the application of performance assessment methods to guide design decisions. This obviously requires careful structuring of the design process itself (Spekkink, 2005: 24–25). A similar view is presented by Becker (2008) in her well-cited paper on the fundamentals of performance-based building design, which emphasizes the following core stages of performance-based design: (i) the identification and formulation of the user needs, (ii) the translation of user needs into quantifiable performance requirements and (iii) the use of design and evaluation tools that are suitable to assess whether the design solutions indeed meet the stated requirements and criteria. Becker positions this in a framework that at its core relates the stakeholders, the building facility and the building process. This core is embedded in the context of four markets: the building market, product market, property market and the capital and insurance market. Yet this approach is mainly focussed on performance analysis, checking whether targets have been met and preventing faults; it leaves open where design solutions come from and how performance evaluation is embedded in the wider creative design process (*ibid.*). A view that gives more attention to the overall design process is presented by Lin and Gerber (2014), who coin the phrase of 'designing-in performance'. These authors distinguish between efforts that aim to enhance the design process with analysis tools and domain knowledge and efforts that support generation, rapid evaluation and search for design alternatives; their solution includes the use of parametric models to generate designs.

Just as other types of design, performance-based building design can be viewed through the lens of design theory. Design theories can typically be grouped in one of the following categories: normative, empirical and 'design as an art'. Normative design theory is concerned with proposing rational approaches towards design, developing standards and guidelines. Empirical design theory notes that normative theories are seldom used in practice and instead aims to describe what actually happens. The 'design as an art' category is based on the belief that design cannot be described or prescribed by a methodology, but requires a high degree of flexibility and reflection (Stempfle and Badke-Schaub, 2002). Most discussions of performance-based building tend to be either normative or related to the arts; there is little – if any – work that takes the empirical approach. This may be due to the fact that description and modelling of design processes requires detailed data; this may lead to an under-representation of more conceptual and creative activities (Almendra and Christiaans, 2011).

Normative approaches towards performance-based design, such as the aforementioned work of Becker (2008), typically emphasize the need for a rational underpinning of design choices and the use of analysis tools. However, different approaches have different degrees of freedom. Within the European PeBBu project and the ASTM Standard on Whole Building Functionality and Serviceability, the performance concept is primarily taken as the comparison between demand and supply, which both are formulated in different languages. The demand side specifies *what* is required and *why*; the supply side specifies *how* demands can be met (Szigeti and Davis, 2005: 11). In the systems engineering domain, this is echoed by Gilb (2005: 350) who defines engineering as 'an iterative process of determining a set of designs, with rigorous attention to quantified and measurable control of their impact on requirements. The design engineering process implies the matching of potential and specified design ideas with quantified performance requirements, quantified resource requirements and defined design and condition constraints'. INCOSE (2003: 88) defines systems engineering as 'a bridging process translating an identified need into a system solution composed of specified implementable hardware and software elements'. Authors like Welle *et al.* (2012) and Lin and Gerber (2014) suggest that multidisciplinary design optimization (MDO) is important in ensuring building performance, where optimization implies a search for the 'best' design solution and thus goes beyond just the match of demand and supply. Shi and Yang (2013) hold that performance-driven architectural design needs to be based on the quantification of building performance over a range of aspects, followed by optimization and integration of findings into the design. Others go even further and try to address the issue of the complexity of the many different criteria, interests and requirements that relate to building performance; for instance Alwaer and Clements-Croome (2010) hold this to be a prime obstacle for the assessment of performance, which they try to overcome by developing a model that helps to identify and build consensus regarding relevant performance indicators.[1] Yet Leatherbarrow (2005: 7) warns that the use of a new approach and of new tools will not necessarily change architectural practice and theory. While it may lead to new understandings, a focus on technical performance alone may lead to a narrow functionalist view that only takes into account those things that can be measured and predicted; yet there are many issues in design that cannot be rationalized.

In literature there are only few empirical explorations of the performance-based building design process. Architectural magazines and publications by engineering associations often present case studies, but mostly these presentations are about the final product and do not provide significant insights into how design decisions were made, or any alternatives that were contemplated at some point but later cast aside. Scientific publications contain an abundance of papers that discuss the building design process in general, often supported by anecdotal evidence; however there are few papers that report on design processes in terms of a detailed sequence of interrelated activities. Exceptions mostly study design in an educational context, which can be explained by academics having both a research and teaching role and thus being able to set students task that fit with their research objectives. For instance, Struck *et al.* (2009) observe the emergence of building design solutions in undergraduate student projects.

---

1 It is unfortunate that Alwaer and Clements-Croome use the term KPI, which mostly occurs in a business and project management context.

Bleil de Souza (2013) describes experiments with novice designers (students), where design decisions were studied using qualitative thematic analysis of design journals. From the work she identified differences in the subprocesses of making design proposals, making design changes and reviewing performance results. Almendra and Christiaans (2011) provide an example that combines the study of building interior and product design. Studies of professional design in actual practice are rare; this is probably due to the fact that following an actual design process takes a significant time investment and because actors may want to keep some of their detailed work processes confidential. The retrospective case studies by de Wilde (2004) fall in this category. Zapata-Lancaster (2014) and Zapata-Lancaster and Tweed (2016) report on ethnographic studies of the design process of non-domestic buildings in the United Kingdom. Zapata-Lancaster (2014) reports that in low-carbon design the design team sets their aspirations higher than the legal requirements and does so early in the design process. She finds that design teams set up a range of energy performance assessment 'gateways' to test how performance progresses. Also, results indicate that knowledge is transferred from project to project. Zapata-Lancaster and Tweed (2016) suggest that architectural design is a type of problem solving, which typically is done in an exploratory and highly flexible manner, and without a rigorous framing of the problem. They also suggest that the social interaction and negotiation with other actors plays an important role. Action research on performance-based design, seeking to enhance the use of simulation during the design process, has been conducted in the United Kingdom by the Energy Design Advice Scheme (EDAS) and the Scottish Energy Systems Group (SESG), as reported by Maver and McElroy (1999) and McElroy (2009: iv–x). Harries *et al.* (2013) give a first-hand account of integrated holistic design – using building performance simulation tools – in the design of the London 2012 Velodrome building.

Approaches that emphasize design as an art start from the premise that the building design process should not be measured, analysed and controlled in all aspects and that tacit knowledge also has a role to play. This assumption resonates with many authors who question the purely technological approach to building performance assessment and tool development. For instance, Zapata-Poveda and Tweed (2014) assert that human activities always need to be considered in a social context and that it is likely that the tools, policies and approaches for achieving building performance will be used in other ways than expected by their developers. For some, it allows to distinguish between routine activities that can be streamlined, while other activities are unique and innovative and can be allowed a degree of freedom (Gann and Whyte, 2003). Along the same lines, Krish (2011) distinguishes between 'routine' and 'creative' design problems, noting that the routine design process is increasingly approached by semi-automated design procedures, whereas creative design remains more difficult to support. This fits well with the view that the traditional architectural design process is mainly focussed on space, form, aesthetics and functions, and that the consideration of performance belongs within the functional domain; see for instance Shi and Yang (2013). Others take a wider view and recognize performance as a guiding principle for the entire building design process that may sit next to, or even replace, the emphasis on form (Kolarevic, 2005a: 195). *Performative architecture* can be defined as architecture that interacts actively with a changing social, cultural and technological context (*ibid.*, 205). Leatherbarrow (2005: 7–8) approaches the concept of performance from the background of architectural theory and practice. Here the focus is on what the architectural work, and by extension a building, does. This leaves open whether that performance is similar to a technical artefact like a car or stereo, or similar to a theatrical or concert performance. This extends existing architectural theory, where the focus is on an object and what that object represents. A similar idea is put forward by Hensel (2013: 10) who presents *performance-oriented architecture* as based on a vision where 'the building reveals what it does'.

De Wit and Augenbroe (2002) view design evolution as a series of design decisions, where each decision rests on the input from a range of domain experts. Yet design is considered by many to be an ill-defined or 'wicked' problem-solving process, on the basis of the fact that one often has to deal with incomplete requirements, contradictions and changing conditions and that solving one issue may lead to other problems (de Groot, 1999; Lin and Gerber, 2014). These issues are amplified in performance-based design. Design for performance requires the deep analysis of a design problem, including the identification of the key criteria, the way the performance for each of these criteria can be quantified, and multi-criterion decision making (Augenbroe, 2011: 15). Hazelrigg (2012: 11) points out that the fundamental concept for a rational choice is the selection of the alternative that gives the outcome that is most preferred by the decision maker. Thus making a rational choice requires that this is selected from a range of options, taking into account known preferences. However, in most cases there is no one way process from requirement to solution but instead a series of iterations with a need to carry out verification and change management along the way. To maintain consistency in such an iterative process, there is a need to manage the steps between stakeholder requirements, proposed technical solutions and selected technical solutions (Huovila, 2005: 22). Foliente (2000) points out that in practice there often is a mix of prescriptive and performance-based requirements. The more performance-based these requirements are stated, the more freedom is given to the designer to come up with new solutions. Further complexity results from the need to look beyond the design process, as it is believed that the development of a feedback loop from actual building performance in operation back to design will lead to better design decisions. This should include information on how buildings are actually used versus how they were designed to be used; this idea is sometimes known as the concept of 'design for operation' (Frankel *et al.*, 2015: 7). Further complexity may rise from the fact that performance-based design often takes more time, is more costly and in its pure form might be less reliable in ensuring long-term performance than reliance on prescribed solutions (Becker, 2008). Moreover, a focus on performance measurement may sometimes obstruct good judgment; attention might be skewed towards that which is being measured, while other aspects get overlooked (Gann and Whyte, 2003). Or, as stated by Neely *et al.* (1997) in the context of process and operations management, 'It has long been recognized that inadequately designed performance measures can result in dysfunctional behaviour. Often because the method of calculating performance – the formula – encourages individuals to pursue inappropriate courses of action'. Performance-based design thus should be used with some caution.

Many authors such as Gann and Whyte (2003), Hensen and Lamberts (2011: 6), Schade *et al.* (2011) or Zanni *et al.* (2016) call for a logical evaluation process to take place during the building design process. Such a process is seen as iteration between making design decisions and evaluating the implications of these decisions on building performance in order to achieve a continuous improvement process

(Choudhary *et al.*, 2005b). Similar positions are taken in systems engineering (Gilb, 2005: 27; Hazelrigg, 2012: 289; SEBoK, 2014: 161–164; INCOSE, 2015: 11–12). Attempts to structure the building design process in a logic-driven way have been the subject of many doctoral research projects in architectural and building science; see for instance Hand (1998), Morbitzer (2003), de Wilde (2004), Hansen (2007), Hopfe (2009), McElroy (2009), Hu (2009) or Struck (2012).

On a general holistic level, the performance-based design process can be described using the stages of the RIBA Plan of Work as discussed in Chapter 2 (RIBA, 2013), with requirements being developed during 'strategic definition' and 'preparation and brief' and with design solutions being generated during the subsequent stages. However, it must be noted that in detail there will be many iterations as both requirements and design co-evolve. Becker (2005: 69) writes that the translation of user needs into building design solutions typically takes one of two pathways. The first pathway is the application of solutions that are known to meet the needs; often these are included in building codes as *Deemed to Satisfy Solutions* or *Approved Documents*. The second pathway is to turn the user needs into performance criteria that can be used as criteria for the evaluation of a range of design solutions. Work along this route of performance requirements often starts with requirements for the complete building, yet at some point in time these needs to start to relate to actual components and products. At this point there often is a need for transition to lower-level dedicated requirements that apply to these components/products, which allows the designer to check whether they are effective in the context of the overall design (CIB Report 64, 1982: 11). Described differently, design may be seen as moving through a hierarchy of ideas where solutions on one level form the constraints at other levels (Barton and Love, 2000). As a consequence, there is likely to be iterations between requirements and solutions and between component requirements and overall building requirements. To some extent, these iterations are desirable; Charles and Thomas (2009) complain that in practice many design processes are overly linear, with a range of early design decisions being made without input of engineers. One should also keep in mind that building design needs to respond to the local conditions and constraints, such as the local climate, which may require further iterations. Geva *et al.* (2014) present a case of a building in Israel designed by a Swiss architect, which is overheating but which would have had good thermal comfort when constructed in Switzerland. Furthermore, uncertainties about building properties, building use and consequently building performance are present throughout most of the design process. A challenging task for designers is the development of designs and details that are robust towards these aspects.

A figure that is shown in various publications, in different guises, depicts the general trend of decreasing design freedom as the design process progresses; see Figure 8.1. This graph is sometimes known as the 'design paradox'; it is used to describe how, as the design becomes more and more established, the freedom for designers to change the design is reduced, whereas knowledge about the detail, cost and performance becomes more established. See for instance Macdonald (2002: 4), Geebelen (2003: 20), Löhnert *et al.* (2003), Cantin *et al.* (2012) or Marsh (2016) for different variants. It reflects the idea that at inception, the canvas for the design is blank and there is no design data, only needs; at the completion of the design process, all details have been specified (Spekkink, 2005: 32–33). Such figures also visualize the idea that there are phases in the design process where the building is not yet completely defined, where information is missing

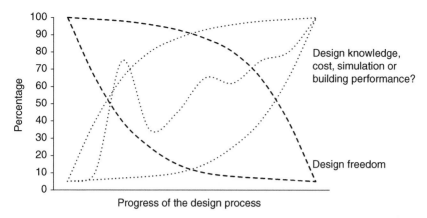

Figure 8.1 Decreasing design freedom graph.

and detail still needs to be filled in. Any performance prediction done during such stages is subject to uncertainties and therefore conditional to assumptions; results should be presented in a way that reflects these uncertainties. However, while this over-all trend may be correct, such figures should be handled with extreme care; for instance there is no hard evidence that these trend lines are as smooth as often depicted or that there is an inverse relationship between one aspect and another; it is also very possible that in reality quantities may be subject to strong fluctuations. Empirical work by Bradbury (2015) shows irregular, stepwise changes in quantified energy performance as building design progresses, with sudden changes in performance corresponding to specific interventions. Other assumptions may also be questioned; for instance site constraints may limit the initial design freedom to less than 100% right at the start of the process, whereas a shell and core design project will leave freedom open at the end. Some researchers make rather sweeping statements about the activities that take place in the design process and the order of these activities. By way of example, Cemesova *et al.* (2015) state that in current practice 'an architect designs a building, and an energy specialist manually recreates it in a building energy (BEM) analysis tool, adding missing materials data such as thermal transfer coefficients (or U-values)'. But as pointed out by Negendahl (2015), there can be a range of relations and partnerships. And in most cases, architectural design and engineering efforts will partly overlap and allow for iterations.

Within the overall design process, there is special attention for the early stage. The terms *early design* and *conceptual design* are often used to indicate the same phase. This interest is due to the fact that this is where the first ideas about the actual building design materialize. Because of this, it sets the context for many later developments; this includes future building performance. Many researchers such as Hansen (2007), Welle *et al.* (2011), Struck (2012), Jrade and Jalaei (2013), Pelken *et al.* (2013) and Østergård *et al.* (2017) therefore recommend that performance aspects should be given attention right from the start of the project; however, the order in which specific design param-eters should be addressed depends on the specifics of each project. Overall, this is a difficult design stage to grasp, as it is typically characterized by uncertainties and ambiguities, making it hard to model and organize (Almendra and Christiaans, 2011).

The boundaries of conceptual design, while firm in the RIBA plan, are less clear in real practice. For instance some work focusses on the design of building systems and within that defines a further conceptual systems design stage; see Méndez Echenagucia *et al.* (2015) for an example on façade design. Others even distinguish an early design stage within building retrofit projects (Shao *et al.*, 2014). The actual content of the early design stage also remains elusive and is highly variable; it has been described as a process 'where many threads of possibilities are developed in parallel; these concepts are then abandoned or re-combined until a satisfactory scheme emerges' (Krish, 2011). Macmillan *et al.* (2002) give an insightful report in mapping of actual early building design work, discussing the iterations that take place, the impact of the experience of the design team and the nature of the design project at hand. Early design is recognized to pose a range of challenges, such as the fact that the design is still subject to rapid change and evolution, conflict of requirements, time and cost constraints, and the need to balance input from a range of design stakeholders (Østergård *et al.*, 2016). Moreover, at the early stage of design, there is no building to measure or occupants to interrogate, thus leaving only simulation tools to provide performance predictions (Jin *et al.*, 2012). To some extent, performance may also be benchmarked against experience; in reality quantification is mostly limited to quick estimations only. Historic data from other projects may be used to underpin assumptions, if used with caution (Zapata-Poveda and Tweed, 2014). It is often stated that early design decisions constrain the remaining design space and that choices made are difficult to change later in the design process (Østergård *et al.*, 2016); however this is not necessarily the case with BIM technology where building designs may be parameterized and thus can be adapted more easily.

While the recognition of overall stages helps to maintain a general overview, actual design processes are almost always complex and dynamic; it is hard to foresee the detail of what will happen. Some describe the design process as a 'sequence of activities that operate on the product of the design process'; the design or product thus evolves over time and different design versions emerge (Roldán *et al.*, 2010). As the essence of design is to create new objects and systems, it is inherent to the design process that the actual actions at some level become uniquely linked to the evolving design. Even design processes that are limited to only a part of the building, such as retrofit of facades or similar, still remain complex and are dependent on the input from many stakeholders, actual building conditions and others; see for instance the process map by Ferreira *et al.* (2013). A similar example is the work by Jin *et al.* (2012) who focus on the design of building façades, positioning the façade design process as a separate entity from the overall building design process. In their view this process consists of a series of decisions on aspects like glazing type, orientation and window–wall ratio, which need to be based on computational assessment of four environmental factors (thermal, air quality, acoustics and visual), which in turn are related to occupant productivity and economic benefit. However, beyond such highly unique activities, at an atomic level, other activities may become repetitive once more. By way of example, even highly individualistic artists such as painters often work with paint and brushes, so actions like 'mix paint' and 'pick up brush' are common among many of them. In order to understand, support and where possible direct some of the various design activities, design process planning and design project management are of interest in both the architectural and the engineering disciplines.

An attempt at a high-resolution process map of the overall design process workflow of an 'integrated design process' (IDP) for building design, with a focus on optimal use of solar energy, was developed in the context of IEA Task 23. This work gives an excellent overview of the complexity of the various process stages, actors, design challenges, methods and tools that are available and how all of these interact in a generic process that moves from analysis to synthesis. Yet it will be clear for anyone studying this process map that it is very hard to apply this to structure a newly initiated process and the actions that are to be taken. This process map as well as some of its underlying issues also feature in the thesis of Hansen (2007).

Gilb (2005) provides an overall approach for competitive engineering, which aims to be at the forefront of design and provide clients with products that meet or even exceed their requirements. However, as noted by Augenbroe (2011: 18), this approach needs to be tuned to each new project and relies strongly on the experience of the actors; it involves reinvention and rediscovery for each new project. Roldán *et al.* (2010) describe the complexity of capturing and tracing design processes in other industries, such as chemical and software engineering. They suggest to use a formal model that captures the design actions as well as the underlying rationale for these actions, noting that reasons for doing things are easily lost over time. Such an approach would be highly interesting for exploring the detailed drivers in performance-based building design; however studies following this suggestion will require significant time and resources.

At a detailed process level, some activities may have clear process logic and need to be dealt with in a certain order. For instance, it makes sense to first define the location of a wall and then to insert a door or window in that wall. However, the subprocess may fit at different places in the overall design project; for instance, one could first design all walls and then all windows, or first design one specific wall, insert the window, and then repeat this process for the next wall. In building performance analysis, the situation is similar, but here one also has to consider physics. For instance, one may want to first analyse solar access to a room in order to establish the need for artificial lighting, which then impacts on the thermal balance of that room. An example is given in Figure 8.2, which concerns the logic of sizing the HVAC system. The activities that do have a logical order can be sequenced in a 'swimming lane'. While this process requires an internal logic, there is no general rule that will force the design team to consider HVAC system sizing only after other activities such as designing the load-bearing structure or considering the acoustics of the project. Other design activities cannot be as easily sequenced; for instance, there may be iterative considerations of system performance in different domains such as lighting and acoustics, while design activities concern both. In general, design tasks can have different relationships in terms of information flow between them, which requires them to be carried out sequentially (if the next task is information dependent), concurrent (if there is no information flow between the two tasks) or overlapping (if there is a limited amount of information flow). Only if these relationships are known in advance can this be used for optimal planning (Joglekar *et al.*, 2001).

Against this backdrop of general high-level and high-resolution views of the design process, different suggestions are made on how to strive towards building performance. At the technical end of the spectrum, Becker (2008) gives a mechanistic list of 10 steps to embed performance thinking in the design process; see Table 8.1. At the opposite end, Bleil de Souza (2012) recommends a training of designers/architects in mathematics and physics to ensure a tacit influx of performance thinking into the design process and

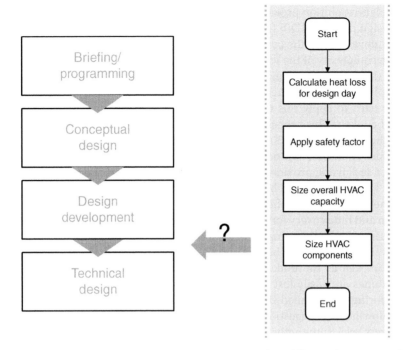

Figure 8.2 Subprocess with logic sequence that may fit at different places in overall design process.

Table 8.1 Core steps of performance-based design.

Step 1	Identify all stakeholders of the future building, their activities and the corresponding stakeholder needs
Step 2	List all conditions and activities that might be a barrier to achieving the stakeholder needs; study these as generalized loads and, where needed, the combinations of loads that might occur
Step 3	For each stakeholder need, identify performance indicators that allow to specify the performance requirements
Step 4	For each performance indicator, define what it is that indicates stakeholder dissatisfaction or performance failure
Step 5	Define accepted levels of failure, and acceptable percentages of dissatisfied stakeholders, for each stakeholder need while taking into account all stakeholder activities that have been identified
Step 6	Define the loads that will be applied to the building in terms of checking performance
Step 7	Set the threshold values (limits) for the performance indicators
Step 8	Apply any safety factors to modify the threshold values into design targets
Step 9	Identify which analysis tools can be used to study the performance of the building under the loads that need to be studied
Step 10	Employ approaches that allow the analysis of the performance of all relevant design options

to drive systematic and scientific exploration in building design. At the same time she suggests that the engineering and physical science community needs to gain a deeper understanding of the needs and typical approaches of building designers. Other efforts are more flexible and aim for the middle ground; these may help to organize different tasks in a convenient way. For instance, the Design Structure Matrix approach may be used to map relationships between building systems or activities, which can be used to structure and manage the design process (Nomaguchi *et al.*, 2015). Similar benefits can be obtained from using a modular approach to the design of buildings, employing approaches from the manufacturing and software industries (Gann *et al.*, 2003). Inoue *et al.* (2012) suggest an approach named Preference Set-based Design (PSD), which, in contrast to investigating a single point solution, works on the basis of narrowing down a set of solutions and preferences. PSD aims to narrow these sets down to a set of more preferred, robust solutions while taking into account design uncertainties. Lin and Gerber (2014) use the concept of 'designing-in performance' to describe design processes that have access to performance feedback, which misses from the conventional design process. Other approaches are limited to specific sectors; for instance, Evidence-Based Design is a form of performance-based design that is increasingly popular in the context of healthcare. It focusses on patient health, staff well-being, increasing safety and reducing stress. Evidence-Based Design relates to Evidence-Based Medicine, a scientific approach for the selection of treatment choices. In the context of buildings, it employs POE approaches, with a focus on environmental checklists, patient surveys and stakeholder focus groups (Thiel *et al.*, 2014). There also are other approaches that are more creative. For instance, one way to find novel designs is by studying analogies; a well-known example is biomimicry,[2] where patterns and solutions from nature are used as inspiration to solve human problems. There are various programming approaches towards finding analogies, such as linguistic matching; however finding analogies may also be based on searches for similar performances (Lucero *et al.*, 2016).

Beyond this, efforts towards performance-based building design typically are grouped under a range of subjects. First of all, there is a large body of literature that aims to achieve good building performance through a process named *integrated design*,[3] which champions collaboration between different disciplines and consideration of various aspects. Other efforts are more analytical in nature and focus on *analysis, evaluation and optimization*. In the area of design process management, efforts towards building performance are often visible in *project and workflow management*.

- **Integrated design**

  Building performance relates to a wide range of aspects and depends on many individual systems as well as their interaction. Therefore many efforts towards performance-based design focus on the need for various disciplines to collaborate; often the resulting process is named integral or integrated design (Spekkink, 2005: 31; Glicksman, 2008; Charles and Thomas, 2009; Pelken *et al.*, 2013). This is not unique to building; some other disciplines use Integrated Product Teams (IPTs) with specific areas of responsibility for the development and operation of a product (INCOSE, 2005: 50). Other terms used in this context are Integrated Design Process (IDP) and

---

2 An interesting website to explore in the context of biomimicry is www.asknature.org.
3 The term integral design is also used.

integrated project delivery (IPD); see Kent and Becerik-Gerber (2010) for a discussion of their application in construction. Further terms may link the integration to optimization, such as in 'multidisciplinary design optimization' (Welle *et al.*, 2011). The need for integrated design seems to accelerate; for instance Cantin *et al.* (2012) suggest that existing practices are more and more unable to meet the increasing demands and complexity of future buildings. They point to the fragmentation of professional and trade responsibilities and segmentation of the design process as key problems, and suggest that an approach that accounts for building structures and systems, building functions and transformation may improve the situation; yet how actors are to achieve this in detail is left open. Hansen (2007: 107) suggests that integrated design can be achieved by bringing together a set of actors from different disciplines into a design team or by giving actors an education in various disciplines. Structured forms of collaboration could ensure that the members of a (project) team or organization share the same goals (Chiu, 2002). In theory most building performance problems have many aspects that are interrelated, and in an ideal case these should be considered concurrently. However, this often is not practical, leading to an approach where such problems are partitioned and delegated to relevant specialists (Choudhary *et al.*, 2005a). The relationship between the different actors then becomes a point of importance. Negendahl (2015) provides an interesting view of collaboration in the design process, discerning between situations where an engineer is an assistant to an architect, where an engineer is a partner to the architect and cases where boundaries between architect and engineer disappear and a hybrid practitioner emerges. Koch and Buhl (2013) compare and contrast an architectural and engineering view and report that a key issue seems to be increased interaction between actors in different disciplines, which requires additional efforts by all parties involved. But whatever the roles, where design is carried out by a range of domain experts that are each responsible for separate performance requirements and solutions, it is important to assess that the final design, in its totality, meets all requirements (Becker, 2008). Unfortunately some practitioners adhere to an 'expert knows best' approach, which often fails to take into account the views of other stakeholders (Gann and Whyte, 2003). Some universities have started to teach integrated design to students in architecture and engineering. There are various modes; for instance experts may be brought into the design studio to give students access to engineering knowledge and tools, academics may provide a similar service, or students may be asked to provide this support to each other (Charles and Thomas, 2009). Reinhart *et al.* (2012) present an example of role playing in architectural education, which the authors suggest does support the development of an energy modelling culture among the participants. A deep review of some of the challenges in collaboration between architects and consultants is given by Alsaadani and Bleil de Souza (2016).

- **Analysis, evaluation and optimization**
  Other efforts aim to achieve performance-based design by emphasizing a rational process, which includes analysis and evaluation and, in some cases, optimization. Zeigler *et al.* (2000: 12) discern three types of system analysis, with differences based on the objective of the analysis: efforts that aims to understand the behavioural characteristics of a system, efforts that aims to infer how the system works and what causes the behaviour, and efforts that aim to design a system that has the behaviour that is desired. Marques *et al.* (2010) view analysis in a different way, based on whether

one looks backwards or forward; they discern (i) analysis of what has happened and how the project meets given requirements and (ii) analysis of what is likely to happen as the outcome of decisions and evolution of the project. The first is *a posteriori*, and the second *a priori* analysis. These positions map to the viewpoint of Bordass *et al.* (2001c), who argue that findings on building performance should be used in two ways: identified problems should lead to diagnosis, and where possible to corrections; success should be used for selling and advocacy of good approaches for subsequent design processes. In the building domain, notable proponents of performance analysis – and especially the use of building performance simulation – are Mahdavi (1998), who observed that there is a need to improve the logic in the design processes that are insufficiently integrated; Clarke (2001: xi), who holds that this will lead to a 'cheaper, better and quicker design process'; and Hensen and Lamberts (2011: 6), who argue that 'predicting and analyzing future building behaviour in advance is far more efficient and economical than fixing problems when the building is in the use phase'. Augenbroe (2011: 15) discusses the role of simulation as key tool for analysis in a rational design decision process. Optimization of building design solutions is, among others, championed by Wright (2016). In analysis and evaluation, and especially optimization, it is important to define what alternatives are explored. In technical terms this set of alternatives is named the *option space* or *search space*; where the evaluation concerns design options, the term *design space* may also be used. Within a design space, INCOSE distinguishes between system architectures that conceptually define a system (in terms of what elements) and the characteristics (attributes, design features and parameters) of these elements (INCOSE, 2015: 64–65). In performance analysis, the option space is typically coupled to an *outcome space*; here the option space describes the design variants, while the outcome space contains the information of performance of these individual design variants. An example of the use of design and outcome spaces, albeit named alternative and objective space, is provided by Gane and Haymaker (2012). In design, parameters that characterize alternatives may represent nominal or characteristic values. It is often assumed that such nominal values represent about 90% of the products that may be encountered. Target values for performance may be stated as design values, which include multiplication with a safety factor γ_p (Becker, 2008). In the discussion of alternatives, it is easy to believe that all design parameters are independent; however this is not necessarily the case (Tian *et al.*, 2015). Similarly, not all parameters vary over a continuous range; for instance HVAC systems such as boilers, furnaces and chillers are typically produced in a series that have a stepwise variation. It is important to prevent engineering design from being constrained by optimization, in terms of only allowing for an exploration of established solutions and the corresponding parameters (Schwitter, 2005: 114); sometimes it is possible to 'think out of the box' and find good design alternatives. And while analysis may help in steering performance-based design, Andersen *et al.* (2013) note that achieving good daylighting is dependent on so many parameters that hard performance targets have only limited value structuring the design process.

- **Project and workflow management**
  As performance-based design involves many aspects and the collaboration of different disciplines, it becomes important to explore the dependencies between different tasks and activities. This is a non-trivial problem as different domain experts may have a different understanding of these dependencies. Addressing such issues, as well

as questions about design collaboration and leadership, is the domain of design management. To some extent this addresses similar challenges as faced by construction project management, which studies the scheduling and timing of activities, cost control and management of risks. Efforts to standardize the design process are typically more limited by the varying nature of building projects; however there is some progress in the area of industrial house building where work processes are more repetitive; see for instance Jansson *et al.* (2016). In his discussion of processes in the context of building information modelling, Eastman (1999: 333–346) points out that while the design process can be seen as static at the top level, lower levels become increasingly project specific and dependent on particular tasks and challenges. Managing and structuring the activities at the lower levels requires design process management, which can be supported at a basic level by Gantt charts, Critical Path Method graphs, Integral Definition models such as IDEFØ, state transition diagrams and Petri nets. Rekola *et al.* (2012) have explored the role of design management in making buildings more sustainable in Finnish design practice; they find that social interaction, leadership and influencing of colleagues play a key role in ensuring that design outcomes meet the objectives.

Project management deals with the process of delivering a unique product. The management involves project initiation, planning, execution, monitoring, completion and overall control. A well-known approach to project management is the PRINCE2 method. Well-known commercial generic project management systems are Microsoft Project, Primavera and RPLan. A general overview of process modelling and monitoring of engineering processes is provided by Heer and Wörzberger (2011). Project management may apply to an overall construction project but can also focus on design.

Workflow Management (WfM) is an approach that models and reuses repetitive stages of processes; the corresponding Workflow Management Systems allow planning, enactment and monitoring of process progress. It may also allow the automation of parts of a process. For a more extensive discussion and application to product engineering and design, see for instance Derks *et al.* (2003). An application of workflow management to collaborative building design is provided by Tuzmen (2002). A further example of the power of workflow is the work by Gil *et al.* (2011) that describes how it can support the careful design of computational experiments by ensuring process logic and efficient use of computational resources. Heer and Wörzberger (2011) point out that workflow management systems can only support subprocesses that can be defined in advance; connecting such a workflow to the overall process remains the task of a design process manager and is likely to vary from project to project. Examples of general workflow management systems are Activiti, Apache Orchestration Director Engine, Aristaflow, Bonita BPM, IBM Websphere Business Process Server, jBPM, WorkflowGen and YAWL.

An early example of advanced process management in the context of building performance analysis was included in the EU COMBINE project, which employed Petri Nets to structure the data exchange in so-called project windows (Augenbroe, 1995: 36–40; Hand, 1998: 152–158; Clarke, 2001: 309–316). The later Design Analysis Interface (DAI) Initiative instead applied a commercial workflow management system to model and enact process logic (Augenbroe *et al.*, 2003: 18–19; Augenbroe *et al.*, 2004). A more recent example of the use of Petri nets to support resource management in the building design process is presented by Cheng *et al.* (2013). In modern BIM systems, underlying processes play an important role in defining what information needs to be exchanged and made available in Model View Definitions (MVD); the underlying workflows are based on industry requirements (Eastman *et al.*, 2011: 120–123). Process management and planning in a wider sense, not limited to design but also including the procurement and construction phases, is discussed by Verheij (2005) and Ren *et al.* (2008).

In spite of all efforts, performance-based design is still a field under development. Becker (2005: 85–86) provided an overview of research challenges for performance based building in general, many of which are still applicable a decade later. Among these are the need for a better understanding of human health, safety and other responses and how these relate to the use of performance thresholds or targets. Support systems that facilitate the handling of stakeholder needs, performance requirements and indicators are still uncommon in building design, with most work still being academic. Development of databases with statistically sound data on stakeholder needs, performance requirements and generalized loads that can be used to develop benchmarked performance indicators has not yet happened, possibly due to the high variability across projects and the complexity of the industry. Some progress has been made with improving POE procedures and risk analysis, as well as with validation and verification of some of the performance assessment tools, but further work is needed in most of these areas. Monitoring and analysis of performance-based demonstration projects is taking place but needs to continue in order to capture the evolution of building technology and changing use patterns. Part of the problem may lie at the root: according to Popova and Sharpanskykh (2010), in spite of all the effort put into the definition of goals, performance analysis and evaluation, in practical management contexts, much of this work is still done in an ad hoc way, without a systematic approach, making it hard to develop standard processes for downstream activities. It should thus not come as a surprise that Burry *et al.* (2001) observe that researchers 'struggle' in identifying what it is that designers actually do, what support they require, and how that support is best provided. Yet on a positive note, there might also be an option to use the new analysis tools in a new, generative manner, helping designers to shape buildings in a novel way (Kolarevic, 2005a: 198).

8.2 Performance-Based Design Decisions

A crucial connection between rational analysis approaches towards building performance and the wider design process is made by focussing on design decisions. Isolating performance-related design decisions allows maximum freedom for the remaining process while strengthening the position of logic in those individual decisions. The focus on performance-related design decisions fits flexibly with various views of the (building) design process while linking to well-established theories on decision making. The notion of design as decision making is well established in building science; the seminal book by Markus *et al.* (1972: 13–28) already includes a full chapter on the topic.

Design thinking often discerns four key cognitive operations: generation, exploration, comparison and selection of design alternatives (Stempfle and Badke-Schaub, 2002).

In this view the selection of alternatives clearly is a design decision moment; this maps well to activities such as the selection of energy conservation measures and other procurement tasks. Zanni *et al.* (2016) report on case studies with design practitioners that were structured around the identification of 'incidents' that had some effect on the building design, which the authors related to the decision points, project timeline, reflection and justification. It must be noted that there are other views with regard to design decisions; for instance Almendra and Christiaans (2011) prefer to name activities in the (conceptual) design stage just 'interactions', as there are many actions that interact with each other and thus make it hard to decide what is a true decision. Some design variables may later be found to be dominant, and their adoption thus becomes a design driver; but the identification as a key design decision can only be made retrospectively. For some authors, especially those with an architectural background, design is mainly about defining geometry (Ercan and Elias-Ozkan, 2015); this is another view where generation, exploration, comparison and selection take place in rapid interaction and pinpointing decisions may be difficult. Support of such highly integrated activities requires tacit knowledge; in terms of ensuring building performance, this maps to the suggestion by Bleil de Souza (2013) that building designers should be trained in the fundamental sciences and physics, allowing them to drive building performance from within the creative process. However, focus on rational decision-making moments is a common approach in many fields. For instance, the discipline of Operations Research (OR) uses decision-making units (DMUs) to identify parts of the decision-making process that can be evaluated and managed (Lim and Zhu, 2013). Keeney and Gregory (2005) argue that while decisions can sometimes be taken intuitively, analysis is required to enhance the quality of decisions and that such analysis must be based on a the review of a set of alternatives and their consequences. Schade *et al.* (2011) mention the use of 'quality gates' that form the entry point to a next stage of a project; these gates can only be passed after an evaluation of the design against key criteria. In this manner, the quality gate enforces a go/no-go decision moment in the process. In a similar way, Zanni *et al.* (2016) view decisions as information gates; they discern 'hard' gates where the design is frozen until a decision is made and 'soft' gates where parallel processes are allowed to proceed. Barton and Love (2000) introduce the notion of a decision chain, where choices made during the phase of conceptual design are linked and have consequences for later decisions about the detail of the design.

Decisions are choices that are made on the basis of careful deliberation (French, 1988: 27). Important stages that precede decisions are the recognition that there actually is a decision problem and the structuring of that problem in terms of gathering data on the state of issue, on the possible actions that may be taken, and on the outcomes of these actions (Gettys *et al.*, 1987). Discussing design decisions, Almendra and Christiaans (2011) categorize a range of design decision activities that map to this overall structure: framing of a design problem, exploring alternatives and actually solving design problems; they categorize idea generation and information management as supporting activities. Hazelrigg (2012: 7–9) identifies the following attributes of decisions: they involve the commitment of resources, they are made in the present, they are made by an individual, they consist of a choice from a set of alternatives, they aim to achieve a more desirable future, they involve an element of risk, and they are based on preferences.

There is a significant body of knowledge on decision making, both generic and with a focus towards design and engineering. Key books on the subject are for instance

Decision Theory (French, 1988), *Decisions with Multiple Objectives* (Keeney and Raiffa, 1993), *Decisions under Uncertainty* (Jordaan, 2005) and *Fundamentals of Decision Making* (Hazelrigg, 2012). Much of this theory on decision making is *normative*: it describes how decisions ought to be made in order to make the best choice possible given the information that is available. However, this may not be what decision makers do in practice; finding out what happens in reality is *descriptive*, which seems to get a lot less attention in the decision-making literature. Hazelrigg (2012: 3–5) warns that this wider literature on decision making roughly falls into three different categories. At one end there is literature that is highly mathematical but rather removed from design and engineering practice. At the other end are 'self-help' approaches that are often lacking in mathematical validity. The middle ground is captured by work that mainly caters for business decisions, with mixed mathematical rigour. Hazelrigg thus stresses the need to review the logic, validity and consistency of decision-making methods.

While one may strive for the best possible decision, it is important to keep in mind that in reality there will almost always be constraints in terms of the time, knowledge and resources that are available to underpin a decision. This situation is described by the term 'bounded rationality' (Simon, 1997: 20). In this context Simon also uses the term 'satisficing', which describes the situation where decision makers are concerned with finding a satisfactory solution, not necessarily an optimal one. Obviously, the amount of effort invested in a decision may vary hugely – for instance, the evaluations that will go into deciding on the type of front door to go into a dwelling will be of a complete different order of magnitude than the evaluations that will underpin the siting of a new airport or nuclear power plant.

Another important notion in studies of decision making is that there is a huge difference between the decision itself and the outcomes of that decision. This is due to the fact that many decisions are made under uncertainty; thus decisions need to be based on predictions while taking into account the fact that the desired outcome may not materialize. Consequently many examples in the decision-making literature discuss bets that involve decisions as well as the risk that the decision maker is willing to take. Hazelrigg (2012: 7) goes one step further and defines a decision strictly as a mental activity, a commitment to action; thus the decision is separated from the action itself, which then in turn may or may not lead to the desired outcome.

Decisions, as a choice made after careful deliberation, require a range of alternatives from which the choice can be made. In technical terms the set of alternatives is named the option space. One aspect that is often overlooked in the literature on decision making is where these options come from. Most theory suggests approaches for exploring a range of given options and processes for selecting the best or a 'good enough' option from this range; yet this overlooks giving an answer to the question 'how do people choose what to choose from?' (Johnson and Raab, 2003). Where the decision is to identify the best, or a good enough option, there is another related problem: the decision maker also needs to decide how to invest his time and efforts and has a choice to make between exploring for better alternatives or in exploiting known alternatives and selecting a good or even the best one. This trade-off is also known as the exploration/ exploitation dilemma (Toyokawa *et al.*, 2014).

Analysis and prediction are key to exploring the prospects of the available alternatives. The operationalization of any performance data is deeply intertwined with the analysis technique that is used. On a fundamental level this goes back to the choice between quantitative and qualitative approaches. Here the advantages of the one approach are often the disadvantages of the other. Where sufficient data, skill and time are available to make this practicable, quantitative analysis is seen to be preferable to qualitative analysis (INCOSE, 2005: 41–42). However, most design decisions involve several performance aspects and thus inherently need to satisfy a set of performance requirements at the same time, and this needs a range of analysis process as well. Mostly this also means that there are several stakeholders involved in the decision (Augenbroe, 2011: 22). Moreover, decisions about a performance level typically come with risk, which in turn might have consequences for the liability of the decision maker (CIB Report 64, 1982: 7). Quantifying risk adds another layer of complexity to analysis efforts.

Another important question is the issue of who actually makes the decision. In some cases there may be a single actor who is the decision maker, but in many cases there actually is a group of people involved: the design team, members of a board of directors or other stakeholders. Decisions made by a group are typically based on some sort of voting system; underlying mechanisms may be unanimity, majority, plurality or dictatorship (PMBOK, 2008: 108). Unfortunately, group decisions are problematic due to Arrow's impossibility theorem, which states that where group members have three or more options, there is no voting system that can convert individual ranked preferences in a fair and consistent ranking for the group (French, 1988: 280–323). Hazelrigg (2012: 8) thus argues that, for reasons of mathematical consistency, decisions are made by individuals only; groups may have emergent behaviour, but there is no such thing as 'group decisions'.

While the concept of design decisions is widely used in the literature on building performance, there is little discussion about the different type of design decisions. However, one may discern the following categories:

- Many authors hold that a key decision in building design is the selection of systems and options that can be incorporated in a developing scheme. Markus *et al.* (1972: 13) describe design as a search for physical systems or hardware as solutions to design problems. CIB Report 64 (1982: 22) emphasizes product selection in design. De Wit (2001: 146–147) provides a deep discussion of a choice that keeps a building as it is or opts to include an active cooling system. De Wilde (2004) hones in on the selection of energy-saving building components. Schade *et al.* (2011) also discuss decisions regarding building energy efficiency, but in a wider framework. Singhaputtangkul *et al.* (2014) focus on decisions about the envelope of sustainable buildings. Ochoa and Capeluto (2015) review decisions on retrofit systems for the façades of residential buildings.
- Less prominent in building performance literature, but equally important, are decisions regarding the building location, orientation and geometry. Decisions on these subjects concern 'form finding' and thus the very essence of architectural design. An example that deals with explicit geometry decisions in the design of a mixed-use building in terms is provided by Lin and Gerber (2014).
- Another decision concerns system and component sizing. In some cases this may be a follow-up decision once a specific system has been chosen; for instance a building services engineer may first select a chiller type and then decide on the specific size. However, this is not always the case; for instance a decision on the amount of glass area to be used in a façade may actually be decisive for the choice of cladding system. Sizing decisions are often linked to optimization efforts, as they are easily parameterized (Nguyen *et al.*, 2014).

- Retrofit decisions represent a particular type of design decisions. In this case, there is an existing building, which means that there is more information than in a newly built project; one may even have information about occupant behaviour and control settings. At the same time, retrofit decisions are challenging, as investments, expected energy savings and other performance aspects require a trade-off (Rysanek and Choudhary, 2013).

Other decisions also play an important role in performance-based design, but might not necessarily be seen as design decisions in the strict sense of shaping the design. Among these are:

- Decisions that define performance criteria, as discussed in Chapters 3 and 5. These drive the evaluation of what is a satisficing or optimal design alternative.
- Decisions that set conditions for the future use of the building, such as number of occupants and operating hours.
- Choices about the quantification methods and tools that will be used in the design process, noting that tools may have underlying assumptions and constraints that may drive design in a certain direction.
- Decisions about building certification and corresponding target levels, such as the choice to aim for LEED gold or BREEAM excellent status.
- Beyond decisions about the selection of hardware and systems, there also are decisions to be made about partnering, collaboration and bidding for projects. Building the team is an important issue in the success of construction projects. Performance data about other actors may be crucial in this respect. In construction, such information may be hard to obtain, but in for instance the medical world, performance indicators are used to measure the quality of care provision of various hospitals so that patients can make an informed choice (Anema *et al.*, 2013).

Beyond these decisions, the design process may also lead to a range of design questions and problems. Questions and problems are not equivalent to decisions. As defined by Hazelrigg (2012: 14), problems can be solved; the solution or answer to the problem is either right or wrong. Decisions are made; the decision can be either good or bad. Whether a decision is good or bad depends on the alternatives, beliefs on outcomes and preferences of the decision maker. Closed questions lead to a predefined outcome; open questions do not. Such questions also map to a generic interest in any system as described by Montgomery (2013: 14–15). The following design questions may appear in performance-based building design:

- Does the building design meet the building regulations?
- Will the resulting building be energy efficient, comfortable or meeting other requirements? Or is there a need to modify the design to meet requirements?
- Factor screening: what are the parameters that have the largest impact on a given performance issue, so that the designer may decide to prioritize working on these?
- Discovery: what is the impact on performance of some unusual system or intervention?
- Which system in a certain category has the best fit with a building design?
- How can this building design be improved in order to get a better performance?
- Robustness: what happens to the performance of the building if it is subjected to adverse conditions?

- Is the design in a good shape in order to progress to a next stage, such as moving from conceptual to technical design? Or is it at a level that may be presented to key stakeholders for a go/no-go decision?

These different issues clearly show that each design decision needs to be seen in a specific context. It puts some doubt on generic terms like 'what-if analysis', a concept used by some authors to motivate an exploration of the relation between building design parameters and building performance, as this lacks specificity; see for instance Hand (1998: 70–72), Hopfe and Hensen (2011) or Østergård *et al.* (2016).

Some of the decisions have been recognized before, and suggestions have been made for making choices. For instance, CIB Report 64 presents the following methods for selecting products that are to be used in buildings: subjective selection (by individual experts or by groups), selection on the basis of the availability of test methods, selection based on functional analysis, selection on the basis of feedback from products in use (reports on complaints and failures, as well as stakeholder surveys) and selection based on the study of stakeholder requirements. It is noted that these methods are not mutually exclusive and that in many cases a combination of these methods will be used (CIB Report 64, 1982: 12).

Optimization, the process of finding the best alternative, clearly has an important role in making design decisions. As noted by Choudhary *et al.* (2005a), the use of numerical optimization techniques can help to ensure consistency in decisions, especially where this is combined with computer analysis. Some authors are focussing the aim of design decision on making the choice of an optimal solution such as a system or material; see for instance Løken (2007), Castro-Lacouture *et al.* (2009), Kämpf *et al.* (2010), Rahman *et al.* (2012), Cedillos Alvarado *et al.* (2016) or Lindberg *et al.* (2016). Others, such as Khatib *et al.* (2011), Fux *et al.* (2013) and Huang and Niu (2016), emphasize the use of optimization to find optimal design parameter values and system sizes, often through the study of performance of alternatives in relation to a reference case. Few authors, such as Evins (2015), consider both system selection and parameter variation within those systems at the same time, which obviously is more computationally intensive. Yet with the large number of design alternatives in building design and the many criteria that may govern decisions, it may be wise to allow satisficing and the adoption of solutions that are 'good enough'. This favours work such as that of Langner *et al.* (2012) who take a more moderate approach and only aim for an understanding of the impact of parameters on performance and of the interactions between parameters rather than on finding an optimal solution. Junghans (2013) points out that many decision makers would prefer a set of alternatives over just one single optimal solution, especially if improvement over alternatives is marginal; this provides them with more flexibility in moving forwards. Hazelrigg (2012: 98) argues that optimizing separate parts of a decision problem by use of multiple objective functions that are optimized separately is not a valid approach and is likely to lead to irrational choices; instead one should define the appropriate overall objective function and use this to find the preferred solution for the complete search space. If objective functions lead to results that seem inappropriate, one should review that objective function instead of trying to solve the issue by imposing constraints.

Where the objective is to select one option out of a range of alternatives, it is worth to spend some time in narrowing down the search space by elimination of infeasible design subspaces. In some cases this is intuitive, as with homebuyers going through a list of houses that are for sale and eliminating those that are over budget or do not have the desired number of bedrooms. For other decisions this elimination may become more complex, and infeasible solutions may have to be identified by comparing a figure of the target and allowable performance with the predicted performance of the solutions (Inoue *et al.*, 2012). Further caution is needed as some of the alternatives that are considered in a decision may interact. For instance, Eisenhower *et al.* (2012) take the view that building design decisions should concern the combination of a building system and its operation; they identify this as DOS (design and operation scenario) and optimize over a range of DOSs rather than building design variations only. Another layer of interaction between decisions can be found in the framing of design decisions; for instance Welle *et al.* (2012) note that in current practice the decision about what rooms to analyse in terms of for instance daylighting is mainly a manual process, driven by the expertise of architects, engineers and consultants. Another pitfall in thinking about design decisions is the assumption that all parameters operate on a continuous domain. In part this may be caused by some of the modelling tools, where for instance window height and wall thickness may look like very similar parameters. However, while glass often is cut to the exact size required for a specific project, wall thickness may in fact be governed by the standardized dimensions of bricks or studwork available on the market and thus operates on a discrete domain.

As should be obvious from the long discourses in the earlier chapters on needs, requirements, goals and targets, the setting of criteria for a decision is a non-trivial task. Yet it has been suggested that financial considerations are the main driving factor for decision making in the building industry (Langston, 2013). Consequently some researchers argue that Cost–Benefit Analysis (CBA) is the main tool to support decision making in construction; the key principle of this approach is to check whether project benefits exceed the costs. Authors such as Jin *et al.* (2012) reduce the underpinning of decisions to a single economic measure, in this particular case for façade decisions. Junghans (2013) emphasizes the monetary side of building renovation decisions and proposes a ranking strategy based on the analysis of investments and expected returns. Again, they suggest one key criterion: Life Cycle Cost (LCC) including Net Present Value (NPV). However, there is a question whether monetized decisions can value environmental goods and services in a proper way (Ding, 2005). Even in extremely price-driven, competitive sectors like the manufacturing industry, efforts to manage performance consider product costs, quality, flexibility and time; typically there also is significant attention for product development, customer satisfaction and the manufacturing process (Girard and Doumeingts, 2004). Monetizing decisions may thus be appropriate in some cases, but a careful deliberation should explore the context of each specific situation; standardized blanket approaches should be distrusted.

Selection of alternatives in the context of a range of performance aspects mostly requires subjective judgment. Sometimes it is useful to aggregate information into a single number using weighting factors (CIB Report 64, 1982: 20). Some of the methods employed in this context are the Weighted Score method, Paired Comparison method and Cost–Benefit Analysis. Many performance assessment schemes that aim to combine various aspects use some form of summation. In its simplest form these just add up scores; these do not take into account the fact that some aspects are more important than others. More advanced schemes use weighing factors; these allow putting more importance on key aspects; however they still only take a linear approach. Also, there is a need to incorporate an effect of 'diminishing return' at the high end of the performance spectrum; this incorporates the utility of performance to the stakeholder rather than a simple straight value. This is, for instance, implemented in the utility profiles named 'contingencies' in the ProMES system (Pritchard *et al.*, 2012: 20–24). Inoue *et al.* (2012) have developed mathematical functions to define one ultimate Preference Robustness Index (PRI) that is an aggregate of some other indices: a Design Preference Index (DPI), a Design Accuracy Index (DAI), a Design Convergence Index (DCI) and a Design Stability Index (DSI). Another approach to the comparison of a range of performances is the use of graphical methods. CIB Report 64 (1982: 20) describes the use of a 'performance profile', where all aspects are depicted in one figure, with one horizontal axis each. Another approach is the use of a 'radar plot'; this is a radial graph used in a range of domains that is useful for the display of multivariate data.

8.2.1 Normative Decision Methods

Normative decision methods give guidance on how to make the best possible decision, assuming an idealized decision maker who operates completely logical. This does not necessarily represent how real people make decisions, but presents the gold standard in how decisions ought to be made. The basic activities that make up a rational decision consist of recognition and diagnosis of the decision problem, the search for and/or design of alternatives, evaluation and screening of the alternatives, and ultimately making a choice.

The best possible choice is to decide on the alternative that gives the highest chance of a desired outcome, once the decision is enacted. To identify what is desired, one needs a standard of measurement; typically the term value is used to represent numerically what is 'good'. Where decisions involve a degree of uncertainty, the term expected value may be used; expected value represents the probability-weighted average of all possible values or $x_1 p_1 + x_2 p_2 + \cdots + x_n p_n$, where x is the expected value of each outcome and p the corresponding probability.

Another closely related term is utility. Utility is a concept that represents the 'degree of liking' or 'measure of preference' that the decision maker has for alternatives. Utility can be used to trade the liking for aspect A against the liking for aspect B, if there is a relation that more of A leads to less B; points that have the same user preference can then be plotted as an indifference or Pareto curve. Just like expected value, one may also calculate expected utility. A difference between value and utility is that utilities may represent the willingness to take risk, as exemplified by the following. Looking at the outcomes of an event, making money is mostly always seen as a good outcome, and one may measure outcomes in for instance pound sterling. Using pounds allows one to measure the value of different outcomes. Yet decision makers typically have a different preference for the amount of consequence that they may obtain: the difference between making £100 on a starting capital of £1 is not the same as making £100 on a starting capital of £1000. Where consequences are described by expected utility, and hence involve an amount of risk, a relation between attribute (such as revenue) and expected utility can be plotted. Where the relation is a straight line, this is risk neutral; other relations indicate an inclination to take or avert risk (Jordaan, 2005: 190).

In formal approaches towards decision making, the target of the decision maker is represented by an objective function. This objective function is a mathematical construct that ranks alternatives according to the preference of a decision maker; using of objective functions allows automating the analysis process and prevents having to rely on the decision maker to be involved in the comparison of each and every pair of options (Hazelrigg, 2012: 45). But preferences are best expressed as directly as possible, without structures and frameworks that may constrain the order. Approaches that carry some risk towards misrepresentation of the actual preferences of the decision maker are for instance linear multi-attribute objective functions and multiplicative objective functions (*ibid.*, 47–50).

For most decisions, the decision maker has a range of criteria. This is reflected in the wide uptake of Multi Criteria Decision Making (MCDM), an approach that helps to select a single preferred solution from a range of alternatives in a situation where there are competing criteria. The literature also uses the term Multi Criteria Decision Analysis (MCDA), which strictly speaking is the preparatory work that precedes the decision. Some authors also use different terms for criteria; French (1988: 105) already notes that 'the terms multi-attribute decision making, multi-criteria decision making and multi-objective decision making are used almost interchangeably'. Augenbroe (2011: 31) points out that the theory of Multi Criteria Decision Analysis (MCDA) structures the process of taking decisions based on performance assessment. The key to MCDA is to look at how each alternative scores for the criteria given and then to consider how these criteria are weighted against each other. In other words, the process is about 'weighting both the differences between options, and how much that difference matters'. This assessment is then used to actually make the decision.

There is a large number of MCDA techniques. The strongest theoretical foundation is provided for Multi-Attribute Utility Theory (MAUT); see for instance Keeney and Raiffa (1993), French (1988), Jordaan (2005) and Hazelrigg (2012). Other methods include the Analytic Hierarchy Process (AHP), Data Envelopment Analysis (DEA), the Dominance-based Rough Set Approach (DRSA), Goal Programming (GP), Multi-Attribute Value Theory (MAVT), Value Analysis, Outranking, Weighted Product Model, Weighted Sum Model and the Technique for Order Preference by Similarity to Ideal Solution (TOPSIS). In general, these MDCA methods can be grouped in three main categories: (i) utility/value measurement methods; (ii) goal, aspiration and reference level models; and (iii) outranking models (Løken, 2007). Mathematically, the first category employs numerical values to identify what alternative is preferred ($a \succ b$). The second category measures the deviation between actual performance and goal using a formula $z_i + \delta_i \geq g_i$ where z_i is the performance attribute, δ_i is the deviation and g_i is the goal value. Outranking models employ pairwise comparison between each possible set of two alternatives. Løken (2007) points out that the choice of a MCDA approach may have a significant impact on the decision and therefore recommends the combination of two or more methods.

Two issues are crucial to the outcomes of all these methods: the selection of what criteria to use and the determination of how these criteria compare to each other. In effect the choice of criteria and their comparison are decisions in themselves. Yet typically in most MCDA methods, this selection of criteria and weights is left to the judgment of the decision maker. Wang *et al.* (2009) suggest that the selection of decision criteria itself should be based on a range of principles: the criteria should be set up in a systemic way, ensuring that they represent the essential issues at stake; the criteria should be consistent with the overall goal of the decision maker; the criteria must be independent in other to prevent double counting; criteria must be measurable; and criteria must be comparable and normalized. There are two principles for setting weights. The first one is direct, where decision makers (typically experts) set the weighting factors, which are then known as *a priori* weightings. The second one is indirect; here decision makers are asked to compare alternatives and weighting factors are computed from the outcomes; these weightings are then named *a posteriori* weightings (Langston, 2013). The indirect method to set weightings may be supported by MCDA methods like the Analytical Hierarchical Process, as exemplified by Hopfe *et al.* (2013). Where there are only a few decision criteria, the importance of the weightings on these criteria increases; this is sometimes seen as favouring a system where there are many criteria, thus reducing the risk that one single criterion becomes dominant (Langston, 2013). However, Wang *et al.* (2009) report that in practice the use of equal weights is the most popular weighting method. It is interesting to note that some weightings in the environmental domain have come under criticism; for instance some argue that energy use, due to the direct relation with climate change, should be given a much higher priority than it typically is given (Campbell, 2009). Yet Langston (2013) found that a change of weighting factors did not significantly change the outcome of a facility management case study decision.

While Hazelrigg emphasizes the idea that decisions are always made by single decision makers, Augenbroe (2011: 31) sees room for different roles in MCDA and paints a more nuanced picture: the decision maker may be supported by a decision analyst; there also may be other stakeholders who influence the decision maker.

There are many publications that discuss the use of MDCA in building design, especially from an academic, normative point of view; see for instance de Wit (2001: 144–156), Soebarto and Williamson (2001), de Wilde and van der Voorden (2002), de Wilde (2004: 74–81), Chinese *et al.* (2011), Welle *et al.* (2011), Akadiri *et al.* (2013), Hopfe *et al.* (2013) and Sun *et al.* (2015). Many of these papers relate to the domain of energy efficiency and the selection of energy conservation measures. A good overview of various methods and their uptake in construction is provided by Jato-Espino *et al.* (2014). Contributions that are focussed on single methods are for instance Iwaro *et al.* (2014), who present an integrated criteria weighting framework based on the Analytical Hierarchy Process that they use to set average weighting factors based on input from a range of stakeholders such as architect, engineers, quantity surveyors and others in support of facade decisions. Løken (2007) presents work in the area of energy planning, which he describes as a clear candidate for the application of MDCA due to the complexity of underlying problems, the involvement of various stakeholders and the need to take into account a range of criteria. Wang *et al.* (2009) categorize these criteria as technical, economic, environmental and social. Frenette *et al.* (2008, 2010) review the complexity of proper multi-criteria assessment of a relatively simple building system: a light-frame wood assembly. Juan *et al.* (2010) presents a system that aims to support the trade-off between renovation costs, improved building quality and environmental impact; the work includes an application to a building reuse case in Taiwan. Mela *et al.* (2012) explore the use of six MCDM methods, including the weighted sum method and weighted product method, in the context of construction, applying them to an office building, hall and residential case study. Langston (2013) studies the use of MCDA in the context of the renovation of an industrial building in Australia, focussing on Facility Management. Hopfe *et al.* (2013) describe the use of MCDM, including AHP, in a choice between two design alternatives during the conceptual design of an office building in the Netherlands. Shao *et al.* (2014) present an approach that combines a process for the design team with numerical optimization and apply this to an office building retrofit in Germany. Wang (2015) presents the application of multi-criteria decision making using the TOPSIS approach in the context of building energy benchmarking. Wang *et al.* (2017b) apply normative theory to retrofit decisions.

8.2.2 Naturalistic Decision Making

Naturalistic Decision Making emerged in the mid-1980s in response to a notion that what many decision makers do in actual practice does not fully align with classic normative decision theory. A key issue is that normative decision theory focusses on *single decision events*; yet much real-life decision making relates to developing scenarios, such as the decisions of a fire crew commander, medical specialist or chief executive (Klein *et al.*, 1993: 3–5). Where normative decision theory is concerned with how to make the best possible decision, naturalistic decision making is deeply based on descriptive studies about what people actually do in reality. Over time, the work in naturalistic decision making has emphasized the way experts think and decide. Rather than prescribe what would be an optimal decision, it studies how the environment impacts on an expert's decision-making process. Moreover, this field also tries to identify and understand the difference between how novices and experts make decisions (Shan and Yang, 2016).

Naturalistic Decision Making is typically found in situations where one or more of the following factors complicate the decision-making situation: problems that are ill-structured and not directly recognizable as options; a context that is dynamic and uncertain, and where the decision maker has to deal with incomplete and imperfect information; where there are shifting, ill-defined or competing goals; where there is a series of events rather than a single decision, with interaction between actions and feedback; where decisions are made under temporal stress; where stakes are high and of real importance to the decision maker; in case of multiple actors; and where there are organizational goals and norms to take into account (Klein *et al.*, 1993: 7–10). Among the areas where naturalistic decision making is found, design engineering design is explicitly named (Klein *et al.*, 1993: 107).

A closely related area is that of heuristics, sometimes also known as rules of thumb. Heuristics are practical methods in decision making that cannot be guaranteed to lead to optimal decisions but where outcomes are typically seen as 'good enough'. While the scientific literature tends to favour normative decision theory, heuristics are also subject of evaluation. For instance, the field of fast and frugal heuristics (FFH) explores when heuristics work well, what principles are used to generate a decision and how heuristics perform in complex real-world situations (Shan and Yang, 2016). In the wider area of naturalistic decision making, different research methodologies can be employed to study what real decision makers do, such as after-action reviews, cognitive task analysis, in-depth interviews, argumentation analysis, functional analysis, observation and ethnography (Gore *et al.*, 2015). There is evidence to suggest that in processes where there are large differences between alternatives and their consequences, a decision maker in the real world may just follow a course of action that is obvious to an expert, without making any formal decision (Shan and Yang, 2016); this renders the decision-making process fluid and makes it hard to recognize design decision moments.

In naturalistic decision making a decision is seen as something that evolves over time rather than one static decision event; this therefore requires different cognitive strategies and processes (Meso *et al.*, 2002). There is a range of models that describe naturalistic decision making (Klein *et al.*, 1993: 138–204), with the following being the most prominent ones:

- Recognition Primed Decisions describe how experts deal with decisions based on their past experience. A situation may be familiar, fairly familiar or complex; for a familiar situation a suitable action can be selected quickly, whereas new and novel situation require deeper analytical processes.
- Image Theory, which describes how goals, beliefs and values and plans for realization interact.
- Decision Ladders, which capture a range of steps that are taken from the very first stage of activating a decision process towards execution of the resulting choices.

Skills of decision makers can be classified in a range of categories of increasing understanding as follows: novice, advanced beginner, competent, proficient and expert (Shan and Yang, 2016). An important related factor in naturalistic decision making is situation awareness; here experience helps experts to recognize cues that are linked to potential actions that may help to steer the situation in the desired direction (Elliott *et al.*, 2007). Situational awareness is especially important in the lead-up to a decision moment (Gore *et al.*, 2006). To some extent, situational awareness may be trained; this for instance is what pilots learn in extensive sessions in flight simulators. In other domains, experience can be gained from training using a simulated micro world in the computer; however setting up sufficiently real-life complex challenges in a virtual environment is a non-trivial task (Elliott *et al.*, 2007).

Naturalistic decision making may be seen by some as inferior to normative decision making, yet key concepts from this field are now used in for instance the training of North American power systems operators, where simulator-based training prepares employees to use recognition-primed decisions, while capturing situation awareness, mental models used and mapping cognitive task analysis (Greitzer *et al.*, 2010). Zannier *et al.* (2007) suggest, on the basis of empirical research with software designers, that both rational decision making and naturalistic decision making may be used in the same design project. These researchers note that problem structuring is an important aspect of design; more structured situations lead to rational decisions, and less structured to naturalistic ones; typically one of the approaches is dominant in a given situation, but aspects of the other are mostly recognizable. Decision making can also be studied on the organizational level, aptly named Organizational Decision Making (ODM). Seminal work in this field is provided by Simon (1997); a brief more recent overview as well as the relation between organizational decision making and naturalistic decision making is given by Lipshitz *et al.* (2006). Jenkins *et al.* (2010) describe a decision ladder that is based on naturalistic decision making as well as Cognitive Work Analysis, an approach stemming from process management in nuclear facilities; this decision ladder starts by setting goals and then has a number of levels: observe that a decision is needed, collection of information, diagnosis of the system state and a cycle that explores options and their consequences. Once a suitable action is identified, this is cascaded downwards to identify tasks and processes that can be executed. Again this work shows integration between normative and naturalistic approaches. In the context of building design, case studies reported by de Wilde (2004: 59) indicate that most decisions on energy-saving measures went straight for one single option; selection from a range of options was the exception. Welle *et al.* (2011) report that the development of a single design option may cost a month and that as a consequence many design processes have very few iterations. From observation of design, Bleil de Souza (2012) sees a process of 'framing' that has similarities to obtaining situational awareness; she does not recognize 'what-if questions' and related decision moments. Jato-Espino *et al.* (2014) present a review into the uptake of multi-criteria decision making methods in construction based on academic literature; they warn that there is a tendency to use the best-known techniques only; however it must be noted that the underlying material is academic and that industry practice probably lags further behind. More recent interviews with industry experts (Jones, 2016: 42) also raise doubt about the use of normative methods in practice.

8.2.3   Decision-Making Challenges

While the normative and naturalistic decision-making theories provide many insights on how decisions ought to be made and are actually made in practice, and how this may support performance based building design, there also remain significant challenges.

As mentioned earlier in the chapter, there is a paucity of information in the scientific literature about what actually happens during performance based building design. Yet there is evidence of a disconnect between choices regarding the selection of energy conservation measures and the use of analysis tools to assess the impact on energy performance (de Wilde, 2004: 56–57) that seems to indicate a lack of exploration of alternatives and high reliance on heuristics (*ibid.*, 59). Practitioners often mention high workload pressures in industry and limited funding to do analysis, employing approaches like building simulation (de Wilde and Prickett, 2009). Other scientists also note that only a limited set of design alternatives is considered and that analysis of how well these alternatives perform often takes place later in the design process and mainly for confirmation purposes. Often lack of time and software issues are cited as a reason for late analysis (Lin and Gerber, 2014). A long-standing concern is that design decisions often miss access to required information (Augenbroe, 2002; Crawford *et al.*, 2011; Zanni *et al.*, 2016). Against this background, claims such as that of Wang *et al.* (2009) that MCDA methods are gaining popularity in energy decision making must be viewed with caution. Targeted interviews with industry experts on how they make decisions in current practice seems to raise questions about the use of normative decision-making theory (Garmston, 2016; Jones, 2016) and looks more aligned with naturalistic decision making. The link between decision making and design remains intricate and complex; already at the start of the century, Soebarto and Williamson (2001) point out that many researchers and tool developers in the domain of building performance assume that performance requirements can be defined before the design process starts, whereas designers often explore requirements and let these evolve together with the design. A similar observation can be made with respect to the criteria for making performance-based design decisions.

Decisions always deal with the future and hence require prediction. Prediction comes with uncertainty and requires the application of probability theory. Augenbroe (2011: 15) thus emphasizes the need to incorporate uncertainty analysis in performance assessments that are to inform design decisions. Williamson (2010) notes that building performance simulation results are sometimes less accurate than hoped for and therefore may lead to decisions that are not as well founded as one would expect. Hazelrigg (2012: 153) warns for the need to use Kolmogorov's probability theory[4] and Bayes' theorem.[5] An application of the inclusion of quantitative risk analysis in the context of building retrofit decisions is presented by Heo *et al.* (2013). This work employs Bayesian calibration in order to quantify, and subsequently update, uncertainties in energy models that are used to study the predicted impact of installation of energy conservation measures. Yet during conceptual building design, uncertainties are high; the confidence interval for early design decisions has been found to be no more than 50% (Rezaee *et al.*, 2015).

Another issue that remains open for debate is how performance-based design decisions should account for the various stakeholders. On the basis of mathematical consistency, Hazelrigg (2012: 8) argues against theories that cater for group decisions and claims that decisions are made by individuals only. Yet it is obvious that building performance requires dialogue between the many stakeholders and actors in the design process. Hazelrigg (2012: 493–511) also refutes a number of popular decision methods on the basis that they do not meet requirements of consistent mathematical logic: fuzzy logic, quality function deployment, the Pugh method, the Analytic Hierarchy Process (AHP), axiomatic design, Taguchi method and Six Sigma. Yet there is a significant number of researchers that continue to use and promote these approaches in the building performance domain.

8.3   Tools for Performance-Based Design

There are many instruments that may support performance-based design, which can range from traditional design tools such as pencil and paper for sketching purposes all the way to advanced computer environments that provide complex functionality. They include the performance quantification tools as discussed in the chapter 6; however typically these are used in conjunction with other instruments. The domain of tools is continuously changing, especially where it comes to the new digital tools that have become a key feature in the efforts to design novel high quality buildings (Schwitter, 2005: 113).

As with the quantification tools, there are different ways of categorizing design tools that all have their own use, without one view necessarily being better than the others. In the modern day and age, most of the tools are based on digital technology

---

4 Kolmogorov's probability theory provides the main foundation of modern statistics. Key are two axioms. The first axiom relates to the definition of a sample space, which is a finite set of all possible outcomes of an event. The second axiom states that the probability that at least one of the events contained in the sample space will occur is always one.
5 Bayes' theorem links the probability of an event to conditions that might be related to that event; it gives a mathematical approach to update beliefs in the outcome if new evidence becomes available.

and computer based; however one should keep in mind the use of traditional sketching, models, catalogues with product information and traditional books. Among the traditional tools are also 'rules of thumb', which are general guidelines that specify what size of a building system, such as beam height or window area, to select, but without explaining why (Donn, 2001). Turning to digital tools, in his classic book on the use of computer environments to support building design, Eastman (1999: 13) discerns tools that help to define geometry and materials, analysis tools and simulation tools, automatic and interactive detailing tools, expert systems, code evaluation tools and site investigation and development tools. Hendricx (2000: 15–16) differentiates between automated and assisted design tools. She then goes on to discuss tools that support space allocation, layout development, shape grammars, case-based reasoning (CBR), artificial intelligence (AI) and expert systems. Raphael and Smith (2003: 3–4) order computer tools by defining the main engineering tasks as simulation (observing behaviour), diagnosis (identification of causes), synthesis (finding physical configurations) and more general interpretation. Hansen (2007: 128) differentiates between tools that help structure and manage the design process in terms of phases, tasks and actors (process tools), tools that help structure technological issues (strategy tools), tools that suggest promising solutions (design tools) and evaluation and analysis tools. Ferreira *et al.* (2013) discern discrete decision analysis tools that assess the performance of a finite and not overly large group of design alternatives, as well as continuous or mixed problem decision analysis tools that assess the performance of a much larger group of options. In many cases there are iterations between different categories. For instance, Becker (2008) makes the distinction between design tools (which provide design solutions) and assessment tools (which predict the performance) but also acknowledges that iteration allows the use of assessment tools to guide design generation. Lin and Gerber (2014) suggest that in order to fit into the constraints of real design processes, tools and support environments should enable a rapid generation of design options, rapid analysis of the performance of these options, support multi-criteria analysis and trade-off, and include optimization and search capabilities. On the basis of observation from the interaction of users with a daylighting expert system tool, Andersen *et al.* (2013) hold that interactive tools are more acceptable to designers than black box tools that generate design solutions without giving the user the ability to intervene.

The development of tools that target design support in themselves has to consider the requirements that their use in the building design process imposes. In general, this seems to be a rather underdeveloped field. Typical approaches in software engineering such as described by Pressman (2005) emphasize requirement elicitation and engineering and employ the development of well-described use cases (*ibid.*, 175–206). Yet there is scant evidence of these approaches being employed in the development of building design tools. Instead, the building research literature contains a range of generic claims. For instance, it is often stated that building design, especially in the early phases, moves at a fast pace and hence requires very fast feedback on building performance; this would be at odds with the relatively long modelling and computation times that many building performance analysis tools require (Geyer and Schlüter, 2014). Negendahl (2015) holds that most tools that target early design support are not sufficiently flexible to accommodate rapid design evolution and do not provide the information sought by designers. A rather generic view seems to be that tools should allow to 'predict the consequences' of design decisions, especially in terms of indoor environment, environmental impact and costs (Østergård *et al.*, 2016); obviously such a view overlooks other design activities such as option generation.

Creative and Generative Tools

A brief overview of some of the tools for performance-based design ought to start with tools that help the design team to create and express ideas. Key aspects in this domain are sketching and visualization; while these were originally paper based, this work can now also be done with computers. Sketching and visualization are essential for creative design, helping the designer to formulate, check and develop ideas; sometimes the process of sketching is named visual reasoning (Krish, 2011). The search for alternatives can be supported by tools that help designers establish an option or search space. Again, there is a wide range of options; for instance tools may convey information of previous projects and case studies, possibly in the form of a database, or may support workshops and conversations with clients. Other support instruments are sustainability cards, project gateways, online fora and blogs, documentation tools, detail retrieval systems or diary entries. Activities like workshops and brainstorm sessions may also be considered to be a tool of some sort (Zapata-Lancaster, 2014). More specific support for identifying potential design alternatives would for instance be a typology of existing facades and the potential refurbishment options for each of these façade types as presented by Cuerda *et al.* (2014).

A closely related category is that of generative tools. Typically these tools use some form of permutation of parameterized archetype designs to create a large range of variations; examples are shape grammars, geometric optimization and genetic algorithms (Kitchley and Srivathsan, 2014). Architectural scripting tools are computer programs that define architectural form. Relationships between elements may be manipulated by means of parameters; this is called parametric scripting (Nembrini *et al.*, 2014). Well-known architectural parametric scripting tool are Maya and Autodesk Revit. Filtering of design alternatives may be done automatically, based on some objectives or search criteria, but it is also possible to leave this filtering to a human expert, especially where some criteria are not computable (Krish, 2011). Some recent work aims to support early design by rapid iterations between different design alternatives and performance evaluation, where possible using parameterization and optimization (Lin and Gerber, 2014). It is worth bearing in mind that while generative tools may yield a wide range of solutions, many of which may be innovative and unexpected to the human mind, mathematically these tools actually operate in a constrained search space, searching for permutations of parameters and geometry that have been set by the tool operator. By nature, tools like genetic algorithms do not 'think outside the box' and are limited to the parameters and parameter ranges that are input to the tool. A popular tool for generative building design is the Rhino 3D environment, combined with the additional Grasshopper visual programming language. Further plug-ins for this system are the Diva system for environmental analysis, the Galapagos evolutionary solver, the Ladybug add-on that links Grasshopper to climate data and the Honeybee system that connects to thermal and lighting simulation software. Shi and Yang (2013) demonstrate the use of these tools to design a non-traditional roof surface; Ercan and Elias-Ozkan (2015) present their application to the design of solar shading devices for a case study building in Cyprus. Turrin *et al.* (2011) present a study into the combination of parametric modelling using ParaGen with computational analysis tools to drive architectural design development. Lin and Gerber (2014) demonstrate the use of another generative system, based on the use of Revit and H.D.S. Beagle, for the design of a mixed-use building in the United States. In other work, Gerber *et al.* (2017) combine agent-based systems with generative approaches, performance simulation, optimization and coordination.

Performance Quantification/Analysis Tools

As quantification is a crucial factor in performance based building design, performance analysis tools obviously play an important role. A generic overview of the various ways to categorize these tools and a discussion of key categories has already been provided in Chapter 6. The reader is reminded of the many different viewpoints, such as classification according to performance aspect, computational algorithm, system resolution or explicitness of the underlying model. Here the discussion focusses on the use of these tools as support instrument for building design.

The development of analysis tools that support the design and operation of buildings that perform well is the 'holy grail' of much of the building performance literature; it is no coincidence that a seminal book like that of Clarke (2001) is titled *Energy simulation in building design* and the work by Hensen and Lamberts (2011) is titled *Building performance simulation for design and operation* – even when the discussion of the role of building simulation in performance-based design is mainly limited to the chapter by Augenbroe (2011).

An influential vision about the role of analysis tools in the design process has been given by Clarke (2001: 4–5) and describes two situations: a 'toolbox metaphor' where design tools are disconnected from the design process and a 'computer-supported design environment approach' as an alternative where tools are fully integrated. In the first situation, the onus is on the designer who must recognize tasks, identify an appropriate tool, run this tool and interpret what the results mean for the design. In the alternative situation, all of this is automated, but at the cost of a clear definition of the roles of all components in the system. It is interesting that in approach one, there is a two-way exchange between designer and tools on the basis of tasks; in the second approach the designer sends decisions to the environment and receives back the implications of these decisions. Along the lines of integrated systems, Mahdavi (1999) introduced the idea of 'bidirectional inference'. This would be a mechanism that not only allows the traditional mapping from a building design to the related performance but also supports the mapping from a preferred performance to design alternatives that will provide that performance. Yet in spite of the fact that 15 years have passed, the realization of these fully integrated system seems to progress only very slowly. Some persistent barriers to the use of analysis tools in design are well documented, such as issues with the availability of appropriate tools and models for specific design situations, lack of trust in computational results, high level of expertise required to use advanced tools, the costs in terms of both time and money of doing analysis work, and data exchange issues (de Wilde, 2004: 28–29). Further complexity in providing design support is caused by the fact that many design decisions relate to multiple performance aspects and stakeholders, the intrinsic properties of various stages of the design process where information needed to do an analysis is unavailable or uncertain, the problem of the performance gap between prediction and measurement once buildings are operational, and issues with the mapping between typical building information models and analysis needs (Augenbroe, 2011: 22–25). Williamson (2010) points out that the accuracy of analysis tools is often discussed, whereas the decision-making problems for which these tools are to be used are often not defined in great detail, if at all.

Analysis tools play an important role in the quantification of building performance, but also have their limitations. The availability of tools and models is crucial for performance-based legislation; these allow the development of the criteria, the design of buildings and building parts that meet these criteria and the verification of performance (through review or audit) once the building is complete (Foliente, 2000). At the same time, as noted by Williamson (2010), 'simulation applications should not be seen as surrogates to reality and interpreted as logical answers to substantive problems'. Some authors such as Jensen *et al.* (2009) claim that building simulation tools are used to analyse the energy consumption of a range of competing design alternatives. Hensen and Lamberts (2011: 5) hold that these tools have gone through a technology hype cycle and have now achieved their productivity plateau, although they acknowledge that the tool users may display different levels of innovativeness and preparedness to use simulation. However, there is a sparsity of research that reports on actual use of the use of tools in building design practice. Bleil de Souza (2013) points out that there is only limited research on the actual use of building simulation tools in real design practice, with methods tending 'to be limited to interviews with building designers, structured on-line surveys, reports of specific case studies and reporting experiences of interactions between specialists and building designers working in collaboration to solve specific problems'. Earlier work by de Wilde (2004) found a disconnect between decisions on the selection of energy-saving measures and the use of analysis tools. A decade later, the case studies reported by Zapata-Poveda and Tweed (2014) still indicate that actual calculations/simulations only take place after the completion of the conceptual design and mainly have compliance as their objective. Similarly, Eisenhower *et al.* (2012) see the main use of building simulation as part of efforts to demonstrate code compliance and to achieve certification with systems such as BREEAM and LEED; they hold that other use is limited to minor studies in some specific companies and to research efforts. Augenbroe (2011: 23) notes that performance quantifications at those early stages are by necessity a conditional statement, as many design features are still undecided; by consequence he states that this may make the outcomes of simulation efforts nearly irrelevant. Computational design support in the early stages therefore should be used with care, taking into account what information is still missing and how that may impact on any results. Frankel *et al.* (2015:1) report that in the United States in 2015, building models are still mostly used to verify and demonstrate compliance; according to their work this covers about 80% of tool use, and only 20% of tool use is aimed at supporting design decisions.

There are various routes that aim to integrate analysis tools into the performance based building design process. On a high level (de Wilde, 2004: 24–27), one may discern between efforts that aim to improve the tools themselves by improving functionality, user interfaces, data exchange capabilities and similar, and efforts that are independent of specific tools such as the Energy Design Advice Scheme EDAS. Bleil de Souza (2012) discerns the following categories in efforts to better integrate analytical tools into the building design process: work that targets specific users such as architects; work that is aimed at specific design stages; work that focusses on creating collaborative environments; and work that tries to position tools as design advisors. Attia *et al.* (2012) suggest that access to analysis tools is driven by the usability of the tool interface, linkage to a design knowledge base, ability to simulate complex components and interoperability. While these may be important practical considerations, it is more likely that tool selection is driven by the particular design decision at hand, combined with expertise and modelling requirements of the analyst. Attempts to define the design decision at hand and to link this to a suitable analysis procedure as defined in an 'Analysis Function' was the underlying premise of the Design Analysis Integration (DAI) Initiative (Augenbroe *et al.*, 2004). Østergård *et al.* (2016) consider single projects in a wider context and differentiate between 'predesign informative' simulation and 'proactive simulation'. Predesign informative simulation is carried out before the design process starts, whereas proactive simulation may be carried out during the process. On the side of tool-independent efforts towards integration, some researchers suggest that the best way to integrate analytical tools into the building design process is by focussing on collaboration in design teams made up of various experts (Savanović, 2009; Keeler and Vaidya, 2016); others believe that there is a need to develop specific tools for designers that match the way designers operate and think (Bleil de Souza, 2012). Schmid (2008) emphasizes the use of simulation tools with architecture students as a way to teach building physics and get students to evaluate the implications of their design decisions, which then will lead to performance-based design once they enter practice. This idea of training is also supported by Reinhart *et al.* (2012) who hold that designers should be equipped with the ability to read and interpret the results of simulation and should have the knowledge of how to adapt their design in response. At the same time, too much emphasis on simulation might be counterproductive and take attention away from understanding the performance targets themselves. It might also undermine the trust in the calculation results as these can be seen as theoretical only. Finally, the emphasis might shift towards getting the simulation right rather than getting the design right (Zapata-Poveda and Tweed, 2014).

Analytical support for design is also linked to the phase of conceptual design. Many authors have pointed out that it is easier to modify a building design early in the design process and conclude that therefore analysis tools should also be used early in the design process; see for instance Donn (2001), Gratia and De Herde (2003), Jensen *et al.* (2009), Kleindienst and Andersen (2012), Andersen *et al.* (2013), Lin and Gerber (2014), Nembrini *et al.* (2014), Reinhart *et al.* (2012), Futrell *et al.* (2015), Negendahl (2015) and Østergård *et al.* (2016). This has led to the development of a range of 'early design tools', such as Energy-10 (Balcomb, 1997), Ecotect (Marsh, 1997), the Building Design Advisor (Papamichael *et al.*, 1997), OPTI (Gratia and De Herde, 2003) and the Building Performance Sketch, a concept developed around the COMFEN tool for fenestration design (Donn *et al.*, 2012). More recently there are some efforts to develop early design tools based on regression analysis of the outcomes of more advanced analysis methods; see for instance Marsh (2016), Hester *et al.* (2017) or Østergård *et al.* (2017). However, Mahdavi (1999) already pointed out that the tool development community sometimes erroneously assumes that early design is equal to simple relationships and configurations and that therefore simplified models will be sufficient for the early design stages; this is typically not the case. This view is now generally accepted, and the use of simplified tools is now considered controversial; the common critique is that tools that simplify geometry and systems are too far removed from the complexity of actual buildings, whereas tools that limit access to only part of the functionality do not make full use of the capability of the underlying calculation engines (Nembrini *et al.*, 2014). A related issue that reappears in many efforts on tools aimed at early design is that the tool developer believes to be able to anticipate what specific evaluation of performance will be of importance during the design process. A classic example is the LT method, which flattens the design space to a few basic parameters such as percentage of glazing, internal heat load and ventilation rate. The original version was developed in the late 1980s, but derivatives have been promoted well into the new century (Baker and Steemers, 1996; Baker *et al.*, 2013). While such tools may be useful in an educational context, where users gain a feel for relations between different factors that impact on a goal such as energy efficiency or thermal comfort, they have little to contribute to actual design challenges. The aforementioned Building Design Advisor is more advanced and combines a central building model with a range of analysis tools in the domains such as lighting, energy and airflow analysis (Papamichael *et al.*, 1997, 1999). It has been promoted as a decision-making tool for the early phases of schematic building design, based on extensive use of default values; later work on inclusion of a code checking facility in the BDA implicitly acknowledges the need to expand the scope and uptake of the program (Reichard and Papamichael, 2005). Interestingly Marsh, who developed Ecotect as a 'simple and intuitive tool for architects, to be used during early design' (Marsh, 1997; Roberts and Marsh, 2001), later went on to believe that the future of simulation lies with a full integration of simulation and BIM in an approach termed BIMS (building information modelling and simulation), which then leads to a building response system that is an integral part of the final building (Marsh and Khan, 2011). An intermediate solution may be the use of metamodels, sometimes named surrogate models. Here the underlying premise is that many advanced simulation tools provide useful results but are too complex and cumbersome for use during the design stages. Instead, the researchers working on metamodels use simulation tools and machine learning to produce algorithms that provide fast access to results that are equivalent to simulation outcomes, provided the machine learning has been trained from valid data. See for instance Eisenhower *et al.* (2012), Geyer and Schlüter (2014) or Østergård *et al.* (2017). While simplified tools may not represent the full complexity of real buildings in real operational conditions, they still may be a useful learning help for students; the simplifications may help to make relationships easier to understand, without the distraction of other factors (Chase, 2005). Furthermore, the limitations of simplified tools should not be used as an argument to move to the opposite end of the tool complexity spectrum. The 'best, most detailed' simulation might not necessarily be the appropriate one for supporting a design decision either, as it carries the risk of being time consuming and computationally heavy. Sometimes a simplified model, or a normative calculation, can do the job just as well; a careful consideration of what is needed is required (Augenbroe, 2011: 22). One way to seek improvement of the usability of analytical tools is by interviewing the users of current tools in order to find out why tools are being selected and what shortcomings these tools are perceived to have; see for instance Attia *et al.* (2012). However, such efforts are good for gradual improvement of existing approaches; they are less suitable for a fundamental review of what is required or possible by thinking 'out of the box'.

In spite of some claims, it is hard to quantify the actual impact of building performance simulation tools on the building design process. It is virtually impossible to design a test where there is a control group or baseline without simulation that can be meaningfully compared with a group that uses simulation. The makers of the EnergyPlus did couple the download of new versions of the simulation engine to a survey of previous use, which required the total floor area of modelled buildings. But even these figures need to be questioned, as downloads by academic users (who were not involved in specifying systems for building developments) must have significantly biased the numbers obtained. Similarly, claims that design tools have been validated as for instance made by Lin and Gerber (2014) need to be viewed with scepticism; there currently exists no formal process for such validation that is generally accepted by the scientific community.

It is interesting to note that there are many academic papers that present complex software and ICT structures for design support by analytical tools that have little, if any, follow-up in practice. Recent examples are the Computerized Partnering Performance Index System by Yeung *et al.* (2009), ThermOpt by Welle *et al.* (2011), the Parametric Systems Modelling approach by Geyer (2012) and Geyer and Buchholz (2012), ZEBO Decision Support Tool for Zero Energy Design (Attia *et al.*, 2013), Virtual Design Studio (Pelken *et al.*, 2013; Zhang *et al.*, 2013), the Evolutionary Energy Performance Feedback for Design (EEPFD) framework by Lin and Gerber (2014), the SCRIPT platform by Petri *et al.* (2014), PassivBIM by Cemesova *et al.* (2015) and the SIPRIUS framework for sustainable urban regeneration (Laprise *et al.*, 2015). The same goes for older and well-cited work such as SEMPER (Mahdavi, 1999; Wong and Mahdavi, 2000) and the DAI Workbench (Augenbroe *et al.*, 2004).

Rating Systems

Rating systems such as BREEAM, LEED and others are fundamentally certification systems, which thus are developed to assess a building design once complete in terms of whether certification criteria are met. This puts them in a different category than analysis tools and real design tools; they are there to endorse completed products (Soebarto and Williamson, 2001). However, given that designers aim to get a good rating score, these systems do have an indirect impact on the design process and can be said to have had a significant influence on the design of some of the certified buildings. Rating systems may have both a positive and negative contribution to the design process. By their presence they focus attention on performance aspects, leading to discussion and thus consideration of these aspects. However, this may be to the detriment of aspects that are not covered by the system. In some cases, this may also lead to 'gaming the system' in order to chase maximum scores rather than a building that performs well (Cole, 2005). Vilcekova and Kridlova-Burdova (2014) provide some insights into the considerations that need to be given to the set-up of this type of schemes. There is evidence that rating schemes especially impact buildings designs that are close to the certification thresholds (Matisoff *et al.*, 2014). Obviously, the impact of these schemes on design depends on the buy-in of the actors in the design stage (Schweber and Haroglu, 2014). Rating schemes for a long time had a focus on operational energy use; life cycle energy use has only been seriously added recently. Some criticism may be levelled at rating systems for missing some aspects or issues, for instance by those who aim for a wider definition of sustainability such as Marjaba and Chidiac (2016) or those who focus on deeper explo-ration of single aspects such as Kabak *et al.* (2014). Shi *et al.* (2016) note the potential conflict between various objectives and how these are weighted. Rating schemes may also be subject to critique due to the fact that they award points for a whole range of performance criteria; this may mean that some of them, such as energy use, may not get the full attention they would do when considered as a separate issue (Campbell, 2009). However, rating and certification are also seen to have marketing benefits, especially among non-profit organizations (Matisoff *et al.*, 2014), and help to make performance an issue in building design. While building rating systems such as BREEAM and LEED are designed to assess a completed building design, the fact that they employ a range of criteria makes it relatively straightforward to tie them into a MCDM approach. For instance Kabak *et al.* (2014) present a system that combines BREEAM criteria with local building regulation aspects in a design approach for Turkey. An overview of the role of rating and certification in the context of building energy efficiency is provided by Pérez-Lombard *et al.* (2009).

Design Decision Support Systems

Several authors have developed design decision support systems (DDSS) for architec-ture and building engineering. Different types of decision support systems are used in various industries; for instance there are dedicated systems to support logistics and decisions on routing and scheduling in the transport sector. Where there is process automation, they can be part of supply chain management (SCM) systems and enter-prise resource planning (ERP) applications (Gayialis and Tatsiopoulos, 2004). In the context of performance based building design, design decision support systems are often focussed on specific performance aspects; for instance de Groot (1999) targets office lighting design; Bazilian and Prasad (2002) focus on building integrated photovol-taic (PV) systems at the residential scale. Zareian and Krawinkler (2012) present a DDSS that addresses seismic design at the building and story level; Singhaputtangkul *et al.* (2013) one that focuses on the design of building envelopes. Lee *et al.* (2013b) present a decision support system that accounts for the impact of cost estimating and scheduling on the design of supertall buildings, and Rahman *et al.* (2012) a decision support system for the selection of optimal roofing materials. Design decision support systems come in many different forms and often include routines that help to establish potential alterna-tives, evaluate these alternatives and support decision making. By way of example, Flourentzou *et al.* (2002), in their development of the TOBUS decision support tool for building retrofit, recognize the need to carry out a diagnosis of the state of all existing building systems and elements, an assessment of the indoor environmental quality in the building and a study of energy efficiency before proceeding to the analysis of refur-bishment scenarios and the associated costs. In general design decision support systems must be flexible and configurable and should not constrain the work of designers (Roldán *et al.*, 2010). There are various descriptions in the literature of tools that sup-port multi-criteria decision making, and application of multi-criteria decision making to case studies; see for instance the work of Kalay (1999), Soebarto and Williamson (2001), Chinese *et al.* (2011), Akadiri *et al.* (2013), Hopfe *et al.* (2013) and Sun *et al.* (2015). There are also digital support environments that emphasize multi-objective optimization but at the same time accommodate the understanding that some aspects are hard to quantify and require human judgment (Shao *et al.*, 2014). Geyer (2012) presents a design support system based on Parametric Systems Modelling (PSM), an approach rooted in systems engineering, and Systems Modelling Language (SysML). Seven main steps of a highly automated approach are highlighted, which link use cases, requirements, activity blocks, internal activities and parametric diagrams to obtain measures of effectiveness. However, it should be clear that the straight jump from SysML diagrams to measures of effectiveness is not as simple as suggested, since it lacks the deep reflection on setting up a virtual experiment that is meaningful for stakeholder needs to get to performance predictions and does not mention how targets, benchmarks and constraints are set up. The analysis process in this work seems to be hardwired into the approach, which makes it hard to flexibly address a different problem than the one studied by the author. In the context of process industry, Wang (2014) describes how multidisciplinary engineering design is supported by tool integration and interoperability and how this requires a system that combines project and workflow management, dedicated geometry modelling and a mix of optimization, sensitivity analysis and search methods. Ochoa and Capeluto (2015) present an expert system[6] to support the selection of appropriate systems for façade retrofit. Their system includes simple decision rules, such as the exclusion of building integrated PV panels on North facades, and the choice to prioritize energy efficiency or return on investment.

Optimization Tools

Another tool that is often used in the context of performance-based design, either in combination with analysis tools or built into a decision support system, is optimization. In general, optimization is the process of *making the best or most effective use of a situation or resource.* In engineering, this is often done by means of mathematical concepts such as minimalization and maximization (Blanchard and Fabrycky, 2011: 253–302); however Hazelrigg (2012: 41–45) warns against taking a narrow view and only focussing on finding the best solution on the basis of perfect information; instead he suggests to think of optimization as making the best possible decision in any given situation. Wright (2016) takes an even wider view and defines optimization as 'a process of exploration that increases understanding and aids decision-making'. Optimization is a key technique in mathematics; however it is also extensively discussed in the context of decision making; see for instance Jordaan (2005: 220–271) or Hazelrigg (2012: 41–100). Like other techniques, there are different ways to classify optimization approaches. An important classification relates to the search space and discerns between local and global search. As implied by the name, local searches only cover a limited subset of alternatives; global searches aim to cover the full search space. Where the search space is defined by parameters, a local optimization may consider one parameter only, whereas a global search would consider all potential permutations of all parameter values. An examination of all possible alternatives is also known as brute force optimization or exhaustive search; however this may not always be

---

6 An expert system is a software tool that uses a database of expert knowledge. This database can help to provide input to support decisions. Expert systems are used in various domains, including engineering.

computationally feasible as the number of options from system variants and parameter permutations tends to explode to extremely large search spaces (Talbourdet *et al.*, 2013). Where local searches consider single parameters, this is also named 'one-at-a-time' optimization. Another split is between derivative-based and derivative-free optimization. The former approach is only possible if one can develop a formal objective function and use the mathematical process of differentiation to find the sensitivity of that function to change. Derivative-free optimization approaches need to be used where the objective function is too complex for differentiation. A closely related view discerns between deterministic optimalization, where a rigorous mathematical approach defines all variables that are to be considered, and stochastic optimization, where random variables are used to explore the search space. Further terms in optimization are heuristics, which is used to indicate partial search algorithms, and meta-heuristics, which typically describes a higher-level procedure that helps to find appropriate heuristics. There is a wide range of optimization techniques, which can broadly be divided in optimization algorithms, iterative methods, global convergence methods and heuristics. More specific methods are for instance simulated annealing (SA), the simplex algorithm, particle swarm optimization (PSO), ant colony optimization (ACO), graph and tree search algorithms, and hill climbing methods.

A review of optimization algorithms for building design is provided by Machairas *et al.* (2014); Nguyen *et al.* (2014) make a similar contribution. Evins (2013) also reviews optimization approaches, but his work includes an in-depth review of application of optimization to building envelope design, building services engineering, whole building design and retrofit. Wetter and Polak (2005) discuss the details of derivative-free optimization in building simulation. Machairas *et al.* (2014) list a range of optimization software that can be used in conjunction with building analysis tools: modeFrontier, Matlab, GenOpt, MultiOpt, BEopt and others. Nguyen *et al.* (2014) give a comprehensive overview of optimization programs that can be applied to building performance analysis, listing Altair HyperStudy, BEopt, BOSS quattro, DAKOTA, GENE-ARCH, GenOpt, GoSUM, iSIGHT, jEPlus + EA, LIONsolver, Matlab, MOBO, modeFrontier, ModelCenter, MultiOpt2, Opt-E-Plus, ParadisEO and TNROPT and discussing some of the key features of these optimizers. Hamdy *et al.* (2016) review the performance of a number of multi-objective optimization algorithms and in doing so give a good overview of the field. Nguyen *et al.* (2014) consider the optimization process as consisting of three main phases: pre-processing, running the optimization and post-processing. Pre-processing involves formulation of the problem, including the setting up of a simulation model, setting objectives and constraints, defining the design variables, choosing an optimization algorithm and linking that algorithm to the simulation model. The actual optimization process is then automated but still requires monitoring; issues to review are convergence, model termination and dealing with errors. Post-processing requires analysis of the results and may include verification efforts, sensitivity analysis and presentation work.

There is a large body of literature on optimization in the building performance domain, which initially focussed on building services but has expanded to cover a wide range of performance aspects and building systems. While a full overview is beyond the scope of this section, Wright *et al.* (2002) discuss multi-criteria decision making and optimization for the design of a single-zone HVAC system. Wetter (2004) introduces the GenOpt system. Choudhary *et al.* (2005a, 2005b) discuss the iteration between optimizations at different system levels for a workshop design problem and an Air Handling Unit sizing problem for a healthcare facility. Kämpf *et al.* (2010) compare the use of global optimization approaches when applied within EnergyPlus thermal simulation. Eisenhower *et al.* (2012) apply optimization to a metamodel that is trained with detailed building simulation outcomes rather than directly linking an optimization routine to a simulation tool. Talbourdet *et al.* (2013) discuss the use of a genetic algorithm in combination with TRNSYS simulation. Lin and Gerber (2014) use what they call multidisciplinary design optimization to support generative design in a system named Evolutionary Energy Performance Feedback for Design; in principle this work combines a genetic algorithm with multi-criteria decision analysis. Bucking *et al.* (2014) present an approach that clusters parameters according to their predicted influence on overall building energy efficiency. Futrell *et al.* (2015) explore the design of facades during early building design, optimizing for thermal and lighting performance. Hamdy and Sirén (2015) discuss the complexity of optimizing for cost-optimal decisions on energy efficiency levels of building, which requires the exploration of a search space that contains more than 1.6^{10} solutions. Harb *et al.* (2015) use integer linear programming to select optimal heating systems for residential buildings in Germany. Azari *et al.* (2016) combine an artificial neural network and a genetic algorithm to optimize a building envelope in terms of life cycle assessment, exploring combinations of window–wall ratio for the South and North orientation, R-value, insulation material, glazing type and framing material. Menberg *et al.* (2016) contribute a study into the performance of sensitivity methods themselves when applied to transient thermal models.

Uncertainty and Sensitivity Analysis Tools

A related category of tools support the techniques of uncertainty analysis (UA) and sensitivity analysis (SA). Uncertainty analysis studies how variation of the input of an experiment influences the outcomes; this is also known as propagation of uncertainties. Sensitivity analysis works in the opposite direction and aims to establish which of the input parameters have the most impact on the experimental outcomes. Within the realm of building design decision making, well-known work that deals with uncertainty analysis are the theses by de Wit (2001) and Macdonald (2002). Further contributions that include sensitivity analysis are made by Hopfe (2009) and Struck (2012). Various studies at Georgia Tech expanded the field towards the analysis of risk in design; see for instance Hu (2009) on the risk in the design of off-grid solar houses or Sanguinetti (2012) on handling risk in façade retrofit design. The tools used to conduct uncertainty analysis and sensitivity analysis typically include a computational environment such as Matlab or Visual Basic, the dedicated SimLab platform that is specifically geared towards uncertainty and sensitivity analysis, and statistical analysis tools such as the R Environment.

Building Information Modelling

Within the context of tools for performance based building design, the concept of Building Information Modelling (BIM) cannot remain undiscussed. BIM concerns the use of a digital representation of all information that pertains to the building as an object as well as to the building life cycle. This digital approach is changing the way information is exchanged within the industry, the way buildings are designed and constructed and even the way buildings look. BIM has a long history, dating back to the emergence of computer-aided design or CAD systems with the Sketchpad application by Sutherland in 1963 (Eastman, 1999: 35). Initially efforts focussed on building product models to capture information about the building as an object and process models to capture the life cycle aspects. Seminal works on the subject are *Building product models: computer environments supporting design and construction* (Eastman, 1999) and the more recent *BIM Handbook* (Eastman *et al.*, 2011). Building information models are now crucial in representing and visualizing building designs; they also are a key source for input for analysis tools. In theory, BIM could also support the efficient sharing of data between various building performance analysis tools, such as structural and energy analysis (Belsky *et al.*, 2016); however in actual practice this information sharing is still challenging. Part of the problem is that experts are selective in analysing only parts and abstractions of buildings, making fully automated data transfer approaches a cumbersome approach. For instance Welle *et al.* (2012) discuss the decomposition of BIM models for daylighting analysis in terms of determining what spaces should be studied in depth. As stated by the authors, the goal of decomposition is 'to systematically decompose the problem into independent or loosely coupled sub-problems', thus increasing speed, accuracy and scalability of the analysis efforts while reducing costs. Yet BIM is very successful in enabling many tasks that relate to data management, such as automatically generating bills of quantities and producing documentations about a design; see for instance Ilhan and Yaman (2016) or Ciribini *et al.* (2016) for more complex challenges in practice. In the realm of performance analysis, there still is a range of issues. For instance, even for the relative simple domain of rule checking where a building model is subjected to tests that lead to a pass, fail, warning or unknown status, expectations are not yet fully met (Pauwels *et al.*, 2011). Data imports from one tool to the other, for instance from Revit to tools like DesignBuilder and similar, often still require manual intervention by the analyst. Cemesova *et al.* (2015) present an extension of the IFC, which allows linking the IFC model in a BIM environment to the Passive House Planning Package (PHPP) tool in Excel. While this work allows a two-directional information exchange from IFC to PHPP and back, it does not give any specific direction on the information sought other than to get the default PHPP output. Solutions that are being explored are for instance the development of specific model view definitions and semantic enrichment of BIM data (Belsky *et al.*, 2016). Becerik-Gerber and Rice (2010) present an overview of the most popular tasks that are conducted using BIM systems; interestingly, the prime uses are still visualization, building assembly and construction sequencing, with building performance analysis-related tasks such as environmental analysis only ranking 12th in the overview and being used in only 19% of the cases observed.

Tools for Naturalistic Decision Making

Typically the development of tools that support naturalistic decision making is challenging, as the underlying models such as RPD are mainly qualitative and descriptive. Capturing events, sequences and their transitions requires a high level of expertise commensurate with that of the decision makers. However, there are some efforts in this field, using neural networks, pattern matching, rule-based systems and agent technology (Nowroozi *et al.*, 2012). Tools in this field look at situational awareness and should pay specific attention to problem formulation and asking the right questions. Moreover, such systems should fit in with teamwork and knowledge organization; they may capitalize on earlier work in expert systems, artificial intelligence and case-based reasoning (Meso *et al.*, 2002). Stanton and Wong (2010) provide an overview of recent efforts to use computing technology to support naturalistic decision making. Overall they suggest that humans and computers have different strengths and that a joint approach is most promising.

8.4   Performance Visualization and Communication

As noted by INCOSE (2003: 245–246), 'measurement by itself does not control or improve process performance. Measurement results must be provided to decision makers in a manner that provides the needed insight for the right decisions to be made'. This is true whether the measurement data is obtained through physical measurement, simulation, expert judgment or stakeholder evaluation. Providing insight is definitely an issue where it comes to conveying measurement results to others who use these to make decisions. In many cases it is also relevant where the person undertaking the measurements is the same as the person making the decision: raw data often does not directly yield insight and may need to be processed and visualized. Yet it is known that decision makers often face problems of information overload, information being isolated and not giving an overall picture, and information being inconclusive (Jain *et al.*, 2011). Visualization is one way of making sense of large amounts of performance data. Visualization typically is based on a range of underlying technologies, such as geometric modelling, perceptual theory, psychology, computer graphics and image processing (Pilgrim, 2003: 4). Traditionally, visualization is used in the building design and construction sector to represent buildings and building systems; this visualization is crucial for collaborative conceptual design, design evolution and development, marketing and the handling of details during the construction process. However, visualization does not have to be limited to physical objects; it can also include abstract building performance data (Bouchlaghem *et al.*, 2005). This visualization of performance also can help in design development, conveying key information to designers and their clients (Kolarevic, 2005a: 197–198). Effectiveness of visualization or perceptualization techniques depends on the application domain, the stakeholders that interact with the visualizations, and the tasks that they are supposed to support (Prazeres, 2006: v). Visualization needs to account for the user perception and human physiology, especially vision; the human eye typically allows discerning differences in geometry, position, orientation, size, colour, transparency and texture (Pilgrim, 2003: 30–31). Tufte (2001: 177–190) provides general guidelines for visualization of information, based on discussion of many good and bad examples. These include the use of an appropriate format and design; synergy between words, numbers and drawings; attention to achieving a sense of balance, proportion and scale; giving access to detail; basing visualizations on a story or narrative; professional production; and preventing the use of any unnecessary decoration. Many researchers hold the notion that data visualization should be available rapidly and that presentation in terms of simple graphics with headline statements will enhance communication (Gann and Whyte, 2003). Harries *et al.* (2013) mention the use of airflow simulation in the context of conveying a story to other design team members in the design of the London Velodrome; clearly this is way of capturing attention and starting discussions.

The concept of communication becomes important as soon as two or more actors are involved in performance-based design. Communication is a transmission of information, typically between someone who sends the information and others who receive it. In most cases it is interpersonal, meaning that feelings, values and opinions do play a role. Communication requires that (design) information is conveyed well, that all parties involved have the same understanding of the information and that no noise is created during the information transmission, that the message intended by the sender gets across to the receiver, and that the right persons are reached within an organization (Chiu, 2002). As noted by Pati *et al.* (2009), the different stakeholders in the building procurement process all have different backgrounds and expectations. Dialogues and negotiations between these stakeholders are made complex by social factors, the specific power and authority of some actors and the lack of a common vocabulary for discussion of building performance issues. A deeper discussion of communication in construction is provided by Dainty *et al.* (2006). It must be noted that visualization is just one tool for communication. Research into the area of remote collaboration confirms the importance of audio and text communication in combination with visual information (Casera and Kropf, 2010). Collaboration itself, the process of working with others to produce something, goes beyond communication – this also requires negotiation, reflection, decision making and consultation (Chiu, 2002). As with all tools, visualization and discussion about building performance must be used in the right manner; within the product development literature, it is recognized that an extreme focus on aspects such as product quality, reducing lead times and cost cutting also carries the risk of increased communication overheads and excessive design iterations (Joglekar *et al.*, 2001). As discussed by Yin *et al.* (2011), the performance of designers may also be studied directly rather than just the performance of the products that result from the design process. Typical criteria in this context are issues such as capacity for collaboration, meeting the design brief, decision-making ability and capability for innovation. When applied with care, such design team performance measurement may be used to identify team strengths and weaknesses and to improve team communication.

The result from performance measurement is data. Data in itself is nothing more than a set of values, typically presented in a numeric, symbolic or other format. Where data is processed, organized, structured and interpreted, it yields information. In other words, data in itself lacks meaning; data with meaning becomes information (Tomić and Milić, 2013). Information in turn can be converted into knowledge: that what human beings know and how they understand the world. The process of extracting information from a large amount of data is called data mining. This may be supported by visualization and computational techniques such as pattern finding, scatter plots, regression analysis, clustering and tree classification (Morbitzer, 2003: 129–155; Yu *et al.*, 2013). In many fields, information technologies lead to the availability of large amounts of data. Sometimes this amount is seen as becoming too large to handle, leading to the use of the term 'data tsunami'. Visualization of this data is one way to manage and interpret it (Few, 2006). Even where data amounts are manageable, just collecting, processing and analysing data is seldom sufficient. To have an impact, in most cases this needs to be complemented with advocacy messages and actions that aim to influence decisions. In spite of many parties sharing data, there is little evidence of this actually impacting management plans and overall programme efficiency. To have an impact, it is often necessary to adjust the message to the recipients and their expertise level (Chiranov, 2014). Gursel *et al.* (2009) point out that building performance data is typically context dependent and semantically rich; the use of different tools and methods leads to fragmentation and redundancy. They suggest an integral reference model in attempt to better handle and communicate such data.

As said, one way to analyse performance data is through visualization. A distinction can be made between generic data visualization, which may apply to any type of data, and scientific visualization, which concerns the representation of data in a way that allows scientists to understand, interpret and illustrate their data; often the latter pertains to physical data and three-dimensional phenomena (Pilgrim, 2003: 5). An example of scientific visualization is the use of contours in maps that indicate flooding risk and associated uncertainties; Lim *et al.* (2016) discuss variations and how these may impact on decision making in terms of the choice for a building location. On a fundamental level, visualization may be based on a range of dimensions; 1D, 2D and 3D representations are common, but some techniques such as radial graphs and matrix representation allow capturing even further dimensions (Pilgrim, 2003: 32). Surface occlusion may be solved by a range of approaches such as reflection, multiple viewpoints, animation, unfolding, exploding or projected information (*ibid.*, 35). The following types of charts are often used to visualize data. To compare different quantities, bar charts, trend lines and pie charts are useful. An alternative that allows comparison of how one or more options score on a range of aspects in a single figure is the radar graph; this is used in other domains, such as the medical sector (Saary, 2008), but also may be used to convey building performance data such as in Pati *et al.* (2009). Time series plots can be used to visualize a chronological sequence of data, scatter or XY plots to help establish interdependency of two variables, carpet plots to reveal patterns in long time series of single variables and box plots to show statistical distributions. To help the visualization, filtering and clustering of data may also be useful (Neumann and Jacob, 2008: 27–28). Sankey diagrams, which represent both the direction and the quantity of flow processes, may be used to address building performance aspects like the flow of resources such as energy and water; for an example at the level of a University Campus, see Abdelalim *et al.* (2015). There are also options to develop visualizations that meet particular objectives; for instance Lee and Rojas (2014) plot performance on concentric circles, with each circle representing the performance of an activity, with the distance from the centre representing cost and the angle representing time. Only active processes are plotted, allowing activities to appear when started and disappear when completed. Such dedicated visualization shows performance over time and is useful in a process management context. A special kind of visualization is the use of information dashboards (Few, 2006), which aims to capture the most important information for a task in a single screen.

There is a range of tools that may be used to visualize data. Spreadsheets such as Microsoft Excel can be used to process data and create a range of graphic representations, such as bar charts, line graphs, pie charts, scatter plots, radar charts and surface plots. However, in this tool the visualization is limited to the predefined formats; typically additional post-processing and data handling is needed to handle domain-specific tasks such as finding the warmest hour in a 24 hour series (Pilgrim, 2003: 19).

Other more advanced environments are for instance Matlab, which is frequently used to plot histograms. Some analysis tools also have predefined visualization back ends; for instance, DesignBuilder produces standard temperature time series. It is important to keep in mind that many tools also have inherent limitations in the amount of data they can handle, especially where one works with big data. Even for tools and models where data visualization is a key objective, such as BIM, computing requirements may require careful considerations. For instance, it may be worth to explicitly cull elements from the visualization that are outside the view, on the backside of an object or occluded by other objects. Johansson *et al.* (2015) provide a good overview of BIM viewers and issues that play in rendering building images. While most visualizations are either single static images or at best a predefined sequence (movie), Packham (2003) introduces the concept of interactive visualization for engineering design. In his system the computer analyses and clusters data, but users can direct further refinements and zoom in on high performance regions that are of interest in some way. An application of this system, showing human-led exploration of solutions for the structural design of masonry wall panels, is described by Rafiq and Sui (2010); de Wilde *et al.* (2008) have combined the same visualization tool with transient thermal simulation to explore the uncertainties in predicting the impact of climate change on the performance of a typical terraced residential building in the United Kingdom.

Building performance visualization efforts are often integrated with the development of measurement and analysis tools. Simulation engines like EnergyPlus typically produce large data files that can be picked up for post-processing by the likes of OpenStudio, DesignBuilder and Sefaira, which all can produce time series of temperatures, load profiles and a range of other visuals; other tools such as TRNSYS, WUFI and Radiance have built-in capabilities. Seminal work that attempted to visualize building performance across a range of domains was carried out by the Energy Systems Research Unit at the University of Strathclyde, which developed the Integrated Performance View or IPV (Morbitzer, 2003: 123–168) and subsequent I^2PV (Prazeres, 2006: 135–209). The idea behind the IPV is to 'collate a range of performance indicators to represent the multi-variant performance of a building' (Prazeres, 2006: 20); see Figure 8.3.

In the area of building measurement and monitoring, the concept of an energy dashboard has gained popularity. Examples are the NREL campus energy dashboard, Honeywell Energy Awareness Dashboard, Siemens Building Technologies Navigator system, Energy Efficiency Education Dashboard, Vabi Building Performance Dashboard and many others. Few (2006: 164–172) gives some general recommendations for good dashboard design, including the need for proper organization of information, consistency for quick and accurate interpretation, the support of meaningful comparisons while discouraging meaningless ones, the application of an aesthetically pleasing design with appropriate colours, resolution and text and providing access to underlying data; Few also stresses the need for proper testing of any new dashboard.

Decision makers at executive level in industry have to deal with information inundation, information isolation and information indecision. In other words, they receive too much information, sources of information are isolated and there is no guidance on how this information might be used to improve performance (Jain *et al.*, 2011). High-level decisions on significant commitment of resources are typically made in the boardroom. Where building performance analysis results are to be used to inform decision making at this level, for instance to get support for a new building development, it is even more

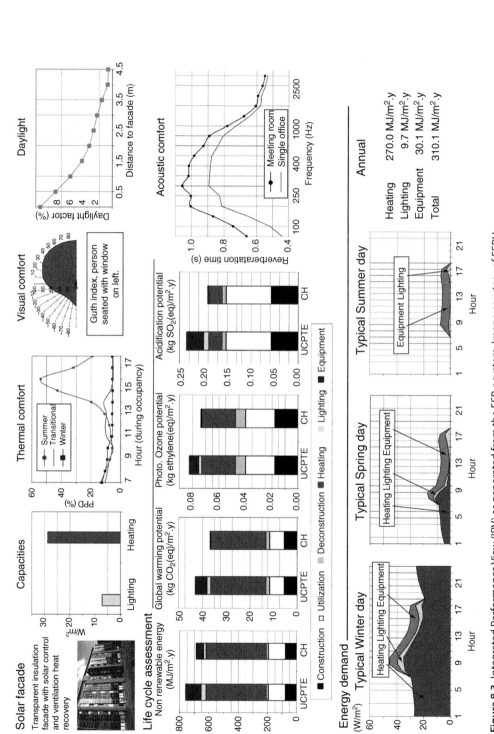

Figure 8.3 Integrated Performance View (IPV) as generated from the ESP-r system. Image courtesy of ESRU.

important that findings are visualized and communicated in an effective manner. Board members typically expect to have access to underlying data in high quality reports (Hamza and de Wilde, 2014). This is based on the fact that decision makers at this level know that there are problems with reports, since data in reports has no meaning unless it is interpreted. As interpretation is a human task, this activity risks issues with the quality of the derived information caused by data volume and presentation and user factors such as concentration, previous knowledge, experience, bias and others. Good executives will try to query these issues. Further problems for executives are 'avalanches of data' and 'being unable to get the information wanted'. Visualization is one way to make data accessible to these decision makers; another might be the use of automated data interpretation using expert systems and domain knowledge (Tomić and Milić, 2013). However, a key concern is that any interpretation can be explained and is open for discussion. This follows the trend that in business, KPI data is often presented using visualization by various charts (time series, pie and bar), balanced scorecards and information dashboards. Sometimes these visualizations involve dials and thermometers, as well as drill-down facility to access and if needed present less summarized data (Tomić and Milić, 2013).

A different level of visualization of building performance can be achieved by immersive building simulation, where humans can interact with a computer-generated environment. Immersive simulation is rooted in augmented reality or virtual reality (VR). A good introduction to this emerging area is provided by Malkawi (2003). Virtual Reality is not just a visualization of the 3D environment or of building performance, but comes with immersion and interactivity: the sensation of being within that environment, the ability to move around in that environment and the capability to manipulate objects (Dunston *et al.*, 2010). VR systems can be linked with BIM to develop virtual prototypes or virtual mock-ups. Such a VR environment would for instance be useful in healthcare facility design, a field that currently uses physical mock-ups. It is noted that the use of physical mock-ups in design is rarely described in the academic literature (*ibid.*)

8.5 Construction for Performance

However good the design, ultimately the performance of an actual building also relates to how well that design is executed and realized. As pointed out by Ahire and Dreyfus (2000), it is important to pay equal attention to both design and process management, as the quality of products depends on both. The building as an artefact is created through a complex construction process, which involves many actors and a large number of activities. The general field that aims to direct this process in the best possible manner is construction management. Within this domain, quality management is a well-covered field, with a range of textbooks covering the subject. See for instance Watson and Howard (2011), who discuss key theories on quality, project and corporate performance, quality management models and management systems, learning in organizations and health and safety aspects. Quality of construction is closely related to the quality provided by the suppliers of materials, assemblies and equipment that goes into buildings. For a review of supplier quality management, see AlMaian *et al.* (2016), who stress the importance of work observation, inspections, effort tracking and testing.

One route towards improving various quality aspects of construction is a move towards increased off-site construction, where building systems are becoming more modular and can be prefabricated in controlled factory conditions rather than on the typical construction site under outdoor conditions (Marjaba and Chidiac, 2016).

A general approach for quality management, applicable to most disciplines, is given by the systems enshrined in ISO 9000 and ISO 9001. ISO 9000 provides the fundamentals of quality management (customer focus, leadership, involvement of people, process management, systems view, continuous improvement, effective decision making and synergy), whereas ISO 9001 spells out the requirements for meeting the standard and becoming certified. Also frequently mentioned is the ISO 14000 standard, which covers environmental management and thus relates to the energy and water use of buildings, as well as emissions. The application of ISO 9001 to construction is discussed by for instance Pheng and Teo (2003) or Lau and Tang (2009).

Unfortunately the construction industry is known for issues with time and cost overruns and with buildings often having defects on completion (Harrington *et al.*, 2012). Tam *et al.* (2006) use non-compliance records, complaints and warnings and fines and penalties as 'performance indicator' for regulatory compliance. However, lack of defects is a prerequisite for performance, not a guarantee. It has been suggested that too much attention has been paid on measuring and improving the process; yet as a consequence the quality of the products has been brought into question (Gann and Whyte, 2003). Different approaches have been developed that aim to deal with these issues, such as the application of Total Quality Management (TQM), Lean Manufacturing, Six Sigma and others; these are well known from for instance the automotive sector. However, these methods seem to gain only limited traction in the building industry. Yet, albeit slowly, construction methods change over time; for instance fast-track construction has gone from a research topic to a mainstream method over the last 15 years. This has significant consequences for the actors and responsibility for quality assurance, with shifting roles (Minchin *et al.*, 2010; Kraft and Molenaar, 2014).

Lizarralde *et al.* (2011) discuss the fact that construction is not carried out by a permanent team, but by a temporary multi-organization where different actors have different skills, interest and perceptions; their work goes on to explore how procurement is influenced by both team-internal and team-external pressure groups. This in turn makes it hard to manage project teams in the construction sector. Moreover, the project management in the sector often has a specific type of decision making, which reviews project progress on a daily basis; key issues are whether progress towards the objectives has been made, what developments may be anticipated and how actions and events may help steer the project to ensure it is completed in time, on budget and with the required quality (Marques *et al.*, 2010). Proper planning is key aspect of ensuring the process performance of construction work; without this, it becomes much harder to meet targets in terms of time, cost, quality and safety. Proper planning prevents tensions between tasks that should be done and tasks that actually can be done (Hamzeh *et al.*, 2015).

In a wider sense, the performance of construction processes can be measured in terms of cost, time, quality, safety, waste reduction and customer satisfaction. Ling (2004) contributes a correlation analysis between 60 explanatory factors and a wider range of 10 project performance metrics such as unit cost, construction speed, workmanship quality and client satisfaction. Explanatory factors taken into account include gross floor area, type of building, form of contract, ownership, bidding procedure and others; the study was based on data of 65 projects. Xiao and Proverbs (2002) have studied customer satisfaction with contractors in Japan, the United Kingdom and the United States. They not only report similar levels of satisfaction but also found fewer defects, longer liability periods and fewer recalls of contractors in Japan. Hoonakker *et al.* (2010) discuss the view of contractors towards quality issues in construction.

There are a number of reasons for the performance challenges faced by the building construction process. Key issues are the low profit margins in construction, which leads to high pressure value engineering and change orders. Often savings are made on systems, since the overall building form and geometry is hard to change and offers fewer options. Another area is workmanship, where contractors may elect to work with cheap labour. The recruitment and selection process of workers in construction is a complex area in itself. Since construction is site specific and project based, staffing has particular dimensions and does not map well to recruitment processes in other industries; see for instance Lockyer and Scholarios (2007) who report on the situation in Scotland. At a basic level, building control is the mechanism for ensuring that buildings meet the requirements of legislation. Unfortunately this also remains a challenging area, especially where the building legislation is complex and performance based (van der Heijden *et al.*, 2007); in this case there often is a lack of enforcement to ensure that construction actually follows the design that has been approved.

Where it comes to building performance, in many cases 'quality' of construction is still measured in terms of number of defects rather than in terms of meeting performance requirements. Numbers of defects often are significant; for the United Kingdom, Sommerville and McCosh (2006) report a peak value of 389 defects in a single domestic property. There are different types of defects: those where a building or building system does not function properly; those where systems or parts are omitted; and those where some part of the building is damaged upon handover (Sommerville and McCosh, 2006). In the United Kingdom the process of checking a building for minor faults is sometimes named 'snagging' and the defects that are found 'snags'. Other terms for these defects are rework, quality failure and repairs. Typically these are identified by way of visual inspection during a walk around the building. The resulting workload and administration is seen as being problematic, although ICT-based technologies are helping to ease the burden (Sommerville *et al.*, 2004) and allow to keep better track of defect positions and attributes (Laofor and Peansupap, 2012). Boukamp and Akinci (2007) discuss how product models, process models and data collected from a construction site may be compared to identify deviations and exceptions, which if severe enough may be escalated to defects. Some defects are, unfortunately, not visible and need more advanced techniques for their identification. For a limited number of construction elements, such as pipelines and ducts, it may be possible to use autonomous robotic vehicles to gain access, whereas most defects in these systems can be anticipated and hence can be detected automatically; see for instance Guo *et al.* (2009). For others, such as underground structures, use of advanced techniques such as ground-penetrating radar may be required (Qin *et al.*, 2016). In another context of invisibility, pressure waves, infrared thermography, X-ray or acoustic laser and imaging techniques may be used to detect problems with bonding in fibre-reinforced bonded concrete systems (Yu *et al.*, 2016). An approach to detect invisible thermal defects is the use of a thermographic camera (Fox *et al.*, 2015, 2016).

On a more positive note, technological developments are slowly making an impact on construction management. For instance, new techniques are making it easier to deliver made-to-measure building components; by way of example, Larsen *et al.* (2011) describe the complex workflow that is available for producing and installing prefabricated façade elements in a retrofit context. A key challenge here is capturing the right dimensions, for which tacheometry, photogrammetry and 3D laser scanning may now be used, helping to increase accuracy over manual measurements. BIM is also seen as a force that may help to improve quality in construction. It may do so by eliminating conflicts through clash detection, thus reducing the need for later rework. BIM typically creates more consistent design data; it can also be used to check for quality compliance (Chen and Luo, 2014). Wang (2008) discusses the use of Radio Frequency Identification (RFID) in the context of construction quality inspection, with a special focus on material tracking, testing and management. New sensing and image recognition techniques also help to track what happens during the construction process (Brilakis *et al.*, 2006, 2011; Dimitrov and Golparvar-Fard, 2014; Teizer, 2015).

The phase of building handover, which completes the construction, is commonly known as commissioning. Commissioning, as well as the variants of retrocommissioning and recommissioning, was already discussed in Section 2.1 of this book. Building commissioning focusses on making sure that all building systems are correctly installed and that these operate as planned in the design stage (Gursel *et al.*, 2009). However, a proper process that goes beyond checking that a building has been completed and actually performs as expected requires resources in terms of time and budget (Noye *et al.*, 2016). Yet Hinge and Winston (2009) express concern that building commissioning is dominated by checklists and misses out on third-party verification. One may discern different levels of commissioning, which answer questions that range from establishing whether systems are there, whether these systems work and are connected, to whether they deliver the required functions and are optimized. Even if all levels are covered, building performance has a tendency to degrade over time; this can be countered by a process called 'continuous commissioning', which involves ongoing measurement, analysis and intervention where needed (Neumann and Jacob, 2008: 1). Djuric and Novakovic (2009) go even further and suggest a need for lifetime commissioning. Potts and Wall (2002) discuss the need to embed commissioning in the complete building cycle, right from the initial feasibility study of a project, noting that clients may not be familiar with the process and recommending that one actor takes specific responsibility for commissioning. POE is often part of commissioning efforts and typically takes place during the first year of building occupancy; it studies the proper operation of building services and control system as well as training of key staff, updating of building information and ensures learning and feedback between the client and the design team (Jaunzens *et al.*, 2003). In the context of continuous commissioning, building performance analysis can be viewed as a four-step process: (i) benchmarking or operational rating, (ii) certification or asset rating, (iii) optimization and (iv) regular inspection (Neumann and Jacob, 2008: 1).

8.6 Design and Construction for Performance Challenges

A good example of the challenges of performance-based design in practice is the winning entry of Team Germany for the Solar Decathlon 2009. The Solar Decathlon is a competition by the US Department of Energy, where teams of university students design and build a solar house. These houses are then brought together at a single location – originally the National Mall in Washington, DC – where they compete in the 10 categories that explain the Decathlon name: architecture, market potential, engineering, communications, innovation, water, health and comfort, appliances, home life and energy. The competition has detailed rules and regulations that in fact provide an excellent statement of requirements to underpin the design process. Beyond the rules and regulations of the competition, the Solar Decathlon also has a specific building code that protects public health and ensures safety, in line with the International Residential Code and the National Electrical Code that apply to domestic construction in the United States. Upon completion of each house ready for the competition, a third party installs instrumentation and monitoring equipment that captures lighting levels, indoor temperature, relative humidity and electricity usage.

The first Solar Decathlon took place in 2002, followed by a second event in 2005 and a third in 2007; since then the competition has taken place every 2 years. International expansion has established separate Solar Decathlons in Europe (since 2010), China (since 2013), Latin America and the Caribbean (since 2015) and the Middle East (to start in 2018). The 2009 competition involved 20 entries including that of the Technische Universität Darmstadt, which was branded Team Germany; the building was named surPLUShome; see Figure 8.4.

From a process point of view, the design of the surPLUShome started with individual designs by all 16 architecture students that were part of the team. After an internal jury round by internal and external assessors, students teamed up in 8 teams that developed good ideas in more detail, providing more depth to architectural and energy properties. A second jury round decided on three projects going forward. Again, these three

Figure 8.4 SurPLUShome by Team Germany, winning entry to the Solar Decathlon 2009. Image © Jim Tetro, courtesy of US Department of Energy.

projects were developed; a third and final jury then decided on the ultimate concept design. This concept then evolved through a series of stages. Stage one consisted of workshops that established spatial arrangements, relations to the outside world and the building image, sculptural design and exploration of a timber frame structure. Stage two explored landscape design, navigation, dynamic space configurations and flexible furniture. Stage three developed a 1:20 scale model for discussion with solar firms. Stage four presented the design to the public at an exhibition in the city of Munich. From there the design was frozen and only modified to adapt to availability of specific products. The final design is described in a detailed set of drawings that is 429 pages long; the building comes with a project manual of 126 pages.

Conceptually, the actual design of the surPLUShome as realized started with a focus on the façade. This was combined with the concept of a single, multifunctional room, making the house a two-storey cube. The surface of this cube is a solar envelope, which contains a 11.1 kW photovoltaic system with crystalline panels on the roof and an additional 250 thin-film solar cells in the façade; these thin-film cells are copper indium gallium diselenide (CIGS) panels. Together these systems produce 200% of the electricity required by the building itself. Automated louvre systems are employed to prevent overheating. The fabric of the building is built to Passivhaus standards; the façade includes vacuum insulation panels. Internally the building employs phase change materials to maintain a comfortable indoor temperature. The building also has a reversible electric heat pump system for the provision of heating, cooling and domestic hot water.

The description of the design process, as provided by the team, shows a clear pathway of evolution and integration of ideas of various participants. However, the key decision moments were driven by jury decisions, basing themselves on expert opinion. Hidden in the design process are a few other decisions that were crucial for the performance of the final design. One of these is the choice to maximize the area of the building envelope in order to create a maximum area to install PV. Similarly a key decision was the choice to risk putting the roof panels at an angle of zero degrees in anticipation of overcast, rainy weather during the competition, instead of using an optimal tilt towards the sun. Other decisions were forced upon the team, for instance by the availability of technology. An example is the façade; one original idea was to weave the building skin from flexible photovoltaic strips, but such a material was not available. Hence the decision was made to apply the traditional principle of shingles. A company was found that could manufacture smaller panels than usually used for PV, allowing a blend of the traditional approach with modern materials. One advantage of the principle of shingles is a slight tilt, which helps to get a better orientation towards the sun.

After inception and testing in Germany, the surPLUShome was disassembled to be transported to Washington, DC, for the competition. The transport took about 4 weeks. After the competition, the house was returned to Germany and put on permanent display on the campus of the TU Darmstadt.

8.7 Reflections on Designing for Performance

Design of buildings that perform well, or performance-based design, has been studied by many authors. However, giving guidance on how to do this type of design well is extremely difficult. Many approaches only give generic directions, lacking specifics.

Yet at the same time one needs to be aware that putting the design in a performance-based straightjacket does not work either. But there are subprocesses that have their own process logic and that can be improved by making sure the design teams follow a proper procedure. Processes that have been demonstrated to work may even be automated to some degree.

The importance of early design stages seems obvious, but using this knowledge to do things differently is hard. Time is not an unlimited resource, and shifting some tasks to the front of the process (prioritization of specific activities in the early design stage) means that other tasks will be delayed, unless extra resources and time are made available. The process of early design is not fully understood, nor are the constraints that govern what happens in this stage.

Integrated design, which brings together all expertise and considers all relevant aspects, is another approach that appears to make sense. But this also requires additional resources to be made available to enable workshops and meetings that span across all disciplines and issues. It is unclear where these resources will come from and whether there are indeed sufficient efficiency gains downstream to cover this with just a time shift of efforts.

Research on performance-based design seems to have only limited impact. There are many reports in the literature about new ICT tools, Design Decision Support Systems, early design tools and integrated environments; however most of these have remained at a proof-of-concept stage. A key problem with many of these tools seems to be a lack of situational awareness – they provide solutions to preconceived design problems, but since design processes are highly unique, these design problems may not be relevant in a given actual context. It is hard to map actual search spaces, design constraints and user knowledge. Without this, it seems dubious that tools can replace expertise and be handled by people that lack proper training and knowledge, or that tools can identify optimal design solutions that are indeed acceptable to the specific design project at hand.

A key assumption in many efforts that aim to support design is that one can identify 'design decision moments' within the design process that require information to make a rational choice. There are some decision moments that seem highly realistic, such as the choice of a facade retrofit solution for an existing building. However, in other cases decisions may not be as clear-cut and may in fact arise from a chain of small steps and choices. Further empirical research on decision making in design practice is required to get a better understanding on this issue and to help match support systems to the actual design processes used in practice. This issue relates to the debate about normative decision making versus naturalistic decision making.

When reviewing design for performance, the tensions between design practice and academia re-emerge. Actors in industry own the actual processes but often repeat the same procedures and use well-known strategies, lacking innovation. Academics may suggest novel ways to do things but often fail to appreciate the pressures and complexities of actual design practice. There are insufficient studies that provide hard empirical data on design process logic and events, while there are too many that stay with generic assumptions and high-level theories. One thing that seems to be overlooked in the scientific literature is the fact that individual building design projects typically are part of a chain of processes from the point of view of the actors; it seems there is insufficient attention on how one project may influence the next or how lessons are carried over along the project chain.

Making good design decisions and designing buildings that will perform well is one thing; however getting executives to accept proposals is a separate task. Communication and visualization play an important role in getting buy-in. Beyond this there are many stories about good designs falling down due to excessive value engineering. Construction remains a separate game, with emphasis on preventing defect as stage one towards performance. Communication with clients and other stakeholders therefore remains of great importance and should not be underestimated.

8.8 Summary

This chapter deals with the process of creating buildings that performs well. It discusses the general process of performance-based design, including integrated (integral) design. From there, it hones in on key design decisions and covers decision-making theory and optimization. A range of tools that aim to support performance-based design are explored. Furthermore, the chapter briefly explores visualization, communication and some key aspects of actual performance that are crucial to achieve the performance goals.

Performance-based design is a topic that gets significant attention but remains hard to pin down. While there is consensus about the importance of design in achieving buildings that perform well, opinions on how to do this range from a view that favours systematic work and analytic approaches to a flexible view where actors rely on knowledge and experience. To some extent this matches more general design theories, which can be categorized as normative (theories on how design ought to take place), descriptive (theories about how design takes place in practice) and design as an art (theories that refute prescription). There are various general descriptions of the development of design freedom and knowledge about the design, cost and performance of a building as the design progresses; while these are useful in stressing the importance of early or conceptual design, detailed trends should only be accepted as indicative and not taken as actual depictions of real processes. Similarly, process maps of the design process should also be handled with care. At a high level, phases as provided by the RIBA Plan of Work are useful to structure and measure general progress. At a lower level however, most design processes are highly individual, and prescriptive roadmaps dubious. However, at yet another level, there are logic sequences of tasks and activities that may be captured in 'swimming lanes' and supported by workflow management systems. Within this context, some authors suggest enacting process logic as a way to ensure building performance; others emphasize the need for proper training or collaboration in the setting of an integrated design process.

One way to connect rational analysis procedures with design is through a focus on design decision moments. This approach isolates key points that can be scrutinized in terms of performance, while the rest of the design process is left free. Key design decisions that would benefit from this approach are the selection of systems, components and options; choices relating to location, orientation and geometry; decisions about system and component sizing; and retrofit decisions. Related choices are the selection of performance criteria, building operation conditions, choice of quantification tools and methods, decisions about certification and design team aspects. Beyond this, one may also raise a number of design questions that would warrant performance analysis, such as questions about meeting regulations, or areas that would benefit from improvement. In some cases decision gates can be built into the design process to ensure performance analysis takes place. Normative decision theory can be used to structure design decisions: it prescribes how decisions ought to be made. Key concepts in normative decision making are value and utility; mathematically the best foundation for decision making is provided by multi attribute utility (MAUT); see Hazelrigg (2012: 493–511) for a critical discussion of some other popular techniques. However, one may also relate to naturalistic decision theory, which emphasizes the use of expertise in decision making, especially where there are chains of decisions; in this context situational awareness is an important issue.

There is a wide range of tools that aim to support performance-based design. This includes, but is not limited to, the building performance analysis tools as discussed in Chapter 6. Others include creative tools that help designers express their ideas, generative tools that help to develop an option space, rating and certification systems that encourage good performance, dedicated design decision support systems, optimization tools, uncertainty and sensitivity analysis tools, building information modelling tools and tools that help to increase situational awareness.

Visualization not only plays an important role in communicating performance information to others but is also a tool in itself to make sense of large amounts of data. Visualization may range from simple time series graphs and scatter plots to advanced immersive visualizations where people can interact with a virtual environment. What visualization is appropriate depends on the use; in some cases it may be used to convey information to executive decision makers. Key visualizations are the Integrated Performance View (Morbitzer, 2003: 123–168; Prazeres, 2006: 135–209) and the more recent range of building performance dashboards.

It is important to keep in mind that building performance depends not only on the building design but also on the construction process that allows these buildings to actually materialize. The domain of construction management is concerned with controlling cost, time, quality, health and safety, waste and customer satisfaction. Quality management systems like ISO 9000 and ISO 14000 are often used in this context. However, current approaches often leave significant defects in newly completed projects, although new (and often digital) approaches are improving the situation. Further improvements can be achieved by a proper commissioning process, which ensures that buildings perform as intended once handed over to the client and operational.

Recommended Further Reading:

- *Fundamentals of decision making for engineering design and systems engineering* (Hazelrigg, 2012) for a solid introduction to normative models and utility theory; note appendix C on non-rigorous methods.
- *Decision making in action: models and methods* (Klein *et al.*, 1993) for the seminal work on naturalistic decision making.
- *The visual display of quantitative information* (Tufte, 2001) for an overview of many of the factors that need consideration when presenting performance information to others.

Activities:

1 Undertake a design process observation as independent, non-participating researcher. Try to identify any design decisions made while you are observing. While you do undertake the observation, make a list of additional information you would like to have from the actors in the process; add this information through interviews after the observation is completed.

2 Carry out an actual design task, preferably in the context of a marked academic assignment or, even better, real professional practice. Reflect on the decisions you are making while carrying out the task and the role of performance in those decisions.

3 Consider a design process where a choice has to be made between double or triple glazing. Set up a normative multi-criteria decision and gather the required information to support a rational choice.

4 Conduct interviews with both architects and engineers, and explore whether they use normative or naturalistic decision making in their daily practice. Are there significant differences between the responses of these two types of professionals?

5 Define three chains of activities (swimming lanes) that may be part of various design processes that each respond to process logic: one regarding a search for product data, one about demonstrating compliance with building regulations, and one on reporting to the client.

6 Gain access to a real construction project. Explore what quality assurance system is in place for this project, what is being done to find defects during construction, and what remedial actions are taken. Critically reflect on the sources of information you may access.

8.9 Key References

Andersen, M., J. Gagne and S. Kleindienst, 2013. Interactive expert support for early stage full-year daylighting design: a user's perspective on Lightsolve. *Automation in Construction*, 35, 338–352.

Augenbroe, G., 2011. The role of simulation in performance based building. In: Hensen, J. and R. Lamberts, eds., *Building performance simulation for design and operation*. Abingdon: Spon Press.

Bakens, W., G. Foliente and M. Jasuja, 2005. Engaging stakeholders in performance-based building: lessons from the Performance-Based Building (PeBBu) Network. *Building Research & Information*, 33 (2), 149–158.

Becerik-Gerber, B. and S. Rice, 2010. The perceived value of building information modeling in the U.S. building industry. *Journal of Information Technology in Construction*, 15, 185–201.

Becker, R., 2008. Fundamentals of performance-based building design. *Building Simulation*, 1 (4), 356–371.

Bleil de Souza, C., 2012. Contrasting paradigms of design thinking: the building thermal simulation tool user vs the building designer. *Automation in Construction*, 22, 112–122.

Bleil de Souza, C., 2013. Studies into the use of building thermal physics to inform design decision making. *Automation in Construction*, 30, 81–93.

Clarke, J., 2001. *Energy simulation in building design*. Oxford: Butterworth-Heinemann, 2nd edition.

Cross, N., 2011. *Design thinking: understanding how designers think and work*. London: Bloomsbury.

Eastman, C., P. Teicholz, R. Sacks and K. Liston, 2011. *BIM Handbook – A guide to building information modeling*. Hoboken, NJ: Wiley, 2nd edition.

Evins, R., 2013. A review of computational optimisation methods applied to sustainable building design. *Renewable and Sustainable Energy Reviews*, 22, 230–245.

Frankel, M., J. Edelson and R. Colker, 2015. *Getting to outcome-based building performance: report from a Seattle Summit on performance outcomes*. Vancouver, WA/Washington, DC: New Buildings Institute/National Institute of Building Sciences.

French, S., 1988. *Decision theory: an introduction to the mathematics of rationality*. Chichester, UK: Ellis Horwood/Halsted Press.

Gilb, T., 2005. *Competitive engineering: a handbook for systems engineering, requirements engineering, and software engineering using planguage*. Oxford: Butterworth-Heinemann.

Glicksman, L., 2008. Energy efficiency in the built environment. *Physics Today*, 61 (7), 35–40.

Hansen, H., 2007. Sensitivity analysis as a methodological approach to the development of design strategies for environmentally sustainable buildings. Ph.D. Thesis. Aalborg: Aalborg University.

Hazelrigg, G.A., 2012. Fundamentals of decision making for engineering design and systems engineering. http://www.engineeringdecisionmaking.com/ [Accessed 21 December 2017].

Heer, T. and R. Wörzberger, 2011. Support for modeling and monitoring of engineering design processes. *Computers and Chemical Engineering*, 35 (4), 709–723.

Hensen, J. and R. Lamberts, eds., 2011. *Building performance simulation for design and operation*. Abingdon: Spon Press.

Jasuja, M., ed., 2005. *PeBBU Final Report - Performance Based Building thematic network 2001–2005*. Rotterdam: CIB.

Jato-Espino, D., E. Castillo-Lopez, J. Rodriguez-Hernandez and J. Canteras-Jordana, 2014. A review of application of multi-criteria decision making methods in construction. *Automation in Construction*, 45, 151–162.

Jordaan, I., 2005. *Decisions under uncertainty: probabilistic analysis for engineering decisions*. Cambridge: Cambridge University Press.

Kalay, Y., 1999. Performance-based design. *Automation in Construction*, 8 (4), 395–409.

Keeney, R. and H. Raiffa, 1993. *Decisions with multiple objectives – preferences and value tradeoffs*. Cambridge: Cambridge University Press.

Klein, G., J. Orasanu, R. Calderwood and C. Zsambok, eds., 1993. *Decision making in action: models and methods*. Norwood, OH: Ablex Publishing Corporation.

Lawson, B., 2005. *How Designers Think: the design process demystified*. London: Architectural Press and Routledge, 4th edition.

Lin, S. and D. Gerber, 2014. Designing-in performance: a framework for evolutionary energy performance feedback in early stage design. *Automation in Construction*, 38, 59–73.

Macdonald, I.A., 2002. Quantifying the Effects of Uncertainty in Building Simulation. Ph.D. Thesis. Glasgow: University of Strathclyde.

McElroy, L., 2009. Embedding Integrated Building Performance Assessment in Design Practice. Ph.D. Thesis. Glasgow: University of Strathclyde.

Montgomery, D.C., 2013. *Design and analysis of experiments*. Hoboken, NJ: Wiley, 8th edition.

Morbitzer, C., 2003. Towards the Integration of Simulation into the Building Design Process. Ph.D. Thesis. Glasgow: University of Strathclyde.

Negendahl, K., 2015. Building performance in the early design stage: an introduction to integrated dynamic models. *Automation in Construction*, 54, 39–53.

Nguyen, A., S. Reiter and P. Rigo, 2014. A review on simulation-based optimization methods applied to building performance analysis. *Applied Energy*, 113, 1043–1058.

Oxman, R., 2006. Theory and design in the first digital age. *Design Studies*, 27 (3), 229–265.

Pérez-Lombard, L., J. Ortiz, R. González and I. Maestre, 2009. A review of benchmarking, rating and labelling concepts with the framework of building energy certification schemes. *Energy and Buildings*, 41 (3), 272–278.

Raphael, B. and I. Smith, 2003. *Fundamentals of Computer-Aided Engineering*. Chichester, UK: Wiley.

Rezaee, R., J. Brown, G. Augenbroe and J. Kim, 2015. Assessment of uncertainty and confidence in building design exploration. *Artificial Intelligence for Engineering Design, Analysis and Manufacturing*, 29 (4), 429–441.

Schade, J., T. Olofsson and M. Schreyer, 2011. Decision-making in a model-based design process. *Construction Management and Economics*, 29 (4), 371–382.

Simon, H.A., 1996. *The sciences of the artificial*. Cambridge, MA: MIT Press, 3rd edition.

Stempfle, J. and P. Badke-Schaub, 2002. Thinking in design teams – an analysis of team communication. *Design Studies*, 23 (5), 473–496.

Struck, C., P. de Wilde, C. Hopfe and J. Hensen, 2009. An investigation of the option space in conceptual building design for advanced building simulation. *Advanced Engineering Informatics*, 23 (4), 386–395.

Tufte, E., 2001. *The visual display of quantitative information*. Cheshire: Graphics Press.

Watson, P. and T. Howard, 2011. *Construction Quality Management: principles and practice*. Abingdon: Spon Press.

de Wilde, P., 2004. Computational Support for the Selection of Energy Saving Building Components. Ph.D. Thesis. Delft: Delft University Press.

de Wit, S., 2001. Uncertainty in predictions of thermal comfort in buildings. Ph.D. Thesis. Delft: Delft University Press.

Yin, Y., S. Qin and R. Holland, 2011. Development of a design performance measurement matrix for improving collaborative design during a design process. *International Journal of Productivity and Performance Management*, 60 (2), 152–184.

Zapata-Lancaster, G. and C. Tweed, 2016. Tools for low-energy building design: an exploratory study of the design process in action. *Architectural Engineering and Design Management*, 12 (4), 279–295.

Zapata-Poveda, G. and C. Tweed, 2014. Official and informal tools to embed performance in the design of low carbon buildings. An ethnographic study in England and Wales. *Automation in Construction*, 37, 38–47.

9

Building Operation, Control and Management

While the previous chapter explored the implementation of building performance theory in the design stage, this chapter focuses on building operation and management. Ultimately, even if a building is designed well and constructed according to plan, how good it performs in practice and under actual use conditions still depends on the interaction between the building, the occupants and a broad range of external factors. Part of this interaction is based on automated responses by the building, often governed by a central building management system (BMS). However, systems may also age and malfunction, and control settings may drift or become inappropriate over time; it therefore is useful to monitor building performance, detect anomalies, explore what causes them and intervene to bring building performance back to the intended levels. Traditionally, this was mostly handled in-house by the owner or local estates department; a new trend is to outsource some of these tasks under (energy) performance contracts.

Building management and control covers a wide range of systems and human activities, which ensure that a building does what it is supposed to be doing. To achieve this, buildings are equipped with a range of sensors that capture the building state and behaviour, control algorithms that evaluate this status and behaviour in order to decide on actions that need to be taken, and actuators that help to enact these control actions. The sensor data collected by building systems and control signals generated can be logged with a time stamp, leading into the domain of building monitoring. The boundaries between control and monitoring systems are not always clear; for instance both approaches may have an interest in observing the development of room temperatures over time. However, it is important to keep in the mind that the prime objective of a control system is to manage the building system; as a consequence, control systems are not typically designed with monitoring in mind. A broader term that covers both aspects is performance tracking. Performance tracking is not just a matter of technical data capture; to find out how the building performs requires an observation of the interaction between people, processes and building behaviour. For it to be effective, people need to buy into the practice of performance tracking; these people may range across various organizational levels, from the service room operators to the senior management. Processes need to be structured so that activities are continuous and followed up where needed; this also requires proper resources and accountability, as well as incentives. Performance monitoring provides the basic information needed to make performance tracking operational (California Commissioning Collaborative, 2011: 9–11). The people who manage buildings (estate departments, control engineers, facility managers) are also

Building Performance Analysis, First Edition. Pieter de Wilde.
© 2018 John Wiley & Sons Ltd. Published 2018 by John Wiley & Sons Ltd.

concerned with building system maintenance. Maintenance has five main objectives: to ensure (i) system functionality, (ii) achieving the system design life, (iii) plant and environmental safety, (iv) cost effectiveness and (v) effective use of resources (Muchiri *et al.*, 2011). Monitoring is an important tool for saving resources; Ahmed *et al.* (2010) suggest that close monitoring and control may lead to energy savings in the order of 10%–40% in modern, commercial buildings. In terms of building quality, monitoring of performance levels allows the prediction of unacceptable levels before they occur; this prediction is a key aspect of preventive maintenance (Djurdjanovic *et al.*, 2003). In this context, the term prognostics is sometimes used for the prediction of system performance degradation, prediction of future failures, and remaining lifetime (Katipamula and Brambley, 2005a). If a building is to perform as intended, then the owners and occupants of this building have an obligation to maintain and not to misuse that building (CIB Report 64, 1982: 7). However, a mismatch between occupant expectations and the conditions produced by building systems can lead to complaints; such complaints may be used as feedback to the operators and help change the buildings management and, in some cases, systems. There are two types of complaints: volunteered or solicited. Volunteered complaints are put forward by occupants because something causes discomfort or could be done better. Solicited complaints are collected on purpose, for instance via an occupant satisfaction survey. Complaints can play a role in marketing but also form a reputational risk. Handling of complaints can be reactive or proactive (Goins and Moezzi, 2013).

In the wider sense of building management, efforts to evaluate the building performance of buildings in use can range from single energy audits and surveys that take place at one point in time, to more longitudinal studies over time, all the way to continuous performance evaluation. One can also distinguish between special evaluations that set goals and establish a baseline, and routine checks throughout the building lifetime; in the latter case it is important to establish regular assessment intervals (Carbon Trust, 2012). For some areas, International Standards such as BS EN ISO 50001:2011 on Energy Management Systems are in place to guide organizations to monitor and, where possible, to improve their energy performance. This recognizes a need to combine monitoring, measurement and analysis in a wider process, which also requires a policy, planning and implementation, actual interventions and audits of this wider process to be in place (BS EN ISO 50001:2011: vi).

Building systems originally were subject to manual control; obviously this is now shifting to automation. For modern buildings, most of the control relates to building services systems. Automation in buildings started with timers; a next stage was sensor-based control such as in the familiar thermostat system. More advanced control systems have many sensors across a range of building and environmental parameters, and some include smart and dynamic control with advanced features such as vision-based systems (Shih, 2014). Classical thermostats may result in frequent changes; a simple solution is to create a 'dead zone' in between on and off settings. However, while this prevents oscillating control, it does not stop overshoots; this can be addressed by Proportional-Integral-Derivative or PID controllers (Dounis and Caraiscos, 2009). The structure of many modern building automation systems can be described in terms of a *field layer*, which involves sensors, actuators and devices; an *automation layer* that involves control loops, data transfer and storage, and a *management layer* that relates to data analysis, archiving and similar (Domingues *et al.*, 2016).

The intelligence in building automation systems is first and foremost based on a set of knowledge-based control rules, typically in the format of if-then-else statements. More advanced systems may have self-learning functions. While performance prediction mostly plays a role during design, it may also be used in the operation and control of buildings. Here models can be used to predict the near-term future in order to find the optimal control setting, allowing anticipation rather than reaction. This is generally known as model predictive control (MPC). Model predictive control works especially well where there are time-dependent effects, such as thermal storage, which can be exploited by checking the predicted and actual state of the building system (Park, 2003; Henze, 2013). Model-based predictive control can also been described as repeated optimization of a control problem, which is carried out at each time step of the controller (Coffey, 2013). For a recent overview, see Rockett and Hathway (2017).

Monitoring is typically concerned with a well-known range of building performance parameters such as indoor temperature, relative humidity, lighting levels and consumption of resources such as gas and electricity. In some cases more details may be captured, such as radiant temperature, indoor air quality, (day)lighting levels and occupancy (Ahmad *et al.*, 2016). However, in special cases there may be other interests and phenomena that are observed. Examples are the monitoring of bacterial growth in historical buildings, with an aim to prevent possible degradation of sensitive wall paintings and materials (Piñar *et al.*, 2001), monitoring for presence of pests (Capella *et al.*, 2011), monitoring of structural strains and 3D acceleration in case of earthquakes (Torfs *et al.*, 2013) or monitoring of the structural load from tourists visiting heritage buildings (Dai *et al.*, 2016).

A special area of control is that of fault detection and diagnostics (FDD). The first element relates to identifying outliers in the typical building behaviour that indicate that there is indeed a fault. This is a non-trivial task as there may be many data streams, and separating actual faults from incidental extremes requires deeper analysis; for instance high-energy consumption in the evening may be due to a system malfunction but could also result from extending building operation beyond the regular opening hours. The second element is about finding the root cause of faults. FDD plays an important role in managing buildings, reducing any systems downtime and making maintenance, prevention and repair efforts as efficient as possible. Efforts in fault detection and diagnostics are required at various stages: during commissioning, operation and maintenance. Typical problems that may be found through FDD at the commissioning stage include issues such as wrong sizing of systems, installation problems, wiring faults and incorrect control settings. Faults found during operation include system degradation and malfunctioning; often these faults may have severe consequences due to accumulation of waste and compensation actions by other systems. Fault detection during maintenance aims at preventing future failures before these faults actually occur (Katipamula and Brambley, 2005a). In many engineering domains, the operation of systems goes hand in hand with a deep interest in RAMS: reliability, availability, maintainability and safety integrity (Smith, 2011: 3). Here reliability is concerned with the probability that a system will perform a function, under stated conditions, for a given period of time. Availability is concerned with the system being able to function when required. Maintainability relates to the probability that a system, once it has failed, can be restored to operational condition through a set of repair actions. Finally, safety relates to the requirement of not harming people or other assets during operation (*ibid.*).

There is a substantial body of knowledge on reliability engineering, maintainability and working with risks; see for instance the books by Smith (2011) or O'Connor and Kleyner (2012).

Building performance management is also leading to new business strategies; this is especially the case for Energy Service Companies (ESCOs) and Energy Performance Contracts (EPCs); in spite of the prominent position of energy in these terms, there are also efforts to undertake similar efforts with a wider view and looking at other resources. The underlying idea of energy performance contracts is to use future energy savings to fund investments in energy conservation measures. For the client, this yields a facility upgrade without having to make a capital investment; for the ESCO it yields a share in savings made through reducing the energy bill of a client. Typically the ESCO takes over the operation and management (O&M) of the facility during the duration of the contract (Yik and Lee, 2004; Deng *et al.*, 2014; Pätäri and Sinkkonen, 2014; Nolden *et al.*, 2016). Energy performance contracts need to take into account the production costs of creating energy savings and transaction costs of setting up the arrangements between client and ESCO. However, ESCOs can offer clients advantages in terms of economies of scale, scope and specialization; often they also provide help in financing energy conservation measures (Nolden *et al.*, 2016). ESCOs may have different backgrounds, with roots in utilities, facility management, engineering, system and control supplies and conversion equipment companies (Sorrell, 2007).

This chapter thus explores building operation, control and management, covering building automation systems, model-based predictive control, monitoring, fault detection and diagnostics, as well as the new world of performance service companies and performance contracts.

9.1   Building Performance Management and Control

Building performance management and control is a subject area that is closely intertwined with the full building life cycle, including the design process that sets out the intention of what a building is supposed to do and how. Performance management and control may be of importance during commissioning, audits, fabric and system testing, during regular operation, or the focus of specific studies and efficiency programs and may be carried out by either internal or external parties (Guerra-Santin and Tweed, 2015). In many cases the evaluation of building performance thus requires collaboration between a range of specialists, including instrumentation vendors and suppliers, measurement analysts, installation contractors and others (Gillespie *et al.*, 2007: iv). There are formal standards for some areas of building performance management, such as ISO 50001:2011, which covers energy management systems and BS EN 16247-1 on energy audits; ISO 16484-5:2012 and EN 15232 cover building automation and control systems. Ahmad *et al.* (2016) give an overview of energy legislation in a number of countries and how this relates to a need for the monitoring of buildings. Chen *et al.* (2016) present a deep discussion of fundamental building control knowledge and its representation.

Management and control may take place at different levels. At the building operation level, decisions can be made about things like defining setpoints and controlling the on/off behaviour of systems; in this context it is worthwhile to distinguish between systems that are *dispatchable*, such as boilers, generators and appliances, and others that are *non-dispatchable*, such as photovoltaic arrays and building heat and cooling loads (Gruber *et al.*, 2015). A different level considers the interaction between the building and the energy supply grid. At the interface between these systems, building management may steer processes in order to allow for efficient scheduling, which can help to balance the overall load on the grid, optimize the use of off-peak tariffs or ensure operation at times when renewable energy is available on site. Key terms in this context are *demand response management* and *load shaping* (*ibid.*). In this management and control context, it makes sense to distinguish processes that are non-deferrable, deferrable and interruptible. Similarly, one can discern processes that have one single energy profile, which allow for alternative profiles or even hybrid profiles (Mauser *et al.*, 2016). At the facility management and building maintenance level, there are decisions about the order of activities and interventions – one could work in the order in which the need for action becomes evident, but typically one may want to prioritize certain tasks that are deemed to be of higher importance. A key target in maintenance is to limit downtime of facilities due to failure as well as for repairs (Djurdjanovic *et al.*, 2003). Yet another level of building performance management relates to performance contracts, and the activities that are required to ensure the terms of such contracts are met by the parties involved.

Building performance, like the performance of most other complex systems, mostly has a tendency to drift downwards. This phenomenon is caused by malfunction of components, wear and tear, systems getting misaligned over time and similar issues. Once a gap opens up between the current performance of the system and the desired performance, this may result in pressure to lower the performance goal instead of taking corrective action with the corresponding costs and efforts. However, a combination of performance reductions and lowered goals will at some point lead to crisis and ultimately collapse of the system (SEBoK, 2014: 129). Maintenance is the key tool to counteract this process of performance reduction. In traditional maintenance systems, interventions take place at the first indication or recognition of failure. An alternative principle is the analysis of monitored data that allows the observation of drift in the building performance signature during typical, normal operation. Most maintenance is reactive or blindly proactive. Reactive maintenance takes places after a system fails and consists of fixing or replacing components; this is often named a 'fail and fix' (FAF) approach. Blind proactive maintenance schedules a replacement of components after a given time, assuming a level of performance degradation without actually measuring the reduced performance level. Predictive maintenance, sometimes named the 'predict and prevent' (PAP) approach, allows to predict unacceptable performance levels before they occur, allowing preventive interventions at the right point in time. By intervening at the right time, systems and components can be used as long as possible, allowing maximum return on investment and postponing replacement costs. Condition-Based Maintenance (CBM) goes even further and adds actual status measurements to the predict and prevent approach (Djurdjanovic *et al.*, 2003). Katipamula and Brambley (2005a) note that building equipment lags behind other fields in terms of condition-based maintenance; although some progress has been made in the decade since this work was published, the general position has not fundamentally changed.

Analysis of building operation and performance is a key concern for the discipline of facility management. The personnel in charge of a building have a range of tasks that requires different information. Depending on circumstances they may be able to work with aggregate performance data; however specific tasks may require refinement towards a higher spatial or temporal resolution. Utility bill analysis is a low-level starting point for the analysis of building energy use. In many cases, this is a good way to access historical gas and electricity use data, but one has to bear in mind that some data might be estimated only (California Commissioning Collaborative, 2011: 27–29). The findings from data analysis may not only trigger follow-up activities (such as maintenance, servicing or replacing of components) but also lead to the identification of a further or different data need, for instance to verify the impact of an intervention (Gursel *et al.*, 2009). Facility management personnel thus needs to follow a systemic and standardized approach to collect data about the building, its state and management; this needs to be combined with analysis of actionable variables, optimization and management policies into a cycle of continuous monitoring and evaluation (Zhang and Gao, 2010). Data for use in facility management may stem from different sources: wired automation networks, wireless sensor networks, digital building information models, cost and financial departments, and proprietary meters (Ahmed *et al.*, 2010). Given these many sources, managing this data in a way that allows advanced analysis becomes more important. Unfortunately, a lot of data lives on 'islands of information' that are difficult to relate to other information. Curry *et al.* (2013) discuss how linked data technology and open data protocols may help to structure this data in a better manner. Ahmed *et al.* (2010) suggest the use of data warehouse (DW) technology to store and analyze building performance data, moving beyond traditional database systems. Data warehousing is used in other industries and has advanced functions for the aggregation, analysis and comparison of data. However, a central issue in the work of building operators and facility managers is to deal with complaints and the potential for complaints; recent work suggest that the majority of effort in the industry still goes to responding to issues rather than prevention (Goins and Moezzi, 2013).

Apart from the obvious benefits in terms of helping owners to manage the resources and costs associated with operating a building, and ensuring occupant satisfaction, building performance monitoring can also help the key stakeholders to position themselves at the forefront of building efficiency and performance. This for instance has benefits in terms of gaining green credentials and demonstrating corporate social responsibility (CSR). Monitoring may also help protect against liabilities such as poor indoor air quality, gas leaks and ventilation problems (California Commissioning Collaborative, 2011: 12–14).

A key concept for building performance management is monitoring and targeting (M&T). Monitoring and targeting is concerned with measurement, detecting and prioritizing exceptions, and intervening to improve performance (Vesma, 2009: 15). M&T can be defined as a 'systematic assessment of actual against expected performance … to assist in target-setting and diagnosis of abnormal performance' (*ibid.*, 76). In essence, M&T is a process that aims to understand building performance as well as to set targets that may be used in reviewing performance in the future.

A related concept is that of measurement and verification (M&V). Measurement and verification also involves the quantification of performance and comparison of the findings with established targets; however the term M&V is typically used in the context of performance contracting and in most cases assumes that binding target values are already in place. A prominent document pertaining to M&V is the International

Performance Measurement and Verification Protocol (IPMVP) for the monitoring of energy and water use by buildings. It defines standard terms and best practice for M&V, while acknowledging that these M&V activities typically need to be tailored to each specific process (Efficiency Valuation Organization, 2014a: iv). Given the contractual and business context, M&V techniques must be trustworthy while at the same time providing a right balance between time, cost and accuracy (Granderson *et al.*, 2016). The Efficiency Valuation Organization (2012: 5) lists the following purposes of M&V: to determine and increase energy savings delivered through energy conservation measures; to document financial transactions related to energy efficiency; to enhance financing for efficiency projects; to improve engineering design, facility operations and maintenance; to manage energy budgets; to enhance the value of emission–reduction credits; to support the evaluation of regional efficiency programs; and to increase public understanding of energy management as a public policy tool. M&V typically involves the following activities: the installation, calibration and maintenance of meters; the gathering and screening of data; the use of modelling and computation to provide estimates; computational analysis on the basis of the measured data; and reporting and quality assurance (Efficiency Valuation Organization, 2012: 4).

Whether the context is building control, facility management, M&T or M&V, building performance tracking consists of a cycle. Stage one of the cycle is concerned with collecting data and actually tracking performance, in terms of how performance changes over time. Stage two involves the analysis of the performance data to find out where anomalies occur. Stage three diagnoses the issues that cause the anomalies and identifies potential solutions. Finally, stage four applies these solutions and verifies the results, closing the loop towards further data collection (California Commissioning Collaborative, 2011: 3).

9.1.1   Building Automation Systems

In this day and age, the operation of many building systems is controlled by building automations systems (BAS) that steer the actions of heating, cooling, lighting and other building services. Data from the BAS can also be accessed to check status, raise alarms and investigate causes for anomalies in building performance. The key to automation systems is that there is a non-manual initiation of a control action; typically these control actions relate to hardware in a range of different rooms or zones (Stein *et al.*, 2006: 1360–1369). Such automation rests on a network of sensors that report on the building state and performance, computerised reasoning about what actions to take, and a control system that actually enacts the selected course of action. To do this, building automation systems typically collect a significant amount of data about the building and environment, such as the outdoor air temperature, zone temperatures, supply air/water temperatures, return air/water temperatures, flow rates, damper positions, valve positions, system on/off status, thermostat settings and many others. While there is a wide variation amongst the various building automation systems, they all have a large number of sensors and measure hundreds or even thousands of data points. The amount of data is increased significantly increased due to the need to capture these parameters for multiple rooms, floors and systems. In building automation systems, historical data is often named 'trend' data (Zibin *et al.*, 2016). A familiar example is the use of building automation in the operation of heating and cooling systems. For these to operate well, proper control of the plant is a key requirement. Typically this involves the setting of correct timings, temperature set points, zoning, frost prevention, boiler sequencing and, in air-based systems, the mixing and recirculation of air flows. The time of heating needs to match the occupancy of the building. In buildings where some time is needed to reach comfort conditions, optimum start-up times need to be set. Temporal overrides to meet non-standard occupancy without altering the basis scheme help to deal with deviations in typical use (Vesma, 2009: 24).

Building automation systems can be found in residential, commercial and public buildings; buildings with such systems are sometimes named 'smart buildings'. Obviously the BAS may control the typical building systems such as heating and ventilation and decide on the use of different energy sources such as the national grid or renewable energy systems; however it may also control the operation of appliances in the buildings, such as domestic washing machines or tumble dryers (Mauser *et al.*, 2016).

Apart from having the advantage that they remove the need for human operation of the systems, building automation systems may also play a role in saving resources. As they control the most active systems in buildings, the automation has a direct impact on the use of utilities like gas, electricity and water; for instance automated systems can prevent overheating and more advanced controls can modulate heat supply to the building. In some advanced systems the automation system may be able to sense that electric equipment is in stand-by mode and further reduce energy use by fully shutting down the related power outlets (Suryadevara *et al.*, 2015). Where smart grid initiatives try to encourage users to reduce energy demand, the options to actually enforce demand reduction are limited; a building automation system however is in direct, real-time control of the building systems (Kang *et al.*, 2014).

Building automation is a complex, multidisciplinary and fast evolving field; as a consequence there are inconsistencies in definitions and approaches that may lead to issues (Domingues *et al.*, 2016). The most prominent and generic term, used as title for this section, is building automation system (BAS). Others that have significant overlap are: Building Automation and Control System (BACS); Building Management System (BMS); Building Energy Management System (BEMS); Energy Management and Control System (EMCS); Energy Management and Information System (EMIS); and Home Energy Management System (HEMS). In terms of software used for building energy management, sometimes the term energy information system (EIS) is used; such a system comprises data acquisition software, performance monitoring software and further components for the storage, analysis and display of building energy information. It must however be noted that most building automation systems react to the state of the building, but do not necessarily archive this status data; capturing and storing this data often is a first step in employing them in building monitoring (California Commissioning Collaborative, 2011: 29–30). On the other hand, typical energy information systems do not have a control function. In terms of formal definition, building automation systems should be kept apart from an Energy Management System (EnMS) as defined by BS EN ISO 50001:2011: 2, which is a more general 'set of interrelated or interacting elements to establish an energy policy and energy objectives, and processes and procedures to achieve those objectives'. See Table 9.1.

Control and optimization of building systems can take place at different levels. At a bottom level, all subsystems require management; however there also is need to see how these systems interact and to ensure integrated building optimization. Viewing this

Table 9.1 Building automation systems synonyms.

Acronym	Meaning
BAS	Building Automation System
BMS	Building Management System
BEMS	Building Energy Management System
EMCS	Energy Management and Control System
EMIS	Energy Management and Information System
HEMS	Home Energy Management System
EnMS	Energy Management System

in the opposite direction, the control of a building involves the management of a range of interdependent problem parts (IPPs) that need to be coordinated (Mauser *et al.*, 2016). Building automation systems normally consist of sensors, actuators and controllers, which are connected by a control network that is often labelled the communication or data bus system (Domingues *et al.*, 2016). In terms of building control systems, it is important to distinguish between systems that control single rooms or devices, such as thermostats, and more advanced systems that coordinate operation across a full building or even a district or campus.

Especially with the developments in digital technologies, building automation is a fast-paced field. There is wide range of commercial building automation systems, such as Becon Manager (LG), Desigo (Siemens), Dupline (Carlo Gavazzi), Metasys (Johnson Controls), SmartStruxure (Schneider Electric) and WEBs-N4 and WEBs-AX (Honeywell). With the split between hardware and software, it has become a practice to discern different data points. A data point may relate to a *physical data point*, such as the input/output ports of a device, or a *virtual data point* that is an abstract object such as a parameter representing a temperature. Data points may also be readable or writable (Domingues *et al.*, 2016). However, the uptake of digital technologies also leads to changes in system functionality; for instance in the recent past, building security systems were often based on a separate wired network of movement and contact sensors linked to a central alarm console; in modern automation systems this function may be integrated into a wider building management system (Bhatt and Verma, 2015). At the same time, many building management systems, especially the slightly more dated versions, are designed to control buildings during normal operation; typically they have limited data storage and analysis capacities, making them less suited for analysis of faults and malfunctions (Painter *et al.*, 2012). Some further challenges with current systems are a tendency towards custom solutions, lack of interoperability with systems from other vendors, closed specifications, complexity that requires specialist personnel and a lack of flexibility beyond very strict specifications (Domingues *et al.*, 2016).

Dounis and Caraiscos (2009) give an overview of some of the computational approaches used inside the reasoning elements of building control systems, such as fuzzy logic, supervisory control and agent-based systems. A general framework for the control of systems is SCADA (Supervisory Control and Data Acquisition), which is used in the process industry and for infrastructure management as well as for building automation (Vesma, 2009: 7). SCADA systems consist of supervisory computers at the top level, remote terminal units (RTUs) that act as sensors and actuators, programmable logic controllers (PLCs) and a Human Machine Interface (HMI). PLCs now also face competition from direct digital controllers (DDC), sometimes also known as distributed digital controllers; these bring control decisions closer to the systems that need to be managed (Domingues *et al.*, 2016).

The central computers of a building automation system need to communicate with the systems they control. The instructions passed on inside the building automation system for changing the state or operation of components are named commands. The messages that report back to the computer on component states or operations are either event or alarm notifications. Essential instructions, messages and information about interactions with users are stored in data logs that can be either temporary or permanent (Domingues *et al.*, 2016).

While advances in ICT make building systems increasingly advanced, ensuring that these systems communicate with each other and with the central building automation system remains challenging and requires an open and flexible network (Jiang *et al.*, 2009). A standard approach is to use the Simple Network Management Protocol (SNMP), which is a standard internet protocol for control of devices. On the commercial building level, an often-used protocol for building automation and control is BACnet. Another protocol for commercial buildings is KNX; further alternatives are ModBus, M-Bus, LonTalk, Z-Wave and EnOcean. Wireless communication may use the ZigBee internet communication protocol, which is mainly used in home automation. For a review of these protocols and their capabilities, see Domingues *et al.* (2016). Increasingly, monitoring systems rely on the internet for data transmission; Zhao *et al.* (2013) present the implementation of this concept for large public buildings.

While the building automation system controls the operation of the building, the BAS itself is also a system in its own right. Therefore, as with other building systems, this will also benefit from commissioning. At a basic level this allows to adjust various setpoints to their best value and to make sure the BAS works as intended (Painter *et al.*, 2012). Most building automation systems will need calibration; this is a non-trivial task as most buildings have bespoke system layouts, and the number of measurements typically is lower than the number of model parameters, leading to undetermined calibrations (Zibin *et al.*, 2016). Like other systems, the BAS itself also requires periodical 'tuning' to ensure proper operation; resetting the controls of HVAC systems in large office buildings to good values, such as adjusting schedules, minimal damper flow and thermostat setpoints, may lead to a reduction of 20% in energy use (Fernandez *et al.*, 2015). A specific point of attention is the sensors. Proper control relies on correct input data; however sensors may get damaged during construction, may be installed improperly or can malfunction during building operation. Additional data collection may help to verify whether a system as installed performs as intended (Painter *et al.*, 2012).

Smart meters are a new phenomenon that should be seen in close relation to building automation. While some authors allow for a loose interpretation of smart meters as systems that can be remotely read by building owners, utilities and third parties, strictly speaking a smart meter should have some sort of optimization function and be able to control building systems and appliances (Suryadevara *et al.*, 2015). In other words, it should not just be a monitoring device but also have a control function. Measurement of building performance by smart meters is sometimes labelled M&V 2.0; the required metering systems may be named Advanced Metering Infrastructure or AMI (Granderson *et al.*, 2016). In wireless networks, power saving is often important as systems may run on batteries; here power use is related to the life of the system (Capella *et al.*, 2011). For some low power monitoring systems, battery life may be as long as 12 years (Torfs *et al.*, 2013).

O'Neill *et al.* (2013) present the application of advanced building energy management to a demonstration building that combines monitoring, benchmarking, predictive modelling, diagnostics and visualization.

9.1.2 Model-Based Predictive Control

Typical control systems obtain the building status data from sensors, compare that status with a desired status and then use some form of rule to decide on appropriate control actions. These rules can be quite advanced; for instance a Proportional-Integral-Derivative (PID) controller decides on proportion (how much), integral (for how long) and differential (how fast) a system like a boiler or furnace needs to operate. However, identifying the desired status may be a challenging task; for instance the desirable air temperature in a building may depend on a range of factors such as individual occupant preferences, activities and outdoor weather conditions. One solution for this problem is the use of self-learning control systems, which register occupant preferences over a period of time to establish a typical preferred pattern. Another challenge may be that finding the preferred status requires some degree of foresight, for instance to anticipate occupancy changes, weather, a slow response of thermal mass or fluctuating energy prices. A typical example is the control of smart façades with their intricate balance of airflow, solar gain and daylighting (Park, 2003: xxi–xxii). This can be addressed by using models to explore the likely future states of the building so that optimal control actions may be identified; this approach is known as model-based predictive control (MPC). General control problems that suggest the use of predictive techniques are the forecasting of energy prices, load and usage patterns, and supply of renewable energies that depend on meteorological conditions (Kang *et al.*, 2014). An important factor in all these prediction challenges is the time horizon for optimization. For some issues, such as temperature control, a short horizon of one hour may be effective; for others such as energy pricing and weather conditions, it may be beneficial to use 24 h predictions (Gruber *et al.*, 2015).

As there is a wide range of models, there is a similar range of approaches to model-based predictive control. At one end of the spectrum, one may conduct a range of simulations with complex models to create a predefined 'lookup table' that is available in advance and that can easily be integrated into control systems; in that case all simulation work has been done beforehand (Coffey, 2013). At the other end, one may do full dynamic simulation within the control loop; however this requires significant simulation speed as control decisions do not allow for long computer runs, with 5 or 10 minutes being the absolute maximum. An intermediate solution is the use of reduced order models, which are developed by training algorithms with the data from more advanced simulations. These may be appropriate for control, as they are easy to scale and less time consuming to create and calibrate; however these models are less accurate and should not be used for extrapolation (O'Neill *et al.*, 2013). While significant scientific work has been done on predictive, optimal and adaptive control, implementation in industrial systems seems to lag behind (Dounis and Caraiscos, 2009). De Coninck *et al.* (2016) present automated inverted modelling, in terms of calibration of applicable simulation models, in order to predict loads in buildings. Their MPC algorithms are demonstrated to work in the context of two well-monitored houses in Germany.

Another interesting interface between building performance simulation and the operation of systems is provided by the Building Controls Virtual Test Bed (BCVTB); rather than bringing simulation into the control loop of buildings, this allows to conduct 'hardware in the loop' simulations that incorporate the behaviour of actual systems (Wetter, 2011). Kwak *et al.* (2015) present a detailed study into real-time model-based predictive control using the BCVTB in order to manage the damper position of an air handling unit.

9.2 Performance Monitoring

Performance monitoring first of all requires the identification of building management and control objectives. These objectives need to be expressed in terms of performance measures, which in turn lead to a need to collect data. Once it is clear what data is needed, measurement equipment needs to be defined that can capture that data. As measurement equipment (sensors) just produces readings, these readings need to be taken through data acquisition, transmitted to a central archive and stored. A final step is data analysis, visualization and storage. Accordingly, at a high level, monitoring systems can be considered to consist of three subsystems: the actual sensors and metering systems; data acquisition, transmission and handling systems; and data management, analysis and reporting systems (Gillespie *et al.*, 2007; Vesma, 2009: 8).

One can monitor the whole building or individual systems within it. Monitoring of the whole facility allows the operator to identify how a building compares in relation to similar buildings elsewhere and whether performance changes over time and in line with expectations. The monitoring of individual systems allows to check whether systems perform as anticipated, to keep track of efficiency, to better understand how systems contribute to the overall building performance and beyond what is visible at the overall level, and to pinpoint where any problems might be located (California Commissioning Collaborative, 2011: 43). Obviously, where a building automation system already gathers data about the building state, this data can be fed into the monitoring system. This is most easily done where the BAS operates digitally. However, it must be noted that many building automation systems are proprietary and that monitoring mostly requires substantially more measurement points than are used in the building control context (Gillespie *et al.*, 2007: 1). Data collected during building performance monitoring for instance concerns outdoor air temperature, indoor temperature, electricity use, gas use, water use, chilled and hot water supply/loop/return temperature, gas flow, electricity peak power, damper fractions, duct pressure, system efficiency and boiler output (Gillespie *et al.*, 2007: 22–30).

Before investing in additional meters and sensors, it is worth carrying out an in-depth operational verification to ensure systems work as intended. Operational verification can include visual inspection, sample spot measurements, short-term testing, data trending and control-logic reviews. Where monitoring is carried out as part of an M&V process, this helps to ensure the original system functions properly before any interventions are carried out and to ensure that the proper interventions are selected in the first place. Operational verification is also useful for making sure interventions themselves work as planned; obviously starting a more complex measurement process would be useless if the intervention is in fact faulty (Efficiency Valuation Organization, 2014a: 2–3).

A monitoring system gives access to both the current and past performance of a building and relevant (sub)systems. The first and foremost reasons for monitoring are thus routine reporting on how a building is performing and troubleshooting if things go wrong. Building monitoring data is also used for benchmarking, where operators are interested in knowing how the performance of a facility compares with peers or historical data of the building itself (California Commissioning Collaborative, 2011: 25). Performance monitoring or performance tracking may help operators to prioritize interventions in single buildings or in a building portfolio. It may also help to establish a baseline, which is useful for identifying future problems and excessive use of resources and which can be used to quantify the impact of interventions (*ibid.*, 8). Where the monitoring concerns the use of resources such as gas, electricity and water, it helps to track typical consumption and peak loads to identify any unnecessary running of equipment, to establish system health as well as deterioration and to review the efficiency of operation strategies (Gillespie *et al.*, 2007: 2). Furthermore, monitoring is also used extensively in a building research context; in this case it is important to keep in mind that building measurement at best is a semi-controlled experiment, where variation in contextual conditions such as the weather, occupant behaviour and control settings also needs to be captured. Performance monitoring does not have to be permanent; with the advent of wireless technology, it is now relatively simple to install a sensor, conduct monitoring until an issue has been resolved and then move the sensor to another location where further benefits can be gained (Jang *et al.*, 2008).

The set-up of a specific building monitoring system depends on the objectives of the monitoring. Beyond the typical wired and wireless meters and sensors, further information may be collected manually, through the use of questionnaires and surveys. Analysis may also require data about the building that is static, such as floor area; one can thus discern between static data that pertains to the building and dynamic data types such as use schedule, occupancy characteristics, weather data and utility use data (Mantha *et al.*, 2016). Gillespie *et al.* (2007: 3) discern three levels of building performance monitoring: basic (covering essential measurement of typical buildings and systems), intermediate (progressive measurement of typical buildings and systems) and advanced (sophisticated measurement of unique and/or critical facilities and systems). They suggest to start simple and build up complexity slowly, testing the monitoring system while it is being put in place. The installation of a monitoring system requires resources in terms of hardware, labour and time. As these resources may be costly, monitoring is best not done on an ad hoc basis, but through a dedicated monitoring brief, embedded into a wider project context and carried out by experts (Gillespie *et al.*, 2007: 8).

Data about building behaviour may stem from a range of sources, such as wired sensors, actuators and meters; originally sensors were wired but these same sources are increasingly available as wireless options. Further data may be obtained from radio frequency identification (RFID) tags or can be obtained from a range of sources: monitoring tools, control systems, diagnosis and maintenance tools, data mining and decision support tools and building simulation software; often data is stored in a building information model (Ahmed *et al.*, 2010). For electricity meters, there typically is a choice between meters that are included in the supply line or indirect 'clamp on' meters. Flow meters allow capturing the amount of water, gas, steam, compressed air, oil and other fluids or gasses that go into or leave the building. However, some issues like expected temperatures and pressures, flow rates and viscosity of the fluid or gas being measured need careful consideration when selecting a meter (Vesma, 2009: 3). For electricity, the IPMVP suggests to capture the voltage, amperage and power factor (Efficiency Valuation Organization, 2014a: 11). For many monitoring purposes, it is important that meters can give a pulse output, which can be counted and converted to a range of flows such as water, gas or electricity (California Commissioning Collaborative, 2011: 42). Where data on individual buildings or systems is difficult to obtain, it may be possible to derive this from wider patterns; for instance a Sankey diagram of a cluster of buildings may help to establish flows that can be used for this purpose (Abdelalim *et al.*, 2015).

Considerations that are important when setting up monitoring efforts are the selection of measurement boundaries, a measurement baseline and the reporting period. One may also want to define on/off tests. The measurement boundaries define the scope of the monitoring efforts; they provide a clear hint on where meters ought to be positioned. These borders also need to be reviewed in terms of potential leakage across the boundaries to prevent measurement errors. The measurement baseline will represent the regular operation conditions of the building and will be used to identify anomalies or, in the case of M&V, the success of any interventions. The baseline thus should represent the full operational cycle over an appropriate length of time. The reporting period should be of sufficient length to once again capture the full operational cycle over an appropriate length of time, so that outliers in the building behaviour can be captured. For M&V it should be selected in such a way that the impact of the interventions becomes quantifiable. On/off tests are useful where systems can be easily controlled, so that their impact on the building measurement can be isolated and thus be recognized in the overall data profile (Efficiency Valuation Organization, 2014a: 3–4).

Many of today's building automation systems also have a lot of sensors that can be employed to capture the system status and performance, yet often this data is not analyzed and used (Djurdjanovic *et al.*, 2003; Yu *et al.*, 2013). A building automation system typically allows to monitor current to electrical equipment, power used by electrical equipment, valve positions, outside air temperature (OAT), supply air temperature (SAT), return air temperatures (RAT), water supply temperature, wet bulb/dry bulb temperatures, speed parameters of fans, coil valve positions and many others. These can be used to identify unnecessary operation, short cycling, leakages, etc. Most building automation systems have alarm capabilities to alert operators if a parameter is outside an expected range, which may also be of interest when analyzing performance (California Commissioning Collaborative, 2011: 30–31).

Important factors in the selection of monitoring equipment are accuracy, ease of deployment, cost, availability, granularity and, if wireless, what communication protocols are used (Ahmad *et al.*, 2016). Beyond what data is to be collected and what sensor captures that best, another important issue is the exact sensor location; within the complex geometry and systems of buildings maintaining an overview of sensor positions is often a non-trivial task, especially where sensors are part of wireless systems (Jang *et al.*, 2008). In this context it is important to develop a unique naming of all measurement points in the monitoring system; there will be many similar points, such as a range of room temperatures, valve positions and others, which can easily be confused. A unique name, which is used across drawings, databases and others, helps to clearly identify each point and its location (Gillespie *et al.*, 2007: 9–10). Where possible, sensors and measurement points should be equipped with a durable tag or label that shows this identifier, as well as what is being measured (Vesma, 2009: 5). Another important issue is to ensure that all data can be related to a valid time stamp (Gillespie *et al.*, 2007: 17).

Modern monitoring projects typically employ Automated Meter Reading (AMR). AMR has the benefit of providing high temporal resolution data, with 30 minute intervals being common. Moreover, AMR allows remote reading; this is helpful where multiple buildings or sites are monitored, access is restricted, readings need to be synchronized and delay in processing time needs to be minimized (Vesma, 2009: 7).

The quality of a building performance monitoring system is dependent on a number of key properties. Assuming the right sensors have been selected and positioned at the correct points in the building, the following issues become important:

- Measurement errors: the difference between observed values and a 'true' values; the measurement error may be random or systematic.
- Repeatability: consistency of repeated measurements of the same value, under the same experimental conditions.
- Scan rate: the frequency at which the sensors and meters undertake measurements.
- Sample interval: the time required by a measurement system for acquiring readings from all data points, placing these in memory, and writing these to a display or file.
- Refresh rate: the frequency of display or data file updates.

Further factors to take into account are accuracy, linearity, drift or stability over time, dynamics, resolution, signal-to-noise ratio, unit conversion, and data storage and retrieval frequency (Gillespie *et al.*, 2007: 13).

Metering errors can never be fully avoided, as no meter is 100% accurate. However, metering errors can be minimized by regular calibration. Other sources of measurement error include poor placement of sensors and data transmission errors (Efficiency Valuation Organization, 2014b: 12–13). Good organization and recording of all activities helps to reduce the risk of confusing sensors, although a mistake in wired systems is easily made. For measuring electricity, the IPMVP points out the need to be vigilant towards distortion of the signal by other devices in the vicinity (Efficiency Valuation Organization, 2014a: 11).

A couple of aspects may cause issues with monitoring. Data confidentiality, privacy and security may become a problem, especially where residential buildings are the subject of the monitoring efforts; many monitoring systems may be relatively open and can lead to unintended vulnerability of building occupants (Ahmad *et al.*, 2016). By way of example, a simple electricity use meter reading may be used to establish and infer activity levels in a home; this may be used in both a positive manner (for instance to keep an eye on elderly people living in assisted living, where a lack of activity triggers an action to check on their well-being) or a negative manner (establishing when a home is empty and timing a burglary). Often permission of the building occupant is required before monitoring can start; this is easier to obtain from a single occupant than from a range of occupants in a multi-tenant building. Where monitoring efforts are foreseen, it may be useful to include a right for monitoring in lease contracts (California Commissioning Collaborative, 2011: 27).

The success of a monitoring campaign is not solely a technical issue but also depends on how results are taken forward in an organization. The California Commissioning Collaborative (2011: 19–23) recommends to use a building performance management framework, which ensures that sufficient resources are available, that team members and their roles are known and that quantifiable goals are set. A performance management framework should also consider incentives to motivate staff, accountability and provide clear definitions for activities and, where relevant, contracts.

It must be noted that traditional monitoring is mostly limited to the sensors and measurement devices that relate to the building systems. However, there may also be 'soft' data available that is not integrated into performance analysis, such as human resources allocations, financial information, data from social media or communications between occupants. Most building automation and monitoring systems are not set up to connect such other sources of data, making correlation between different data difficult (Curry *et al.*, 2013; Corry *et al.*, 2014).

As with all systems, sensors and data transmission pathways may malfunction, leading to data gaps. Identification of such gaps is important, as it allows the quick repair of equipment where needed in order to minimize the number of readings that are missed and in order to understand that any analysis results over the period with data gaps show issues pertaining to the monitoring system rather than to the performance of the building being monitored. Brown *et al.* (2010) describe the handling of suspicious entries in the data on energy consumption of 300 buildings that were monitored over 7 years. In some cases it may be possible to fill in missing data; for instance Hu *et al.* (2014) discuss how interpolation can be used to restore missing temperature entries. Problems with faulty data are not unique to the building discipline; van den Broeck *et al.* (2005) discuss the process of data cleaning in medical research, describing a process that consists of data screening, diagnosis and editing. The screening identifies missing and excess data, outliers and inconsistencies, strange patterns and suspect analysis results. Diagnosis points out where data is missing or must be faulty, what outliers are in fact true extremes and true normal, and what data cannot be diagnosed but remains suspect. Finally, the editing process may correct some data, delete it or decide to leave it unchanged. Identifying faulty data is not always straightforward, as problems can be hidden in large volumes; finding irregularities becomes increasingly difficult if changes in a process take place over time (Au *et al.*, 2010). The amount of data obtained from monitoring can be significant; Dong *et al.* (2014) report 3000 data points for a multifunctional case study building with three storeys while noting that this number is actually limited due to the bandwidth of the BEMS of the building. These data points are monitored at a sampling frequency of 5 minutes, thus creating a large data set over time.

Once data has been collected, and possibly cleaned, it needs to be analyzed. Guerra-Santin and Tweed (2015) distinguish different levels of data analysis: core, advanced and bespoke analysis. These may also be seen as indicative, diagnostic and investigative analysis. At a basic level there is direct evaluation, where important parameter values are captured directly and have a meaning that can be assessed in itself, such as the air temperature in a room. However, for a lot of data beyond this basic level, some aggregation is required; for instance this is the case where temperature data is used to calculate heating or cooling degree days. When analyzing measured building data, it is not always clear what the driving factors are and what are dependent variables. This therefore needs careful consideration. For instance, in terms of building energy consumption, the outside temperature typically is the main driver; however there are other situations where a production process may dominate or where the key factor is hours of darkness that drives electricity use for lighting (Carbon Trust, 2012: 8).

Analysis efforts make use of a range of techniques to process monitoring data. Gillespie *et al.* (2007: 15) give a list of data aggregation calculations that are standard for many energy performance monitoring systems, which include math and data functions (addition, subtraction, multiplication, square root, minimum, maximum, total/integration over time, standard deviation, running average, sum and average, logic-triggered sampling), data resolution functions (that group data according to time or space) and standard functions (power conversion, load calculation, wet bulb temperature calculation, establishing the heat balance quotient). In M&V efforts analysis often takes the form of linear regression. For instance, a linear regression formula might relate monthly energy use to heating degree days, cooling degree days and building occupancy, using an empirical formula (Efficiency Valuation Organization, 2014b: 5). Other often-used terms are load profiling, trend analysis and pattern recognition. This includes the use of clustering algorithms that allow the automated identification of similar profiles in a large dataset; some of the mathematical approaches that may be applied for this purpose such as Fast Fourier Transform (FFT), the K-means algorithm, self-organizing maps (SOM) and support vector clustering are discussed by Panapakidis *et al.* (2014).

Analysis of data on the actual use of resources is only useful if it can be compared with what would have been an appropriate amount. Advanced analysis systems thus not only review recent data but also compare this with expectancies. Defining what is appropriate requires an understanding of the patterns of resource consumption in relation to driving factors such as the weather, activities, level of production and similar. A good approach to predict consumption is the use of regression analysis. A key process is to establish an expected trend on the basis of historical data; in the world of M&V, this is often called modelling, which may lead to confusion with first-principle modelling as used in the building simulation domain as discussed in Chapter 6. An example of such a model would be a trend line that gives the correlation between outdoor air temperature and expected energy use of a facility (Vesma, 2009: 8–14). This type of analysis does require a baseline or benchmarks, which are used to specify within what range the observed parameters are expected to vary (Gillespie *et al.*, 2007: 1). The importance of an accurate baseline is also emphasized by Granderson and Price (2014), who explored the impact of length and temporal resolution of the training period on energy use predictions. Typical models used to model thermal performance baselines are mean-week models, which compare values for the same day and time in the week; change-point models that relate heating and cooling loads to the outdoor temperature and where the change point indicates where heating and cooling processes commence; and day-time-temperature regression models that correlate heating to the time of day, day of the week and temperature variables (Granderson and Price, 2014). Correlations are often captured by means of multivariate linear regression (MLR), artificial neural networks (ANN) and Fourier series. Other techniques that are often used are CUSUM analysis and exponentially weighted moving average charts. Srivastav *et al.* (2013) also stress the importance of baselines but add the quantification of uncertainties by using Gaussian regression. In some systems, an alert may be generated if the difference between actual and predicted use exceeds a certain threshold (California Commissioning Collaborative, 2011: 45–47). It is important to realize that these analysis techniques may be able to detect faults and errors but can seldom diagnose them; mostly they need human operators for find out what causes the faults (*ibid.*, 32).

Analysis errors may stem from a range of sources. First of all one needs to be sure that the data that is processed is correct and does not stem from faulty sensors, that the measurement process is carried out correctly and that there are no sampling errors: if measurements are taken at the wrong point in time or from a wrong process, the sensors and process may be right but readings will still be of little use. In general, sampling errors can be caused by inhomogeneous populations, insufficient precision of the measurements, errors in signal aggregation/disaggregation, and sample size. However, even if all data is of good quality, the analysis process itself may also lead to erroneous findings. For instance, key variables may be left out of a correlation analysis or the wrong type of relationship may be assumed; a classical example is the assumption of linearity in a relation that in fact may be polynomial. Other typical errors within regression models are the use of out of range data, omission of relevant variables, inclusion of irrelevant variables, use of the wrong function (not all relations are linear) and data shortage (Efficiency Valuation Organization, 2014b: 7–8). When used with care, statistics and uncertainty analysis may help to quantify and control these errors. Some typical statistical terms that are useful in this context are sample variance (S^2), sample standard deviation (s), sample standard error (SE), sample standard deviation of the total (S_{tot}), coefficient of variation (cv), precision, absolute precision and relative precision; for further discussion see for instance the statistics and uncertainty section of the IMPVP (*ibid.*, 2–5).

There is a range of tools and systems that supports the analysis of monitoring data; some of these have dedicated visualization capabilities. At this point monitoring becomes part of a performance information system, which may be used for a range of purposes such as facility management and maintenance. In the world of building energy monitoring, which dominates much of the measurement domain, Energy Information Systems (EIS) have been defined as tools that 'store, analyze and display current and historical energy use, typically displaying hour-by-hour data for each meter. EIS provide the capability for the user to analyze energy use patterns using a variety of graphical formats. This hourly tracking enables system improvements to be viewed at the meter level and problems to be more easily and quickly identified'. An EIS consists of data acquisition systems, data transmission systems, an EIS server for data storage and web interfaces. Many EIS include advanced energy analysis algorithms that will predict energy use from a range of variables, as well as alarms that will be triggered if some predicted level of energy use is exceeded (California Commissioning Collaborative, 2011: 36–38). Results of analysis of the use of resources like energy may be reported in the form of an overspend league table, which lists the most important overspends at the top of the first page and thus clearly identifies where action should be directed (Vesma, 2009: 16; Carbon Trust, 2012: 5–6). A typical CBM system includes sensors, signal processing, condition monitoring, health assessment, prognostics, decision support and presentation (Djurdjanovic *et al.*, 2003). Hitchcock (2003: 10) suggest the use of scalars, vectors (bar charts), time series, tables (2D XY plots), graphs (3D XYZ plots) and distributions to represent building performance data. Gillespie *et al.* (2007: 63–72) recommend a range of graphical representations, including a floor-level graphic for the identification of sensor locations and systems, plant graphs that show key components and status, tables with data for specific points, time series plots for the current last 24h, group trends for appropriate periods (24 h, 7 days, 1 month, 1 year) and XY graphs to study correlations. Some systems allow the real-time collection and display of building performance; this enables better interaction of the occupants and operators with the building. Another level can be reached by including automated alerts generated by advanced software (California Commissioning Collaborative, 2011: 34). Lawrence Berkeley National Laboratory in the United States has developed a prototype tool named Metracker that captures the building performance objectives and their metrics over the full building life cycle. Metracker employs an underlying IFC model to capture the building and performance data. Conceptually this means that the performance data is stored in an IFC model, whereas Metracker is the interface to access and visualize this data (Hitchcock, 2003: 8–9). Some of the commercial data analysis tools for building performance analysis, mostly with a focus on energy management, are Cepenergy, C3ntinel, e-Bench, EnergyPeriscope, eSight, Measurabl, Optima and PowerLogic.

9.2.1 Specialized Monitoring Techniques

Beyond the traditional monitoring efforts that are common to building automation and building science projects, there are some areas that are seeing high interest and rapid developments. Two of these are discussed here: non-intrusive electricity load monitoring and the monitoring of occupant behaviour.

1) **Non-intrusive load monitoring (NILM)**
   The particulars of alternating current (AC) systems allow more advanced analysis techniques to be applied to electricity use. Rather than detailed sub-metering, this has enabled the development of a specialist research area on non-intrusive load monitoring (NILM).[1] Instead of measuring the electricity usage of single appliances, which would require a large number of meters that need to be set up within a building, this technique analyzes the current and frequency changes in the overall electricity supply in detail, capturing changes in the consumption and attributing these to appliances (Giri and Bergés, 2015). Due to the rapid changes in the state of electric appliances, a high frequency of measurement is essential (Norford and Leeb, 1996). Recognition of load signatures is also an issue for the development of smart homes (Figueiredo *et al.*, 2012), activity recognition for homes of reduced-autonomy residents (Belley *et al.*, 2014) and smart grid applications (Bouhouras *et al.*, 2014). NILM relies on machine learning techniques that can be trained to recognize typical appliance behaviour. The usual process consists of training of the algorithm, event detection, extraction of event features and attribution to appliances; sometimes additional calculations are undertaken in the background to keep track of energy use. Appliances can be categorized in terms of electricity use as on-off appliance, appliance with continuously variable load, appliance with permanent load, and finite state machine (Giri and Bergés, 2015). Algorithms used are for instance Nearest Neighbor Rule (NNR), Support Vector Machines and Artificial Neural Networks (Figueiredo *et al.*, 2012; Tsai and Lin, 2012). Advanced approaches may use spectral analysis of the waveform of individual loads (Bouhouras *et al.*, 2014). Training of NILM algorithms remains complex, as there are many different appliances in most buildings, which

---

1 Some authors also use the term non-intrusive appliance load monitoring (NIALM).

often also are subject to rapid technological development. Marceau and Zmeaureanu (2000) present work that disaggregates the electricity consumption of a residential building, showing that the error in attributing loads to the three main end uses (heating, hot water and refrigerator) is below 10%. Training of algorithms may, to some extent, be helped by positioning electromagnetic field sensors close to appliances (Giri *et al.*, 2013). Research on NILM uses publicly available data sets that allow the testing of analysis approaches on data for which ground truth data on actual appliance use is available (Giri and Bergés, 2015).

2) **Monitoring of occupant behaviour**
Another special category of building monitoring concerns occupant behaviour; this is of interest as the interaction between building and humans is key to the use of resources, perception of building quality and many processes that take place inside buildings. Occupant behaviour can be split in two main aspects: human presence and human actions that relate to the building operation. When conducting monitoring of occupant behaviour, care needs to be taken with privacy issues, as data might be considered sensitive. Different approaches have been developed, including manual methods such as observation by researchers, diary studies, automated presence detection by passive infrared (PIR) sensors, monitoring of proxies such as CO_2 levels in rooms, all the way to vision-based approaches; for a concise overview, see the section on occupant monitoring and data collection by Yan *et al.* (2015) or the work of Gilani and O'Brien (2016). Further reports about specific types of occupant behaviour monitoring can be found in Li *et al.* (2012b), who explore the use of RFID based occupancy detection to detect what occupants are in a thermal zone. Correia da Silva *et al.* (2013) have monitored the interaction between building occupants and lighting and shading systems in offices, using the lighting and shading system status as proxy for occupant action. Naghiyev *et al.* (2014) present a qualitative review of the PIR- and CO_2-based approaches, as well as the emerging device-free localization technique that employs measurements of radio signal strength, exploiting absorption of this signal by human bodies. Shih (2014) discusses an advanced system that combines a programmable camera that can pan, tilt and zoom and that is combined with a support vector machine to recognize activities and occupant identification. Ahn and Park (2016) used webcams to study occupancy in an academic laboratory. Guerra-Santin *et al.* (2016) employ a mixed method that combines both quantitative methods (monitoring of indoor parameters and building operation) and qualitative methods (interviews to understand daily routines, preferences and attitudes); their work also gives an account of further work on occupancy monitoring. Jeong *et al.* (2016) study the manual control of windows, correlating window state with indoor and outdoor conditions. Newsham *et al.* (2017) have studied the use of sensors added to a PC in order to capture occupancy in an office, reporting better accuracy than a ceiling-based PIR sensor. Wang and Shao (2017) exploit modern ICT systems by using the detection of Wi-Fi-enabled systems to study occupancy of a library building. Foulds *et al.* (2013) explore how building monitoring links to social practices research and how this combination can increase the understanding of the everyday life of the occupants. However, occupant behaviour is a complex area; for instance Wei *et al.* (2014a) describe no less than 27 underlying factors that all have an impact on how occupants regulate the heating in homes, which is only one of the systems that interact with the building occupants.

9.2.2 International Performance Measurement and Verification Protocol

A monitoring guideline that requires separate discussion in the context of building operation and control is the International Performance Measurement and Verification Protocol (IPMPV). This protocol is central to many monitoring efforts in the context of M&V projects. The IPMVP distinguishes between four main options in carrying out measurement and verification efforts around an intervention. These are Option A, where only key parameters are measured to assess the impact of an isolated retrofit intervention; Option B, where all relevant parameters are measured to assess the impact of an isolated retrofit intervention; Option C, which aims at measurement of the performance of a complete facility; and Option D, which concerns a calibrated simulation of a full facility or a sub-facility. IPMVP options A and B (retrofit isolation) limit the measurement to a subsystem of the building, especially an energy conservation measure (ECM). Typically energy flows are measured at the boundary of the ECM. Depending on circumstances, it might be sufficient to measure only a limited set of parameters, possibly also over a shorter period, which is covered by option A. In other cases a more comprehensive and long-term testing is required, as per option B. IPMVP option C covers a whole facility or building. Given the scale of the measurement this clearly is a more costly measurement effort, which may need justification in terms of savings made. In some cases existing utility meters and billing information can be used; however it must be noted that billing information sometimes includes estimated values. Whole building metering also requires the capture of independent variables such as the weather, building occupancy and production volume. IPMVP option D employs calibrated building simulation software to predict savings. This is useful when there are multiple interventions, and consequently it is hard to single out the impact from each single intervention by actual metering, where baseline data is missing or where the monitoring pertains to a building that is still being designed and hence that does not yet exist. For more details on the IPMVP, see Efficiency Valuation Organization (2012, 2014a, 2014b).

The IPMVP has a particular terminology for some aspects of monitoring; it defines the following concepts in specific context of M&V projects:

- Baseline: energy performance as measured before the implementation of any energy performance actions or interventions.
- Energy conservation measures (ECM): measures that have the specific aim of improving the energy efficiency of a building and to reduce energy demand; interestingly this term is also applied to water usage and water demand management.
- Operational verification: the confirmation, from measurements, of the potential to achieve savings by certain interventions.
- Adjustments: increments in the consumption or demand of the baseline to make sure the baseline used for establishing the potential interventions is a fair representation of the intervention.
- Independent variable: a parameter that impacts on the energy use of a building, which is subject to regular change.

The IPMVP recommends the use of an M&V plan for measurement efforts. This should state the intent, what IPMVP option is to be used, where measurement boundaries will be positioned and what baseline period will be used. It will also define the reporting period, conditions for making adjustments, analysis procedure and energy prices used in calculations. An M&V plan should contain detailed meter specifications, monitoring responsibilities, expected accuracy, budget, reporting format and details on any quality assurance procedures that are in place.

The IPMVP makes generic recommendations in terms of ensuring that accuracy of measurements will be balanced against costs. Furthermore, it encourages work to be as complete as possible (considering all effects of an intervention), the use of conservative values when making estimates, and efforts to ensure consistency across different projects, staff, measurement periods and both demand reduction and energy generation projects. On a general level, it aspires to ensure that work done as per IPMVP will be both relevant and transparent (Efficiency Valuation Organization, 2014a: 2).

9.3   Fault Detection and Diagnostics

Monitoring and data analysis may identify anomalies in building performance, yet the identification of such issues does not always lead to easy understanding of what the underlying problem is. Buildings are large and complex, making it mostly difficult – if not impossible – for operators to visually identify faults and errors; often an intensive search process is required to fully understand causes and effects. Fault Detection and Diagnostic (FDD) tools have been developed to support building engineering staff with this search process; the fault detection part of FDD tools helps with finding the anomalies in the large amounts of building performance data, whereas the diagnostic part provides functionalities that help to identify the location or root cause of these problems (California Commissioning Collaborative, 2011: 50). FDD may address different building levels, such as the system level containing AHUs, fans and compressors, or higher levels such as a complete HVAC system; at the top level it concerns the complete building (Magoulès *et al.*, 2013). However, the focus of FDD efforts seem to be mainly on AHUs, with less work on the zone or whole building level (Gunay *et al.*, 2017).

Diagnostics aim to find faults in the building system. Capozzoli *et al.* (2015) consider all abnormal consumption of resources a fault, but some irregular meter readings may have a good reason: for instance additional amounts of energy and water will be used during an event that takes place in an office or educational building outside regular opening hours. True faults that lead to performance anomalies are caused by issues such as system degradation (wear and tear), improper operation or lack of maintenance. Faults may manifest themselves in different ways; some issues lead to higher energy consumption, while others may mean that targets such as a heating setpoint may not be achieved. The onset of problems may be sudden, for instance when a valve gets stuck, but in other cases changes may take place gradually, making the problem harder to spot (Magoulès *et al.*, 2013). It may be useful to differentiate between partial failure (soft faults) and complete failure (hard faults); in partial faults maintenance may still allow to prevent the occurrence of a hard fault (Padilla and Choinière, 2015). It is generally accepted that buildings lose part of their designed (energy) efficiency over time after being commissioned or re-commissioned (Srivastav *et al.*, 2013). In this context it then becomes important to decide how much deviation of measured performance from initial performance is allowed and when the expected slow degradation becomes a fault; to do this requires the definition of some threshold of allowable deviation that should not be exceeded (Magoulès *et al.*, 2013). In a similar vein, Katipamula and Brambley (2005a) comment that performance of a building system can be captured in a performance measure named figure of merit (FOM); when faults occur, this FOM decreases until a level performance is reached that is no longer acceptable and intervention becomes necessary. Djurdjanovic *et al.* (2003) mention the use of a Performance Confidence Value (PCV), ranging between zero and one, which quantifies the similarity between a performance signature and normal operation (values close to one) and faulty behaviour (values close to zero). The underlying data for PCV are often (Gaussian) distributions.

Typical problems that are encountered in the operation of buildings, especially HVAC systems, present a long list. Examples are scheduling problems, simultaneous heating and cooling, faulty controls and sensors, deactivated or incorrectly set control, lack of calibration of sensors, lack of maintenance, improper or lacking hydraulic balancing, improper setpoints and resets, wrong staging and sequencing of equipment, malfunction and leakage of dampers and valves and over-/under-sizing of equipment (Neumann and Jacob, 2008: 22–23; California Commissioning Collaborative, 2011: 56). Katipamula and Brambley (2005b) discuss the symptom patterns for a number of selected faults in HVAC systems and components, demonstrating that a deep understanding of these is required for successful fault identification. Some problems may require significant expertise with system operation; for instance Veronica (2013) discusses the detection of faulty control settings in HVAC controls, which for instance may lead to excessive oscillations in control variables. A good overview of some of the problems that may occur in HVAC systems is the BSRIA HVAC Troubleshooting guide by Deramchi and Bell (2014). Zhang and Hong (2017) group problems in VAV systems in the following categories: fouling faults, sensor offsets, control faults, performance degradation and stuck valves. Another good overview is the paper by Gunay *et al.* (2017).

While finding a discrepancy between a historical or computed baseline and actual measurements may help to detect an issue, this does not help to identify what causes the discrepancy; this is the domain of diagnosis (Wu and Sun, 2011b). Diagnosis can be thought of as the process of isolating the symptoms that relate to a problem, so that the issues that may cause this problem can be evaluated. Deep understanding between cause and effect then allows identification of the issue at hand. Diagnosis also involves decisions on whether the system can be allowed to continue to operate, an alarm needs to be raised, control settings need to be changed and whether repair or maintenance needs to be requested (Katipamula and Brambley, 2005a). The two key questions that are asked in diagnostics are (i) why is performance degraded? and (ii) when may an observed system or process fail? The first question relates to finding causes; the second question relates to forecasting and prevention. Diagnosis thus pertains to both monitoring and forecasting of facility health. It is often based on the matching between actual data profiles and existing profiles that either describe healthy or faulty operation modes (Djurdjanovic *et al.*, 2003). FDD for buildings has its roots in other fields such as the aeronautics, space, nuclear, defence and process industries, where main efforts started in the 1970s. Work in the HVAC sector started only in the late 1980s and initially focussed on systems like compressors, chillers and heat pumps, with air handling units following shortly thereafter (Katipamula and Brambley, 2005a).

Most FDD tools either work on the basis of an expert (knowledge-based) rule or on the basis of performance data comparison (Capozzoli *et al.*, 2015). The expert rules are typical conditional statements (IF condition THEN consequent, ELSE alternative) that mimic the thinking of human experts. Performance data comparison is based on the comparison of measured values against either an historical trend or otherwise expected values, for instance, results from intentionally recreating or modelling typical faults, which allows the FDD system to recognize the corresponding patterns. Data signatures can be established at different building system levels, such as fan or damper, air handling unit or zone level. Finding issues at the more detailed levels requires corresponding detail in the reference data (Wu and Sun, 2011b; Hygh *et al.*, 2012). There are many types of FFD; some indicate the location of where a fault takes place and others even suggest a range of potential causes for the fault (California Commissioning Collaborative, 2011: 57). Fault detection and diagnosis may run in sequence, where a first system identifies faults and a second system then tries to identify the cause of any faults that have been found; in other systems, detection and diagnosis are integrated (Katipamula and Brambley, 2005a). Schein *et al.* (2006) discuss the detailed workings of a rule-based system that is able to identify 5 main faults that may occur in air handling units on the basis of 28 rules; Schein and Bushby (2006) apply the same approach to the wider HVAC system level covering 19 faults.

The basis of a diagnostic system is the comparison of measured performance metrics with a set of reference performance metrics. Where measured values deviate from the reference values, further data mining is needed to establish the nature of these anomalies and to come up with recommended courses of action. Data mining may employ techniques such as principal component analysis, T^2 and Q statistics (O'Neill *et al.*, 2014). However, before resorting to computational strategies, it is worthwhile to review the system configuration and employ a spatial and temporal partition strategy, which aims to find out whether issues are local or system wide and, where possible, to identify the location of problems so that analysis efforts can be directed at possible problem areas (Wu and Sun, 2011a). Another helpful strategy that may help to diagnose the cause of the problem is the process of Active Functional Testing (AFT). This process creates very specific control actions that lead to special system behaviour, which in turn makes it easier to recognize typical error patterns (Padilla and Choinière, 2015).

The potential reasons for anomalies in monitored data may be formally represented in the form of a 'fault tree', which then yields hypotheses for the identification of the causes of any outliers during the diagnostic stage (Capozzoli *et al.*, 2015). The corresponding investigation is named fault tree analysis (FTA); although it can also be used to identify regular states such as whether a system is turned on or off. Fault tree analysis is used in risk and reliability analysis in domains such as nuclear and aerospace engineering. It helps to graphically describe the logic that leads to an issue. Key elements in fault trees are OR gates, where any of the inputs may lead to the top issue or AND gates where all inputs need to be present to lead to the top issue. For an application of FTA to building performance and recognition of faults, especially in the structural domain, see Pan (2006) or Yu *et al.* (2013).

FDD may apply statistical methods to compare the current system status with that of a past situation that was assessed as representing the normal condition or baseline. Alternatively, the system status may be compared to outcomes of a simulation effort (Wu and Sun, 2011b; Pavlak *et al.*, 2016; Boxer *et al.*, 2017). The statistical methods used are similar to those used for regular monitoring. Jacob *et al.* (2010) discuss the use of black-box models for fault detection, emphasizing linear regression and change point models and the need to identify and remove outliers in the training data. Typical tools for signal processing and feature extraction are (auto)regression models, Fourier transforms, wavelets transforms, time-frequency analysis and domain-specific models and knowledge. For performance analysis in terms of building health monitoring, the signals and features can be assessed using statistical analysis, feature maps, logistic regression, neural networks, particle filters and Hidden Markov Models. For performance prediction, signals and features can be processed using autoregressive moving average prediction, Elman recurrent neural networks or Match Matrix prediction. Finally, for condition diagnosis use can be made of statistical matching or Support Vector Machines (Djurdjanovic *et al.*, 2003). A deep review of methods for fault detection, diagnostics and prognostics in building services is provided by Katipamula and Brambley (2005a, 2005b), who relate this to FDD theory from other disciplines.

Modern FDD systems monitor a wide range of building performance measures and apply a large number of rules to detect issues; for instance the Diagnostic Agent for Building Operation (DABO) system is reported to track 275 indicators and apply 800 rules to supervise the proper operation of mechanical and electrical building systems (Ferretti *et al.*, 2015). O'Neill *et al.* (2014) propose a building performance monitoring and diagnostics system built on the basis of the BCVTB and EnergyPlus simulation models; similar work is presented by Boxer *et al.* (2017) or Zhang and Hong (2017). In contrast to other approaches, the use of models has the advantage of not needing any training data; however this comes at the price of requiring knowledge of how systems ought to behave in terms of physical principles. Unfortunately there are limitations to the systems representations[2] in EnergyPlus and similar simulation programs that put constraints to their use in FDD. An advanced system like the Watchdog Agent (Djurdjanovic *et al.*, 2003) takes data from sensors as key input, combining this with historical information, expert knowledge and engineering models. This is used for three purposes: performance assessment, performance prediction and performance diagnosis; these in turn feed into proactive maintenance decisions, possible product reuse and product redesign. Liang and Du (2007) employ a Support Vector Machine for a single zone HVAC system, which is trained with data from a lumped parameter model in order to capture three potential problems: a stuck damper, blocking of a coil and issues with the fan. Capozzoli *et al.* (2015) demonstrate the use of FDD at the level of a small cluster of eight buildings, instead of the usual application to single buildings.

As with design, building monitoring and analysis results also need to be communicated to users. O'Neill *et al.* (2014) present an energy monitoring visualization dashboard; Pavlak *et al.* (2016) and Boxer *et al.* (2017) employ traffic light signals to represent status, deviation and fault occurrence. Further detail on information dashboard design is provided by Few (2006). Beyond the status and development of building performance, any notable exceptions must be flagged up to the client or operator. This may take the form of alarms. Some software systems allow users to set alerts that are triggered when some predefined thresholds are exceeded; in most cases expert facility managers will then proceed to investigate the causes and correct any issues found (Carbon Trust, 2012: 18). In some cases, a simple observation of a measured value like inlet air

---

2 In many simulation tools, systems may be simplified to a relationship between parameters; for instance, an air handling unit may be captured by a mathematical representation between air speed, mixing and air temperatures, without explicit representation of a fan and fan speed.

temperature is sufficient to monitor system performance; an alarm can be set to warn the operator if this value goes outside the acceptable range. However, in many FFD systems, underlying principles and their assessment are much more complex. For instance, in an air handling unit, there may be a stuck valve that limits the amount of cold air that enters the unit, but this may be compensated by a cooling coil working harder. To safeguard against such issues, one can set up smart alarms which monitor a range of sensors (California Commissioning Collaborative, 2011: 54). In setting up FDD tools, users must define the tolerance level for alarms. This requires finding a proper balance: tight tolerances result in many issues being flagged and needing follow-up; wider tolerances reduces the number of issues needing attention but means that some faults may remain undetected. A common approach is to start with a wider tolerance and progressively tighten this as the operation of the building is becoming better (*ibid.*, 61). Some systems also may provide a league table that orders systems or at a higher level orders buildings in terms of status or efficiency so that operators can direct their attention to the least efficient entries.

While FDD tools help to note the occurrence of an issue and to identify what is the cause, which allows fast interventions and short response times by the building operator, it is important to note that in the end human intervention is needed to follow up and actually fix the problem. This requires the existence of a process that guarantees that the FDD tools are actually used, and any alarms are followed up (California Commissioning Collaborative, 2011: 60; Costa *et al.*, 2013). While in building management monitoring may be a permanent process, dedicated fault detection is often carried out during system downtime. Downtime starts with realization time, which is the time between fault occurrence and the operator realizing there is a fault, and ends once the system is back to normal operation and has been verified to work properly. Once the occurrence of a fault has been established, access time starts, for instance to remove covers or to connect measurement systems. This is followed by the period of actual diagnosis: finding the fault and what causes this. Next is part procurement if any systems or components need replacement, followed by actual replacement time, checking, and if needed system alignment (Smith, 2011: 22–23); see Figure 9.1.

While FDD approaches can be very helpful, it is also worthwhile to remember some drawbacks and limitations. Many systems are only based on relatively simple rules; if that is the case one should not expect them to be able to respond to the more complex interactions that may take place at the full building level (O'Neill *et al.*, 2014). And while analysis of monitored data can make use of various computational approaches, such as fuzzy logics, stochastic models, vector support machines and others, typically the

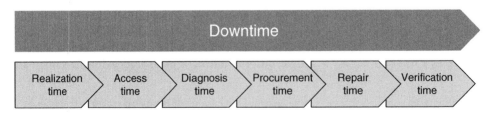

Figure 9.1 Downtime and its components.

identification of failure modes and use of models for prediction of failure times requires deep knowledge of the assessed system or process (Djurdjanovic *et al.*, 2003). In terms of commercial systems, fault detection and diagnosis modules are often provided by a separate platform, which leads to issues of integration and communication between the FDD module and the central building automation system (Dong *et al.*, 2014). O'Neill *et al.* (2014) state that there is no FDD system that is fully integrated yet and that the existing offerings are limited add-on systems. Another problem with advanced building management systems for fault detection is that these have limited transferability, as systems are highly specific for a given building; this means that software needs to be scaled, calibrated and tested to fit new buildings (O'Neill *et al.*, 2013). While BIM technology may benefit FDD analysis, in practice there still is a disconnect between the design and operational stage. Standards such as COBie (Construction Operations Building information exchange) and HVACie (HVAC information exchange) aim to support a better link between design and facility management; researchers are working on BIM infrastructure that combines static building information, prediction by way of simulation, monitored data and FDD (Dong *et al.*, 2014).

9.4 Performance Service Companies and Contracts

Building operation, control and management is not purely a technical issue: since buildings require important investments to construct and cash flows to operate, there also is a clear business perspective on building performance. Traditionally this is the domain of Facility Management, which covers maintenance, service provision, policy and procedures for provision of services and contract management (Atkin and Brooks, 2009). Similar efforts on system operation and management can be found in the domain of systems engineering, which supports processes like life cycle management (LCM) and service life management (SLM) through the monitoring of systems using key performance indicators, technical performance parameters and service performance measures (SEBoK, 2014: 569). In the built environment, there is an increasing business focus on efforts that relate to the management of energy consumption: energy is a key resource for building operation that is subject to increasing costs, significant price volatility and environmental regulations. As a consequence there is an emerging energy efficiency business that not only supplies renewable energy to buildings but also invests in retrofits to upgrade existing facilities.

From a business point of view, building performance is never a prime objective; business is concerned with generating revenue. However, to operate and make such revenues, many companies require premises, and these lock up significant amounts of money in buildings that thus are strategic assets. On the operational side, part of the business cash flow is tied up in operating these buildings and thus is not available for other activities in the organization or marketplace. However, some specialist companies are making this very fact their core business proposition, again with the energy domain leading the field: here Energy Service Companies (ESCOs) are offering Energy Performance Contracts (EPCs) to building owners on the basis of them taking on the investments in buildings in return for a share in the savings that will be achieved. Business decisions are framed in typical economical terms. This means that building performance management needs to relate to the required investment, annual projected savings, net present value (NPV),[3] internal rate of return (IRR),[4] return on investment (ROI)[5] and payback period.[6] Business decisions are made at a high level, typically by a Chief Financial Officer (CFO) who has to balance business priorities, available resources, activities of the competition, and impact on cash flow against any investment opportunities.

The finance sector is the part of business community that makes investments; financing is the process of finding a source of capital investment for a procurement activity. There is a range of options to improve building performance, from monitoring to the installation of new systems and retrofit, that come at different prices. The cost of making a strategic investment in better performance must be balanced against the cost of inaction; in some situations the hand of the decision maker is forced because a system is broken and requires immediate replacement. Building owners have different options to finance performance upgrades: they may be able to provide capital themselves or may gain access to a commercial loan (traditional borrowing of money), operational lease (where the equipment remains property of the lessor) or a capital lease (where the lessee gains ownership of the equipment at the end of the lease). In some cases there are special schemes for government or municipality financing. In many countries there are also specific programmes, such as the property assessed clean energy (PACE) program in the United States. Further complexity is added by local tax and rebate systems. Where finance involves a loan, one must consider the interest rate, monthly payment and deferred payment, which all have an impact on cash flow operation; for loans the building or equipment installed in the building may become a lien to secure payment. In all these structures, it is important to establish ownership of the building and building systems, who has the benefit of any performance upgrades and reduced operating costs, and who has the responsibilities if targets are not met.

Energy efficient operation of buildings is tied into the operation of (inter)national power grids, which goes beyond the boundaries of single facilities. Building management thus is related to (smart) grid operation; power utility companies may want to balance power demand from the grid and thus provide financial incentives for building owners to reduce energy demand for a finite period (Kang *et al.*, 2014) or increase their use during demand valleys (Gruber *et al.*, 2015). The energy market is complex; building owners for instance may opt to hedge against future energy price rise by buying energy at fully fixed prices, or they may opt to go for fully indexed pricing where they buy energy at spot market price; they can also purchase some energy through block pricing and top this up with fully indexed energy. What is appropriate depends on the specific operation, access to capital and willingness to take risks.

The current Energy Service Company (ESCO) concept emerged in the United States in the 1970s, and the United States still is the most mature ESCO market; Larsen *et al.* (2012) estimate that the US ESCO industry has generated around US\$23 billion between 1990 and 2008. In the United Kingdom, ESCOs became established in the 1980s, but in many other countries the concept was only taken up in the mid-1990s (Fang *et al.*, 2012a).

---

3 NPV: cash inflow minus cash outflow, with adjustment of future cash flows to the present time.
4 IRR: budgeting measure on the basis of NPV, which represents the rate at which the breakeven point is reached for any given period of time.
5 ROI: measure for the efficiency of an investment, calculated by ROI = (gain - cost)/cost.
6 Payback period: length of time needed to recover the initial investment.

Marino *et al.* (2011) report that there are somewhere between 650 and 1040 ESCOs in Europe in 2010, especially in Germany, Italy and France. However, ESCOs operate in particular market segments only; for instance Suhonen and Okkonen (2013) discuss the potential of ESCOs to enter the residential market through collaboration with housing associations in Finland, but overall find that the ESCO model is not really attractive in this sector.

An energy performance contract (EPC) is an agreement where an ESCO guarantees specific energy savings to a building owner, in return for a share of the profits that will result from these energy savings. In this context, ESCOs may also be named EPC provider or contractor. EPCs are sometimes called energy saving performance contracts (ESPCs) or energy saving agreement (ESA). Fundamentally, EPCs transfer the authority over the energy management of a building, including decisions on investments, improvement and maintenance, from the client to the ESCO for the duration of the contract. In terms of business, the use of EPCs and ESCOs is similar to outsourcing. EPCs are specific to a certain site, the facilities on that site, and relate to human assets and expertise; obviously they vary significantly in depth, scope and method of finance. Simple EPCs just cover electricity and hot water supply; more advanced contracts include further performance aspects such as lighting, thermal comfort and relative humidity (Sorrell, 2007). In contrast, a traditional energy supply contract just governs the provision of electricity, gas, steam/hot water or coolant, but not final energy services (*ibid.*); energy suppliers do not bear the performance risks that comes with EPCs (Pätäri and Sinkkonen, 2014). EPCs specify energy savings (results) rather than the specific technology to be deployed (Nolden *et al.*, 2016). Examples of the use of EPCs are provided by Murphy *et al.* (2013) who discuss actual savings through an EPC in a governmental lab in the United States or Polzin *et al.* (2016) who explore the experiences with the use of EPCs by local authorities in Germany. Hufen and de Bruijn (2016) report on the successful implementation use of EPCs for the maintenance and management of nine swimming pools in Rotterdam in the Netherlands.

For the client, the key advantage of an EPC is that it shifts investment costs and risks to the ESCO, allowing the client to concentrate on their core activities. For the ESCO, the EPC gives access to a share in the profits of energy use reduction in the building of the client. Obviously, due to the financial stakes in an EPC, it is important that there is trust between client and ESCO; this can be supported by professional operation, transparent and shared data, and clear M&V processes; use of an independent third party to carry out that M&V is recommended. Waltz (2002) discusses the role of M&V in energy performance contracting, with an emphasis on the US context – such as the development of the IPMVP from the earlier North American Energy Measurement and Verification Protocol (NEMVP) and some typical M&V terminology such as 'energy savings' (which include 'cost avoidance' measures), stipulated values and calculations and estimates (which can be used even when a measure has already been applied). The paper also outlines the main options for quantifying energy savings: on the basis of utility bills at the whole building level (option C), on the basis of agreed but fixed calculations (option A) or on the basis of specific measurement (option B).

In a traditional energy efficiency upgrade, all that the client may get in terms of ensured operation of newly installed systems is a warranty. In an EPC, the ESCOs income is tied into the savings that are to be made, so the ESCO has a vested interest in ensuring all systems work properly and that operation and management is carried out well (Lee *et al.*, 2015b). The working of EPCs is obviously impacted by taxes, rebates and subsidies; Lu and Shao (2016) have explored the influence of government subsidies on the performance levels and pricing in the EPCs offered by ESCOs, demonstrating significant effects of subsidies on ESCO demand and profits. Performance contracts are still under development; strictly EPCs deal with energy but sometimes ESCOs also include other utilities such as water (Murphy *et al.*, 2013). Saari and Aalto (2006) discuss the development of performance contracts that would cover indoor environmental quality.

The process of setting up an EPC consists of typical procurement stages: budgeting, auditing the existing facility, putting out a request for proposals, evaluating bids received, organizing project financing, contracting, and measurement and verification once the project is implemented (Lee *et al.*, 2015b). A good relationship of mutual trust between a client and ESCO helps to prevent issues; in this context it is good if the client occasionally looks at energy monitoring data or if a third party is brought in to check and quantify savings (Waltz, 2002). A crucial aspect of setting up an EPC is the agreement on a mutually acceptable baseline performance that can be used for determining any energy savings achieved (Yik and Lee, 2004). Another one is the decision on the contract period; this has a direct relation to the competiveness of bids as well as the potential revenue (Deng *et al.*, 2014). Furthermore, the client and the ESCO need to agree on how to divide the profits as well as the risks that stem from the project (Qian and Guo, 2014). These issues are non-trivial, as the related decisions need to be made under uncertainties such as the actual performance of any energy conservation measures and the future fluctuation of energy prices (Deng *et al.*, 2014; Qian and Guo, 2014). Beyond these basic uncertainties, there are further risks: for instance one of the parties involved in an EPC may default, or installation costs of energy conservation measures may be higher than anticipated. Clients may also be concerned about long payback periods, the complexity of the arrangement and their own repayment ability. In general, EPC project risks can be classified as economic risk, finance risk (involving a third party), project design risk, installation risk, technology risk, operational risk and measurement and verification risk (Lee *et al.*, 2015b). In starting a move towards setting up an EPC, it is also worth bearing in mind that something as simple as who is asked for guidance may have a significant impact on the outcome. For instance, engagement with lawyers and policy makers may lead towards use of an EPC, whereas engagement with an energy consultant may tend to point clients towards keeping the facility upgrade in-house (Polzin *et al.*, 2016). Wang *et al.* (2017b) discuss how normative decision theory may support the handling of uncertainties in energy performance contracts.

Problems in working with EPCs may stem from clients being reluctant to enter into this type of contract, financial institutions being reluctant to offer loans, difficulties in measuring energy performance and unforeseen fluctuations in contextual factors such as energy prices, weather or the building use pattern (Yik and Lee, 2004). There also may be legal barriers, low governmental support, a general lack of awareness, information and understanding and a lack of expertise (Pätäri and Sinkkonen, 2014). EPCs are meant to create benefits for both the client and the ESCO, yet in reality a mismatch between expected and actual energy savings may cause problems. Yik and Lee (2004) discuss cooperation between client and ESCO, suggesting a partnership may be a way to improve on conventional contracts. EPCs may not be suitable for all organizations and situations (Sorrell, 2007). Nolden *et al.* (2016) suggest that they works best in medium size and complexity facilities, where the expertise of the ESCO is likely to outweigh in-house expertise but where the facility leaves room for the ESCO to make sufficient efficiency gains to make a profit.

9.5 Building Operation, Control and Management Challenges

A building that exemplifies the challenges of dealing with special requirements in terms of building operation, control and management is the Test Centre of the European Space Agency (ESA) at the European Space Research and Technology Centre (ESTEC) in Noordwijkerhout in the Netherlands. ESA is an intergovernmental organization that has been established in 1975 and that operates on an annual budget of €5.25 billion. Within this centre, ESA tests satellites; this has to happen in a space that is extremely clean, as satellites are designed to operate in space where there is no dust. Contamination with dust may cause spacecraft electronics to fail and can have a negative impact on the mirrors and lenses used in some satellites. The Test Centre has facilities that allow to tests various aspects of the satellite life cycle, which includes being the subject of severe vibrations and acoustic noise during launch. Satellites are also tested under vacuum and low temperatures conditions for prolonged periods of time. To support such testing, the ESTEC Test Centre includes facilities such as the Large Space Simulator (LSS), a Thermal Vacuum Test Chamber (Phenix), a multi-axis hydraulic shaker (Hydro), the Large European Acoustic Facility (LEAF), the Compact Payload Test Range (CPTR), an Electromagnetic Test Chamber (Maxwell) and a large cleanroom (Rosetta). See Figure 9.2. Beyond comprehensive testing ESA also provides cleaning, detection,

Figure 9.2 'Rosetta' Clean room at ESTEC, Noordwijk, the Netherlands. Image courtesy of European Space Agency – ESA.

disinfection and sterilization services for spacecraft. This paragraph focuses on the clean room facilities.

The ESTEC Cleanrooms are ISO certified. The ISO classes give limits for the maximum number of particles per cubic meter for a range of particle sizes. Four of the ESTEC cleanrooms are ISO Class 8, which means they have less than 3 520 000 particles of 0.5 µm or smaller per cubic meter; a tent is available that offers ISO Class 5, which provides an environment with less than 100 000 particles of 0.1 µm or smaller per cubic meter. The total floor space of the four interlinked clean rooms is 1309 m². Operations are supported by dedicated air locks and checkout rooms. The clean rooms have overhead cranes that can carry equipment and spacecraft of up to 32 ton. ESTEC also operates one ISO class 1 clean room of 35 m² surface; this is the highest category in the ISO classification, requiring an environment with less than 10 particles of 0.1 µm or smaller per cubic meter. This latter clean room requires extreme filtering of ventilation air, with two-stage HEPA (High-Efficiency Particulate Arrestance) filters on fresh air, fan filer units (FFU) that remove particles from recirculated air using Ultra Low Penetration Air (ULPA) filters and plus Airborne Molecular Contamination (AMC) filters of classes A, B, C and D. HEPA filters have an efficiency between 99.97% and 99.995%, while ULPA filters are 99.9995% efficient.

To achieve higher rates, clean rooms are typically nested, with critical areas contained in clean areas and clean areas contained in transition zones. In this manner, access is via zones that are increasingly clean and act as buffer. Beyond contamination from the outside, further sources may be inside the facility: there may be small contributions from the walls, floors, ceilings, paint and coatings of the clean room; equipment may emit particles from wear and tear; cleaning equipment in itself may be a source of contaminants; and people may introduce skin flakes, oils, hairs and clothing debris.

The operation of clean rooms also requires a strict operational regime to be in place. Personnel entering the facilities walk over a special floor mat that removes any dirt from their shoes. They then have to wear special equipment: a dust-free suit, covers over their head, gloves, a facemask and covers over their shoes. Any equipment and spacecraft entering the clean room facility has to be cleaned. To facilitate operations, the clean-rooms in ESTEC are interlinked, so that they may be moved from one test chamber to another without a need for repackaging and the associated risk of contamination. Personnel inside the clean rooms are not allowed to walk fast, run, sit on equipment and work surfaces or remove items from beneath their clean room garments. Only specially approved clean room paper and ball points are allowed inside clean rooms.

Cleaning is an essential element of operating clean rooms. To prevent cleaning equipment from being counterproductive, special materials are used, which give of minimal amounts of particles; disinfectants are filtered, and only clean room approved wipes can be used. There is a prescribed cleaning routine that goes from top to bottom, following the direction of the airflow.

There is a range of ISO standards that govern the design, classification and operation of clean rooms: ISO-16644-1 (classification by airborne particles), ISO-16644-2 (monitoring for compliance), ISO-16644-3 (measurement and testing), ISO-16644-4 (design, construction and start-up) and ISO-16644-5 (clean room operations). Clean rooms are measured for particulate count in three different situations: as built, at rest and operational. Measurement takes place via an advanced particle counter that needs to be calibrated; measurement locations are to be distributed evenly throughout the area of the clean room and take place at height of work level. Further testing involves airflow and velocity ratings and filter integrity tests based on introducing aerosols in the air system and then scanning the filter face.

9.6   Reflections on Building Automation and Monitoring

Building automation and professional monitoring are areas that are rather distinct from the worlds of physical testing, building simulation and stakeholder evaluation. Many systems that control buildings employ quite different approaches and logic. Moreover, many of these systems are proprietary for commercial reasons. Terminology used in the domains of M&T and M&V is noticeable different.

The pace of development in automation and monitoring is high. AMR is continuously capturing more aspects of buildings, while also increasing temporal resolution, creating a sea of data. However, this data only has use if it is properly analyzed. Machine learning may go far in detecting trends and correlations in this data, but ultimately one needs to reach an understanding of what the data means and how it relates to the key human concerns about building performance. This requires alignment with the other approaches to building performance, which at present seems underdeveloped. There are interesting tensions. In the past, measured data often consisted of monthly energy bills; simulation operated with much higher hourly resolution. With AMR, building data may now be available for intervals of 30 minutes of less. While it is possible to reduce the simulation time step to match this, some fundamental inputs such as weather data are still only provided on an hourly basis. It would be wise not to embark on an arms race for higher resolution, but to search for a time step that aligns with the underlying analysis needs, whether the tool is hardware in a real building or software simulating a virtual building.

Some actors see monitored data as the real thing and the ultimate truth. However, it is worth keeping in mind that systems and their settings may be imperfect from the start; and even if they once were good, they will typically deteriorate and drift over time. There may be many sources of error, such as sensor position, malfunction and misinterpretation. Some of the capabilities of working with data may be overrated; actual delivery by BMS and FDD systems needs to be carefully checked, and in the end often requires human intervention. Reactive maintenance is still quite acceptable in buildings, whereas it is not in the automotive or aerospace industries. Future developments in the area need to improve on that status, for instance through self-learning BMS and FDD, and training of these systems on the basis of simulated data even before a building is constructed.

While overall building automation has good prospects, there are also some challenging areas where other solutions may be required. For instance, Passivhaus buildings rely strongly on a fabric first approach and need only very limited use of building services, which renders automation less important. A challenge to all automation efforts is the variability in human behaviour. To work well building automation systems need to be aware of changes to human patterns and therefore changing needs. In this context there also is an unanswered question: is it better to automate buildings and give systems intelligent control, or is it better to educate the building occupants and let them manage the building?

Many recent trends in building management business are related to energy efficiency, such as the new approaches of ESCOs and EPCs. It will be interesting to explore whether these can also be applied to a wider area of performance aspects.

9.7 Summary

This chapter deals with building performance during the operational stage. It looks at building performance management, maintenance, building automation systems and controls, including model-based predictive control. It covers building monitoring in terms of the process of sensing and measuring, data handling and data analysis, paying special attention to M&T and M&V. The chapter discusses fault detection and diagnosis, and closes with a section on the business that is emerging around building performance, especially the concept of energy performance contracts.

Building control systems manage the building in terms of handling the on/off status and detailed behaviour of building components. This enables the building to meet setpoint values; advanced systems may also react to requests by the grid for demand response management and load shaping. In general, building performance has a tendency to drift downward; maintenance aims to counteract this tendency. There are different types of maintenance: fail-and-fix, predict-and-prevent and condition-based. In many cases, monitoring of performance and comparison of measured values with a baseline or benchmark is a driving force in decisions about when to intervene. The systematic assessment of actual performance against expected performance is named monitoring and targeting (M&T). A similar concept is measurement and verification (M&V); however this is mostly used in the context of interventions in buildings and performance contracts, with the aim of establishing the savings made by such interventions.

Modern building control is provided by building automation systems (BAS); sometimes this may also be named building management system (BMS), building energy management system (BEMS), energy management and control system (EMCS) or similar. These systems employ a network of sensors to capture the building state and performance, computing capacity to decide on the best action to take, and control systems to enact this course of action. In digital automation systems, messages about the building state or performance are event or alarm notifications; instructions to systems and components are commands. There is a range of communication protocols to handle these messages, such as the general Simple Network Management Protocol (SNMP), BacNet, LonTalk and others. Building automation systems have the advantage of removing the need for human operation of the building systems; if operated well they may also save resources like gas, electricity and water. There is a wide range of commercial building automation systems; like all systems these benefit from proper commissioning and maintenance. Buildings with advanced automation systems are often named 'smart'. Many modern buildings employ automated meter reading (AMR). Meters that interact with the building are often named smart meters. Model-based predictive control employs computation to predict future states, which then can be used to decide on the best control action. The underlying models may be very simple (look-up tables based on advanced computation), intermediate (reduced order models that stem from observations or calculations) or advanced (full transient representation of the relevant physics). However, it is crucial that input is provided at the time scale required for the control, which still limits some of the more advanced approaches.

In essence, building performance monitoring is measurement over time. There is a wide range of data that may be collected, combined with different spatial and temporal resolution. Sometimes data can be collected from the building automation system, but it must be noted that many automation systems are proprietary and that their prime objective is control, not data collection. Most monitoring will involve the installation of dedicated meters and sensors, data acquisition systems such as loggers and transmission equipment, and data collection and archiving systems. Often static data about the building, such as floor areas, also needs to be gathered. Monitoring may be augmented with data from questionnaires, surveys and billing. For performance management, measured data needs to be compared with benchmark or baseline values, which represent the way the building ought to perform. A key document that defines approaches in this area is the International Performance Measurement and Verification Protocol (IPMVP). One should take into account the accuracy of the measurement equipment and any potential errors; regular calibration may help reduce problems. A careful review of data quality, and where needed a process of data cleansing, is recommended before moving on to the analysis process. Analysis of monitored data may be done visually by human operators; however mathematical techniques such as regression analysis are useful and may be used to automate the process.

Fault detection and diagnosis (FDD) reviews the anomalies that are found during monitoring and aims to identify any underlying problems. This requires deep understanding of how systems work; some anomalies may be outliers but not necessarily a fault. Many diagnosis tools work on the basis of expert rules, which are structured in the form of a fault tree; this allows identification of problems using 'if-then-else' logic. More advanced systems employ pattern matching, comparing measured data with the data signature that can be expected when certain faults occur. FDD systems need an appropriate user interface to communicate with humans; in the end human operatives are required to correct faults and carry out preventive maintenance.

Building performance is not just a technical issue but also a business opportunity. There is an increasing interest in outsourcing building operation, control and management. Key aspects to consider in this context are Energy Service Companies (ESCOs) and Energy Performance Contracts (EPCs).

Recommended Further Reading:

- The *Building Performance Tracking Handbook* (California Commissioning Collaborative, 2011) for a good overview of the many reasons to monitor building performance and the main approaches available.
- The *Specification Guide for Performance Monitoring Systems* (Gillespie *et al.*, 2007) for a more in-depth review of some of the considerations that play a role when selecting monitoring equipment.
- The paper on building automation by Domingues *et al.* (2016) for a deeper review of the concepts and technology of these systems.
- The report on Core Concepts of the International Performance Measurement and Verification Protocol (IPMVP) for an introduction to M&V and the required data analysis (Efficiency Valuation Organization, 2014a).

- The papers by Katipamula and Brambley (2005a, 2005b) for a good introduction to FDD in the building services domain.
- The work by Lee *et al.* (2015b) for a discussion of the risks involved in Energy Performance Contracting projects.

Activities:

1   Compare and contrast the working of a thermostatic radiator valve and that of a classic room thermostat in terms of the interaction with the rest of the HVAC system.

2   Select a building that allows you gain access to documentation in terms of use of resources. Draw a section of this building, and overlay a Sankey diagram that shows the resources entering that building, as well as how rest products of those resources leave. Annotate the flows with annual values.

3   Conduct a small monitoring campaign. If you have access to a full building, such as an apartment or house, take daily readings of the consumption for utilities such as electricity, gas and water. If you only have access to a room, install a couple of plug-in power and energy meters to capture the electricity consumption of key equipment like your PC, fridge and TV. After you have collected data for a week, analyze the data and try to find out what the main drivers for your consumption are. In the following week, try to bring down the consumption.

4   Study the fault detection and diagnosis processes of a medical doctor (GP) and a car mechanic. What are the parallels and differences between these processes and FDD in buildings?

5   Find a room with a digital thermostat that allows you to read the setpoint as well as the actual temperature. Use an accurate digital handheld thermometer to take detailed temperature readings in the room. Draw a floor plan and a section of the room with isotherms, showing spatial differences and stratification. Then review the accuracy of the thermostat reading and associated control actions.

6   Find out what companies are active in your area that might offer Energy Performance Contracts to local schools and universities. Review the structure of these companies; are they pure ESCOs, or are they larger companies that also take on the ESCO function?

9.8   Key References

Ahmed, A., J. Ploennigs, K. Menzel and B. Cahill, 2010. Multi-dimensional building performance data management for continuous commissioning. *Advanced Engineering Informatics*, 24 (4), 466–475.

Brown, N., A. Wright, S. Shukla and G. Stuart, 2010. Longitudinal analysis of energy metering data from non-domestic buildings. *Building Research & Information*, 38 (1), 80–91.

California Commissioning Collaborative, 2011. *The Building Performance Tracking Handbook*. California Energy Committee.

Capozzoli, A., F. Lauro and I. Khan, 2015. Fault detection analysis using data mining techniques for a cluster of smart office buildings. *Expert Systems with Applications*, 42 (9), 4324–4338.

Corry, E., J. O'Donnell, E. Curry, D. Coakley, P. Pauwels and M. Keane, 2014. Using semantic web technologies to access soft AEC data. *Advanced Engineering Informatics*, 28 (4), 370–380.

Costa, A., M. Keane, J. Torrens and E. Corry, 2013. Building operation and energy performance: monitoring, analysis and optimisation toolkit. *Applied Energy*, 101, 310–316.

Curry, E., J. O'Donnell, E. Corry, S. Hasan, M. Keane and S. O'Riain, 2013. Linking building data in the cloud: integrating cross-domain building data using linked data. *Advanced Engineering Informatics*, 27 (2), 206–219.

Djurdjanovic, D., J. Lee and J. Ni, 2003. Watchdog Agent – an infotronics-based prognostic approach for product performance degradation assessment and prediction. *Advanced Engineering Informatics*, 17 (3/4), 109–125.

Domingues, P., P. Carreira, R. Vierira and W. Kastner, 2016. Building automation systems: concepts and technology review. *Computer Standards & Interfaces*, 45, 1–12.

Dong, B., Z. O'Neill and Z. Li, 2014. A BIM-enabled information infrastructure for building energy fault detection and diagnostics. *Automation in Construction*, 44, 197–211.

Efficiency Valuation Organization, 2012. *International Performance Measurement and Verification Protocol: Concepts and options for determining energy and water savings, Volume 1*. Washington, DC: Efficiency Valuation Organization.

Efficiency Valuation Organization, 2014a. *International Performance Measurement and Verification Protocol: Core concepts*. Washington, DC: Efficiency Valuation Organization.

Efficiency Valuation Organization, 2014b. *International Performance Measurement and Verification Protocol: Statistics and uncertainty for IPMVP*. Washington, DC: Efficiency Valuation Organization.

Few, S., 2006. *Information Dashboard Design – The Effective Visual Communication of Data*. Sebastopol, CA: O'Reilly Media.

Gillespie, K., P. Haves, R. Hitchcock, J. Deringer and K. Kinney, 2007. *A specification guide for performance monitoring systems*. Berkeley, CA: Lawrence Berkeley National Laboratory.

Giri, S. and M. Bergés, 2015. An energy estimation framework for event-based methods in non-intrusive load monitoring. *Energy Conversion and Management*, 90, 488–498.

Granderson, J. and P. Price, 2014. Development and application of a statistical methodology to evaluate the predictive accuracy of building energy baseline models. *Energy*, 66, 981–990.

Granderson, J., S. Touzani, C. Custodio, M. Sohn, D. Jump and S. Fernandes, 2016. Accuracy of automated measurement and verification (M&V) techniques for energy savings in commercial buildings. *Applied Energy*, 173, 296–308.

Gruber, J., F. Huerta, P. Matatagui and M. Prodanović, 2015. Advanced building energy management based on a two-stage receding horizon optimization. *Applied Energy*, 160, 194–205.

Hu, J., O. Ogunsola, L. Song, R. McPherson, M. Zhu, Y. Hong and S. Chen, 2014. Restoration of 1–24 hour dry-bulb temperature gaps for use in building performance monitoring and analysis – Part I. *HVAC&R Research*, 20 (6), 594–605.

Jang, W., W. Healy and M. Skibniewski, 2008. Wireless sensor networks as part of a web-based building environmental monitoring system. *Automation in Construction*, 17 (6), 729–736.

Jiang, Z., J. Xia and Y. Jiang, 2009. An information sharing building automation system. *Intelligent Buildings International*, 1 (3), 195–208.

Katipamula, S. and M. Brambley, 2005a. Methods for fault detection, diagnostics and prognostics for building systems: a review, part I. *HVAC&R Research*, 11 (1), 3–25.

Katipamula, S. and M. Brambley, 2005b. Methods for fault detection, diagnostics and prognostics for building systems: a review, part II. *HVAC&R Research*, 11 (2), 169–187.

Lee, P., P. Lam and W. Lee, 2015b. Risks in energy performance contracting (EPC) projects. *Energy and Buildings*, 92, 116–127.

O'Neill, Z., X. Pang, M. Shashanka, P. Haves and T. Bailey, 2014. Model-based real-time whole building energy performance and diagnostics. *Journal of Building Performance Simulation*, 7 (2), 83–99.

Painter, B., N. Brown and M. Cook, 2012. Practical application of a sensor overlay system for building monitoring and commissioning. *Energy and Buildings*, 48, 29–39.

Park, C., 2003. *Occupant responsive optimal control of smart façade systems.* Ph.D. Thesis. Atlanta: Georgia Institute of Technology.

Shih, H., 2014. A robust occupancy detection and tracking algorithm for the automatic monitoring and commissioning of a building. *Energy and Buildings*, 77, 270–280.

Sorrell, S., 2007. The economics of energy service contracts. *Energy Policy*, 35 (1), 507–521.

Veronica, D., 2013. Automatically detecting faulty regulation in HVAC controls. *HVAC&R Research*, 19 (4), 412–422.

Vesma, V., 2009. *Energy management principles and practice.* London: British Standards Institute.

Wetter, M., 2011. Co-simulation of building energy and control systems with the Building Control Virtual Test Bed. *Journal of Building Performance Simulation*, 4 (3), 185–203.

Wu, S. and J. Sun, 2011b. Cross-level fault detection and diagnosis of building HVAC systems. *Building and Environment*, 46 (8), 1558–1566.

Yan, D., W. O'Brien, T. Hong, X. Feng, H. Gunay, F. Tahmasebi and A. Mahdavi, 2015. Occupant behavior modeling for building performance simulation: current state and future challenges. *Energy and Buildings*, 107, 264–278.

Yik, F. and W. Lee, 2004. Partnership in building energy performance contracting. *Building Research & Information*, 32 (3), 235–243.

10

High Performance Buildings

While all buildings can be assessed in terms of their performance, there is one category of buildings that makes performance a key characteristic and aspires to be state of the art with regard to this particular aspect: high performance buildings. However, this is another concept that is poorly defined and at times appears to be reduced to a buzzword; for instance many authors use the terms high performance building, green building and sustainable building interchangeably, without proper distinction (Prum, 2010). One root cause for this situation is the lack of definition of the term building performance itself, as addressed in the previous chapters. Beyond that, 'high' performance is a relative term; it positions specific buildings relative to their peers and the wider building stock. One can compare the situation with the automotive sector, where a high performance car may have a different interpretation in the contexts of off-roading or Formula One racing. Similarly, the term high performance building seems more applicable to certain categories of buildings like offices and public buildings, while there is some tension in applying it to other categories such as family homes (Trubiano, 2013: 3).

The ambivalent use of the term high performance buildings is clearly demonstrated by Erhorn and Erhorn-Kluttig (2011), who give a long list of terms that may be seen by some as equivalent to high performance buildings in the member states of the European Union: low-energy house, eco-building, energy saving house, ultra-low-energy house, 3-litre-house, passive house, zero-heating energy house, zero energy house, plus-energy house, zero-emission house, zero carbon house, emission-free house, carbon-free house, energy self-sufficient house, energy autarkic house, triple-zero house, BREEAM building, green building, code for sustainable homes house, bioclimatic house, very low-energy house and climate-active house. These authors discern three main clusters: terms that relate to low-energy consumption, terms that relate to low greenhouse gas emissions and terms that relate to green building aspects. However, this ambivalence can lead to issues. Prum (2010) discusses the use of terms like 'high performance buildings', 'green buildings' and 'sustainable construction' from a legal point of view, noting that this is a problem in contracts; where disputes arise it is often left to the court or arbitrator to interpret these terms and whether a breach of contract has taken place. Further increasing complexity, some authors use compound expressions such as 'high performance green building', where high performance is a class within a wider set of green buildings. But one has to wonder how this differs from a 'green high performance building', which indicates a green class of buildings in a wider set of high performance buildings. Given the prominence of the topic of energy efficiency in the building performance literature,

Building Performance Analysis, First Edition. Pieter de Wilde.
© 2018 John Wiley & Sons Ltd. Published 2018 by John Wiley & Sons Ltd.

it should come as no surprise that some authors take this as the main criterion for high performance buildings, albeit one that needs to be balanced with occupant health and productivity; see for instance Day and Gunderson (2015). Torcellini *et al.* (2006: 1) see zero energy buildings (ZEB) as a type of high performance building that is designed in such a way as to balance energy generation and use on the building site over the course of 1 year, but that interpretation still leaves open what high performance buildings in general are. So the development of a precise definition is overdue.

At the same time, high performance buildings are associated with state-of-the-art design and construction. Some leading architects and engineers are employing principles of building performance and innovation to define building form, and as a key ingredient towards innovative and groundbreaking design (Loftness *et al.*, 2005). Thus many buildings in this category employ novel, innovative technologies. This lead to an interesting conundrum: these techniques may not yet be proven and may suffer from teething troubles, in which case they are detrimental to building performance. Torcellini *et al.* (2006: 1) note that designers and owners may be hesitant to adopt such technologies. Yet at the same time, innovation often requires the acceptance of a certain degree of risk, and emerging technology certainly has a significant pull on architects, building services engineers and their clients.

As will be obvious from the preceding chapter, realization of good building performance is closely related to building operation, automation and control. In this context, other concepts that require further examination are smart and intelligent buildings. Coffey (2013) suggests that high performance buildings may require more advanced operation, for instance by means of model predictive control. But again this area is one of change and developing concepts, especially so as building intelligence is closely related to ICT developments and the latest trends in computing and automation.

Given the interest in high performance buildings, it is no surprise that there are universities that offer programmes on the subject. In the United States, the Georgia Institute of Technology has been running a Building Performance Lab with students at both MSc and PhD levels since 2009; the present offering is a Master of Sciences that consists of a general major in Architecture with a concentration in High Performance Buildings. In Asia, Hong Kong Polytechnic University offers both an MSc and Postgraduate Diploma (PgD) in High Performance Buildings since 2013. In the United Kingdom, Plymouth University opened an MSc programme in High Performance Buildings in 2016. The University of Sydney in Australia offers an MSc in Architectural Science with a High Performance Buildings pathway option. In the United States, ASHRAE offers High Performance Building Design Professional (HPBD) Certification (ASHRAE, 2016). The content of this certification includes building sustainability concepts, HVAC systems and processes, energy analysis (preliminary analysis, envelope design, massing, orientation, ventilation, code compliance), indoor environment analysis (lighting, air quality, thermal comfort), controls and monitoring (optimization strategies and hardware), benchmarking of projects, water conservation, commissioning and operation and management.

Leading the field of construction requires extra efforts and resources. It is often stated that the design of high performance buildings requires an integrated design approach (Hopfe *et al.*, 2013). Authors such as Korkmaz *et al.* (2010) and Swarup *et al.* (2011) point out that the realization of high performance buildings puts tension on the finances and time available for a construction project and thus requires additional efforts in terms of planning, design and actual construction processes. Yet over the last decennia buildings have seen significant change due to the increased capabilities for air conditioning, artificial lighting, telecommunication and information technologies and power supply. These developments in turn lead to a growth in the demands and desires of building occupants and owners, which is unlikely to be reversible (Braham, 2005: 57). Political goals are adding further pressures. In the United States, the Architecture 2030 Challenge aims to reduce the energy use by new buildings with 50% in the near term, while the long-term goal is to achieve net-ZEB by 2030 (Frankel *et al.*, 2015: 3). In the EU, the European Commission has set a target of a 40% cut in greenhouse gas emissions for 2030 as compared with those of 1990 and also aims for a share of at least 27% of renewable energy in energy production. Worldwide, such policies not only are complemented at local city level with further initiatives that increase friendly competition and capacity building but may also include stricter standards, benchmarking and audits (Trencher *et al.*, 2016). Wang *et al.* (2017a) discuss some of the challenges that face buildings over the next century.

This chapter explores the definition of high performance buildings and the criteria for qualifying as such a building. It reviews some of the emergent technologies that are providing the industry with novel systems and approaches that may be employed in state-of-the-art projects, and which push the performance boundaries of tomorrow's building stock. Finally, it explores the category of smart and intelligent buildings, as a special kind of buildings that in many ways is leading the field.

10.1   Existing Definitions for High Performance Buildings

The concept of high performance buildings originates from North America, and this appears to be the area where the idea still finds the most traction. The term came to prominence with the publication of the *High Performance Building Guidelines* by the City of New York (NYC DDC, 1999). This defined high performance buildings as buildings that go beyond the typical construction projects in terms of being energy and resource efficient, specifying the objectives of high performance buildings as 'maximizing operational energy savings, providing healthy interiors, and limiting the detrimental environmental impacts of the buildings' construction and operation' (*ibid.*, 13–14). From this local start, the term went nationwide in 2005 with the introduction of the US Energy Policy Act, which also covers buildings and which defines a high performance building as 'a building that integrates and optimizes all major high performance building attributes, including energy efficiency, durability, life-cycle performance and occupant productivity' (NIBS, 2008: 1). As stipulated in the act, this led to the formation in 2007 of the High Performance Building Council, which sits in the National Institute of Building Sciences (*ibid.*, 5). In 2007, this US Energy Policy Act was complemented by the Energy Independence and Security Act. Here the definition of a high performance building is extended with more explicit terms, now becoming 'a building that integrates and optimizes on a life cycle basis all major high performance attributes including energy conservation, environment, safety, security, durability, accessibility, cost-benefit, productivity, sustainability, functionality, and operational considerations' (NIBS, 2011: 4). Later, the High Performance Building Council of the National Institute of Building Sciences redefined high performance buildings as buildings 'which address human, environmental, economic and total societal impact, and which are the result of the application of the highest level design, construction and maintenance principles' (NIBS, 2008: 6). They highlight the importance of cost effectiveness, safety and security, sustainability, accessibility, functionality, productivity, historic preservation and aesthetics (*ibid.*, 9–19).

In other nations, the uptake of the concept seems a lot less significant. In the United Kingdom, there was a push for high performance buildings for Further and Higher Education (HEEPI, 2008); however this seems to have been superseded with the national drive towards zero carbon buildings. In Europe, Erhorn and Erhorn-Kluttig (2011) published their report on terms and definitions for high performance buildings that exemplifies the wide range of terms used on the continent, as well as a strong focus on energy and carbon emissions. There also was a workshop on High Performance Buildings in 2013, organized in collaboration between the Joint Research Centre (JRC) of the European Commission, the International Network for Information on Ventilation and Energy Performance (INIVE) and the Italian National Agency for New Technologies, Energy and Sustainable Economic Development (ENEA). The call for the meeting defined a high performance building as 'a building that consumes as little as possible energy during a whole year of heating, cooling, ventilation, light, hot water and copes with the presence of people and domestic appliances'; the programme indicates a firm focus on energy and the Energy Performance of Buildings Directive. In Asia, many publications on high performance building seem to use it as a synonym for green building.

Various authors in the academic literature have attempted a description or definition of high performance buildings:

- On the basis of case studies into six building design projects, Torcellini *et al.* (2006: 118–121) conclude that the following aspects represent the best practice in designing high performance buildings: the use of a whole building project for the design, construction and operation; planning to include both commissioning and POE; implementation of monitoring processes; good use of daylighting, energy recovery ventilation, natural ventilation, evaporative cooling, demand-responsive control systems, on-site energy production and storage, and management of loads.
- Kibert (2016: 491–492) defines high performance buildings as 'facilities designed, built, operated, renovated and disposed of using ecological principles for the purpose of promoting occupant health and resource efficiency and minimizing the impact on the built environment'.
- Prum (2010) defines high performance buildings as buildings 'which address human, environmental, economic and total societal impact, are the result of the application of the highest level design, construction, operation and maintenance principles', adding that such buildings require a paradigm shift for the built environment.
- Abaza (2012) represents the view that high performance buildings are those that save resources, especially energy, compared with a baseline building that is designed to just meet the building codes. In this approach, the key determinants are an integrated design process in which all actors participate, the use of modelling tools to quantify energy consumption, the use of life cycle modelling to select materials and systems and the application of some sort of quality assurance process. However, this view does not specify any requirements for the high performance building itself; it is suggested that the key measure for identifying high performance buildings is the energy use intensity.

- Ren (2013) defines a high performance building as 'a sustainable building with better environmental, economic and socio-cultural features and performance than standard practices', but which also requires aesthetical attractiveness, safety, health and comfort and economic efficiency throughout the life cycle. He notes that inspiration for high performance buildings may stem from previous buildings, natural objects and phenomena, or advanced sciences and technologies (*ibid.*, 100–103).
- Hyde (2014) considers high performance buildings a trend that implies a reliance on technology to overcome problems in the built environment. He suggests that this reliance on technology, or the 'technical fix', needs to be accompanied by a change in owner and occupant values, which he labels an 'attitudinal fix'. On this basis he suggest to change the definition of a high performance building to a building that 'integrates and optimizes all major attributes, including asset and environmental management, energy efficiency, life-cycle performance, and occupant productivity'.

These descriptions and definitions vary widely, with many of them attempting to establish a wide range of performance aspects that need to be taken into account. A similar conclusion was reached by Trubiano (2013: 4), who also points out the lack of a commonly accepted definition of high performance buildings, stating that 'for some, the term is loosely associated with sustainable design practices that result in "green" buildings and carbon-neutral designs. For others, it is aligned with the specific gains of prefabrication and new materials. And for others still, "high performance" describes building systems that operate more efficiently by using less energy for supplying light, air and heat. Rarely are these positions reconciled in one all encompassing definition of high performance, at the center of which is the equal importance of metrics, ethics and design'. The qualification of 'high' is mostly interpreted by terms like going beyond the typical, optimal, best practice and use of highest principles. Erhorn and Erhorn-Kluttig (2011), while providing a rather wide view of building categories that may be considered a high performance building, also observe that many of these concepts are implicit and descriptive only, and that only a few cases include benchmarks or target values.

There are many buildings that are presented as high performance buildings; for instance many buildings that are certified under the LEED, BREEAM and Passivhaus schemes are seen to fit the category. ASHRAE publishes the *High Performance Buildings* magazine[1] that highlights the measured performance of exemplary buildings, products and technology and discusses integrated design, operation and maintenance practice. Specific case studies of high performance buildings are published in this magazine that highlight innovations, challenges that have been overcome and lessons learned from the project. These exemplary buildings typically perform well in the areas of energy and water use, recycling, environmental education, new technology or products, operation and maintenance. Case study presentations follow a generic template that requires a building description, discussion of building performance data (at least 12 months of measured data to be included) and photos of the actual building. Other high performance buildings can be identified on the basis of prizes they have received. For instance, the Chartered Institution of Building Services Engineers in the United Kingdom has been running Building Performance Awards for a decade in order to 'recognize people, products and projects that demonstrate engineering excellence in the built environment'.

---

1 Available online from www.hpbmagazine.org.

Categories include Building Performance Champion, Building Performance Consultancy, Building Performance Training Programme, Collaborative Working Partnership, Energy Management Initiative, Facilities Management Team, Lighting for Building Performance, Product and Innovations and Projects of the Year for seven distinct sectors (residential, public use, leisure, commercial/industrial, international, energy-efficient product or innovation, and energy saving product or innovation).

Grumman and Hinge (2012) report on a review of the self-selected sample of buildings that were presented in ASHRAE's *High Performance Buildings*. They observe a high variation in building functions, energy use and cost. While typically the buildings reported in the magazine include an energy-efficient envelope, use of daylighting, lighting controls, renewable energy systems, rain and waste water systems, there is 'no one formula or set menu of features and design techniques that applies to each project'. The editors of the *High Performance Buildings* magazine note that it is hard to get performance information for buildings that are showcased in the publication. They suggest that there might be technical issues, such as insufficient metering, but that another key problem is that there are complicated relationships between building designers and owner/occupants that prevents performance information from being shared (Hinge and Winston, 2009).

On paper many buildings exceed requirements and thus are considered high performance buildings; yet this performance is seldom validated in use (Gill *et al.*, 2010). Addressing this issue is complex and relates to the problem of the 'performance gap' (Carbon Trust, 2011; Menezes *et al.*, 2012; de Wilde, 2014; Fedoruk *et al.*, 2015; van Dronkelaar *et al.*, 2016). Furthermore, current data on high performance buildings is often isolated from data that pertains to the generic stock. Cross-comparison would help to ensure more realistic assessment of the high-profile buildings (Hinge and Winston, 2009). In 2011, the National Institute of Building Sciences identified a need to increase stakeholder trust in high performance buildings and suggested that data collection for this type of buildings would help to build that trust (NIBS, 2011: 4). This has led to the US Department of Energy maintaining a catalogue of case studies of high performance buildings.[2]

Beyond this, further challenges exist. According to Nguyen *et al.* (2014), 'high performance buildings require an efficient performance-based design process which forces the implementation/integration of optimizations techniques into building performance simulation programs'. However, where high performance buildings employ innovative technology these buildings may have features and systems that are not represented in all available simulation tools. Tian *et al.* (2009) present a case study of a LEED Platinum building in Canada and discuss some of the issues when modelling this building in DOE-2 or EnergyPlus. Day and Gunderson (2015) point out that many buildings require occupant engagement to deliver on their promises and deliver benefits to their stakeholders; this is especially so for the high performing ones. And high performance buildings are (by definition) pushing the boundaries of performance. By consequence, the design and construction of such buildings might not always go right, so the lessons learned are often the most interesting aspect of reporting on these buildings (Grumman and Hinge, 2012).

---

2 Available from https://buildingdata.energy.gov.

10.2   Emerging Technologies

In order to achieve a high performance, many high performance buildings employ emerging technologies. Emerging technologies are a moving target, so any description risks become outdated quickly as new ideas and systems emerge and others become obsolete. This section gives a brief overview of some developments that have been notable in recent years. However, keeping on top of emerging technology is an area where students and professionals need to embark on a lifelong journey of continuous professional development (CPD).

Construction technology is typically not an area that sees disruptive innovation; many systems have been in use for a long time and have evolved over time. For instance, concrete was already used in the Roman Empire. Similarly, a wide range of energy saving technologies for buildings has been developed since the energy crisis of the 1970s; many of these are still somewhat experimental and niche products, whereas others have gone mainstream. Examples of such energy saving building features and components are advanced glazing systems, ambient heat sources, atria/courtyards, aquifers, black attics, chemical energy storage systems, clerestories/skylights, cogeneration units, concentrating solar collectors, cooling ceilings, energy piles, geothermal heat sources, heat pipes, heat pumps, holographic optical elements, light shelves, movable insulation systems, roof ponds and Trombe walls (de Wilde, 2004: 165–172). One area that shows increased rates of development due to the uptake of ICT is that of environmental control and automation systems. Traditionally these consisted of physical devices such as thermostatic radiator valves (TRV) and control panels such as timers and thermostats. More recently these have been expanded with web-based interfaces and apps that allow the interfacing with building systems through mobile devices. A deep review of some existing products in the market space is provided by Chien and Mahdavi (2008), discussing control options and functional coverage, provision of information, mobility and reconfigurability and cognitive user requirements. Chew *et al.* (2017) review the prospects of emerging smart lighting systems.

It is hard to define boundaries for the concept of emerging technologies. At one end, there are subtle changes in architectural design that still have significant impacts. For instance, a move away from the typical US cubicle office layout towards a more flexible layout where spaces are designed for the type of work to be done rather than for an individual worker can, in appropriate cases, reduce the floor area, furniture cost, capital cost, as well as hours lost (due to noise, visitors and waiting for others) per employee, and thus help to increase productivity (Laing *et al.*, 2011); yet in other cases it may alienate the workforce and lead to a fall in output. At the other end of the spectrum are novel of-the-shelf systems, which are self-contained and can be added to both new and existing buildings, such as the latest building automation systems. A different category of emergent technology manifests itself in the construction process, where automation and ideas like modular construction change the way things are done.

Advances in material sciences create a strong basis for novel products. For instance, ACMA (2016) gives an overview of novel fibre-reinforced polymer (FRP) architectural products, discussing their characteristics and impact on building design and construction. The complex forms that can be achieved with FRP makes them suitable to easily replicate traditional ornaments while also offering a wide range for new form finding. This material thus links into techniques such as 3D scanning and computer numerical control (CNC) milling. In general, there is a long tradition on materials that self-repair, especially when subject to cracking and corrosion (Dry, 1997). Insulation material is another field within material science that sees continuous developments. Traditional materials such as mineral wool may be improved, but there also are novel ones such as aerogels and dynamic insulation materials; for an in-depth review of this wide field, see Jelle (2011). For a review of aerogel insulation, see Baetens *et al.* (2011); for a study of the prospects of dynamic insulation materials, see Menyhart and Krarti (2017). On the interface between material science and system development, a novel approach for the development of a thin but highly efficient insulation layer is the use of a vacuum insulation panel (VIP). This technology is expensive but has a niche market in retrofit, where thick solutions may be infeasible due to historical and aesthetic constraints (Johansson, 2014). Most VIP systems consist of an evacuated inner core board that sits inside an envelope that helps to maintain the vacuum; typical core materials are fumed silica, expanded perlite, fibreglass and polyurethane foam. For an analysis of the potential space heating savings and payback times, see Alam *et al.* (2017). In the field of thermal building engineering, phase-change materials (PCMs) are an interesting alternative to traditional thermal mass, allowing the capture and release of heat near the melting point of the PCM. PCMs can be incorporate in both active and passive systems; a typical technique is encapsulation (Akeiber *et al.*, 2016). Kenisarin and Mahkamov (2016) review the use of PCM for passive thermal control in dwellings.

Glazing materials are subject to continuous developments. While good insulating properties can be achieved by double and triple glazing in combination with a cavity and an inert gas, a lot of development is taking place in terms of coatings that may be applied to all faces of the glass in the window system. Emerging technologies include thermochromic films, which reduce solar transmission at higher temperatures, and electrochromic films, which allow control of both the visible and thermal properties (Granqvist, 2016). Windows that change their properties are sometimes named 'smart glazing'; if the change can be controlled, this may also be named 'switchable glazing'. For a study of the prospects of thermochromic systems in terms of energy efficiency and daylight access, see Costanzo *et al.* (2016). Alternatives for glass are also emerging, such as ETFE (ethylene tetrafluoroethylene), which has good light transmission and thermal properties and has been used for very distinct architectural projects such as the National Aquatics Center in Beijing (China) or the Eden Project Biomes in Cornwall (UK). ETFE is produced as foil, with ETFE cushions being the product that is typically used in buildings (Hu *et al.*, 2017). Modern technology also allows the development of complete virtual windows, which use LED technology and static images to create both the light and the view which would be provided by a real window and which can be positioned in an internal wall that otherwise would just be a blind wall (Mangkuto *et al.*, 2014).

Other emerging technology exists at the building component and system level. One key area is that of façades; Aksamija (2013) gives an overview of façades that she labels 'high performance building envelopes'. There are many types of façades. Double skin façades have a history that dates back to the early 1900s; however there still is a fascination in architecture with the concept. With the many options for the inner and outer layers and the space between them, the possibilities are endless; control variations further augment the complexity (He *et al.*, 2011; Barbosa and Ip, 2014; Pomponi *et al.*, 2016). A number of new technologies is available for use in building façades, which enhance solar control, daylighting and thermal behaviour. Amongst these are spectrally selective glazing, angular selective systems such as prismatic glazing and between-pane louvres, solar filters such as light shelves and secondary skins, sunlight redirection systems such as laser-cut panels and holographic optical elements, and double skin façade and active façade systems (Lee *et al.*, 2002a). Adaptive solar façades allow for control, for instance to open and close ventilation openings, to change louvre position and tilts or to optimize the position of building integrated photovoltaic (BIPV) panels. For a recent example, see the work by Nagy *et al.* (2016). Loonen *et al.* (2013) review the closely related emerging field of climate adaptive building shells.

The use of plants and trees is a traditional approach for passive cooling in for instance Mediterranean climates (Kontoleon and Eumorfopoulou, 2010). Recently this is also used more widely and integrated with modern construction technology. Plants may be used on 'green roofs'; another application is on vertical surfaces, which is typically named 'living wall' or 'green façade' or more generic a vertical greenery system (VGS). Advantages of this approach can be the shading by leaves, cooling effects through evapotranspiration, capturing of particles in the air, rainwater retention and provision of a wind barrier (Pulselli *et al.*, 2014; Pérez *et al.*, 2017; Riley, 2017).

Existing HVAC systems are continuously improved, with new components and systems being added. Recent trends focus on electrical systems such as heat pump technology. Another example is the development of personalized ventilation systems, which deliver clean air directly to building occupants and reduce the exposure of these occupants to any indoor contaminants. Personal ventilation systems require the use of air terminal devices, typically in the form of nozzles, close to the occupant; these are also named ductless personalized ventilation systems (Khalifa *et al.*, 2009; Dalewski *et al.*, 2014). While there has been significant research and development of systems that use typical renewable energy sources such as solar, wind and hydro, there now also is interest in systems that harvest indoor energy. A prime example is the use of piezoelectric floors, which generate electricity from the footfall of pedestrians; trials with such systems have been conducted in railway stations, dance floors and academic facilities, capitalizing on areas of high traffic movement (Li and Strezov, 2014). Advanced technology from other domains is also making inroads to the building services domain. For instance, face recognition (Bajwa *et al.*, 2013) can be included in building security systems to ensure that only people with authorization are able to access a building.

New construction processes are also appearing. A notable example is the 3D printing of buildings, with remarkable developments in the working with concrete that combines novel computational design with the use of robotic arms to guide extrusion (Gosselin *et al.*, 2016). 3D printing is sometimes also known as 'additive manufacturing' or 'contour crafting'; its success depends on advances in computing as well as materials science (Labonnote *et al.*, 2016). Initial developments took place for extreme environments, such as the construction of bases on the moon or Mars; however the technology is now finding wider application. An overview of the use of 3D printing in the construction sector is provided by Wu *et al.* (2016), who discuss the 3D printing of construction parts, architectural models and entire building projects. Labonnote *et al.* (2016) give examples of the actual application of this technology in architectural projects. Other contributions on the subject of additive manufacturing in construction are Perkins and Skitmore (2015) and Bos *et al.* (2016). For an example of additive manufacturing on a remote construction site, see Figure 10.1.

Figure 10.1 Contour crafting at a space base (Image courtesy of NASA/B.Khoshnevis/USC/NIAC).

Nanotechnology is also making inroads into construction; products available to the market are for instance nano-polymer bonds in steel coating that helps prevent corrosion, glass coatings that render windows self-cleaning, nanotubes in ceramics that improve resistance to stress while reducing volume and weight, nanotubes that replace steel reinforcing bars in concrete and nanopores that increase insulation (Arora *et al.*, 2014). Each of these products involves a research area in itself; for instance Pacheco-Torgal and Jalali (2011) explore the role of nanotechnology in improving Portland cement products.

Vähä *et al.* (2013) discuss the increased role of automation in the building construction process. Bock (2015) gives an interesting account of the prospective of robotics for construction. He extrapolates labour productivity curves and development curves, giving an outlook for the next 35 years of construction and outlining how this may involve robot-oriented design, robotic industrialization, use of construction robots and site automation culminating in ambient robotics. Some technology, such as floor-cleaning robots, window-cleaning robots and logistics robots are already available today. Another interesting development is the increasing interaction between smart buildings and electric vehicles, which explores the integration of power supply, reliability and flexibility of both systems (Wang *et al.*, 2012).

10.3   Smart and Intelligent Buildings

As the conditions under which buildings have to achieve their performance vary due to changes in occupancy, weather or prices of electricity on the grid, buildings that are aiming for high performance are helped by automation, anticipation and an ability to adapt to changes in their circumstances. Leatherbarrow (2005: 15) states that 'the true measure of a building's preparations is their capacity to respond to both foreseen and unforeseen developments. Stated in reverse, bad buildings are those that cannot respond to unexpected conditions because they have been so rigidly attuned to environmental norms'. The category of buildings that can adapt and react is often known as smart or intelligent buildings. To some extent these terms fit in the category of buzzwords, along with concepts such as connectivity, open architecture and interoperability (Hoffmann, 2009). However, as pointed out by Barrett *et al.* (2013), there is still a major challenge to design and construct buildings that correspond much better to human life stages, how people understand built spaces and their sensory experience; and buildings designed with such ideas in mind may have significant implications for schools, offices or homes for the elderly constructed in the future. And with the advent of the smartphone, work progressing on self-driving vehicles and similar developments, there clearly is an interest in smart and intelligent buildings. But as with high performance buildings, a spread of definitions of intelligent and smart buildings can be found in the literature.

An early discussion of various definitions of intelligent buildings is provided by Clements-Croome (1997), who concludes that intelligent buildings are those that 'can cope with social and technological change and are adaptable to short-term and long-term human needs'. Chen *et al.* (2006) define intelligent buildings as buildings that provide a cost-effective and productive environment while supporting people, products and processes. Wong *et al.* (2008) present an approach to capture the intelligence of a building through a system of 69 'key intelligent indicators' that are aggregated into a 'system intelligent score' (SIS). The evaluation of these 'indicators' is done by a survey amongst 81 experts, yielding relative weighting by this group; for buildings most of these 'indicators' can only be assessed in a binary manner, checking weather a certain feature is present or not. Perumbal *et al.* (2010) broadly define an intelligent building as 'an entity that maximizes the efficiency of building services and at the same time ensures effective resource management with minimum life-time costs'. They go on to state that intelligent buildings often are shaped by a range of subsystems but that these subsystems require intelligent integration for the building to be more than a sum of the parts; this requires subsystem interoperability.

Definitions of smart buildings show subtle differences between smart buildings in general, smart homes that are limited to the domestic context and smart buildings as elements in a wider smart utility grid.

Hoffmann (2009) defines smart buildings as buildings that employ 'intelligent systems that help to run a facility more efficiently and deliver leaps in productivity'. This requires that the buildings and systems have embedded intelligence and can connect to – and thus communicate with – other devices. The Royal Academy of Engineering identifies a need to ensure that smart buildings communicate with their occupant, allowing people to engage with what happens rather than feel at the mercy of a building control system. At the same time, smart buildings should anticipate the needs of the occupants (RAEng, 2013: 10–13).

Smart homes may include telecare systems for people with extra needs such as the elderly or people with disabilities and cognitive impairment. These telecare systems may for instance include fall detectors or medication management, or computerized prompts to make sure occupants undertake some key activities like taking medication in order to assist people with memory issues (Swann, 2008). Smart homes may employ a dedicated sensor network that includes motion detectors, RFID tags and non-intrusive load monitoring to analyse what happens in the house and to enable the home to intervene and assist the occupant when needed (Belley *et al.*, 2014). The development of smart homes goes hand in hand with the evolution of home area networks (HANs), the wired and wireless systems that are available within a dwelling and which allow devices to communicate (Bhatt and Verma, 2015). Jabłoński (2015) includes other features in his view of smart homes, including the provision of convenience, communication, comfort and security through technology such as audio instructions, digital and automated control of building light and HVAC, video communication facilities and intruder alarms. A deep discussion about the potential roles of smart home technologies is provided by Wong *et al.* (2017).

Lawrence *et al.* (2016) define smart systems as systems with advanced control systems and technologies, which allow them to respond to other systems. In this view, a smart building is a building that can respond to demand management requests from the power grid through digital data streams; this allows the building to minimize electricity cost, while it helps the grid operator to balance the network. A smart grid is an electricity distribution network that includes digital data streams, allowing this network to produce, transmit and distribute electricity in an efficient manner. Sinopoli (2016: xi) views smart buildings as those buildings that use advanced technology in order to improve performance, operation, occupant satisfaction and financial returns while reducing the need for service and maintenance as well as other costs.

Obviously, the concepts of smart and intelligent buildings are closely related. On a general level, there may be a tendency to label buildings in general 'intelligent', whereas residential buildings may be named 'smart'. For instance, the Institution of Engineering and Technology sees intelligent buildings as part of an increasingly integrated built environment that combines smart homes, intelligent transport, smart grid and smart cities. They consider an intelligent building to be a building that combines technologies and interconnected systems in order to support the building occupant, responding to changes in use while ensuring efficient operation of the building (IET, 2016). Another way to look at things is expressed by Velikov and Thün (2013: 855), who hold the view that smart systems depend on intrinsic system properties to drive their behaviour, whereas intelligent systems rely on computation and automation. An alternative may be to revert to more specific definitions. For instance, in the context of building automation control, the term 'sentient building' has been coined by Mahdavi (İçoğlu and Mahdavi, 2007; Chien and Mahdavi, 2008); this refers to a building that has the ability to feel and perceive and react accordingly through a network of sensors, a dynamic and self-updating representation of the building and system status, and self-regulation.

The development of smart grids is an area that receives significant effort. Grid operators are increasingly interested in forecasting electricity loads; this helps them to balance demand and supply and prevent the power outages that may result from imbalances. In the domestic sector, loads relate to appliance usage, domestic hot water consumption and space heating; prediction relates these to weather conditions, time of day, day of the week and time of year. Modelling may involve Markov chain models, a stochastic state representation. Sandels *et al.* (2014) present work on forecasting the electricity demand of households in Sweden based on studies of the use of appliances, domestic hot water and space heating; Sandels *et al.* (2015) discuss the forecasting of electricity demand in offices in the same country. Such information is important for smart building control as well as grid management. Different scale levels are involved in the communication between a smart building and a smart grid; these involve the level of systems within the building, the level of the building itself, a level of nano- or microgrids such as in a campus or urban grid, the general operational grid level as managed by a utility company or state, the national level and the international level. Smart operation requires communication between all these levels (Lawrence *et al.*, 2016).

Increasingly, it is also becoming clear that smart and intelligent buildings and systems may also lead to additional risks. Such risk may stem from different issues, such as overreliance on technology, novel failure modes in interconnected systems and cyber security problems (RAEng, 2013: 16; IET, 2016). Future developments of smart and intelligent buildings will interact with the development of the Internet of things (IoT), the concept where devices will be interconnected and anticipate the needs of the building occupants in a wide range of domains, way beyond the current focus of building automation systems in thermal comfort, lighting, ventilation and energy efficiency (Lilis *et al.*, 2017).

10.4   High Performance Building Challenges

A building that demonstrates the complexity of designing, constructing and operating a high performance building is the Belgian Princess Elisabeth Station, a research facility at Utsteinen Nunatak, Antarctica. The building went into service in spring 2009. It can provide living accommodation for up to 48 people working at or in the vicinity of the station, as well as research facilities for a range of research programmes in the fields of glaciology, earth and atmospheric sciences, terrestrial microbiology, geophysics and meteorology. A typical (summer) operating crew consists of the station leader, a doctor, electrical/telecom/network and bio-engineers, a mechanic, field guides and a chef. The station acts as a hub for exploration in the 20°–30°E sector of the Antarctic, with operations taking place within a radius of 300 km around the station. The building has to withstand the extreme Antarctic climate; at the same time it is the only zero-emissions research station on this special continent. It is anchored in a granite ridge with steel rods that go 6 m deep into the base rock and are held in place by resin. The station can sleep 16 expedition members. Additional accommodation, technical and garage space are constructed on the leeward side of the ridge and are cantilevered on one side with the other side being supported by sliding foundations to compensate for ice movement. See Figure 10.2. After the Princess Elisabeth Station, a number of new generation research stations has been built following the International Polar Year 2007–2009: the Neumayer III Station (Germany, 2009), Halley VI Station (Britain, 2012), Bharati Station (India, 2013), Jang Bogo Station (South Korea, 2014) and the Comandate Ferraz Station (Brazil, 2018).

The Princess Elisabeth Station project was launched in 2004, when Belgium (an original signatory of the Antarctic Treaty) responded to the invitation of the International Polar Foundation to build a new Station in the Antarctic during the International Polar Year

Figure 10.2 Princess Elisabeth Research Station, Utsteinen Nunatak, Antarctica (Image courtesy of International Polar Foundation).

2007–2009. It was decided that the new building was to be built on rock and to make it a technological showcase, demonstrating that it is possible to have an Antarctic research station that is self-supporting in terms of energy. The new facility was designed to operate in the Antarctic summer, which runs from November to February. The building has to withstand the hostile polar environmental conditions, specifically extreme wind speeds, subzero temperatures, very dry atmosphere, snow erosion and accumulation, and potential blasting by grit and stone projectiles. Air temperatures at the site vary between –40°C and –5°C, while wind gusts of up to 150 km/h have been recorded. Wind conditions at the Utsteinen ridge are special, involving Katabatic winds (downslope winds that come from the ice-covered Antarctic plateau at the South of the Sør Rondane Mountains) but slightly tempered by the surrounding mountains so that conditions are better for wind turbines. The building is located 431 km away from the nearest Antarctic base, the Russian Novolazarevskaya Station, meaning that it has to be self-sufficient and cannot rely on any neighbours. The isolated site and extreme climate conditions mean that the building will have to do way better than research stations in more temperate climates – the station must be autonomous, reliable and able to cope with the local weather. On top of this the building also has high environmental ambitions.

The following partners were key stakeholders in the design and construction of the Princess Elisabeth Station: The International Polar Foundation, as client, supervisor and project coordinator; Philippe Samyn and Partners, architects and engineers, for the building core, skin and reactive support structures; Prefalux for the building concept; 3E for the pre-construction systems assessment; the von Karman Institute for aerodynamics; GDF Suez subsidiary Laborelec for the grid design and construction; Schneider

Electric for the intelligent control systems and power electronics; Consolar for the solar thermal systems; and EPAS and Laborelec for the bioreactor design and assembly. The Belgian construction giant Besix coordinated the construction work in Antarctica.

After the decision to build the station was taken in 2004, a number of expeditions was launched to find a suitable location, a safe route to the coast and to transport construction material to the site. In September 2007, the station shell was put together for a dry run of the assembly and was exhibited to the public in Brussels. It was then dismantled and packed and taken by ship to Breid Bay on the coast of Antarctica. From here it was transported overland for another 220 km to its final destination: Utsteinen. A temporary base camp was put in place during the construction of the station. Station personnel and scientists are transported to the station by twin-engine ski planes. Annual provisioning of the station is by ship and land transport. The formal inauguration took place in February 2009. The station has theoretical design life of about 25 years but will probably last longer as the environmental factors are gradually brought under optimal management.

With the extreme climate, the design of the building involved some special performance studies. Given the special wind conditions, aerodynamic studies were carried out using CFD modelling and wind tunnel experiments, validated by field measurements. These studies aimed to limit snow accumulation in the lee or upwind of the building, control forces acting on the building, reduce erosion effects, noise and vibrations, and to assess the best turbine positions for wind power production (Rodrigo *et al.*, 2012).

In terms of energy efficiency, the first design principle was to aim for as much of a passive solar building as was feasible, with a compact shape, highly insulated envelope, windows designed to make maximal use of solar gains while limiting heat losses, and an advanced ventilation system with heat recovery. The windows make the most of the natural light during the Antarctic summer and give a beautiful view of the surrounding landscape. The building shell includes, from outside to inside, 1.5 mm thick stainless steel plates to protect against the high winds and any windblown rocks and debris, a 2 mm EPDM waterproof membrane, 42 mm multiplex panels, 400 mm of lightweight expanded polystyrene insulation, 74 mm multiplex panel with heavy-duty Kraft paper with thick aluminium vapour barrier, and wool-felt wall covering. The second design principle is the optimal use of renewable energy. To this end, the station employs $22 \, m^2$ thermal solar panels (providing 12% of the energy supply), $380 \, m^2$ of PV panels (40%) and 9 wind turbines (48%); some PV panels are integrated into the building, but others are placed next to the building on the ridge. Electrical energy is stored in a battery room containing 15 tonnes of sealed lead acid batteries, whereas thermal heat is stored in special heat exchange tanks with phase change materials. As backup for safety reasons, the station has two conventional 110 KVa diesel generators. In due course, hydrogen fuel cells will be installed to capitalize on the excess energy that cannot currently be stored due to limited electric storage. The ventilation capacity of the station is a maximum of $600 \, m^3$/hour; the U-value of the building shell is $0.07 \, W/m^2 K$. Energy systems were selected on the basis of the use of proven technology; even so, the systems run independently so that a failure in one system does not compromise another. Redundancy studies were carried out by using Failure Mode Effect Analysis (FMEA). System design was based on a projected estimation of day-to-day energy needs of the station occupants. However, this is continuously refined to take into account real use. The use of renewable energy reduces the need to bring fuel to the station, limiting the risk of spillage. The building has extensive Environmental Management Plans in operation, and impact on local fauna and flora is monitored.

The station has an advanced building automation system that takes the form of a micro smart grid; it can be managed remotely via satellite control if the station is unoccupied during the winter. The station start-up phase after winter can also be managed remotely. The building automation system has 700 digital inputs, 500 digital outputs, 300 analogue inputs and 50 analogue outputs. The control and energy management is based on prioritization of users, with lower levels being denied power if there is insufficient energy to support the higher functions. Priority 1 is for security and medical emergencies; priority 2 is station technical operations; priority 3 is scientific research instruments; priority 4 is daily life; and priority 5 is entertainment.

The water systems have been designed to allow 100% recycling of both grey and black water, although drinking water for the occupants is sourced only from melted snow. The station employs membrane bioreactor technology and has two bio-filtration units. In the bioreactors, bacteria help breakdown waste in a first step; further systems involve filters, a chlorine unit and UV treatment. The filtering system has a capacity of 1500l/day; water storage tanks include a fresh water tank of 1500l; a recycled water tank of 1500l; and three hot water tanks of 500l each.

An extensive report with details about the construction and operation of the Princess Elisabeth Station is provided by the Belgian Science Policy and the International Polar Foundation (2007).

10.5 Reflection: A Novel Definition for High Performance Buildings

Where the concept of building performance is left undefined, it is no surprise that definitions for high performance buildings are mainly aspirational. Emerging technology not only is often included to achieve a cutting-edge design but also carries risks as emerging technology has not yet been proven and may be less reliable than existing solutions. Smartness and intelligence may be components of high performance but mainly mean that a building is able to adapt or to anticipate; and the concepts of smart buildings or intelligent buildings also may have different interpretations.

A different definition of high performance buildings can be built from the review of performance earlier in this book. The starting point is the overview of building performance and related key attributes as presented in Table 3.3, which discerns the engineering view, process view and aesthetic view. Since high performance buildings implies that the discussion is limited to the building as an object, this allows for some simplification. Leaving aside the specific interpretation of the adjective high for a moment, a high performance building can then be said in general to be: a building that performs (a) task(s) or function(s) better than normal, in terms of the quality, resource saving, workload capacity, timeliness, readiness, or aesthetic attributes of that building. Note that in this definition, the attributes that pertain to a high performance construction process fit within the object view (cost, time, quality, safety, waste fit within resource saving, whereas customer satisfaction fits within quality). Also note that this definition allows for one or more attributes to be considered: it allows not only for a highly specialized category (for instance a high performance building that is able to process an extreme number of people, as in a sport arena or airport terminal) but also for a multi-functional view (for instance a high performance hospital).

However, further attention is needed for the qualification 'high'. In the definition this has been reworded as 'better than normal'. This implies a relation to a group of buildings and then a ranking where some buildings are seen to be better than others. This question is not answered in any of the literature on high performance buildings, so one has to look elsewhere for an answer. Within many sets of objects, one can group objects into clusters in such a way that the objects within the cluster are more similar to each other than to objects outside the clustering. There is a range of mathematical techniques that supports clustering, for instance single linkage and K-means. The methods are based on quantifying the difference or similarity between object in terms of distance for a range of aspects of interest. Within clusters one may identify sub-clusters that are a subset of another cluster and independent clusters that are not related to any others (Kim *et al.*, 2012). A similar approach is required to establish which buildings are high performance buildings and which are not.

Setting boundaries is not an easy task. By way of example, in education there is deep discussion about performance level descriptions and the 'cut scores' that define when students pass or fail a subject or grade. Education uses performance levels descriptors (PLDs) that define levels of proficiency (Perie, 2008). Typical levels may be 'below basic, basic, proficient, accelerated and advanced', or 'unsatisfactory, limited, satisfactory, exceeding' or similar. PLDs set the standard for assessing in what category students fit. Good practice in the field is to first define the numbers and names of the various categories that will be used, then set out the intention or policy of the categories and only then progress to fully develop descriptors for each subject and grade level (*ibid.*). A similar issue arises with high performance buildings; clients of buildings that are found to be just underneath a threshold level for high performance buildings will not be pleased with that outcome. At the same time, rising above average performance may carry some risk: for instance in psychology there is work that indicates that there may be tendencies in the workforce to punish low and high performers and that co-workers are averse to violation of performance norms (Jensen *et al.*, 2014). In the world of high performance buildings, such an effect may have parallels in higher insurance premiums, taxes or rents.

To identify high performance buildings, one first has to define clusters. At the absolute minimum, one could split the building stock in two groups: high and low performance buildings. As there are only two categories, one would expect 50% of buildings in each category. A categorization that allows for more differentiation would have a category of 'regular performance' in the middle. While such an approach may sound reasonable, it leaves a question of where one puts the thresholds for each category. This is to some degree arbitrary; for instance one could set thresholds at 25% for low performance and 75% for high performance. Approaches used in education may also give a way forward. For instance, student grading systems also identify categories that allow the identification of the top students. In the UK undergraduate grading approach, a first-class degree requires a mark of 70% or higher, an upper second class a mark between 60% and 69%, a lower second class a mark between 50% and 59% and a third class mark is awarded for a mark between 40% and 49%, whereas marks below 40% are a fail. Applying this to high performance buildings, the cut grade would be 70%. Yet another logical way to handle the issue would be to assume that building performance of a group of buildings follows a normal distribution. One could then proceed to define buildings within one standard deviation (SD) of the mean to be regular performing buildings, which would account for 68% of the population. Beyond this one could identify the top 16% as high performance buildings and the low 16% as low performing buildings. One could also use the limits of two standard deviation of the mean as boundaries, with the top and bottom 2.3% as ultra-low and ultra-high performance, thus introducing five classes. While the exact cut rate will remain subjective and open for discussion, this leads to the conclusion that roughly speaking high performance buildings are the 16%–30% best performing buildings in their specific category.

10.6   Summary

This chapter explores the concept of high performance buildings, a category that by its very name aspires to be leading the industry in terms of performance. It includes a review of definitions found in literature and then develops a new way to look at the criteria for claiming to be a high performance building that is better aligned with the discussion of building performance in the preceding chapters. Since high performance buildings often employ innovations to achieve their performance, the chapter also reviews emergent technology. Finally, it explores the related concepts of smart and intelligent buildings.

Many definitions of high performance buildings have been suggested, with most of these stemming from North America. In the United States, high performance buildings are defined in the Energy Policy Act of 2005 and the Energy Independence and Security act of 2007; however the corresponding definitions seem more concerned with defining all performance aspects that need to be taken into account than with a distinction of how 'high performance' is discerned from 'typical performance'. Definitions from the scientific literature also try to specify performance aspects to be covered; many emphasize efficient use of resources. Many authors see high performance buildings as a synonym for terms like green, sustainable and zero carbon buildings or as buildings that employ advanced technology. Case studies and examples of high performance buildings can be found in the ASHRAE *High Performance Buildings* magazine, or identified by means of the annual CIBSE building performance awards.

Another way to define high performance buildings is as a building that performs (a) task(s) or function(s) better than normal, in terms of the quality, resource saving, workload capacity, timeliness, readiness, or aesthetic attributes of that building. Note that this allows for focus on a single task or function, as well as for a wider set. Boundaries for what consists 'better than normal' are by nature subjective; however a review of a range of approaches suggests that 'high performance' is mostly taken to apply to something like the top 16%–30% of a certain category.

High performance buildings are often associated with innovation and thus frequently employ emerging technologies. This is by nature a fast-developing field, where students and professionals need to remain informed of the latest trends through continuous professional development. Emerging technology may cover advances in material science, systems and construction processes. Technologies to watch are, amongst others, the wide range of energy conservation measures, building automation systems, architectural design concepts, novel materials such as fibre-reinforced polymers, self-repairing materials, aerogels, dynamic insulation, vacuum insulation panels, phase-change materials, thermochromic and electrochromic glazing, ethylene tetrafluoroethylene foils, double skin façades, active façades, climate adaptive building shells, living walls and green façades, personalized ventilation systems, piezoelectric floors, face recognition, additive manufacturing or 3D printing, advances in robotics and the fusion of building systems with other products such as automotive systems.

Smart and intelligent buildings may achieve a better performance by their capability to anticipate and adapt to changing conditions. There are various definitions that try to pin down what exactly makes a building smart or intelligent. Broadly speaking, buildings in general tend to be named intelligent, whereas homes are often named smart. Smart homes are sometimes associated with telecare. In the context of the interaction between buildings and the utility grid, the terms 'smart building' and 'smart grid' are often employed in parallel; here load forecasting plays a significant role. Future smart and intelligent buildings will relate to developments like the Internet of Things.

Recommended Further Reading:

- *High Performance Building Guidelines* (NYC DDC, 1999) and the report of the National Institute of Buildings to the US Congress on high performance buildings (NIBS, 2008) for a good understanding of the emergence of the concept in the United States.
- The online *HPB Magazine* of ASHRAE (www.hpbmagazine.org) for a range of self-reported high performance building case studies.
- The paper by Bock (2015) on the emergence of robots in constructions, for a vision of how many aspects in building may change in the future.
- The paper by Lawrence *et al.* (2016) on smart buildings and smart grids, for a discussion of the increased integration of buildings into the utility grid.

Activities:

1 Select a building that is claimed to be a high performance building. Collect data about what the building set out to achieve in terms of performance, as well as data about how it actually performs. Compare and contrasts aims and achievements, and critically discuss whether the building indeed qualifies as high performance building.

2 Review the report by Erhorn and Erhorn-Kluttig (2011) on terms that are closely related to the concept of high performance buildings. Then attempt to establish proper definitions for each of these terms that allow you to distinguish between each of these.

3 Review the literature to establish how the criteria for energy-efficient buildings developed over the past half century. What was a reasonable threshold in 1970, 1980, 1990, 2000 and 2010?

4 Develop a plan for yourself to stay informed about emerging construction technology in terms of new materials, systems and construction methods. Set a date in your diary a full 365 days ahead, and check back how you have been doing in terms of staying on top of developments.

5 Review the smart assistance and similar functions offered in modern automobiles, using company websites, sales brochures and scientific publications. Relate these capabilities to the smart control options offered in a modern office.

6 Review the concept of the 'Turing Test' as a standard for computer intelligence. Relate the test to building automation, and explore how this may become a benchmark for testing building intelligence in some point in the future.

10.7   Key References

Barrett, P., L. Barrett and F. Davies, 2013. Achieving a step change in the optimal sensory design of buildings for users at all life-stages. *Building and Environment*, 67, 97–104.

Bock, T., 2015. The future of construction automation: technological disruption and the upcoming ubiquity of robotics. *Automation in Construction*, 59, 113–121.

Clements-Croome, D., 1997. What do we mean by intelligent buildings? *Automation in Construction*, 6 (5/6), 395–400.

Grumman, D. and A. Hinge, 2012. What makes buildings high performing. *High Performance Buildings*, 46–54.

Hinge, A. and D. Winston, 2009. Documenting performance: does it need to be so hard? *High Performance Buildings*, 18–22.

Hyde, R., 2014. The technology and attitudinal fix. Redefining the definition of high performance building. *Architectural Science Review*, 57 (3), 155–158.

IET, 2016. *Intelligent buildings: understanding and managing security risks*. Stevenage: The Institution of Engineering and Technology.

Labonnote, N., A. Rønnquist, B. Manum and P. Rüther, 2016. Additive construction: state-of-the-art, challenges and opportunities. *Automation in Construction*, 72, 347–366.

Lawrence, T., M. Boudreau, L. Helsen, G. Henze, J. Mohammadpour, D. Noonan, D. Patteeuw, S. Pless and R. Watson, 2016. Ten questions concerning integrating smart buildings into the smart grid. *Building and Environment*, 108, 273–283.

Lilis, G., G. Conus, N. Asadi and M. Kayal, 2017. Towards the next generation of intelligent building: an assessment study of current automation and future Internet of Things based systems with a proposal for transitional design. *Sustainable Cities and Society*, 28, 473–481.

NIBS, 2008. *Assessment to the US Congress and US Department of Energy on high performance buildings*. Washington, DC: National Institute of Building Science.

NIBS, 2011. *Data needs for achieving high-performance buildings*. Washington, DC: National Institute of Building Science.

NYC DDC, 1999. *High Performance Building Guidelines*. New York: New York City Department of Design and Construction.

Prum, D., 2010. Green buildings, high performance buildings, and sustainable construction: does it really matter what we call them? *Villanova Environmental Law Journal*, 21 (1), 1.

RAEng, 2013. *Smart buildings: people and performance.* London: Royal Academy of Engineering.

Torcellini, P., S. Ples, M. Deru, B. Griffith, N. Long and R. Judkoff, 2006. *Lessons learned from case studies of six high-performance buildings.* Golden, CO: National Renewable Energy Laboratory.

Trubiano, F., 2013. *Design and construction of high-performance homes.* Abingdon: Routledge.

Wong, J., J. Leung, M. Skitmore and L. Buys, 2017. Technical requirements of age-friendly smart home technologies in high-rise residential buildings: a system intelligence analytical approach. *Automation in Construction*, 73, 12–19.

Epilogue

11

Emergent Theory of Building Performance Analysis

The first 10 chapters of this book have introduced the concept of building performance and then explored different dimensions: the context in which building performance is placed, the demand side that defines what performance is desired and how this demand can be captured, the inner workings of the concept of building performance itself and how performance is expressed, the criteria that can be used to assess performance, the wide range of performance quantification tools, the application of building performance analysis in the full complexity of the context, the application of the concept of building performance in design and operations, and finally the aspirational category of high performance buildings. Given the large number of partial contributions made by other researchers, these 10 chapters are predominantly descriptive in nature, giving a wide overview of the state-of-the-art on building performance analysis and how the different efforts fit together.

This chapter builds on these previous chapters and distils an emergent theory of building performance analysis. The adjective 'emergent' has been added deliberately: while it is possible to draw conclusions from the previous work, and to make recommendations, a proven theory requires further testing and discussion within the field.

The emergent theory is presented as follows: this chapter present series of facts and observations on building performance analysis, with comments and explanations for these facts and observations. Where possible these are followed by principles and hypotheses; hypotheses are more tentative and may require further research. Future work may be able to develop stronger statements of causality and laws, but that is felt to be outside the realm of an emergent theory.

Further sections give an initial guideline for conducting building performance analysis in practice and identify some R&D challenges that the field needs to explore. The chapter concludes with an epilogue that brings together the different strands of the book.

11.1 Observations, Explanations, Principles and Hypotheses

Observations on the field of building performance analysis:

- There is a large body of knowledge on building performance in the scientific literature, but so far it consists mainly of many individual contributions on aspects of building performance only. It is hard to get an overview of the overall field in its full complexity.

- Given the large number of contributions on building performance, some of the older seminal contributions such as the book by Markus *et al.* (1972) and CIB W60 Report 64 (1982) are at risk of fading into history and being forgotten.
- Contributions to the development of building performance analysis mainly stem from inside the construction discipline, leading to a rather introvert body of knowledge.
- There are many other disciplines that have developed ideas and theories on the subject of performance. Surely some of these ideas and theories may benefit the field of building performance analysis.

Comment: Building performance analysis is still a relatively young discipline. Given the complexity of the subject people work on many frontiers, but a comprehensive overview of the field has been missing so far.

Observations on the ownership of building performance:

- There sometimes is tension between different disciplines, such as architecture, building services engineering and facility management, in terms of who should take the lead in ensuring building performance.
- Most buildings are complex and require teamwork across various disciplines to achieve good performance.

Explanation: Being a complex subject, building performance relates to many disciplines. Actors from each of these disciplines may show leadership on the subject. Who becomes project lead may depend more on interpersonal factors than background.

Principle: *There is no reason to assign the ownership of the concept of building performance to one specific discipline, as long as each design project or existing building in operation has a 'champion' that guards it.*

Observations on the complexity of building performance:

- Buildings are highly individual, bespoke and complex products, the performance of which sometimes is difficult to control.
- The context in which building performance lives is often underestimated. One needs to appreciate the strong complexity of the many systems and parts that make up buildings, the many stakeholders that relate to even one single building, and the fact that many relations change over time.
- Reduction of building stakeholders to stereotypes with preconceived attributes and capabilities is dangerous; many 'stakeholders' are in fact categories of people, and those people cover a wide spectrum of human beings.
- Most stakeholders are involved with various projects that relate to building performance in one way or another and bring experience and learning from one project to the next.

Comment 1: The commonality and daily use of buildings makes that many people take buildings for granted.

Comment 2: Stereotyping of building stakeholders is one way to simplify the complexity of the many organizations and people that relate to buildings.

Observation on industry–academia disconnect:

- There is a degree of disconnect between industry and academia where it comes to views on building performance analysis.

  Explanation: Industry has low profit margins and maintains production by repeating well-proven processes and solutions, but this inhibits innovation. Academia is concerned with how things may be progressed, but may underestimate the complexity of the real-world practical context.

Observation on transfer of building performance concepts from one building to the other:

- Technologies and systems that work well in one building may not work in a different building.

  Explanation: As buildings are highly customized and individual products, with performance depending on the interaction between many systems and components, it is difficult to extrapolate findings from one design to another.

Observations on terms and definitions in the building performance analysis field:

- The concept of 'building performance' has not properly been defined in previous work.
- The building performance literature does not employ a common terminology; this leads to misunderstandings and makes it difficult to understand some aspects of the field.
- There is no fixed terminology for the expression of building performance, with different authors using performance metrics, performance indicators and performance measures for the same concept.
- The loose use of the concept of 'building performance' in combination with terms like green buildings, eco buildings or sustainable buildings risks to distract from the idea of building performance as a measurable, quantifiable building attribute.

  Explanation: As there is no generally accepted theory and terminology of building performance analysis, all contributors are free to use their own preferred terms; however this hinders progress and is evidence of lack of coherence.

  Principle 1: *In principle the systems engineering terminology of Measures of Effectiveness (MOEs), Key Performance Parameters (KPPs), Measures of Performance (MOPs) and Technical Performance Measures (TPMs) is applicable to the building performance domain. A good generic 'catch all' term is Performance Measure.*

  Principle 2: *Key Performance Indicators (KPIs) belong to the domains of business and management. They should not be used in the context of building performance, except for the specific subdomain of construction management.*

  <u>Definition</u>: *Building performance* is a concept that describes, in a quantifiable way, how well a building and its systems provide the tasks and functions expected of that building. Requirements may stem from three main views: an engineering view of buildings as an object, a process view of building as a construction activity and an aesthetic view. Important performance requirements in the engineering view pertain to building quality, resource savings, workload capacity, timeliness and readiness.

Observation on some errors in working with the concept of building performance:

- Sometimes building attributes are mixed up. Building properties like size and materials are not performance attributes; performance depends on building behaviour and is a function of excitation, observation, quantification and sometimes aggregation.
- There often is confusion between functional requirements and performance requirements.
- The decomposition of a building in systems, subsystems, parts and components does not map one-to-one with a functional decomposition of what the building does; for instance a window has many different functions. However, the mistake of assuming a one-to-one relation ('hardware classification trap') is a common one in the building discipline.

Explanation: The lack of a common frame of reference allows for easy mistakes.

Hypothesis: *There is a need for teaching building performance analysis terms and principles, ensuring a common understanding and frame of discussion.*

Observations on the body of knowledge required for building performance analysis:

- A deep understanding of building performance requires a broad knowledge of many domains, including architecture, engineering, material science, computing and others.
- Efforts to bring concepts from other domains, such as the automotive and aerospace industries, into construction often fail to come to full fruition.
- Without a generally accepted theory on building performance analysis, it is difficult to ascertain whether new observations, understandings and suggestions fit with existent domain knowledge. This makes it difficult to test and review the state-of-the-art in the domain and to appreciate new contributions that may cause a paradigm shift.

Explanation: Building performance analysis deals with a complex subject. Furthering the field should be a balanced effort where experts in the specific area of building performance gather and expand the domain knowledge. These domain experts should also carefully examine extraneous concepts from other fields such as systems engineering, which may advance the domain in some areas. Building performance analysis should neither be xenophile nor xenophobic.

Principle: *The field of building performance analysis lacks a common knowledge base (Building Performance Analysis Body of Knowledge – BPABoK?), which makes it difficult to learn from past efforts and experience.*

Observations on the nature of building performance analysis:

- Building performance analysis is in essence about studying the match between demand (needs, requirements) and supply (behaviour offered by a building).
- Building performance analysis requires the quantification of both demand and supply so that these can be compared.
- Building performance is not a static attribute, but changes with context, loading of the building and use conditions.

Explanation: Analysis implies a detailed examination. Building performance analysis thus needs to explore the needs and requirements that a building must meet and how well it actually meets those needs and requirements.

Hypothesis: *The building performance for every single performance aspect of a building can be quantified, although some aspects have more advanced quantification methods available than others.*

Observations on performance requirements:

- Building design briefs only contain a limited set of performance requirements; most performance requirements need to be established by the design team during the design process and carefully elicited from the various stakeholders.
- Performance criteria are by nature subjective; however whether the building performance meets a certain criterion can be measured objectively.
- Formal requirements, defined in terms of diagrams, structured language templates or mathematical expressions, make it easier to analyze building performance.
- There are five main reasons for performance analysis: factor screening (finding out what factors have the most impact), optimization (finding the best configurations and parameter values), confirmation (checking that performance is as expected), discovery (establishing the performance of innovations) and study of robustness (finding out how a building responds to adverse conditions). These drivers need to be made explicit together with the requirements as they may impact on the appropriate analysis process.

Explanation: The development of a detailed set of requirements requires hard work. Where the focus is developing a building design, formal writing of requirements may take a back seat.

Hypothesis 1:   *As requirements and design proposals interact, a certain vagueness in terms of requirements may yield more design freedom. Retrospective development or updating of a set of design requirements is not seen to be a value-adding activity. However, for performance analysis, having proper requirements is an essential prerequisite.*

Hypothesis 2:   *The requirement engineering process of systems engineering is applicable to building design performance analysis.*

Observation on taxonomies of building performance aspects:

- Concepts like 'Total Building Performance' and 'Whole Building Functionality', which try to give an ordered overview of all relevant performance aspects of buildings, only find limited acceptance in the discipline.
- The strong focus in building performance literature on energy efficiency and mitigation of climate change distracts from some other pressing challenges faced by the building sector.

Comment: The relevance of building performance aspects varies from case to case and is highly context dependent. Priorities also change over time, with stakeholders and with the functions that buildings are to provide.

Explanation: Humans can only keep a limited number of factors in mind at the same time. The use of categories and mental maps helps to cover more aspects. Similarly, prioritization of a limited number of issues helps to deal with information overload.

Principle: *Categorizing building performance requirement as relating to quality, work capacity, saving of resources, timeliness and readiness is a workable structure, which also helps to focus the mind on a wider range of performance aspects.*

Hypothesis: *It is unlikely that one can define one single overarching taxonomy or hierarchy of building performance aspects that applies to buildings in general.*

Observations on specification of building performance analysis efforts:

- Building performance studies seldom make use of the formulation of a proper hypothesis (H_1/H_A) and experiments to contrast this hypothesis to a null hypothesis (H_0).
- Short-hand statements of used performance measures by unit, such as EUI or daylight autonomy, do not describe the full experiment needed to arrive at quantification of performance.
- The work on defining performance quantification experiments by way of PTMs, PAMs and AFs is just a first step; none of these has yet reached the status of industry standard.

Explanation: The complexity of buildings, the context in which they operate and the nature of the various phenomena that work on buildings make it difficult to give a comprehensive description of a performance quantification experiment. However, that should not be an excuse not to attempt such a description.

Hypothesis: *There is a need for fundamental studies that explore how to unequivocally quantify building performance through excitation, observation of states and inputs/outputs, and how to deal with the reality of semi-controlled experiments in actual buildings.*

Observations on building performance criteria:

- Many building performance criteria are stated in a binary way, with a threshold value that can either be achieved or not.
- It is not true that for all criteria more is better; there are some (such as use of resources) where less is better and yet others where the best value is a midpoint in a range (such as comfort temperature). Moreover, for some criteria, there is a diminishing return at some ranges.
- The use of performance bands implies targets, as performance is grouped within each band.

Explanation: Many performance requirements are rooted in legal requirements. By nature, building codes and regulations are based on enforcing a certain minimum performance and state this in a way that allows assessment of conformity.

Principle: *The use of performance targets rather than goals signposts a desired outcome but leaves room to 'get close'. Targets thus are a better tool to incentivize working towards performance while allowing trade-off between conflicting targets.*

Hypothesis 1: *The use of performance bands may lead to inactivity in the middle of the band and increased activity near the bandwidth boundaries.*

Hypothesis 2: *Development of building performance criteria requires hard work that reflects the specific context, loads, needs and interest while also taking into account what evaluation method may be applied.*

Observations on building performance quantification methods:

- There is a large number of different performance quantification methods, which in general can be classified as physical measurement methods, building performance analysis software tools, expert judgment and stakeholder perception surveys.
- There is no one-to-one mapping between the many performance quantification experiments (from the key categories of physical measurement, building simulation, expert judgment and stakeholder perception survey) unless individual researchers set up this mapping for a very specific context.
- Most performance quantification methods have their own distinct strengths; there is some overlap but little redundancy.
- There is relatively little literature on the use of expert judgment in the construction sector.
- Post-Occupancy Evaluation is a term that risks misinterpretation; its name relates to the start of the building use phase, but POE is often taken as synonym for stakeholder perception or stakeholder satisfaction surveys.

Comment: The existent building performance quantification methods have been developed from different backgrounds and traditions.

Principle: *The more building performance data becomes digitized, the easier comparison between various building performance quantification methods becomes. However, proper alignment between these methods, in a way that ensures that they all quantify the same aspect using the same experiment, requires hard work.*

Hypothesis: *Ultimately there always remains a need to configure performance quantification methods to account for the context, needs and dynamics of individual cases.*

Observations on software tools for building performance quantification during design:

- Analysis tools that aim for a reduction of the modelling process by limiting the input to a predefined search space mostly fail to connect to design processes, as there are few real design situations where this limited search space represents the exact interest of the design team. In other words, such 'simplified' tools are 'solutions in search of a problem that fits that solution'.
- The building performance analysis tools that are regularly used in practice mostly span a wide range of design options, giving them a high degree of flexibility to accommodate different designs.
- Tools that allow for only a limited set of design alternatives may not fit well with actual design processes; however, they have an important role in teaching novices the interaction between selected design parameters and performance and helping these novices to gain expertise that may support tacit design in later projects.

Explanation: Design and operation of buildings are complex and highly specific processes; one needs to find a tool that matches the actual project and performance issues that are of interest to the team working on the building.

Hypothesis 1: *In education, there is a rationale for progressing from simplified tools to more complex software. Directly starting with complex tools may make it more difficult for learners to establish a feel for the impact of key input parameters on the associated performance outcomes.*

Hypothesis 2: *The building performance analysis field lacks a methodology that helps analysts to find the proper tool for a certain analysis task.*

Observations on automated building performance analysis and BIM:

- Automated building performance analysis is only possible if there is a predefined workflow that enacts an agreed process.
- Just pushing large amounts of building data to a range of digital performance analysis tools does not guarantee meaningful results; therefore the use of BIM requires dedicated mappings that send relevant information to appropriate tools.
- Existing mappings between BIM tools and building performance analysis tools often require manual intervention by the user to work.

Explanation 1: Building performance analysis needs more than only a description of the building: it also requires the definition of excitation, observation of building states and inputs/outputs and how these observations need to be aggregated.

Explanation 2: Mappings between BIM tools and analysis tools often miss user assumptions and are typically not set up to handle expert simplifications like grouping rooms into zones or simplifying geometry. The attempt to transfer all data from BIM into analysis tools often leaves redundant data that needs to be removed or reinterpreted.

Hypothesis: *The prospects of BIM to support building performance analysis by 'brute force' data parsing are limited. What is required is a rethink driven by analysis needs and where data is gathered from BIM systems in a selective manner. IFC 'model views' are a first step in that direction; however building informatics specialists need to team up with building performance analysis experts to take this to the next level.*

Observations on building performance gaps:

- If building performance is not as expected, this is too often accepted and does not have serious repercussions within the industry. There is no 'performance blacklist' such as is used in the aviation industry, where some companies that fail to meet standards are precluded from operations.
- While it has become fashionable in literature to talk about 'the' performance gap, there are in fact many such gaps. There are gaps for different aspects (energy use, thermal comfort, lighting, acoustics) and between different quantification methods. Furthermore, each of these performance gaps is time dependent, as various design and life cycle stages are likely to be involved.
- Occupant behaviour and actual use of a building may be very different from what was anticipated by the design team, which has a significant influence on building performance.

Explanation 1: There is no standard format for describing the experiments and observations that one can adhere to when quantifying building performance, which hinders replicability and comparability.

Explanation 2: Performance varies over time, with context and with use conditions; these are not easily established with precision. A first step towards bridging any performance gap is to define very accurately what performance aspect is to be quantified at what point in the building life cycle, including the specific testing conditions and context.

Explanation 3: Even if it has been established exactly when and how performance prediction and measurement are to be compared, one must expect some variation at both ends due to a range of uncertainties in the building itself, in the prediction method and in measurement approaches.

Hypothesis: *One may reduce performance gaps, but different performance quantification methods are unlikely to ever give identical results. What matters is the understanding and limiting differences to acceptable variations.*

Observations on performance-based design decisions:

- Almost all literature on making performance-based design decisions stems from academics, who follow the path of normative decision making. This causes a gap with design practice, where expertise plays a large role and decisions are made 'on the go'.
- Ensuring building performance during the building design process can take place via two different routes: (i) through the application of tacit knowledge and expertise, following the theory of naturalistic decision making or (ii) by means of explicit decisions at key decision moments, which follow the normative theory on decisions.
- There are very few empirical studies about design decision making on performance-related issues in actual building design in the public domain.

Explanation: Similar to the history of decision-making studies in general, scientists in the building performance domain are interested in studying how decisions ought to be made; industry experts are more concerned with satisficing and making design progress with limited time and resources.

Principle: *It is futile to try to develop a detailed generic process map for performance-based building design. Process logic only works for parts of the process.*

Hypothesis 1: *If normative decision-making is to play a role in performance-based building design, it needs to do so at decision moments that can be isolated from the rest of the process and that are seen as sufficiently important to warrant performance analysis in order to underpin decisions. Further work is required to recognize such decision moments.*

Postulate: Key design decisions are (i) the selection of systems, (ii) choices pertaining to building location, orientation and geometry, (iii) decisions on system and component sizing and (iv) retrofit decisions.

Hypothesis 2: *The tacit, expertise-based approach to support performance-based design will benefit from inclusion of building performance analysis theory in education of the various professions that may deliver future building performance champions.*

Hypothesis 3: *There still is a large distance between building performance analysis capabilities and the complexity of actual design projects. An approach that ensures that building design decisions are made in a rational manner and that ensures that future buildings will perform as expected needs to be developed. This probably will have to be done by experts in design methodology.*

Observations on early/conceptual and integral design:

- The importance of early and conceptual design towards later building performance is often emphasized, but there are very few actionable guidelines that help designers act differently in this phase.
- Integral design, in terms of consideration of many performance aspects by a wide range of stakeholders and experts, appears to be a logical approach. Yet application seems to be limited to high profile flagship projects only.
- Designing a building that will perform well is one thing; getting clients and other stakeholders to accept this design, and to construct and operate it as intended, is another challenge.

Comment: Additional efforts on early design or integrated design require extra resources in terms of time and manpower.

Principle: *Without access to additional resources, a rethinking of the design process that emphasizes early or integrated design needs to make sufficient downstream savings to allow upfront investments.*

Observations on design questions:

- Not all building performance analysis efforts are necessarily related to decisions that have to be made.
- Building performance analysis may also be driven by genuine questions by design team members and other building stakeholders throughout the building life cycle.

Hypothesis: *Recognition of design questions is another starting point for the streamlining of building performance analysis efforts.*

Postulate: Key design questions pertain to (i) whether or not a building will be meeting the regulations, (ii) whether or not a building will meet other requirements, (iii) which parameters will have the largest impact on performance, (iv) discovery of the impact of unusual systems or designs, (v) directions on how to improve on a design, (vi) robustness towards averse conditions and (vii) check before moving to next stage.

Observations on reporting and visualization of performance:

- Representation of building performance, whether in numbers or graphs, needs to be tailored to the analysis effort. Often this may require comparison between different alternatives or comparison with peers.
- Radar plots are often used in academic literature, but seem to be seldom used in building design practice and the wider construction industry.

Explanation: Radar plots fit well with the concept of multi-criteria decision making and comparison across a range of alternatives. Industry may be more concerned with tracking the performance of single options.

Hypothesis: *Performance profiles may be preferred over radar plots as they do not suggest a rounded, complete overview with a central focus point, but a list that always can be extended.*

Observations on digital support environments:

- Researchers have developed a range of digital design support environments that offer advanced features, including interoperability, optimization, uncertainty propagation and workflow management. Uptake of these environments in practice is limited.
- In the building performance literature, there are many examples of work on optimization; there is significantly less work on satisficing. Again, there are some optimization frameworks that are mostly used in an academic context.

Explanation: Many 'support environments' are complex proof-of-concepts systems. They demonstrate what might be possible, but are not necessarily designed to meet the demands of daily industry practice.

Hypothesis: *Efforts to support design should have situational awareness in terms of the search space, design drivers, user knowledge and context.*

Observations on the building construction process:

- The construction process is known to often lead to a significant number of defects; this means buildings are not constructed as per design specifications.
- Construction management has a different view of performance and often measures performance in terms of business KPIs.

Comment: The construction process is a difficult to control, in-situ manufacturing process, with low profit margins.

Hypothesis: *Due to tight budgets, time pressures and the issues with working with a varying workforce under open-sky conditions, construction management often prioritizes process control over product performance.*

Observations on performance measurement and monitoring:

- Only few aspects of building performance can be measured directly, without a need to aggregate data.
- Savings cannot be measured directly, but always require a benchmark value or baseline for reference.
- Building performance data is only useful if it is provided together with details on the conditions under which this data was collected, in terms of building excitation, operation, observation and possibly aggregation of measurements.
- The balance between monitored and simulated building performance data is shifting. In the past, monitored building data was relatively coarse, whereas simulation data on the basis of an hourly timestep was seen to be detailed. Recent trends reverse this, with monitoring timesteps of 30 minutes or less being the norm.
- There is still a surprising lack of publicly available building performance benchmark data.

Explanation: Measured/monitored building performance data needs analysis to yield actionable information, either for building operation or for decisions about maintenance and refurbishment.

Observations on building automation, metering and FDD:

- AMR creates big data, which needs to be analyzed to bring benefits.
- Metered data is not always correct, but may be faulty due to incorrect measurement positions, sensor errors and similar.
- Control settings of any automation system may drift or become outdated because of changes in the use pattern.
- Building monitoring systems may raise alerts, but mostly human expertise is called for to confirm performance issues and to intervene when and where this is needed.

Comment: Identifying outliers and system errors may be done by mathematical routines built on machine learning approaches; however identifying the causes of these outliers and errors is a much more complex matter.

Hypothesis: *Building automation systems and FDD systems may benefit from being trained on the basis of virtual data as generated through simulation. This could allow system preparation before the building is physically completed, a set-up that recognizes the theoretical behaviour, as well as training to recognize errors and faults that have a low occurrence rate.*

Observations on high performance buildings:

- Definitions of high performance buildings thus far are mainly aspirational.
- Many attempts at a definition of high performance buildings mainly give an overview of a large number of performance aspects that must be addressed.

Explanation: To define a high performance building, one first needs to define and quantify building performance but this was observed to be left undefined before.

Definition: A *high performance building* is a building that performs (a) task(s) or function(s) better than 70%–84% of its peers, in terms of the quality, resource saving, workload capacity, timeliness, readiness, representativeness or entertainment attributes of that building.

11.2 Suggested Guidelines for Building Performance Analysis

As buildings are complex, highly customized, individual products, there is no standard roadmap on how to carry out building performance analysis in detail; instead there is a need to establish the needs and requirements, criteria and appropriate performance quantification methods for each specific case. However, as starting point, this section provides some generic guidance for the common cases of analysis in a design, operational and research context. These should be used in combination with the template for the development of performance criteria as provided in Appendix B and the checklist for tool/instrument configuration provided in Appendix C.

11.2.1 Building Performance Analysis During Design

For a performance-based building design project, the following steps should help to work towards a design that meets a range of well-defined performance requirements:

1) Appoint a 'building performance champion' early in the design process. This champion must not necessarily have a specific role such as architect or building services engineer; however it is important that someone takes ownership of the issue of building performance.

2) Invest resources (time, manpower) in establishing building performance requirements. This effort can be part of the briefing process; it requires interaction with most stakeholders in the project. Make sure to move beyond functional requirements. Ensure that the performance requirements qualify under the categories of building quality, workload capacity, resource saving, timeliness and readiness.

3) Properly develop the performance requirements and make them as explicit as possible by using graphical methods such as UML or IDEF diagrams. Then establish criteria in terms of goals, targets and benchmarks, and formulate all performance requirements in terms of a mathematical 'performance measurement expression'.

4) Define 'performance gateway' points in the design process where performance will be verified. These may be design decisions that are deemed critical, such as choice of the façade system or choice of HVAC components, or transition points in the process such as where the design moves from conceptual to preliminary design.

5) For any performance gateway, decide what performance aspects need to be reviewed. Some gateways may be passed by reviewing a small subset, others may require a full performance review across a wide range of requirements. But note that it is unlikely that there will be only one singe aspect involved with any one gateway, as almost all decisions relate to a range of interrelated issues.

6) Where a performance gateway relates to a choice between options, make sure that the search space is properly defined. Review whether the options considered are the only ones or whether a larger set of alternatives needs to be considered. Setting the search space is crucial for the success of choices, and in setting up the search space, one may want to 'think outside the box'.

7) Once the search space and criteria have been defined, decide how performance of each option for each criterion is going to be quantified. This may require a review of the 'performance measurement expressions' that have been defined and matching these with software and expertise that is available to support the analysis. Where possible, use a template that describes the performance quantification that can be used by the four main approaches (measurement, simulation, expert judgment, stakeholder evaluations) later in the building life cycle to limit the development of a performance gap due to incompatibility between methods.

8) Consider whether there are significant uncertainties in options, criteria and quantification methods, and decide how these will be handled.

9) Quantify the performance of each option for each performance criterion using the performance quantification methods that have been selected.

10) Present quantification results in an overview, for instance using a matrix structure or performance profile. Evaluate whether all design options are feasible or whether some do not meet criteria and thus need to be rejected as alternative.

11) Discuss and evaluate findings with the stakeholders and decision makers.

12) If appropriate, use multi-attribute utility theory to rank options and find the options that most closely meet the requirements. Keep in mind that weighting factors are subjective and should be set with stakeholder involvement. Be careful with group decisions, and remember Arrow's impossibility theorem.

13) Carry out a review of the analysis process and ensure proper documentation of all activities, assumptions, findings and recommendations.
14) Consider what lessons may be drawn from the analysis effort and whether these can be carried over to future efforts.

11.2.2 Building Performance Analysis During Operation

Many of the steps recommended for building performance analysis during design also apply to analysis of an existing building that is in operation. However, there are key differences in the analysis methods available and also in the control that analysts can exert over real-life experiments. The following points deserve additional attention:

1) Invest time in establishing what performance aspects are to be analyzed and why. There may be refurbishment decisions to be made, complaints to be addressed or a genuine interest in finding out how the facility is doing with regard to certain aspects. Consider doing an in-depth stakeholder analysis and requirement elicitation.
2) Review whether performance is to be related to a group of similar buildings (peers), to a historical trend, or to interventions. Explore what benchmarks and baselines should be used as reference.
3) Establish the detailed performance requirements and criteria that will be used to assess the building performance for the aspects of interest.
4) Define the 'performance measurement expressions' that will govern the analysis efforts.
5) Decide on performance quantification method; for buildings that actually exist, options include physical measurement, building simulation, expert judgment as well as stakeholder perception studies. Consider whether different methods can be used in combination to allow for triangulation of findings.
6) Choose the analysis timeframe, and define relevant interventions. Note that in software, parameters of the analysis can easily be managed. This is not the case in real-life measurement, where semi-controlled experiments are typical; for instance weather conditions cannot be set by the analyst. However, there may be some prospects to 'force' behaviour, for instance by setting extreme control setpoint values, directed interventions in occupancy and others. Where factors are not controllable, these need to be accounted for, and the duration of analysis needs to be selected in such a way as to allow the building to cover the most common situations.
7) Capitalize on the theory on Design of Experiments (DOE) to set up measurement and experimentation in an efficient manner, ensuring that data is collected in such a way as to enable the use of statistical analysis.
8) Where measurement involves analysis during occupation, ensure that proper ethics protocols are followed and that there are no breaches or risks in terms of security and privacy.
9) In building stakeholder perception studies, be vigilant about the potential impact of external factors (such as working pressure, health) on outcomes of a study that should principally reflect on the facility.
10) Include error analysis in the review of all performance quantifications.

11.2.3   Building Performance Analysis in Research

The use of building performance analysis in a research context puts additional emphasis on methodological aspects: science requires experiments to be reproducible, reliable, valid and transferable. This leads to the following:

1) In research, significant attention should be paid to defining analysis efforts in such a way that they can be repeated by other scientists. Given the complexity of most building performance analysis efforts, this is a non-trivial challenge. Documentation should precisely describe the type and location of sensors, sampling of any stakeholder groups, and selection of experts. Where software is used, models should be made available for review.
2) Another area that needs additional efforts is the validity of building performance analysis studies, in terms of making sure that measurements and quantification actually capture the issues of interest. This is of particular importance where analysis is based on semi-controlled experiments; given the complexity of buildings and their use, contaminating signals are easily overlooked.
3) Selection of the appropriate tools and methods carries extra weight in research. The choice for specific sensors, software, survey techniques or questionnaires thus needs to be motivated.
4) Measurements and quantifications should be repeated, so that conclusions are not drawn from single observations. Where possible, data triangulation should be attempted.
5) It is of importance to consider the transferability of any building performance analysis effort and to explore the applicability to other buildings and situations. Keeping in mind the highly individual, bespoke and complex nature of most buildings as well as the variations in context and use, generalization of findings needs special attention.
6) Building performance analysis research has an extra responsibility to consider uncertainties and errors in any studies.

11.3   Future Challenges

A generally accepted theory of building performance analysis is slowly emerging, with the previous sections attempting to build a framework that can be discussed, tested, expanded and modified as necessary. The hypotheses that were developed in Section 11.1 would be a good starting point for further work. Specific challenges are summarized as follows:

- Terminology on the subject of building performance analysis has been noted in this work to be diverse and often lacking. This text attempts to give a balanced discussion and move towards common terms and understanding. However, debate amongst industry and academia is required to make progress and actually change current practice.
- Concepts from other disciplines, such as systems engineering, process management and manufacturing, may help to strengthen the field of building performance analysis. However, to evolve as a scientific discipline, the field building performance analysis needs to develop self-confidence and to embark on a journey where such concepts are critically reviewed for their prospects when adapted.

- Significant work is required to better align the four main approaches for quantifying building performance (physical measurement, building simulation, expert judgment and stakeholder perception studies). This requires efforts by domain experts to develop performance quantification procedures that are the same across all four approaches. Without this, 'performance gaps' will continue to exist for all performance aspects across the board. Such performance quantification procedures must consider building excitation, observation of states and inputs/outputs, and how to deal with the reality of semi-controlled experiments in actual buildings.
- Software for the support of performance-based design needs a deep rethink:
  - The process of performance-based design is still poorly understood. There is a need to establish those parts of the process that require adherence to process logic so that these can be supported and possibly automated while leaving other parts free to respond flexibly to the needs of creative activities.
  - While BIM provides a novel way of creating and manipulating 3D objects in the computer, an established process for linking these emerging and changing 3D objects to an evaluation of how a building (which consists of a set of such objects) performs does not exist yet. There are two issues to be addressed: the specific virtual experiment used for performance analysis (which goes beyond just definition of the objects) needs to be specified; there also is a need to define when the virtual experiment becomes appropriate (for instance, there is no use in doing an energy efficiency analysis before the façade has been added to the design).
  - There remains a need to find the proper models and tools to conduct an analysis, rather than fit the design process to analysis tools that happen to be available.
- The large amount of data generated by the many sensors in modern buildings provides a range of challenges. Approaches are needed to analyze this data and to distil actionable information to better manage buildings and to identify errors and faults. At the same time, this should not infringe on security and privacy.
- The teaching of building performance analysis needs a deep review. At present, this is a relatively disjointed area, with separate courses in traditional disciplines (structural engineering, mechanical engineering), systems (façade engineering, HVAC systems) and specialist areas (building simulation) covering parts of the same domain but failing to provide a coherent approach.
- There are deep questions to be asked about the directions that future building technology should take. For instance, one may opt to educate building occupants about their influence on building performance but allow them a large amount of control, or one may decide to automate buildings and develop them to a level that occupant intervention becomes almost negligible. A similar issue is the choice between centralized and local systems; the traditional system boundary around buildings may require rethinking and extend both outwards (smart grid and city) and inwards (internet of things). Here building performance interacts with both the performance of a wider urban fabric as well as with that of the systems and appliances contained inside the building.
- The building construction process dimension of building performance is strongly associated with work in management and manufacturing. Key challenges in this domain are better process control in order to manage on-time and on-budget delivery, while maintaining health and safety on site, reducing waste and customer satisfaction. In terms of buildings as a product, the current emphasis in construction is still

on preventing defects. The core issues about the performance of the resulting product, in terms of building quality, resource saving, workload capacity, timeliness and readiness, are slowly gaining interest. However, deeper integration between product and process management is also required.

- The aesthetic view of building performance is only just emerging. Significant efforts will be required to develop this view and create a strong position that equals the product and process dimensions.

11.4 Closure

This book sets out to bring together the existent body of knowledge on the subject of building performance analysis. It explored four key issues: the concept of building performance itself, the measurement and quantification of building performance, the way that building performance can be used to guide the improvement of buildings, and what the building domain may learn from other disciplines.

Building performance has been found to be mostly left as an implicit concept. A review of the background of building performance, key uses in the field of building design and construction, and a review of building performance requirements has been used to develop the following working definition: *Building performance is a concept that describes, in a quantifiable way, how well a building and its systems provide the tasks and functions expected of that building. Requirements may stem from three main views: an engineering view of buildings as an object, a process view of building as a construction activity, and an aesthetic view concerned with the notions of form and appreciation. Important performance requirements in the engineering view pertain to building quality, resource savings, workload capacity, timeliness and readiness.*

Building performance can be measured and quantified using four main approaches: physical measurement, building simulation, expert judgment and stakeholder evaluation studies. In principle, all these approaches observe how building states, input and output change when the building is subjected to some form of excitation that leads to a response and thus some sort of behaviour. Work in the area of performance quantification is complicated due to the fact that buildings are highly individual, bespoke and complex systems. Given their size and costs, analysis often has to take place in real buildings, where full control of experimental conditions (climate, occupant behaviour and similar), is not possible.

Using building performance analysis to improve buildings and the built environment can be achieved in all parts of the building life cycle, but especially in the phases of building design and building operation. Embedding a performance-based approach in the design process requires a balancing of logic-based analysis processes with more open and flexible form finding activities. Once a building is operational, performance analysis may be used to review how a building performs in relation to its own historical track record or in relation to a group of peers. Analysis may help to point to faults and errors that can be remedied to reinstate the original performance levels.

Building performance analysis may benefit from studying concepts from related fields, such as systems engineering, process management and manufacturing. However, given the complexity and particularities of buildings and their use, there is a need for domain experts to test which concepts will work in the building context and which will not.

Moving forward requires self-confidence of the building performance analysis domain to take on this role.

High performance buildings represent the cutting edge in terms of buildings that meet a set of particular performance requirements. Thus far the term has been mainly used in a generic aspirational manner. From the point of view of building performance analysis, it makes sense to ensure that these criteria are well defined and fit under the building performance requirement categories of quality, resource saving, workload capacity, timeliness, readiness, representativeness or aesthetics; note that it is not necessary to cover all categories. However, high performance buildings should be amongst the best 30%–16% of peers in their category.

The chapters of this book provide an overview of the background and history of building performance analysis and the relation to other disciplines. They discuss the context of building performance, and the complexity of the life cycle, systems and stakeholders involved. Performance criteria in terms of stakeholder needs, building functions and requirements are investigated and contrasted with the building and its systems in order to develop the concept of building performance as a comparison of 'demand' and 'supply'. Attention is paid to goals, targets, constraints, baselines and benchmark that are needed to evaluate the fit between demand and supply. Different quantification methods, covering the domains of physical measurement, simulation and computation, expert judgment and stakeholder perception studies, are discussed. The hard work of setting performance criteria, finding suitable quantification methods and adjusting tools to the analysis of specific cases is explored. The focus then shifts to the application of building performance analysis in the building design and construction process, as well as building operation and management. Special attention is also given to the concept of high performance buildings.

The deep review of the existing knowledge on building performance analysis as presented in this book is then used to develop an emergent theory for this field, which is formulated in terms of observations, comments, explanations, principles and hypothesis. Furthermore this is used to give broad guidelines for carrying out building performance analysis during design, or building operation or in a research context. Some issues that require further work are described. This is offered to the building performance community for further exploration and debate.

Appendix A

Overview of Building Performance Aspects

This appendix gives an indicative list of building performance aspects while acknowledging that it is not possible to develop a definitive list. The overview includes a list of verbs that help to turn the building performance aspect into a function. It also indicates what type of performance requirement would be asked for if the function is demanded from a building.

Aspect	Verb (aspect ▶ function)	Performance requirement
1. Occupant satisfaction	provide	quality
2. Continuity of service	provide	quality
3. Thermal comfort	maintain	quality
(a) Air temperature	control	quality
(b) Radiant temperature	control	quality
(c) Air velocity	control	quality
(d) Relative humidity	control	quality
(e) Air speed	control	quality
(f) Overheating	prevent	quality
(g) Undercooling	prevent	quality
(h) Wind chill	prevent	quality
4. Acoustical comfort	maintain	quality
(a) Speech intelligibility	provide	quality
(b) Reverberation times	control	quality
5. Visual comfort	maintain	quality
(a) Glare	prevent	quality
(b) Flickering	prevent	quality
6. Olfactory comfort	maintain	quality
(a) Odour	control	quality
7. Indoor air quality	maintain	quality
(a) Smoke, fumes, stale air	dispose of	quality
(b) Fresh air	provide	quality
8. Structural integrity	maintain	quality

(Continued)

Building Performance Analysis, First Edition. Pieter de Wilde.
© 2018 John Wiley & Sons Ltd. Published 2018 by John Wiley & Sons Ltd.

Aspect	Verb (aspect ▶ function)	Performance requirement
9. View to the outside	provide	quality
(a) Outside world	provide connection with	quality
(b) Circadian rhythm	support	quality
10. Identity	provide	quality
11. Privacy	provide	quality
12. Inclusivity	support	quality
13. Relative humidity	control	quality
14. Vibration	protect from/limit	quality
15. Noise	protect from/limit	quality
16. Glare	protect from/limit	quality
17. Precipitation	keep out	quality
18. Ground/surface water	keep out	quality
19. Unwanted visitors/vermin	keep out	quality
20. Outdoor pollutants	keep out	quality
21. Electricity/gas/water	provide uninterrupted supply	quality
22. Drainage/sewerage	provide safe and adequate	quality
23. Wayfinding	support	quality
24. Wind flow around building	control	quality
25. Condensation	prevent	quality
26. Contamination	prevent	quality
27. Complaints	minimize number of	quality
28. Fire ignition	prevent	quality
29. Fire spread	prevent	quality
30. Congestion, crowding	prevent	quality
31. Community	provide sense of	quality
32. Historical significance	have	quality
33. Local and national heritage	contribute to	quality
34. Income/revenue	generate	workload capacity
35. Key processes/work	enable	workload capacity
36. Productivity	enable	workload capacity
37. Ease of movement/ circulation	provide	workload capacity
38. Structural loading	carry	workload capacity
(a) Dead load (own weight)	resist	workload capacity
(b) Live load (occupants, furniture)	resist	workload capacity
(c) Live load (wind, precipitation)	resist	workload capacity
(d) Cycling loads (fatigue)	resist	workload capacity
39. Heating/cooling	provide	workload capacity

Aspect	Verb (aspect ▶ function)	Performance requirement
40. Ventilation/fresh air	provide	workload capacity
41. Daylight/sunlight	provide	workload capacity
42. Hot and cold water	supply	workload capacity
43. Artificial lighting	provide	workload capacity
44. ICT connectivity	provide	workload capacity
45. Safety	ensure	resource saving
(a) Falling risk	mitigate	resource saving
(b) Cutting risk	mitigate	resource saving
(c) Risk from machines	mitigate	resource saving
(d) Electrocution risk	mitigate	resource saving
46. Energy	make efficient use of	resource saving
47. Water	make efficient use of	resource saving
48. Material	make efficient use of	resource saving
49. Renewable energy	generate	resource saving
50. Rainwater	harvest	resource saving
51. Waste	minimize	resource saving
52. Local ecosystem	protect	resource saving
53. Rare and endangered species	protect	resource saving
54. Wear and tear	resist	resource saving
55. Decay and rot	resist	resource saving
56. Corrosion	resist	resource saving
57. Construction costs	control	resource saving
58. Construction time	control	resource saving
59. Operational costs	control	resource saving
60. Cleanability	provide	resource saving
61. Maintenance and repair	efficiently provide	resource saving
62. Greenhouse gas emissions	limit	resource saving
63. Access control	provide	responsiveness
64. HVAC control	provide	responsiveness
65. Lighting control	provide	responsiveness
66. Darkness	provide	responsiveness
67. Solar radiation	control	responsiveness
68. Urban context	respond to	responsiveness
69. Site conditions	respond to	responsiveness
70. Outside hours access	allow	responsiveness
71. Modifications to building	allow	responsiveness
72. Service life	manage	responsiveness
73. Fire/smoke alarm	raise	readiness
74. Intrusion alarm	raise	readiness
75. Evacuation	allow	readiness
(a) Evacuation route	provide	readiness
(b) Evacuation time	allow	readiness
(c) Survival time in refuges	guarantee	readiness

(Continued)

Aspect	Verb (aspect ▶ function)	Performance requirement
76. Burglary	resist	readiness
77. Vandalism	resist	readiness
78. Extreme events	resist	readiness
(a) Fire (smoke and heat)	minimize impact of	readiness
(b) Explosion	minimize impact of	readiness
(c) Radioactivity spread	minimize impact of	readiness
(d) Poisonous substance spread	minimize impact of	readiness
(e) Heat waves	cope with	readiness
(f) Cold spells	cope with	readiness
(g) Natural disasters	resist	readiness
(h) Human-made disasters	resist	readiness
79. Disease and infection	stop spreading of	readiness
80. Buildability	provide	readiness
81. Flexibility	possess	readiness
82. Disposability	provide	readiness
83. Aesthetics	consider	aesthetics
(a) Architectural statement	make	aesthetics
(b) Creativity	demonstrate	aesthetics
(c) Interpretation	require	aesthetics
(d) Communication	engage in	aesthetics
(e) Embodiment	represent	aesthetics
(f) Image	portray	aesthetics
(g) Eloquence in composition	demonstrate	aesthetics
(h) Enchantment	instil	aesthetics
(i) Movement	suggest	aesthetics
(j) Structural elegance	express	aesthetics

Appendix B

Criterion Development Template

Name: _____

Identifier: _____

Author: _____

Version: _____

Status: _____

Type of performance aspect: (   ) Workload capacity

(   ) Resource saving

(   ) Timeliness

(   ) Readiness

(   ) Aesthetics

(   ) Other: _____

Purpose of assessment: _____

_____

_____

_____

_____

Stakeholders: _____

_____

_____

Building Performance Analysis, First Edition. Pieter de Wilde.
© 2018 John Wiley & Sons Ltd. Published 2018 by John Wiley & Sons Ltd.

Room for negotiation: _

_ _

_ _

Essential building features, systems, components and parts that must be included in analysis:

_ _

_ _

_ _

_ _

Specific parameters of interest: _

_ _

_ _

Time horizon: _

Frequency: _

Spatial resolution: ( ) System level

( ) Room

( ) Zone

( ) Floor

( ) Whole building

( ) Other: _ _ _ _ _ _ _ _ _ _ _ _ _ _ _ _ _ _

Appendix C

Tool/Instrument Configuration Checklist

Name: _ _ _ _ _ _ _ _ _ _ _ _ _ _ _ _ _

Identifier: _ _ _ _ _ _ _ _ _ _ _ _ _ _ _ _ _

Author: _ _ _ _ _ _ _ _ _ _ _ _ _ _ _ _ _

Version: _ _ _ _ _ _ _ _ _ _ _ _ _ _ _ _ _

Status: _ _ _ _ _ _ _ _ _ _ _ _ _ _ _ _ _

Relates to the quantification of performance criterion: _ _ _ _ _ _ _ _ _ _ _ _ _ _ _ _ _ _ _

Fundamental measurement purpose: ( ) Establish response to stimuli/excitation

( ) Create input–output pair time series

( ) Reproduce system behaviour

( ) Other: _

Building Performance Analysis, First Edition. Pieter de Wilde.
© 2018 John Wiley & Sons Ltd. Published 2018 by John Wiley & Sons Ltd.

Checklist points:

1. Does the tool/instrument capture all building features and systems of interest?   Yes / No

   List key systems, system boundaries and variables that have been considered:

   _____

   _____

   _____

   _____

   _____

2. Does the tool/instrument enable to apply the load, stimulus and excitation of interest? Yes / No

   List loads, stimuli and excitation of relevance:

   _____

   _____

   _____

   _____

   _____

3. Does the tool/instrument measure the relevant building response?   Yes / No

   List key system states and outputs that need to be captured:

   _____

   _____

   _____

   _____

   _____

4. Is any aggregation processes needed to obtain performance measure values?   Yes /No

    Detail of any required aggregation process:

                           _____

                           _____

                           _____

                           _____

                           _____

5. Have any test runs been conducted to check the measurement works?   Yes / No

    Detail of any test runs:

                           _____

                           _____

                           _____

                           _____

                           _____

    Any corrections made:

                           _____

                           _____

                           _____

                           _____

                           _____

6. Has the validity of the tool/method been reviewed?   Yes /No

    Validation details:

                           _____

                           _____

                           _____

                           _____

                           _____

7. Is anything known about reliability of the tool/method?   Yes / No

   Reliability details:

   $$_____$$

   $$_____$$

   $$_____$$

   $$_____$$

   $$_____$$

8. Is anything known about reproducibility of the measurements?   Yes / No

   Reproducibility details:

   $$_____$$

   $$_____$$

   $$_____$$

   $$_____$$

   $$_____$$

Appendix D

Measurement Instruments

acoustic camera	microphone array that is used to localize and identify sound sources; can be two- or three-dimensional
albedometer	device that measures the albedo, which is the proportion of light reflected by the Earth's surface (%); typically consisting of two pyranometers, with one measuring the direct solar radiation and the other measuring the reflected radiation
anemometer	device to capture wind speed (m/s); may be a cup-type, vane or hot-wire anemometer
balances and scales	instruments that measure the weight of an object (kg)
carbon dioxide meter	device that captures the presence and concentration of CO_2; may be a non-dispersive infrared (NDIR) or chemical sensor
carbon monoxide meter	instrument that detects the presence of CO; typical sensors are of the opto-chemical, biomimetic, electro-chemical or semiconductor type
chromameter	a device that measures the colour of surfaces
clamp meter	a meter that has two arms that 'clamp' around an electrical wire, allowing it to measure the electric current in that wire
electrical multimeter	device that measures electrical voltage (V or U), resistance (Ω) and current (I)
flow meter	device that establishes the flow rate of a liquid or gas (m^3/s)
gloss meter	instrument that measures the specular reflection of a surface in Gloss Units (GU)
heat flux sensor	instrument that captures the heat flow in watts (W) through a surface
illuminance meter	instrument that captures the incident luminous flux in lux (lx) on a surface
infrared thermometer	thermometer that uses thermal radiation and assumed emissivity of an object to measure the temperature (°C, °F or K)

Building Performance Analysis, First Edition. Pieter de Wilde.
© 2018 John Wiley & Sons Ltd. Published 2018 by John Wiley & Sons Ltd.

integrating sound level meter	device that measures sound energy in terms of the sound pressure level (SPL) over a period of time and then establishes average the sound energy over that time in terms of an equivalent continuous sound level (L_{eq})
laser distance meter	rangefinder that uses a laser beam to measure a distance in meter (m)
luminance meter	instrument that measures the luminance of an object for a given direction and solid angle in candela per square meter (cd/m^2)
manometer	instrument that measures pressure using a column of liquid in pascal (Pa)
moisture meter	device that measures the percentage of water in a sample (%), typically on the basis of establishing electric resistance
motion sensor	device that detects movement, typically of people in around buildings
multimeter	instrument that combines several measurement functions; often used in the context of electronics for a device that measures voltage (V), current (I) and resistance (R); however also used for other combinations such as an instrument that captures light levels, temperature, humidity and sound levels
noise logger	instrument that uses spectral noise logging (SNL) to test basins, reservoirs and ductwork for integrity
occupancy sensor	device that detects the presence of human beings; can be based on different principles such as passive infrared detection, ultrasonic or microwave technology, audio detection, CO_2 measurement, electromagnetic switches on doors or windows or key card technology
oscilloscope	device that measures the regular change of states of something, typically the variation in signal voltage, sound and vibration, allowing to observe waveforms in terms of amplitude, frequency and similar
particle counter	instrument that detects and counts physical particles through light obstruction, scattering or direct imaging
personal noise dosimeter	portable sound level meter that captures the noise exposure of a person over time
pressure sensor	instrument that measures atmospheric pressure (p), pressure in a vessel (p) or simply the presence or absence of pressure (on /off)
pyranometer	instrument that measures solar irradiance on a planar surface (W/m^2)
pyrgeometer	instrument that measures near-surface infrared radiation
pyrheliometer	instrument that measures direct beam solar irradiance

pyrometer	instrument that measures high temperatures such as in furnaces; used in meteorology and climate studies
radiometer	instrument that measures radiant flux of electromagnetic radiation
spectrophotometer	instrument that measures the intensity of light in a part of the spectrum
spectroradiometer	device that measures the spectral power distribution of a light source
sound intensity meter	equipment for measuring sound pressure and particle velocity, used to establish sound intensity (I); typically consisting of two microphones
sound level meter	instrument that measures sound pressure level (SPL)
thermal camera	device that captures radiation in the infrared spectrum (0.75–300 µm) and produces an image of that radiation; software may be used to convert the radiation measurement to apparent temperature
thermal conductivity probe	device for the measurement of conductivity; typically in the form of a needle probe that can be inserted into a sample
thermal imager	handheld thermal camera
thermocouple	electrical device that consists of two different conductors that produce temperature dependent voltage, which can be used to measure temperature (°C, °F or K)
thermometer	instrument that measures temperature (°C, °F or K); the type may be expansion thermometer (liquid in glass, bi-metal), electrical resistance thermometer, thermocouple or infrared
dry-bulb	measures the air temperature with a thermometer that is shielded from radiation and moisture
wet-bulb	measures the lowest temperature that can be achieved in a space through the evaporation of water
infrared	thermometer that uses thermal radiation and assumed emissivity of an object to establish temperature
globe	captures the combined effect of radiation, air temperature and air velocity by measuring the temperature inside hollow copper sphere that is painted black; typically the diameter of the sphere is 150 mm
vibration meter	instrument that measures the frequency (Hz) of a vibration
water level meter	instrument that indicates the depth (d) in wells, boreholes, tanks, rivers, ponds, etc.
water run-off gauge	instrument that measures the amount of water (V) that is discharged from a surface, typically as a result of rainfall
weather station	a platform that measures air temperature (°C, °F or K), atmospheric pressure (Pa), relative humidity (%RH), wind speed (m/s) and wind direction, solar radiation (W/m^2) and rainfall (mm/day).

Glossary

absorption process by which radiation is intercepted by matter and converted into internal thermal energy

acausal not governed by a cause–effect relationship

accuracy degree of closeness of a measurement to a true value

additive manufacturing production of items by gradually adding materials; often used as synonym for 3D printing

aesthetics principles concerned with the nature and appreciation of beauty

affordance possibility of action on an object; often used to describe what a product offers to a user and that user is aware of

aggregation formation of something by bringing things together

ambition specified performance level that is being sought

architecture the art and practice of designing and constructing buildings, as well as the complex structure of something as in 'software architecture'

architectural programming the process of briefing in American English

auralization process of making a theoretical sound field audible

automated meter reading technology that captures meter readings without need for human intervention, typically at a high frequency; in most cases data is transmitted to a central database for analysis and storage

bandwidth range of frequencies or performance levels

baseline starting point for comparison; often used to denote a historical trend that is captured before some sort of intervention takes place

benchmark standardized situation or test case that can be used for evaluation and comparison

bias systematic deviation of an observed value from a known true parameter

biomimicry design based on emulating examples from nature (biology)

brief document that captures the client's business case and requirements for a building design and development project

building automation the use of technology to replace human control and operation of buildings

building code regulations, imposed by government, that contain requirements that buildings must meet

building information modelling the process of creating and managing data about a building, its construction process and performance across the full building life cycle

Building Performance Analysis, First Edition. Pieter de Wilde.
© 2018 John Wiley & Sons Ltd. Published 2018 by John Wiley & Sons Ltd.

building performance attribute that relates to an engineering, process or aesthetic view of the building; typically expresses, in a quantifiable way, how well a building meets performance requirements

building services the systems installed in buildings to make them comfortable, functional, efficient and safe; predominantly used in British English (more or less equivalent to HVAC systems in American English)

calculation mathematical determination of a quantity or extent

calibration process of adjusting parameters in a computational model to improve agreement with experimental data

capillary motion moisture flow in narrow spaces

circadian rhythm daily cycle of physical, mental and behavioural change in response to changes between light and dark

climate general weather conditions prevailing in a geographic location over a long period

cognitive bias systematic deviation from normative judgment

communication the process of transferring information, which requires sending and receiving

complexity the quality of consisting of many different and connected parts, leading to challenges in description, computation and understanding

constraint limitation or restriction

construction management the activities that oversee the design, planning and construction of a (building) project

coupled simulation simulation that combines models of different physical aspects such as lighting and thermal

criterion (plural: criteria) principle by which something may be judged or decided

data point physical input or output of a system, or abstract parameter used in system controls

data logger device that records data over time

data triangulation approach that employs data from two or more research methods to allow cross-comparison and validation

decision making cognitive process of selecting of an option from a set of alternatives, resulting in a choice

design decision support system IT system that helps actors in making design choices, typically supporting the generation and ranking of design options

design of experiments (DoE) the process of planning experiments in such a way that the data can be analysed by statistical methods in an efficient manner, leading to valid conclusions

design paradox paradox that relates to the fact that as design progresses, design freedom decreases, while knowledge about the object that is being designed increases

diffusion movement of heat or a substance from a region of higher concentration to one of lower concentration

discretization process of splitting up continuous models and functions into discrete parts, as step towards numerical evaluation and computing

downtime the time that passes from the moment a system fails to perform as expected to performance being restored

efficacy ability to produce the desired or intended performance

efficiency ability to be productive with a minimum of wasted effort or expense

elicitation process of collecting input (especially pertaining to needs and requirements) from stakeholders

energy performance contract contract that outsources the finance, installation and operation of energy conservation measures to a third party

energy service company provider of energy performance contracts

ethnography research methodology that studies subjects in their natural setting

experiment process that is carried out with the aim of gaining knowledge about some sort of phenomenon, often in a way that allows the testing of a hypothesis

expert judgment assessment carried out by persons with specific knowledge and experience in a certain area of interest

fault detection and diagnostics identification of outliers in system behaviour and the underlying cause

function that what a system or object is expected to do

functional requirement specification of some action or function that a system must be able to provide; not be confused with a performance requirement, which specifies how well that function must be provided

generative tool tool that helps with the creation of design alternatives, typically on the basis of permutation of parameterized archetypes

goal objective towards which efforts are directed, with strict binary judgment: either achieved or missed

heuristics approach to problem solving that is based on experience and trial and error

high performance building a building that performs certain tasks or functions better than a significant number of its peers

hydronic heating heating system that is based on the use of water as a transport medium ('wet system')

hysteresis process where the properties of a system depend on its history

indicator thing that indicates a state or level

integral (integrated) design design process that aims to bring together different expertise in order to develop a design that performs well across a range of aspects

intelligent building building that can respond to changing conditions, such as the preferences of occupants and the energy grid operator

iron triangle management model that represents the need for a trade-off between time, cost and quality

key performance indicator (KPI) measure for the achievement of organizational goals in a business or construction process context

Likert scale ordinal ranking scale used for representing human perception of a subject

limit terminal point or boundary beyond which something cannot pass

loss prevention standard fire and security standard developed by the BRE (United Kingdom)

luminaire lighting unit, consisting of lamp, reflector and other parts

maintenance the process of keeping a system in good condition

measure standard unit to express size, amount, degree, extent or quality

mereology the study of the relations between parts and wholes

metric system or standard of measurement

metrology the scientific study of measurement

model-based predictive control application of some sort of model to predict near future states of the building, which is used to identify optimal control actions at the present point in time

modelling the specification of a system that is to be simulated by way of a set of instructions, rules, equations and constraints

monitoring measurement over time

multi-attribute utility theory a (mathematical) decision theory that helps to establish the preferences of a decision maker who is completely rational

naturalistic decision theory a decision theory that represents how decisions are made in practice, usually emphasizing decision making by experts

norm description of a required performance level; synonym for standard

normative decision theory theory that describes how a decision should be made to realize the best possible outcome

normative measurement method method that designates some outcomes as permitted or desired and other outcomes as not permitted or undesirable

observation the activity of closely monitoring something, allowing one to note significant details

ontology definition of a set of concepts in a subject area and the relation between these concepts

optimization the process of searching for the best possible alternative

option space (search space) the set of all possible design alternatives, typically defined by system configurations and parameter permutations

outcome space (solution space) the set that represents the performance of all elements of the option space across all performance aspects of interest

performance-based design (PBD) design process that explicitly targets good building performance, often through the use of performance quantification methods

performance gap discrepancy between predicted performance and measured performance

performance profile graphical representation that allows the comparison of different building alternatives using parallel axes (contrast with radar plot, where the axes are brought together in a centre point); see CIB Report 64 for example

performance quantification measurement of building performance using physical experiments, simulation, expert judgment or stakeholder assessment

performative architecture architecture that interacts actively with a changing social, cultural and technological context

photometry the science of the measurement of light

post-occupancy evaluation (POE) examination of building performance post-handover, mostly on the basis of stakeholder assessment but sometimes also including extensive monitoring efforts

precision measure for the repeatability of a measurement

project management the process of managing the initiation, planning, execution, monitoring, completion and overall control of some sort of project

prognostics the process of predicting when a system will no longer perform as desired

prototype early version of a product that is tested before the production of a larger series of final products

proxy a figure that is used to represent something else, either because that other thing cannot be measured directly or for reasons of simplification

quality degree of excellence

quantification the process of measurement, expressing the quantity of something

random error a measurement error that is not systematic

rating scheme evaluation system leading to the classification of something in terms of quality, performance and other attributes

refraction bending of a wave when entering a medium where the speed of that wave is different

requirement characteristic task, action or function that must be performed to achieve a desired outcome

requirements engineering systematic approach to the specification and management of requirements

resilience capacity of a system to recover from difficulties and malfunctions

robustness ability to withstand or overcome adverse conditions

round-robin test inter-laboratory test where the same experiment is carried out independently by various laboratories

safety factor an additional factor that allows a system to deal with more than the expected load and thus cope with emergency situations

satisficing the process of searching for an alternative that is 'good enough'

sensitivity analysis the process that establishes which of the input parameters of an experiment have the most impact on the outcomes

sick building syndrome situation where symptoms of poor health are related to being in a certain building, but without a specific cause having been identified

simulation use of the computer to imitate reality, employing mathematical models that account for physical principles and good engineering practice

smart grid energy supply network that employs digital communication technology that allows this network to respond to local changes in energy usage

smart home intelligent domestic building, often including an element of telecare for the elderly and other people who may require some sort of assistance

stakeholder a person or party who has an interest in something

standard document that sets rules and processes for certain activities, such as experimentation and construction

system a set of elements in interaction, a whole of parts that interconnects

systems engineering interdisciplinary field of science that focuses on the design and management of systems

tacit knowledge knowledge that remains uncodified, as opposed to explicit knowledge that can be written down and transmitted to others

target objective towards which efforts are directed, but allowing for deviation

thermochromic system that undergoes a change of colour when heated or cooled

threshold level or point at which something starts or ceases to happen

uncertainty analysis the process that studies how variation of the input of an experiment influences the outcomes

user/stakeholder evaluation assessment by the user/stakeholder of how well a product performs

user/stakeholder satisfaction degree to which a product meets the expectations of the user/stakeholder

validation process that determines how well a model represents the real world

verification the process of checking that a model or tool does what the developer wants it to do

weather the state of the atmosphere in terms of temperature, relative humidity, wind speed, precipitation and similar at a given location at a specific point in time

workflow management the modelling of repetitive processes in order to support planning, enactment and process control

Building Performance Abbreviations

AF	Analysis Function
AFT	Active Functional Testing
AMI	Advanced Metering Infrastructure
AMR	Automated Meter Reading
ASET	Available Safe Egress Time
BDT	Blower Door Test
BPE	Building Performance Evaluation
BPI	Building Performance Index
COP	Coefficient of Performance
CSF	Critical Success Factor
DEA	Data Envelopment Analysis
DEC	Display Energy Certificate
DPM	Design Performance Measure
DQI	Design Quality Indicator
EDP	Engineering Demand Parameter
EE	Expert Elicitation
EPC	Energy Performance Certificate
EPC	Energy Performance Contract
ESCo	Energy Service Company
EUI	Energy Use Intensity
FC	Functional Concept
FDD	Fault Detection and Diagnosis
FPT	Functional Performance Test
IM	Intensity Measure
IPMVP	International Performance Measurement and Verification Protocol
IPT	Integrated Product Team
IPV	Integrated Performance View
KPI	Key Performance Indicator
KPP	Key Performance Parameter
MOE	Measure of Effectiveness
MOP	Measure of Performance
MRV	Measurement, Reporting and Verification
MSA	Measurement System Analysis
M&T	Monitoring and Targeting
M&V	Measurement and Verification

Building Performance Analysis, First Edition. Pieter de Wilde.
© 2018 John Wiley & Sons Ltd. Published 2018 by John Wiley & Sons Ltd.

NDT	Non-destructive Technique
NIALM	Non-intrusive Appliance Load Monitoring
NILM	Non-intrusive Load Monitoring
O&M	Operation and Maintenance
P4P	Pay for Performance
PAM	Performance Assessment Method
PBB	Performance-Based Building
PBD	Performance-Based Design
PI	Performance Indicator
PMS	Performance Measurement System
POE	Post-Occupancy Evaluation
PR	Performance Requirement
PTM	Performance Test Method
P4P	Pay for Performance
RE	Requirements Engineering
RSET	Required Safe Escape Time
SC	Solution Concept
SCADA	Supervisory Control and Data Acquisition
SEJ	Structured Expert Judgment
SoR	Statement of Requirements
SPM	Service Performance Measure
TPM	Technical Performance Measure
TRL	Technology Readiness Level
UN	User Needs
UTM	Universal Testing Machine
VFA	Video Fire Analysis
VORD	Viewpoint-Oriented Requirements Definition
WBS	Work Breakdown Structure

Generic Abbreviations

ABM	agent-based model
aDRI	annual driving rain index
AEC	architecture, engineering and construction
AHP	analytic hierarchy process
AHU	air handling unit
AIA	American Institute of Architects
ANN	artificial neural network
ASHRAE	American Society of Heating, Refrigerating and Air-Conditioning Engineers (USA)
BAS	building automation system
BCVTB	Building Controls Virtual Test Bed
BEMS	building energy management system
BES	building energy simulation
BIPV	building integrated photovoltaics
BMS	building management system
BRE	Building Research Establishment (UK)
BRI	building-related illness
CA	cellular automata
CAD	computer-aided design
CBM	condition-based maintenance
CBR	case-based reasoning
CDD	cooling degree day
CEN	Comité Européen de Normalisation/European Committee for Standardization
CFD	computational fluid dynamics
CHP	combined heat and power
CIB	Conseil International du Bâtiment/International Council for Research and Innovation
CIBSE	Chartered Institution of Building Services Engineers (UK)
CSTB	Centre Scientifique et Technique du Bâtiment (France)
CFRP	carbon fibre reinforced polymer
CPD	continuous professional development
DDC	direct digital control
dDRI	directional driving rain index
DDSS	design decision support system

Building Performance Analysis, First Edition. Pieter de Wilde.
© 2018 John Wiley & Sons Ltd. Published 2018 by John Wiley & Sons Ltd.

DEA	data envelopment analysis
DF	daylight factor
DPV	ductless personalized ventilation
DV	displacement ventilation
EBCx	Existing Building Commissioning
EBD	evidence-based design
ECM	energy conservation measure
EDT	early decay time
EIS	energy information system
EKE	expert knowledge elicitation
EMIS	energy management and information system
EMS	energy management system
EMCS	energy management and control system
EPC	energy performance coefficient
EOL	end-of-life
ERP	enterprise resource planning
ETFE	ethylene tetrafluoroethylene
FAF	fail and fix (maintenance practice)
FEA	finite element analysis
FM	facility management
FSE	fire safety engineering
GDP	gross domestic product
GIS	geographic information system
GUI	graphical user interface
HAM	heat and moisture (or heat, air and moisture)
HCI	human–computer interaction
HDD	heating degree day
HRR	heat release rate
HVAC	heating, ventilation and air conditioning
IAQ	indoor air quality
IBC	International Building Code
IBPSA	International Building Performance Simulation Association
ICC	International Code Council
IDEF-0	Integral DEFinition process modelling method
IDM	Integrated Data Model, product model as developed in COMBINE
IEA	International Energy Agency
IEQ	indoor environmental quality
IFC	Industry Foundation Classes
IoT	Internet of Things
IPD	integrated project delivery
IRR	internal rate of return
ISO	International Organization for Standardization
LCA	life cycle assessment
MAS	multi-agent system
MCDA	multi-criteria decision analysis
MCDM	multi-criteria decision making
MOO	multi-objective optimization

MPC	model-based predictive control
MPC	model predictive control
NPV	net present value
nZEB	net Zero Energy Building
nZED	net Zero Energy District
OAT	outside air temperature
O&M	operation and maintenance
PAP	predict and prevent (maintenance practice)
PCM	phase change material
PID	passive infrared detector
PLM	product life cycle management
PMV	predicted mean vote
PPD	predicted percentage of dissatisfied
PPM	parts per million
PPP	public–private partnership
QFD	quality function deployment
RAMS	reliability, availability, maintainability and safety-integrity
RAT	return air temperature
RE	requirements engineering
RIBA	Royal Institute of British Architects (UK)
ROI	return on investment
RSA	response surface approximation
SAT	supply air temperature
SBS	sick building syndrome
sDA	spatial daylight autonomy
TGT	tracer gas technique
TMY	test meteorological year
TQM	total quality management
TRY	test reference year
UML	Unified Modeling Language
VAV	variable air volume
VIP	vacuum insulation panel
VOC	volatile organic compound
WDR	wind-driven rain
ZCB	Zero Carbon Building

List of Figures and Tables

Figures

Building Performance Analysis, First Edition. Pieter de Wilde.
© 2018 John Wiley & Sons Ltd. Published 2018 by John Wiley & Sons Ltd.

Tables

Symbols and Units

SI base units

Measure	Name	Abbreviation or symbol
length	metre	m
mass	kilogram	kg
time	second	s
electric current	ampere	A
temperature	kelvin	K
amount of substance	mole	mol
luminous intensity	candela	cd

Derived units

Measure	Name	Abbreviation or symbol	Equivalent
frequency	hertz	Hz	$1/s$
angle	radian	rad	m/m
solid angle	steradian	sr	m^2/m^2
force	newton	N	$kg{\cdot}m/s^2$
pressure, stress	pascal	Pa	N/m^2
energy, work, heat	joule	J	$N{\cdot}m$, $W{\cdot}s$
power, radiant flux	watt	W	J/s
electric charge	coulomb	C	$s{\cdot}A$
electromotive force, potential	volt	V	W/A
electrical resistance	ohm	Ω	$1/S$, V/A
electrical conductance	siemens	S	$1/\Omega$, V/A
electrical capacitance	farad	F	C/V, s/Ω
magnetic flux	weber	Wb	J/A
inductance	henry	H	$V{\cdot}s/A$, $\Omega{\cdot}s$
magnetic flux density	tesla	T	$V{\cdot}s/m^2$
luminous flux	lumen	lm	$cd{\cdot}sr$
illumination	lux	lx	lm/m^2

Building Performance Analysis, First Edition. Pieter de Wilde.
© 2018 John Wiley & Sons Ltd. Published 2018 by John Wiley & Sons Ltd.

Symbols often used in the context of buildings

Measure	Symbol	Unit
area	A	m^2
density	ρ	kg/m^3
specific heat capacity	c_p	J/kgK
heat transfer coefficient	h	W/m^2K
mass flow rate	$\dot{m}$	kg/s
heat transfer	Q	J
temperature	T	K (Kelvin)
		°C (degree Celcius)
		°F (degree Fahrenheit)
power: Kilowatt	kW	$10^3 J/s$
energy: Kilowatt-hour	kWh	$10^9 J$
energy use intensity	EUI	$kWh/m^2 year$
rounds per minute	rpm	–
shape Coefficient	C_f	–
south Exposure Coefficient	C_s	–
coefficient of performance	COP	%
relative Compactness	RC	–
window-to-Wall Ratio	WWR	%
window-to-Floor Ratio	WFR	%
window-to-Surface Ratio	WSR	%

SI prefixes

	Abbreviation or symbol	Factor
atto-	a	10^{-18}
femto-	f	10^{-15}
pico-	p	10^{-12}
nano-	n	10^{-9}
micro-	μ	10^{-6}
milli-	m	10^{-3}
centi-	c	10^{-2}
deci-	d	10^{-1}
deca-	da	10
hecto-	h	10^2
kilo-	k	10^3
mega-	M	10^6
giga-	G	10^9
tera-	T	10^{12}
exa-	E	10^{18}

Imperial units

Measure	Name	Abbreviation or symbol	Conversion to SI units
length	inch	in	1 in = 25.4 millimetres
	foot	ft	1 ft = 12 in = 0.3048 metre
	yard	yd	1 yd = 3 feet = 0.9144 metre
	mile	ml	1 ml = 1.609 kilometres
volume	gallon	gal	1 gl = 4.546 litres
mass	pound	lb	1 lb = 0.4536 kilogram
	ton	t	1 t = 1016.05 kilograms
flow	cubic foot per minute	cfm	1 cfm = 0.472 litres/second
heat	British thermal unit	Btu	1 Btu = 1054 joules

About the Author

Pieter de Wilde holds a chair in Building Performance Analysis at Plymouth University in the United Kingdom. He is an expert in building science, with a focus on thermal aspects. His particular interest is thermal building performance simulation and its use in the building services and engineering community. He also has an interest in building performance monitoring in their actual day-to-day context which enables critical reflection on the validity of computational results and their meaning in building operation and facility management practice. He is secretary of IBPSA-World, chair of IBPSA-England and EG-ICE, Fellow of both IBPSA and CIBSE, and member of ASHRAE.

Contact data:

Prof. Dr. Ir. Pieter de Wilde
Professor of Building Performance Analysis
Plymouth University
Room 301A, Roland Levinsky Building
Drake Circus
Devon, PL4 8AA
UK
T +44 (0)1752 586115
F +44 (0)1752 585155
W www.bldg-perf.org
E pieter.dewilde@plymouth.ac.uk
E pieter@bldg-perf.org

References: Longlist and Secondary Sources

Abahri, K., B. Belarbi and A. Trabelsi, 2011. Contribution to analytical and numerical study of combined heat and moisture transfers in porous building materials. *Building and Environment*, 46 (7), 1354–1360.

Abaza, M., 2012. High performance buildings using whole building integrated design approach. *Strategic Planning for Energy and the Environment*, 31 (4), 19–34.

Abdelalim, A., W. O'Brien and Z. Shi, 2015. Visualization of energy and water consumption and GHG emissions: a case study of a Canadian university campus. *Energy and Buildings*, 109, 334–352.

Abernethy, M., M. Horne, A. Lillis, M. Malina and F. Selto, 2005. A multi-method approach to building causal performance maps from expert knowledge. *Management Accounting Research*, 16 (2), 135–155.

ABPMP, 2015. Association of Business Process Management Professionals homepage [online]. Saint Paul, MN. Available from www.abpmp.org [Accessed 28 April 2015].

Abran, A., J. Desharnais and J. Cuadrado-Gallego, 2012. Measurement and quantification are not the same: ISO 15939 and ISO 9126. *Journal of Software: Evolution and Process*, 24, 585–601.

Abuku, M., B. Blocken and S. Roels, 2009. Moisture response of building facades to wind-driven rain: field measurements compared with numerical simulations. *Journal of Wind Engineering and Industrial Aerodynamics*, 97 (5/6), 197–207.

Ackoff, R., 1971. Towards a system of systems concepts. *Management Science*, 17 (11), 661–671.

ACMA, 2016. *Guidelines and recommended practices for fiber-reinforced polymer (FRP) architectural products*. Arlington, VA: American Composites Manufacturers Association.

Action Energy, 2000. *Energy Consumption Guide 19: Energy use in offices*. London: Department of the Environment, Transport and the Regions.

Afonso, C., 2015. Tracer gas technique for measurement of air infiltration and natural ventilation: case studies and new devices for measurement of mechanical air ventilation in ducts. *International Journal of Low-Carbon Technologies*, 10 (3), 188–204.

Aguwa, C., L. Monplaisir and O. Turgut, 2012. Voice of the customer: customer satisfaction ratio based analysis. *Expert Systems with Applications*, 39 (11), 10112–10119.

Ahire, S. and P. Dreyfus, 2000. The impact of design management and process management on quality: an empirical investigation. *Journal of Operations Management*, 18 (5), 549–575.

Building Performance Analysis, First Edition. Pieter de Wilde.
© 2018 John Wiley & Sons Ltd. Published 2018 by John Wiley & Sons Ltd.

Ahmad, M., M. Mourshed, D. Mundow, M. Sisinni and Y. Rezgui, 2016. Building energy metering and environmental monitoring: a state-of-the-art review and directions for future research. *Energy and Buildings*, 120, 85–102.

Ahmed, A., J. Ploennigs, K. Menzel and B. Cahill, 2010. Multi-dimensional building performance data management for continuous commissioning. *Advanced Engineering Informatics*, 24 (4), 466–475.

Ahn, K. and C. Park, 2016. Correlation between occupants and energy consumption. *Energy and Buildings*, 116, 420–433.

Ahn, K., D. Kim, Y. Kim, S. Yoon and C. Park, 2016. Issues to be solved for energy simulation of an existing office building. *Sustainability*, 8 (4), 345.

Ahn, K., Y. Kim, D. Kim, S. Yoon and C. Park, 2013. Difficulties and issues in simulation of a high-rise office building. In: Wurtz, E., J. Roux, and M. Woloszyn, eds., *Building Simulation '13, 13th International IBPSA Conference*, Chambéry, France, 25–30 August 2013 (USB).

Ahuja, V., J. Yang and R. Shankar, 2010. Benchmarking framework to measure extent of ICT adoption for building project management. *Journal of Construction Engineering and Management*, 136 (5), 538–545.

Ahzahar, N., N. Karim, S. Hassan and J. Eman, 2011. A study of contribution factors to building failures and defects in construction industry. *Procedia Engineering*, 20, 249–255.

Ai, Z. and C. Mak, 2013. CFD simulation of flow and dispersion around an isolated building: effects of inhomogeneous ABL and near-wall treatment. *Atmospheric Environment*, 77, 568–578.

Ajayi, S., L. Oyedele, M. Bilal, O. Akinade, H. Alaka, H. Owolabi and K. Kadiri, 2015. Waste effectiveness of the construction industry: understanding the impediments and requisites for improvements. *Resources, Conservation and Recycling*, 102, 101–112.

Akadiri, P., P. Olomolaiye and E. Chinyio, 2013. Multi-criteria evaluation model for the selection of sustainable materials for building projects. *Automation in Construction*, 30, 113–125.

Akeiber, H., P. Nejat, M. Majid, M. Wahid, F. Jomehzadeh, I. Famileh, J. Calautit, B. Hughes and S. Zaki, 2016. A review on phase change material (PCM) for sustainable passive cooling in building envelopes. *Renewable and Sustainable Energy Reviews*, 60, 1470–1497.

Aksamija, A., 2013. *Sustainable facades – design methods for high-performance building envelopes*. Hoboken, NJ: Wiley.

Alam, M., H. Singh, S. Suresh and D. Redpath, 2017. Energy and economic analysis of vacuum insulation panels (VIPs) used in non-domestic buildings. *Applied Energy*, 188, 1–8.

Alcamo, G. and M. De Lucia, 2014. A new test cell for the evaluation of thermo-physical performance of facade building components. *International Journal of Sustainable Energy*, 33 (4), 954–962.

Ale, B., 2009. *Risk: an introduction. The concepts of risk, danger and chance*. London: Routledge.

Alha, M. and V. Pohjola, 1995. Eliciting expert knowledge: a study on connectionist approach for building a cognitive model of process design. *Computers & Chemical Engineering*, 19, 809–814.

Ali, M., V. Vukovic, M. Hussain Sahir and G. Fontanella, 2013. Energy analysis of chilled water system configurations using simulation-based optimization. *Energy and Buildings*, 59, 111–122.

Aliverdi, R., L. Naeni and A. Salehipour, 2013. Monitoring project duration and cost in a construction project by applying statistical quality control charts. *International Journal of Project Management*, 31 (3), 411–423.

Al-Jibouri, S., 2003. Monitoring systems and their effectiveness for project cost control in construction. *International Journal of Project Management*, 21 (2), 145–154.

Allacker, K., D. Maia de Souza and S. Sala, 2014. Land use impact assessment in the construction sector: an analysis of LCIA models and case study application. *International Journal for Life Cycle Assesment*, 19 (11), 1799–1809.

Allen, E., 2005. *How buildings work: the natural order of architecture*. Oxford: Oxford University Press, 3rd edition.

AlMaian, R., K. LaScola Needy, K. Walsh, T. Alves and N. Scala, 2016. Analyzing supplier quality management practices in the construction industry. *Quality Engineering*, 28 (2), 175–183.

Almeida, N., V. Sousa, L. Alves Dias and F. Branco, 2010. A framework for combining risk-management and performance-based building approaches. *Building Research & Information*, 38 (2), 157–174.

Almendra, R. and H. Christiaans, 2011. Decision making in design: an experiment with Dutch and Portuguese students. *Design Principles and Practices: an International Journal*, 5 (3), 65–77.

Alonso, A. and F. Martellotta, 2015. Room acoustic modelling of textile materials hung freely in space: from the reverberation chamber to ancient churches. *Journal of Building Performance Simulation*, 9 (5), 469–486.

Alonso, A., J. Sendra, R. Suárez and T. Zamarreño, 2014. Acoustic evaluation of the cathedral of Seville as a concert hall and proposals for improving the acoustic quality perceived by listeners. *Journal of Building Performance Simulation*, 7 (5), 360–378.

AlQahtany, A., Y. Rezgui and H. Li, 2013. A proposed model for sustainable urban planning development for environmentally friendly communities. *Architectural Engineering and Design Management*, 9 (3), 176–194.

Alsaadani, S. and C. Bleil de Souza, 2016. Of collaboration or condemnation? Exploring the promise and pitfalls of architect-consultant collaborations for building performance simulation. *Energy Research & Social Science*, 19, 21–36.

Álvarez-Morales, L., T. Zamarreño, S. Girón and M. Galindo, 2014. A methodology for the study f the acoustic environment of Catholic cathedrals: application to the cathedral of Malaga. *Building and Environment*, 72, 102–115.

Alwaer, H. and D. Clements-Croome, 2010. Key performance indicators (KPIs) and priority setting in using the multi-attribute approach for assessing sustainable intelligent buildings. *Building and Environment*, 45 (4), 799–807.

d'Ambrosio Alfano, F., M. Dell'Isola,G. Ficco and F. Tassini, 2012. Experimental analysis of air tightness in Mediterranean buildings using the fan pressurization method. *Building and Environment*, 53, 16–25.

Amecke, H., 2012. The impact of energy performance certificates: a survey of German home owners. *Energy Policy*, 46, 4–14.

Amin, M., 2013. The smart-grid solution. *Nature*, 499 (7457), 145–147.

Amore, M., M. Rossoni, M. Marengo and P. de Wilde, 2016. Impact of occupancy on energy consumption by an educational facility. In: Ślusarczyk, S. and P. de Wilde, eds. *EG-ICE 2016, Conference on Intelligent Computing in Engineering*, Krakow, Poland, 29 June–1 July 2016.

Amundadottir, M., S. Rockcastle, M. Khanie and M. Andersen, 2017. A humn-centric approach to assess daylight in buildings for non-visual health potential, visual interest and gaze behavior. *Building and Environment*, 113, 5–21.

Andersen, M., 2015. Unweaving the human response in daylighting design. *Building and Environment*, 91, 101–117.

Andersen, M., J. Gagne and S. Kleindienst, 2013. Interactive expert support for early stage full-year daylighting design: a user's perspective on Lightsolve. *Automation in Construction*, 35, 338–352.

Anderson, K., 2014. *Design energy simulation for architects*. New York: Routledge.

Andrée, K., D. Nilsson and J. Eriksson, 2016. Evacuation experiments in a virtual reality high-rise building: exit choice and waiting time for evacuation elevators. *Fire and Materials*, 40 (4), 554–567.

Anema, H., S. van der Veer, J. Kievit, E. Krol-Warmerdam, C. Fischer, E. Steyerberg, D. Dongelmans, A. Reidinga, N. Klazinga and N. de Keizer, 2013. Influences of definition ambiguity on hospital performance indicator scores: examples from The Netherlands. *European Journal of Public Health*, 24 (1), 73–78.

Ang, G., M. Groosman and N. Scholten, 2005. Dutch performance-based approach to building regulations and public procurement. *Building Research & Information*, 33 (2), 107–119.

Angrisani, G., F. Minichiello, C. Roseli and M. Sasso, 2010. Desiccant HVAC system driven by a micro-CHP: experimental analysis. *Energy and Buildings*, 42 (11), 2028–2035.

Antretter, F., M. Pazold, J. Radon and H. Künzel, 2013. Kopplung von dynamischer Wärmebrückenberechnung mit hygrothermischer Gebäudesimulation. *Bauphysik*, 35 (3), 181–192.

Appelfeld, D. and S. Svendsen, 2011. Experimental analysis of energy performance of a ventilated window for heat recovery under controlled conditions. *Energy and Buildings*, 43 (11), 3200–3207.

Appleby, P., 2013. *Sustainable retrofit and facilities management*. Abingdon: Routledge.

Arendt, K. and M. Krzaczek, 2014. Co-simulation strategy of transient CFD and heat transfer in building thermal envelope based on calibrated heat transfer coefficients. *International Journal of Thermal Sciences*, 85, 1–11.

Arjunan, A., C. Wang, K. Yahiaoui, D. Mynors, T. Morgan, V. Nguyen and M. English, 2014. Development of a 3D finite element acoustic model to predict the sound reduction index of stud based double-leaf walls. *Journal of Sound and Vibration*, 333 (23), 6140–6155.

Armbrust, M., A. Fox, R. Griffith, A. Joseph, R. Katz, A. Konwinski, G. Lee, D. Patterson, A. Rabkin, I. Stoica and M. Zaharia, 2010. A view of cloud computing. *Communications of the ACM*, 53 (4), 50–58.

Arora, S., R. Foley, J. Youtie, P. Shapira and A. Wiek, 2014. Drivers of technology adoption – the case of nanomaterials in building construction. *Technological Forecasting & Social Change*, 87, 232–244.

Arrunda, M., J. Firmo, J. Correia and C. Tiago, 2016. Numerical modelling of the bond between concrete and CFRP laminates at elevated temperatures. *Engineering Structures*, 110, 233–243.

Ascione, F., R. De Masi, F. De Rossi, S. Ruggiero and G. Vanoli, 2016. MATRIX, a multi activity test-room for evaluating the energy performances of 'building/HVAC' systems in Mediterranean climate: experimental set-up and CFF/BPS numerical modeling. *Energy and Buildings*, 126, 424–446.

Asdrubali, F. and G. Baldinelli, 2011. Thermal transmittance measurements with the hot box method: calibration, experimental procedures, and uncertainty analyses of three different approaches. *Energy and Buildings*, 43 (7), 1618–1626.

ASHRAE, 2002. Measurement of energy demand and savings – ASHRAE Guideline 14-2002. Atlanta, GA: American Society of Heating, Refrigerating and Air-Conditioning Engineers.

ASHRAE, 2015. American Society of Heating, Refrigerating and Air-Conditioning Engineers homepage [online]. Atlanta, GA. Available from www.ashrae.org [Accessed 26 November 2015].

ASHRAE, 2016. ASHRAE High-Performance Building Design Professional (HPBD) Certification Candidate Guidebook. Atlanta, GA: American Society of Heating, Refrigerating and Air-Conditioning Engineers.

ASHRAE, 2017. *Handbook of fundamentals*. Atlanta, GA: American Society of Heating, Refrigerating and Air-Conditioning Engineers.

Assefa, G., M. Glaumann, T. Malmqvist, B. Kindembe, H. Hult, U. Myhr and O. Eriksson, 2007. Environmental assessment of building properties – where natural and social sciences meet: the case of EcoEffect. *Building and Environment*, 42 (3), 1458–1464.

Assum, T. and M. Sørensen, 2010. Safety performance indicator for alcohol in road accidents – international comparison, validity and data quality. *Accident Analysis and Prevention*, 42 (2), 595–603.

Atkin, B. and A. Brooks, 2009. *Total Facilities Management*. Chichester: Wiley-Blackwell, 3rd edition.

Atkinson, R., 1999. Project management: cost, time and quality, two best guesses and a phenomenon, its time to accept other success criteria. *International Journal of Project Management*, 17 (6), 337–342.

Attia, S., J. Hensen, L. Beltrán and A. De Herde, 2012. Selection criteria for building performance simulation tools: contrasting architects' and engineers' needs. *Journal of Building Performance Simulation*, 5 (3), 155–169.

Attia, S., A. De Herde, E. Gratia and J. Hensen, 2013. Achieving informed decision-making for net zero energy buildings design using building performance simulation tools. *Building Simulation*, 6 (1), 3–21.

Au, S., R. Duan. S. Hesar and W. Jiang, 2010. A framework of irregularity enlightenment for data pre-processing in data mining. *Annals of Operation Research*, 174 (1), 47–66.

Augenbroe, G., 1994. An overview of the COMBINE project. In: *ECPPM Conference*, Dresden, Germany, 5–7 October 1994, pp. 547–554.

Augenbroe, G., 1995. COMBINE 2, Final Report. Brussels: Commission of the European Communities.

Augenbroe, G., 2002. Trends in building simulation. *Building and Environment*, 37 (8/9), 891–902.

Augenbroe, G., 2003. Trends in building simulation. In: Malkawi, A.M. and G. Augenbroe, eds., *Advanced building simulation*. New York: Spon Press.

Augenbroe, G., 2011. The role of simulation in performance based building. In: Hensen, J. and R. Lamberts, eds., *Building performance simulation for design and operation*. Abingdon: Spon Press.

Augenbroe, G., 2017. Personal communication, 1 May 2017.

Augenbroe, G. and C. Park, 2005. Quantification methods of technical building performance. *Building Research & Information*, 33 (2), 159–172.

Augenbroe, G., P. de Wilde, H. Moon, A. Malkawi, R. Choudhary, A. Mahdavi and R. Brame, 2003. Design Analysis Interface (DAI) Final Report. Atlanta, GA: Georgia Institute of Technology.

Augenbroe, G., P. de Wilde, H. J. Moon and A. Malkawi, 2004. An interoperability workbench for design analysis integration. *Energy and Buildings*, 36 (8), 737–748.

Augusztinovicz, F., 2012. Vibro-acoustic analysis of the stage floor of a concert hall – a case study. *Applied Acoustics*, 73 (6/7), 648–658.

Aurum, A. and C. Wohlin, 2003. The fundamental nature of requirements engineering activities as a decision-making process. *Information and Software Technology*, 45 (14), 945–954.

Awbi, H., 2003. *Ventilation of buildings*. London: Spon Press, 2nd edition.

Azadegan, A., K. Papamichail and P. Sampaio, 2013. Applying collaborative process design to user requirements elicitation: a case study. *Computers in Industry*, 64 (7), 798–812.

Azadi, H., P. Ho, E. Hafni, K. Zarafshani and F. Witlox, 2011. Multi-stakeholder involvement and urban green space performance. *Journal of Environmental Planning and Management*, 54 (6), 785–811.

Azari, R., S. Garshasbi, P. Amini, H. Rashed-Ali and Y. Mohammadi, 2016. Multi-objective optimization of building envelope design for life cycle environmental performance. *Energy and Buildings*, 126, 524–534.

Azzopardi, E. and R. Nash, 2013. A critical evaluation of importance-performance analysis. *Tourism Management*, 35, 222–233.

Bachman, L. R., 2003. *Integrated buildings: the systems basis of architecture*. Hoboken, NJ: Wiley.

Bae, S., H. Song, I. Nam, G. Kim, J. Lee and J. Yi, 2014. Quantitative performance analysis of graphite-LiFePO$_4$ battery working at low temperature. *Chemical Engineering Science*, 118 (18), 74–82.

Baetens, R., B. Jelle and A. Gustavsen, 2011. Aerogel insulation for building applications: a state-of-the-art review. *Energy and Buildings*, 43 (4), 761–769.

Bajwa, U., I. Taj, M. Anwar and X. Wang, 2013. A multifaceted independent performance analysis of facial subspace recognition algorithms. *PLoS ONE*, 8 (2), e56510.

Bakens, W., G. Foliente and M. Jasuja, 2005. Engaging stakeholders in performance-based building: lessons from the Performance-Based Building (PeBBu) Network. *Building Research & Information*, 33 (2), 149–158.

Baker, P., 2008. Evaluation of round-robin testing using the PASLINK test facilities. *Building and Environment*, 43 (2), 181–188.

Baker, N. and K. Steemers, 1996. LT-method 3.0 – a strategic energy-design tool for Southern Europe. *Energy and Buildings*, 23 (3), 251–256.

Baker, P. and H. van Dijk, 2008. PASLINK and dynamic outdoor testing of building components. *Building and Environment*, 43 (2), 143–151.

Baker, N., M. Guedes, N. Shaikh, L. Calixto and R. Aguiar, 2013. The LT-Portugal software: a design tool for architects. *Renewable Energy*, 49, 156–160.

Balaras, C. and A. Argiriou, 2002. Infrared thermography for building diagnostics. *Energy and Buildings*, 34 (2), 171–183.

Balcomb, J., 1997. Energy-10: a design-tool computer program. In: Spitler, J. and J. Hensen, eds., *Building Simulation '97, 5th International IBPSA Conference*, Prague, Czech Republic, 8–10 September 1997, Volume I, pp. 49–56.

Ballesteros, M., M. Fernández, S. Quintana, J. Ballesteros and I. González, 2010. Noise emission evolution on construction sites. Measurement for controlling and assessing its impact on the people and environment. *Building and Environment*, 45 (3), 711–717.

Balocco, C., G. Petrone and G. Cammarata, 2012. Assessing the effects of sliding doors on an operating theatre climate. *Building Simulation*, 5 (1), 73–83.

Barazzetti, L., F. Banfi, R. Brumana, G. Gusmeroli, M. Previtali and G. Schiantarelli, 2015. Cloud-to-BIM-to-FEM: structural simulation with accurate historic BIM from laser scans. *Simulation Modelling Practice and Theory*, 57, 71–87.

Barbason, M. and S. Reiter, 2014. Coupling building energy simulation and computational fluid dynamics: application to a two-storey house in a temperate climate. *Building and Environment*, 75, 30–39.

Barbera, G., G. Barone, P. Mazzoleni and A. Scandurra, 2012. Laboratory measurement of ultrasound velocity during accelerated aging tests: implication for the determination of limestone durability. *Construction and Building Materials*, 36, 977–983.

Barbosa, S. and K. Ip, 2014. Perspectives of double skin façades for naturally ventilated buildings: a review. *Renewable and Sustainable Energy Reviews*, 40, 1019–1029.

Barbosa, R. and N. Mendes, 2008. Combined simulation of central HVAC systems with a whole-building hygrothermal model. *Energy and Buildings*, 40 (3), 276–288.

Baroghel-Bouny, V., K. Kinomura, M. Thierry and S. Moscardelli, 2011. Easy assessment of durability indicators for service life prediction or quality control of concretes with high volumes of supplementary cementitious materials. *Cement & Concrete Composites*, 33 (8), 832–847.

Bar-On, D. and R. Oxman, 2002. Context over content: ICPD, a conceptual schema for the building technology domain. *Automation in Construction*, 11 (4), 467–493.

Barreira, E., R. Almeida and J. Delgado, 2016. Infrared thermography for assessing moisture related phenomena in building components. *Construction and Building Materials*, 110, 251–269.

Barrett, P., J. Hudson and C. Stanley, 1999. Good practice in briefing: the limits of rationality. *Automation in Construction*, 8 (6), 663–642.

Barrett, P., L. Barrett and F. Davies, 2013. Achieving a step change in the optimal sensory design of buildings for users at all life-stages. *Building and Environment*, 67, 97–104.

Barrios, G., G. Huelsz and J. Rojas, 2014. Ener-Habitat: a cloud computing numerical tool to evaluate the thermal performance of walls/roofs. *Energy Procedia*, 57, 2042–2051.

Bartak, M., I. Beausoleil-Morrison, J. Clarke, J. Denev, F. Drkal, M. Lain, I. Macdonald, A. Melikov, Z. Popiolek and P. Stankov, 2002. Integrating CFD and building simulation. *Building and Environment*, 37 (8/9), 865–871.

Barton, J. and D. Love, 2000. Design decision chains as a basis for design analysis. *Journal of Engineering Design*, 3 (11), 283–297.

Basmadjian, D., 2003. *Mathematical modeling of physical systems: an introduction*. Oxford: Oxford University Press.

Bastl, K., U. Berger and M. Kmenta, 2016. Ten questions about pollen and symptom load and the need for indoor measurements in built environment. *Building and Environment*, 98, 200–208.

Batterman, S., C. Jia, G. Hatzivasilis and C. Godwin, 2006. Simultaneous measurement of ventilation using tracer gas techniques and VOC concentrations in homes, garages and vehicles. *Journal of Environmental Monitoring*, 8 (2), 249–256.

Bauer, E., E. Pavón, E. Barreira and E. Kraus de Castro, 2016. Analysis of building facade defects using infrared thermography: laboratory studies. *Journal of Building Engineering*, 6, 93–104.

Bauwens, G. and S. Roels, 2014. Co-heating test: a state-of-the-art. *Energy and Buildings*, 82, 163–172.

Bayazit, N., 2004. Investigating design: a review of forty years of design research. *Design Issues*, 20 (1), 16–29.

Bazilian, M. and D. Prasad, 2002. Modelling of a photovoltaic heat recovery system and its role in a design decision support tool for building professionals. *Renewable Energy*, 27 (1), 57–68.

BCO, 2009. *2009 Guide to Specification*. London: British Council for Offices.

Beach, T., O. Rana, Y. Rezgui and M. Parashar, 2015. Governance model for cloud computing in building information management. *IEEE Transactions on Services Computing*, 8 (2), 314–327.

Beausoleil-Morrison, I., B. Griffith, T. Vesanen and A. Weber, 2009. A demonstration of the effectiveness of inter-program comparative testing for diagnosing and repairing solution and coding errors in building simulation programs. *Journal of Building Performance Simulation*, 2 (1), 63–73.

Beausoleil-Morrison, I., F. Macdonald, M. Kummert, T. McDowell and R. Jost, 2014. Co-simulation between ESP-r and TRNSYS. *Journal of Building Performance Simulation*, 7 (2), 133–151.

Becerik-Gerber, B. and S. Rice, 2010. The perceived value of building information modeling in the U.S. building industry. *Journal of Information Technology in Construction*, 15, 185–201.

Becker, R., ed., 2005. Performance Based Building international state of the art – final report. Rotterdam: CIB.

Becker, R., 2008. Fundamentals of performance-based building design. *Building Simulation*, 1 (4), 356–371.

Beersma, B., J. Hollenbeck, S. Humphrey, H. Moon, D. Conlon and D. Ilgen, 2003. Cooperation, competition and team performance: toward a contingency approach. *Academy of Management Journal*, 46 (5), 572–590.

Behnia, A., H. Chai and T. Shiotani, 2014. Advanced structural health monitoring of concrete structures with the aid of acoustic emission. *Construction and Building Materials*, 65, 282–302.

Beji, T., S. Verstockt, R. Van de Walle and B. Merci, 2015. Global analysis of multi-compartment full-scale fire tests ('Rabot2012'). *Fire Safety Journal*, 76, 9–18.

Belgian Science Policy and International Polar Foundation, 2007. Construction and operation of the new Belgian research station, Dronning Maud Land, Antarctica. Brussels: Final Comprehensive Environmental Evaluation (CEE).

Belley, C., S. Gaboury, B. Bouchard and A. Bouzouane, 2014. An efficient and inexpensive method for activity recognition within a smart home based on load signatures of appliances. *Pervasive and Mobile Computing*, 12, 58–78.

Belsky, M., R. Sacks and I. Brilakis, 2016. Semantic enrichment for building information modeling. *Computer-Aided Civil and Infrastructure Engineering*, 31 (4), 261–274.

Berardi, U., 2013. The position of the instruments for the sound insulation measurement of building façades: from ISO 140-5 to ISO 16283-3. *Noise Control Engineering*, 61 (1), 70–80.

Berardi, U., B. Meacham, N. Dembsey and Y. You, 2014. Fire performance assessment of a fiber reinforced polymer wall panel used in a single family dwelling. *Fire Technology*, 50 (6), 1607–1617.

Berendonk, C., R. Stalmeijer and L. Schuwirth, 2013. Expertise in performance assessment: assessors' perspectives. *Advances in Health Sciences Education*, 18 (4), 559–571.

Berger, J., S. Guernouti, M. Woloszyn and F. Chinesta, 2015. Proper generalised decomposition for heat and moisture multizone modelling. *Energy and Buildings*, 105, 334–351.

Berggren, B., M. Hall and M. Wall, 2013. LCE analysis of buildings – taking the step towards net zero energy buildings. *Energy and Buildings*, 62, 381–391.

von Bertalanffy, L., 1968. *General Systems Theory – Foundations, Development, Applications*. New York: George Braziller Ltd.

Bertelsen, N., A. Frandsen, K. Haugbølle, P. Huovila, B. Hansson and O. Karud, 2010. *CREDIT Performance Indicator Framework*. Hørsholm: Danish Building Research Institute/Aalborg University.

Bewoor and Kulkarni, 2009. *Metrology & Measurement*. Chennai, India: McGraw Hill Education.

Bhatt, J. and H. Verma, 2015. Design and development of wired building automation systems. *Energy and Buildings*, 103, 396–413.

Bibby, C. and M. Hodgson, 2013. Field measurement of the acoustical and airflow performance of interior natural-ventilation openings and silencers. *Buildings and Environment*, 67, 265–273.

Birkland, T., 2009. Disasters, lessons learned, and fantasy documents. *Journal of Contingencies and Crisis Management*, 17 (3), 146–156.

Björkman, J. and E. Mikkola, 2001. Risk assessment of a timber frame building by using CRISP simulation. *Fire and Materials*, 25 (5), 185–192.

Blanchard, B.S. and W.J. Fabrycky, 2011. *Systems engineering and analysis*. Upper Saddle River, NJ: Prentice Hall, 5th edition.

Bleil de Souza, C., 2012. Contrasting paradigms of design thinking: the building thermal simulation tool user vs the building designer. *Automation in Construction*, 22, 112–122.

Bleil de Souza, C., 2013. Studies into the use of building thermal physics to inform design decision making. *Automation in Construction*, 30, 81–93.

Blocken, B., 2015. New initiative: 'ten questions' paper series in building & environment. *Building and Environment*, 94, 325–326.

Blocken, B. and J. Carmeliet, 2004. A review of wind-driven rain research in building science. *Journal of Wind Engineering and Industrial Aerodynamics*, 92 (13), 1079–1130.

Blocken, B., T. Stathopoulos, J. Carmeliet and J. Hensen, 2011. Application of computational fluid dynamics in building performance simulation for the outdoor environment: an overview. *Journal of Building Performance Simulation*, 4 (2), 157–184.

Blocken, B., D. Derome and J. Carmeliet, 2013. Rainwater runoff from building facades: a review. *Building and Environment*, 60, 339–361.

Blocken, B., T. Stathopoulos and J. van Beeck, 2016. Pedestrian-level wind conditions around buildings: review of wind-tunnel and CFD techniques and their accuracy for wind comfort assessment. *Building and Environment*, 100, 50–81.

Bluyssen, P., 2010. Towards new methods and ways to create healthy and comfortable buildings. *Building and Environment*, 45 (4), 808–818.

Blyth, A. and J. Worthington, 2010. *Managing the brief for better design*. Abingdon: Routledge, 2nd edition.

Boardman, B., 2010. *Fixing fuel poverty – challenges and solutions*. London: Earthscan.

Bobran, H., 1967. *Handbuch der Bauphysik*. Berlin: Verlag Ullstein.

Bock, T., 2015. The future of construction automation: technological disruption and the upcoming ubiquity of robotics. *Automation in Construction*, 59, 113–121.

Bodart, M., A. Deneyer, A. De Herde and P. Wouters, 2007. A guide for building daylight scale models. *Architectural Science Review*, 50 (1), 31–36.

Bodart, M., R. De Peñaranda, A. Deneyer and G. Flamant, 2008. Photometry and colorimetry characterisation of materials in daylighting evaluation tools. *Building and Environment*, 43 (12), 2046–2058.

Bojórquez, J. and S. Ruiz, 2014. An efficient approach to obtain optimal load factors for structural design. *The Scientific World Journal*, 2014, 456826.

Bolchini, C., A. Geronazzo and E. Quintarelli, 2017. Smart buildings: a monitoring and data analysis methodological framework. *Building and Environment*, 121, 93–105.

Bolger, F. and G. Rowe, 2015. The aggregation of expert judgment: do good things come to those who weight? *Risk Analysis*, 35 (1), 5–11.

Bonvini, M., A. Leva and E. Zavaglio, 2012. Object-oriented quasi-3D sub-zonal airflow models for energy-related system-level building simulation. *Simulation Modelling Practice and Theory*, 22, 1–12.

Bordass, B. and A. Leaman, 2005. Making feedback and post-occupancy evaluation routine 1: a portfolio of feedback techniques. *Building Research & Information*, 33 (4), 347–352.

Bordass, B., R. Cohen, M. Standeven and A. Leaman, 2001a. Assessing building performance in use 2: technical performance of the Probe buildings. *Building Research & Information*, 29 (2), 103–113.

Bordass, B., R. Cohen, M. Standeven and A. Leaman, 2001b. Assessing building performance in use 3: energy performance of the Probe buildings. *Building Research & Information*, 29 (2), 114–128.

Bordass, B., A. Leaman and P. Ruyssevelt, 2001c. Assessing building performance in use 5: conclusions and implications. *Building Research & Information*, 29 (2), 144–157.

Borrmann, A., A. Kneidl, G. Köster, S. Ruzika and M. Thiemann, 2012. Bidirectional coupling of macroscopic and microscopic evacuation models. *Safety Science*, 50 (8), 1695–1703.

Bos, F., R. Wolfs, Z. Ahmed and T. Salet, 2016. Additive manufacturing of concrete in construction: potentials and challenges of 3D concrete printing. *Virtual and Physical Prototyping*, 11 (3), 209–225.

Van Den Bossche, N. and A. Janssens, 2016. Airtightness and watertightness of window frames: comparison of performance and requirements. *Building and Environment*, 110, 129–139.

Bottoms, S., 2008. Performatics: against definition. *Performance Research: A Journal of the Performing Arts*, 13 (2), 147–148.

Bouchlaghem, D., H. Shang, J. Whyte and A. Ganah, 2005. Visualisation in architecture, engineering and construction (AEC). *Automation in Construction*, 14 (3), 287–295.

Bouhouras, A., A. Milioudis and D. Labridis, 2014. Development of distinct load signatures for higher efficiency of NILM algorithms. *Electric Power Systems Research*, 117, 163–171.

Boukamp, F. and B. Akinci, 2007. Automated processing of construction specifications to support inspection and quality control. *Automation in Construction*, 17 (1), 90–106.

Boxer, E., G. Henze and A. Hirsch, 2017. A model-based decision support tool for building portfolios under uncertainty. *Automation in Construction*, 78, 34–50.

Boyce, P. and P. Raynham, 2009. *The SLL Lighting Handbook*. London: CIBSE Society of Light and Lighting.

Bradbury, S., 2015. Interrogating the dynamics of regulations on the design of energy performance. *Housing, Architecture and Culture*, 3 (3), 337–354.

Bradley, J., 2011. Review of objective room acoustic measures and future needs. *Applied Acoustics*, 72 (10), 713–720.

Braham, W., 2005. Biotechniques: remarks on the intensity of conditioning. In: Kolarevic, B. and A. Malkawi, eds., *Performative architecture: beyond instrumentality*. New York: Spon Press.

BRE, 2011. BREEM New Construction, Non-Domestic Buildings. Watford: Building Research Establishment, Technical Manual SD5073-2.0:2011.

BRE, 2014. Loss Prevention Standard LPS 1280 – Testing procedures for the LPCB approval and listing of duct smoke detectors using point smoke detectors. Watford: Building Research Establishment.

Bribián, I., A. Usón and S. Scarpellini, 2010. Life cycle assessment in buildings: state-of-the-art and simplified LCA methodology as a complement for building certification. *Building and Environment*, 44 (12), 2510–2520.

Brilakis, I., L. Soibelman and Y. Shinagawa, 2006. Construction site image retrieval based on material cluser recognition. *Advanced Engineering Informatics*, 20 (4), 443–452.

Brilakis, I., M. Park and G. Jog, 2011. Automated vision tracking of project related entities. *Advanced Engineering Informatics*, 25 (4), 713–724.

Brinks, P., O. Kornadt and R. Oly, 2015. Air infiltration assessment for industrial buildings. *Energy and Buildings*, 86, 663–676.

Brischke, C., A. Rapp, R. Bayerbach, N. Morsing, P. Fynholm and C. Welzbacher, 2008. Monitoring the 'material climate' of wood to predict the potential for decay: results from in situ measurements on buildings. *Building and Environment*, 43 (10), 1575–1582.

British History Online, 2015. The Assize of Buildings. Available from www.british-history. ac.uk [Accessed 30 December 2015].

van den Broeck, J., S. Cunningham, R. Eeckels and K. Herbst, 2005. Data cleaning: detecting, diagnosing and editing data abnormalities. *PLoS Medicine*, 2 (1), 0966–0970.

Bronzo, M., M. Valaderes de Oliviera and K. McCormack, 2012. Planning, capabilities and performance: an integrated value approach. *Management Decision*, 50 (6), 1001–1021.

Brown, N., A. Wright, S. Shukla and G. Stuart, 2010. Longitudinal analysis of energy metering data from non-domestic buildings. *Building Research & Information*, 38 (1), 80–91.

de Bruijn, H. and P. Herder, 2010. System and actor perspectives on sociotechnical systems. *IEEE Transactions of Systems, Man and Cybernetics*, 39 (5), 981–992.

Brundtland, G.H. et al., 1987. *Our common future*. Oxford: Oxford University Press.

Brunesi, E., R. Nascimbene, F. Parisi and N. Augenti, 2015. Progressive collapse fragility of reinforced concrete framed structures through incremental dynamics analysis. *Engineering Structures*, 104, 65–79.

Brunsgaard, C., P. Dvořáková, A. Wyckmans, W. Stutterecker, M. Laskari, M. Almeida, K. Kabele, Z. Magyar, P. Bartkiewicz and P. Op't Veld, 2014. Integrated energy design – education and training in cross-disciplinary teams implementing energy performance of buildings directive (EPBD). *Building and Environment*, 72, 1–14.

Bryan, T., 2010. *Construction technology: analysis and choice.* Oxford: Wiley-Blackwell, 2nd edition.

Bryson, B., 2003. *A short history of nearly everything.* London: Black Swan.

Bryson, B., 2011. *At Home – a short history of private life.* London: Black Swan/ Random House.

BS EN 15221-7:2012. Facility management. Part 7: guidelines for performance benchmarking. London: British Standards Institute.

BS EN ISO 50001:2011. Energy management systems – requirements with guidance for use. London: British Standards Institute.

BS ISO 15686-10:2010. Buildings and constructed assets – service life planning. Part 10: when to assess functional performance. London: British Standards Institute.

BSRIA, 2016. Product testing overview [online]. Bracknell. Available from www.bsria. co.uk/services/test/product-testing/ [Accessed 11 July 2016].

Bu, Z., S. Kato and T. Takahashi, 2010. Wind tunnel experiments on wind-induced natural ventilation rate in residential basements with areaway space. *Building and Environment*, 45 (10), 2263–2272.

Bucking, S., R. Zmeureanu and A. Athienitis, 2014. A methodology for identifying the influence of design variations on building energy performance. *Journal of Building Performance Simulation*, 7 (6), 411–426.

Bueno, B., J. Wienold, A. Katsifaraki and T. Kuhn, 2015. Fener: a Radiance-based modelling approach to assess the thermal and daylighting performance of complex fenestration systems in office spaces. *Energy and Buildings*, 94, 10–20.

Bunse, K., M. Vodicka, P. Schönsleben, M. Brülhart and F. Ernst, 2011. Integrating energy efficiency performance in production management – gap analysis between industrial needs and scientific literature. *Journal of Cleaner Production*, 19 (6/7), 667–679.

Buonomano, A., 2016. Code-to-code validation and application of a dynamic simulation too for the building energy performance analysis. *Energies*, 9 (4), 301.

Burge, P., 2004. Sick building syndrome. *Occupational and Environmental Medicine*, 61 (2), 185–190.

Burman, E., 2016. Assessing the operational performance of educational buildings against design expectations – a case study approach. EngD thesis. London: University College London.

Burpee, H. and E. McDade, 2013. *Comparative analysis of hospital energy use: Pacific Northwest and Scandinavian.* Seattle, WA: University of Washington.

Burry, M., J. Coulson, J. Preston and E. Rutherford, 2001. Computer-aided design decision support: interfacing knowledge and information. *Automation in Construction*, 10 (2), 203–215.

Butler, D., 2008. Architects of a low-energy future. *Nature*, 452 (3), 520–523.

Byström, A., X. Cheng, U. Wickström and M. Veljkovic, 2012. Full-scale experimental and numerical studies on compartment fire under low ambient temperature. *Building and Environment*, 51, 255–262.

Cabeza, L., L. Rincón, V. Vilariño, G. Pérez and A. Castell, 2014. Life cycle assessment (LCA) and life cycle energy analysis (LCEA) of buildings and the building sector: a review. *Renewable and Sustainable Energy Reviews*, 29, 394–416.

California Commissioning Collaborative, 2011. *The Building Performance Tracking Handbook.* Sacramento, CA: California Energy Commission.

Cameron, W., 1963. *Informal sociology: a casual introduction to sociologic thinking.* New York: Random House.

Campbell, P., 2009. Overrated ratings: criteria for 'green buildings' need to make energy performance a priority – as do universities. *Nature,* 461 (10), 146.

Candido, C., J. Kim, R. de Dear and L. Thomas, 2016. BOSSA: a multidimensional post-occupancy evaluation tool. *Building Research & Information,* 44 (2), 214–228.

Cantin, R., A. Kindinis and P. Michel, 2012. New approaches for overcoming the complexity of future buildings impacted by energy constraints. *Futures,* 44 (8), 735–745.

Capehart, B., H. Indig and L. Capehart, 2004. If buildings were built like cars: the potential for information and control systems technology in new buildings. *Strategic Planning for Energy and the Environment,* 24 (2), 7–27.

Capella, J., A. Perles, A. Bonastre and J. Serrano, 2011. Historical building monitoring using an energy-efficient scalable wireless sensor network architecture. *Sensors,* 11 (11), 10074–10093.

Capozzoli, A., F. Lauro and I. Khan, 2015. Fault detection analysis using data mining techniques for a cluster of smart office buildings. *Expert Systems with Applications,* 42 (9), 4324–4338.

Caragliu, A. and C. Del Bo, 2012. Smartness and European urban performance: assessing the local impacts of smart urban attributes. *Innovation – The European Journal of Social Science Research,* 25 (2), 97–113.

Carbon Trust, 2011. *Closing the gap: lessons learned on realising the potential of low carbon building design.* London: Carbon Trust.

Carbon Trust, 2012. *Monitoring and targeting: techniques to help organisations control and manage their energy use.* London: Carbon Trust.

CarbonBuzz, 2016. CarbonBuzz homepage [online]. London. Available from www.carbonbuzz.org [Accessed 26 January–June 2016].

Carfrae, J., 2011. The moisture performance of straw bale construction in a temperate maritime climate. PhD thesis. Plymouth: Plymouth University.

Carfrae, J., P. de Wilde, S. Goodhew, P. Walker and J. Littlewood, 2010. A cost effective probe for the long term monitoring of straw bale buildings. *Building and Environment,* 46 (1), 156–164.

Carletti, C. F. Sciurpi, L. Pierangioli, F. Asdrubali, A. Pisello, F. Bianci, S. Sambuco and C. Guattari, 2016. Thermal and lighting effects of an external venetian blind: experimental analysis in a full scale test room. *Building and Environment,* 106, 45–56.

Carlon, E., V. Kumar Verma, M. Schwarz, L. Golicza, A. Prada, M. Baratieri, W. Haslinger and C. Schmidl, 2015. Experimental validation of a thermodynamic boiler model under steady state and dynamic conditions. *Applied Energy,* 138, 505–516.

Carpio, M., M. Martín-Morales and M. Zamorano, 2015. Comparative study by an expert panel of documents recognized for energy efficiency certification of buildings in Spain. *Energy and Buildings,* 99, 98–103.

Carrillo de Gea, J., J. Nicolás, J. Fernández Alemán, A. Toval, C. Ebert and A. Vizcaíno, 2012. Requirements engineering tools: capabilities, survey and assessment. *Information and Software Technology,* 54 (10), 1142–1157.

Carroll, T. and F. Rachford, 2011. Non-linear dynamics method for target identification. *IET Radar, Sonar and Navigation,* 5 (7), 741–746.

Casera, S. and P. Kropf, 2010. Collaboration in scientific visualization. *Advanced Engineering Informatics*, 24 (2), 188–195.

Castro-Lacouture, D., J. Sefair, L. Flórez and A. Medaglia, 2009. Optimization model for the selection of materials using a LEED-based green building rating system in Colombia. *Building and Environment*, 44 (6), 1162–1170.

Cedillos Alvarado, D., S. Acha, N. Shah and C. Markides, 2016. A Technology Selection and Operation (TSO) optimisation model for distributed energy systems: mathematical formulation and case study. *Applied Energy*, 180, 491–503.

Çelebi, M., M. Huang, A. Shakal, J. Hooper and R. Klemencic, 2013. Ambient response of a unique performance-based design tall building with dynamic response modification features. *The Structural Design of Tall and Special Buildings*, 22 (10), 816–829.

Cemesova, A., C. Hopfe and R. McLeod, 2015. PassivBIM: enhancing interoperability between BIM and low energy design software. *Automation in Construction*, 57, 17–32.

Cerdá, S., A. Giménez, J. Romero, R. Cibrián and J. Miralles, 2009. Room acoustical parameters: a factor analysis approach. *Applied Acoustics*, 70 (1), 97–109.

Cerkez, B., 2014. Perceptual literacy and the construction of significant meanings within art education. *International Journal of Art & Design Education*, 33 (2), 272–285.

Cha, H., J. Kim and J. Han, 2009. Identifying and assessing influence factors on improving waste management performance for building construction projects. *Journal of Construction Engineering and Management*, 135 (7), 647–656.

Chambers, D. and E. LaBarre, 2014. Why professional judgment is better than objective description in dental faculty evaluations of student performance. *Journal of Dental Education*, 78 (5), 681–693.

Chan, W., J. Joh and M. Sherman, 2013. Analysis of air leakage measurements of US houses. *Energy and Buildings*, 66, 616–625.

Chandon, P., R. Smith, V. Morwitz, E. Spangenberg and D. Sprott, 2011. When does the past repeat itself? The interplay of behavior prediction and personal norms. *Journal of Consumer Research*, 38 (3), 420–430.

Chandra, V. and M. Loosemore, 2010. Mapping stakeholders' cultural learning in the hospital briefing process. *Construction Management and Economics*, 28 (7), 761–760.

Charles, P. and C. Thomas, 2009. Four approaches to teaching with building performance simulation tools in undergraduate architecture and engineering education. *Journal of Building Performance Simulation*, 2 (2), 95–114.

Chase, S., 2005. Generative design tools for novice designers: issues for selection. *Automation in Construction*, 14 (6), 689–698.

Chatillon, J., 2007. Influence of sound directivity on noise levels in industrial halls: simulation and experiments. *Applied Acoustics*, 68 (6), 682–698.

Chen, W. and H. Hao, 2014. Experimental and numerical study of composite lightweight structural insulated panel with expanded polystyrene core against windborne debris impact. *Materials and Design*, 60, 409–423.

Chen, L. and H. Luo, 2014. A BIM-based construction quality management model and its applications. *Automation in Construction*, 46, 64–73.

Chen, Y. and S. Treado, 2014. Development of a simulation platform based on dynamic models for HVAC control analysis. *Energy and Buildings*, 68, 376–386.

Chen, Z., D. Clements-Croome, J. Hong, H. Li and Q. Xu, 2006. A multicriteria lifespan energy efficiency approach to intelligent building assessment. *Energy and Buildings*, 38 (5), 393–409.

Chen, C., W. Hsieh, W. Hu, C. Lai and T. Lin, 2010. Experimental investigation and numerical simulation of a furnished office fire. *Building and Environment*, 45 (12), 2735–2742.

Chen, W., P. Shen and Z. Shui, 2012. Determination of water content in fresh concrete mix based on relative dielectric constant measurement. *Construction and Building Materials*, 34, 306–312

Chen, W., H. Hao and H. Du, 2014. Failure analysis of corrugated panel subjected to windborne debris impacts. *Engineering Failure Analysis*, 44, 229–249.

Chen, C., M. Delmas and M. Lieberman, 2015. Production frontier methodologies and efficiency as a performance measure in strategic management research. *Strategic Management Journal*, 36 (1), 19–36.

Chen, Y., S. Treado and J. Messner, 2016. Building HVAC control knowledge data schema – towards a unified representation of control system knowledge. *Automation in Construction*, 72, 174–186.

Cheng, M., H. Tsai and Y. Lai, 2009. Construction management process reengineering performance measurement. *Automation in Construction*, 18 (2), 183–193.

Cheng, E., N. Ryan and S. Kelly, 2012. Exploring the perceived influence of safety management practices on project performance in the construction industry. *Safety Science*, 50 (2), 363–369.

Cheng, F., H. Li, Y. Wang, M. Skitmore and P. Forsythe, 2013. Modeling resource management in the building design process by information constraint Petri nets. *Automation in Construction*, 29, 92–99.

Chesné, L., T. Duforestel, J. Roux and G. Rusaouën, 2012. Energy saving and environmental resources potentials: towards new methods of building design. *Building and Environment*, 58, 199–207.

Chew, M. and N. De Silva, 2004. Factorial method for performance assessment of building facades. *Journal of Construction Engineering and Management*, 130 (4), 525–533.

Chew, I., D. Karunatilaka, C. Tan and V. Kalavally, 2017. Smart lighting: the way forward? Reviewing the past to shape the future. *Energy and Buildings*, 149, 180–191.

Chi, J., S. Wu and C. Shu, 2011. A fire risk simulation system for multi-purpose building based fire statistics. *Simulation Modelling Practice and Theory*, 19 (4), 1243–1250.

Chien, S. and A. Mahdavi, 2008. Evaluating interface designs for user-system interaction media in buildings. *Advanced Engineering Informatics*, 22 (4), 484–492.

Chinese, D., G. Nardin and O. Saro, 2011. Multi-criteria analysis for the selection of space heating systems in an industrial building. *Energy*, 36 (1), 556–565.

Chiranov, M., 2014. Creating measurement addiction – a tool for better advocacy and improved management. *Performance Measurement and Metrics*, 15 (3), 99–111.

Chiu, M., 2002. An organizational view of design communication in design collaboration. *Design Studies*, 23 (2), 187–210.

Chiu, L., R. Lowe, R. Raslan, H. Altamirano-Medina and J. Wingfield, 2014. A socio-technical approach to post-occupancy evaluation: interactive adaptability in domestic retrofit. *Building Research & Information*, 42 (5), 574–590.

Choi, D., 2011. Reactive goal management in a cognitive architecture. *Cognitive Systems Research*, 12 (3/4), 293–308.

Choi, J., V. Loftness and A. Aziz, 2012. Post-occupancy evaluation of 20 office buildings as basis for future IEQ standards and guidelines. *Energy and Buildings*, 46, 167–175.

Choudhary, R., 2012. Energy analysis of the non-domestic building stock of Greater London. *Building and Environment*, 51, 243–254.

Choudhary, R., P. Papalambros and A. Malkawi, 2005a. A hierarchical design optimization approach for meeting building performance targets. *Architectural Engineering and Design Management*, 1 (1), 57–76.

Choudhary, R., P. Papalambros and A. Malkawi, 2005b. Analytic target cascading in simulation-based building design. *Automation in Construction*, 14 (4), 551–568.

Chow, C. and W. Chow, 2010. Heat release rate of accidental fire in a supertall building residential flat. *Building and Environment*, 45 (7), 1632–1640.

Chow, L. and T. Ng, 2007. A fuzzy gap analysis model for evaluating the performance of engineering consultants. *Automation in Construction*, 16 (4), 425–435.

Chow, W. and G. Zou, 2009. Numerical simulation of pressure changes in closed chamber fires. *Building and Environment*, 44 (6), 1261–1275.

Christoph, M., M. Vis, L. Rackliff and H. Stipdonk, 2013. A road safety performance indicator of vehicle fleet compatibility. *Accident Analysis and Prevention*, 60, 396–401.

Chua, K. and S. Chou, 2011. A performance-based method for energy efficiency improvement of buildings. *Energy Conversion and Management*, 52 (4), 1829–1839.

Chuang, M., K. Tsai, P. Lin and A. Wu, 2015. Critical limit states in seismic buckling-restrained brace and connection designs. *Earthquake Engineering & Structural Dynamics*, 44 (10), 1559–1579.

Chung, W., 2011. Review of building energy-use performance benchmarking methodologies. *Applied Energy*, 88 (5), 1470–1479.

Chung, J., M. Kumaraswamy and E. Palaneeswaran, 2009. Improving megaproject briefing through enhanced collaboration with ICT. *Automation in Construction*, 18 (7), 966–974.

Chytas, P., M. Glykas and G. Valiris, 2011. A proactive balanced scorecard. *International Journal of Information Management*, 31 (5), 460–468.

CIB Working Commission W60, 1982. CIB Report 64: Working with the Performance Approach to Building. Rotterdam: CIB.

CIBSE, 2006. *Commissioning Code A: 1996 (2006)*. London: The Charted Institution of Building Services Engineers.

CIBSE, 2013. *TM54 Evaluating operational energy performance of buildings at the design stage*. London: The Charted Institution of Building Services Engineers.

CIBSE, 2015a. The Chartered Institution of Building Services Engineers homepage [online]. London. Available from www.cibse.org [Accessed 26 November 2015].

CIBSE, 2015b. *CIBSE Guide A: Environmental Design*. London: The Chartered Institution of Building Services Engineers, 8th edition.

CIBSE, 2015c. *CIBSE AM11: Building performance modelling*. London: The Chartered Institution of Building Services Engineers, 8th edition.

Ciribini, A., S. Mastrolembo Ventura and M. Paneroni, 2016. Implementation of an interoperable process to optimise design and construction phases of a residential building: a BIM pilot project. *Automation in Construction*, 71 (1), 62–73.

Citherlet, S., J. Clarke and J. Hand, 2001. Integration in building physics simulation. *Energy and Buildings*, 33, 451–461.

Clarke, J., 2001. *Energy simulation in building design*. Oxford: Butterworth-Heinemann, 2nd edition.

Clarke, J., 2015. A vision for building performance simulation: a position paper prepared on behalf of the IBPSA Board. *Journal of Building Performance Simulation*, 8 (2), 39–43.

Clarke, J. and J. Hensen, 2015. Integrated building performance simulation: progress, prospects and requirements. *Building and Environment*, 91, 294–306.

Clements-Croome, D., 1997. What do we mean by intelligent buildings? *Automation in Construction*, 6 (5/6), 395–400.

Clements-Croome, D., ed., 2006. *Creating the productive workplace.* Abingdon: Taylor & Francis, 2nd edition.

Clifford, M., R. Brooks, A. Howe, A. Kennedy, S. McWilliam, S. Pickering, P. Shayler and P. Shipway, 2009. *An introduction to mechanical engineering.* Abingdon: Hodder Education.

Coakley, D., P. Raftery and M. Keane, 2014. A review of methods to match building energy simulation models to measured data. *Renewable and Sustainable Energy Reviews*, 37, 123–141.

Coffey, B., 2013. Approximating model predictive control with existing building simulation tools and offline optimization. *Journal of Building Performance Simulation*, 6 (3), 220–235.

Cohen, R., M. Standeven, B. Bordass and A. Leaman, 2001. Assessing building performance in use 1: the Probe process. *Building Research & Information*, 29 (2), 85–102.

Cole, R., 2005. Building environmental assessment methods: redefining intentions and roles. *Building Research & Information*, 33 (5), 455–467.

Cole, R., P. Busby, R. Guenter, L. Briney, A. Blaviesciunaite and T. Alencar, 2012. A regenerative design framework: setting new aspirations and initiating new discussions. *Building Research & Information*, 40 (1), 95–111.

Collinge, W. and C. Harty, 2014. Stakeholder interpretations of design: semiotic insights into the briefing process. *Construction Management and Economics*, 32 (7/8), 760–772.

Collinge, W., A. Landis, A. Jones, L. Schaefer and M. Bilec, 2014. Productivity metrics in dynamic LCA for whole buildings: using a post-occupancy evaluation of energy and indoor environmental quality tradeoffs. *Building and Environment*, 82, 339–348.

de Coninck, R., F. Magnusson, J. Åkesson and L. Helsen, 2016. Toolbox for development and validation of grey-box building models for forecasting and control. *Journal of Building Performance Simulation*, 9 (3), 288–303.

Cook, M., 2007. *The design quality manual – improving building performance.* Oxford: Blackwell.

Cooke, R., 2013. Expert judgment assessment: quantifying uncertainty on thin ice. *Nature Climate Change*, 3, 311–312.

Cooke, R. and L. Goossens, 2000. Procedures guide for structured expert judgment in accident consequence modelling. *Radiation Protection Dosimetry*, 90 (3), 303–309.

Cooke, R. and L. Goossens, 2008. TU Delft expert judgment data base. *Reliability Engineering and System Safety*, 93 (5), 657–674.

Cooper, I., 2001. Post-occupancy evaluation – where are you? *Building Research & Information*, 29 (2), 58–163.

Corani, G. and A. Mignatti, 2015. Credal model averaging for classification: representing prior ignorance and expert opinions. *International Journal of Approximate Reasoning*, 56, 264–277.

Corgnati, S., M. Perino and V. Serra, 2007. Experimental assessment of the performance of an active transparent façade during actual operating conditions. *Solar Energy*, 81 (8), 993–1013.

Correia da Silva, P., V. Leal and M. Andersen, 2013. Occupants interaction with electric lighting and shading systems in real single-occupied offices: results from a monitoring campaign. *Building and Environment*, 64, 152–168.

Corry, E., J. O'Donnell, E. Curry, D. Coakley, P. Pauwels and M. Keane, 2014. Using semantic web technologies to access soft AEC data. *Advanced Engineering Informatics*, 28 (4), 370–380.

Corry, E., P. Pauwels, S. Hu, M. Keane and J. O'Donnell, 2015. A performance assessment ontology for the environmental and energy management of buildings. *Automation in Construction*, 57, 249–259.

Costa, A., M. Keane, J. Torrens and E. Corry, 2013. Building operation and energy performance: monitoring, analysis and optimisation toolkit. *Applied Energy*, 101, 310–316.

Costanzo, V., G. Evola and L. Marletta, 2016. Thermal and visual performance of real and theoretical thermochromic glazing solutions for office buildings. *Solar Energy Materials & Solar Cells*, 149, 110–120.

Counsell, C. and L. Wolf, 2001. *Performance analysis: an introductory coursebook.* Abingdon: Routledge.

Coussement, K., D. Benoit and M. Antioco, 2015. A Bayesian approach for incorporating expert opinions into decision-support systems: a case study of online customer-satisfaction detection. *Decision Support Systems*, 79, 24–32.

Cracolici, M., M. Cuffaro and P. Nijkamp, 2010. The measurement of economical, social and environmental performance of countries: a novel approach. *Social Indicators Research*, 95 (2), 339–356.

Crawford, R., I. Czerniakowski and R. Fuller, 2011. A comprehensive model for streamlining low-energy building design. *Energy and Buildings*, 43 (7), 1748–1756.

Crawley, D., L. Lawrie, F. Winkelmann, W. Buhl, Y. Huang, C. Pedersen, R. Strand, R. Liesen, D. Fisher, M. Witte and J. Glazer, 2001. EnergyPlus: creating a new-generation building energy simulation program. *Energy and Buildings*, 33 (4), 319–331.

Crawley, D., L. Lawrie, C. Pedersen, F. Winkelmann, M. Witte, R. Strand, R. Liesen, W. Buhl, Y. Huang, R. Henninger, J. Glazer, D. Fisher, D. Shirley, B. Griffith, P. Ellis and L. Gu, 2004. EnergyPlus: new, capable, and linked. *Journal of Architectural Planning Research*, 21 (4), 292–302.

Crawley, D., J. Hand, M. Kummert and B. Griffith, 2008. Contrasting the capabilities of building energy performance simulation programs. *Building and Environment*, 43 (4), 661–673.

Cronemberger, J., M. Almagro Corpas, I. Cerón, E. Caamaño-Martín and S. Vega Sánchez, 2014. BIPV technology application: highlighting advances, tendencies and solutions through Solar Decathlon Europe Houses. *Energy and Buildings*, 83, 44–56.

Crook, B. and N. Burton, 2010. Indoor moulds, sick building syndrome and building related illness. *Fungal Biology Review*, 24 (3/4), 106–113.

Cropper, P., T. Yang, M. Cook, D. Fiala and R. Yousaf, 2010. Coupling of a model of human thermoregulation with computational fluid dynamics for predicting human-environment interaction. *Journal of Building Performance Simulation*, 3 (3), 233–243.

Cross, N., 2011. *Design thinking: understanding how designers think and work.* London: Bloomsbury.

Cuce, E., C. Yong and S. Riffat, 2015. Thermal performance investigation of heat insulation solar glass: a comparative experimental study. *Energy and Buildings*, 86, 595–600.

Cuerda, E., M. Pérez and J. Neila, 2014. Facade typologies as a tool for selecting refurbishment measures for the Spanish residential building stock. *Energy and Buildings*, 76, 119–129.

Cui, W. and L. Caracoglia, 2015. Simulation and analysis of intervention costs due to wind-induced damage on tall buildings. *Engineering Structures*, 87, 183–197.

Cui, W., G. Cao, J. Park, Q. Qoyang and Y. Zhu, 2013. Influence of indoor air temperature on human thermal comfort, motivation and performance. *Building and Environment*, 68, 114–122.

Curry, E., J. O'Donnell, E. Corry, S. Hasan, M. Keane and S. O'Riain, 2013. Linking building data in the cloud: integrating cross-domain building data using linked data. *Advanced Engineering Informatics*, 27 (2), 206–219.

Dai, L., N. Yang, L. Zhang, Q. Yang and S. Law, 2016. Monitoring crowd load effect of typical ancient Tibetan building. *Structural Control and Health Monitoring*, 23 (7), 998–1014.

Dainty, A., M. Cheng and D. Moore, 2003. Redefining performance measures for construction project managers: an empirical evaluation. *Construction Management and Economics*, 21 (2), 209–218.

Dainty, A., D. Moore and M. Murray, 2006. *Communication in construction: theory and practice*. London: Taylor & Francis.

Dalewski, M., A. Melikov and M. Vesely, 2014. Performance of ductless personalised ventilation in conjunction with displacement ventilation: physical environment and human response. *Building and Environment*, 81, 354–364.

Dasgupta, P., 2007. *Economics – a very short introduction*. Oxford: Oxford University Press.

Daum, D. and N. Morel, 2009. Assessing the total energy impact of manual and optimized blind control in combination with different lighting schedules in a building simulation environment. *Journal of Building Performance Simulation*, 3 (1), 1–16.

Davidich, D. and G. Köster, 2013. Predicting pedestrian flow: a methodology and proof of concept based on real-life data. *Plos One*, 8 (12), e83355.

Day, C., 2015. *The eco-home design guide – principles and practice for new build and retrofit*. Cambridge: Green Books.

Day, J. and D. Gunderson, 2015. Understanding high performance buildings: the link between occupant knowledge of passive design systems, corresponding behaviors, occupant comfort and environmental satisfaction. *Building and Environment*, 84, 114–124.

de Dear, R. and G. Brager, 1998. Developing an adaptive model of thermal comfort and preference. *ASHRAE Transactions*, 104 (1), 145–167.

de Dear, R., T. Akimoto, E. Arens, G. Brager, C. Candido, K. Cheong, B. Li, N. Nishihara, S. Sekhar, S. Tanabe, J. Toftum, H. Zhang and Y. Zhu, 2013. Progress in thermal comfort research over the last twenty years. *Indoor Air*, 23 (6), 442–461.

Defraeye, T., B. Blocken and J. Carmeliet, 2011. Convective heat transfer coefficients for exterior building surfaces: existing correlations and CFD modelling. *Energy Conversion and Management*, 52 (1), 512–522.

Deiterding, R. and S. Wood, 2013. Parallel adaptive fluid-structure interaction simulation of explosions impacting on building structures. *Computers & Fluids*, 88, 719–729.

Delhomme, F., G. Debicki and Z. Chaib, 2010. Experimental behaviour of anchor bolts under pullout and relaxation tests. *Construction and Building Materials*, 24 (3), 266–274.

Delile, H., J. Blichert-Toft, J. Goiran, S. Keay and F. Albarède, 2014. Lead in ancient Rome's city waters. *Proceedings of the National Academy of Sciences of the United States*, 111 (18), 6594–6599.

Della Crociata, S., F. Martellotta and A. Simone, 2012. A measurement procedure to assess indoor environment quality for hypermarket workers. *Building and Environment*, 47, 288–299.

De Marco, A., D. Briccarello and C. Rafele, 2009. Cost and schedule monitoring of industrial building projects: case study. *Journal of Construction Engineering and Management*, 135 (9), 853–862.

Demyanyk, Y. and O. van Hemert, 2011. Understanding the subprime mortgage crisis. *The Review of Financial Studies*, 24 (6), 1848–1880.

Deng, Q., L. Zhang, Q. Cui and X. Jiang, 2014. A simulation-based decision model for designing contract period in building energy performance contracting. *Building and Environment*, 71, 71–80.

Dent, R. and G. Alwani-Starr, 2001. Performance measurement of design activities – a summary report and key performance indicators. London: CIRIA, Project Report 85.

Deramchi, S. and R. Bell, eds., 2014. HVAC Troubleshooting. Bracknell: BSRIA, Guide BG 25/2014.

Derks, W., R. Weston, A. West, R. Harrison and D. Shorter, 2003. Role of workflow management systems in product engineering. *International Journal of Production Research*, 41 (15), 3393–3418.

Deru, M. and P. Torcellini, 2005. Performance metrics research project – final report. Golden, CO: National Renewable Energy Laboratory.

Dibley, M., H. Li, J. Miles and Y. Rezgui, 2011. Towards intelligent agent based software for building related decision support. *Advanced Engineering Informatics*, 25 (2), 311–329.

Van Dijk, H. and G. van der Linden, 1993. The PASSYS method for testing passive solar components. *Building and Environment*, 28 (2), 115–126.

Dijkman, R., B. Sprenkels, T. Peeters and A. Janssen, 2015. Business models for the Internet of Things. *International Journal of Information Management*, 35 (6), 672–678.

Dimitri, R. and F. Tornabene, 2015. A parametric investigation of the seismic capacity for masonry arches and portals of different shapes. *Engineering Failure Analysis*, 52, 1–34.

Dimitrov, A. and M. Golparvar-Fard, 2014. Vision-based material recognition for automated monitoring of construction progress and generating building information modeling from unordered site image collections. *Advanced Engineering Informatics*, 28 (1), 37–49.

Ding, G., 2005. Developing a multicriteria approach for the measurement of sustainable performance. *Building Research & Information*, 33 (1), 3–16.

Djurdjanovic, D., J. Lee and J. Ni, 2003. Watchdog Agent – an infotronics-based prognostic approach for product performance degradation assessment and prediction. *Advanced Engineering Informatics*, 17 (3/4), 109–125.

Djuric, N. and V. Novakovic, 2009. Review of possibilities and necessities for building lifetime commissioning. *Renewable & Sustainable Energy Reviews*, 13 (2), 486–492.

DoE, 1994. *Rethinking the Team ('Latham Report')*. London: Department of the Environment.

Doerr, K. and K. Gue, 2013. A performance metric and goal-setting procedure for deadline-oriented processes. *Production and Operations Management*, 22 (3), 726–738.

Dolce, M., A. Kappos, A. Masi, G. Penelis and M. Vona, 2006. Vulnerability assessment and earthquake damage scenarios of the building stock of Potenza (Southern Italy) using Italian and Greek methodologies. *Engineering Structures*, 28 (3), 357–371.

Dols, W., L. Wang, S. Emmerich and B. Polidoro, 2015. Development and application of an updated whole-building coupled thermal, airflow and contaminant transport simulation program (TRNSYS/CONTAM). *Journal of Building Performance Simulation*, 8 (5), 326–337.

Domingues, P., P. Carreira, R. Vierira and W. Kastner, 2016. Building automation systems: concepts and technology review. *Computer Standards & Interfaces*, 45, 1–12.

Dong, A., M. Maher, M. Kim, N. Gu and X. Wang, 2009. Construction defect management using a telematic digital workbench. *Automation in Construction*, 18 (6), 814–824.

Dong, B., Z. O'Neill and Z. Li, 2014. A BIM-enabled information infrastructure for building energy fault detection and diagnostics. *Automation in Construction*, 44, 197–211.

Donn, M., 2001. Tools for quality control in simulation. *Building and Environment*, 36 (6), 673–680.

Donn, M., S. Selkowitz and B. Bordass, 2012. The building performance sketch. *Building Research & Information*, 40 (2), 186–208.

Dorer, V. and A. Weber, 2009. Energy and CO_2 emissions performance assessment of residential micro-cogeneration systems with dynamic whole-building simulation programs. *Energy Conversion and Management*, 50 (3), 648–657.

Dorst, K. and P. Vermaas, 2005. John Gero's Function-Behaviour-Structure model of designing: a critical analysis. *Research in Engineering Design*, 16 (1), 17–26.

Douglas, J., 1996. Building performance and its relevance to facilities management. *Facilities*, 14 (3/4), 23–32.

Dounis, A. and C. Caraiscos, 2009. Advanced control systems engineering for energy and comfort management in a building environment – a review. *Renewable and Sustainable Energy Reviews*, 13 (6/7), 1246–1261.

van Dronkelaar, C., D. Cóstolo, R. Mangkuto and J. Hensen, 2014. Heating and cooling energy demand in underground buildings: potential for saving in various climates and functions. *Energy and Buildings*, 71, 129–136.

van Dronkelaar, C., M. Dowson, C. Spataru and D. Mumovic, 2016. A review of the regulatory energy performance gap and its underlying causes in non-domestic buildings. *Frontiers in Mechanical Engineering*, 1, 17.

Dror, I., 2013. The ambition to be scientific: human expert performance and objectivity. *Science and Justice*, 53 (2), 81–82.

Drust, B., 2010. Performance analysis research: meeting the challenge. *Journal of Sports Sciences*, 28 (9), 921–922.

Dry, C., 1997. Building materials that self repair. *Architectural Science Review*, 40 (2), 49–52.

DTI, 1998. *Rethinking Construction ('Egan Report')*. London: Department of Trade and Industry.

Du, J. and S. Sharples, 2010. Analysing the impact of reflectance distributions and well geometries on vertical surface daylight levels in atria for overcast skies. *Building and Environment*, 45 (7), 1733–1745.

Du, Y., X. Mao, D. Xu and F. Ren, 2012. Analyzing the effects of failure on fire equipment in building by FAST. *Procedia Engineering*, 45, 655–662.

Du, Z., P. Xu, X. Jin and Q. Liu, 2015. Temperature sensor placement optimization for VAV control using CFD-BES co-simulation strategy. *Building and Environment*, 85, 104–113.

Duarte, C., R. Budwig and K. Van den Wymelenberg, 2015. Energy and demand implication of using recommended practice occupancy diversity factors compared to real occupancy data in whole building energy simulation. *Journal of Building Performance Simulation*, 8 (6), 408–423.

Dubois, M., F. Cantin and K. Johnsen, 2007. The effect of coated glazing on visual perception: a pilot study using scale models. *Lighting Research and Technology*, 39 (3), 283–304.

Duives, D., W. Daamen and S. Hoogendoorn, 2013. State-of-the-art crowd motion simulation models. *Transport Research Part C*, 37, 193–209.

Duncan, J., 2005. Performance-based building: lessons from implementation in New Zealand. *Building Research & Information*, 33 (2), 120–127.

Dunston, P., L. Arns and J. McGlothlin, 2010. Virtual reality mock-ups for healthcare facility design and a model for technology hub collaboration. *Journal of Building Performance Simulation*, 3 (3), 185–195.

Duprat, F., N. Tru Vu and A. Sellier, 2014. Accelerated carbonation tests for the probabilistic prediction of the durability of concrete structures. *Construction and Building Materials*, 66, 597–605.

Durrani, F., M. Cook and J. McGuirk, 2015. Evaluation of LES and RANS CFD modelling of multiple states in natural ventilation. *Building and Environment*, 92, 167–181.

Dwyre, C. and C. Perry, 2015. Expanded fields: architecture/landscape/performance. *Journal of Performance and Art*, 37 (1), 1–7.

Eastman, C.M., 1999. *Building product models: computer environments supporting design and construction.* Baton Roca, FL: CRC Press.

Eastman, C., P. Teicholz, R. Sacks and K. Liston, 2011. *BIM Handbook – A guide to building information modeling.* Hoboken, NJ: Wiley, 2nd edition.

van Eck, D., 2011. Supporting design knowledge exchange by converting models of functional decomposition. *Journal of Engineering Design*, 22 (11/12), 839–858.

Eckert, C., A. Ruckpaul, T. Alink and A. Albers, 2012. Variations in functional decomposition for an existing product: experimental results. *Artificial Intelligence for Engineering Design, Analysis and Manufacturing*, 26 (2), 107–128.

Edwards, N., 2008. Performance-based building codes: a call for injury prevention indicators that bridge health and building sectors. *Injury Prevention*, 14 (5), 329–332.

Edwards, M., 2014. Fetal death and reduced birth rates associated with exposure to lead-contaminated drinking water. *Environmental Science & Technology*, 48 (1), 739–746.

Efficiency Valuation Organization, 2012. *International Performance Measurement and Verification Protocol: Concepts and options for determining energy and water savings*, Volume 1. Washington, DC: Efficiency Valuation Organization.

Efficiency Valuation Organization, 2014a. *International Performance Measurement and Verification Protocol: Core concepts.* Washington, DC: Efficiency Valuation Organization.

Efficiency Valuation Organization, 2014b. *International Performance Measurement and Verification Protocol: Statistics and uncertainty for IPMVP.* Washington, DC: Efficiency Valuation Organization.

EIA, 2016. US Energy Information Administration homepage [online]. Washington, DC. Available from www.eia.gov/consumption/ [Accessed 29 April 2016].

Eisenhower, B., Z. O'Neill, S. Narayanan, V. Fonoberov and I. Mezić, 2012. A methodology for meta-model based optimization in building energy models. *Energy and Buildings*, 47, 292–301.

El Dien, H. and P. Woloszyn, 2004. Prediction of the sound field into high-rise building facades due to its balcony ceiling form. *Applied Acoustics*, 64 (4), 431–440.

El-Diraby, T., 2013. Domain ontology for construction knowledge. *Journal of Construction Engineering and Management*, 139 (7), 768–784.

Elliott, T., M. Welsh and T. Nettelbeck, 2007. Investigating naturalistic decision making in a simulated microworld: what questions should we ask? *Behavior Research Methods*, 39 (4), 901–910.

El Mankibi, M., Z. Zhai, S. Al-Saadi and A. Zoubir, 2015. Numerical modeling of thermal behaviors of active multi-layer living wall. *Energy and Buildings*, 106, 96–110.

El-Omari, S. and O. Moselhi, 2011. Integrated automated data acquisition techniques for progress reporting of construction projects. *Automation in Construction*, 20 (6), 699–705.

Entrop, A. and H. Brouwers, 2010. Assessing the sustainability of buildings using a framework of triad approaches. *Journal of Building Appraisal*, 5 (4), 293–310.

Ercan, B. and S. Elias-Ozkan, 2015. Performance-based parametric design explorations: a method for generating appropriate building components. *Design Studies*, 38, 33–53.

Erden, M., H. Komoto, T. van Beek, V. D'Amelio, E. Echavarria and T. Tomiyama, 2008. A review of function modeling: approaches and applications. *Artificial Intelligence for Engineering Design, Analysis and Manufacturing*, 22 (2), 147–169.

Erhorn, H. and H. Erhorn-Kluttig, 2011. Terms and definitions for high performance buildings. Concerted Action: Energy Performance of Buildings. Detailed Report. EPBD.

Evans, J., 2004. An exploratory study of performance measurement systems and relationships with performance results. *Journal of Operations Management*, 22 (3), 219–232.

Evins, R., 2013. A review of computational optimisation methods applied to sustainable building design. *Renewable and Sustainable Energy Reviews*, 22, 230–245.

Evins, R., 2015. Multi-level optimization of building design, energy system sizing and operation. *Energy*, 90, 1775–1789.

Fallah, A. and T. Taghikhany, 2015. Sliding mode fault detection and fault-tolerant control of smart dampers in semi-active control of building structures. *Smart Materials and Structures*, 24 (12), 125030.

Famighetti, T., K. Cunefare and E. Muhlberger, 2006. Qualification and performance of a reverberation chamber equipped with lightweight diffusers. *Noise Control Engineering Journal*, 54 (3), 201–211.

Fan, Y. and K. Ito, 2012. Energy consumption analysis intended for real office space with energy recovery ventilator by integrating BES and CFD approaches. *Building and Environment*, 52, 57–67.

Fang, W., S. Miller and C. Yeh, 2012a. The effect of ESCOs on energy use. *Energy Policy*, 51, 558–568.

Fang, Z., W. Song, Z. Li, T. Tian, W. Lv, J. Ma and X. Xiao, 2012b. Experimental study on evacuation process in a stairwell of a high-rise building. *Building and Environment*, 47, 316–321.

Fanger, P., 1970. *Thermal comfort*. Copenhagen: Danish Technical Press.

Farmer, D., D. Johnston and D. Miles-Shenton, 2016. Obtaining the heat loss coefficient of a dwelling using its heating system (integrated coheating). *Energy and Buildings*, 117, 1–10.

Farrell, R. and C. Hooker, 2013. Design, science and wicked problems. *Design Studies*, 34 (6), 681–705.

Farris, J., E. van Aken, G. Letens, P. Chearskul and G. Coleman, 2010. Improving the performance review process. *International Journal of Operations & Production Management*, 31 (4), 376–404.

Federspiel, C., Q. Zhang and E. Arens, 2002. Model-based benchmarking with application to laboratory buildings. *Energy and Buildings*, 34 (3), 203–214.

Fedoruk, L., R. Cole, J. Robinson and A. Cayuela, 2015. Learning from failure: understanding the anticipated-achieved building energy performance gap. *Building Research & Information*, 43 (6), 750–763.

Felgner, F., R. Merz and L. Litz, 2006. Modular modelling of thermal building behaviour using Modelica. *Mathematical and Computer Modelling of Dynamical Systems*, 12 (1), 35–49.

Fellows, R. and A. Liu, 2008. *Research Methods for Construction*. Chichester: Wiley-Blackwell, 3rd edition.

Feng, W., J. Grunewald, A. Nicolai, C. Zhang and J. Zhang, 2012. CHAMPS-Mulitzone – a combined heat, air, moisture and pollutant simulation environment for whole-building performance analysis. *HVAC&R Research*, 18 (1), 233–251.

Féral, J., 2008. Introduction: towards a genetic study of performance – Take 2. *Theatre Research International*, 33 (3), 223–233.

Fernández, M., S. Quintana, N. Chavarría and J. Ballesteros, 2009. Noise exposure of workers in the construction sector. *Applied Acoustics*, 70 (5), 753–760.

Fernandez, N., S. Katipamula, W. Wang, Y. Huang and G. Liu, 2015. Energy saving modelling of re-tuning energy conservation measures in large office buildings. *Journal of Building Performance Simulation*, 8 (6), 391–407.

Fernie, S., R. Leiringer and T. Thorpe, 2006. Change in construction: a critical perspective. *Building Research & Information*, 34 (2), 91–103.

Ferreira, J., M. Duarte Pinheiro and J. de Brito, 2013. Refurbishment decision support tools: a review from a Portuguese user's perspective. *Construction and Building Materials*, 49, 425–447.

Ferretti, N., M. Galler, S. Bushby and D. Choinière, 2015. Evaluating the performance of diagnostic agent for building operation (DABO) and HVAC-Cx tools using the virtual cybernetic building testbed. *Science and Technology for the Built Environment*, 21 (8), 1154–1164.

Few, S., 2006. *Information Dashboard Design – The Effective Visual Communication of Data*. Sebastopol, CA: O'Reilly Media.

Figueiredo, M., A. de Almeida and B. Ribeiro, 2012. Home electrical signal disaggregation for non-intrusive load monitoring (NILM) systems. *Neurocomputing*, 96, 66–73.

Flourentzou, F., J. Genre and C. Roulet, 2002. TOBUS software – an interactive decision aid tool for building retrofit studies. *Energy and Buildings*, 34 (2), 193–202.

Foliente, C., 2000. Developments in performance-based building codes and standards. *Forest Products Journal*, 50 (7/8), 12–21.

Foliente, G., 2005a. Incentives, barriers and PBB implementation. In: Becker, R., ed., *Performance Based Building international state of the art – final report*. Rotterdam: CIB.

Foliente, G., 2005b. PBB Research and development roadmap summary. In: Becker, R., ed., *Performance Based Building international state of the art – final report*. Rotterdam: CIB.

Foliente, G., R. Leicester and L. Pham, 1998. Development of the CIB proactive program on Performance Based Building codes and standards. Victoria: CSIRO.

Forcada, N., M. Macarulla, M. Gangolells and M. Casals, 2014. Assessment of construction defects in residential buildings in Spain. *Building Research & Information*, 42 (5), 629–640.

Foucquier, A., S. Robert, F. Suard, L. Stéphan and A. Jay, 2013. State of the art in building modelling and energy performances prediction: a review. *Renewable and Sustainable Energy Reviews*, 23, 272–288.

Foulds, C., J. Powell and G. Seyfang, 2013. Investigating the performance of everyday domestic practices using building monitoring. *Building Research & Information*, 41 (6), 622–636.

Fox, A., 2011. Cloud computing – what's in it for me as a scientist? *Science*, 331 (6016), 406–407.

Fox, M., 2015. Thermography for thermal building assessment and improvement. PhD thesis. Plymouth: Plymouth University.

Fox, M., D. Coley, S. Goodhew and P. de Wilde, 2014. Thermography methodologies for detecting energy related building defects. *Renewable and Sustainable Energy Reviews*, 40, 296–310.

Fox, M., D. Coley, S. Goodhew and P. de Wilde, 2015. Time-lapse thermography for building defect detection. *Energy and Buildings*, 92, 95–106.

Fox, M., S. Goodhew and P. de Wilde, 2016. Building defect detection: external versus internal thermography. *Building and Environment*, 105, 317–331.

Frankel, M., J. Edelson and R. Colker, 2015. Getting to outcome-based building performance: report from a Seattle Summit on performance outcomes. Vancouver, WA/Washington, DC: New Buildings Institute/National Institute of Building Sciences.

Fraunhofer Institute for Building Physics, 2016. Measuring and testing facilities overview [online]. Holzkirchen. Available from www.pruefstellen.ibp.fraunhofer.de/en/measuring-testing-facilities.html [Accessed 11 July 2016].

French, S., 1988. *Decision theory: an introduction to the mathematics of rationality*. Chichester: Ellis Horwood/Halsted Press.

Frenette, C., D. Derome, R. Beauregard and A. Salenikovich, 2008. Identification of multiple criteria for the evaluation of light-frame wood wall assemblies. *Journal of Building Performance Simulation*, 1 (4), 221–236.

Frenette, C., R. Beauregard, I. Abi-Zeid, D. Derome and A. Salenikovich, 2010. Multicriteria decision analysis applied to the design of light-frame wood wall assemblies. *Journal of Building Performance Simulation*, 3 (1), 33–52.

Frontczak, M., S. Schiavon, J. Goins, E. Arens, H. Zhang and P. Wargocki, 2012. Quantitative relationships between occupant satisfaction and satisfaction aspects of indoor environmental quality and building design. *Indoor Air*, 22 (2), 119–131.

Fuchs, M., J. Teichmann, M. Lauster, P. Remmen, R. Streblow and D. Müller, 2016. Workflow automation for combined modeling of buildings and district energy systems. *Energy*, 117 (2), 478–484.

Fulford, R. and C. Standing, 2014. Construction industry productivity and the potential for collaborative practice. *International Journal of Project Management*, 32 (2), 315–326.

Fumo, N., 2014. A review of the basics of building energy estimation. *Renewable and Sustainable Energy Reviews*, 31, 53–60.

Futrell, B., E. Ozelkan and D. Bentrup, 2015. Bi-objective optimization of building enclosure design for thermal and lighting performance. *Building and Environment*, 92, 591–602.

Fux, S., M. Benz and L. Guzzella, 2013. Economic and environmental aspects of the component sizing for a stand-alone building energy system: a case study. *Renewable Energy*, 55, 438–447

Galea, E., 2012. Evacuation and pedestrian dynamics guest editorial – 21st century grand challenges in evacuation and pedestrian dynamics. *Safety Science*, 50 (8), 1653–1654.

Galiana, M., C. Llinares and Á. Page, 2012. Subjective evaluation of music hall acoustics: response of expert and non-expert users. *Building and Environment*, 58, 1–13.

Galle, P., 2009. The ontology of Gero's FBS model of designing. *Design Studies*, 30 (4), 321–339.

Gane, V. and J. Haymaker, 2012. Design scenarios: enabling transparent design spaces. *Advanced Engineering Informatics*, 26 (3), 618–640.

Gann, D. and J. Whyte, Guest eds., 2003. Design quality, its measurement and management in the built environment. *Building Research & Information*, 31 (5), 314–317.

Gann, D., A. Salter and J. Whyte, 2003. Design Quality Indicator as a tool for thinking. *Building Research & Information*, 31 (5), 318–333.

Gao, X. and A. Malkawi, 2014. A new methodology for building energy performance benchmarking: an approach based on intelligent clustering algorithm. *Energy and Buildings*, 84, 607–616.

Gärling, T., 2014. Past and present environmental psychology. *European Psychologist*, 19 (2), 127–131.

Garmston, H., 2016. Decision-making in the selection of retrofit façades for non-domestic buildings. PhD thesis. Plymouth: Plymouth University.

Garrido, M., P. Paulo and F. Branco, 2012. Service life prediction of façade paint coatings in old buildings. *Construction and Building Materials*, 29, 394–402.

Gayialis, S. and I. Tatsiopoulos, 2004. Design of an IT-driven decision support system for vehicle routing and scheduling. *European Journal of Operation Research*, 152 (2), 382–398.

Gaylord, E., C. Gaylord and J. Stallmeyer, 1997. *Structural Engineering Handbook*. New York: McGraw-Hill, 4th edition.

Geebelen, B., 2003. Daylight availability prediction in the early stages of the building design process. PhD thesis. Leuven: Katholieke Universiteit Leuven.

van der Geer, E., H. van Tuijl and C. Rutte, 2009. Performance management in healthcare: performance indicator development, task uncertainty, and types of performance indicators. *Social Science and Medicine*, 69 (10), 1523–1530.

Gelenbe, E. and F. Wu, 2012. Large scale simulation for human evacuation and rescue. *Computers and Mathematics with Applications*, 64 (12), 3869–3880.

Gendron, Y., D. Cooper and B. Townley, 2007. The construction of auditing expertise in measuring government performance. *Accounting, Organizations and Society*, 32 (1/2), 101–129.

Georgiou, J., 2010. Verification of a building defect classification system for housing. *Structural Survey*, 28 (5), 370–383.

Gerber, D., E. Pantazis and A. Wang, 2017. A multi-agent approach for performance based architecture: design exploring geometry, user, and environmental agencies in façades. *Automation in Construction*, 76, 45–58.

Gere, J. and B. Goodno, 2013. *Mechanics of Materials*. Stamford, CT: Cengage Learning, 8th edition.

Gerlich, V., K. Sulovská and M. Záležák, 2013. COMSOL Multiphysics validation as simulation software for heat transfer calculation in buildings: building simulation software validation. *Measurement*, 46 (6), 2003–2012.

Gernay, T. and J. Franssen, 2015. A performance indicator for structures under natural fire. *Engineering Structures*, 100, 94–103.

Gero, J., 1990. Design prototypes: a knowledge representation schema for design. *AI Magazine*, 11 (4), 26–36.

Gero, J. and U. Kannengiesser, 2003. The situated function-behaviour-structure framework. *Design Studies*, 25 (4), 373–391.

Gero, J. and U. Kannengiesser, 2007. A function-behaviour-structure ontology of processes. *Artificial Intelligence for Engineering Design, Analysis and Manufacturing*, 21 (4), 379–391.

Gettys, C., R. Pliske, C. Manning and J. Casey, 1987. An evaluation of human act generation performance. *Organizational Behavior and Human Decision Processes*, 39 (1), 23–51.

Geva, A., H. Saaroni and J. Morris, 2014. Measurements and simulations of thermal comfort: a synagogue in Tel Aviv, Israel. *Journal of Building Performance Analysis*, 7 (3), 233–250.

Geyer, P., 2012. Systems modelling for sustainable building design. *Advanced Engineering Informatics*, 26 (4), 656–668.

Geyer, P. and M. Buchholz, 2012. Parametric systems modeling for sustainable energy and resource flows in buildings and their urban environment. *Automation in Construction*, 22, 70–80.

Geyer, P. and A. Schlüter, 2014. Automated metamodel generation for design space exploration and decision-making: a novel method supporting performance-oriented building design and retrofitting. *Applied Energy*, 119, 537–556.

Gielingh, W., 1988. General AEC reference model (GARM). *CIB W74+W78 Workshop on Conceptual Modelling of Buildings*, Lund, Sweden, 25–27 October 1988, pp. 165–178.

Gielingh, W., 2005. Improving the Performance of Construction by the Acquisition, Organization and Use of Knowledge. PhD thesis. Delft: Delft University of Technology.

Gil, Y., V. Ratnakar, J. Kim, J. Moody and E. Deelman, 2011. Wings: intelligent workflow-based design of computational experiments. *IEEE Intelligent Systems*, 26 (1), 62–72.

Gilani, S. and W. O'Brien, 2016. Review of current methods, opportunities, and challenges for in-situ monitoring to support occupant modelling in office spaces. *Journal of Building Performance Simulation*, 10 (5/6), 1–27.

Gilb, T., 2005. *Competitive engineering: a handbook for systems engineering, requirements engineering, and software engineering using planguage*. Oxford: Butterworth-Heinemann.

Gill, Z., M. Tierney, I. Pegg and N. Allan, 2010. Low-energy dwellings: the contribution of behaviours to actual performance. *Building Research & Information*, 38 (5), 491–508.

Gillespie, K., P. Haves, R. Hitchcock, J. Deringer and K. Kinney, 2007. *A specification guide for performance monitoring systems*. Berkeley, CA: Lawrence Berkeley National Laboratory.

Girard, P. and G. Doumeingts, 2004. Modelling the engineering design system to improve performance. *Computers & Industrial Engineering*, 46 (1), 43–67.

Giri, S. and M. Bergés, 2015. An energy estimation framework for event-based methods in non-intrusive load monitoring. *Energy Conversion and Management*, 90, 488–498.

Giri, S., M. Bergés and A. Rowe, 2013. Towards automated appliance recognition using an EMF sensor in NILM platforms. *Advanced Engineering Informatics*, 27 (4), 447–485.

Glaser, D., O. Feng, J. Voung and L. Xiao, 2004. Towards an algebra for lighting simulation. *Building and Environment*, 39 (8), 895–903.

Gleckler, P., K. Taylor and C. Doutriaux, 2008. Performance metrics for climate models. *Journal of Geophysical Research*, 113, D06104.

Glicksman, L., 2008. Energy efficiency in the built environment. *Physics Today*, 61 (7), 35–40.

Gloriant, F., P. Tittelein, A. Joulin and S. Lassue, 2015. Modeling a triple-glazed supply-air window. *Building and Environment*, 84, 1–9.

Göçer, Ö., Y. Hua and K. Göçer, 2015. Completing the missing link in building design process: enhancing post-occupancy evaluation method for effective feedback for building performance. *Building and Environment*, 89, 14–27.

Goel, S., J. Hofman, S. Lehaie, D. Pennock and D. Watts, 2010. Predicting consumer behavior with web search. *Proceedings of the National Academy of Sciences of the United States*, 107 (41), 17486–17490.

Gogotsi, Y. and P. Simon, 2011. True performance metrics in electrochemical energy storage. *Science*, 334 (6058), 917–918.

Goins, J. and M. Moezzi, 2013. Linking occupant complaints to building performance. *Building Research & Information*, 41 (3), 361–372.

Goldsworthy, M., 2012. Dynamic coupling of the transient system simulation and fire dynamics programs. *Journal of Building Performance Simulation*, 5 (2), 105–114.

Gong, J. and C. Caldas, 2011. An object recognition, tracking, and contextual reasoning-based video interpretation method for rapid productivity analysis of construction operations. *Automation in Construction*, 20 (8), 1211–1226.

Goodhew, S., 2016. *Sustainable construction processes: a resource text*. Chichester: Wiley Blackwell.

Goossens, L., R. Cooke, A. Hale and L. Rodić-Wiersma, 2008. Fifteen years of expert judgment at TU Delft. *Safety Science*, 46 (2), 234–244.

Gordon, S. and K. Gallo, 2011. Structuring expert input for a knowledge-based approach to watershed condition assessment for the Northwest Forest Plan, USA. *Environmental Monitoring & Assessment*, 172 (1/4), 643–661.

Gore, J., A. Banks, L. Millward and O. Kyriakidou, 2006. Naturalistic decision making and organisations: reviewing pragmatic science. *Organization Studies*, 27 (7), 925–942.

Gore, J., R. Flin, N. Stanton and B. Wong, 2015. Applications for naturalistic decision-making. *Journal of Occupational and Organizational Psychology*, 88 (2), 223–230.

Gossauer, E. and A. Wagner, 2007. Post-occupancy evaluation of thermal comfort: state of the art and new approaches. *Advances in Building Energy Research*, 1 (1), 151–175.

Gosselin, C., R. Duballet, P. Roux, N. Gaudillière, J. Dirrenberger and P. Morel, 2016. Large-scale 3D printing of ultra-high performance concrete – a new processing route for architects and buildings. *Materials and Design*, 100, 102–109.

Gousseau, P., B. Blocken, T. Stathopoulos and G. van Heijst, 2011. CFD simulation of near-field dispersion on a high-resolution grid: a case study by LES and RANS for a building group in downtown Montreal. *Atmospheric Environment*, 45 (2), 428–438.

Gowreesunker, B., S. Tassou and M. Kolokotroni, 2013. Coupled TRNSYS-CFD simulations evaluating the performance of PCM plate heat exchangers in an airport terminal building displacement conditioning system. *Building and Environment*, 65, 132–145.

Granadeiro, V., L. Pina, J. Duarte, J. Correia and V. Leal, 2013. A general indirect representation for optimization of generative design systems by genetic algorithms: application to a shape grammar-based design system. *Automation in Construction*, 35, 374–382.

Granderson, J. and P. Price, 2014. Development and application of a statistical methodology to evaluate the predictive accuracy of building energy baseline models. *Energy*, 66, 981–990.

Granderson, J., S. Touzani, C. Custodio, M. Sohn, D. Jump and S. Fernandes, 2016. Accuracy of automated measurement and verification (M&V) techniques for energy savings in commercial buildings. *Applied Energy*, 173, 296–308.

Granqvist, C., 2016. Recent progress in thermochromics and electrochromics: a brief survey. *Thin Solid Films*, 614, 90–96.

Gratia, E. and A. De Herde, 2003. Design of low energy office buildings. *Energy and Buildings*, 35 (5), 473–491.

Greitzer, F., R. Podmore, M. Robinson and P. Ey, 2010. Naturalistic decision making for power system operators. *International Journal of Human-Computer Interaction*, 26 (2/3), 278–291.

Groat, L. and D. Wang, 2013. *Architectural Research Methods*. Hoboken, NJ: Wiley, 2nd edition.

de Groot, E., 1999. Integrated Lighting System Assistant. PhD thesis. Eindhoven: Eindhoven University Press.

Gross, J., 1996. Developments in the application of the performance concept in building. In: Becker, R. and M. Paciuk, eds., *CIB-ASTM-ISO-RILEM 3rd International Symposium*, Tel-Aviv, Israel, 9–12 December 1996.

Gruber, J., F. Huerta, P. Matatagui and M. Prodanović, 2015. Advanced building energy management based on a two-stage receding horizon optimization. *Applied Energy*, 160, 194–205.

Grumman, D. and A. Hinge, 2012. What makes buildings high performing. *High Performance Buildings*, 5, 46–54.

Grynning, S., F. Goia, E. Rognvik and B. Time, 2013. Possibilities for characterization of a PCM window system using large scale measurements. *International Journal of Sustainable Built Environment*, 2 (1), 56–64.

Gudmundsson, A., J. Löndahl and M. Bohgard, 2007. Methodology for identifying particle sources in indoor environments. *Journal of Environmental Monitoring*, 9 (8), 831–838.

Guerra-Santin, O. and C. Tweed, 2015. In-use monitoring of buildings: an overview and classification of methods. *Energy and Buildings*, 86, 176–189.

Guerra-Santin, O., N. Herrera, E. Cuerda and D. Keyson, 2016. Mixed methods approach to determine occupants' behaviour – analysis of two case studies. *Energy and Buildings*, 130, 546–566.

Gulvanessian, H., 2009. EN 1990 Eurocode 'basis of structural design' – the innovative head Eurocode. *Steel Construction*, 2 (4), 222–227.

Gunay, B., W. Shen and C. Yang, 2017. Characterization of a building's operation using automation data: a review and case study. *Building and Environment*, 118, 196–210.

Guo, X. and S. Silva, 2008. High-performance transistors by design. *Science*, 320 (5876), 618–619.

Guo, W., L. Soibelman and J. Garrett, 2009. Automated defect detection for sewer pipeline inspection and condition assessment. *Automation in Construction*, 18 (5), 587–596.

Guo, B., T. Yiu and V. González, 2015. Identifying behaviour patterns of constructions safety using system archetypes. *Accident Analysis and Prevention*, 80, 125–141.

Gupta, R. and M. Gregg, 2012. Using UK climate change projections to adapt existing English homes for a warming climate. *Building and Environment*, 55, 20–42.

Gupta, A. and P. Yadav, 2004. SAFE-R: a new model to study the evacuation profile of a building. *Fire Safety Journal*, 39 (7), 539–556.

Gursel, I., S. Sariyildiz, Ö. Akin and R. Stouffs, 2009. Modeling and visualization of lifecycle building performance assessment. *Advanced Engineering Informatics*, 23 (4), 396–417.

Gursel Dino, I. and R. Stouffs, 2014. Evaluation of reference modeling for building performance assessment. *Automation in Construction*, 40, 44–59.

Gustafsson, M., G. Dermentzis, J. Myhren, C. Bales, F. Ochs, S. Holmberg and W. Feist, 2014. Energy performance comparison of three innovative HVAC systems for renovation through dynamic simulation. *Energy and Buildings*, 82, 512–519.

Gutiérrez-Montes, C., E. Sanmiguel-Rojas, A. Kaiser and A. Viedma, 2008. Numerical model and validation experiments of atrium enclosure fire in a new fire test facility. *Building and Environment*, 43 (11), 1912–1928.

Habert, G., Y. Bouzidi, C. Chen and A. Jullien, 2010. Development of a depletion indicator for natural resources used in concrete. *Resources, Conservation and Recycling*, 54 (6), 364–376.

Hadjri, K. and C. Crozier, 2009. Post-occupancy evaluation: purpose, benefits and barriers. *Facilities*, 27 (1/2), 21–33.

Hafner, I., M. Rößler, B. Heinzl, A. Körner, M. Landsiedl and F. Breitenecker, 2014. Investigation communication and step-size behaviour for co-simulation of hybrid physical systems. *Journal of Computational Science*, 5 (3), 427–438.

Ham, Y. and M. Golparvar-Fard, 2013. An automated vision-based method for rapid 3D energy performance modeling of existing buildings using thermal and digital imagery. *Advanced Engineering Informatics*, 27 (3), 395–409.

Ham, Y. and M. Golparvar-Fard, 2014. 3D visualisation of thermal resistance and condensation problems using infrared thermography for building energy diagnostics. *Visualisation in Engineering*, 2, 12.

Hamdy, M. and K. Sirén, 2015. A multi-aid optimization scheme for large-scale investigation of cost-optimality and energy performance of buildings. *Journal of Building Performance Simulation*, 9 (4), 411–430.

Hamdy, M., A. Nguyen and J. Hensen, 2016. A performance comparison of multi-objective optimization algorithms for solving nearly-zero-energy-building design problems. *Energy and Buildings*, 121, 57–71.

Hammond, D., J. Dempsey, F. Szigeti and G. Davis, 2005. Integrating a performance-based approach into practice: a case study. *Building Research & Information*, 33 (2), 128–141.

Hamza, N. and P. de Wilde, 2014. Building simulation for the boardroom: an explanatory study. *Journal of Building Performance Simulation*, 7 (1), 52–67.

Hamzeh, F., I. Saab, I. Tommelein and G. Ballard, 2015. Understanding the role of 'tasks anticipated' in lookahead planning through simulation. *Automation in Construction*, 49, 18–26.

Han, S. and S. Hong, 2003. A systematic approach for coupling user satisfaction with product design. *Ergonomics*, 46 (13/14), 1441–1461.

Han, S., M. Chae, K. Im and H. Ryu, 2008. Six Sigma approach to improve performance in construction operations. *Journal of Management in Engineering*, 24 (1), 21–31.

Han, H., C. Shin, I. Lee and K. Kwon, 2011. Tracer gas experiments for local mean ages of air from individual supply inlets in a space with multiple inlets. *Building and Environment*, 46 (12), 2462–2471.

Han, G., J. Srebric and E. Enache-Pommer, 2014. Variability of optimal solutions for building components based on comprehensive life cycle cost analysis. *Energy and Buildings*, 79, 223–231.

Hanby, V., M. Cook, D. Infield, Y. Ji, D. Loveday, L. Mei and M. Holmes, 2008. Nodal network and CFD simulation of airflow and heat transfer in double skin facades with blinds. *Building Services Engineering Research and Technology*, 29 (1), 45–59.

Hand, J., 1998. Removing Barriers to the Use of Simulation in the Building Design Professions. PhD thesis. Glasgow: University of Strathclyde.

Hand, J., 2008. Improving non-geometric data available to simulation programs. *Building and Environment*, 43 (4), 674–685.

Hann, R., 2012. Blurred architecture: duration and performance in the work of Diller Scofidio + Renfro. *Performance Research*, 17 (5), 9–18.

Hannah, D. and O. Kahn, 2008. Performance/Architecture: Introduction. *Journal of Architectural Education*, 61 (4), 4–5.

Hansen, H., 2007. Sensitivity analysis as a methodological approach to the development of design strategies for environmentally sustainable buildings. PhD thesis. Aalborg: Aalborg University.

Hansen, K. and J. Vanegas, 2003. Improving design quality through briefing automation. *Building Research and Information*, 31 (5), 379–386.

Hantouche, E., N. Abboud, M. Morovat and M. Engelhardt, 2016. Analysis of steel bolted double angle connections at elevated temperatures. *Fire Safety Journal*, 83, 79–89.

Harb, H., J. Reinhardt, R. Streblow and D. Müller, 2015. MIP approach for designing heating systems in residential buildings and neighbourhoods. *Journal of Building Performance Simulation*, 9 (3), 316–330.

Harfield, S., 2007. On design 'problematization': theorising differences in designed outcomes. *Design Studies*, 28 (2), 159–173.

Harries, A., G. Brunelli and I. Rizos, 2013. London 2012 Velodrome – integrating advanced simulation into the design process. *Journal of Building Performance Simulation*, 6 (6), 401–419.

Harrington, H., F. Voehl and H. Wiggin, 2012. Applying TQM to the construction industry. *The TQM Journal*, 24 (4), 352–363.

Hartkopf, V. and V. Loftness, 1999. Global relevance of total building performance. *Automation in Construction*, 8 (4), 377–393.

Hartkopf, V., V. Loftness and P. Mill, 1986a. Integration for performance. In: Rush, R., ed., *The building systems integration handbook*. Boston, MA: The American Institute of Architects.

Hartkopf, V., V. Loftness and P. Mill, 1986b. The concept of total building performance and building diagnostics. In: Davis, G. ed., *Building performance: function, preservation and rehabilitation*. Philadelphia, PA: American Society for Testing and Materials.

Hasegawa, T., 2004. Climate change, adaptation and government policy for the building sector. *Building Research & Information*, 32 (1), 61–64.

Hasegawa, K., H. Yoshino, U. Yanagi, K. Azuma, H. Osawa, N. Kagi, N. Shinohara and A. Hasegawa, 2015. Indoor environmental problems and healt status in water-damaged homes due to tsunami disaster in Japan. *Buildings and Environment*, 93, 24–34.

Hassanain, M., H. Mathar and A. Aker, 2016. Post-occupancy evaluation of a university student cafeteria. *Architectural Engineering and Design Management*, 12 (1), 67–77.

Haverinen, U., M. Vahteristo, D. Moschandreas, A. Nevalainen, T. Husman and J. Pekkanen, 2003. Knowledge-based and statistically modeled relationships between residential moisture damage and occupant reported health symptoms. *Atmospheric Environment*, 37 (4), 577–585.

Hayashi, Y., K. Tamura, M. Mori and I. Takahashi, 1999. Simulation analysis of buildings damaged in the 1995 Kobe, Japan, earthquake considering soil-structure interaction. *Earthquake Engineering & Structural Dynamics*, 28 (4), 371–391.

Hazelrigg, G.A., 2012. Fundamentals of decision making for engineering design and systems engineering. http://www.engineeringdecisionmaking.com/ [Accessed 22 December 2017].

He, G., L. Shu and S. Zhang, 2011. Double skin facades in the hot summer and cold winter zone in China: cavity open or closed? *Building Simulation*, 4 (4), 283–291.

He, C., H. Salonen, X. Ling, L. Crilley, N. Jayasundara, H. Cheung, M. Hargreaves, F. Huygens, L. Knibbs, G. Ayoko and L. Morawska, 2014. The impact of flood and post-flood cleaning on airborne microbiological and particle contamination in residential houses. *Environment International*, 69, 9–17.

HEEPI, 2008. Building a sustainable legacy: the benefits of high performance buildings to further and higher education, and how they can be achieved. Brighouse: Higher Education Environmental Performance Improvement.

Heer, T. and R. Wörzberger, 2011. Support for modeling and monitoring of engineering design processes. *Computers and Chemical Engineering*, 35 (4), 709–723.

Heffernan, E., 2015. Delivering Zero Carbon Homes and sustainable communities: the potential of group self-build housing in England. PhD thesis. Plymouth: Plymouth University.

Heffernan, E., W. Pan, X. Liang and P. de Wilde, 2015. Zero carbon homes: perceptions from the UK construction industry. *Energy Policy*, 79, 23–36.

van der Heijden, J., H. Visscher and F. Meijer, 2007. Problems in enforcing Dutch building regulations. *Structural Survey*, 25 (3/4), 319–329.

Hendricx, A., 2000. A core object model for architectural design. PhD thesis. Leuven: Katholieke Universiteit Leuven.

Hens, H., 2015. Combined heat, air, moisture modelling: a look back, how, of help? *Building and Environment*, 91, 138–151.

Hensel, M., 2013. *Performance-oriented architecture: rethinking architectural design and the built environment*. Chichester: Wiley.

Hensen, J. and R. Lamberts, eds., 2011. *Building performance simulation for design and operation*. Abingdon: Spon Press.

Henze, G., 2013. Model predictive control for buildings: a quantum leap? *Journal of Building Performance Simulation*, 6 (3), 157–158.

Heo, Y., R. Choudhary and G. Augenbroe, 2012. Calibration of building energy models for retrofit analysis under uncertainty. *Energy and Buildings*, 47, 550–560.

Heo, Y., G. Augenbroe and R. Choudhary, 2013. Quantitative risk management for energy retrofit projects. *Journal of Building Performance Simulation*, 6 (4), 257–268.

Herbert, G., 1999. Architect-engineer relationships: overlappings and interactions. *Architectural Science Review*, 42 (2), 107–110.

Hernandez, P. and P. Kenny, 2010. From net energy to zero energy buildings: defining life cycle zero energy buildings (LC-ZEB). *Energy and Buildings*, 42 (6), 815–821.

Hesaraki, A., E. Bourdakis, A. Ploskić and S. Holmberg, 2015. Experimental study of energy performance in low-temperature hydronic heating systems. *Energy and Buildings*, 109, 108–114.

Hester, J., J. Gregory and R. Kirchain, 2017. Sequential early-design guidance for residential single-family buildings using a probabilistic metamodel of energy consumption. *Energy and Buildings*, 134, 202–211.

Hinge, A. and D. Winston, Winter 2009. Documenting performance: does it need to be so hard? *High Performance Buildings*, 18–22.

Hinze, J., S. Thurman and A. Wehle, 2013. Leading indicators of construction safety performance. *Safety Science*, 51 (1), 23–28.

Hitchcock, R., 2003. Standardized Buildings Performance Metrics – Final Report. Berkeley, CA: Lawrence Berkeley National Laboratory.

Hitchcock, R., 2004. Software interoperability in support of whole-building performance assurance. *Journal of Architectural and Planning Research*, 21 (4), 303–311.

Hoc, J., 2000. From human-machine interaction to human-machine cooperation. *Ergonomics*, 43 (7), 833–843.

Hoegl, M. and K. Parboteeah, 2003. Goal setting and team performance in innovative projects: on the moderating role of teamwork quality. *Small Group Research*, 34 (1), 3–19.

Hoffmann, T., 2009. *Smart Buildings*. Milwaukee, WI: Johnson Controls Inc. (white paper).

Hojem, T., K. Sørensen and V. Lagesen, 2014. Designing a green building: expanding ambitions through social learning. *Building Research & Information*, 42 (5), 591–601.

Holt, L., M. Ninneman and M. Kaphushion, 2007. *Building code adaptation, compliance and enforcement*. Denver, CO: The Holt Group LLC.

Hölttä, V. and H. Koivo, 2011. Performance metrics for web-forming processes. *Journal of Process Control*, 21 (6), 885–892.

Honfi, D., A. Mårtensson and S. Thelandersson, 2012. Reliability of beams according to Eurocodes in serviceability limit state. *Engineering Structures*, 35, 48–53.

Hong, T., L. Yang, D. Hill and W. Feng, 2014. Data and analytics to inform energy retrofit of high performance buildings. *Applied Energy*, 126, 90–106.

Hong, T., H. Sun, Y. Chen, S. Taylor-Lange and D. Yan, 2016. An occupant behavior modeling tool for co-simulation. *Energy and Buildings*, 117, 272–281.

Hoonakker, P., P. Carayon and T. Loushine, 2010. Barriers and benefits of quality management in the construction industry: an empirical study. *Total Quality Management & Business Excellence*, 21 (9), 953–969.

Hopfe, C., 2009. Uncertainty and sensitivity analysis in building performance simulation for decision support and design optimization. PhD thesis. Eindhoven: Eindhoven University Press.

Hopfe, C. and J. Hensen, 2011. Uncertainty analysis in building performance simulation for design support. *Energy and Buildings*, 43 (10), 2798–2805.

Hopfe, C. and R. McLeod, 2015. *The Passivhaus designer's manual – a technical guide to low and zero energy buildings*. Abingdon: Routledge.

Hopfe, C., G. Augenbroe and J. Hensen, 2013. Multi-criteria decision making under uncertainty in building performance assessment. *Building and Environment*, 69, 81–90.

Hornikx, M., 2015. Acoustic modelling for indoor and outdoor spaces. *Journal of Building Performance Simulation*, 8 (1), 1–2.

Hornikx, M., C. Hak and R. Wenmaekers, 2015. Acoustic modelling of sports halls, two case studies. *Journal of Building Performance Simulation*, 8 (1), 26–38.

Horonjeff, R. and F. McKelvey, 1994. *Planning & design of airports*. Boston, MA: McGraw Hill.

Horová, K., T. Jána and F. Wald, 2013. Temperature heterogeneity during travelling fire on experimental building. *Advances in Engineering Software*, 62/63, 119–130.

Hoyle, C., W. Chen, N. Wang and G. Gomez-Levi, 2011. Understanding and modelling heterogeneity of human preferences for engineering design. *Journal of Engineering Design*, 22 (8), 583–601.

Hrivnak, J., 2007. Is relative sustainability relevant? *Architectural Research Quarterly*, 11 (2), 167–176.

Hronsky, J. and D. Groves, 2008. Science of targeting: definition, strategies, targeting and performance measurement. *Australian Journal of Earth Sciences*, 55 (1), 3–12.

Hsiao, S., C. Hsu and Y. Lee, 2012. An online affordance evaluation model for product design. *Design Studies*, 33 (2), 126–159.

Hsu, D., 2014. Improving energy benchmarking with self-reported data. *Building Research & Information*, 42 (5), 641–656.

Hu, H., 2009. Risk-conscious Design of Off-grid Solar Energy Houses. PhD thesis. Atlanta, GA: Georgia Institute of Technology.

Hu, H. and G. Augenbroe, 2012. A stochastic model based energy management system for off-grid solar houses. *Building and Environment*, 50, 90–103.

Hu, J., O. Ogunsola, L. Song, R. McPherson, M. Zhu, Y. Hong and S. Chen, 2014. Restoration of 1–24 hour dry-bulb temperature gaps for use in building performance monitoring and analysis – Part I. *HVAC&R Research*, 20 (6), 594–605.

Hu, J., W. Chen, B. Zhao and D. Yang, 2017. Buildings with ETFE foils: a review on material properties, architectural performance and structural behavior. *Construction and Building Materials*, 131, 411–422.

Hua, J., J. Wang and K. Kumar, 2005. Development of a hybrid field and zone model for fire smoke propagation simulation in buildings. *Fire Safety Journal*, 40 (2), 99–119.

Huang, Y. and J. Niu, 2016. Optimal building envelope design based on simulated performance: history, current status and new potentials. *Energy and Buildings*, 117, 387–398.

Huang, S., Q. Li and S. Xu, 2007. Numerical evaluation of wind effects on a tall steel building by CFD. *Journal of Constructional Steel Research*, 63 (5), 612–627.

Hufen, H. and H. de Bruijn, 2016. Getting the incentives right. Energy performance contracts as a tool for property management by local government. *Journal of Cleaner Production*, 112 (4), 2717–2729.

Hughes, M. and R. Bartlett, 2010. Performance analysis. *Journal of Sport Sciences*, 20 (10), 735–737.

Hull, E., K. Jackson and J. Dick, 2011. *Requirements Engineering*. London: Springer.

Huo, F., W. Song, L. Chen, C. Liu and K. Liew, 2016. Experimental study on characteristics of pedestration evacuation on stairs in a high-rise building. *Safety Science*, 86, 165–173.

Huovila, P., 2005. *Decision Support Toolkit*. Rotterdam: CIB.

Hurley, W. and D. Lior, 2002. Combining expert judgment: on the performance of trimmed mean vote aggregation procedures in the presence of strategic voting. *European Journal of Operational Research*, 140 (1), 142–147.

Hyde, R., 2014. The technology and attitudinal fix. Redefining the definition of high performance building. *Architectural Science Review*, 57 (3), 155–158.

Hygh, J., J. DeCarolis, D. Hill and R. Ranjithan, 2012. Multivariate regression as an energy assessment tool in early building design. *Building and Environment*, 57, 165–175.

IBPSA, 2015. International Building Performance Simulation Association homepage [online]. Toronto, ON. Available from www.ibpsa.org [Accessed 26 November 2015].

IBPSA-USA, 2016. Building Energy Software Tool Directory [online]. Available from http://www.buildingenergysoftwaretools.com [Accessed 15 February 2016].

İçoğlu, O. and A. Mahdavi, 2007. VIOLAS: a vision-based sensing system for sentient building models. *Automation in Construction*, 16 (5), 685–712.

IEA, 1994. Annex 21: Calculation of energy and environmental performance of buildings. Subtask B: appropriate use of programs. Volume 1: executive summary. Watford: Building Research Establishment.

IEA, 2015. *World Energy Outlook 2015*. Paris: International Energy Agency.

IEA, 2016a. Annex 60: New generation computational tools for building and community energy systems based on the Modelica and Functional Mockup Interface standards homepage [online]. Berkeley, CA. Available from www.iea-annex60.org [Accessed 24 August 2016].

IEA, 2016b. Annex 66: Definition and simulation of occupant behavior in buildings homepage [online]. Berkeley, CA. Available from www.annex66.org [Accessed 16 June 2016].

IET, 2016. Intelligent buildings: understanding and managing security risks. Stevenage: The Institution of Engineering and Technology.

Igos, E., E. Benetto, I. Baudin, L. Tiruta-Barna, Y. Mery and D. Arbault, 2013. Cost-performance indicator for comparative environmental assessment of water treatment plants. *Science of the Total Environment*, 443, 367–374.

Ikeda, Y., 2016. Verification of system identification utilizing shaking table tests of a full-scale 4-story steel building. *Earthquake Engineering & Structural Dynamics*, 45 (4), 543–562.

Ilhan, B. and H. Yaman, 2016. Green building assessment tool (GBAT) for integrated BIM-based design decisions. *Automation in Construction*, 70, 26–37.

Illampas, R., D. Charmpis and I. Ioannou, 2014a. Laboratory testing and finite element simulation of the structural response of an adobe masonry building under horizontal loading. *Engineering Structures*, 80, 362–376.

Illampas, R., I. Ioannou and C. Charmpis, 2014b. Adobe bricks under compression: experimental investigation and derivation of stress-strain equation. *Construction and Building Material*, 53, 83–90.

Imam, S., D. Coley and I. Walker, 2017. The building performance gap: are modellers literate? *Building Services Engineering Research & Technology*, 38 (3), 351–375.

INCOSE, 2003. *Systems Engineering Handbook: a guide for system life cycle processes and activities*. San Diego, CA: International Council on Systems Engineering.

INCOSE, 2005. Technical Measurement. San Diego, CA: Practical Software & Systems Measurement (PSM) and International Council on Systems Engineering, Report INCOSE-TP-2003-020-01.

INCOSE, 2015. *Systems Engineering Handbook: a guide for system life cycle processes and activities*. San Diego, CA: Wiley.

INCOSE, 2016. The International Council on Systems Engineering homepage [online]. San Diego, CA. Available from www.incose.org [Accessed 28 April 2016].

Incropera, F.P., D.P. DeWitt, T.L. Bergman and A.S. Lavine, 2007. *Fundamentals of heat and mass transfer*. Hoboken, NJ: Wiley, 6th edition.

Inoue, M., K. Lindow, R. Stark, K. Tanaka, E. Nahm and H. Ishikawa, 2012. Decision-making support for sustainable product creation. *Advanced Engineering Informatics*, 26 (4), 782–792.

Ismael, M. and D. Oldham, 2005. A scale model investigation of sound reflection from building façades. *Applied Acoustics*, 66 (2), 123–147.

ISO 6241:1984. Performance standards in building – principles for their preparation and factors to be considered. Geneva: International Organisation for Standardization.

ISO 7162:1992. Performance standards in building – content and format of standards for evaluation performance. Geneva: International Organisation for Standardization.

ISO 15939:2007. Systems and software engineering – Measurement process. Geneva: International Organisation for Standardization.

ISO 15686-10:2010. Buildings and constructed assets – service life planning. Part 10: when to assess functional performance. Geneva: International Organisation for Standardization.

ISO 14031:2013. Environmental management – environmental performance evaluation – guidelines. Geneva: International Organisation for Standardization.

ISO 37120:2014. Sustainable development of communities – indicators for city services and quality of life. Geneva: International Organisation for Standardization.

ISO/IEC JTC 1, 2014. Smart Cities – Preliminary Report. Geneva: International Organisation for Standardization.

Iwaro, J., A. Mwasha, G. Williams and R. Zico, 2014. An integrated criteria weighting framework for the sustainable performance assessment and design of building envelope. *Renewable and Sustainable Energy Reviews*, 29, 417–434.

Jabłoński, I., 2015. Integrated living environment: measurements in modern energy efficient smart building with implemented the functionality of telemedicine. *Measurement*, 101, 211–235.

Jacob, D., S. Dietz, S. Komhard, C. Neumann and S. Herkel, 2010. Black-box models for fault detection and performance monitoring of buildings. *Journal of Building Performance Simulation*, 3 (1), 53–62.

Jain, S., K. Triantis and S. Liu, 2011. Manufacturing performance measurement and target setting: a data envelopment analysis approach. *European Journal of Operational Research*, 214 (3), 616–626.

Jang, W., W. Healy and M. Skibniewski, 2008. Wireless sensor networks as part of a web-based building environmental monitoring system. *Automation in Construction*, 17 (6), 729–736.

Jang, I., J. Park and P. Seong, 2012. An empirical study on the relationship between functional performance measure and task performance measure in NPP MCR. *Annals of Nuclear Energy*, 42, 96–103.

Janssen, R., 2010. EU energy performance of buildings directive gets tough (briefing). *Proceedings of the Institution of Civil Engineers*, 163 (2), 53.

Janssen, H., B. Blocken and J. Carmeliet, 2007. Conservative modelling of the moisture and heat transfer in building components under atmospheric excitation. *International Journal of Heat and Mass Transfer*, 50 (5/6), 1128–1140.

Jansson, G., E. Viklund and H. Lidelöw, 2016. Design management using knowledge innovation and visual planning. *Automation in Construction*, 72 (3), 330–337.

Januševičius, T., J. Mažuolis and D. Butkus, 2016. Sound reduction in samples of environmentally friendly building materials and their compositions. *Applied Acoustics*, 113, 132–136.

Jasuja, M., ed., 2005. PeBBU Final Report – Performance Based Building thematic network 2001–2005. Rotterdam: CIB.

Jato-Espino, D., E. Castillo-Lopez, J. Rodriguez-Hernandez and J. Canteras-Jordana, 2014. A review of application of multi-criteria decision making methods in construction. *Automation in Construction*, 45, 151–162.

Jaunzens, D., R. Grigg, R. Cohen, M. Watson and E. Picton, 2003. *Building performance feedback: getting started*. BRE Digest 478. Garston: Building Research Establishment.

JCGM, 2012. *International vocabulary of metrology – basic and general concepts and associated terms (VIM)*. Sèvres: Joint Committee for Guides in Metrology, 3rd edition.

Jelle, B., 2011. Traditional, state-of-the-art and future thermal building insulation materials and solutions – properties, requirements and possibilities. *Energy and Buildings*, 43 (10), 2549–2563.

Jenkins, D., N. Stanton, P. Salmon, G. Walker and L. Rafferty, 2010. Using the decision-ladder to add a formative element to naturalistic decision-making research. *International Journal of Human-Computer Interaction*, 26 (2/3), 132–146.

Jensen, S., 1995. Validation of building energy simulation programs: a methodology. *Energy and Buildings*, 22 (2), 133–144.

Jensen, K., J. Toftum and P. Friis-Hansen, 2009. A Bayesian Network approach to the evaluation of building design and its consequences for employee performance and operational costs. *Building and Environment*, 44 (3), 456–462.

Jensen, J., P. Patel and J. Raver, 2014. Is it better to be average? High and low performance as predictors of employee victimization. *Journal of Applied Psychology*, 99 (2), 296–309.

Jeon, G., J. Kim, W. Hong and G. Augenbroe, 2011. Evacuation performance of individuals in different visibility conditions. *Building and Environment*, 46 (5), 1094–1103.

Jeong, W., J. Kim, M. Clayton, J. Haberl and W. Yan, 2014. Translating building information modeling to building energy modeling using model view definition. *The Scientific World Journal*, 2014, 638276.

Jeong, B., J. Jeong and J. Park, 2016. Occupant behavior regarding the manual control of windows in residential buildings. *Energy and Buildings*, 127, 206–216.

Jewell, C. and R. Flanagan, 2012. Measuring construction professional services exports: a case for change. *Building Research & Information*, 40 (3), 337–347.

Ji, Y. and M. Cook, 2007. Numerical studies of displacement natural ventilation in multi-storey buildings connected to an atrium. *Building Services Engineering Research and Technology*, 28 (3), 207–222.

Ji, Y. and P. Xu, 2015. A bottom-up and procedural calibration method for building energy simulation models based on hourly electricity submetering data. *Energy*, 93, 2337–2350.

Ji, Y., M. Cook and V. Hanby, 2007. CFD modelling of natural displacement ventilation in an enclosure connected to an atrium. *Building and Environment*, 42 (3), 1158–1172.

Ji, L., H. Tan, S. Kato, Z. Bu and T. Takahashi, 2011. Wind tunnel investigation on influence of fluctuating wind direction on cross natural ventilation. *Building and Environment*, 46 (12), 2490–2499.

Jiang, Z., J. Xia and Y. Jiang, 2009. An information sharing building automation system. *Intelligent Buildings International*, 1 (3), 195–208.

Jin, Q., M. Overend and P. Thompson, 2012. Towards productivity indicators for performance-based façade design in commercial buildings. *Building and Environment*, 57, 271–281.

Joas, F., M. Pahle, C. Flachsland and A. Joas, 2016. Which goals are driving the Energiewende? Making sense of the German energy transformation. *Energy Policy*, 95, 42–51.

Joglekar, N., A. Yassine, S. Eppinger and D. Whitney, 2001. Performance of coupled product development activities with a deadline. *Management Science*, 47 (12), 1605–1620.

Johansson, P., 2014. Building retrofit using Vacuum Insulation Panels – hygrothermal performance and durability. PhD thesis. Gothenburg: Chalmers University of Technology.

Johansson, M., M. Roupé and P. Bosch-Sijtsema, 2015. Real-time visualization of building information models (BIM). *Automation in Construction*, 54, 69–82.

Johns, C., 1903. The code of laws promulgated by Hammurabi, King of Babylon. Edinburgh: T. &T. Clark, Project Gutenberg eBook 2005 edition.

Johnson, J. and M. Raab, 2003. Take the first: option-generation and resulting choices. *Organizational Behavior and Human Decision Processes*, 91 (2), 215–229.

Jones, L., 2016. A critical appraisal of a theoretical approach to decision making in the construction industry. BSc dissertation. Plymouth: Plymouth University.

Joo, J., N. Kim, R. Wysk, L. Rothrock, Y. Son, Y. Oh and S. Lee, 2013. Agent-based simulation of affordance-based human behaviors in emergency evacuation. *Simulation Modelling Practice and Theory*, 32, 99–115.

Jordaan, I., 2005. *Decisions under uncertainty: probabilistic analysis for engineering decisions*. Cambridge: Cambridge University Press.

Joseph, A. and D. Hamilton, 2008. The Pebble Projects: coordinated evidence-based case studies. *Building Research & Information*, 36 (2), 129–145.

Josephson, P. and Y. Hammarlund, 1999. The causes and costs of defects in construction: a study of seven building projects. *Automation in Construction*, 8 (6), 681–687.

Jrade, A. and F. Jalaei, 2013. Integrating building information modelling with sustainability to design building projects at the conceptual stage. *Building Simulation*, 6 (4), 429–444.

Juan, Y., P. Gao and J. Wang, 2010. A hybrid decision support system for sustainable office building renovation and energy performance improvement. *Energy and Buildings*, 42 (3), 290–297.

Judge, T. and J. Kammeyer-Mueller, 2012. On the value of aiming high: the causes and consequences of ambition. *Journal of Applied Psychology*, 97 (4), 758–775.

Judkoff, R., 2008. Testing and validation of building energy simulation tools – IEA final task management report. Golden, CO: National Renewable Energy Laboratory.

Judkoff, R. and J. Neymark, 1995. Building energy simulation test (BESTEST) and diagnostic method. Golden, CO: National Renewable Energy Laboratory.

Jung, J., C. Schneider and J. Vlacich, 2010. Enhancing the motivational affordance of information systems: the effects of real-time performance feedback and goal setting in group collaboration environments. *Management Science*, 56 (4), 724–742.

Junghans, L., 2013. Sequential equi-marginal optimization method for ranking strategies for thermal building renovation. *Energy and Buildings*, 65, 10–18.

Kääriäinen, H., M. Rudolph, D. Schaurich, K. Tulla and H. Wiggenhauser, 2001. Moisture measurement in building materials with microwaves. *Nondestructive Testing & Evaluation*, 34 (6), 389–394.

Kabak, M., E. Köse, O. Kirilmaz and S. Burmaoğlu, 2014. A fuzzy multi-criteria decision making approach to assess building energy performance. *Energy and Buildings*, 72, 382–389.

Kalay, Y., 1999. Performance-based design. *Automation in Construction*, 8 (4), 395–409.

Kalz, D., J. Pfafferott, S. Herkel and A. Wagner, 2009. Building signatures correlating thermal comfort and low-energy cooling: in-use performance. *Building Research & Information*, 37 (4), 413–432.

Kamara, J., C. Anumba and N. Evbuomwan, 2001. Assessing the suitability of current briefing practices in construction within a concurrent engineering framework. *International Journal of Project Management*, 19 (6), 337–351.

Kamara, J., G. Augenbroe, C. Anumba and P. Carrillo, 2002. Knowledge management in the architecture, engineering and construction industry. *Construction Innovation*, 2 (1), 53–67.

Kämpf, J., M. Wetter and D. Robinson, 2010. A comparison of global optimization algorithms with standard benchmark functions and real-world applications using EnergyPlus. *Journal of Building Performance Simulation*, 3 (2), 103–120.

Kandil, A. and J. Love, 2014. Signature analysis calibration of a school energy model using hourly data. *Journal of Building Performance Simulation*, 7 (5), 326–345.

Kanfer, R., M. Wolf, T. Kantrowitz and P. Ackerman, 2010. Ability and trait complex predictors of academic and job performance: a person-situation approach. *Applied Psychology*, 59 (1), 40–69.

Kang, C., 2011. Performance measure of residual vibration control. *Journal of Dynamic Systems, Measurement, and Control*, 133 (4), 044501/1–044501/6.

Kang, S., J. Park, K. Oh, J. Noh and H. Park, 2014. Schedule-based real time energy flow control strategy for building energy management system. *Energy and Buildings*, 75, 239–248.

Kansara, T. and I. Ridley (2012). Post occupancy evaluation of buildings in a zero carbon city. *Sustainable Cities and Society*, 5, 23–25.

Kaplan, R. and D. Norton, 1992. The balanced scorecard – measures that drive performance. *Harvard Business Review*, 70 (1), 71–79.

Karlsen, L., P. Heiselberg, I. Bryn and H. Johra, 2015. Verification of simple illuminance based measures for indication of discomfort glare from windows. *Building and Environment*, 92, 615–626.

Kassimali, A., 2015. *Structural Analysis*. Stamford, CT: Cengage Learning.

Katipamula, S. and M. Brambley, 2005a. Methods for fault detection, diagnostics and prognostics for building systems: a review, part I. *HVAC&R Research*, 11 (1), 3–25.

Katipamula, S. and M. Brambley, 2005b. Methods for fault detection, diagnostics and prognostics for building systems: a review, part II. *HVAC&R Research*, 11 (2), 169–187.

Kavousian, A. and R. Rajagopal, 2014. Data-driven benchmarking of building energy efficiency utilizing statistical frontier models. *Journal of Computing in Civil Engineering*, 28 (1), 79–88.

Kazaz, A. and M. Birgonul, 2005. The evidence of poor quality in high rise and medium rise housing units: a case study of mass housing projects in Turkey. *Building and Environment*, 40 (11), 1548–1556.

Ke, M., C. Yeh and J. Jian, 2013. Analysis of building energy consumption parameters and energy savings measurement and verification by applying eQuest software. *Energy and Buildings*, 61, 100–107.

Keeler, M. and P. Vaidya, 2016. *Fundamentals of integrated design for sustainable building*. Hoboken, NJ: Wiley, 2nd edition.

Keeney, R. and R. Gregory, 2005. Selecting attributes to measure the achievement of objectives. *Operations Research*, 53 (1), 1–11.

Keeney, R. and H. Raiffa, 1993. *Decisions with multiple objectives – preferences and value tradeoffs*. Cambridge: Cambridge University Press.

Keller, E., 2015. Being an expert witness in geomorphology. *Geomorphology*, 231, 383–389.

Kelliher, D. and K. Sutton-Swaby, 2012. Stochastic representation of blast load damage in a reinforced concrete building. *Structural Safety*, 34 (1), 407–417.

Kelly, J., K. Hunter, G. Shen and A. Yu, 2005. Briefing from a facilities management perspective. *Facilities*, 23 (7/8), 356–367.

Kenisarin, M. and K. Mahkamov, 2016. Passive thermal control in residential buildings using phase change materials. *Renewable and Sustainable Energy Reviews*, 55, 371–398.

Kent, D. and B. Becerik-Gerber, 2010. Understanding construction industry experience and attitudes towards integrated project delivery. *Journal of Construction Engineering and Management*, 136 (8), 815–825.

Kersten, W., M. Crul, D. Geelen, S. Meijer and V. Franken, 2015. Engaging beneficiaries of sustainable renovation – exploration of design-led participatory approaches. *Journal of Cleaner Production*, 106, 690–699.

Kesic, T., 2015. Vital signs: towards meaningful building performance indicators. Green paper. Toronto, ON: University of Toronto.

Kesten, D., S. Fiedler, F. Thumm, A. Löffler and U. Eicker, 2010. Evaluation of daylight performance in scale models and a full-scale mock-up office. *International Journal of Low-Carbon Technologies*, 5 (3), 158–165.

de Keyser, R. and C. Ionescu, 2010. Modelling and simulation of a lighting control system. *Simulation Modelling Practice and Theory*, 18 (2), 165–176.

Khalifa, H., M. Janos and J. Dannenhoffer III, 2009. Experimental investigation of reduced-mixing personal ventilation jets. *Building and Environment*, 44 (8), 1551–1558.

Khatib, T., A. Mohamed, K. Sopia and M. Mahmoud, 2011. Optimal sizing of building integrated hybrid PV/diesel generator system for zero load rejection of Malaysia. *Energy and Buildings*, 43 (12), 3430–3435.

Khazaii, J., 2016. *Advanced decision making for HVAC engineering: creating energy efficient smart buildings*. New York: Springer.

Khodyakov, D., S. Hempel, L. Rubenstein, P. Shekelle, R. Foy, S. Salem-Schatz, S. O'Nell, M. Danz and S. Dalal, 2011. Conducting online expert panels: a feasibility and experimental replicability study. *BMC Medical Research Methodology*, 11, 174.

Khosrowshahi, F. and A. Alani, 2011. Visualisation of impact of time on the internal lighting of a building. *Automation in Construction*, 20 (2), 145–154.

Kibert, C., 2016. *Sustainable construction: green building design and delivery*. Hoboken, NJ: Wiley, 4th edition.

Kibert, C. and M. Fard, 2012. Differentiating among low-energy, low-carbon and net-zero-energy building strategies for policy formulation. *Building Research & Information*, 40 (5), 625–637.

Kim, J., 2016. The impact of occupant modeling on energy outcomes of building energy simulation. PhD thesis. Atlanta, GA: Georgia Institute of Technology.

Kim, K. and J. Haberl, 2015. Development of methodology for calibrated simulation in single-family residential buildings using three-parameter change-point regression model. *Energy and Buildings*, 99, 140–152.

Kim, D. and C. Park, 2011. Difficulties and limitations in performance simulation of a double skin façade with EnergyPlus. *Energy and Buildings*, 43 (12), 3635–3645.

Kim, C., C. Lee, M. Lehto and M. Yun, 2011. Affective evaluation of user impressions using virtual product prototyping. *Human Factors and Ergonomics in Manufacturing & Service Industries*, 21 (1), 1–13.

Kim, B., B. Chitturi and N. Grishin, 2012. Self consistency grouping: a stringent clustering method. *BMC Bioinformatics*, 13, S3.

Kim, I., R. Galiza and L. Ferreira, 2013. Modeling pedestrian queuing using micro-simulation. *Transportation Research Part A*, 49, 232–240.

Kim, E., G. Plessis, J. Hubert and J. Roux, 2014. Urban energy simulation: simplification and reduction of building envelope models. *Energy and Buildings*, 84, 193–202.

Kinney, P., H. Roman, K. Walker, H. Richmond, L. Conner and B. Hubbell, 2010. On the use of expert judgment to characterize uncertainties in the health benefits of regulatory controls of particulate matter. *Environmental Science & Policy*, 13 (5), 434–443.

Kirschner, K., J. Braspenning, J. Jacobs and R. Grol, 2012. Design choices made by target users for a pay-for-performance program in primary care: an action research approach. *BMC Family Practice*, 12, 25.

Kitchley, J. and A. Srivathsan, 2014. Generative models and the design process: a design tool for conceptual settlement planning. *Applied Soft Computing*, 14, 634–652.

Kittler, R., 2007. Daylight prediction and assessment: theory and design practice. *Architectural Science Review*, 50 (2), 94–99.

Klein, G., J. Orasanu, R. Calderwood and C. Zsambok, eds., 1993. *Decision making in action: models and methods*. Norwood, NJ: Ablex Publishing Corporation.

Kleindienst, S. and M. Andersen, 2012. Comprehensive annual daylight design through a goal-based approach. *Building Research & Information*, 40 (2), 154–173.

Kleingeld, A., H. van Mierlo and L. Arends, 2011. The effect of goal setting on group performance: a meta-analysis. *Journal of Applied Psychology*, 96 (6), 1289–1304.

Klotz, L., D. Mack, B. Klapthor, C. Tunstall and J. Harrison, 2010. Unintended anchors: building rating systems and energy performance goals for U.S. buildings. *Energy Policy*, 38 (7), 3557–3566.

Kneidl, A., A. Borrmann and D. Hartmann, 2012. Generation and use of sparse navigation graphs for microscopic pedestrian simulation models. *Advanced Engineering Informatics*, 26 (4), 669–680.

Kneidl, A., D. Hartmann and A. Borrmann, 2013. A hybrid multi-scale approach for simulation of pedestrian dynamics. *Transport Research Part C*, 37, 223–237.

Koch, C. and H. Buhl, 2013. 'Integrated Design Process' a concept for green energy engineering. *Engineering*, 5 (3), 292–298.

Koch, A., S. Girard and K. McKoen, 2012. Towards a neighbourhood scale for low- or zero-carbon building projects. *Building Research & Information*, 40 (4), 527–537.

Kokogiannakis, G., P. Strachan and J. Clarke, 2008. Comparison of the simplified methods of the ISO 13790 standard and detailed modelling programs in a regulatory context. *Journal of Building Performance Analysis*, 1 (4), 209–219.

Kolarevic, B., 2005a. Computing the performative. In: Kolarevic, B. and A. Malkawi, eds., *Performative architecture: beyond instrumentality*. New York: Spon Press.

Kolarevic, B., 2005b. Towards the performative in architecture. In: Kolarevic, B. and A. Malkawi, eds., *Performative architecture: beyond instrumentality*. New York: Spon Press.

Kolarevic, B. and A. Malkawi, eds., 2005. *Performative architecture: beyond instrumentality*. New York: Spon Press.

Kong, F. and H. Wang, 2011. Heat and mass coupled transfer combined with freezing process in building materials: modeling and experimental verification. *Energy and Buildings*, 43 (10), 2850–2859.

Konis, K., 2013. Evaluating daylighting effectiveness and occupant visual comfort in a side-lit open-plan office building in San Francisco, California. *Building and Environment*, 59, 662–677.

Konis, K., 2017. A novel circadian daylight metric for building design and evaluation. *Building and Environment*, 113, 22–38.

Kontokosta, C. and C. Tull, 2017. A data-driven predictive model of city-scale energy use in buildings. *Applied Energy*, 197, 303–317.

Kontoleon, K. and E. Eumorfopoulou, 2010. The effect of the orientation and proportion of a plant-covered wall layer on the thermal performance of a building zone. *Building and Environment*, 45 (5), 1287–1303.

Koo, J., Y. Kim and B. Kim, 2012. Estimating the impact of residents with disabilities on the evacuation in a high-rise building: a simulation study. *Simulation Modelling Practice and Theory*, 24, 71–83.

Korkmaz, S., D. Riley and M. Horman, 2010. Piloting evaluation metrics for sustainable high-performance building project delivery. *Journal of Construction Engineering and Management*, 136 (8), 877–885.

Kortuem, G., F. Kawsar, D. Fitton and V. Sundramoorthy, 2010. Smart objects as building blocks for the Internet of Things. *IEEE Internet Computing*, 14 (1), 44–51.

Kota, S., J. Haberl, M. Clayton and W. Yan, 2014. Building information modeling (BIM)-based daylighting simulation and analysis. *Energy and Buildings*, 81, 391–403.

Kountouriotis, V., S. Thomopoulos and Y. Papelis, 2014. An agent-based crowd behaviour model for real time crowd behaviour simulation. *Pattern Recognition Letters*, 44, 30–38.

Kovacic, I. and V. Zoller, 2015. Building life cycle optimization for early design phases. *Energy*, 92, 409–419.

Koziol, K., J. Vilatela, A. Moisala, M. Motta, P. Cunniff, M. Sennett and A. Windle, 2007. High-performance carbon nanotube fiber. *Science*, 318 (5858), 1892–1895.

Kozłowski, M., M. Dorn and E. Serrano, 2015. Experimental testing of load-bearing timber-glass composite shear walls and beams. *Wood Material Science & Engineering*, 10 (3), 276–286.

Kraft, E. and K. Molenaar, 2014. Fundamental project quality assurance organizations in highway design and construction. *Journal of Management in Engineering*, 30 (4), 04014015.

Krainer, A., 2008. Passivhaus contra bioclimatic design. *Bauphysik*, 30 (6), 393–404.

Kreibich, H. and A. Thieken, 2009. Coping with floods in the city of Dresden, Germany. *Natural Hazards*, 51 (3), 423–436.

Krieger, F., 1957. A casebook on Soviet Astronautics, Part II. Santa Monica, CA: The RAND Corporation, Research Memorandum RM-1922, 21 June 1957.

Krish, S., 2011. A practical generative design method. *Computer-Aided Design*, 43 (1), 88–100.

Kroeker, K., 2011. A new benchmark for artificial intelligence. *Communications of the ACM*, 54 (8), 13–15.

Krogerus, M. and R. Tschäppeler, 2011. *The decision book: fifty models for strategic thinking*. London: Profile Books.

Kuliga, S., T. Thrash, R. Dalton and C. Hölscher, 2015. Virtual reality as an empirical research tool – exploring user experience in a real building and a corresponding virtual model. *Computers, Environment and Urban Systems*, 54, 363–375.

Kuligowski, E. and R. Peacock, 2005. A review of building evacuation models. Washington, DC: National Institute of Standards and Technology, Technical Note 1471.

Kumar, N., V. Dayal and P. Sarkar, 2012. Failure of wood-framed low-rise buildings under tornado wind loads. *Engineering Structures*, 39, 79–88.

Künzel, H. and D. Zirkelbach, 2013. Advances in hygrothermal building component simulation: modelling moisture sources likely to occur due to rainwater leakage. *Journal of Building Performance Simulation*, 6 (5), 346–353.

Künzel, H., A. Holm, D. Zirkelbach and A. Karagiozis, 2005. Simulation of indoor temperature and humidity conditions including hygrothermal interactions with the building envelope. *Solar Energy*, 78 (4), 554–561.

Kus, H., E. Özkan, Ö. Göcer and E. Edis, 2013. Hot box measurements of pumice aggregate concrete hollow block walls. *Construction and Building Materials*, 38, 837–845.

Kwak, Y., J. Huh and C. Jang, 2015. Development of a model predictive control framework through real-time building energy management system data. *Applied Energy*, 155, 1–13.

Kwasniewski, L., 2010. Nonlinear dynamic simulations of progressive collapse for a multistory building. *Engineering Structures*, 32 (5), 1223–1235.

Kwon, O., E. Lee and H. Bahn, 2014. Sensor-aware elevator scheduling for smart building environments. *Building and Environment*, 72, 332–342.

Kylili, A., P. Fokaides, P. Christou and S. Kalogirou, 2014. Infrared thermography (IRT) applications for building diagnostics: a review. *Applied Energy*, 134, 531–549.

Labate, 2016. Metrics, measures and Indicators [online]. Available from https://thecarebot. github.io/metrics-measures-and-indicators/ [Accessed 20 June 2017].

Labonnote, N., A. Rønnquist, B. Manum and P. Rüther, 2016. Additive construction: state-of-the-art, challenges and opportunities. *Automation in Construction*, 72, 347–366.

Lai, J., 2013. Gap theory based analysis of user expectation and satisfaction: the case of a hostel building. *Building and Environment*, 69, 183–193.

Lai, J. and F. Yik, 2009. Perception of importance and performance of the indoor environmental quality of high-rise residential buildings. *Building and Environment*, 44 (2), 352–360.

Lai, M., Y. Li and Y. Liu, 2010. Determining the optimal scale width for a rating scale using an integrated discrimination function. *Measurement*, 43 (10), 1458–1471.

Laing, A., D. Craigh and A. White, 2011. Vision statement: High-performance office space. *Harvard Business Review*, 1271, 32–33.

Lam, D., A. Chan and D. Chan, 2010. Benchmarking success of building maintenance projects. *Facilities*, 28 (5/6), 290–305.

Lan, L. and Z. Lian, 2016. Ten questions concerning thermal environment and sleep quality. *Building and Environment*, 99, 252–259.

Lan, L., P. Wargocki, D. Wyon and Z. Lian, 2011. Effects of thermal discomfort in an office on perceived air quality, SBS symptoms, physiological responses, and human performance. *Indoor Air*, 21 (5), 376–390.

Landlin, A. and C. Nilsson, 2001. Do quality systems really make a difference? *Building Research & Information*, 29 (1), 12–20.

Langmans, J. and S. Roels, 2015. Experimental analysis of cavity ventilation behind rainscreen cladding systems: a comparison of four measuring techniques. *Building and Environment*, 87, 177–192.

Langmans, J., A. Nicolai, R. Klein and S. Roels, 2012. A quasi-steady state implementation of air convection in a transient heat and moisture building component model. *Building and Environment*, 58, 208–218.

Langner, M., G. Henze, C. Corbin and M. Brandemuehl, 2012. An investigation of design parameters that affect commercial high-rise office building energy consumption and demand. *Journal of Building Performance Simulation*, 5 (5), 313–328.

Langston, C., 2013. The impact of criterion weights in facilities management decision making: an Australian case study. *Facilities*, 31 (7/8), 270–289.

Laofor, C. and V. Peansupap, 2012. Defect detection and quantification system to support subjective visual quality inspection via digital image processing: a tiling work case study. *Automation in Construction*, 24, 160–174.

Laporti, V., M. Borges and V. Braganholo, 2009. Athena: a collaborative approach to requirements elicitation. *Computers in Industry*, 60 (6), 367–380.

Lappegard Hauge, Å., J. Thomsen and T. Berker, 2011. User evaluations of energy efficient buildings: literature review and further research. *Advances in Building Energy Research*, 5 (1), 109–127.

Laprise, M., S. Lufkin and E. Rey, 2015. An indicator system for the assessment of sustainability integrated into the project dynamics of regeneration of disused urban areas. *Building and Environment*, 86, 29–38.

Larsen, K., F. Lattke, S. Ott and S. Winter, 2011. Surveying and digital workflow in energy performance retrofit projects using prefabricated elements. *Automation in Construction*, 20 (8), 999–1011.

Larsen, P., C. Goldman and A. Satchwell, 2012. Evolution of the U.S. energy service company industry: market size and project performance from 1990–2008. *Energy Policy*, 2012, 802–820.

Larsson, K. and C. Simmons, 2011. Measurements of structure-borne sound from building service equipment by a substitution method – round robin comparisons. *Noise Control Engineering Journal*, 59 (1), 75–86.

Latif, E., S. Tucker, M. Ciupala, D. Wijeyesekera, D. Newport and M. Pruteanu, 2016. Quasi steady state and dynamic hygrothermal performance of fibrous hemp and stone wool insulations: two innovative laboratory based investigations. *Building and Environment*, 95, 391–404.

Lau, A. and S. Tang, 2009. A survey on the advancement of QA (quality assurance) to TQM (total quality management) for construction contractors in Hong Kong. *International Journal of Quality & Reliability Management*, 26 (5), 410–425.

Law, A., E. Choi and R. Britter, 2004. Re-entrainment around a low-rise industrial building: 2D versus 3D wind tunnel study. *Atmospheric Environment*, 38 (23), 3817–3825.

Lawrence, A., 1970. *Architectural Acoustics*. Barking: Elsevier.

Lawrence, T., M. Boudreau, L. Helsen, G. Henze, J. Mohammadpour, D. Noonan, D. Patteeuw, S. Pless and R. Watson, 2016. Ten questions concerning integrating smart buildings into the smart grid. *Building and Environment*, 108, 273–283.

Lawson, B., 2005. *How Designers Think: the design process demystified*. London: Architectural Press and Routledge, 4th edition.

Leaman, A. and B. Bordass, 2001. Assessing building performance in use 4: the Probe occupant surveys and their implications. *Building Research & Information*, 29 (2), 129–143.

Leaman, A., F. Stevenson and B. Bordass, 2010. Building evaluation: practice and principles. *Building Research & Information*, 38 (5), 564–577.

Leatherbarrow, D., 2005. Architecture's unscripted performance. In: Kolarevic, B. and A. Malkawi, eds., *Performative architecture: beyond instrumentality*. New York: Spon Press.

Lee, W., 2010. Benchmarking the energy performance for cooling purposes in buildings using a novel index – total performance of energy for cooling purposes. *Energy*, 35 (1), 50–54.

Lee, T., 2015. Performance metrics as drivers of quality – getting to second gear. *Circulation*, 131 (11), 967–968.

Lee, S. and R. Davidson, 2010. Physics-based simulation model of post-earthquake fire spread. *Journal of Earthquake Engineering*, 14 (5), 670–687.

Lee, W. and K. Lee, 2009. Benchmarking the performance of building energy management using data envelopment analysis. *Applied Thermal Engineering*, 29 (16), 3269–3273.

Lee, N. and E. Rojas, 2014. Activity Gazer: a multi-dimensional visual representation of project performance. *Automation in Construction*, 44, 25–32.

Lee, E., S. Selkowitz, V. Bazjanac, I. Vorapat and C. Kohler, 2002a. *High-performance commercial building façades*. Berkeley, CA: Lawrence Berkeley National Laboratory.

Lee, W., C. Chau, F. Yik, J. Burnett and M. Tse, 2002b. On the study of the credit-weighting scale in a building environmental assessment scheme. *Building and Environment*, 37 (12), 1385–1396.

Lee, C., J. Lee and Y. Kim, 2008. Demand forecasting for new technology with a short history in a competitive environment: the case of the home networking market in South Korea. *Technological Forecasting and Social Change*, 75 (1), 91–106.

Lee, M., K. Mui, L. Wong, W. Chan, E. Lee and C. Cheung, 2012. Student learning performance and indoor environmental quality (IEQ) in air-conditioned university teaching rooms. *Building and Environment*, 49, 238–244.

Lee, E., X. Pang, S. Hoffmann, H. Goudey and A. Thanachareonkit, 2013a. An empirical study of a full-scale polymer thermochromic window and its implications on material science development objectives. *Solar Energy Materials & Solar Cells*, 116, 14–26.

Lee, J., H. Lee, M. Park and S. Ji, 2013b. A decision support system for super tall building development. *The Structural Design of Tall and Special Buildings*, 22 (16), 1230–1247.

Lee, S., T. Kamada, S. Uchida and D. Linzell, 2014. Imaging defects in concrete structures using accumulated SIBIE. *Construction and Building Materials*, 67, 180–185.

Lee, H., D. Tommelein and G. Ballard, 2015a. Target-setting practice for loans for commercial energy-retrofit projects. *Journal of Management in Engineering*, 31 (3), 04014046.

Lee, P., P. Lam and W. Lee, 2015b. Risks in energy performance contracting (EPC) projects. *Energy and Buildings*, 92, 116–127.

Lehmann, B., H. Güttinger, V. Dorer, S. van Vlesen, A. Thieman and T. Frank, 2010. Eawag Forum Chriesbach – simulation and measurement of energy performance and comfort in a sustainable office building. *Energy and Buildings*, 42, 1958–1967.

Leidinger, M., T. Sauerwald, W. Reimringer, G. Ventura and A. Schütze, 2014. Selective detection of hazardous VOCs for indoor air applications using a virtual gas sensor array. *Journal of Sensors and Sensor Systems*, 3 (2), 253–263.

Lengsfeld, K. and A. Holm, 2007. Entwicklung und Validierung einer hygrothermischen Raumklima-Simulationssoftware WUFI®-Plus. *Bauphysik*, 29 (3), 178–186.

Lenoir, A., G. Baird and F. Garde, 2012. Post-occupancy evaluation and experimental feedback of a net zero-energy building in a tropical climate. *Architectural Science Review*, 55 (3), 156–168.

Leung, M., J. Yu, C. Dongyu and T. Yuan, 2014. A case study exploring FM components for elderly in care and attention homes using post occupancy evaluation. *Facilities*, 32 (11/12), 685–708.

Leyten, J. and S. Kurvers, 2013. Limitations of climate chamber studies into thermal comfort and workers' performance – letter to the editor. *Indoor Air*, 23 (5), 439–440.

Li, D. and B. Han, 2015. Behavioral effect on pedestrian evacuation simulation using cellular automata. *Safety Science*, 80, 41–55.

Li, X. and V. Strezov, 2014. Modelling piezoelectric harvesting potential in an educational building. *Energy Conversion and Management*, 85, 435–442.

Li, Y. and J. Van de Lindt, 2012. Loss-based formulation for multiple hazards with application to residential buildings. *Engineering Structures*, 38, 123–133.

Li, D., E. Tsang, K. Cheung and C. Tam, 2010. An analysis of light-pipe system via full-scale measurements. *Applied Energy*, 87 (3), 799–805.

Li, L., H. Zhang, Q. Xie, L. Chen and C. Xu, 2012a. Experimental study on fire hazard of typical curtain materials in ISO 9705 fire test room. *Fire and Materials*, 36 (2), 85–96.

Li, N., G. Calis and B. Becerik-Gerber, 2012b. Measuring and monitoring occupancy with an RFID based system for demand-driven HVAC operations. *Automation in Construction*, 24, 89–99.

Li, D., T. Chau and K. Wan, 2014a. A review of the CIE general sky classification approaches. *Renewable and Sustainable Energy Reviews*, 31, 563–574.

Li, P., Y. Li, J. Seem, H. Qiao, X. Li and J. Winkler, 2014b. Recent advances in dynamic modeling of HVAC equipment. Part 2: Modelica-based modeling. *HVAC&R Research*, 20 (1), 150–161.

Li, Z., Y. Han and P. Xu, 2014c. Methods for benchmarking building energy consumption against its past or intended performance: an overview. *Applied Energy*, 124, 325–334.

Li, B., Y. Dong and D. Zhang, 2015. Fire behaviour of continuous reinforced concrete slabs in full-scale multi-storey steel-framed building. *Fire Safety Journal*, 71, 226–237.

Li, Y., X. Zhang, G. Ding and Z. Feng, 2016. Developing a quantitative construction waste estimation model for building construction projects. *Resources, Conservation and Recycling*, 106, 9–20.

Liang, J. and R. Du, 2007. Model-based fault detection and diagnosis of HVAC systems using support vector method. *International Journal of Refrigeration*, 30 (6), 1104–1114.

Liang, Y., Y. Hu, J. Chen, Y. Shen and J. Du, 2014. A transient thermal model for full-size vehicle climate chamber. *Energy and Buildings*, 85, 256–264.

Lilis, G., G. Conus, N. Asadi and M. Kayal, 2017. Towards the next generation of intelligent building: an assessment study of current automation and future Internet of Things based systems with a proposal for transitional design. *Sustainable Cities and Society*, 28, 473–481.

Lim, S. and J. Zhu, 2013. Incorporating performance measures with target levels in data envelopment analysis. *European Journal of Operational Research*, 230 (3), 634–642.

Lim, S., S. Park, H. Chung, M. Kim, J. Baik and S. Shin, 2015. Dynamic modeling of building heat network system using Simulink. *Applied Thermal Engineering*, 84, 375–389.

Lim, N., S. Brandt and S. Seipel, 2016. Visualisation and evaluation of flood uncertainties based on ensemble modelling. *International Journal of Geographical Information Science*, 30 (2), 240–262.

Limbachiya, M., M. Medday and S. Fotiadou, 2012. Performance of granulated foam glass concrete. *Construction and Building Materials*, 28 (1), 759–768.

Lin, E., 2006. Performance practice and theatrical privilege: rethinking Weinmann's concepts of locus and platea. *New Theatre Quarterly*, 22 (3), 283–298.

Lin, J., 2012. Exceedance probability of extensive damage limit for general buildings in Taiwan. *Disaster Prevention and Management: an International Journal*, 21 (3), 386–397.

Lin, S. and D. Gerber, 2014. Designing-in performance: a framework for evolutionary energy performance feedback in early stage design. *Automation in Construction*, 38, 59–73.

Lin, C., E. Chu, L. Ku and J. Liu, 2014. Active disaster response system for a smart building. *Sensors*, 14, 17451–17470.

Lindberg, K., G. Doorman, D. Fischer, M. Korpås, A. Ånestad and I. Sartori, 2016. Methodology for optimal energy system design of Zero Energy Building used mixed-integer linear programming. *Energy and Buildings*, 127, 194–205.

Lindsey, S., 2005. A matter of fact. *The Architects' Journal*, 221 (25), 44.

Ling, F., 2004. Key determinants of performance of design-bid-build projects in Singapore. *Building Research & Information*, 32 (2), 128–139.

Lipshitz, R., G. Klein and J. Carroll, 2006. Naturalistic decision making and organizational decision making: exploring the intersections. *Organization Studies*, 27 (7), 917–923.

Lirola, J., E. Castañeda, B. Lauret and M. Khayet, 2017. A review on experimental research using scale models for buildings: application and methodologies. *Energy and Buildings*, 142, 72–110.

Liu, Z. and W. Hansen, 2016. Freeze-thaw durability of high-strength concrete under deicer salt exposure. *Construction and Building Materials*, 102, 478–485.

Liu, J. and J. Niu, 2016. CFD simulation of the wind environment around an isolated high-rise building: an evaluation of SRANS, LES and DES models. *Building and Environment*, 96, 91–106.

Liu, S. and A. Novoselac, 2015. Air Diffusion Performance Index (ADPI) of diffusers for heating mode. *Building and Environment*, 87, 215–223.

Liu, M., D. Claridge and W. Turner, 2003. Continuous commissioning of building energy systems. *Journal of Solar Energy Engineering*, 125 (3), 275–281.

Liu, J., H. Aizawa and H. Yoshino, 2004. CFD prediction of surface condensation on walls and its experimental validation. *Building and Environment*, 39 (8), 905–911.

Liu, T., Y. Luan and W. Zhong, 2012. Earthquake responses of clusters of building structures caused by a near-field thrust fault. *Soil Dynamics and Earthquake Engineering*, 42, 56–70.

Liu, M., K. Wittchen and P. Heiselberg, 2014. Verification of a simplified method for intelligent glazed façade design under different control strategies in a full-scale façade test facility – preliminary results of a south facing single zone experiment for a limited summer period. *Building and Environment*, 82, 400–407.

Lizarralde, G., M. de Blois and I. Latunova, 2011. Structuring of temporary multi-organizations: contingency theory in the building sector. *Project Management Journal*, 42 (4), 16–36.

Llau, A., L. Jason, F. Dufour and J. Baroth, 2015. Adaptive zooming method for the analysis of large structures with localized nonlinearities. *Finite Elements in Analysis and Design*, 106, 73–84.

Lo, T. and K. Choi, 2004. Building defects diagnosis by infrared thermography. *Structural Survey*, 22 (5), 259–263.

Lo, L. and A. Novoselac, 2012. Cross ventilation with small openings: measurements in a multi-zone test building. *Building and Environment*, 57, 377–386.

Locke, E. and G. Latham, 2002. Building a practically useful theory of goal setting and task motivation. *American Psychologist*, 57 (9), 705–717.

Lockyer, C. and D. Scholarios, 2007. The 'rain dance' of selection in construction: rationality as ritual and the logic of informality. *Personnel Review*, 36 (4), 528–548.

Loffler, G., 2008. Perception of contours and shapes: low and intermediate stage mechanisms. *Vision Research*, 48 (2), 2106–2127.

Loftness, V., K. Lam and V. Hartkopf, 2005. Education and environmental performance-based design: a Carnegie Mellon perspective. *Building Research & Information*, 33 (2), 196–203.

Loftness, V., B. Hakkinen, O. Adan and A. Nevalainen, 2007. Elements that contribute to healthy building design. *Environmental Health Perspectives*, 115 (6), 965–970.

Lohman, C., L. Fortuin and M. Wouters, 2004. Designing a performance measurement system: a case study. *European Journal of Operation Research*, 156 (2), 267–286.

Löhnert, G., A. Dalkowski and W. Sutter, 2003. Integrated Design Process: a guideline for sustainable and solar-optimised building design. Berlin/Zug: International Energy Agency, Solar Heating & Cooling Programme, Task 23.

Løken, E., 2007. Use of multicriteria decision analysis methods for energy planning problems. *Renewable and Sustainable Energy Reviews*, 11 (7), 1584–1595.

Lomas, K., H. Eppel, C. Martin and D. Bloomfield, 1997. Empirical validation of building energy simulation programs. *Energy and Buildings*, 26 (3), 253–275.

Lombardi, P., S. Giorano, H. Faroud and W. Yousef, 2012. Modelling the smart city performance. *Innovation – The European Journal of Social Science Research*, 25 (2), 137–149.

Loonen, R., M. Trčka, D. Cóstola and J. Hensen, 2013. Climate adaptive building shells: state-of-the-art and future challenges. *Renewable and Sustainable Energy Reviews*, 25, 483–493.

Love, P., J. Liu, J. Matthews, C. Sing and J. Smith, 2015. Future proofing PPPs: life-cycle performance measurement and building information modelling. *Automation in Construction*, 56, 26–35.

Low, S., J. Liu and K. Oh, 2008. Influence of total building performance, spatial and acoustic concepts on buildability scores of facilities. *Facilities*, 26 (1/2), 85–104.

Lozano, S. and E. Gutiérrez, 2011. Efficiency analysis and target setting of Spanish airports. *Networks and Spatial Economics*, 11 (1), 139–157.

Lozano, S. and G. Villa, 2009. Multiobjective target setting in data envelopment analysis using AHP. *Computers & Operations Research*, 36 (2), 549–564.

Lu, Z. and S. Shao, 2016. Impact of government subsidies on pricing and performance level choice in energy performance contracting: a two-step optimal decision model. *Applied Energy*, 184, 1176–1183.

Lu, X., X. Lu, H. Guan and L. Ye, 2013. Collapse simulation of reinforced concrete high-rise building induced by extreme earthquakes. *Earthquake Engineering & Structural Dynamics*, 42 (5), 705–723.

Lu, Z., X. Chen, X. Lu and Z. Yang, 2016. Shaking table test and numerical simulation of an RC frame-core tube structure for earthquake-induced collapse. *Earthquake Engineering & Structural Dynamics*, 45 (9), 1537–1556.

Luccioni, B., R. Ambrosini and R. Danesi, 2004. Analysis of building collapse under blast loads. *Engineering Structures*, 26 (1), 63–71.

Luce, R., 1996. The ongoing dialog between empirical science and measurement theory. *Journal of Mathematical Psychology*, 40 (1), 78–98.

Lucero, B., J. Linsey and C. Turner, 2016. Frameworks for organising design performance metrics. *Journal of Engineering Design*, 27 (4/6), 1–30.

Ludovico-Marques, M. and C. Chastre, 2012. Effect of salt crystallization ageing on the compressive behavior of standstone blocks in historical buildings. *Engineering Failure Analysis*, 26, 247–257.

Lützkendorf, T. and D. Lorenz, 2006. Using an integrated performance approach in building assessment tools. *Building Research & Information*, 34 (4), 334–356.

Ma, J., W. Song, W. Tian, S. Lo and G. Liao, 2012. Experimental study on an ultra high-rise building evacuation in China. *Safety Science*, 50 (8), 1665–1674.

Ma, J., S. Lo, W. Song, W. Wang, J. Zhang and G. Liao, 2013. Modeling pedestrian space in complex building for efficient pedestrian traffic simulation. *Automation in Construction*, 30, 25–36.

Maamari, F., M. Andersen, J. de Boer, W. Carroll, D. Dumortier and P. Greenup, 2006. Experimental validation of simulation methods for bi-directional transmission properties at the daylighting performance level. *Energy and Buildings*, 38 (7), 878–889.

MacdonaldI., 2002. Quantifying the Effects of Uncertainty in Building Simulation. PhD thesis. Glasgow: University of Strathclyde.

Machairas, V., A. Tsangrassoulis and K. Axarli, 2014. Algorithms for optimization of building design: a review. *Renewable and Sustainable Energy Reviews*, 31, 101–112.

Maciejewska, M. and A. Szczurek, 2015. Representativeness of shorter measurement sessions in long-term indoor air monitoring. *Environmental Science: Processes & Impacts*, 17, 381–388.

Macilwain, C., 2013. Halt the avalanche of performance metrics. *Nature*, 500 (7462), 255.

Macmillan, S., J. Steele, P. Kirby, R. Spence and S. Austin, 2002. Mapping the design process during the conceptual phase of building projects. *Engineering Construction and Architectural Management*, 9 (3), 174–180.

Maghareh, A., S. Dyke, A. Prakash and G. Bunting, 2016. Establishing a predictive performance indicator for real-time hybrid simulation. *Earthquake Engineering & Structural Dynamics*, 43 (15), 2299–2318.

Magoulès, F., H. Zhao and D. Elizondo, 2013. Development of an RDP neural network for building energy consumption fault detection and diagnosis. *Energy and Buildings*, 62, 133–138.

Mahbub, A., H. Kua and S. Lee, 2010. A total building performance approach to evaluating building acoustics performance. *Architectural Science Review*, 35 (2), 213–223.

Mahdavi, A., 1998. Computational decision support and the building delivery process: a necessary dialogue. *Automation in Construction*, 7 (2/3), 205–211.

Mahdavi, A., 1999. A comprehensive computational environment for performance based reasoning in building design and evaluation. *Automation in Construction*, 8 (4), 427–435.

Mahdavi, A., 2011. People in building performance simulation. In: Hensen, J. and R. Lamberts, eds., *Building performance simulation for design and operation*. Abingdon: Spon Press.

Mahdavi, A. and S. Dervishi, 2010. Approaches to computing irradiance on building surfaces. *Journal of Building Performance Simulation*, 3 (2), 129–134.

Mahdavi, A. and R. Ries, 1998. Towards computational eco-analysis of building designs. *Computers and Structures*, 67 (5), 375–387.

Mahdavi, A., P. Mathew, S. Kumar and N. Wong, 1997a. Bi-directional computational design support in the SEMPER environment. *Automation in Construction*, 6 (4), 353–373.

Mahdavi, A., P. Mathew and K. Lam, 1997b. Aggregate space-time performance indicators for simulation-based building evaluation procedures. In: Spitler, J. and J. Hensen, eds., *Building Simulation '97, Fifth International IBPSA Conference*, Prague, Czech Republic, 8–10 September 1997, pp. II-293–II-298.

Mahdavi, A., J. Lechleitner and J. Pak, 2008. Measurements and predictions of room acoustics in atria. *Journal of Building Performance Simulation*, 1 (2), 67–74.

Mak, C. and Z. Wang, 2015. Recent advances in building acoustics: an overview of prediction methods and their applications. *Building and Environment*, 91, 118–126.

Malkawi, A., 2003. Immersive building simulation. In: Malkawi, A.M. and G. Augenbroe, eds., *Advanced building simulation*. New York: Spon Press.

Malkawi, A.M. and G. Augenbroe, eds., 2003. *Advanced building simulation*. New York: Spon Press.

Mallor, F., C. García-Olaverri, S. Gómez-Elvira and P. Mateo-Collazas, 2008. Expert judgment-based risk assessment using statistical scenario analysis: a case study – running the bulls in Pamplona (Spain). *Risk Analysis*, 28 (4), 1103–1019.

Malmqvist, T., M. Glaumann, S. Scarpellini, I. Zabalza, A. Aranda, E. Llera and S. Díaz, 2011. Life cycle assessment in buildings: the ENSLIC simplified method and guidelines. *Energy*, 36 (4), 1900–1907.

Mangkuto, R., S. Wang, B. Meerbeek, M. Aries and E. van Loenen, 2014. Lighting performance and electrical energy consumption of a virtual window prototype. *Applied Energy*, 135, 261–273.

Manley, M. and Y. Kim, 2012. Modeling emergency evacuation of individuals with disabilities (exitus): an agent-based public decision support system. *Expert Systems with Applications*, 39 (9), 8300–8311.

Mann, D., 2011a. Capturing the voice of the customer before the customer knows what they want: TRIZ, spiral dynamics, and the fourth turning. *Procedia Engineering*, 9, 573–581.

Mann, S., 2011b. *Sustainable Lens: A Visual Guide*. Dunedin: New Splash.

Mansour, O. and S. Radford, 2016. Rethinking the environmental and experiential categories of sustainable building design: a conjoint analysis. *Building and Environment*, 98, 47–54.

Mantha, B., C. Menassa and V. Kamat, 2016. A taxonomy of data types and data collection methods for building energy monitoring and performance simulation. *Advances in Building Energy Research*, 10 (2), 263–293.

Mantzouratos, N., D. Gardiklis, V. Dedoussis and P. Kerhoulas, 2004. Concise exterior lighting simulation methodology. *Building Research & Information*, 32 (1), 42–47.

Maravas, A. and J. Pantouvakis, 2012. Project cash flow analysis in the presence of uncertainty in activity duration and cost. *International Journal of Project Management*, 30 (3), 374–384.

Marć, M., J. Namieśnik and B. Zabiegała, 2017. The miniaturised emission chamber system and home-made passive flux sampler studies of monoaromatic hydrocarbon emissions from selected commercially-available floor coverings. *Building and Environment*, 123, 1–12.

Marceau, M. and R. Zmeaureanu, 2000. Nonintrusive load disaggregation computer program to estimate the energy consumption of major end uses in residential buildings. *Energy Conversion and Management*, 41 (13), 1389–1403.

Mardaljevic, J., L. Heschong and E. Lee, 2009. Daylight metrics and energy savings. *Lighting Research and Technology*, 41 (3), 261–283.

Marino, A., P. Bertoldi, S. Rezessy and B. Boza-Kiss, 2011. A snapshot of the European energy service market in 2010 and policy recommendations to foster a further market development. *Energy Policy*, 39 (10), 6190–6198.

Marjaba, G. and S. Chidiac, 2016. Sustainability and resiliency metrics for buildings – critical review. *Building and Environment*, 101, 116–125.

Markus, T., P. Whyman, J. Morgan, D. Whitton, T. Maver, D. Canter and J. Fleming, 1972. *Building Performance*. London: Applied Science Publishers.

Marmot, A., J. Eley and S. Bradley, 2005. Phase 2: programming/briefing – programme review. In: Preiser, W. and J. Vischer, eds., *Assessing building performance*. Oxford: Butterworth-Heinemann.

Marques, G., D. Gourc and M. Lauras, 2010. Multi-criteria performance analysis for decision making in project management. *International Journal of Project Management*, 29 (8), 1057–1069.

Marsh, A. 1997. Performance analysis and conceptual design. PhD thesis. Perth: University of Western Australia.

Marsh, G., 2008. Blue sky thinking for green build. *Renewable Energy Focus*, 9 (6), 4–11.

Marsh, R., 2016. LCA profiles for building components: strategies for the early design process. *Building Research & Information*, 44 (4), 358–375.

Marsh, A. and A. Khan, 2011. Simulation and the future of design tools for ecological research. *Architectural Design*, 81 (6), 82–91.

Marszal, A., P. Heiselberg, J. Bourrelle, E. Musall, K. Voss, I. Sartori and A. Napolitano, 2011. Zero energy building – a review of definitions and calculation methodologies. *Energy and Buildings*, 43 (4), 971–979.

Martin, K., A. Campos-Celador, C. Escudero, I. Gómez and J. Sala, 2012. Analysis of a thermal bridge in a guarded hot box testing facility. *Energy and Buildings*, 50, 139–149.

Maslow, A., 1943. A theory of human motivation. *Psychological Review*, 50 (4), 370–396.

Mathew, P., L. Dunn, M. Sohn, A. Mercado, C. Custudio and T. Walter, 2015. Big-data for building energy performance: lessons from assembling a very large national database of building energy use. *Applied Energy*, 140, 85–93.

Matisoff, D., D. Noonan and A. Mazzolini, 2014. Performance or marketing benefits? The case of LEED certification. *Environmental Science & Technology*, 48 (3), 2001–2007.

Matson, N. and M. Piette, 2005. High Performance commercial building systems: review of California and National benchmarking methods. Berkeley, CA: Lawrence Berkeley National Laboratory.

Matusiak, B. and C. Klöckner, 2016. How we evaluate the view out through the window. *Architectural Science Review*, 59 (3), 203–211.

Mauser, I., J. Müller, F. Allerding and H. Schmeck, 2016. Adaptive building energy management with multiple commodities and flexible evolutionary optimization. *Renewable Energy*, 87, 911–921.

Maver, T. and L. McElroy, 1999. Information technology and building performance. *Automation in Construction*, 8 (4), 411–415.

Mavroidis, I., S. Andronopoulos and J. Bartzis, 2012. Computational simulation of the residence of air pollutants in the wake of a 3-dimensional cubical building. The effect of atmospheric stability. *Atmospheric Environment*, 63, 189–202.

Mawdesley, M. and S. Al-Jibouri, 2009. Modelling construction project productivity using systems dynamics approach. *International Journal of Productivity and Performance Management*, 59 (1), 18–36.

McAllister, T., F. Sadek, J. Gross, S. Kirkpatrick, R. MacNeil, R. Bocchieri, M. Zarghamee, O. Erbay and A. Sarawit, 2013. Structural analysis of impact damage to World Trade Center Buildings 1, 2 and 7. *Fire Technology*, 49 (3), 615–642.

McCleary, P., 2005. Performance (and performers) in search of direction (and a director). In: Kolarevic, B. and A. Malkawi, eds., *Performative architecture: beyond instrumentality.* New York: Spon Press.

McCuskey Shepley, M., K. Zimmerman and M. Boggess, 2009. Architectural office post-occupancy evaluation. *Journal of Interior Design*, 34 (3), 17–29.

McElroy, L., 2009. Embedding Integrated Building Performance Assessment in Design Practice. PhD thesis. Glasgow: University of Strathclyde.

McGrath, P. and M. Horton, 2011. A post-occupancy evaluation (POE) study of student accommodation in an MMC/modular building. *Structural Survey*, 29 (3), 244–252.

McLean, A., H. Bulkeley and M. Crang, 2015. Negotiating the urban smart grid: socio-technical experimentation in the city of Austin. *Urban Studies*, 53 (15), 3246–3263.

McLeod, R., C. Hopfe and Y. Rezgui, 2012. An investigation into recent proposals for a revised definition of zero carbon homes in the UK. *Energy Policy*, 46, 25–35.

McMullan, R., 2012. *Environmental science in building.* Basingstoke: Palgrave Macmillan, 7th edition.

McNeil, A. and E. Lee, 2013. A validation of the Radiance three-phase simulation method for modelling annual daylight performance of optically complex fenestration systems. *Journal of Building Performance Simulation*, 6 (1), 24–37.

Meir, I., Y. Garb, D. Jiao and A. Cicelsky, 2009. Post-occupancy evaluation: an inevitable step towards sustainability. *Advances in Building Energy Research*, 3 (1), 189–219.

Meissner, P., C. Brands and T. Wulf, 2017. Quantifying blind spots and weak signals in executive judgment: a structured integration of expert judgment into the scenario development process. *International Journal of Forecasting*, 33 (1), 244–253.

Mela, K., T. Tiainen and M. Heinisuo, 2012. Comparative study of multiple criteria decision making methods for building design. *Advanced Engineering Informatics*, 26 (4), 716–726.

Melnyk, S., D. Steward and M. Swink, 2004. Metrics and performance measurement in operations management: dealing with the metrics maze. *Journal of Operations Management*, 22 (3), 209–217.

Menberg, K., Y. Heo and R. Choudhary, 2016. Sensitivity analysis methods for building energy models: comparing computational costs and extractable information. *Energy and Buildings*, 133, 433–445.

Méndez Echenagucia, T., A. Capozzoli, Y. Cascone and M. Sassone, 2015. The early design stage of a building envelope: multi-objective search through heating, cooling and lighting energy performance analysis. *Applied Energy*, 154, 577–591.

Méndez Fernández, D. and S. Wagner, 2015. Naming the pain in requirements engineering: a design for a global family of surveys and first results from Germany. *Information and Software Technology*, 57, 616–643.

Menezes, C., A. Cripps, D. Bouchlaghem and R. Buswell, 2012. Predicted vs. actual energy performance of non-domestic buildings: Using post-occupancy evaluation data to reduce the performance gap. *Applied Energy*, 97, 355–364.

Meng, X., 2012. The effect of relationship management on project performance in construction. *International Journal of Project Management*, 30 (2), 188–198.

Meng, Q., H. Hao and W. Chen, 2016. Laboratory test and numerical study of structural insulated panels strengthened with glass fibre laminate against windborne debris impact. *Construction and Building Materials*, 114, 434–446.

Mentese, S., N. Mirici, M. Otkun, C. Bakar, E. Palaz, D. Tasdibi, S. Cevizci and O. Cotuker, 2015. Association between respiratory health and indoor air pollution exposure in Canakkale, Turkey. *Building and Environment*, 93, 72–83.

Menyhart, K. and M. Krarti, 2017. Potential energy savings from deployment of dynamic insulation materials for US residential buildings. *Building and Environment*, 114, 203–218.

Meroney, R., 2016. Ten questions concerning hybrid computational/physical model simulation of wind flow in the built environment. *Building and Environment*, 96, 12–21.

Meso, P., M. Troutt and J. Rudnicka, 2002. A review of naturalistic decision making research with some implications for knowledge management. *Journal of Knowledge Management*, 6 (1), 63–73.

Messer, M., J. Panchal, J. Allen and F. Mistree, 2011. Model refinement decisions using the process performance indicator. *Engineering Optimization*, 43 (7), 741–762.

Meth, H., M. Brhel and A. Maedche, 2013. The state of the art in automated requirements elicitation. *Information and Software Technology*, 55 (10), 1695–1709.

Mfula, A., V. Kukadia, R. Griffiths and D. Hall, 2005. Wind tunnel modelling of urban building exposure to outdoor pollution. *Atmospheric Environment*, 39 (15), 2737–2745.

Michalak, P., 2014. The simple hourly method of EN ISO 13790 standard in Matlab/ Simulink: a comparative study for the climatic conditions of Poland. *Energy*, 75, 568–578.

Miles, E. and E. Clenney, 2012. Extremely difficult negotiator goals: do they follow the predictions of goal-setting theory? *Organizational Behavior and Human Decision Processes*, 118 (2), 108–115.

Miller, E. and L. Buys, 2008. Retrofitting commercial office buildings for sustainability: tenants' perspectives. *Journal of Property Investment & Finance*, 26 (6), 552–561.

Mills, A., P. Love and P. Williams, 2009. Defect costs in residential construction. *Journal of Construction Engineering and Management*, 135 (1), 12–16.

Minchin, R., N. Hendriquez, A. King and D. Lewis, 2010. Owners respond: preference for task performance, delivery systems and quality management. *Journal of Construction Engineering and Management*, 136 (3), 283–293.

Mlecnik, E., T. Schütze, S. Jansen, G. de Vries, H. Visscher and A. van Hal, 2012. End-user experiences in nearly zero-energy houses. *Energy and Buildings*, 49, 471–478.

Moers, F., 2006. Performance measure properties and delegation. *The Accounting Review*, 81 (4), 897–924.

Mohamed, A., A. Hasan and K. Sirén, 2014. Fulfillment of net-zero energy building (NZEB) with four metrics in a single family house with different heating alternatives. *Applied Energy*, 114, 385–399.

Mommertz, E., 2008. *Acoustics and sound insulation*. Berlin: Berkhäuser.

Montgomery, D.C., 2013. *Design and Analysis of Experiments*. Hoboken, NJ: Wiley, 8th edition.

Moore, J., C. Davis, M. Coplan and S. Greer, 2009. *Building Scientific Apparatus*. Cambridge: Cambridge University Press, 4th edition.

Morbitzer, C., 2003. Towards the Integration of Simulation into the Building Design Process. PhD thesis. Glasgow: University of Strathclyde.

Morgan, M., 1960. *Vitruvius: the ten books on architecture*. English translation. New York: Dover Publications.

Morrison, G., R. Shaughnessy and S. Shu, 2011. Setting maximum emission rates for ozone emitting consumer appliances in the United States and Canada. *Atmospheric Environment*, 45 (11), 2009–2016.

Muchiri, P., L. Pintelon, L. Gelders and H. Martin, 2011. Development of maintenance function performance framework and indicators. *International Journal of Production Economics*, 131 (1), 295–302.

Muller, A., F. Demouge, M. Jeguirim, P. Fromy and J. Vantelon, 2013. The use of Petri nets and a two-zone model for fire scene reconstruction. *Fire Safety Journal*, 55, 139–151.

Mulvey, D., P. Redding, C. Robertson, C. Woodall, P. Kingsmore, D. Bedwell and S. Dancer, 2011. Finding a benchmark for monitoring hospital cleanliness. *Journal of Hospital Infection*, 77 (1), 25–30.

Mumovic, D. and M. Santamouris, eds., 2009. *A handbook of sustainable building design & engineering*. London: Earthscan.

Mumovic, D., M. Davies, I. Ridley, H. Altamirano-Medina and T. Oreszczyn, 2009. A methodology for post-occupancy evaluation of ventilation rates in schools. *Building Services Engineering Research and Technology*, 30 (2), 143–152.

Mundo-Hernández, J., M. Valerdi-Nochebuena and J. Sosa-Oliver, 2015. Post-occupancy evaluation of a restored industrial building: a contemporary art and design gallery in Mexico. *Frontiers of Architectural Research*, 4 (4), 330–340.

Murphy, L., 2014. The influence of the Energy Performance Certificate: the Dutch case. *Energy Policy*, 67, 664–672.

Murphy, L., S. Morgan, M. Ossi, J. Ohlson and K. Linthicum, 2013. Reductions in energy and water use in a government laboratory through an energy savings performance contract (ESPC). *Strategic Planning for Energy and the Environment*, 32 (4), 65–76.

Murray, J. and M. Sasani, 2016. Near-collapse response of existing RC building under severe pulse-type ground motion using hybrid simulation. *Earthquake Engineering & Structural Dynamics*, 45 (7), 1109–1127.

Mustafaraj, G., D. Marini, A. Costa and M. Keane, 2014. Model calibration for building energy efficiency simulation. *Applied Energy*, 130, 72–85.

Mylopoulos, M. and G. Regehr, 2011. Putting the expert together again. *Medical Education*, 45 (9), 920–926.

Nabil, A. and J. Mardaljevic, 2005. Useful daylight illuminance: a new paradigm for assessing daylight in buildings. *Lighting Research and Technology*, 37 (1), 41–59.

Nagel, R., R. Hutcheson, D. McAdams and R. Stone, 2011. Process and event modelling for conceptual design. *Journal of Engineering Design*, 22 (3), 145–164.

Naghiyev, E., M. Gillott and R. Wilson, 2014. Three unobtrusive domestic occupancy measurement technologies under qualitative review. *Energy and Buildings*, 69, 507–514.

Nagy, Z., B. Svetozarevic, P. Jayathissa, M. Begle, J. Hofer, G. Lydon, A. Willmann and A. Schlueter, 2016. The adaptive solar facade: from concept to prototypes. *Frontiers of Architectural Research*, 5 (2), 143–156.

Nah, K. and J. Kreifeldt, 1996. The use of computer-generated bivariate charts in anthropometric design. *Applied Ergonomics*, 27 (6), 397–409.

Narmashiri, K., N. Sulong and M. Jumaat, 2012. Failure analysis and structural behaviour of CFRP strengthened steel I-beams. *Construction and Building Materials*, 30, 1–9.

Nassar, K., 2010. A model for assessing occupant flow in building spaces. *Automation in Construction*, 19 (8), 1027–1036.

Nastar, N., J. Anderson, C. Brandow and R. Nigbor, 2010. Effects of low-cycle fatigue on a 10-storey steel building. *The Structural Design of Tall and Special Buildings*, 19 (1/2), 95–113.

Navarro, J. and J. Escolano, 2015. Simulation of building indoor acoustics using an acoustic diffusion equation model. *Journal of Building Performance Simulation*, 8 (1), 3–14.

Navarro, I., Á. Gutiérrez, C. Montero, E. Rodríguez-Ubiñas, E. Matallanas, M. Castillo-Cagigal, M. Porteros, J. Solórzano, E. Caamaño-Martín, M. Egido, J. Páez and S. Vego, 2014. Experiences and methodology in a multidisciplinary energy and architecture competition: Solar Decathlon Europe 2012. *Energy and Buildings*, 83, 3–9.

Navon, R., 2007. Research in automated measurement of project performance indicators. *Automation in Construction*, 16 (2), 176–188.

Neely, A., 2005. The evolution of performance measurement research. *International Journal of Operations & Production Management*, 25 (12), 1264–1277.

Neely, A., H. Richards, J. Mills, K. Platts and M. Bourne, 1997. Designing performance measures: a structured approach. *International Journal of Operations & Production Management*, 17 (11), 1131–1152.

Negendahl, K., 2015. Building performance in the early design stage: an introduction to integrated dynamic models. *Automation in Construction*, 54, 39–53.

Nembrini, J., S. Samberger and G. Labelle, 2014. Parametric scripting for early design performance simulation. *Energy and Buildings*, 68, 786–798.

Neumann, C. and D. Jacob, eds., 2008. Guidelines for the evaluation of building performance. Report of the Building EQ SAVE action. Freiburg: Fraunhofer Institute for Solar Energy Systems.

Newsham, G., B. Birt, C. Arsenault, L. Thompson, J. Veitch, S. Mancini, A. Galasiu, B. Gover, I. Macdonald and G. Burns, 2012. Do green buildings outperform conventional buildings? Indoor environment and energy performance in North American offices. Ottawa: National Research Council Canada.

Newsham, G., H. Xue, C. Arsenault, J. Valdes, G. Burns, E. Scarlett, S. Kruithof and W. Shen, 2017. Testing the accuracy of low-cost data streams for determining single-person office occupancy and their use for energy reduction of building services. *Energy and Buildings*, 135, 137–147.

Newton, C., S. Wilks, D. Hes, A. Aibinu, R. Crawford, K. Goodwin, C. Jensen, D. Chambers, T. Chan and L. Aye, 2012. More than a survey: an interdisciplinary post-occupancy tracking of BER schools. *Architectural Science Review*, 55 (3), 196–205.

Neymark, J., R. Judkoff, G. Knabe, H. Le, M. Dürig, A. Glass and G. Zweifel, 2002. Applying the building energy simulation test (BESTEST) diagnostic method to verification of space conditioning equipment models used in whole-building energy simulation programs. *Energy and Buildings*, 34 (9), 917–931.

Ng, E., K. Lam, W. Wu and T. Nagakura, 2001. Advanced lighting simulation in architectural design in the tropics. *Automation in Construction*, 10 (3), 365–379.

Nguyen, A., S. Reiter and P. Rigo, 2014. A review on simulation-based optimization methods applied to building performance analysis. *Applied Energy*, 113, 1043–1058.

NIBS, 2008. *Assessment to the US Congress and US Department of Energy on high performance buildings.* Washington, DC: National Institute of Building Science.

NIBS, 2011. Data needs for achieving high-performance buildings. Washington, DC: National Institute of Building Science.

Nicol, F. and M. Humpreys, 2007. Maximum temperatures in European office buildings to avoid heat discomfort. *Solar Energy*, 81 (3), 295–304.

Nicol, F. and M. Humpreys, 2010. Derivation of the adaptive equations for thermal comfort in free-running buildings in European standard EN15251. *Building and Environment*, 45 (1), 11–17.

Nicol, F. and S. Roaf, 2005. Post-occupancy evaluation and field studies of thermal comfort. *Building Research & Information*, 33 (4), 338–346.

Nikolaou, T., I. Skias, D. Kolokotsa and G. Stavrakakis, 2009. Virtual Building Dataset for energy and indoor thermal comfort benchmarking of office buildings in Greece. *Energy and Buildings*, 41 (12), 1409–1416.

Nikolaou, T., D. Kolokotsa and G. Stavrakakis, 2011. Review on methodologies for energy benchmarking, rating and classification of buildings. *Advances in Building Energy Research*, 5 (1), 53–70.

Nikolopoulos, N., A. Nikolopoulos, T. Larsen, K. Stefanos and P. Nikas, 2012. Experimental and numerical investigation of the tracer gas methodology in the case of a naturally cross-ventilated building. *Building and Environment*, 56, 379–388.

Nilpueng, K. and S. Wongwises, 2010. Two-phase gas-liquid flow characteristics inside a plate heat exchanger. *Experimental Thermal and Fluid Science*, 34 (8), 1217–1229.

Nilsson, D. and A. Johansson, 2009. Social influence during the initial phase of a fire evacuation – analysis of evacuation experiments in a cinema theatre. *Fire Safety Journal*, 44 (1), 71–79.

Nilsson, M., H. Frantzich and P. van Hees, 2013. Selection and evaluation of fire related scenarios in multifunctional buildings considering antagonistic attacks. *Fire Science Reviews*, 2 (3), 1–20.

Nolden, C., S. Sorrell and F. Polzin, 2016. Catalysing the energy service market: the role of intermediaries. *Energy Policy*, 98, 420–430.

Nomaguchi, Y., M. Saito and K. Fujita, 2015. Multi-domain DSM method for design process management of complex system. *Journal of Industrial and Production Engineering*, 32 (7), 465–472.

Noon, S. et al., 2012. *A street through time: a 12,000-year walk through history*. London: Dorling Kindersley.

Nordin, S., E. Lidén and A. Gidlöf-Gunnarsson, 2009. Development and evaluation of a category ratio scale with semantic descriptors: the Environmental Annoyance Scale. *Scandinavian Journal of Psychology*, 50 (2), 93–100.

Norford, L. and S. Leeb, 1996. Non-intrusive electrical load monitoring in commercial buildings based on steady-state and transient load-detection algorithms. *Energy and Buildings*, 24 (1), 51–64.

Norford, L., R. Socolow, E. Hsieh and G. Spadaro, 1994. Two-to-one discrepancy between measured and predicted performance of a 'low-energy' office building: insights from a reconciliation based on the DOE-2 model. *Energy and Buildings*, 21 (2), 121–131.

Norman, G. and D. Streiner, 2003. *PDQ Statistics*. Shelton: People's Medical Publishing House, 3rd edition.

Nouidui, T., M. Wetter and W. Zuo, 2014. Functional mock-up unit for co-simulation import in EnergyPlus. *Journal of Building Performance Simulation*, 7 (3), 192–202.

Nowroozi, A., M. Shiri, A. Aslanian and C. Lucas, 2012. A general computational recognition primed decision model with multi-agent rescue simulation benchmark. *Information Sciences*, 187, 52–71.

Noye, S., R. North and D. Fisk, 2016. Smart systems commissioning for energy efficient buildings. *Building Services Engineering Research & Technology*, 37 (2), 194–204.

NYC DDC, 1999. *High Performance Building Guidelines*. New York: City of New York, Department of Design and Construction.

Nytsch-Geusen, C., J. Huber, M. Ljubijankic and J. Rädler, 2013. Modelica BuildingSystems – eine Modellbibliothek zur Simulation komplexer energietechnischer Gebäudesysteme. *Bauphysik*, 35 (1), 21–29.

O'Brien, W., I. Gaetani, S. Carlucci, P. Hoes and J. Hensen, 2017. On occupant-centric building performance metrics. *Building and Environment*, 122, 373–385.

O'Connor, P. and A. Kleyner, 2012. *Practical reliability engineering*. Chichester: Wiley.

O'Donnell, J., 2009. Specification of optimum holistic building environmental and energy performance information to support informed decision making. PhD thesis. Cork: University College Cork.

O'Grady, M. and G. O'Hare, 2012. How smart is your city? *Science*, 335 (6076), 1581–1582.

O'Kelly, M., M. Walter and J. Rowland, 2014. Simulated hygrothermal performance of a Passivhaus in a mixed humid climate under dynamic load. *Energy and Buildings*, 81, 211–218.

O'Neill, Z., T. Bailey, B. Dong, M. Shashanka and D. Luo, 2013. Advanced building energy management system demonstration for Department of Defence buildings. *Annals of the New York Academy of Sciences*, 1295, 44–53.

O'Neill, Z., X. Pang, M. Shashanka, P. Haves and T. Bailey, 2014. Model-based real-time whole building energy performance and diagnostics. *Journal of Building Performance Simulation*, 7 (2), 83–99.

Oberkampf, W. and C. Roy, 2010. *Verification and Validation in Scientific Computing*. Cambridge: Cambridge University Press.

Obrecht, C., 2012. High Performance Lattice Bolzmann Solvers on Massively Parallel Architectures with Applications to Building Aeraulics. PhD thesis. Lyon: Institut National des Sciences Appliquées.

Ochoa, J., 2014. Reducing plan variations in delivering sustainable building projects. *Journal of Cleaner Production*, 85, 276–288.

Ochoa, C. and I. Capeluto, 2015. Decision methodology for the development of an expert system applied in an adaptable energy retrofit façade system for residential buildings. *Renewable Energy*, 78, 498–508.

Ochoa, C., M. Aries and J. Hensen, 2012. State of the art in lighting simulation for building science: a literature review. *Journal of Building Performance Simulation*, 5 (4), 209–233.

Ogunlana, S., H. Li and F. Sukhera, 2003. System dynamics approach to exploring performance enhancement in a construction organisation. *Journal of Construction Engineering and Management*, 129 (5), 528–536.

Oğuz, O., A. Akaydin, T. Yilmaz and U. Güdükbay, 2010. Emergency crowd simulation for outdoor environments. *Computers and Graphics*, 34 (2), 136–144.

Oh, S. and J. Haberl, 2016. Origins of analysis methods used to design high-performance commercial buildings: whole-building energy simulation. *Science and Technology for the Built Environment*, 22 (1), 118–137.

Ohlsson, K. and T. Olofsson, 2014. Quantitative infrared thermography imaging of the density of heat flow rate through a building element surface. *Applied Energy*, 134, 499–505.

Ohnesorge, D., K. Richter and G. Becker, 2010. Influence of wood properties and bonding parameters on bond durability of European Beech (*Fagus sylvatica* L.) glulams. *Annals of Forest Science*, 67 (6): 601.

Olenick, S. and D. Carpenter, 2003. An updated international survey of computer models for fire and smoke. *Journal of Fire Protection Engineering*, 13 (2), 87–110.

Olesen, B., 2005. The European Energy Performance of Buildings Directive EPBD (editorial). *HVAC&R Research*, 11 (4), 505–509.

Olesen, B., 2007. The philosophy behind EN15251: Indoor environmental criteria for design and calculation of energy performance of buildings. *Energy and Buildings*, 39 (7), 740–749.

OMG, 2016. Unified Modelling Language (UML) resource page [online]. Needham, MA: Object Management Group. Available from www.uml.org [Accessed 28 February 2016].

Orehounig, K., R. Evins and V. Dorer, 2015. Integration of decentralized energy systems in neighbourhoods using the energy hub approach. *Applied Energy*, 154, 277–289.

Oreszczyn, T. and R. Lowe, 2010. Challenges for energy and buildings research: objectives, methods and funding mechanisms. *Building Research & Information*, 38 (1), 107–122.

Ornelas, J., A. Silva Júnior and J. Barros Fernandes, 2012. Yes, the choice of performance measure does matter for ranking of US mutual funds. *International Journal of Finance and Economics*, 17 (1), 61–72.

Osborne, A., S. Baur and K. Grantham, 2010. Simulation prototyping of an experimental solar house. *Energies*, 3 (6), 1251–1262.

van Ostaeyen, J., A. van Horenbeek, L. Pintelon and J. Duflou, 2013. A refined typology of product-service systems based on functional hierarchy modeling. *Journal of Cleaner Production*, 51, 261–276.

Østergård, T., R. Jensen and S. Maagaard, 2016. Building simulations supporting decision making in early design – a review. *Renewable and Sustainable Energy Reviews*, 61, 187–201.

Østergård, T., R. Jensen and S. Maagaard, 2017. Early building design: informed decision-making by exploring multidimensional design space using sensitivity analysis. *Energy and Buildings*, 142, 8–22.

Oxford Dictionary of English, 2010. Oxford: Oxford University Press, 3rd edition.

Oxman, R., 2006. Theory and design in the first digital age. *Design Studies*, 27 (3), 229–265.

Oyedele, L., K. Tham, M. Fadeyi and B. Jaiyeoba, 2012. Total building performance approach in building evaluation: case study of an office building in Singapore. *Journal of Energy Engineering*, 138 (1), 25–30.

Özcan, E. and M. Arentsen, 2014. Nonconformity of policy ambitions with biomass potential in regional bioenergy transition: a Dutch example. *Energy Policy*, 65, 212–222.

Ozer, M., 2008. Improving the accuracy of expert predictions of the future success of new internet services. *European Journal of Operational Research*, 184 (3), 1085–1099.

Ozorhon, B., D. Arditi, I. Dimen and T. Birgonul, 2011. Toward a multidimensional performance measure for international joint ventures in construction. *Journal of Construction Engineering and Management*, 137 (6), 403–411.

Pacheco, C. and I. Garcia, 2012. A systematic literature review of stakeholder identification methods in requirements elicitation. *The Journal of Systems and Software*, 85 (9), 2171–2181.

Pacheco-Torgal, F. and S. Jalali, 2011. Nanotechnology: advantages and drawbacks in the field of construction and building materials. *Construction and Building Materials*, 25 (2), 582–290.

Packham, I., 2003. An interactive visualisation system for engineering design using evolutionary computing. PhD thesis. Plymouth: University of Plymouth.

Padilla, M. and D. Choinière, 2015. A combined passive-active sensor fault detection and isolation approach for air handling units. *Energy and Buildings*, 99, 214–219.

Painter, B., N. Brown and M. Cook, 2012. Practical application of a sensor overlay system for building monitoring and commissioning. *Energy and Buildings*, 48, 29–39.

Palme, D., J. Agudo, H. Sánchez and M. Macías, 2014. An Internet of Things example: classrooms access control over near field communication. *Sensors*, 14 (4), 6998–7012.

Pan, N., 2006. Evaluation of building performance using fuzzy FTA. *Construction Management and Economics*, 24 (12), 1241–1252.

Pan, W., 2014. System boundaries of zero carbon buildings. *Renewable and Sustainable Energy Reviews*, 37, 424–434.

Pan, Y., Z. Huang and G. Wu, 2007. Calibrated building energy simulation and its application in a high-rise commercial building in Shanghai. *Energy and Buildings*, 39 (6), 651–657.

Panapakidis, I., T. Papadopoulos, G. Christoforidis and G. Papagiannis, 2014. Pattern recognition algorithms for electricity load curve analysis of buildings. *Energy and Buildings*, 73, 137–145.

Pang, X., M. Wetter, P. Bhattacharya and P. Haves, 2012. A framework for simulation-based real-time whole building performance assessment. *Building and Environment*, 54, 100–108.

Papadimitriou, E., G. Yannis and J. Golias, 2009. A critical assessment of pedestrian behaviour models. *Transportation Research Part F*, 12 (3), 242–255.

Papamichael, K., J. LaPorta and H. Chauvet, 1997. Building Design Advisor: automated integration of multiple simulation tools. *Automation in Construction*, 6 (4), 341–352.

Papamichael, K., H. Chauvet, J. LaPorta and R. Dandridge, 1999. Product modeling for computer-aided decision-making. *Automation in Construction*, 8 (3), 339–350.

Park, C., 2003. Occupant responsive optimal control of smart façade systems. PhD thesis. Atlanta, GA: Georgia Institute of Technology.

Park, C., D. Lee, O. Kwon and X. Wang, 2013. A framework for proactive construction defect management using BIM, augmented reality and ontology-based data collection template. *Automation in Construction*, 33, 61–71.

Park, H., B. Meacham, N. Dembsey and M. Goulthorpe, 2015. Conceptual model development for holistic fire safety performance analysis. *Fire Technology*, 51 (1), 173–193.

Pätäri, S. and K. Sinkkonen, 2014. Energy service companies and energy performance contracting: is there a need to renew the business model? Insights from a Delphi study. *Journal of Cleaner Production*, 66, 264–271.

Pati, D. and G. Augenbroe, 2007. Integrating formalized user experience within building design models. *Computer-Aided Civil and Infrastructure Engineering*, 22 (2), 117–132.

Pati, D., C. Park and G. Augenbroe, 2009. Roles of quantified expressions of building performance assessment in facility procurement and management. *Building and Environment*, 44 (4), 773–784.

Pauwels, P., D. van Deursen, R. Verstraeten, J. de Roo, R. de Meyer, R. van de Walle and J. van Campenhout, 2011. A semantic rule checking environment for building performance checking. *Automation in Construction*, 20 (5), 506–518.

Pavlak, G., G. Henze, A. Hirsch, A. Florita and R. Dodier, 2016. Experimental verification of an energy consumption signal tool for operational decision support in an office building. *Automation in Construction*, 72, 75–92.

Peacock, R., B. Hoskins and E. Kuligowski, 2012. Overall and local movement speed during fire drill evacuations in buildings up to 31 stories. *Safety Science*, 50 (8), 1655–1664.

Pedro, J., F. Meijer and H. Visscher, 2010. Technical building regulations in EU countries: a comparison of their organization and formulation. *CIB World Congress 2010*, Salford, UK, 10–13 May 2010.

Pelechano, N. and A. Malkawi, 2008. Evacuation simulation models: challenges in modeling high rise building evacuation with cellular automata approaches. *Automation in Construction*, 17 (4), 377–385.

Pelken, P., J. Zhang, Y. Chen, D. Rice, Z. Meng, S. Semahegn, L. Gu, H. Henderson, W. Fing and F. Ling, 2013. 'Virtual Design Studio' – Part 1: Interdisciplinary design processes. *Building Simulation*, 6 (3), 235–251.

Pellegrini-Masini, G. and C. Leishman, 2011. The role of corporate reputation and employees' values in the uptake of energy efficiency in office buildings. *Energy Policy*, 39 (9), 5409–5419.

Peng, C., 2016. Calculation of a building's life cycle carbon emissions based on Ecotect and building information modeling. *Journal of Cleaner Production*, 112, 453–465.

Peng, C., L. Huang, J. Liu and Y. Huang, 2015. Energy performance evaluation of a marketable net-zero-energy house: Solark I at Solar Decathlon China 2013. *Renewable Energy*, 81, 136–149.

Peng, L., P. Nielsen, X. Wang, S. Sadrizadeh, L. Liu and Y. Li, 2016. Possible user-dependent CFD predictions of transitional flow in building ventilation. *Building and Environment*, 99, 130–141.

Perén, J., T. van Hooff, B. Leite and B. Blocken, 2016. CFD simulation of wind-driven upward cross ventilation and its enhancement in long buildings: impact of single-span versus double-span leeward sawtooth roof and opening ratio. *Building and Environment*, 96, 142–156.

Perera, D., D. Winkler and N. Skeie, 2016. Multi-floor building heating models in Matlab and Modelica environments. *Applied Energy*, 171, 46–57.

Perez, C., J. Donoso and L. Medina, 2010. A critical experimental study of the classical tactile threshold theory. *BMC Neuroscience*, 11, 76.

Pérez, G., J. Coma, S. Sol and L. Cabeza, 2017. Green facade for energy savings in buildings: the influence of leaf area index and facade orientation on the shadow effect. *Applied Energy*, 187, 424–437.

Pérez-Bella, J., J Domingues-Hernández, B. Rodriguez-Soria, J. del Coz-Díaz, E. Cano-Suñen and A. Navarro-Manso, 2013. An extended method for comparing watertightness tests for facades. *Building Research & Information*, 41 (6), 706–721.

Pérez-Bella, J., J Domingues-Hernández, E. Cano-Suñen, J. del Coz-Díaz and F. Suárez-Domínguez, 2014. A comparison of methods for determining watertightness test parameters of building facades. *Building and Environment*, 78, 145–154.

Pérez-Lombard, L., J. Ortiz, R. González and I. Maestre, 2009. A review of benchmarking, rating and labelling concepts with the framework of building energy certification schemes. *Energy and Buildings*, 41 (3), 272–278.

Perie, M., 2008. A guide to understanding and developing performance-level descriptors. *Educational Measurement: Issues and Practice*, 27 (4), 15–29.

Perkins, I. and M. Skitmore, 2015. Three-dimensional printing in the construction industry: a review. *International Journal of Construction Management*, 15 (1), 1–9.

Perumbal, T., A. Ramli, C. Leong, K. Samsudin and S. Mansor, 2010. Middleware for heterogeneous subsystem interoperability in intelligent buildings. *Automation in Construction*, 19 (2), 160–168.

Petri, I., T. Beach, Y. Rezgui, I. Wilson and H. Li, 2014. Engaging construction stakeholders with sustainability through a knowledge harvesting platform. *Computers in Industry*, 65 (3), 449–469.

Petrone, F., L. Shan and S. Kunnath, 2016. Modeling of RC frame buildings for progressive collapse analysis. *International Journal of Concrete Structures and Materials*, 10 (1), 1–13.

Pheng, L. and J. Teo, 2003. Implementing Total Quality Management in construction through ISO 9001:2000. *Architectural Science Review*, 46 (2), 159–165.

Pickard, P., ed., 2016. *Toilets: a spotter's guide*. Carlton: Lonely Planet Publications.

Pijpers, F., 2006. Performance metrics. *Astronomy and Geophysics*, 47 (6), 17–18.

Pilgrim, M., 2003. The application of visualisation techniques to the process of building performance analysis. EngD thesis. Loughborough: Loughborough University.

Piñar, G., C. Ramos, S. Rölleke, C. Schabereiter-Gurtner, D. Vybiral, W. Lubitz and E. Denner, 2001. Detection of indigenous Halobacillus populations in damaged ancient wall paintings and building materials: molecular monitoring and cultivation. *Applied and Environmental Microbiology*, 67 (10), 4891–4895.

Pinho, P., M. Pinto, R. Almeida, S. Lopes and L. Lemos, 2016. Aspects concerning the acoustical performance of school buildings in Portugal. *Applied Acoustics*, 106, 129–134.

Pirsig, R., 1974. *Zen and the art of motorcycle maintenance*. London: Transworld Publishers.

Pless, S. and P. Torcellini, 2010. *Net-Zero Energy Buildings: a classification system based on renewable energy supply options*. Golden, CO: National Renewable Energy Laboratory.

PMBOK, 2008. Project Management Body of Knowledge. Newton Square, PA: Project Management Institute Inc.

Podofillini, L. and V. Dang, 2013. A Bayesian approach to treat expert-elicited probabilities in human reliability analysis model construction. *Reliability Engineering and System Safety*, 117, 52–64.

Pohl, K. and C. Rupp, 2015. *Requirement Engineering Fundamentals*. Santa Barbara, CA: Rocky Nook Inc.

Polycarpou, P. and P. Komodromos, 2010. Earthquake-induced poundings of a seismically isolated building with adjacent structures. *Engineering Structures*, 32 (7), 1937–1951.

Polzin, F., P. von Flotow and C. Nolden, 2016. What encourages local authorities to engage with energy performance contracting for retrofitting? Evidence from German municipalities. *Energy Policy*, 94, 317–330.

Pomponi, F., P. Piroozfar, R. Southall, P. Ashton and E. Farr, 2016. Energy performance of double-skin façades in temperate climates: a systematic review and meta-analysis. *Renewable and Sustainable Energy Reviews*, 54, 1525–1536.

Popova, V. and A. Sharpanskykh, 2010. Modeling organisational performance indicators. *Information Systems*, 35 (4), 505–527.

Portal, N., A. van Schijndel and A. Sasic Kalagasidis, 2013. The multiphysics modeling of heat and moisture induced stress and strain of historic building materials and artefacts. *Building Simulation: An International Journal*, 7 (3), 217–227.

Potts, K. and M. Wall, 2002. Managing the commissioning of building services. *Engineering Construction and Architectural Management*, 9 (4), 336–344.

Power, A., 2008. Does demolition or refurbishment of old and inefficient homes help to increase our environmental, social and economic viability? *Energy Policy*, 36 (12), 4487–4501.

Prada, A., F. Cappelletti, P. Baggio and A. Gasparella, 2014. On the effect of material uncertainties in envelope heat transfer simulations. *Energy and Buildings*, 71, 53–60.

Pradhananga, N. and J. Teizer, 2013. Automatic spatio-temporal analysis of construction site equipment operations using GPS data. *Automation in Construction*, 29, 107–122.

Prazeres, L., 2006. An Exploratory Study about the Benefits of Targeted Data Perceptualisation Techniques and Rules in Building Simulation. PhD thesis. Glasgow: University of Strathclyde.

Preiser, W., 2001. Feedback, feedforward and control: post-occupancy evaluation to the rescue. *Building Research & Information*, 29 (6), 456–459.

Preiser, W. and J. Vischer, eds., 2005. *Assessing building performance*. Oxford: Butterworth-Heinemann.

Pressman, R., 2005. *Software Engineering: a practitioner's approach*. New York: McGraw-Hill, 6th edition.

Pretlove, S. and S. Kade, 2016. Post occupancy evaluation of social housing designed and built to Code for Sustainable Homes levels 3, 4 and 5. *Energy and Buildings*, 110, 120–134.

Prior, J. and F. Szigeti, 2003. Why all the fuss about Performance Based Building? Rotterdam: PeBBu Network.

Pritchard, R., S. Weaver and E. Ashwood, 2012. *Evidence-based productivity improvement: a practical guide to the productivity measurement and enhancement system (ProMES)*. New York: Routledge.

Prum, D., 2010. Green buildings, high performance buildings, and sustainable construction: does it really matter what we call them? *Villanova Environmental Law Journal*, 21 (1), 1.

Pulselli, R., F. Pulselli, U. Mazzali, F. Peron and S. Bastianoni, 2014. Emergy based evaluation of environmental performances of living wall and grass wall systems. *Energy and Buildings*, 73, 200–211.

Pursals, S. and F. Garzón, 2009. Optimal building evacuation time considering evacuation routes. *European Journal of Operation Research*, 192 (2), 692–699.

Qian, D. and J. Guo, 2014. Research on the energy-saving and revenue sharing strategy of ESCOs under the uncertainty of the value of energy performance contracting projects. *Energy Policy*, 73, 710–721.

Qin, T., Y. Guo, C. Chan and W. Lin, 2009a. Numerical simulation of the spread of smoke in an atrium under fire scenario. *Building and Environment*, 44 (1), 56–65.

Qin, M., R. Belarbi, A. Aït-Mokhtar and F. Allard, 2009b. Simulation of coupled heat and moisture transfer in air-conditioned buildings. *Automation in Construction*, 18 (5), 614–631.

Qin, Y., Q. Sheng and E. Curry, 2015. Matching over linked data streams in the Internet of Things. *IEEE Internet Computing*, 19 (3), 21–27.

Qin, H., X. Xie, J. Vrugt, K. Zeng and G. Hong, 2016. Underground structure defect detection and reconstruction using crosshole GPR and Bayesian waveform inversion. *Automation in Construction*, 68, 156–169.

Queiroz de Sant'Ana, D. and P. Trobetta Zannin, 2011. Acoustic evaluation of a contemporary church based on in situ measurements of reverberation time, definition, and computer-predicted speech transmission index. *Building and Environment*, 46 (2), 511–517.

Rablen, M., 2010. Performance targets, effort and risk-taking. *Journal of Economic Psychology*, 31 (4), 687–697.

RAEng, 2013. Smart buildings: people and performance. London: Royal Academy of Engineering.

Rafiq, Y. and C. Sui, 2010. Interactive visualisation: a support for system identification. *Advanced Engineering Informatics*, 24 (3), 355–366.

Raftery, P., M. Keane and J. O'Donnell, 2011. Calibrating whole building energy models: an evidence-based methodology. *Energy and Buildings*, 43 (9), 2356–2364.

Rahim, A., 2005. Performativity: beyond efficiency and optimization in architecture. In: Kolarevic, B. and A. Malkawi, eds., *Performative architecture: beyond instrumentality*. New York: Spon Press.

Rahman, S. H. Odeyinka, S. Perera and Y. Bi, 2012. Product-cost modelling approach for the development of a decision support system for optimal roofing material selection. *Expert Systems with Applications*, 39 (8), 6857–6871.

Rahman, A., A. Smith and N. Fumo, 2016. Performance modeling and parametric study of a stratified water thermal storage tank. *Applied Thermal Engineering*, 100, 668–679.

Rajasekar, E., A. Udaykumar, R. Soumya and R. Venkateswaran, 2015. Towards dynamic thermal performance benchmarks for naturally ventilated buildings in a hot-dry climate. *Building and Environment*, 88, 129–141.

Ramesh, T., R. Prakash, K. Shukla, 2010. Life cycle energy analysis of buildings: an overview. *Energy and Buildings*, 42 (10), 1592–1600.

Ramponi, R. and B. Blocken, 2012. CFD simulation of cross-ventilation for a generic isolated building: impact of computational parameters. *Building and Environment*, 53, 34–48.

Raphael, B. and I. Smith, 2003. *Fundamentals of Computer-Aided Engineering*. Chichester: Wiley.

Raslan, R. and M. Davies, 2010. Results variability in accredited building energy performance compliance demonstration software in the UK: an inter-model comparative study. *Journal of Building Performance Simulation*, 3 (1), 63–85.

Raslan, R. and M. Davies, 2012. Legislating building energy performance: putting EU policy into practice. *Building Research & Information*, 40 (3), 305–316.

Ratchev, S., E. Urwin, D. Muller, K. Pawar and I. Moulek, 2003. Knowledge based requirement engineering for one-of-a-kind complex systems. *Knowledge-Based Systems*, 16 (1), 1–5.

Raynor, P. and S. Chae, 2003. Dust loading on electrostatically charged filters in a standard test and a real HVAC system. *Filter + Separation*, 40 (2), 35–39.

Rebaño-Edwards, S., 2007. Modelling perceptions of building quality – a neural network approach. *Building and Environment*, 42 (7), 2762–2777.

Reddy, T., I. Maor and C. Panjapornpon, 2007a. Calibrating detailed building energy simulation programs with measured data – Part I: general methodology. *HVAC&R Research*, 13 (2), 221–241.

Reddy, T., I. Maor and C. Panjapornpon, 2007b. Calibrating detailed building energy simulation programs with measured data – Part II: Application to three case study buildings. *HVAC&R Research*, 13 (2), 243–265.

Redlich, C., J. Sparer and M. Cullen, 1997. Sick-building syndrome. *The Lancet*, 349 (9057), 1013–1016.

Reeves, D., T. Doran, J. Valderas, E. Kontopantelis, P. Trueman, M. Sutton, S. Campbell and H. Lester, 2010. How to identify when a performance indicator has run its course. *British Medical Journal*, 340 (7752), 899–901.

Reichard, G. and K. Papamichael, 2005. Decision-making through performance simulation and code compliance from the early schematic phases of building design. *Automation in Construction*, 14 (2), 173–180.

Reinhart, C. and S. Selkowitz, 2006. Daylighting – light, form, and people. *Energy and Buildings*, 38 (7), 715–717.

Reinhart, C., 2004. Lightswitch-2002: a model for manual and automated control of electric lighting and blinds. *Solar Energy*, 77 (1), 15–28.

Reinhart, C. and C. Davila, 2016. Urban building energy modeling – a review of a nascent field. *Building and Environment*, 97, 196–202.

Reinhart, C. and A. Fitz, 2006. Findings from a survey on the current use of daylight simulations in building design. *Energy and Buildings*, 38 (7), 824–835.

Reinhart, C. and O. Walkenhorst, 2001. Validation of dynamic radiance-based daylight simulations for a test office with external blinds. *Energy and Buildings*, 33 (7), 683–697.

Reinhart, C. and J. Wienold, 2011. The daylight design dashboard – a simulation-based design analysis for daylit spaces. *Building and Environment*, 46 (2), 386–396.

Reinhart, C., T. Dogan, D. Ibarra and H. Wasilowski Samuelson, 2012. Learning by playing – teaching energy simulation as a game. *Journal of Building Performance Simulation*, 5 (6), 359–368.

Rekola, M., T. Mäkeläinen and T. Häkkinen, 2012. The role of design management in the sustainable building process. *Architectural Engineering and Design Management*, 8 (2), 78–89.

Remillieux, M., C. Corcoran, T. Haac, R. Burdisso and U. Svensson, 2012. Experimental and numerical study on the propagation of impulsive sound around buildings. *Applied Acoustics*, 73 (10), 1029–1044.

Ren, J., 2013. High-performance building design and decision-making support for architects in the early design phases. Licentiate thesis. Stockholm: The Royal Institute of Technology.

Ren, Z., C. Anumba, G. Augenbroe and T. Hassan, 2008. A functional architecture for an e-Engineering hub. *Automation in Construction*, 17 (8), 930–939.

Ren, Z., C. Anumba and F. Yang, 2013. Development of CDPM matrix for the measurement of collaborative design performance in construction. *Automation in Construction*, 32, 14–23.

Rezaee, R., J. Brown, G. Augenbroe and J. Kim, 2015. Assessment of uncertainty and confidence in building design exploration. *Artificial Intelligence for Engineering Design, Analysis and Manufacturing*, 29 (4), 429–441.

RIBA, 2013. RIBA Plan of Work 2013. Bristol: Royal Institute of British Architects. Available from www.ribaplanofwork.com [Accessed 31 January 2016].

Rietbergen, M., A. Van Rheede and K. Blok, 2015. The target-setting process in the CO_2 performance ladder: does it lead to ambitious goals for carbon dioxide emission reduction? *Journal of Cleaner Production*, 103, 549–561.

Riley, B., 2017. The state of the art of living walls: lessons learned. *Building and Environment*, 114, 219–232.

Riley, M. and A. Cotgrave, 2009. *Construction technology 2: industrial and commercial building*. Basingstoke: Palgrave Macmillan, 2nd edition.

Riley, M. and A. Cotgrave, 2013. *Construction technology 1: house construction*. Basingstoke: Palgrave Macmillan, 3rd edition.

Rittel, H. and M. Webber, 1973. Dilemmas in a general theory of planning. *Policy Science*, 4 (2), 155–169.

Roaf, S., M. Fuentes and S. Thomas, 2003. *Ecohouse – a design guide*. Oxford: Architectural Press.

Roberts, A. and A. Marsh, 2001. Ecotect: environmental prediction in architectural education. In: Pentilla, H., ed., *Architectural Information Management: 19th eCAADe Conference*, Helsinki, Finland, 29–31 August 2001.

Robertson, S. and J. Robertson, 2013. *Mastering the requirement process: getting requirements right*. Upper Saddle River, NJ: Addison-Wesley (Pearson Education), 3rd edition.

Robinson, O., S. Kemp and I. Williams, 2015. Carbon management at universities: a reality check. *Journal of Cleaner Production*, 106, 109–118.

Roca, P., M. Cervera, L. Pelà, R. Clemente and M. Chiumenti, 2013. Continuum FE models for the analysis of Mallorca Cathedral. *Engineering Structures*, 46, 653–670.

Rockett, P. and E. Hathway, 2017. Model-predictive control for non-domestic buildings: a critical review and prospects. *Building Research & Information*, 45 (5), 556–571.

Rodrigo, J., J. van Beeck and J. Buchlin, 2012. Wind engineering in the integrated design of process Elisabeth Antarctic base. *Building and Environment*, 52, 1–18.

Rodrigues, E., A. Amaral, A. Gaspar and A. Gomes, 2015. How reliable are geometry-based building indices as thermal performance indicators? *Energy Conversion and Management*, 101, 561–578.

Roh, S., Z. Aziz and F. Peña-Mora, 2011. An object-based 3D walk-through model for interior construction progress monitoring. *Automation in Construction*, 20 (1), 66–75.

Roldán, M., S. Gonnet and H. Leone, 2010. TracED: a tool for capturing and tracing engineering design processes. *Advances in Engineering Software*, 41 (9), 1087–1109.

Roozen, N., L. Labelle, M. Rychtáriková and C. Glorieux, 2015. Determining radiated sound power of building structures by means of laser Doppler vibrometry. *Journal of Sound and Vibration*, 346, 81–99.

Rose, H., 2014. Teaching rethorical performance in the humanities. *Text and Performance Quarterly*, 34 (1), 118–119.

Rosenman, M. and J. Gero, 1998. Purpose and function in design: from the socio-cultural to the techno-physical. *Design Studies*, 19 (2), 161–186.

Roth, A., D. Goldwasser and A. Parker, 2016. There's a measure for that! *Energy and Buildings*, 117, 321–331.

Rouault, F., D. Bruneau, P. Sebastian and J. Nadeau, 2016. Use of a latent heat thermal energy storage system for cooling a light-weight building: experimentation and co-simulation. *Energy and Buildings*, 127, 479–487.

Royapoor, M. and T. Roskilly, 2015. Building model calibration using energy and environmental data. *Energy and Buildings*, 94, 109–120.

Royon, L., L. Karim and A. Bontemps, 2013. Thermal energy storage and release of a new component with PCM for integration in floors for thermal management of buildings. *Energy and Buildings*, 63, 29–35.

Ruiz, J., J. Segura and I. Sirvent, 2015. Benchmarking and target setting with expert preferences: an application to the evaluation of educational performance of Spanish universities. *European Journal of Operational Research*, 242 (2), 594–605.

Rüppel, U. and K. Schatz, 2011. Designing a BIM-based serious game for fire safety evacuations. *Advanced Engineering Information*, 25 (4), 600–611.

Rush, R., ed., 1986.*The building systems integration handbook*. Boston, MA: The American Institute of Architects.

Russell-Smith, S., M. Lepech, R. Fruchter and A. Littman, 2015a. Impact of progressive sustainable target value assessment on building design decisions. *Building and Environment*, 85, 52–60.

Russell-Smith, S., M. Lepech, R. Fruchter and Y. Meyer, 2015b. Sustainable target value design: integrating life cycle assessment and target value design to improve building energy and environmental performance. *Journal of Cleaner Production*, 88, 43–51.

Ryan, E. and T. Sanquist, 2012. Validation of building energy modeling tools under idealized and realistic conditions. *Energy and Buildings*, 47, 375–382.

Rysanek, A. and R. Choudhary, 2012. A decoupled whole-building simulation engine for rapid exhaustive search of low-carbon and low-energy building refurbishment options. *Building and Environment*, 50, 21–33.

Rysanek, A. and R. Choudhary, 2013. Optimum building energy retrofits under technical and economic uncertainty. *Energy and Buildings*, 57, 324–337.

Saari, A. and L. Aalto, 2006. Indoor environment quality contracts in building projects. *Building Research & Information*, 34 (1), 66–74.

Saary, M., 2008. Radar plots: a useful way for presenting multivariate health care data. *Journal of Clinical Epidemiology*, 61 (4), 311–317.

Sabol, T. and D. Nishi, 2011. Application of performance-based design to an eccentrically braced frame structure. *The Structural Design of Tall and Special Buildings*, 20, 76–84.

Sagun, A., D. Bouchlaghem and C. Anumba, 2011. Computer simulations vs. building guidance to enhance evacuation performance of buildings during emergency events. *Simulation Modelling Practice and Theory*, 19 (3), 1007–1019.

Saha, S., A. Guha and S. Roy, 2012. Experimental and computational investigation of indoor air quality inside several community kitchens in a large campus. *Building and Environment*, 52, 177–190.

Sahal, N. and M. Lacasse, 2008. Proposed method for calculating water penetration test parameters of wall assemblies as applied to Istanbul, Turkey. *Building and Environment*, 43 (7), 1250–1260.

Sahlin, P., L. Eriksson, P. Grozman, H. Johnsson, A. Shapovalov and M. Vuolle, 2004. Whole-building simulation with symbolic DAE equations and general purpose solvers. *Building and Environment*, 39 (8), 949–958.

Saïd, M., 2007. Measurement methods of moisture in building envelopes – a literature review. *International Journal of Architectural Heritage*, 1 (3), 293–310.

Sameni, S., M. Gaterell, A. Montazami and A. Ahmed, 2015. Overheating investigation in UK social housing flats built to the Passivhaus standard. *Building and Environment*, 92, 222–235.

Samer, M., H. Müller, M. Fiedler, C. Ammon, M. Gläser, W. Berg, P. Sanftleben and R. Brunsch, 2011. Developing the ^{85}Kr tracer gas technique for air exchange rate measurements in naturally ventilated animal buildings. *Biosystems Engineering*, 109 (4), 276–287.

Samer, M., C. Ammon, C. Loebsin, M. Fiedler, W. Berg, P. Sanftleben and R. Brunsch, 2012. Moisture balance and tracer gas technique for ventilation rates measurement and greenhouse gases and ammonia emissions quantification in naturally ventilated buildings. *Building and Environment*, 50, 10–20.

Samuelson, H., A. Ghorayshi and C. Reinhart, 2016. Analysis of a simplified calibration procedure for 18 design-phase building energy models. *Journal of Building Performance Simulation*, 9 (1), 17–29.

Sánchez de Rojas, M., F. Marín, F. Frías, E. Valenzuela and O. Rodríguez, 2011. Influence of freezing test methods, composition and microstructure on frost durability assessment of clay roofing tiles. *Construction and Building Materials*, 25 (6), 2888–2897.

Sandels, C., J. Widén and L. Nordström, 2014. Forecasting household consumer electricity load profiles with a combined physical and behavioral approach. *Applied Energy*, 131, 267–278.

Sandels, C., J. Widén, L. Nordström and E. Andersson, 2015. Day-ahead predictions of electricity consumption in a Swedish office building from weather, occupancy and temporal data. *Energy and Buildings*, 108, 279–290.

Sangi, R., M. Baranski, J. Oltmanns, R. Streblow and D. Müller, 2016a. Modeling and simulation of the heating circuit of a multi-functional building. *Energy and Buildings*, 110, 13–22.

Sangi, R., P. Martínez Martín and D. Müller, 2016b. Thermoeconomic analysis of a building heating system. *Energy*, 111, 351–363.

Sanguinetti, P., 2012. Integrated performance framework to guide façade retrofit. PhD thesis. Atlanta, GA: Georgia Institute of Technology.

dos Santos, G. and N. Mendes, 2006. Simultaneous heat and moisture transfer in soils combined with building simulation. *Energy and Buildings*, 38 (4), 303–314.

dos Santos, J., C. Heuser, V. Moreira and L. Wives, 2011. Automatic threshold estimation for data matching applications. *Information Sciences*, 181 (13), 2685–2699.

Saraiji, R., M. Al Safadi, N. Al Ghaithi and R. Mistrick, 2015. A comparison of scale-model photometry and computer simulation in day-lit spaces using a normalized daylight performance index. *Energy and Buildings*, 89, 76–86.

Sartori, I., A. Napolitano and K. Voss, 2012. Net zero energy buildings: a consistent definition framework. *Energy and Buildings*, 48, 220–232.

Sasic Kalagasidis, A., 2004. HAM-Tools: An integrated simulation tool for heat, air and moisture transfer analyses in building physics. PhD thesis. Gothenburg: Chalmers University of Technology.

Sasic Kalagasidis, A., 2014. A multi-level modelling and evaluation of thermal performance of phase-change materials in buildings. *Journal of Building Performance Simulation*, 7 (4), 289–308.

Savanović, P., 2009. Integral design method in the context of sustainable building design. PhD thesis. Eindhoven: Eindhoven University Press.

Schade, J., T. Olofsson and M. Schreyer, 2011. Decision-making in a model-based design process. *Construction Management and Economics*, 29 (4), 371–382.

Scheck, J. and B. Gibbs, 2015. Impacted lightweight stairs as structure-borne sound sources. *Applied Acoustics*, 90, 9–20.

Schein, J. and S. Bushby, 2006. A hierarchical rule-based fault detection and diagnostic method for HVAC systems. *HVAC&R Research*, 12 (1), 111–125.

Schein, J., S. Bushby, N. Castro and J. House, 2006. A rule-based fault detection method for air handling units. *Energy and Buildings*, 38 (12), 1485–1492.

Schiavi, A., A. Prato and A. Pavoni Belli, 2015. The 'dust spring effect' on the impact sound reduction measurement accuracy of floor coverings in laboratory. *Applied Acoustics*, 97, 115–120.

van Schijndel, A., 2014. A review of the application of Simulink S-functions to multi-domain modelling and building simulation. *Journal of Building Performance Simulation*, 7 (3), 165–178.

Schleibinger, H. and H. Rüden, 1999. Air filters from HVAC systems as a possible source of volatile organic compounds (VOC) – laboratory and field assays. *Atmospheric Environment*, 33 (28), 4571–4577.

Schmid, A., 2008. The introduction of building simulation tools into an architectural faculty: preliminary findings. *Journal of Building Performance Simulation*, 1 (3), 197–208.

Schneider, F. and B. Berenbach, 2013. A literature survey on international standards for systems requirements engineering. *Procedia Computer Science*, 16, 796–805.

Scholz, R. and R. Hansmann, 2007. Combining experts' risk judgments on technology performance of phytoremediation: self-confidence ratings, averaging procedures, and formative consensus building. *Risk Analysis*, 27 (1), 225–240.

Schramm, U., 2005. Phase 1: strategic planning – effectiveness review. In: Preiser, W. and J. Vischer, eds., *Assessing building performance*. Oxford: Butterworth-Heinemann.

Schröter, K., H. Kreibich, K. Vogel, C. Riggelsen, F. Scherbaum and B. Merz, 2014. How useful are complex flood damage models? *Water Resources Research*, 50 (4), 3378–3395.

Schweber, L. and H. Haroglu, 2014. Comparing the fit between BREEAM assessment and design processes. *Building Research & Information*, 42 (3), 300–317.

Schweder, A., 2012. Performance Architecture. *Le Journal Spéciale'Z*, 4, 102–129.

Schwitter, C., 2005. Engineering complexity: performance-based design in use. In: Kolarevic, B. and A. Malkawi, eds., *Performative architecture: beyond instrumentality*. New York: Spon Press.

Scofield, J., 2002. Early performance of a green academic building. *ASHRAE Transactions*, 108, 1214–1230.

SEBoK, 2014. *The Guide to the Systems Engineering Body of Knowledge, v.1.3*. Hoboken, NJ: Body of Knowledge and Curriculum to Advance Systems Engineering/Trustees of the Stevens Institute of Technology.

Sequeira, M. and V. Cortínez, 2016. Optimal acoustic design of multi-source industrial buildings by means of a simplified acoustic diffusion model. *Applied Acoustics*, 103, 71–81.

Sergent, J., E. Zuck, S. Terriah and B. MacDonald, 1992. Distributed neural network underlying musical sight-reading and keyboard performance. *Science*, 257 (5066), 106–109.

Serra, V., F. Zanghirella and M. Perino, 2010. Experimental evaluation of a climate facade: energy efficiency and thermal comfort performance. *Energy and Buildings*, 42 (1), 50–62.

Sesana, M., M. Grecchi, G. Salvalai and C. Rasica, 2016. Methodology of energy efficient building refurbishment: application on two university campus-building case studies in Italy with engineering students. *Journal of Building Engineering*, 6, 54–64.

Sexton, M. and P. Barrett, 2005. Performance-based building and innovation: balancing client and industry needs. *Building Research & Information*, 33 (2), 142–148.

Shabunko, V., C. Lim, S. Brahim and S. Mathew, 2014. Developing building benchmarking for Brunei Darussalam. *Energy and Buildings*, 85, 79–85.

Shah, A. and Y. Ribakov, 2008. Non-destructive measurements of crack assessment and defect detection in concrete structures. *Materials and Design*, 29 (1), 61–69.

Shan, Y. and L. Yang, 2016. Fast and frugal heuristics and naturalistic decision making: a review of their commonalities and differences. *Thinking & Reasoning*, 23 (1), 10–32.

Shao, Y., P. Geyer and W. Lang, 2014. Integrating requirement analysis and multi-objective optimization for office building retrofit strategies. *Energy and Buildings*, 82, 356–368.

Shea, K., R. Aish and M. Gourtovaia, 2005. Towards integrated performance-driven generative design tools. *Automation in Construction*, 14 (2), 253–264.

Shea, A., M. Lawrence and P. Walker, 2012. Hygrothermal performance of an experimental hemp-lime building. *Construction and Building Materials*, 36, 270–275.

Shehata, M., A. Eberlein and A. Fapojuwo, 2007. A taxonomy for identifying requirement interactions in software systems. *Computer Networks*, 51 (2), 398–425.

Shen, T., 2005. ESM: a building evacuation simulation model. *Building and Environment*, 40 (5), 671–680.

Shen, T., Y. Huang and S. Chien, 2008. Using fire dynamic simulation (FDS) to reconstruct an arson fire scene. *Building and Environment*, 43 (6), 1036–1045.

Shen, W., Q. Hao, H. Mak, J. Neelamkavil, H. Xie, J. Dickinson, R. Thomas, A. Pardasani and H. Xue, 2010. Systems integration and collaboration in architecture, engineering, construction and facilities management: a review. *Advanced Engineering Informatics*, 24 (2), 196–207.

Shen, W., Q. Shen and Z. Xialong, 2012a. A user pre-occupancy evaluation method for facilitating the designer-client communication. *Facilities*, 30 (7/8), 302–323.

Shen, W., Q. Shen and Q. Sun, 2012b. Building information modeling-based user activitiy simulation and evaluation method for improving designer-user communications. *Automation in Construction*, 21, 148–160.

Shen, X., C. Zong and G. Zhang, 2012c. Optimization of sampling positions for measuring ventilation rates in naturally ventilated buildings using tracer gas. *Sensors*, 12 (9), 11966–11988.

Shendarkar, A., K. Vasudevan, S. Lee and Y. Son, 2008. Crowd simulation for emergency response using BDI agents based on immersive virtual reality. *Simulation Modelling Practice and Theory*, 16 (9), 1415–1429.

Shi, X. and W. Yang, 2013. Performance-driven architectural design and optimization technique from a perspective of architects. *Automation in Construction*, 32, 125–135.

Shi, J., A. Ren and C. Chen, 2009a. Agent-based evacuation model of large public buildings under fire conditions. *Automation in Construction*, 18 (3), 338–347.

Shi, L., Q. Xie, X. Cheng, L. Chen, Y. Zhou and R. Zhang, 2009b. Developing a database for emergency evacuation model. *Building and Environment*, 44 (8), 1724–1729.

Shi, X., N. Xie, K. Fortune and J. Gong, 2012. Durability of steel reinforced concrete in chloride environments: a review. *Construction and Building Materials*, 30, 125–138.

Shi, Q., Y. Yan, J. Zuo and T. Yu, 2016. Objective conflicts in green buildings projects: a critical analysis. *Building and Environment*, 96, 107–117.

Shields, T. and K. Boyce, 2000. A study of evacuation from large retail stores. *Fire Safety Journal*, 35 (1), 25–49.

Shiflet, A. and G. Shiflet, 2014. *Introduction to computational science: modeling and simulation for the sciences*. Princeton, NJ/Oxford: Princeton University Press, 2nd edition.

Shih, H., 2014. A robust occupancy detection and tracking algorithm for the automatic monitoring and commissioning of a building. *Energy and Buildings*, 77, 270–280.

Shinkle, G., 2012. Organizational aspirations, reference points, and goals: building on the past and aiming for the future. *Journal of Management*, 38 (1), 415–455.

Shohet, I., S. Lavy-Leibovich and D. Bar-On, 2003. Integrated maintenance monitoring of hospital buildings. *Construction Management and Economics*, 21 (2), 219–228.

Siebert, S. and J. Teizer, 2014. Mobile 3D mapping for surveying earthwork projects using an unmanned aerial vehicle (UAV) system. *Automation in Construction*, 41, 1–14.

Siefert, A. and F. Henkel, 2014. Nonlinear analysis of commercial aircraft impact on a reactor building – comparison between integral and decoupled crash simulation. *Nuclear Engineering and Design*, 269, 130–135.

Silva, A., J. de Brito and P. Gaspar, 2012. Application of the factor method to maintenance decision support for stone cladding. *Automation in Construction*, 22, 165–174.

Simon, H.A., 1996. *The sciences of the artificial*. Cambridge, MA: MIT Press, 3rd edition.

Simon, H.A., 1997. *Administrative behavior: a study of decision-making processes in administrative organizations*. New York: The Free Press, 4th edition.

Simon, T., 2010. Just who is at risk? The ethics of environmental regulation. *Human and Experimental Toxicology*, 30 (8), 795–819.

Singh, T., D. Page and J. van der Walls, 2014. The development of accelerated test methods to evaluate the durability of framing timber. *International Biodeterioration & Biodegradation*, 94, 63–68.

Singhaputtangkul, N., S. Low, A. Teo and B. Hwang, 2013. Knowledge-based decision support system quality function deployment (KBDSS-QFD) tool for assessment of building envelopes. *Automation in Construction*, 35, 314–328.

Singhaputtangkul, N., S. Low, A. Teo and B. Hwang, 2014. Analysis of criteria for decision making to achieve sustainability and buildability in building envelope design. *Architectural Science Review*, 57 (1), 20–30.

Sinnot, D. and M. Dyer, 2012. Air-tightness field data for dwellings in Ireland. *Building and Environment*, 51, 269–275.

Sinopoli, J., 2016. *Advanced technology for smart buildings*. Norwood, MA: Artech House.

Skibniewski, M. and S. Gosh, 2009. Determination of key performance indicators with enterprise resource planning systems in engineering construction firms. *Journal of Construction Engineering and Management*, 135 (1), 965–978.

Skylaris, C., 2016. A benchmark for materials simulation. *Science*, 352 (6280), 1394–1395.

Smith, D., 2011. *Reliability, maintainability and risk*. Oxford: Butterworth-Heinemann, 8th edition.

Smythe, K., 2014. A historian's critique of sustainability. *Culture Unbound*, 6 (5), 913–929.

Soderberg, M., S. Kalagnanam, N. Sheehan and G. Vaidyanathanm, 2011. When is a balanced scorecard not a balanced scorecard? *International Journal of Productivity and Performance Management*, 60 (7), 688–708.

Sodja, A. and B. Zupančič, 2009. Modelling thermal processes in buildings using an object-oriented approach and Modelica. *Simulation Modelling Practice and Theory*, 17 (6), 1143–1159.

Soebarto, V. and T. Williamson, 2001. Multi-criteria assessment of building performance: theory and implementation. *Building and Environment*, 36 (6), 681–690.

Sommerville, J., N. Craigh and S. Bowden, 2004. The standardisation of construction snagging. *Structural Survey*, 22 (5), 251–258.

Sommerville, J. and J. McCosh, 2006. Defects in new homes: an analysis of data on 1.696 new UK houses. *Structural Survey*, 24 (1), 6–21.

Song, S. and J. Haberl, 2013. Analysis of the impact of using synthetic data correlated with measured data on the calibrated as-built simulation of a commercial building. *Energy and Buildings*, 67, 97–107.

Song, S., J. Yang and N. Kim, 2012. Development of a BIM-based structural framework optimization and simulation system for building construction. *Computers in Industry*, 63 (9), 895–912.

Song, Y., J. Gong, L. Niu, Y. Li, Y. Yiang, W. Zhang and T. Cui, 2013. A grid-based spatial data model for the simulation and analysis of individual behaviours in micro-spatial environments. *Simulation Modelling Practice and Theory*, 38, 58–68.

Sörqvist, P., 2016. Grand challenges in environmental psychology. *Frontiers in Psychology*, 7 (754), 583.

Sorrell, S., 2007. The economics of energy service contracts. *Energy Policy*, 35 (1), 507–521.

Spekkink, D., 2005. Performance based design of buildings. Rotterdam: CIB.

Spitler, J., 2006. Building performance simulation: the now and not yet. *HVAC&R Research*, 12 (3a), 549–551.

Spodek, J. and E. Rosina, 2009. Application of infrared thermography to historic building investigation. *Journal of Architectural Conservation*, 15 (1), 65–81.

Spyridis, P., S. Konstantis and A. Gakis, 2016. Performance indicator of tunnel linings under geotechnical uncertainty. *Geomechanics and Tunneling*, 9 (2), 158–164.

Srebric, J., 2010. Computational Fluid Dynamics (CFD) challenges in simulation building airflows. *HVAC&R Research*, 16 (6), 729–730.

Srivastav, A., A. Tewari and B. Dong, 2013. Baseline building energy modeling and localized uncertainty quantification using Gaussian mixture models. *Energy and Buildings*, 65, 438–447.

Stafoggia, M., A. Lallo, D. Fusco, A. Barone, M. D'Ovidio, C. Sorge and A. Perucci, 2011. Spie charts, target plots, and radar plots for displaying comparative outcomes of health care. *Journal of Clinical Epidemiology*, 64 (7), 770–778.

Stanton, N. and B. Wong, 2010. Explorations into naturalistic decision making with computers. *International Journal of Human-Computer Interaction*, 26 (2/3), 99–107.

Steeman, H., M. van Belleghem, A. Janssens and M. de Paepe, 2009. Coupled simulation of heat and moisture transport in air and porous materials for the assessment of moisture related damage. *Building and Environment*, 44 (10), 2176–2184.

Steer, K., A. Wirth and S. Halgamuge, 2011. Control period selection for improved operating performance in district heating networks. *Energy and Buildings*, 43 (2/3), 605–613.

Stein, B., J. Reynolds, W. Grondzik and A. Kwok, 2006. *Mechanical and electric equipment for buildings*. Hoboken, NJ: John Wiley & Sons, 10th edition.

Stempfle, J. and P. Badke-Schaub, 2002. Thinking in design teams – an analysis of team communication. *Design Studies*, 23 (5), 473–496.

Stephens, B. and J. Siegel, 2012. Penetration of ambient submicron particles into single-family residences and associations with building characteristics. *Indoor Air*, 22 (6), 501–513.

Stern, P., 2000. Psychology and the science of human-environment interactions. *American Psychologist*, 55 (5), 523–530.

Stoppel, M. and F. Leite, 2013. Evaluating building energy model performance of LEED buildings: identifying potential sources of error through aggregate analysis. *Energy and Buildings*, 65, 185–196.

Stossel, Z., M. Kissinger and A. Meir, 2015. Assessing the state of environmental quality in cities – a multi-component urban performance (EMCUP) index. *Environmental Pollution*, 206, 679–687.

van Straaten, J., 1967. *Thermal Performance of Buildings*. Amsterdam: Elsevier.

Strachan, P., 1993. Model validation using the PASSYS test cells. *Building and Environment*, 28 (2), 153–165.

Strachan, P., 2008. Simulation support for performance assessment of building components. *Building and Environment*, 43 (2), 228–236.

Strachan, N., 2011. UK energy policy ambition and UK energy modelling – fit for purpose? *Energy Policy*, 39 (3), 1037–1040.

Strachan, M. and P. Banfill, 2017. Energy-led refurbishment of non-domestic buildings: ranking measures by attributes. *Facilities*, 35 (5/6), 286–302.

Strachan, P., G. Kokogiannakis and I. Macdonald, 2008. History and development of validation with the ESP-r simulation program. *Building and Environment*, 43 (4), 601–609.

Strachan, P., K. Svehla, I. Heusler and M. Kersken, 2016. Whole model empirical validation on a full-scale building. *Journal of Building Performance Simulation*, 9 (4), 331–350.

Streiner, D., 2013. *A Guide for the Statistically Perplexed – Selected Readings for Clinical Researchers*. Toronto, ON: University of Toronto Press.

Stroud, K.A. and D.J. Booth, 2007. *Engineering mathematics*. Basingstoke: Palgrave MacMillan, 6th edition.

Struck, C., 2012. Uncertainty propagation and sensitivity analysis techniques in building performance simulation to support conceptual building and system design. PhD thesis. Eindhoven: Eindhoven University Press.

Struck, C., P. de Wilde, C. Hopfe and J. Hensen, 2009. An investigation of the option space in conceptual building design for advanced building simulation. *Advanced Engineering Informatics*, 23 (4), 386–395.

Stryer, D. and C. Clancy, 2003. Boosting performance measure for measure. *British Medical Journal*, 326 (7402), 1278–1279.

Sueyoshi, T., M. Goto and T. Ueno, 2010. Performance analysis of US coal-fired power plants by measuring three DEA efficiencies. *Energy Policy*, 38 (4), 1675–1688.

Suhonen, N. and L. Okkonen, 2013. The energy services company (ESCo) as business model for heat entrepreneurship – a case study of North Karelia, Finland. *Energy Policy*, 61, 783–787.

Sun, Y., 2014. Closing the building energy performance gap by improving our predictions. PhD thesis. Atlanta, GA: Georgia Institute of Technology.

Sun, D., C. Jin, A. van Schaik and D. Cabrera, 2009. The design and evaluation of an economically constructed anechoic chamber. *Architectural Science Review*, 52 (4), 312–319.

Sun, Y., P. Huang and G. Huang, 2015. A multi-criteria system design optimization for net zero energy buildings under uncertainties. *Energy and Buildings*, 97, 196–204.

Suryadevara, N., S. Mukhopadhyay, S. Kelly and S. Gill, 2015. WSN-based smart sensors and actuator for power management in intelligent buildings. *IEEE/ASME Transactions on Mechatronics*, 20 (2), 564–571.

Swann, J. 2008. Smart homes: intelligent buildings. *International Journal of Therapy and Rehabilitation*, 15 (6), 273–278.

Swarup, L., S. Korkmaz and D. Riley, 2011. Project delivery metrics for sustainable, high-performance buildings. *Journal of Construction Engineering and Management*, 137 (12), 1043–1051.

Sweets, 2016. Sweets™ Product Catalogs for Building Products [online]. New York: Dodge Data & Analytics, Inc. Available from http://sweets.construction/com [Accessed 26 June 2016].

Szigeti, F. and G. Davis, 2002. Using the ASTM/ANSI standards for whole building functionality and serviceability for major asset and portfolio decisions. *CIB W070 Global Symposium*, Glasgow, UK, 18–20 September 2002, pp. 507–521.

Szigeti, F. and G. Davis, 2005. Performance Based Building: conceptual framework. Rotterdam: CIB.

Szigeti, F., G. Davis and D. Hammond, 2005. Introducing the ASTM facilities evaluation method. In: Preiser, W. and J. Vischer, eds., *Assessing building performance*. Oxford: Butterworth-Heinemann.

Tabares-Velasco, P. and B. Griffith, 2012. Diagnostic test cases for verifying surface heat transfer algorithms and boundary conditions in building energy simulation programs. *Journal of Building Performance Simulation*, 5 (5), 329–346.

Tahmasebi, F. and A. Mahdavi, 2016. An inquiry into the reliability of window operation models in building performance simulation. *Building and Environment*, 105, 343–357.

Talbourdet, F., P. Michel, F. Andrieux, J. Millet, M. El Mankibi and B. Vinot, 2013. A knowledge-aid approach for designing high-performance buildings. *Building Simulation*, 6 (4), 337–350.

Talib, Y., P. Rajagopalan and R. Yang, 2013. Evaluation of building performance for strategic facilities management in healthcare: a case study of a public hospital in Australia, *Facilities*, 31 (13/14), 681–701.

Tam, V., C. Tam, S. Zeng and K. Chan, 2006. Environmental performance measurement indicators in construction. *Building and Environment*, 41 (2), 164–173.

Tan, L., M. Hu and H. Lin, 2015. Agent-based simulation of building evacuation: combining human behavior with predictable spatial accessibility in a fire emergency. *Information Sciences*, 295, 53–66.

Tang, S., 2010. Scale model study of balcony insertion losses on a building façade with non-parallel line sources. *Applied Acoustics*, 71 (10), 947–954.

Tang, L. and Q. Shen, 2013. Factors affecting effectiveness and efficiency of analyzing stakeholders' needs at the briefing stage of public private partnership projects. *International Journal of Project Management*, 31 (4), 513–521.

Tang, C., C. Lin and Y. Hsu, 2008. Exploratory research on reading cognition and escape-route planning using building evacuation plan diagrams. *Applied Ergonomics*, 39 (2), 209–217.

Taplin, R., 2012. Competitive importance-performance analysis of an Australian wildlife park. *Tourism Management*, 33 (1), 29–37.

Tariku, F., K. Kumaran and P. Fazio, 2010. Integrated analysis of whole building heat, air and moisture transfer. *International Journal of Heat and Mass Transfer*, 53 (15/16), 3111–3120.

Tavares, R. and E. Galea, 2009. Evacuation modelling analysis within the operational research context: a combined approach for improving enclosure designs. *Building and Environment*, 44 (5), 1005–1016.

Taylor, T., J. Counsell and S. Gill, 2014. Combining thermography and computer simulation to identify and assess insulation defects in the construction of building facades. *Energy and Buildings*, 76, 130–142.

Tchidi, M., Z. He and Y. Li, 2012. Process and quality improvement using Six Sigma in construction industry. *Journal of Civil Engineering and Management*, 18 (2), 158–172.

Teizer, J., 2015. Status quo and open challenges in vision-based sensing and tracking of temporary resources on infrastructure construction sites. *Advanced Engineering Informatics*, 29 (2), 225–238.

Tham, K. and J. Pantelic, 2010. Performance evaluation of the coupling of a desktop personalized ventilation air terminal device and desk mounted fans. *Building and Environment*, 45 (9), 1941–1950.

Thanachareonkit, A. and J. Scartezzini, 2010. Modelling complex fenestration systems using physical and virtual models. *Solar Energy*, 84 (4), 563–586.

Then, D., 2005. Phase 6: Adaptive reuse/recycling – market needs assessment. In: Preiser, W. and J. Vischer, eds., *Assessing building performance*. Oxford: Butterworth-Heinemann.

Thiel, C., K. LaScola Needy, R. Ries, D. Hupp and M. Bilec, 2014. Building design and performance: a comparative longitudinal assessment of a children's hospital. *Building and Environment*, 78, 130–136.

Thompson, A., 2002. *Architectural Design Procedures*. Oxford: Architectural Press.

Thompson, B. and L. Bank, 2010. Use of system dynamics as a decision-making tool in building design and operation. *Building and Environment*, 45 (4), 1006–1015.

Thompson, P., D. Nilsson, K. Boyce and D. McGrath, 2015. Evacuation models are running out of time. *Fire Safety Journal*, 78, 251–261.

Thomson, W., 1889. *Popular lectures and addresses*. London: Macmillan.

Tian, W. and P. de Wilde, 2011. Thermal building simulation using the UKCP09 probabilistic climate projections. *Journal of Building Performance Simulation*, 4 (2), 105–124.

Tian, Z., J. Love and W. Tian, 2009. Applying quality control in building energy modelling: comparative simulation of a high performance building. *Journal of Building Performance Simulation*, 2 (3), 163–178.

Tian, W., R. Choudhary, G. Augenbroe and S. Lee, 2015. Importance analysis and meta-model construction with correlated variables in evaluation of thermal performance of campus buildings. *Building and Environment*, 92, 61–74.

Tingvall, C., H. Stigson, L. Eriksson, R. Johansson, M. Krafft and A. Lie, 2010. The properties of Safety Performance Indicators in target setting, projections and safety design of the road transport system. *Accident Analysis and Prevention*, 42 (2), 372–376.

Todorovic, M. and J. Kim, 2012. Buildings energy sustainability and health research via interdisciplinarity and harmony. *Energy and Buildings*, 47, 12–18.

Tomažič, S., V. Logar, Ž. Kristl, A. Krainer, I. Škrjanc and M. Košir, 2013. Indoor-environment simulator for control design purposes. *Building and Environment*, 70, 60–72.

Tomić, B. and T. Milić, 2013. Automated interpretation of key performance indicator values and its application in education. *Knowledge-Based Systems*, 37, 250–260.

Tominaga, Y. and B. Blocken, 2016. Wind tunnel analysis of flow and dispersion in cross-ventilated isolated buildings: impact of opening positions. *Journal of Wind Engineering and Industrial Aerodynamics*, 155, 74–88.

Tong, Y., S. Tang, J. Kang, A. Fung and M. Yeung, 2015. Full scale field study of sound transmission across plenum windows. *Applied Acoustics*, 89, 244–253.

Toor, S. and S. Ogunlana, 2010. Beyond the 'iron triangle': stakeholder perception of key performance indicators (KPIs) for large-scale public sector development projects. *International Journal of Project Management*, 28 (3), 228–236.

Torcellini, P., S. Ples, M. Deru, B. Griffith, N. Long and R. Judkoff, 2006. *Lessons learned from case studies of six high-performance buildings*. Golden, CO: National Renewable Energy Laboratory.

Torfs, T., T. Sterken, S. Brebels, J. Santana, R. van den Hoven, V. Spiering, N. Bertsch, D. Trapani and D. Zonta, 2013. Low power wireless sensor network for building monitoring. *IEEE Sensors Journal*, 13 (3), 909–915.

Toyokawa, W., H. Kim and T. Kameda, 2014. Human collective intelligence under dual exploration-exploitation dilemmas. *PLoS ONE*, 9 (4), e95789.

Trčka, M., J. Hensen and M. Wetter, 2009. Co-simulation of innovative integrated HVAC systems in buildings. *Journal of Building Performance Simulation*, 2 (3), 209–230.

Trčka, M., J. Hensen and M. Wetter, 2010. Co-simulation for performance prediction of integrated building and HVAC systems – an analysis of solution characteristics using a two-body system. *Simulation Modelling Practice and Theory*, 18 (7), 957–970.

van Treeck, C., J. Frisch, M. Pfaffinger, E. Rank, S. Paulke, I. Schweinfurth, R. Schwab, R. Hellwig and A. Holm, 2009. Integrated thermal comfort analysis using a parametric manikin model for interactive real-time simulation. *Journal of Building Performance Simulation*, 2 (4), 233–250.

Trencher, G., V. Broto, T. Takagi, Z. Sprigings, Y. Nishida and M. Yarime, 2016. Innovative policy practices to advance building energy efficiency and retrofitting: approaches, impacts and challenges in ten C40 cities. *Environmental Science & Policy*, 66, 353–365.

Trinius, W. and C. Sjöström, 2005. Service life planning and performance requirements. *Building Research and Information*, 33 (2), 173–181.

Trubiano, F., 2013. *Design and construction of high-performance homes*. Abingdon: Routledge.

Tsai, M., 2012. Evaluation of different loading simulation approaches for progressive collapse analysis of regular building frames. *Structure and Infrastructure Engineering*, 8 (8), 765–779.

Tsai, Y. and Y. Cheng, 2012. Analyzing key performance indicators (KPIs) for e-commerce and internet marketing of elderly products: a review. *Archives of Gerontology and Geriatrics*, 55 (1), 126–132.

Tsai, M. and Y. Lin, 2012. Modern development of an adaptive non-intrusive appliance load monitoring system in electricity energy conservation. *Applied Energy*, 96, 55–73.

Tsang, A., A. Jardine and H. Kolodny, 1999. Measuring maintenance performance: a holistic approach. *International Journal of Operations & Production Management*, 19 (7), 691–715.

Tsitsifli, S. and V. Kanakoudis, 2010. Predicting the behaviour of a pipe network using the 'critical Z-score' as its performance indicator. *Desalination*, 250 (1), 258–265.

Tufte, E., 2001. *The visual display of quantitative information*. Cheshire: Graphics Press.

Tuohy, P. and G. Murphy, 2015. Closing the gap in building performance: learning from BIM benchmark industries. *Architectural Science Review*, 58 (1), 47–56.

Turk, Ž., 2006. Construction informatics: definition and ontology. *Advanced Engineering Informatics*, 20 (2), 187–199.

Turkan, Y., F. Bosche, C. Haas and R. Haas, 2012. Automated progress tracking using 4D schedule and 3D sensing technologies. *Automation in Construction*, 22, 414–421.

Turner, W. and H. Awbi, 2015. Experimental investigation into the thermal performance of a residential hybrid ventilation system. *Applied Thermal Engineering*, 77, 142–152.

Turner, C. and M. Frankel, 2008. *Energy performance of LEED for new construction buildings*. White Salmon, WA: New Buildings Institute.

Turpin-Brooks, S. and G. Viccars, 2006. The development of robust methods of post occupancy evaluation. *Facilities*, 24 (5/6), 177–196.

Turrin, M., P. von Buelow and R. Stouffs, 2011. Design explorations of performance driven geometry in architectural design using parametric modeling and genetic algorithms. *Advanced Engineering Informatics*, 25 (4), 656–675.

Tuzmen, A., 2002. A distributed process management system for collaborative building design. *Engineering Construction and Architectural Management*, 9 (3), 209–221.

Ucar, A. and M. Inalli, 2008. Thermal and economic comparisons of solar heating systems with seasonal storage used in building heating. *Renewable Energy*, 33 (12), 2532–2539.

Ugwu, O. and T. Haupt, 2007. Key performance indicators and assessment methods for infrastructure sustainability – a South African construction industry perspective. *Building and Environment*, 42 (2), 665–680.

Ugwu, O., M. Kumaraswamy, A. Wong and S. Ng, 2006. Sustainability appraisal in infrastructure projects (SUSAIP) Part 1: development of indicators and computational methods. *Automation in Construction*, 15 (2), 239–251.

Uihlein, M., 2016. Ove Arup's total design, integrated project delivery, and the role of the architect. *Architectural Science Review*, 59 (2), 102–113.

Underwood, C.P. and F.W.H. Yik, 2004. *Modelling methods for energy in buildings*. Oxford: Blackwell.

Ünver, R., N. Akdağ, G. Gedik, L. Öztürk and Z. Karabiber, 2004. Prediction of building envelope performance in the design stage: an application for office buildings. *Building and Environment*, 39 (2), 143–152.

Uribe, O., J. San Martin, M. Garcia-Alegre, M. Santos and D. Guinea, 2015. Smart building: decision making architecture for thermal energy management. *Sensors*, 15 (11), 27543–27568.

US Department of Energy, 2016. EnergyPlus™ Version 8.7 Documentation: Engineering Reference. Washington, DC: US Department of Energy.

Utkin, L., 2006. A method for processing the unreliable expert judgments about parameters and probability distributions. *European Journal of Operational Research*, 175 (1), 385–398.

Vähä, P., T. Heikkilä, P. Kilpeläinen, M. Järviluoma and E. Gambao, 2013. Extending automation of building construction – survey on potential sensor technologies and robotic applications. *Automation in Construction*, 36, 168–178.

Vanhoucke, M., 2012. Measuring the efficiency of project control using fictitious and empirical project data. *International Journal of Project Management*, 30 (2), 252–263.

Vanier, D., M. Lacasess and A. Parsons, 1996. Using product models to represent user requirements. *Construction on the Information Highway, Proceeding of CIB Workshop W78*, Bled, Slovenia, 10–12 June 1996, pp. 511–524.

Vasilyev, G., V. Lichman, N. Peskov, M. Brodach, Y. Tabunshchikov and M. Kolesova, 2015. Simulation of heat and moisture transfer in a multiplex structure. *Energy and Buildings*, 86, 803–807.

Velikov, K. and G. Thün, 2013. Responsive building envelopes: characteristics and evolving paradigms. In: Trubiano, F., ed., *Design and construction of high-performance homes.* Abingdon: Routledge.

Velis, A., H. Giuliano and A. Méndez, 1995. The anechoic chamber at the Laboratory de acústica y luminotecnia CIC. *Applied Acoustics*, 44 (1), 79–94.

Ventura, C., A. Bebamzadeh and M. Fairhurst, 2015. Efficient performance-based design using parallel and cloud computing. *The Structural Design of Tall and Special Buildings*, 24 (17), 989–1001.

Verheij, J., 2005. Process-mediated Planning of AEC Projects through Structured Dialogues. PhD thesis. Atlanta, GA: Georgia Institute of Technology.

Vermaas, P., 2013. On the formal impossibility of analysing subfunctions as part of functions in design methodology. *Research in Engineering Design*, 24 (1), 19–32.

Vermaas, P. and K. Dorst, 2007. On the conceptual framework of John Gero's FBS-model and the prescriptive aims of design methodology. *Design Studies*, 28 (2), 133–157.

Veronica, D., 2013. Automatically detecting faulty regulation in HVAC controls. *HVAC&R Research*, 19 (4), 412–422.

Vesma, V., 2009. *Energy management principles and practice.* London: British Standards Institute.

Vieira, S., A. Silva, J. Sousa, J. de Brito and P. Gaspar, 2015. Modelling the service life of rendered facades using fuzzy systems. *Automation in Construction*, 51, 1–7.

Vigo, M. and G. Brajnik, 2011. Automatic web accessibility metrics: where we are and where we can go. *Interacting with Computers*, 23 (2), 137–155.

Vilcekova, S. and E. Kridlova-Burdova, 2014. Multi-criteria analysis of building assessment regarding energy performance using a life-cycle approach. *International Journal of Energy and Environmental Engineering*, 5 (2/3), 1–9.

Villain, G., Z. Sbartaï, X. Dérobert, V. Garnier and J. Balayssac, 2012. Durability analysis of a concrete structure in a tidal zone by combining NDT methods: laboratory tests and case study. *Construction and Building Materials*, 37, 893–903.

Visa, I., M. Moldovan, M. Comsit and A. Duta, 2014. Improving the renewable energy mix in a building toward the nearly zero energy status. *Energy and Buildings*, 68, 72–78.

Viswanathan, M., S. Sudman and M. Johnson, 2004. Maximum versus meaningful discrimination in scale response: implications for validity of measurement of consumer perceptions about products. *Journal of Business Research*, 57 (2), 108–124.

Vitali, F., F. Mulas, P. Marini and R. Bellazzi, 2013. Network-based target ranking for polypharmacological therapies. *Journal of Biomedical Informatics*, 46 (5), 876–881.

Volk, R., J. Stengel and F. Schultmann, 2014. Building Information Modeling (BIM) for existing buildings: literature review and future needs. *Automation in Construction*, 38, 109–127.

Vrána, T. and F. Björk, 2008. A laboratory equipment for the study of moisture processes in thermal insulation materials when placed in a temperature field. *Construction and Building Materials*, 22 (12), 2335–2344.

Wagner, N. and V. Agrawal, 2014. An agent-based simulation system for concert venue crowd evacuation modeling in the presence of a fire disaster. *Expert Systems with Applications*, 41 (6), 2807–2815.

Wagner, A., E. Gossauer, C. Moosmann, T. Gropp and R. Leonhart, 2007. Thermal comfort and workplace occupant satisfaction – results of field studies in German low energy office buildings. *Energy and Buildings*, 39 (7), 758–769.

Walker, C., G. Tan and L. Glicksman, 2011. Reduced-scale building model and numerical investigations to buoyancy-driven natural ventilation. *Energy and Buildings*, 43 (9), 2404–2413.

Waltz, J., 2002. Measuring and verifying your energy performance contracts: What are your options? How do you choose? *Strategic Planning for Energy and the Environment*, 21 (4), 47–64.

Wang, Z., 2006. A field study of the thermal comfort in residential buildings in Harbin. *Building and Environment*, 41 (8), 1034–1039.

Wang, L., 2008. Enhancing construction quality inspection and management using RFID technology. *Automation in Construction*, 17 (4), 467–479.

Wang, C., 2014. Insights from developing a multidisciplinary design and analysis environment. *Computers in Industry*, 65 (4), 786–795.

Wang, E., 2015. Benchmarking whole-building energy performance with multi-criteria technique for order preference by similarity to ideal solution using a selective objective-weighting approach. *Applied Energy*, 146, 92–103.

Wang, L. and Q. Chen, 2007a. Theoretical and numerical studies of coupling multizone and CFD models for building air distribution simulations. *Indoor Air*, 17 (5), 348–361.

Wang, L. and Q. Chen, 2007b. Validation of a coupled multizone-CFD program for building airflow and contaminant transport simulations. *HVAC&R Research*, 13 (2), 267–281.

Wang, Y. and L. Shao, 2017. Understanding occupancy pattern and improving building energy efficiency through Wi-Fi based indoor positioning. *Building and Environment*, 114, 106–117.

Wang, Y. and J. Su, 2014. Automated defect and contaminant inspection of HVAC duct. *Automation and Construction*, 41, 15–24.

Wang, H. and Z. Zhai, 2016. Advances in building simulation and computational techniques: a review between 1987 and 2014. *Energy and Buildings*, 128, 319–335.

Wang, S., C. Lin and C. Yu, 2008. Dynamic simulation of backdraft phenomena in a townhouse building fire. *Heat Transfer – Asian Research*, 37 (3), 153–164.

Wang, J., Y. Jing, C. Zhang and J. Zhao, 2009. Review on multi-criteria decision analysis aid in sustainable energy decision-making. *Renewable and Sustainable Energy Reviews*, 13 (9), 2263–2278.

Wang, Z., L. Wang, A. Dounis and R. Yang, 2012. Integration of plug-in hybrid electric vehicles to energy and comfort management for smart building. *Energy and Buildings*, 47, 260–266.

Wang, Y., Y. Dong, B. Li and G. Zhou, 2013. A fire test on continuous reinforced concrete slabs in a full-scale multi-story steel-framed building. *Fire Safety Journal*, 61, 232–242.

Wang, E., Z. Shen and K. Grosskopf, 2014. Benchmarking energy performance of building envelopes through a selective residual-clustering approach using high dimensional dataset. *Energy and Buildings*, 75, 10–22.

Wang, X., R. Astroza, T. Hutchinson, J. Conte and J. Restrepo, 2015a. Dynamic characteristics and seismic behavior of prefabricated steel stairs in a full-scale five-story building shake table test program. *Earthquake Engineering & Structural Dynamics*, 44 (4), 2507–2527.

Wang, Z., H. Zhao, B. Lin, Y. Zhu, Q. Ouyang and J. Yu, 2015b. Investigation of indoor environment quality of Chinese large-hub airport thermal buildings through longitudinal field measurement and subjective survey. *Building and Environment*, 94, 593–605.

Wang, N., P. Phelan, J. Gonzalez, C. Harris, G. Henze, R. Hutchinson, J. Langevin, M. Lazarus, B. Nelson, C. Pyke, K. Roth, D. Rouse, K. Sawyer and S. Selkowitz, 2017a. Ten questions concerning future buildings beyond zero energy and carbon neutrality. *Building and Environment*, 119, 169–182.

Wang, Q., B. Lee, G. Augenbroe and C. Paredis, 2017b. An application of normative decision theory to the valuation of energy efficiency investments under uncertainty. *Automation in Construction*, 73, 78–87.

Ward, I., 2004. *Energy and environmental issues for the practising architect – a guide to help at the initial design stage*. London: Thomas Telford Publishing.

Watson, P. and T. Howard, 2011. *Construction Quality Management: principles and practice*. Abingdon: Spon Press.

Watts, T. and C. McNair-Connolly, 2012. New performance measurement and management control systems. *Journal of Applied Accounting Research*, 13 (3), 226–241.

Way, M. and B. Bordass, 2005. Making feedback and post-occupancy evaluation routine 2: Soft Landings – involving design and building teams in improving performance. *Building Research & Information*, 33 (4), 353–360.

WBDG, 2016. Whole Building Design Guide [online]. Washington, DC: National Institute of Building Science. Available from www.wdbg.org [Accessed 18 March 2016].

Webb, D., G. Soutar, T. Mazzarol and P. Saldaris, 2013. Self-determination theory and consumer behavioural change: evidence from a household energy-saving behaviour study. *Journal of Environmental Psychology*, 35, 59–66.

Weber, R., 2015. Internet of Things: privacy issues revisited. *Computer Law & Security Review*, 31 (5), 618–627.

Wei, L. and L. Qing-Ning, 2012. Performance-based seismic design of complicated tall building structures beyond the code specification. *The Structural Design of Tall and Special Buildings*, 21 (8), 578–591.

Wei, S., R. Jones and P. de Wilde, 2014a. Driving factors for occupant-controlled space heating in residential buildings. *Energy and Buildings*, 70, 36–44.

Wei, W., J. Xiong, W. Zhao and Y. Zhang, 2014b. A framework and experimental study of an improved VOC/formaldehyde emission reference for environmental chamber tests. *Atmospheric Environment*, 82, 327–334.

Weinberg, G., 1975. *An Introduction to General Systems Thinking*. New York: Dorset House Publishing, Silver Anniversary Edition.

Weinberg, G., 1988. *Rethinking Systems Analysis and Design*. New York: Dorset House Publishing.

Weisheit, S., S. Unterberger, T. Bader and R. Lackner, 2016. Assessment of test methods for characterizing the hydrophobic nature of surface-treated high performance concrete. *Construction and Building Materials*, 110, 145–153.

Welle, B., J. Haymaker and Z. Rogers, 2011. ThermalOpt: a methodology for automated BIM-based multidisciplinary thermal simulation for use in optimization environments. *Building Simulation*, 4 (4), 293–313.

Welle, B., Z. Rogers and M. Fischer, 2012. BIM-centric daylight profiler for simulation (BDP4SIM): a methodology for automated product model decomposition and recomposition for climate-based daylighting simulation. *Building and Environment*, 58, 114–134.

Welsh, D. and L. Ordóñex, 2014. The dark side of consecutive high performance goals: linking goal setting, depletion, and unethical behavior. *Organizational Behavior and Human Decision Processes*, 123 (2), 79–89.

Wetter, M., 2004. GenOpt, generic optimization program, user manual. Berkeley, CA: Lawrence Berkeley National Laboratory, Technical report LBNL-54199.

Wetter, M., 2009. Modelica-based modelling and simulation to support research and development in building energy and control systems. *Journal of Building Performance Simulation*, 2 (2), 143–161.

Wetter, M., 2011. Co-simulation of building energy and control systems with the Building Control Virtual Test Bed. *Journal of Building Performance Simulation*, 4 (3), 185–203.

Wetter, M. and E. Polak, 2005. Building design optimization using a convergent pattern search algorithm with adaptive precision simulations. *Energy and Buildings*, 37 (6), 603–612.

Wetter, M., W. Zuo, T. Nouidui and X. Pang, 2014. Modelica Buildings library. *Journal of Building Performance Simulation*, 7 (4), 253–270.

Wetter, M., M. Bonvini and T. Nouidui, 2016. Equation-based languages – a new paradigm for building energy modeling, simulation and optimization. *Energy and Buildings*, 117, 290–300.

Wikberg, F., T. Olofsson and A. Ekholm, 2014. Design configuration with architectural objects: linking customer requirements with system capabilities in industrialized house-building platforms. *Construction Management and Economics*, 32 (1/2), 196–207.

de Wilde, P., 2004. Computational Support for the Selection of Energy Saving Building Components. PhD thesis. Delft: Delft University Press.

de Wilde, P., 2014. The gap between predicted and measured energy performance of buildings: a framework for investigation. *Automation in Construction*, 41, 40–49.

de Wilde, P. and G. Augenbroe, 2009. Energy modelling. In: Mumovic, D. and M. Santamouris, eds., *A Handbook of Sustainable Building Design and Engineering*. London: Earthscan (James & James) Publishers.

de Wilde, P. and G. Augenbroe, 2018. Energy modelling. In: Mumovic, D. and M. Santamouris, eds., *A Handbook of Sustainable Building Design and Engineering*. Abingdon: Taylor & Francis., 2nd edition.

de Wilde, P. and D. Coley, 2012. The implications of a changing climate for buildings. *Building and Environment*, 55, 1–7.

de Wilde, P. and D. Prickett, 2009. Preconditions for the use of simulation in M&E engineering. In: Strachan, P., N. Kelly and M. Kummert, eds., *Building Simulation '09, 11th International IBPSA Conference*, Glasgow, UK, 27–30 July 2009, pp. 414–419.

de Wilde, P. and W. Tian, 2012. Management of thermal performance risks in buildings subject to climate change. *Building and Environment*, 55, 167–177.

de Wilde, P. and M. van der Voorden, 2002. Design analysis integration: supporting the selection on energy saving building components. *Building and Environment*, 37 (8/9), 807–816.

de Wilde, P., Y. Rafiq and M. Beck, 2008. Uncertainties in predicting the impact of climate change on thermal performance of domestic buildings in the UK. *Building Services Engineering Research and Technology*, 29 (1), 7–26.

de Wilde, P., W. Tian and G. Augenbroe, 2011. Longitudinal prediction of the operational energy use of buildings. *Building and Environment*, 46 (8), 1670–1680.

Williams, B., 2006. Building performance: the value management approach. In: Clements-Croome, D., ed., *Creating the productive workplace*. Abingdon: Taylor & Francis, 2nd edition, pp. 434–457.

Williamson, T., 2010. Predicting building performance: the ethics of computer simulation. *Building Research & Information*, 38 (4), 401–410.

Wilson, A., 2013. What performance gap? [online]. London. Available from www.building. co.uk [Accessed 21 June 2016].

Wilson, K., 2016. An investigation of dependence in expert judgment studies with multiple experts. *International Journal of Forecasting*, 33 (1), 325–336.

de Wit, S., 2001. Uncertainty in predictions of thermal comfort in buildings. PhD thesis. Delft: Delft University Press.

de Wit, S. and Augenbroe, G., 2002. Analysis of uncertainty in building design evaluations and its implications. *Energy and Buildings*, 34 (9), 951–958.

Wolf, A., 2008. RILEM TC190-SBJ: development of recommendations on novel durability test methods for wet-applied curtain-wall sealants. *Materials and Structures*, 41 (9), 1473–1486.

Wöllner, C., 2013. How to quantify individuality in music performance? Studying artistic expression with averaging procedures. *Frontiers in Psychology*, 19 (4), 361.

Wong, N. and A. Mahdavi, 2000. Automated generation of nodal representations for complex building geometries in the SEMPER environment. *Automation in Construction*, 10 (1), 141–153.

Wong, J., H. Li and S. Wang, 2005. Intelligent building research: a review. *Automation in Construction*, 14 (1), 143–159.

Wong, J., H. Li and J. Lai, 2008. Evaluating the system intelligence of the intelligent building systems Part I: development of key intelligent indicators and conceptual analytical framework. *Automation in Construction*, 17 (3), 284–302.

Wong, I., H. Choi and H. Yang, 2012. Simulation and experimental studies on natural lighting in enclosed lift lobbies of highrise residential buildings by remote solar lighting. *Applied Energy*, 92, 705–713.

Wong, J., J. Leung, M. Skitmore and L. Buys, 2017. Technical requirements of age-friendly smart home technologies in high-rise residential buildings: a system intelligence analytical approach. *Automation in Construction*, 73, 12–19.

Wongbumru, T. and B. Dewancker, 2016. Post-occupancy evaluation of user satisfaction: a case study of 'old' and 'new' public housing schemes in Bangkok. *Architectural Engineering and Design Management*, 12 (2), 107–124.

Wouters, P., L. Vandaele, P. Voit and N. Fisch, 1993. The use of outdoor test cells for thermal and solar building research within the PASSYS project. *Building and Environment*, 28 (2), 107–113.

Woxenius, J., 2012. Directness as a key performance indicator for freight transport chains. *Research in Transport Economics*, 36 (1), 63–72.

Wright, J., 2016. Personal communication, 10 September 2016.

Wright, J., H. Loosemore and R. Farmani, 2002. Optimization of building thermal design and control by multi-criterion genetic algorithm. *Energy and Buildings*, 34 (9), 959–972.

Wu, C. and M. Hamada, 2002. *Experiments – Planning, Analysis, and Parameter Design Optimization*. Chichester: Wiley, 2nd edition.

Wu, S. and J. Sun, 2011a. A top-down strategy with temporal and spatial partition for fault detection and diagnosis of building HVAC systems. *Energy and Buildings*, 43 (9), 2134–2139.

Wu, S. and J. Sun, 2011b. Cross-level fault detection and diagnosis of building HVAC systems. *Building and Environment*, 46 (8), 1558–1566.

Wu, P., J. Wang and X. Wang, 2016. A critical review of the use of 3-D printing in the construction industry. *Automation in Construction*, 68, 21–31.

Xia, Y., 2014. Simulation on spread of fire smoke in the elevator shaft for a high-rise building. *Theoretical & Applied Mechanics Letters*, 4 (3), 034007.

Xiao, Y. and J. Ma, 2012. Fire simulation test and analysis of laminated bamboo frame building. *Construction and Building Materials*, 34, 257–266.

Xiao, H. and D. Proverbs, 2002. The performance of contractors in Japan, the UK and the USA: an evaluation of construction quality. *International Journal of Quality & Reliability Management*, 19 (6), 672–687.

Xie, H., L. Filippidis, E. Galea, D. Blackshields and P. Lawrence, 2012. Experimental analysis of the effectiveness of emergency signage and its implementation in evacuation simulation. *Fire and Materials*, 36 (5/6), 367–382.

Xiong, J. and A. Tzempelikos, 2016. Model-based shading and lighting controls considering visual comfort and energy use. *Solar Energy*, 134, 416–428.

Xu, Z., K. Wang, Y. Guo, M. Wu and M. Xu, 2016. Hybrid test on building structures using electrodynamic test machine. *Nondestructive Testing and Evaluation*, 32 (1), 90–102.

Yalcintas, M., 2006. An energy benchmarking model based on artificial neural network method with a case example for tropical climates. *International Journal of Energy Research*, 30 (14), 1158–1174.

Yan, D., W. O'Brien, T. Hong, X. Feng, H. Gunay, F. Tahmasebi and A. Mahdavi, 2015. Occupant behavior modeling for building performance simulation: current state and future challenges. *Energy and Buildings*, 107, 264–278.

Yanagisawa, D., Y. Suma, A. Tomoeda, A. Miura, K. Ohtsuka and K. Nishinari, 2013. Walking-distance introduced queuing model for pedestrian queuing system: theoretical analysis and experimental validation. *Transport Research Part C*, 37, 238–259.

Yang, P., X. Tan and W. Xin, 2011. Experimental study and numerical simulation for a storehouse fire accident. *Building and Environment*, 46 (7), 1445–1459.

Yang, T., P. Moehle, Y. Bozorgnia, F. Zareian and J. Wallace, 2012. Performance assessment of tall concrete-core wall building designed using two alternative approaches. *Earthquake Engineering & Structural Dynamics*, 41 (11), 1515–1531.

Yao, J., 2014. Determining the energy performance of manually controlled solar shades: a stochastic model based co-simulation analysis. *Applied Energy*, 127, 64–80.

Ye, H., L. Long, H. Zhang and Y. Gao, 2014. The energy saving index and the performance evaluation of thermochromic windows in passive buildings. *Renewable Energy*, 66, 215–221.

Yeung, J., A. Chan and D. Chan, 2009. A computerized model for measuring and benchmarking the partnering performance of construction projects. *Automation in Construction*, 18 (8), 1099–1113.

Yigitcanlar T. and A. Lönnqvist, 2013. Benchmarking knowledge-based urban development performance: results from the international comparison of Helsinki. *Cities*, 31, 357–369.

Yik, F. and W. Lee, 2004. Partnership in building energy performance contracting. *Building Research & Information*, 32 (3), 235–243.

Yildiz, A., I. Dikmen and M. Birgonul, 2014. Using expert opinion for risk assessment: a case study of a construction project utilizing a risk mapping tool. *Procedia – Social and Behavioural Sciences*, 119, 519–528.

Yin, Y., S. Qin and R. Holland, 2011. Development of a design performance measurement matrix for improving collaborative design during a design process. *International Journal of Productivity and Performance Management*, 60 (2), 152–184.

Youm, K., J. Moon, J. Cho and J. Kim, 2016. Experimental study on strength and durability of lightweight aggregate concrete containing silica fume. *Construction and Building Materials*, 114, 517–527.

Younes, C. and C. Abi Shdid, 2013. A methodology for 3-D multiphysics CFD simulation of air leakage in building envelopes. *Energy and Buildings*, 65, 146–158.

Yu, A., Q. Shen, J. Kelly and K. Hunter, 2005. Application of value management in project briefing. *Facilities*, 23 (7/8), 330–342.

Yu, A., Q. Shen, J. Kelly and K. Hunter, 2007. An empirical study of the variables affecting construction project briefing/architectural programming. *International Journal of Project Management*, 25 (2), 198–212.

Yu, Z., B. Fung and F. Haghighat, 2013. Extracting knowledge from building-related data – a data mining framework. *Building Simulation*, 6 (2), 207–222.

Yu, X., Y. Su and X. Chen, 2014. Application of Relux simulation to investigate energy saving potential from daylighting in a new educational building in UK. *Energy and Buildings*, 74, 191–202.

Yu, T., T. Cheng, A. Zhou and D. Lau, 2016. Remote defect detection of FRP-bonded concrete system using acoustic-laser and imaging radar techniques. *Construction and Building Materials*, 109, 146–155.

Yuan, H., 2012. A model for evaluating the social performance of construction waste management. *Waste Management*, 32 (6), 1218–1228.

Yuan, S. and J. Zhang, 2009. Large eddy simulation of compartment fire with solid combustibles. *Fire Safety Journal*, 44 (3), 349–362.

Yuan, J., Z. Fang, Y. Wang, S. Lo and P. Wang, 2009. Integrated network approach of evacuation simulation for large complex buildings. *Fire Safety Journal*, 44 (2), 266–275.

Yun, M., S. Han, S. Hong and J. Kim, 2003. Incorporating user satisfaction into the look-and-feel of mobile phone design. *Ergonomics*, 46 (13/14), 1423–1440.

Yusoff, W. and M. Sulaiman, 2014. Sustainable campus: indoor environmental quality (IEQ) performance measurement for Malaysian public universities. *European Journal of Sustainable Development*, 3 (4), 323–338.

Zalok, E. and G. Hadjisophocleous, 2011. Assessment of the use of fire dynamics simulator in performance-based design. *Fire Technology*, 47 (4), 1081–1100.

Zambrano, A., J. Fabry and S. Gordillo, 2012. Expressing aspectual interactions in requirements engineering: experiences, problems and solutions. *Science of Computer Programming*, 78 (1), 65–92.

Zanghirella, F., M. Perino and V. Serra, 2011. A numerical model to evaluate the thermal behaviour of active transparent façades. *Energy and Buildings*, 43 (5), 1123–1138.

Zanni, M., R. Soetanto and K. Ruikar, 2016. Towards a BIM-enabled sustainable building design process: roles, responsibilities, and requirements. *Architectural Engineering and Design Management*, 13 (2), 101–129.

Zannier, C., M. Chiasson and F. Maurer, 2007. A model of design decision making based on empirical results of interviews with software designers. *Information and Software Technology*, 49 (6), 637–653.

Zapata-Lancaster, G., 2014. Low carbon non-domestic building design process. An ethnographic comparison of design in Wales and England. *Structural Survey*, 32 (2), 140–157.

Zapata-Lancaster, G. and C. Tweed, 2016. Tools for low-energy building design: an exploratory study of the design process in action. *Architectural Engineering and Design Management*, 12 (4), 279–295.

Zapata-Poveda, G. and C. Tweed, 2014. Official and informal tools to embed performance in the design of low carbon buildings. An ethnographic study in England and Wales. *Automation in Construction*, 37, 38–47.

Zareian, F. and H. Krawinkler, 2012. Conceptual performance-based seismic design using building-level and story-level decision support system. *Earthquake Engineering & Structural Dynamics*, 41 (11), 1439–1453.

Zareian, F., H. Krawinkler, L. Ibarra and D. Lignos, 2010. Basic concepts and performance measures in prediction of collapse of buildings under earthquake ground motions. *The Structural Design of Tall and Special Buildings*, 19 (1/2), 167–181.

Zawidzki, M., M. Chraibi and K. Nishinari, 2014. Crowd-Z: the user-friendly framework for crowd simulation on an architectural floor plan. *Pattern Recognition Letters*, 44, 88–97.

Zeigler, B., H. Praehofer and T. Kim, 2000. *Theory of modeling and simulation – integrating discrete event and continuous complex dynamic systems*. London/San Diego, CA: Academic Press, 2nd edition.

Zhai, Z. and Q. Chen, 2005. Performance of coupled building energy and CFD simulations. *Energy and Buildings*, 37 (4), 333–344.

Zhai, Z. and Q. Chen, 2006. Sensitivity analysis and application guides for integrated building energy and CFD simulation. *Energy and Buildings*, 38 (9), 1060–1068.

Zhang, Y. and P. Barrett, 2010. Findings from a post-occupancy evaluation in the UK primary schools sector. *Facilities*, 28 (13/14), 641–656.

Zhang, X. and H. Gao, 2010. Optimal performance-based building facility management. *Computer-Aided Civil and Infrastructure Engineering*, 25 (4), 269–284.

Zhang, R. and T. Hong, 2017. Modeling of HVAC operational faults in building performance simulation. *Applied Energy*, 202, 178–188.

Zhang, Z. and B. Wachenfeldt, 2009. Numerical study on the heat storing capacity of concrete walls with air cavities. *Energy and Buildings*, 41 (7), 769–773.

Zhang, J., M. Delichatsios and M. Colobert, 2010. Assessment of fire dynamics simulator for heat flux and flame heights predictions from fires in SBI tests. *Fire Technology*, 46 (2), 291–306.

Zhang, H., E. Arens and W. Pasut, 2011. Air temperature thresholds for indoor comfort and perceived air quality. *Building Research & Information*, 39 (2), 134–144.

Zhang, J., P. Pelken, Y. Chen, D. Rice, Z. Meng, S. Semahegn, L. Gu, H. Henderson, W. Feng and F. Ling, 2013. 'Virtual Design Studio' – Part 2: Introduction to overall and software framework. *Building Simulation*, 6 (3), 253–268.

Zhang, L., Y. Feng, Q. Meng and Y. Zhang, 2015a. Experimental study on the building evaporative cooling by using the climatic wind tunnel. *Energy and Buildings*, 104, 360–368.

Zhang, Y., Z. O'Neill, B. Dong and G. Augenbroe, 2015b. Comparisons of inverse modeling approaches from predicting building energy performance. *Building and Environment*, 86, 177–190.

Zhang, X., P. Wargocki and Z. Lian, 2016. Human responses to carbon dioxide, a follow-up study at recommended exposure limits in non-industrial environments. *Building and Environment*, 100, 162–171.

Zhang, Z., L. Gong, Y. Jin, J. Xie and J. Hao, 2017. A quantitative approach to design alternative evaluation based on data-driven performance prediction. *Advanced Engineering Informatics*, 32, 52–65.

Zhao, S., 2011. Simulation of mass fire-spread in urban densely built areas based on irregular coarse cellular automata. *Fire Technology*, 47 (3), 721–749.

Zhao, H. and F. Magoulès, 2012. A review on the prediction of building energy consumption. *Renewable and Sustainable Energy Reviews*, 16 (6), 3586–3592.

Zhao, L., J. Zhang and R. Liang, 2013. Development of an energy monitoring system for large public buildings. *Energy and Buildings*, 66, 41–48.

Zheng, X., T. Zhong and M. Liu, 2009. Modeling crowd evacuation of a building based on seven methodological approaches. *Building and Environment*, 44 (3), 437–445.

Zhong, C., X. Huang, S. Müller Arisona, G. Schmitt and M. Batty, 2014. Inferring building functions from a probabilistic model using public transport data. *Computers, Environment and Urban Systems*, 48, 124–137.

Zhou, X., T. Hong and D. Yan, 2014. Comparison of HVAC system modeling in EnergyPlus, DeST and DOE-2.1E. *Building Simulation*, 7 (1), 21–33.

Zibin, N., R. Zmeureanu and J. Love, 2016. Automatic assisted calibration tool for coupling building automation system trend data with commissioning. *Automation in Construction*, 61, 124–133.

Zimring, C. and J. Reizenstein, 1980. Post-occupancy evaluation – an overview. *Environment and Behavior*, 12 (4), 429–450.

Zingeser, J., 2001. Recommended minimum requirements for small dwelling construction. In: Lide, D., ed., *A century of excellence in measurements, standards and technology*. Boca Raton, FL: CRC Press. NIST Special Publication 958.

Zisis, I., T. Stathopoulos, I. Smith and G. Doudak, 2011. Cladding pressures and primary structural system forces of a wood building exposed to strong winds. *Canadian Journal of Civil Engineering*, 38 (9), 974–983.

Zuo, Q., W. Leonard and E. MaloneBeach, 2010. Integrating performance-based design in beginning interior design education: an interactive dialog between the built environment and its context. *Design Studies*, 31 (3), 268–287.

Zuo, W., A. McNeil, M. Wetter and E. Lee, 2014. Acceleration of the matrix multiplication of Radiance three phase daylighting simulations with parallel computing on heterogeneous hardware of personal computer. *Journal of Building Performance Simulation*, 7 (2), 152–163.

Zuo, W., M. Wetter, W. Tian, D. Li, M. Jin and Q. Chen, 2016. Coupling indoor airflow, HVAC, control and building envelope heat transfer in the Modelica Buildings library. *Journal of Building Performance Simulation*, 9 (4), 366–381.

Index